f_d = stress due to unfactored dead load, at extreme fiber of section where tensile stress is caused by externally applied loads, psi.

f_{pc} = compressive stress in concrete due to effective prestress forces only (after allowance for all prestress losses) at extreme fiber of section where tensile stress is caused by externally applied loads, psi.

f_{ps} = stress in prestressed reinforcement at nominal strength.

f_{pu} = specified tensile strength of prestressing tendons, psi.

f_{py} = specified yield strength of prestressing tendons, psi.

f_r = modulus of rupture of concrete, psi.

f_t = tensile strength of concrete, psi.

f_y = specified yield strength of nonprestressed reinforcment, psi.

f_{yh} = specified yield strength of transverse, reinforcement, psi.

h = overall thickness of member, in.

I = moment of inertia of section resisting externally applied factored loads, in.⁴.

I_b = moment of inertia about centroidal axis of gross section of beam, in.⁴.

I_{cr} = moment of inertia of cracked section transformed to concrete, in.⁴.

I_e = effective moment of inertia for computation of deflection, in.⁴.

I_g = moment of inertia of gross concrete section about centroidal axis, neglecting reinforcement, in.⁴.

k = effective length factor for compression memebers.

K_b = flexural stiffness of beam; moment per unit rotation.

K_c = flexural stiffness of column; moment per unit rotation.

K_{ec} = flexural stiffness of equivalent column; moment per unit rotation.

K_s = flexural stiffness of slab; moment per unit rotation.

K_t = torsional stiffness of torsional member; moment per unit rotation.

l_{db} = basic development length, in.

l_{dh} = development length of standard hook in tension, measured from critical section to outside end of hook (straight embedment length between critical section and start of hook [point of tangency] plus radius of bend and one bar diameter). in.

= l_{hb} × applicable modification factors.

M_a = maximum moment in member at stage deflection is computed.

M_c = facto[r]... comp...

M_d = mom...

M_{cr} = crack...

M_n = nominal moment st...

M_{max} = maximum factored moment at section due to externally applied loads.

M_u = factored moment at section.

n = modular ratio of elasticity.

= E_s/E_c or E_{ps}/E_c.

N_u = factored axial load normal to cross section occurring simultaneously with V_u; to be taken as positive for compression, negative for tension, and to include effects of tension due to creep and shrinkage.

N_{uc} = factored tensile force applied at top of bracket or corbel acting simultaneously with V_u, to be taken as positive for tension.

P_b = nominal axial load strength at balanced strain conditions.

P_c = critical load.

P_n = nominal axial load strength at given eccentricity.

p_{cp} = outside perimeter of the concrete cross-section A_{cp}; in.

p_h = perimeter of centerline of outermost closed transverse torsional reinforcement, in.

r = radius of gyration of cross section of a compression member.

s = spacing of shear or torsion reinforcement in direct parallel to longitudinal reinforcement, in.

t = thickness of a wall of a hollow section, in.

T_u = factored torsional moment at section.

V_e = nominal shear strength provided by concrete.

V_{ci} = nominal shear strength provided by concrete when diagonal cracking results from combined shear and moment.

V_{cw} = nominal shear strength provided by concrete when diagonal cracking results from excessive principal tensile stress in web.

V_d = shear force at section due to unfactored dead load.

V_p = vertical component of effective prestress force at section.

V_s = nominal shear strength provided by shear reinforcement.

V_u = factored shear force at section.

Third edition

REINFORCED CONCRETE

A Fundamental Approach

Dr. Edward G. Nawy, P.E.
Distinguished Professor
Department of Civil and Environmental Engineering
Rutgers University
The State University of New Jersey

PRENTICE HALL, *Upper Saddle River, New Jersey 07458*

Library of Congress Cataloging-in-Publication Data

Nawy, Edward G.
  Reinforced concrete : a fundamental approach / by Edward G. Nawy.
  3rd ed.
      p.   cm. -- (Prentice-Hall international series in civil
engineering and engineering mechanics)
  Includes bibliographical references and index.
  ISBN 0-13-123498-6
  1. Reinforced concrete. 2. Reinforced concrete construction.
I. Title. II. Series.
TA444.N38 1995
624.1'8341--dc20   95-13792
                                                                    CIP

Editorial/production supervision TKM Productions
Manufacturing buyer: Donna Sullivan
Cover design: Bruce Kenselaar
Cover Photo:
Minneapolis City Center Multi-Foods Tower: A 52-story building with 400,000 ft^2 of sandblasted precast
  concrete (Courtesy American Concrete Institute)

The author and publisher of this book have used their best efforts in preparing this
book. These efforts include the development, research, and testing of the theories
and programs to determine their effectiveness. The author and publisher make no
warranty of any kind, expressed or implied, with regard to these programs or the
documentation contained in this book. The author and publisher shall not be liable
in any event for incidental or consequential damages in connection with, or arising
out of, the furnishing, performance, or use of these programs.

 © 1996, 1990, 1985 by Prentice-Hall, Inc.
Simon & Schuster/A Viacom Company
Upper Saddle River, New Jersey 07458

Printed in the United States of America
10 9 8 7 6 5 4 3 2 1

ISBN 0-13-123498-6

Prentice-Hall International (UK) Limited, *London*
Prentice-Hall of Australia Pty. Limited, *Sydney*
Prentice-Hall Canada Inc., *Toronto*
Prentice-Hall Hispanoamericana, S.A., *Mexico*
Prentice-Hall of India Private Limited, *New Delhi*
Prentice-Hall of Japan, Inc., *Tokyo*
Simon & Schuster Asia Pte. Ltd., *Singapore*
Editora Prentice-Hall do Brasil, Ltda., *Rio de Janeiro*

PRENTICE HALL INTERNATIONAL SERIES
IN CIVIL ENGINEERING AND ENGINEERING MECHANICS

William J. Hall, Editor

AU AND CHRISTIANO, *Structural Analysis*
BARSOM AND ROLFE, *Fracture and Fatigue Control in Structures, 2/E*
BATHE, *Finite Element Procedures in Engineering Analysis*
BERG, *Elements of Structural Dynamics*
BIGGS, *Introduction to Structural Engineering*
CHAJES, *Structural Analysis, 2/E*
CHOPRA, *Dynamics of Structures*
COOPER AND CHEN, *Designing Steel Structures*
CORDING, ET AL., *The Art and Science of Geotechnical Engineering*
GALLAGHER, *Finite Element Analysis*
HENDRICKSON AND AU, *Project Management for Construction*
HIGDON ET AL., *Engineering Mechanics, 2nd Vector Edition*
HOLTZ AND KOVACS, *Introduction to Geotechnical Engineering*
HUMAR, *Dynamics of Structures*
JOHNSTON, LIN AND GALAMBOS, *Basic Steel Design, 3/E*
KELKAR AND SEWELL, *Fundamentals of the Analysis and Design of Shell Structures*
MACGREGOR, *Reinforced Concrete: Mechanics and Design, 2/E*
MEHTA, *Concrete: Structure, Properties and Materials*
MELOSH, *Structural Engineering Analysis by Finite Elements*
MEREDITH ET AL., *Design and Planning of Engineering Systems*
MINDESS AND YOUNG, *Concrete*
NAWY, *Prestressed Concrete, 2E*
NAWY, *Reinforced Concrete: A Fundamental Approach, 3/E*
POPOV, *Engineering Mechanics of Solids*
POPOV, *Introduction to the Mechanics of Solids*
POPOV, *Mechanics of Materials, 2/E*
SENNETT, *Matrix Analysis of Structures*
SCHNEIDER AND DICKEY, *Reinforced Masonry Design, 2/E*
WANG AND SALMON, *Introductory Structural Analysis*
WEAVER AND JOHNSON, *Finite Elements for Structural Analysis*
WEAVER AND JOHNSON, *Structural Dynamics by Finite Elements*
WOLF, *Dynamic Soil-Structure Interaction*
WRAY, *Measuring Engineering Properties of Soils*
YANG, *Finite Element Structural Analysis*

"Reflections"—High strength polymer concrete sculpture at Rutgers University. Work by R. H. Karol, the civil engineering class of 1982, and the author.

To
RACHEL E. NAWY

For her high-limit state of endurance over the years,
which made the writing of this book a reality.

CONTENTS

4

REINFORCED CONCRETE   65

5

FLEXURE IN BEAMS   90

6

SHEAR AND DIAGONAL TENSION IN BEAMS   155

7

TORSION **212**

8

SERVICEABILITY OF BEAMS AND ONE-WAY SLABS **273**

9

COMBINED COMPRESSION AND BENDING: COLUMNS 327

10

BOND DEVELOPMENT OF REINFORCING BARS 413

14

INTRODUCTION TO PRESTRESSED CONCRETE **668**

15

SEISMIC DESIGN OF CONCRETE STRUCTURES **721**

APPENDIX A   COMPUTER PROGRAMS IN BASIC **767**

APPENDIX B   TABLES AND NOMOGRAMS **799**

INDEX   **827**

PREFACE

Reinforced concrete is a widely used material for constructed systems. Hence, graduates of every civil engineering program must have, as a minimum requirement, a basic understanding of the fundamentals of reinforced concrete. Additionally, design of the members of a total structure is achieved only by trial and adjustment: assuming a section and then analyzing it. Consequently, design and analysis have been combined to make it simpler for the student when first introduced to the subject of reinforced concrete design.

This third edition of this book revises the previous text so as to conform to the ACI 318-89 Code and also to include major additional topics, such as in Chapter 7 on torsion and Chapter 10 on development length. These have been completely revised because of the new code approaches to these two areas. A new Chapter 15 on seismic design of concrete structures was added, conforming to the ACI 318-95 and the latest Uniform Building Code (1994) provisions. Chapter 9 on compression members was also revamped to accommodate the new ACI Code provisions on the stability of slender columns and discussions of the second-order analysis due to the $P-\Delta$ effects. Another significant feature of this edition is the inclusion of examples in SI units in most of the chapters and a listing of the relevant equation in the SI format. In this manner, students as well as practicing engineers can avail themselves of the tools for transition from the pound-inch (lb-in.) (PI) system to the future System International (SI) that is gaining prevalence in the United States.

The text is an outgrowth of the author's lecture notes evolved in teaching the subject at Rutgers University over the past thirty-five years and the experience accumulated over the years in teaching and research in the areas of reinforced and prestressed concrete inclusive of the Ph.D. level. The material is presented in such a manner that the student can become familiarized with the properties of plain concrete and its components prior to embarking on the study of structural behavior. The book is uniquely different from other textbooks that a good segment of its contents can be covered in one semester in spite of the in-depth discussions of some of its major topics.

The concise discussion presented in Chapters 1 through 4 on the historical development of concrete, the proportioning of the constituent materials, long-term basic behavior, and the development of safety factors should give an adequate introduction to the subject of reinforced concrete. It should also aid in developing fundamental laboratory experiments and essential knowledge of mix proportioning, strength and behavioral requirements, and the concepts of reliability of performance of structures, to which every engineering student should be exposed. The discussion of quality assurance should also give the reader a good introduction to the systematic approach needed to administer the development of concrete structural systems from conception to turnkey use.

Since concrete is a nonelastic material, with the nonlinearity of its behavior starting at a very early stage of loading, only the ultimate-strength approach, or what is sometimes termed the limit state at failure approach, is given in this book. Adequate coverage is given of the serviceability checks in terms of cracking and deflection behavior, as well as long-term effects. In this manner, the design should satisfy all the service-load-level requirements, while ensuring that the theory used in the analysis (design) truly describes the actual behavior of the designed components.

Chapters 5 through 8 cover the flexural, diagonal tension, torsion, and serviceability behavior of one-dimensional members: beams and one-way slabs. Full emphasis has been placed on giving the student and the engineer a feeling for the internal strain distribution in structural reinforced concrete elements and a basic understanding of the reserve strength and the safety factors inherent in the design expressions. Chapter 9, on the analysis and design of columns and other compression members, treats the subject of strain compatibility and strain distribution in a similar manner to that in Chapter 5, on flexural analysis and design of beams. It includes a detailed discussion of how to construct interaction diagrams for columns, as well as how to proportion columns subjected to biaxial bending and buckling. With revised Chapter 7 on torsion, Chapter 9 on compression members, and Chapter 10 on bond and development length in reinforcement, conforming to ACI 318-95, the sequence of design steps of all elements except two-way floors is complete.

It is important to mention that Chapter 6, on diagonal tension, also contains detailed coverage of the behavior of deep beams, corbels, and brackets, with sufficient design examples to supplement the theory. This topic has been included in view of the increased use of precast construction, the wider understanding of the effects of induced horizontal loads on floors, and the frequent need for including shear walls and deep beams in today's multilevel structures. Additionally, Chapter 7 treats the topic of torsion in some detail, considering the space constraints of the book. The discussion ranges from the basic fundamentals of pure torsion in elastic and plastic materials to the design of reinforced concrete members subjected to combined torsion, shear, and bending. The material presented and the accompanying illustrative examples should give the background necessary for pursuing more advanced studies in this area, as listed in the selected references.

Chapter 11 presents an extensive coverage of the subject of analysis and design of two-way slab and plate floor systems. Following a discussion of fundamental behavior, it gives detailed design examples using both the ACI procedures and yield-line theory for the flexural design of reinforced concrete floors. It also includes ultimate-load solutions to most floor shapes and possible gravity loading patterns. Detailed discussion of the deflection behavior and evaluation of two-way panels, as well as the cracking mechanism of such panels, with appropriate analysis examples makes this chapter another unique feature of this concise textbook.

Chapter 13 deals with continuous reinforced concrete structures. It presents a review of the various methods of analysis for continuity of multispan beams and portals and gives relevant examples, including those on the topics of limit theory and plastic hinging. Chapter 14 is an introduction to prestressed concrete. It should serve as a brief treatment of the subject in order to illustrate the fundamental differences between reinforced and prestressed concrete.

Chapter 15 is a new chapter dealing with the seismic behavior of concrete structures. It presents an introduction to the mechanism of earthquakes, the fundamental period of vibration, degrees of freedom, and the Uniform Building Code expressions for the seismic design of concrete elements and shear walls. It also contains several examples on seismic design and a detailed example both in PI and SI units for the design of a high-rise building shear wall and the necessary confining reinforcement.

It is important to emphasize that in this field the use of computers prevails today. Access to transportable personal computers, due to their affordable cost, has made it possible for almost every student to be equipped with such a tool. Hence, extensive flow charts have been presented throughout the book to aid the students in writing their own computer programs. Also, Appendix A contains eight computer programs written in BASIC for IBM and compatible computers covering the topics of flexure, shear, torsion, combined loading, brackets and corbels, and deep beams. As a result, the use of handbook charts was kept to a minimum. Computer program $5\frac{1}{4}$-in. and $3\frac{1}{2}$-in. diskettes can be purchased from the author as indicated in Appendix A. The numerous flow charts for every topic presented in this book should aid the user in developing the logic and step-by-step thinking for easily comprehending the analysis and design procedures for efficient reinforced concrete systems.

Selected photographs of various areas of structural behavior of concrete elements at failure are included in all the chapters. They are taken from the published research work by the author with many of his M.S. and Ph.D. students at Rutgers University over the past three decades. Additionally, photographs of landmark structures, mainly in the United States, are included throughout the book to illustrate the versatility of design in reinforced concrete.

The textbook conforms to the provisions of ACI 318-95 with an eye to stressing the basics, rather than trying every step to the code, which changes once every six years. Consequently, no attempt was made to tie any design or analysis steps to the particular equation numbers in the code; rather, the student is expected to gain the habit of getting familiar with the provisions and section numbers of the ACI Code as a dynamic, ever-changing document. Conversions to SI units are included in the illustrative examples throughout the book; in addition, separate solutions in SI units have been added to most chapters in this edition.

The various topics have been presented in as concise a manner as possible, but without sacrificing the instructional details needed by students first exposed to reinforced concrete design. Hence, the topic of prestressed concrete has been only briefly covered in Chapter 14, and the reader is left to pursue for more advanced works, such as the author's book *Prestressed Concrete: A Fundamental Approach* (2nd ed., 1996), which also conforms to the ACI 318-95 Code.

Portions of this book are intended for a first course at the junior or senior level of the standard college or university curriculum in civil engineering, while the advanced topics can be adequately covered at the graduate level. The contents should also serve as a valuable guideline to the practicing engineer who has to keep abreast of the state of the art in concrete, as well as the designer who is interested in a concise treatment of the fundamentals.

ACKNOWLEDGMENTS

Grateful acknowledgment is due to the American Concrete Institute for contribution to the author's accomplishments and for permitting generous quotations of its ACI 318 Code and the illustrations from other ACI publications. Special mention is made of his original mentor, the late Professor A. L. L. Baker of London University's Imperial College of Science and Technology, who inspired him with the affection that he had developed for systems constructed of concrete. Grateful acknowledgment is also made to the author's many students, both undergraduate and graduate, who have had much to do with generating the writing of this book; to the many who assisted in his research activities over the years, shown in the various photographs of laboratory tests throughout the book; to Dr. P. N. Balaguru at Rutgers, for his input to sections of the first edition, and to the many colleagues at other universities who have continuously used the book and have given advice and suggestions for modifications and additions.

Thanks are due to the panel of reviewers of the first edition: Professor William J. Hall of the University of Illinois at Urbana, Engineering Editor of the Prentice Hall Advanced Series; Professor Vitelmo V. Bertero of the University of California at Berkeley; Professor Dan E. Branson of the University of Iowa; Professor Thomas T. C. Hsu of the University of Houston; and Mr. Gerald B. Neville of the Portland Cement Association, for their suggestions and advice.

Thanks to engineers Mark J. Cipollone for his input to the original manuscript; to Regina Silviera Rocha Souza for her ideas and computational review of the first edition; to Robert M. Nawy, Rutgers engineering class of 1983, for his extensive work on the solutions; and to engineers Abe Daly and Lily Sehayek, Ph.D., for their input to some of the first-edition computer programs in BASIC.

Special gratitude and thanks for this edition are due to Dr. Basile G. Rabbat, PCA Manager of Structural Codes, for his overall input on the new ACI Code provisions; to Professor M. J. Heo of Inchon University for his computational review of the manuscript; to Professor Thomas T. C. Hsu of the University of Houston for his advice and review of Chapter 7 on torsion; to Dr. Douglas D. Lee, President of Douglas D. Lee Associates for reviewing the revised Chapter 10 on development length; to Dr. S. K. Ghosh, Director of Engineering, Standards and Codes, PCA, and Professor Murat Saatcioglu of the University of Ottawa for their review of the newly added Chapter 15 on seismic design of structures.

Thanks also to the Prentice Hall staff and associate editor Alice Dworkin for their commendable efforts in producing this enlarged third edition of the book. Last but not least, the author is deeply indebted to Ms. Kristi A. Latimer, M.S. candidate at Rutgers University, for her diligence, dedication, and assistance in reviewing and processing the changes and the new chapters included in this edition and to her overall work on the text.

Edward G. Nawy
Rutgers University
The State University of New Jersey
New Brunswick, New Jersey

INTRODUCTION

1.1 HISTORICAL DEVELOPMENT OF STRUCTURAL CONCRETE

Use of concrete and its cementatious (volcanic) constituents, such as pozzolanic ash, has been made since the days of the Greeks, the Romans, and possibly earlier ancient civilizations. However, the early part of the nineteenth century marks the start of more intensive use of the material. In 1801, F. Coignet published his statement of principles of construction, recognizing the weakness of the material in tension. J. L. Lambot in 1850 constructed for the first time a small cement boat for exhibition in the 1855 World's Fair in Paris. J. Monier, a French gardener, patented in 1867 metal frames as reinforcement for concrete garden plant containers, and Koenen in 1886 published the first manuscript on the theory and design of concrete structures. In 1906, C. A. P. Turner developed the first flat slab without beams.

Thereafter, considerable progress occurred in this field such that by 1910 the German Committee for Reinforced Concrete, the Austrian Concrete Committee, the American Concrete Institute, and the British Concrete Institute were already established. Many buildings, bridges, and liquid containers of reinforced concrete

Photo 2 Chaochow Bridge on Hsiaoho River, China (A.D. 605–617).

1

Photo 3   Felix Candela's Xochimilco Restaurant, Mexico.

were already constructed by 1920, and the era of linear and circular prestressing began.

The rapid developments in the art and science of reinforced and prestressed concrete analysis, design, and construction have resulted in unique structural systems, such as the Kresge Auditorium, Boston; the 1951 Festival of Britain Dome; Marina Towers and Lake Point Tower, Chicago; the Trump Tower, New York; Two Union Square Towers, Seattle; and many, many others.

Ultimate-strength theories were codified in 1938 in the USSR and in 1956 in England and the United States. Limit theories have also become a part of codes of several countries throughout the world. New constituent materials and composites of concrete have become prevalent, including the high-strength concretes of a strength in compression up to 20,000 psi (137.9 MPa) and 1800 psi (12.41 MPa) in tension. Steel reinforcing bars of strength in excess of 60,000 psi (413.7 MPa) and high-strength welded wire fabric in excess of 100,000 psi (689.5 MPa) ultimate strength are being used. Additionally, deformed bars of various forms have been produced. Such deformations help develop the maximum possible bond between the reinforcing bars and the surrounding concrete as a requisite for the viability of concrete as a structural medium. Prestressing steel of ultimate strengths in excess of 300,000 psi (2068 MPa) is available.

All these developments and the massive experimental and theoretical research that has been conducted, particularly in the last two decades, have resulted in rigorous theories and codes of practice. Consequently, a simplified approach has

Photo 4   Afrikaans Languages Monument, Stellenbosch, South Africa (height of the main dynamically designed hollow columns, 186 ft).

become necessary to an understanding of the fundamental structural behavior of reinforced concrete elements.

1.2 BASIC HYPOTHESIS OF REINFORCED CONCRETE

Plain concrete is formed from a hardened mixture of cement, water, fine aggregate, coarse aggregate (crushed stone or gravel), air, and often other admixtures. The plastic mix is placed and consolidated in the formwork and, then cured to facilitate the acceleration of the chemical hydration reaction of the cement–water mix, resulting in hardened concrete. The finished product has high compressive strength and low resistance to tension, such that its tensile strength is approximately one-tenth of its compressive strength. Consequently, tensile and shear reinforcement in the tensile regions of sections has to be provided to compensate for the weak-tension regions in the reinforced concrete element.

It is this deviation in the composition of a reinforced concrete section from the homogeneity of standard wood or steel sections that requires a modified approach to the basic principles of structural design, as will be explained in subsequent chapters of this book. The two components of the heterogeneous reinforced

Photo 5   Rockefeller Empire State Plaza, Albany, New York—Ammann & Whitney design. (Courtesy of New York Office of General Services.)

Photo 6   Empire State Performing Arts Center, Albany, New York—Ammann & Whitney design.   (Courtesy of New York Office of General Services.)

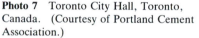

Photo 7   Toronto City Hall, Toronto, Canada.   (Courtesy of Portland Cement Association.)

concrete section are to be so arranged and proportioned that optimal use is made of the materials involved.   This is possible because concrete can easily be given any desired shape by placing and compacting the wet mixture of the constituent ingredients into suitable forms in which the plastic mass hardens.   If the various ingredients are properly proportioned, the finished product becomes strong, durable, and, in combination with the reinforcing bars, adaptable for use as main members of any structural system.

1.3 ANALYSIS VERSUS DESIGN OF SECTIONS

From the foregoing discussion, it is clear that a large number of parameters have to be dealt with in proportioning a reinforced concrete element, such as geometrical width, depth, area of reinforcement, steel strain, concrete strain, and steel stress. Consequently, trial and adjustment are necessary in the choice of concrete sections, with assumptions based on conditions at site, availability of the constituent materials, particular demands of the owners, architectural and headroom requirements, applicable codes, and environmental conditions.   Such an array of parameters has to be considered because of the fact that reinforced concrete is often a site-constructed composite, in contrast to the standard mill-fabricated beam and column sections in steel structures.

Photo 8 Two Union Square Towers, Seattle, Washington; 62 stories and 759 ft high. Concrete strength is 20,000 psi. Design by the NBBJ Group, Architects, Seattle, Washington. (Courtesy of Dr. Weston Hester and Turner Construction Company.)

A trial section has to be chosen for each critical location in a structural system. The trial section has to be analyzed to determine if its nominal resisting strength is adequate to carry the applied factored load. Since more than one trial is often necessary to arrive at the required section, the first design input step generates a series of trial-and-adjustment analyses.

The trial-and-adjustment procedures for the choice of a concrete section lead to the convergence of analysis and design. Hence every design is an analysis once a trial section is chosen. The availability of handbooks, charts, desktop and hand-held personal computers and programs supports this approach as a more efficient, compact, and speedy instructional method, compared with the traditional approach of treating the analysis of reinforced concrete separately from pure design.

Photo 9   The Trump Towers, Fifth Avenue, New York City; concrete strength in excess of 8000 psi.   (Courtesy of Concrete Industry Board.)

CONCRETE-PRODUCING MATERIALS

2

2.1 INTRODUCTION

To understand and interpret the total behavior of a composite element requires a knowledge of the characteristics of its components. Concrete is produced by the collective mechanical and chemical interaction of a large number of constituent materials. Hence a discussion of the functions of each of these components is vital prior to studying concrete as a finished product. In this manner, the designer and the materials engineer can develop skills for the choice of the proper ingredients and so proportion them as to obtain an efficient and desirable concrete satisfying the designer's strength and serviceability requirements.

This chapter presents a brief account of the concrete-producing materials: cement, fine and coarse aggregate, water, air, and admixtures. The cement manufacturing process, the composition of cement, the type and gradation of fine and coarse aggregate, and the function and importance of the water, air, and admixtures are reviewed. The reader is referred to books on concrete, such as the selected references at the end of this chapter, for further information.

Photo 10 LaGuardia Airport parking garage ramps, New York.

Photo 11   North Shore Synagogue, Glencoe, Illinois.   (Courtesy of Portland Cement Association.)

2.2 PORTLAND CEMENT

2.2.1 Manufacture

Portland cement is made of finely powdered crystalline minerals composed primarily of calcium and aluminum silicates.   The addition of water to these minerals produces a paste that, when hardened, becomes of stonelike strength.   Its specific gravity ranges between 3.12 and 3.16 and it weighs 94 lb/ft³, which is the unit weight of a commercial sack or bag of cement.

The raw materials that make cement are:

1. Lime (CaO), from limestone
2. Silica (SiO_2), from clay
3. Alumina (Al_2O_3), from clay

(with very small percentages of magnesia: MgO and sometimes some alkalis).   Iron oxide is occasionally added to the mixture to aid in controlling its composition.

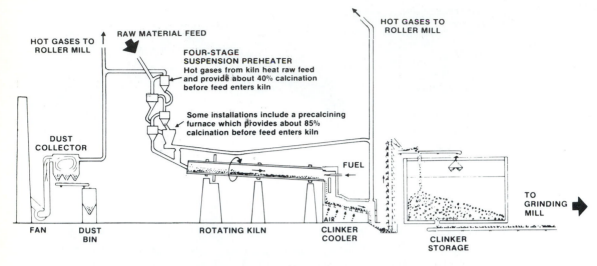

Figure 2.1 Portland cement manufacturing process. (From Ref. 2.5.)

The process of manufacture can be summarized as follows:

1. The raw mix of CaO, SiO$_2$, and Al$_2$O$_3$ is ground with other added minor ingredients either in dry or wet form. The wet form is called a *slurry*.
2. The mixture is fed into the upper end of a slightly inclined rotary kiln.
3. As the heated kiln operates, the material passes from its upper to its lower end at a predetermined, controlled rate.
4. The temperature of the mixture is raised to the point of incipient fusion, that is, the *clinkering temperature*. It is kept at that temperature until the ingredients combine to form at 2700°F the portland cement pellet product. The pellets, which range in size from $\frac{1}{16}$ to 2 in., are called *clinkers*.
5. The clinkers are cooled and ground to a powdery form.
6. A small percentage of gypsum is added during grinding to control or retard the setting time of the cement in the field.
7. Most of the final portland cement goes into silos for bulk shipment; some is packed in 94-lb bags for retail marketing.

Figure 2.1 illustrates schematically the manufacturing process of portland cement. The form and properties of the manufactured compound are described in the following sections.

2.2.2 Strength

The strength of cement is the result of a process of hydration. This chemical process results in recrystallization in the form of interlocking crystals producing the cement gel, which has high compressive strength when it hardens. Table 2.1

TABLE 2.1   PROPERTIES OF CEMENTS

Component	Rate of reaction	Heat liberated	Ultimate cementing value
Tricalcium silicate, C_3S	Medium	Medium	Good
Dicalcium silicate, C_2S	Slow	Small	Good
Tricalcium aluminate, C_3A	Fast	Large	Poor
Tetracalcium aluminoferrate, C_4AF	Slow	Small	Poor

shows the relative contribution of each component of the cement toward the rate of gain in strength. The early strength of portland cement is higher with higher percentages of C_3S. If moist curing is continuous, later strength levels will be greater, with higher percentages of C_2S. C_3A contributes to the strength developed during the first day after placing the concrete because it is the earliest to hydrate.

When portland cement combines with water during setting and hardening, lime is liberated from some of the compounds. The amount of lime liberated is approximately 20% by weight of the cement. Under unfavorable conditions, this might cause disintegration of a structure owing to leaching of the lime from the cement. Such a situation should be prevented by addition to the cement of a silicious mineral such as pozzolan. The added mineral reacts with the lime in the presence of moisture to produce strong calcium silicate.

2.2.3 Average Percentage Composition

Since there are different types of cement for various needs, it is necessary to study the percentage variation in the chemical composition of each type in order to interpret the reasons for variation in behavior. Table 2.2, studied in conjunction with Table 2.1, gives concise reasons for the difference in reaction of each type of cement when in contact with water.

2.2.4 Influence of Fineness of Cement on Strength Development

The size of the cement particles has a strong influence on the rate of reaction of cement with water. For a given weight of finely ground cement, the surface area of the particles is greater than that of the coarsely ground cement. This results in a greater rate of reaction with water and a more rapid hardening process for larger surface areas. This is one reason for the high early strength type III cement giving in 3 days a strength that type I gives in 7 days and a strength in 7 days that type I gives in 28 days.

TABLE 2.2 PERCENTAGE COMPOSITION OF PORTLAND CEMENTS

Type of cement	Component (%)							General characteristics
	C_3S	C_2S	C_3A	C_4AF	$CaSO_4$	CaO	MgO	
Normal: I	49	25	12	8	2.9	0.8	2.4	All-purpose cement
Modified: II	45	29	6	12	2.8	0.6	3.0	Comparative low heat liberation; used in large structures
High early strength: III	56	15	12	8	3.9	1.4	2.6	High strength in 3 days
Low heat: IV	30	46	5	13	2.9	0.3	2.7	Used in mass concrete dams
Sulfate resisting: V	43	36	4	12	2.7	0.4	1.6	Used in sewers and structures exposed to sulfates

2.2.5 Influence of Cement on the Durability of Concrete

Disintegration of concrete due to cycles of wetting, freezing, thawing, and drying and the propagation of resulting cracks is a matter of great importance. The presence of minute air voids throughout the cement paste increases the resistance of concrete to disintegration. This can be achieved by the addition of air-entraining admixtures to the concrete while mixing.

Disintegration due to chemicals in contact with the structure, such as in the case of port structures and substructures, can also be slowed down or prevented. Since the concrete in such cases is exposed to chlorides and sometimes sulfates of magnesium and sodium, it is sometimes necessary to specify sulfate-resisting cements. Usually, type II cement will be adequate for use in seawater structures.

2.2.6 Heat Generation during Initial Set

Since the different types of cement generate different degrees of heat at different rates, the type of structure governs the type of cement to be used. The bulkier and heavier in cross section the structure is, the less the generation of heat of hydration that is desired. In massive structures such as dams, piers, and caissons, type IV cement is more advantageous to use. From our discussion it is seen that the type of structure, the weather, and other conditions under which it is built and will exist are the governing factors in the choice of the type of cement that should be used.

2.3 WATER AND AIR

2.3.1 Water

Water is required in the production of concrete in order to precipitate chemical reaction with the cement, to wet the aggregate, and to lubricate the mixture for easy workability. Normally, drinking water can be used in mixing. Water having harmful ingredients, contamination, silt, oil, sugar, or chemicals is destructive to the strength and setting properties of cement. It can disrupt the affinity between the aggregate and the cement paste and can adversely affect the workability of a mix.

Since the character of the colloidal gel or cement paste is the result only of the chemical reaction between cement and water, it is not the proportion of water relative to the whole of the mixture of dry materials that is of concern, only the proportion of water relative to the cement. Excessive water leaves an uneven honeycombed skeleton in the finished product after hydration has taken place, while too little water prevents complete chemical reaction with the cement. The product in both cases is a concrete that is weaker than and inferior to normal concrete.

2.3.2 Entrained Air

With the gradual evaporation of excess water from the mix, pores are produced in the hardened concrete. If evenly distributed, these could give improved characteristics to the product. Very even distribution of pores by artificial introduction of finely divided, uniformly distributed air bubbles throughout the product is possible by adding air-entraining agents such as vinsol resin. Air entrainment increases workability, decreases density, increases durability, reduces bleeding and segregation, and reduces the required sand content in the mix. For these reasons, the percentage of entrained air should be kept at the required optimum value for the desired quality of the concrete. The optimum air content is 9% of the mortar fraction of the concrete. Air entraining in excess of 5% to 6% of the total mix proportionally reduces the concrete strength.

2.3.3 Water/Cement Ratio

To summarize the preceding discussion, strict control has to be maintained on the water/cement ratio and the percentage of air in the mix. As the water/cement ratio is the real measure of the strength of the concrete, it should be the principal criterion governing the design of most structural concretes. It is usually given as the ratio of weight of water to the weight of cement in the mix.

2.4 AGGREGATES

Aggregates are those parts of the concrete that constitute the bulk of the finished product. They comprise 60% to 80% of the volume of the concrete and have to be so graded that the whole mass of concrete acts as a relatively solid, homogeneous, dense combination, with the smaller sizes acting as an inert filler of the voids that exist between the larger particles.

Aggregates are of two types:

1. *Coarse aggregate:* gravel, crushed stone, or blast-furnace slag
2. *Fine aggregate:* natural or manufactured sand

Since the aggregate constitutes the major part of the mix, the more aggregate in the mix, the cheaper is the cost of the concrete, provided that the mix is of reasonable workability for the specific job for which it is used.

2.4.1 Coarse Aggregate

Coarse aggregate is classified as such if the smallest size of the particle is greater than $\frac{1}{4}$ in. (6 mm). Properties of the coarse aggregate affect the final strength of the hardened concrete and its resistance to disintegration, weathering, and other destructive effects. The mineral coarse aggregate must be clean of organic impurities and must bond well with the cement gel.

The common types of coarse aggregate are:

1. *Natural crushed stone.* This is produced by crushing natural stone or rock from quarries. The rock could be of igneous, sedimentary, or metamorphic type. Although crushed rock gives higher concrete strength, it is less workable in mixing and placing than are the other types.
2. *Natural gravel.* This is produced by the weathering action of running water on the beds and banks of streams. It gives less strength than crushed rock but is more workable.
3. *Artificial coarse aggregates.* These are mainly slag and expanded shale and are frequently used to produce lightweight concrete. They are by-products of other manufacturing processes, such as blast-furnace slag or expanded shale, or pumice for lightweight concrete.
4. *Heavyweight and nuclear-shielding aggregates.* With the specific demands of our atomic age and the hazards of nuclear radiation due to the increasing number of atomic reactors and nuclear power stations, special concretes have had to be produced to shield against x-rays, gamma rays, and neutrons. In such concretes, economic and workability considerations are not of prime importance. The main heavy, coarse aggregate types are steel punchings, barites, magnatites, and limonites.

Whereas concrete with ordinary aggregate weighs about 144 lb/ft³, concrete made with these heavy aggregates weighs from 225 to 330 lb/ft³. The property of heavyweight radiation-shielding concrete depends on the density of the compact product rather than primarily on the water/cement ratio criterion. In certain cases, high density is the only consideration, whereas in others both density and strength govern.

2.4.2 Fine Aggregate

Fine aggregate is a smaller filler made of sand. It ranges in size from No. 4 to No. 100 U.S. standard sieve sizes. A good fine aggregate should always be free of organic impurities, clay, or any deleterious material or excessive filler of size smaller than No. 100 sieve. It should preferably have a well-graded combination conforming to the American Society for Testing and Materials (ASTM) sieve analysis standards. For radiation-shielding concrete, fine steel shot and crushed iron ore are used as fine aggregate.

2.4.3 Grading of Normal-weight Concrete Mixes

The recommended grading of coarse and fine aggregates for normal-weight concretes is presented in Table 2.3.

TABLE 2.3 GRADING REQUIREMENTS FOR AGGREGATES IN NORMAL-WEIGHT CONCRETE (ASTM C-33)

U.S. standard sieve size	Percent passing				
	Coarse aggregate				Fine aggregate
	No. 4 to 2 in.	No. 4 to $1\frac{1}{2}$ in.	No. 4 to 1 in.	No. 4 to $\frac{3}{4}$ in.	
2 in.	95–100	100	—	—	—
$1\frac{1}{2}$ in.	—	95–100	100	—	—
1 in.	25–70	—	95–100	100	—
$\frac{3}{4}$ in.	—	35–70	—	90–100	—
$\frac{1}{2}$ in.	10–30	—	25–60	—	—
$\frac{3}{8}$ in.	—	10–30	—	20–55	100
No. 4	0–5	0–5	0–10	0–10	95–100
No. 8	0	0	0–5	0–5	80–100
No. 16	0	0	0	0	50–85
No. 30	0	0	0	0	25–60
No. 50	0	0	0	0	10–30
No. 100	0	0	0	0	2–10

TABLE 2.4 GRADING REQUIREMENTS FOR AGGREGATES IN LIGHTWEIGHT STRUCTURAL CONCRETE (ASTM C-330).

Size designation	Percentages (by weight) passing sieves having square openings								
	1 in. (25.0 mm)	$\frac{3}{4}$ in. (19.0 mm)	$\frac{1}{2}$ in. (12.5 mm)	$\frac{3}{8}$ in. (9.5 mm)	No. 4 (4.75 mm)	No. 8 (2.36 mm)	No. 16 (1.18 mm)	No. 50 (300 μm)	No. 100 (150 μm)
Fine aggregate No. 4 to 0	—	—	—	100	85–100	—	40–80	10–35	5–25
Coarse aggregate									
1 in. to No. 4	95–100	—	25–60	—	0–10	—	—	—	—
$\frac{3}{4}$ in. to No. 4	100	90–100	—	10–50	0–15	—	—	—	—
$\frac{1}{2}$ in. to No. 4	—	100	90–100	40–80	0–20	0–10	—	—	—
$\frac{3}{8}$ in. to No. 8	—	—	100	80–100	5–40	0–20	0–10	—	—
Combined fine and coarse aggregate									
$\frac{1}{2}$ in. to 0	—	100	95–100	—	50–80	—	—	5–20	2–15
$\frac{3}{8}$ in. to 0	—	—	100	90–100	65–90	35–65	—	10–25	5–15

2.4.4 Grading of Lightweight Concrete Mixes

The grading requirements for lightweight aggregate for structural concrete are given in Table 2.4.

2.4.5 Grading of Heavyweight and Nuclear-shielding Aggregates

The grading requirements to ensure heavyweight concrete are given in Table 2.5.

2.4.6 Unit Weights of Aggregates

The unit weight of the concrete depends on the unit weight of the aggregate, which in turn depends on the type of aggregate: whether it is normal, lightweight, or heavyweight (for radiation shielding). Table 2.6 gives the unit weights of the various aggregates and the corresponding unit weight of the concrete.

TABLE 2.5 GRADING REQUIREMENTS FOR COARSE AGGREGATE FOR AGGREGATE CONCRETE (ASTM C-637)

	Percentage passing	
Sieve size	Grading 1: for $1\frac{1}{2}$ in. (37.5 mm) maximum-size aggregate	Grading 2: for $\frac{3}{4}$ in. (19.0 mm) maximum-size aggregate
Coarse Aggregate		
2 in. (50 mm)	100	—
$1\frac{1}{2}$ in. (37.5 mm)	95–100	100
1 in. (25.0 mm)	40–80	95–100
$\frac{3}{4}$ in. (19.0 mm)	20–45	40–80
$\frac{1}{2}$ in. (12.5 mm)	0–10	0–15
$\frac{3}{8}$ in. (9.5 mm)	0–2	0–2
Fine Aggregate		
No. 8 (2.36 mm)	100	—
No. 16 (1.18 mm)	95–100	100
No. 30 (600 μm)	55–80	75–95
No. 50 (300 μm)	30–55	45–65
No. 100 (150 μm)	10–30	20–40
No. 200 (75 μm)	0–10	0–10
Fineness modulus	1.30–2.10	1.00–1.60

Data in Tables 2.3 to 2.5 reprinted with permission from the American Society for Testing and Materials, Philadelphia, Pa.

TABLE 2.6   UNIT WEIGHT OF AGGREGATES

Type	Unit weight of dry-rodded aggregate (lb/ft^3)[a]	Unit weight of concrete (lb/ft^3)[a]
Insulating concretes (perlite, vermiculite, etc.)	15–50	20–90
Structural lightweight	40–70	90–110
Normal weight	70–110	130–160
Heavyweight	>135	180–380

[a] 1 lb/ft^3 = 16.02 kg/m^3.

2.5 ADMIXTURES

Admixtures are materials other than water, aggregate, or hydraulic cement that are used as ingredients of concrete and that are added to the batch immediately before or during the mixing. Their function is to modify the properties of the concrete so as "to make it more suitable for the work at hand, or for economy, or for other purposes such as saving energy" (Ref. 2.6). The major types of admixtures can be summarized as follows:

1. Accelerating admixtures
2. Air-entraining admixtures
3. Water-reducing admixtures and set-controlling admixtures
4. Finely divided mineral admixtures
5. Admixtures for no-slump concretes
6. Polymers
7. Superplasticizers

2.5.1 Accelerating Admixtures

These admixtures are added to the concrete mix to reduce the time of setting and accelerate early strength development. The best known are calcium chlorides. Other accelerating chemicals include a wide range of soluble salts, such as chlorides, bromides, carbonates, and silicates, and some other organic compounds, such as triethanolamine.

It must be stressed that calcium chlorides should not be used where progressive corrosion of steel reinforcement can occur. The maximum dosage is 1% by weight of the portland cement.

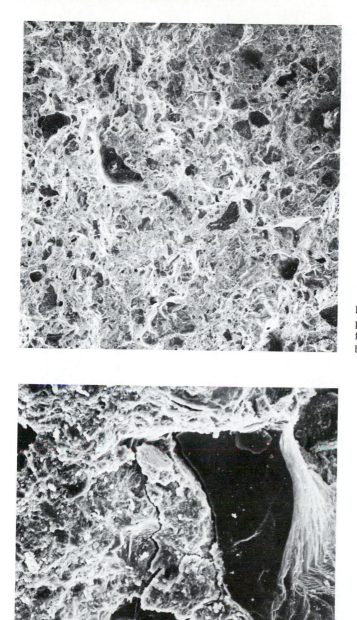

Photo 12 Scanning electron microscope photograph of polymer–cement mortar fracture surface under tension. (Tests by Nawy, Sun, and Sauer.)

Photo 13 Scanning electron microscope photograph of concrete fracture surface. (Tests by Nawy, Sun, and Sauer.)

2.5.2 Air-entraining Admixtures

These admixtures form minute bubbles 1 mm in diameter or smaller in the concrete or mortar during mixing and are used to increase the workability of the mix during placing and the frost resistance of the finished product.   Most air-entraining admixtures are in liquid form, although a few are powders, flakes, or semisolids. The amount of the admixture required to obtain a given air content depends on the shape and the grading of the aggregate used.   The finer the size of the aggregate, the larger is the percentage of admixture needed.   It is also governed by several other factors, such as type and condition of the mixer, use of fly ash or other pozzolans, and the degree of agitation of the mix.   It can be expected that air entrainment reduces the strength of the concrete.   Maintaining cement content and workability, however, offsets the partial reduction of strength because of the resulting reduction in the water/cement ratio.

2.5.3 Water-reducing and Set-controlling Admixtures

These admixtures increase the strength of the concrete.   They also enable reducing the cement content in proportion to the reduction in the water content.

Most admixtures of the water-reducing type are water soluble.   The water they contain becomes part of the mixing water in the concrete and is added to the total weight of water in the design of the mix.   It has to be emphasized that the proportion of the mortar to the coarse aggregate should always remain the same. Changes in the water content, air content, or cement content are compensated for by corresponding changes in the fine aggregate content so that the volume of the mortar remains the same.

2.5.4 Finely Divided Admixtures

These are mineral admixtures used to rectify deficiencies in concrete mix by providing missing fines from the fine aggregate; improving one or more qualities of the concrete, such as reducing permeability or expansion; and reducing the cost of concrete-making materials.   Such admixtures include hydraulic lime, slag cement, fly ash, and raw or calcined natural pozzolan.

2.5.5 Admixtures for No-slump Concrete

No-slump concrete is defined in Ref. 2.6 as a concrete with a slump of 1 in. (25 mm) or less immediately after mixing.   The choice of the admixture depends on the desired properties of the finished product, such as its effect on the plasticity, setting time and strength development, freeze–thaw effects, and strength and cost.

2.5.6 Polymers

These are new types of admixtures that enable producing concretes of very high strength up to a compressive strength of 15,000 psi or higher and a tensile splitting strength of 1500 psi or higher. Such concretes are generally produced using a polymerizing material through (1) modification of the concrete property through water reduction in the field or (2) impregnation and irradiation under elevated temperature in laboratory environment.

Polymer-modified concrete (PMC) is concrete made through the addition of resin and hardener as an "admixture." The principle is to replace part of the mixing water by the polymer so as to attain the high compressive strength and other qualities reported in detail in Ref. 2.7. The optimum polymer/concrete ratio by weight seems to lie within the range of 0.3 to 0.45 to achieve such high compressive strengths.

2.5.7 Superplasticizers

These are also new types of admixtures, which can be termed "high-range, water-reducing chemical admixtures." There are four types of plasticizers:

1. Sulfonated melamine formaldehyde condensates, with a chloride content of 0.005% (MSF)
2. Sulfonated naphthalene formaldehyde condensates, with negligible chloride content (NSF)
3. Modified lignosulfonates, which contain no chlorides

These admixtures are made from organic sulfonates and are termed superplasticizers in view of their considerable ability to facilitate reducing the water content in a concrete mix while simultaneously increasing the slump up to 8 in. (206 mm) or more. A dosage of 1% to 2% by weight of cement is advisable. Higher dosages can result in a reduction in compressive strength.

4. Other superplasticizers, such as sulfonic acid esters or other carbohydrate esters

A dosage of 1% to $2\frac{1}{2}$% by weight of cement is advisable. Higher dosages can result in a reduction in compressive strength unless the cement content is increased to balance this reduction effect. It should be noted that the superplasticizers exert their action by decreasing the surface tension of water and by equidirectional charging of the cement particles. These properties, coupled with the addition of silica fume, help the concrete to achieve high strength and water reduction without loss of workability.

2.5.8 Silica-fume Admixture Use in High-strength Concrete

Silica fume is generally accepted as an efficient admixture for high-strength concrete mixes. It is a new pozzolanic material that has received considerable attention recently in both research and application. Silica fume is a by-product resulting from the use of high-purity quartz with coal in the electric arc furnace in the production of silicon and ferrosilicon alloys. Its main constituent, fine spherical particles of silicon dioxide, makes it an ideal cement replacement, simultaneously raising the concrete strength. Being a waste product with relative ease of collection as compared to fly ash or slag, silica fume is gaining rapid popularity. Norway first experimented with this product, followed by other Scandinavian countries in the 1970s. Canada and the United States have embarked on extensive use of this product since the early 1980s.

Proportions of silica fume in concrete mixes vary from 5% to 30% by weight of the cement depending on strength and workability requirements. However, water demand is greatly increased with increasing proportion of silica fume, and high-range water reducers are essential to keep the water/cement ratio low in order to produce higher-strength, yet workable, concrete. Silica fume seems to attain a high early strength in about 3 to 7 days with relatively less increase in strength at 28 days. The strength-development pattern of flexural and tensile splitting strengths is similar to that of compressive strength gain for silica-fume-added concrete. The addition of silica fume to the mix can produce significant increase in strength, increased modulus of elasticity, and increased flexural strength.

SELECTED REFERENCES

2.1. American Society for Testing and Materials, *Annual Book of ASTM Standards,* Part 14, *Concrete and Mineral Aggregates,* ASTM, Philadelphia, 1993, 834 pp.

2.2. Popovices, S., *Concrete-Making Materials,* McGraw-Hill, New York, 1979, 370 pp.

2.3. ACI Committee 221, "Selection and Use of Aggregate for Concrete," *Journal of the American Concrete Institute,* Proc. Vol. 58, No. 5, 1961, pp. 513–542.

2.4. American Concrete Institute, *ACI Manual of Concrete Practice 1994*, Part I, *Materials*, ACI, Detroit, 1994.

2.5. Portland Cement Association, *Design and Control of Concrete Mixtures*, 12th ed., PCA, Skokie, Ill., 1979, 140 pp.

2.6. ACI Committee 212, "Admixtures for Concrete," in *ACI Manual of Concrete Practice 1983*, ACI 212.1 R-81, ACI, Detroit, 1983, 29 pp.

2.7. Nawy, E. G., Ukadike, M. M., and Sauer, J. A., "High Strength Field Modified Concretes," *Journal of the Structural Division, ASCE,* Vol. 103, No. ST12, December 1977, pp. 2307–2322.

2.8. American Concrete Institute, *Super-plasticizers in Concrete*, Special Publication SP-62, ACI, Detroit, 1979, 427 pp.

2.9. Nawy, E. G., *Fundamentals of High Strength Performance Concrete*, Longman U.K., London, 1995, 400 pp.

CONCRETE

3

3.1 INTRODUCTION

The general knowledge gained from Chapter 2 can now be utilized to design and obtain a concrete of characteristics and functions to suit a definite purpose. As should be realized by now, the proportioning and types of ingredients establish in part the quality of the concrete and hence the quality of the total structural system. Not only must good materials be chosen, but uniformity must be maintained in the whole product. The general characteristics of good concrete are summarized in the following sections.

3.1.1 Compactness

The space occupied by the concrete should, as much as possible, be filled with solid aggregate and cement gel free of honeycombing. Compactness may be the primary criterion for those types of concrete that intercept nuclear radiation.

Photo 14 Lake Point Tower, Chicago. (Courtesy of Portland Cement Association.)

Photo 15   Terminal building, Dulles Airport, Washington, D.C.   (Courtesy of Ammann & Whitney.)

3.1.2 Strength

Concrete should always have sufficient strength and internal resistance to the various types of failure.

3.1.3 Water/Cement Ratio

The water/cement ratio should be suitably controlled to give the required design strength.

3.1.4 Texture

Exposed concrete surfaces should have a dense and hard texture that can withstand adverse weather conditions.

3.1.5 Parameters Affecting Concrete Quality

To achieve the aforementioned properties, good quality control has to be exercised on the factors shown in Fig. 3.1.   The following are the most important parameters:

1. Quality of cement
2. Proportion of cement in relation to water in the mix
3. Strength and cleanliness of aggregate

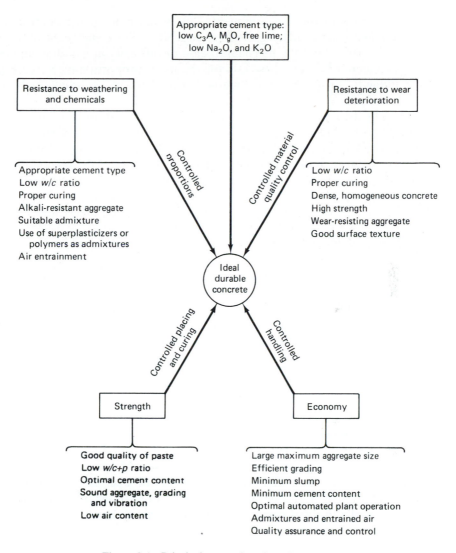

Figure 3.1   Principal properties of good concrete.

4. Interaction or adhesion between cement paste and aggregate
5. Adequate mixing of the ingredients
6. Proper placing, finishing, and compaction of the fresh concrete
7. Curing at a temperature not below 50°F while the placed concrete gains strength
8. Chloride content not to exceed 0.15% in reinforced concrete exposed to chlorides in service and 1% for dry protected concrete

A study of these requirements shows that most of the control actions have to be taken prior to placing the fresh concrete. Since such control is governed by the proportions and the mechanical ease or difficulty in handling and placing, the development of criteria based on the theory of proportioning for each mix should be studied. Most mix design methods have become essentially only of historical and academic value.

The two universally accepted methods for mix proportioning for normal-weight and lightweight concrete are the American Concrete Institute's methods of proportioning, described in the recommended practice for selecting proportions for normal-weight, heavyweight, and mass concrete and the recommended practice for selecting proportions for structural lightweight concrete (Refs. 3.1 and 3.2).

3.2 PROPORTIONING THEORY

Water/cement ratio (w/c ratio) theory states that for a given combination of materials and as long as workable consistency is obtained, the strength of concrete at a given age depends on the ratio of the weight of mixing water to the weight of cement. In other words, if the ratio of water to cement is fixed, the strength of concrete at a certain age is also essentially fixed, as long as the mixture is plastic and workable and the aggregate sound, durable, and free of deleterious materials. Whereas strength depends on the w/c ratio, economy depends on the percentage of aggregate present that would still give a workable mix. The aim of the designer should always be to get concrete mixtures of optimum strength at minimum cement content and acceptable workability. The lower the w/c ratio is, the higher the concrete strength.

Once the w/c ratio is established and the workability or consistency needed for the specific design is chosen, the rest should be simple manipulation with diagrams and tables based on large numbers of trial mixes. Such diagrams and tables allow an estimate of the required mix proportions for various conditions and permit predetermination on small unrepresentative batches.

3.2.1 ACI Method of Mix Design

The flow chart in Fig. 3.2 and the following design example best illustrate the mix design process using the ACI mix design method. One aim of the mix design is to produce workable concrete that is easy to place in the forms. A measure of the degree of consistency and extent of workability is the *slump*. In the slump test, the plastic concrete specimen is formed into a conical metal mold as described in ASTM Standard C-143. The mold is lifted, leaving the concrete to "slump," that is, to spread or drop in height. This drop in height is the slump measure of the degree of workability of the mix.

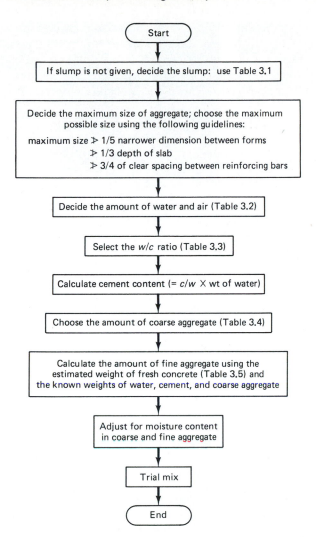

Figure 3.2 Flow chart for normal-weight concrete mix design.

3.2.2 Example 3.1: Mix Design of Normal-weight Concrete

Design a concrete mix using the following details:

Required strength: 4000 psi (27.6 MPa)

Type of structure: beam

Maximum size of aggregate $= \frac{3}{4}$ in. (18 mm)

Fineness modulus of sand $= 2.6$

Dry-rodded weight of coarse aggregate $= 100$ lb/ft^3

Moisture absorption 3% for coarse aggregate and 2% for fine aggregate

Photo 16   Left, $4\frac{1}{2}$-in. slump mix; right, $1\frac{1}{2}$-in. slump mix.

Solution
Required slump for beams (Table 3.1) = 3 in.

$$\text{maximum aggregate size (given)} = \tfrac{3}{4} \text{ in.}$$

For a slump between 3 and 4 in. and a maximum aggregate size of $\frac{3}{4}$ in.,

$$\text{weight of water required per cubic yard of concrete (Table 3.2)} = 340 \text{ lb/yd}^3$$

For the specified compression strength $f'_c = 4000$ psi,

$$w/c \text{ ratio (Table 3.3)} = 0.57$$

Table 3.4 is also needed if volumes instead of weights are used in the mix design calculations. Therefore,

$$\text{amount of cement required per cubic yard of concrete} = \frac{340}{0.57} = 596.5 \text{ lb/yd}^3$$

Using a sand fineness value of 2.6 and Table 3.4,

$$\text{volume of coarse aggregate} = 0.64 \text{ yd}^3$$

TABLE 3.1   RECOMMENDED SLUMPS FOR VARIOUS TYPES OF CONSTRUCTION

	Slump (in.)[a]	
Types of construction	Maximum[b]	Minimum
Reinforced foundation walls and footings	3	1
Plain footings, caissons, and substructure walls	3	1
Beams and reinforced walls	4	1
Building columns	4	1
Pavements and slabs	3	1
Mass concrete	2	1

[a]1 in. = 25.4 mm.
[b]May be increased 1 in. for methods of consolidation other than vibration.

TABLE 3.2   APPROXIMATE MIXING WATER AND AIR CONTENT REQUIREMENTS FOR DIFFERENT SLUMPS AND NOMINAL MAXIMUM SIZES OF AGGREGATES

Slump (in.)	Water (lb/yd³ of concrete for indicated nominal maximum sizes of aggregate)							
	$\frac{3}{8}$ in.[a]	$\frac{1}{2}$ in.[a]	$\frac{3}{4}$ in.[a]	1 in.[a]	$1\frac{1}{2}$ in.[a]	2 in.[a,b]	3 in.[b,c]	6 in.[b,c]
	Nonair-Entrained Concrete							
1 to 2	350	335	315	300	275	260	220	190
3 to 4	385	365	340	325	300	285	245	210
6 to 7	410	385	360	340	315	300	270	—
Approximate amount of entrapped air in nonair-entrained concrete (%)	3	2.5	2	1.5	1	0.5	0.3	0.2
	Air-Entrained Concrete							
1 to 2	305	295	280	270	250	240	205	180
3 to 4	340	325	305	295	275	265	225	200
6 to 7	365	345	325	310	290	280	260	—
Recommended average total air content[d] (percent for level of exposure)								
Mild exposure	4.5	4.0	3.5	3.0	2.5	2.0	1.5[e,f]	1.0[e,f]
Moderate exposure	6.0	5.5	5.0	4.5	4.5	4.0	3.5[e,f]	3.0[e,f]
Extreme exposure[g]	7.5	7.0	6.0	6.0	5.5	5.0	4.5[e,f]	4.0[e,f]

[a]These quantities of mixing water are for use in computing cement factors for trial batches. They are maximal for reasonably well shaped angular coarse aggregates graded within limits of accepted specifications.

[b]The slump values for concrete containing aggregate larger than 1½ in. are based on slump tests made after removal of particles larger than 1½ in. by wet screening.

[c]These quantities of mixing water are for use in computing cement factors for trial batches when 3-in. or 6-in. nominal maximum-size aggregate is used. They are average for reasonably well shaped coarse aggregates, well graded from coarse to fine.

[d]Additional recommendations for air content and necessary tolerances on air content for control in the field are given in a number of ACI documents, including ACI 201, 345, 318, 301, and 302. ASTM C-94 for ready-mixed concrete also gives air-content limits. The requirements in other documents may not always agree exactly, so in proportioning concrete consideration must be given to selecting an air content that will meet the needs of the job and also meet the applicable specifications.

[e]For concrete containing large aggregates that will be wet screened over the 1½-in. sieve prior to testing for air content, the percentage of air expected in the 1½-in.-minus material should be tabulated in the 1½-in. column. However, initial proportioning calculations should include the air content as a percent of the whole.

[f]When using large aggregate in low-cement-factor concrete, air entrainment need not be detrimental to strength. In most cases the mixing water requirement is reduced sufficiently to improve the water/cement ratio and thus to compensate for the strength-reducing effect of entrained-air concrete. Generally, therefore, for these large maximum sizes of aggregate, air contents recommended for extreme exposure should be considered even though there may be little or no exposure to moisture and freezing.

[g]These values are based on the criteria that 9% air is needed in the mortar phase of the concrete. If the mortar volume will be substantially different from that determined in this recommended practice, it may be desirable to calculate the needed air content by taking 9% of the actual mortar volume.

TABLE 3.3 RELATIONSHIP BETWEEN WATER/CEMENT RATIO
AND COMPRESSIVE STRENGTH OF CONCRETE

Compressive strength at 28 days[a] (psi)[b]	Water/cement ratio, by weight	
	Nonair-entrained concrete	Air-entrained concrete
6000	0.41	—
5000	0.48	0.40
4000	0.57	0.48
3000	0.68	0.59
2000	0.82	0.74

[a]Values are estimated average strengths for concrete containing not more than the percentage of air shown in Table 3.2. For a constant water/cement ratio, the strength of concrete is reduced as the air content is increased.

Strength is based on 6 in. × 12 in. cylinders moist-cured 28 days at 73.4 ± 3°F (23 ± 1.7°C) in accordance with Section 9(b) of ASTM C-31, "Making and Curing Concrete Compression and Flexure Test Specimens in the Field."

Relationship assumes maximum size of aggregate about $\frac{3}{4}$ to 1 in.; for a given source, strength produced for a given water/cement ratio will increase as maximum size of aggregate decreases.

[b]1000 psi = 6.9 MPa.

TABLE 3.4 VOLUME OF COARSE AGGREGATE PER UNIT OF VOLUME
OF CONCRETE

Maximum size of aggregate (in.)[a]	Volume of dry-rodded coarse aggregate[b] per unit volume of concrete for different fineness moduli of sand			
	2.40	2.60	2.80	3.00
$\frac{3}{8}$	0.50	0.48	0.46	0.44
$\frac{1}{2}$	0.59	0.57	0.55	0.53
$\frac{3}{4}$	0.66	0.64	0.62	0.60
1	0.71	0.69	0.67	0.65
$1\frac{1}{2}$	0.75	0.73	0.71	0.69
2	0.78	0.76	0.74	0.72
3	0.82	0.80	0.78	0.76
6	0.87	0.85	0.83	0.81

[a]1 in. = 25.4 mm.

[b]Volumes are based on aggregates in dry-rodded condition as described in ASTM C-29, "Unit Weight of Aggregate." These volumes are selected from empirical relationships to produce concrete with a degree of workability suitable for usual reinforced construction. For less workable concrete, such as that required for concrete pavement construction, they may be increased about 10%. For more workable concrete, the coarse aggregate content may be decreased up to 10%, provided that the slump and water/cement ratio requirements are satisfied.

Using the dry-rodded weight of 100 lb/ft³ for coarse aggregate,

$$\text{weight of coarse aggregate} = (0.64) \times (27 \text{ ft}^3/\text{yd}^3) \times 100$$
$$= 1728 \text{ lb/yd}^3$$

estimated weight of fresh concrete for aggregate of $\frac{3}{4}$-in. maximum size
(Table 3.5) = 3960 lb/yd³

TABLE 3.5   FIRST ESTIMATE OF WEIGHT OF FRESH CONCRETE

Maximum size of aggregate (in.)[a]	First estimate of concrete weight[b] (lb/yd³)[c]	
	Nonair-entrained concrete	Air-entrained concrete
$\frac{3}{8}$	3840	3690
$\frac{1}{2}$	3890	3760
$\frac{3}{4}$	3960	3840
1	4010	3900
$1\frac{1}{2}$	4070	3960
2	4120	4000
3	4160	4040
6	4230	4120

[a]1 in. = 25.4 mm.

[b]Values calculated and presented below are for concrete of medium richness (550 lb of cement per cubic yard) and medium slump with aggregate specific gravity of 2.7. Water requirements are based on values for 3- to 4-in. slump in Table 5.3.2 of ASTM C-143. If desired, the estimated weight may be refined as follows if necessary information is available: for each 10-lb difference in mixing water from Table 5.3.2, values for 3- to 4-in. slump, correct the weight per cubic yard 15 lb in the opposite direction; for each 100-lb difference in cement content from 550 lb, correct the weight per cubic yard 15 lb in the same direction; for each 0.1 by which aggregate specific gravity deviates from 2.7, correct the concrete weight 100 lb in the same direction.

weight of fresh concrete per cubic yard, lb

$$= 16.85 G_a(100 - A) + C\left(1 - \frac{G_a}{G_c}\right) - W(G_a - 1)$$

where G_a = weighted average specific gravity of combined fine and coarse aggregate, bulk saturated surface dry density

G_c = specific gravity of cement (generally 3.15)

A = air content, %

W = mixing water requirement, lb/yd³

C = cement requirement, lb/yd³

[c]1 lb/yd³ = 0.6 kg/m³.

$$\text{weight of sand} = [\text{weight of fresh concrete} - \text{weights of (water} +$$
$$\text{cement} + \text{coarse aggregate)}]$$
$$= 3960 - 340 - 596.5 - 1728 = 1295.5 \text{ lb}$$
$$\text{net weight of sand to be taken} = 1.02 \times 1295.5$$
$$(\text{moisture absorption 2\%}) = 1321.41 \text{ lb}$$
$$\text{net weight of gravel} = 1.03 \times 1728$$
$$(\text{moisture absorption 3\%}) = 1779.84 \text{ lb}$$
$$\text{net weight of water} = 340 - 0.02 \times 1295.5 - 0.03 \times 1728$$
$$= 262.25 \text{ lb}$$

For 1 yd³ of concrete:

$$\text{cement} = 596.5 \text{ lb} \simeq 600 \text{ lb (273 kg)}$$
$$\text{sand} = 1321.41 \text{ lb} \simeq 1320 \text{ lb (600 kg)}$$
$$\text{gravel} = 1779.84 \text{ lb} \simeq 1780 \text{ lb (810 kg)}$$
$$\text{water} = 262.25 \text{ lb} \simeq 260 \text{ lb (120 kg)}$$

3.3 PCA METHOD OF MIX DESIGN

The mix design method proposed by the Portland Cement Association (PCA) is essentially similar to the ACI method. Generally, results would be very close once trial batches are prepared in the laboratory. The PCA publication listed in the references gives the details of the method as well as other information on properties of the ingredients.

3.4 MIX DESIGN FOR STRUCTURAL LIGHTWEIGHT CONCRETE

Structural lightweight concrete can best be defined as concrete having a 28-day compressive strength in excess of 2000 psi and an air-dry unit weight less than 115 lb/ft³. The coarse aggregate used is primarily expanded shale, slate, slags, and so on, and the same principles and procedures used in normal-weight concrete are applicable to this type of concrete. Air entrainment is very desirable, if not mandatory. A recommended percentage of air-entraining agents of at least 6% is necessary to give the product acceptable weathering qualities.

3.5 ESTIMATING COMPRESSIVE STRENGTH OF A TRIAL MIX USING THE SPECIFIED COMPRESSIVE STRENGTH

The compressive strength for which the trial mix is designed is not the strength specified by the designer. The mix should be overdesigned to assure that the actual structure has concrete with specified minimum compressive strength. The

Photo 17   Cylinder compression test.

extent of mix overdesign depends on the degree of quality control available in the mixing plant.

ACI Committee 318 specifies a systematic way of determining the compressive strength for mix designs using the specified compressive strength, f_c'. The procedure is presented in a self-explanatory flow-chart form in Fig. 3.3. The cylinder compressive strength f_c' (see Section 3.7) is the test result at 28 days after casting normal-weight concrete. Mix design has to be based on an adjusted higher value f_{cr}'. This adjusted cylinder compressive strength f_{cr}' for which a trial mix design is calculated depends on the extent of field data available.

1. *No cylinder test records available.* If field-strength test records for the specified class (or within 1000 psi of the specified class) of concrete are not available, the trial mix strength f_{cr}' can be calculated by increasing the cylinder compressive strength f_c' by a reasonable value depending on the extent of spread in values expected in the supplied concrete. Such a spread can be quantified by the standard deviation values represented by the values in excess on f_c' in Table 3.6. Table 3.7 can then be used to obtain the water/cement ratio needed for the required cylinder strength value f_c'.

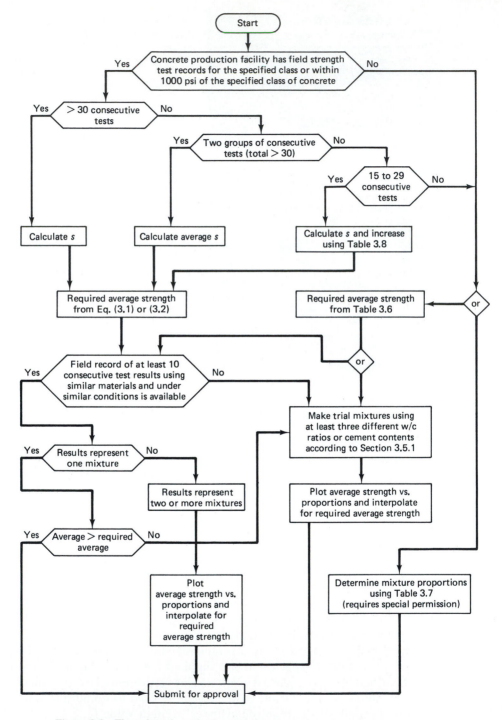

Figure 3.3 Flow chart for selection and documentation of concrete proportions.

TABLE 3.6 REQUIRED AVERAGE COMPRESSIVE STRENGTH WHEN DATA ARE NOT AVAILABLE TO ESTABLISH A STANDARD DEVIATION

Specified compressive strength, f'_c (psi)	Required average compressive strength, f'_{cr} (psi)[a]
Less than 3000	$f'_c + 1000$
3000–5000	$f'_c + 1200$
More than 5000	$f'_c + 1400$

[a]1000 psi = 6.9 MPa.

TABLE 3.7 MAXIMUM PERMISSIBLE WATER/CEMENT RATIOS FOR CONCRETE WHEN STRENGTH DATA FROM FIELD EXPERIENCE OR TRIAL MIXTURES ARE NOT AVAILABLE

Specified compressive strength,[a] f'_c (psi)[b]	Absolute water/cement ratio by weight	
	Nonair-entrained concrete	Air-entrained concrete
2500	0.67	0.54
3000	0.58	0.46
3500	0.51	0.40
4000	0.44	0.35
4500	0.38	c
5000	c	c

[a]28-day strength.   With most materials, the water/cement ratios shown will provide average strengths greater than those calculated using Eqs. 3.1 and 3.2.

[b]1000 psi = 6.9 MPa.

[c]For strengths above 4500 psi for nonair-entrained concrete and 4000 psi for air-entrained concrete, mix proportions should be established using trial mixes.

2. *Data available on more than 30 consecutive cylinder tests.*   If more than 30 consecutive test results are available, Eqs. 3.1, 3.2, and 3.3a in Section 3.5.2 can be used to establish the required mix strength, f'_{cr}, from f'_c.   If two groups of consecutive test results with a total of more than 30 are available, f'_{cr} can be obtained using Eqs. 3.1, 3.2, and 3.3b.

3. *Data available on fewer than 30 consecutive cylinder tests.*   If the number of consecutive test results available is fewer than 30 and more than 15, Eqs. 3.1, 3.2, and 3.3a should be used in conjunction with Table 3.8.   Essentially, the designer should calculate the standard deviation s using Eq. 3.3a, multiply the s value by a magnification factor provided in Table 3.8, and use the magnified s in Eqs. 3.1 and 3.2.   In this manner, the expected degree of spread of cylinder test values as measured by the standard deviation s is well accounted for.

TABLE 3.8 MODIFICATION FACTOR
FOR STANDARD DEVIATION WHEN FEWER
THAN 30 TESTS ARE AVAILABLE

Number of tests[a]	Modification factor for standard deviation[b]
Less than 15	Use Table 3.6
15	1.16
20	1.08
25	1.03
30 or more	1.00

[a]Interpolate for intermediate number of tests.
[b]Modified standard deviation to be used to determine required average strength f'_{cr} in Eqs. 3.1 and 3.2.

3.5.1 Recommended Proportions for Concrete Strength f'_{cr}

Once the required average strength f'_{cr} for mix design is determined, the actual mix can be established to obtain this strength using either existing field data or a basic trial mix design.

1. *Use of field data.* Field records of existing f'_{cr} values can be used if at least 10 consecutive test results are available. The test records should cover a period of time of at least 45 days. The materials and conditions of the existing field mix data should be the same as the ones to be used in the proposed work.

2. *Trial mix design.* If the field data are not available, trial mixes should be used to establish the maximum water/cement ratio or minimum cement content for designing a mix that produces a 28-day f'_{cr} value. In this procedure, the following requirements have to be met:
   (a) Materials used and age of testing should be the same for the trial mix and the concrete used in the structure.
   (b) At least three water/cement ratios or three cement contents should be tried in the mix design. The trial mixes should result in the required f'_{cr}. Three cylinders should be tested for each *w/c* ratio and each cement content tried.
   (c) The slump and air content should be within ± 0.75 in. and 0.5% of the permissible limits.
   (d) A plot is constructed of the compressive strength at the designated age versus the cement content or water/cement ratio, from which one can then choose the *w/c* ratio or the cement content that can give the average f'_{cr} value required.

3.5.2 Trial Mix Design for Average Strength
When Prior Field-strength Data Are Available

If field test data are available for more than 30 consecutive tests, the trial mix should be designed for compressive strength f'_{cr} calculated from

$$f'_{cr} = f'_c + 1.34s \qquad (3.1)$$

or
$$f'_{cr} = f'_c + 2.33s - 500 \qquad (3.2)$$

The larger value of f'_{cr} from Eqs. 3.1 and 3.2 should be used in designing the mix, with the expectation of attaining the minimum f'_c specified design compressive strength. The standard deviation s is defined by the expression

$$s = \left[\frac{\sum (f_{ci} - \bar{f}_c)^2}{n-1} \right]^{1/2} \qquad (3.3a)$$

where f_{ci} = individual strength
$\bar{f}_c$ = average of the n specimens

If two test records are used to determine the average strength, the standard deviation becomes

$$s = \left[\frac{(n_1 - 1)s_1^2 + (n_2 - 1)s_2^2}{n_1 + n_2 - 2} \right]^{1/2} \qquad (3.3b)$$

where s_1, s_2 = standard deviations calculated from two test records, 1 and 2, respectively
n_1, n_2 = number of tests in each test record, respectively

If the number of test results available is fewer than 30 and more than 15, the value of s used in Eqs. 3.1 and 3.2 should be multiplied by the appropriate modification factor value given in Table 3.8.

3.5.3 Example 3.2: Calculation of Design Strength
for Trial Mix

Calculate the average compressive strengths f'_{cr} for the design of a concrete mix if the specified compressive strength f'_c is 5000 psi (34.5 MPa) such that (a) the standard deviation obtained using more than 30 consecutive tests is 500 psi (3.45 MPa); (b) the standard deviation obtained using 15 consecutive tests is 450 psi (3.11 MPa); (c) records of prior cylinder test results are not available.

Solution   (a) Using Eq. 3.1,

$$f'_{cr} = 5000 + 1.34 \times 500$$

$$= 5670 \text{ psi}$$

Using Eq. 3.2,

$$f'_{cr} = 5000 + 2.33 \times 500 - 500$$
$$= 5665 \text{ psi}$$

Hence the required trial mix strength $f'_{cr} = 5670$ psi (39.12 MPa).

(b) $s = 450$ psi in 15 tests. From Table 3.8, the modification factor for s is 1.16. Hence the value of standard deviation to be used in Eqs. 3.1 and 3.2 is 1.16 × 450 = 522 psi (3.6 MPa). Using Eq. 3.1,

$$f'_{cr} = 5000 + 1.34 \times 522$$
$$= 5700 \text{ psi}$$

Using Eq. 3.2,

$$f'_{cr} = 5000 + 2.33 \times 522 - 500$$
$$= 5716 \text{ psi}$$

Hence the required trial mix strength $f'_{cr} = 5716$ psi (39.44 MPa).

(c) Records of prior test results are not available. Using Table 3.6,

$$f'_{cr} = f'_c + 1200 \quad \text{for 5000-psi concrete}$$

Hence the trial mix strength = 5000 + 1200 = 6200 psi (42.78 MPa).

If the mixing plant keeps good records of its cylinder test results over a long period, the required trial mix strength f'_{cr} can be reduced as a result of such quality control, hence reducing costs for the owner.

3.6 MIX DESIGNS FOR NUCLEAR-SHIELDING CONCRETE

Whereas from the foregoing discussion it is seen that the design criterion was the water/cement ratio, in concrete used for shielding against x-rays, gamma rays, and neutrons, the criterion is compactness or density of mix, regardless of workability. To achieve maximum density, tests have been conducted on various mixes using crushed magnatite ore or fine steel shot instead of sand and steel punchings, magnatites, barites, or limonites instead of stone, as discussed previously. Results of these tests for both compactness and strength have shown that the *w/c* ratio has to be limited to 3.5 to 4.0 gal of water per bag of cement.

3.7 QUALITY TESTS ON CONCRETE

3.7.1 Workability or Consistency

Possible tests for workability or consistency include:

1. Slump test by means of the standard ASTM Code. The slump in inches recorded in the mix indicates its workability.

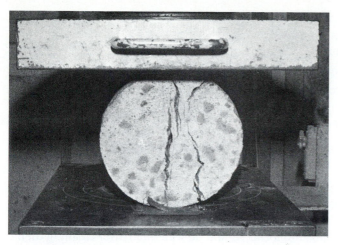

Photo 18   Tensile splitting test.

    2. Remolding tests using Power's flow table.
    3. Kelley's ball apparatus.

The first method is the accepted ASTM standard.

3.7.2 Air Content

Measurement of the air content in fresh concrete is always necessary, especially when air-entraining agents are used.

3.7.3 Compressive Strength of Hardened Concrete

This is done by loading cylinders 6 in. in diameter and 12 in. high in compression perpendicular to the axis of the cylinder.

3.7.4 Flexural Strength of Plain Concrete Beams

This test is performed by three-point loading of plain concrete beams of size 6 in. × 6 in. × 18 in. that have spans three times their depth.

3.7.5 Tensile Splitting Tests

These tests are performed by loading the standard 6 in. × 12 in. cylinder by a line load perpendicular to its longitudinal axis, with the cylinder placed horizontally on the testing machine platten.   The tensile splitting strength can be defined as

$$f'_t = \frac{2P}{\pi DL} \tag{3.4}$$

where P = total value of the line load registered by the testing machine

D = diameter of the concrete cylinder

L = cylinder height

The results of all these tests give the designer a measure of the expected strength of the designed concrete in the built structure.

3.8 PLACING AND CURING OF CONCRETE

3.8.1 Placing

The techniques necessary for placing concrete depend on the type of member to be cast; that is, whether it is a column, a beam, a wall, a slab, a foundation, a mass concrete dam, or an extension of previously placed and hardened concrete. For beams, columns, and walls, the forms should be well oiled after cleaning them, and the reinforcement should be cleared of rust and other harmful materials. In foundations, the earth should be compacted and thoroughly moistened to about 6 in. in depth to avoid absorption of the moisture present in the wet concrete. Concrete should always be placed in horizontal layers that are compacted by means of high-frequency, power-driven vibrators of either the immersion or external type, as the case requires, unless it is placed by pumping. Keep in mind, however, that overvibration can be harmful since it could cause segregation of the aggregate and bleeding of the concrete.

3.8.2 Curing

As seen in Chapter 2, hydration of the cement takes place in the presence of moisture at temperatures above 50°F. It is necessary to maintain such a condition in order that the chemical hydration reaction can take place. If drying is too rapid, surface cracking takes place. This would result in reduction of concrete strength due to cracking, as well as failure to attain full chemical hydration.

To facilitate good curing conditions, any of the following methods can be used:

1. Continuously sprinkling with water.
2. Ponding with water.
3. Covering the concrete with wet burlap, plastic film, or waterproof curing paper.
4. Using liquid membrane-forming curing compounds to retain the original moisture in the wet concrete.
5. Steam curing in cases where the concrete member is manufactured under factory conditions, such as in cases of precast beams and pipes and prestressed girders and poles. Steam-curing temperatures are about 150°F. Curing time is usually 1 day, compared to the 5 to 7 days necessary when using the other methods.

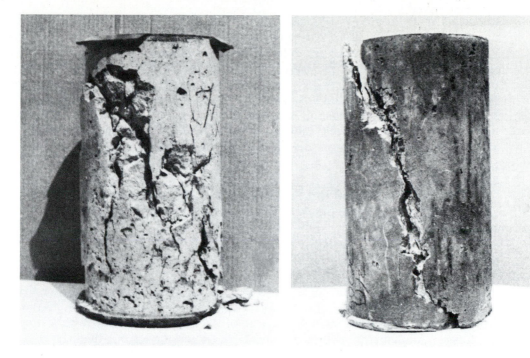

Photo 19   Concrete cylinders tested to failure in compression.   Left, low-epoxy-cement content; right, high-epoxy-cement content.   (Tests by Nawy, Sun, and Sauer.)

3.9 PROPERTIES OF HARDENED CONCRETE

The mechanical properties of hardened concrete can be classified as (1) short-term or instantaneous properties and (2) long-term properties.   The short-term properties can be enumerated as (1) strength in compression, tension, and shear and (2) stiffness measured by modulus of elasticity.   The long-term properties can be classified in terms of creep and shrinkage.   The following sections present some details of the aforementioned properties.

3.9.1 Compressive Strength

Depending on the type of mix, the properties of aggregate, and the time and quality of curing, compressive strengths of concrete can be obtained up to 15,000 psi or more.   Commercial production of concrete with ordinary aggregate is usually in the range from 3000 to 10,000 psi, with the most common concrete strengths in the range from 3000 to 6000 psi.

The compressive strength, f'_c, is based on standard 6 in. $\times$ 12 in. cylinders cured under standard laboratory conditions and tested at a specified rate of loading at 28 days of age.   The standard specifications used in the United States are usually taken from ASTM C-39.   It should be mentioned that the strength of concrete in the actual structure may not be the same as that of the cylinder because of the difference in compaction and curing conditions.

The ACI Code specifies for a strength test the average of two cylinders from the same sample tested at the same age, which is usually 28 days. As for the frequency of testing, the Code specifies that the strength level of an individual class of concrete can be considered as satisfactory if (1) the average of all sets of three consecutive strength tests equal or exceed required f'_c, and (2) no individual strength test (average of two cylinders) falls below the required f'_c by more than 500 psi. The average concrete strength for which a concrete mix must be designed should exceed f'_c by an amount that depends on the uniformity of plant production, as explained in Section 3.5.

It must be emphasized that the design f'_c should not be the average cylinder strength. The design value should be chosen as the conceivable minimum cylinder strength.

3.9.2 Tensile Strength

The tensile strength of concrete is relatively low. A good approximation for the tensile strength f_{ct} is $0.10f'_c < f_{ct} < 0.20f'_c$. It is more difficult to measure tensile strength than compressive strength because of the gripping problems with testing machines. A number of methods are available for tension testing, the most commonly used method being the cylinder splitting test or Brazilian test.

For members subjected to bending, the value of the modulus of rupture f_r rather than tensile splitting strength f'_t is used in design. The modulus of rupture is measured by testing to failure plain concrete beams 6 in. square in cross section, having a span of 18 in. and loaded at its third points (ASTM C-78). The modulus of rupture has a higher value than the tensile splitting strength. The ACI specifies a value of $7.5\sqrt{f'_c}$ for the modulus of rupture of normal-weight concrete.

In most cases, lightweight concrete has a lower tensile strength than does normal-weight concrete. Following are the ACI Code stipulations for lightweight concrete.

1. If the splitting tensile strength f_{ct} is specified,

$$f_r = 1.09f_{ct} \leq 7.5\sqrt{f'_c}$$

2. If f_{ct} is not specified, use a factor of 0.75 for all lightweight concrete and 0.85 for sand–lightweight concrete. Linear interpolation may be used for mixtures of natural sand and lightweight fine aggregate.

3.9.3 Shear Strength

Shear strength is more difficult to determine experimentally than the tests discussed previously because of the difficulty in isolating shear from other stresses. This is one reason for the large variation in shear-strength values reported in the literature,

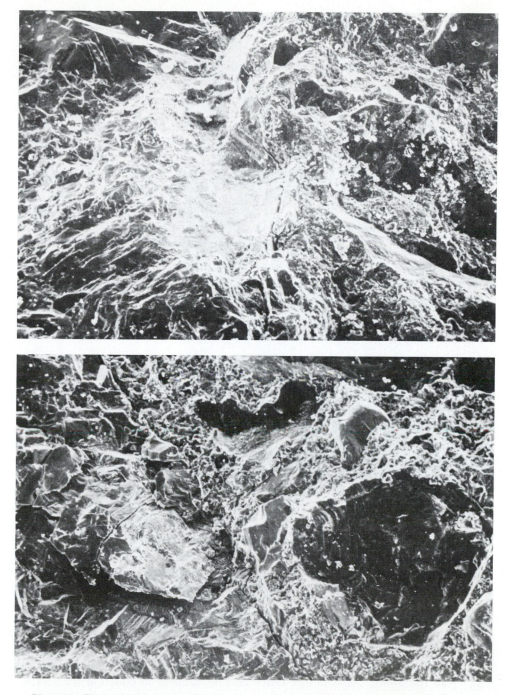

Photo 20   Electron microscope photographs of concrete from specimens in the preceding photographs.   (Tests by Nawy et al.)

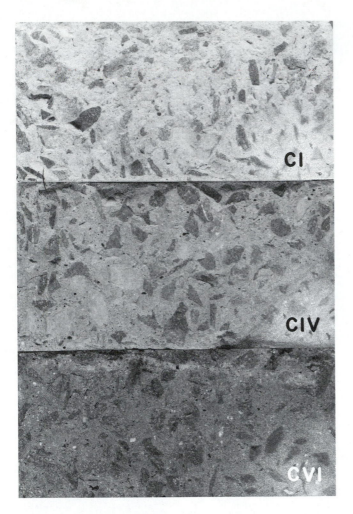

Photo 21 Fracture surfaces in tensile splitting tests of concretes with different *w/c* contents. Specimens CI and CIV have higher *w/c* content, hence more bond failures than specimen CVI. (Tests by Nawy et al.)

varying from 20% of the compressive strength in normal loading to a considerably higher percentage of up to 85% of the compressive strength in cases where direct shear exists in combination with compression. Control of a structural design by shear strength is significant only in rare cases, since shear stresses must ordinarily be limited to continually lower values in order to protect the concrete from failure in diagonal tension.

3.9.4 Stress–Strain Curve

Knowledge of the stress–strain relationship of concrete is essential for developing all the analysis and design terms and procedures in concrete structures. Figure 3.4a shows a typical stress–strain curve obtained from tests using cylindrical concrete specimens loaded in uniaxial compression over several minutes. The first portion of the curve, to about 40% of the ultimate strength f'_c, can be considered essentially linear for all practical purposes.   After approximately 70% of the failure stress, the material loses a large portion of its stiffness, thereby increasing the curvilinearity of the diagram.   At ultimate load, cracks parallel to the direction of loading become distinctly visible, and most concrete cylinders (except those with very low strengths) fail suddenly shortly thereafter.   Figure 3.4b shows the stress–strain curves of concrete of various strengths reported by the Portland Cement Association.   It can be observed that (1) the lower the strength of concrete, the higher the failure strain; (2) the length of the initial relatively linear portion increases with the increase in the compressive strength of concrete; and (3) there is an apparent reduction in ductility with increased strength.

3.9.5 Modulus of Elasticity

Since the stress–strain curve shown in Fig. 3.5 is curvilinear at a very early stage of its loading history, Young's modulus of elasticity can be applied only to the tangent of the curve at the origin.   The initial slope of the tangent to the curve is defined as the initial tangent modulus, and it is also possible to construct a tangent modulus at any point of the curve.   The slope of the straight line that connects the origin to a given stress (about $0.4f'_c$) determines the secant modulus of elasticity of concrete.   This value, termed in design calculation the *modulus of elasticity*, satisfies the practical assumption that strains occurring during loading can be considered basically elastic (completely recoverable on unloading) and that any subsequent strain due to the load is regarded as creep.

The ACI Code gives the following expressions for calculating the secant modulus of elasticity of concrete (E_c).

$$E_c = 33w_c^{1.5}\sqrt{f'_c} \qquad \text{for } 90 < w_c < 155 \text{ lb/ft}^3$$

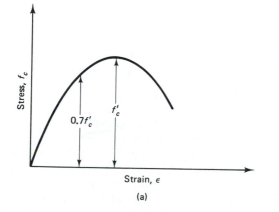

(a)

Figure 3.4   (a) Typical stress–strain curve of concrete; (b) stress–strain curves for various concrete strengths.

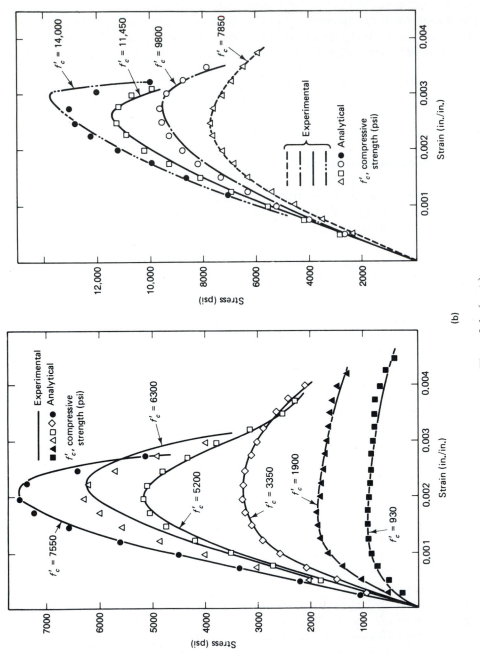

Figure 3.4 (cont.)

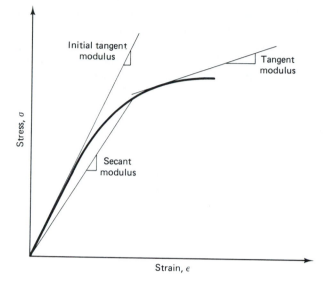

Stress, σ

Strain, ϵ

Initial tangent modulus

Tangent modulus

Secant modulus

Figure 3.5   Tangent and secant moduli of concrete.

where w_c is the density of concrete in pounds per cubic foot (1 lb/ft³ = 16.02 kg/m³) and f_c' is the compressive cylinder strength in psi.   For normal-weight concrete,

$$E_c = 57,000\sqrt{f_c'} \text{ psi} \quad \text{or} \quad E_c = 4730\sqrt{f_c'} \text{ N/mm}^2$$

These expressions are valid only in general terms, since the value of the modulus of elasticity is also affected by factors other than loads, such as moisture in the concrete specimen, the water/cement ratio, age of the concrete, and temperature.   Therefore, for special structures such as arches, tunnels, and tanks, the modulus of elasticity needs to be determined from test results.

Limited work exists on the determination of the modulus of elasticity in tension because the low tensile strength of concrete is normally disregarded in calculations.   It is, however, valid to assume within those limitations that the value of the modulus in tension is equal to that in compression.

3.9.6 Shrinkage

Basically, there are two types of shrinkage: plastic shrinkage and drying shrinkage. *Plastic shrinkage* occurs during the first few hours after placing fresh concrete in the forms.   Exposed surfaces such as floor slabs are more easily affected by exposure to dry air because of their large contact surface.   In such cases, moisture evaporates faster from the concrete surface than it is replaced by the bleed water from the lower layers of the concrete elements.

Drying shrinkage, on the other hand, occurs after the concrete has already attained its final set and a good portion of the chemical hydration process in the cement gel has been accomplished. Drying shrinkage is the decrease in the volume of a concrete element when it loses moisture by evaporation. The opposite phenomenon, that is, volume increase through water absorption, is termed *swelling*. In other words, shrinkage and swelling represent water movement out of or into the gel structure of a concrete specimen due to the difference in humidity or saturation levels between the specimen and the surroundings irrespective of the external load.

Shrinkage is not a completely reversible process. If a concrete unit is saturated with water after having fully shrunk, it will not expand to its original volume. Figure 3.6 relates the increase in shrinkage strain ϵ_{sh} with time. The rate decreases with time since older concretes are more resistant to stress and consequently undergo less shrinkage, such that the shrinkage strain becomes almost asymptotic with time.

Several factors affect the magnitude of drying shrinkage:

1. *Aggregate*. The aggregate acts to restrain the shrinkage of the cement paste; hence concretes with high aggregate content are less vulnerable to shrinkage. In addition, the degree of restraint of a given concrete is determined by the properties of aggregates; those with high modulus of elasticity or with rough surfaces are more resistant to the shrinkage process.

2. *Water/cement ratio*. The higher the water/cement ratio, the higher the shrinkage effects. Figure 3.7 is a typical plot relating aggregate content to water/cement ratio.

3. *Size of the concrete element*. Both the rate and total magnitude of shrinkage decrease with an increase in the volume of the concrete element. However, the duration of shrinkage is longer for larger members since more time is needed for drying to reach the internal regions. It is possible that 1 year may be needed for the drying process to begin at a depth of 10 in. from the exposed surface and 10 years to begin at 24 in. below the external surface.

4. *Medium ambient conditions*. The relative humidity of the medium affects greatly the magnitude of shrinkage; the rate of shrinkage is lower at high states of relative humidity. The environment temperature is another factor, in that shrinkage becomes stabilized at low temperatures.

5. *Amount of reinforcement*. Reinforced concrete shrinks less than plain concrete; the relative difference is a function of the reinforcement percentage.

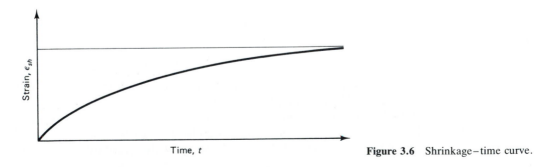

Time, t

Figure 3.6 Shrinkage–time curve.

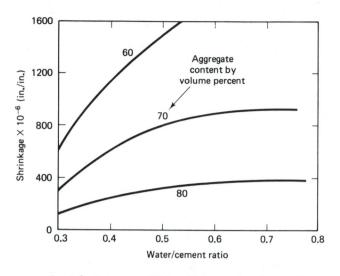

Figure 3.7   *w/c* ratio and aggregate content effect on shrinkage.

6. *Admixtures*. This effect varies depending on the type of admixture. An accelerator such as calcium chloride, used to accelerate the hardening and setting of the concrete, increases the shrinkage. Pozzolans can also increase the drying shrinkage, whereas air-entraining agents have little effect.

7. *Type of cement*. Rapid-hardening cement shrinks somewhat more than other types, while shrinkage-compensating cements minimize or eliminate shrinkage cracking if used with restraining reinforcement.

8. *Carbonation*. Carbonation shrinkage is caused by the reaction between carbon dioxide (CO_2) present in the atmosphere and that present in the cement paste. The amount of the combined shrinkage varies according to the sequence of occurrence of carbonation and drying processes. If both phenomena take place simultaneously, less shrinkage develops. The process of carbonation, however, is dramatically reduced at relative humidities below 50%.

3.9.7 Creep

Creep, or lateral material flow, is the increase in strain with time due to a sustained load. Initial deformation due to load is the *elastic strain*, while the additional strain due to the same sustained load is the *creep strain*. This practical assumption is acceptable since the initial recorded deformation includes few time-dependent effects.

Figure 3.8 illustrates the increase in creep strain with time, and as in the case of shrinkage, it can be seen that creep decreases with time. Creep cannot be observed directly and can be determined only by deducting elastic strain and shrinkage strain from the total deformation. Although shrinkage and creep are not independent phenomena, it can be assumed that superposition of strains is valid; hence

$$\text{total strain } (\epsilon_t) = \text{elastic strain } (\epsilon_e) + \text{creep } (\epsilon_c) + \text{shrinkage } (\epsilon_{sh})$$

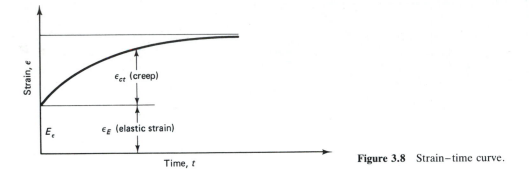

Figure 3.8   Strain–time curve.

An example of the relative numerical values of strain due to the foregoing three factors is presented for a normal concrete specimen subjected to 900 psi in compression:

$$\text{Immediate elastic strain, } \epsilon_e = 250 \times 10^{-6} \text{ in.-in.}$$
$$\text{Shrinkage strain after 1 year, } \epsilon_{sh} = 500 \times 10^{-6} \text{ in.-in.}$$
$$\text{Creep strain after 1 year } \epsilon_c = \underline{750 \times 10^{-6}} \text{ in.-in.}$$
$$\epsilon_t = 1500 \times 10^{-6} \text{ in./in.}$$

These relative values illustrate that stress–strain relationships for short-term loading lose their significance and long-term loadings become dominant in their effect on the behavior of a structure.

Figure 3.9 qualitatively shows in a three-dimensional model the three types of strain discussed resulting from sustained compressive stress and shrinkage.   Since

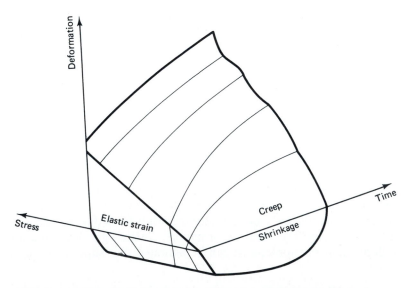

Figure 3.9   Three-dimensional model of time-dependent structural behavior.

creep is time dependent, this model has to be such that its orthogonal axes are deformation, stress, and time.

Numerous tests have indicated that creep deformation is proportional to the applied stress, but the proportionality is valid only for low stress levels. The upper limit of the relationship cannot be determined accurately, but can vary between 0.2 and 0.5 of the ultimate strength f'_c. This range in the limit of the proportionality is expected due to the large extent of microcracks at about 40% of the ultimate load.

Figure 3.10a shows a section of the three-dimensional model in Fig. 3.9 parallel to the plane containing the stress and deformation axes at time t_1. It indicates that both elastic and creep strains are linearly proportional to the applied stress. In a similar manner, Fig. 3.10b illustrates a section parallel to the plane containing the time and strain axes at a stress f_1; hence it shows the familiar creep–time and shrinkage–time relationships.

As in the case of shrinkage, creep is not completely reversible. If a specimen is unloaded after a period under a sustained load, an immediate elastic recovery is obtained that is less than the strain precipitated on loading. The instantaneous recovery is followed by a gradual decrease in strain, called *creep recovery*. The extent of the recovery depends on the age of the concrete when loaded with older concretes presenting higher creep recoveries, while residual strains or deformations become frozen in the structural element (see Fig. 3.11).

Creep is closely related to shrinkage and, as a general rule, a concrete that is resistant to shrinkage also presents a low creep tendency, as both phenomena are related to the hydrated cement paste. Hence creep is influenced by the composition of the concrete, the environmental conditions, and the size of the specimen, but principally creep depends on loading as a function of time.

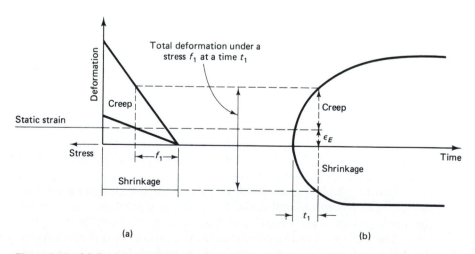

Figure 3.10 (a) Section parallel to the stress–deformation plane; (b) section parallel to the deformation–time plane.

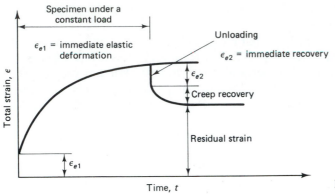

Figure 3.11 Creep recovery versus time.

The composition of a concrete specimen can be essentially defined by the water/cement ratio, aggregate and cement types, and aggregate and cement contents. Therefore, like shrinkage, an increase in the water/cement ratio and in the cement content increases creep. Also, as in shrinkage, the aggregate induces a restraining effect such that an increase in aggregate content reduces creep.

3.9.8 Creep Effects

As in shrinkage, creep increases the deflection of beams and slabs and causes loss of prestress. In addition, the initial eccentricity of a reinforced concrete column increases with time due to creep, resulting in the transfer of the compressive load from the concrete to the steel in the section.

Once the steel yields, additional load has to be carried by the concrete. Consequently, the resisting capacity of the column is reduced and the curvature of the column increases further, resulting in overstress in the concrete and leading to failure.

3.9.9 Rheological Models

Rheological models are mechanical devices that portray the general deformation behavior and flow of materials under stress. A model is basically composed of elastic springs and ideal dashpots denoting stress, elastic strain, delayed elastic strain, irrecoverable strain, and time. The springs represent the proportionality between stress and strain, and the dashpots represent the proportionality of stress to the rate of strain. A spring and a dashpot in parallel form a Kelvin unit, and in series they form a Maxwell unit.

Two rheological models will be discussed: the Burgers model and the Ross model. The Burgers model in Fig. 3.12 is shown since it can approximately simulate the stress–strain–time behavior of concrete at the limit of proportionality with some limitations. This model simulates the instantaneous recoverable strain (a); the delayed recoverable elastic strain in the spring (b); and the irrecoverable time-dependent strain in dashpots (c and d). The weakness in this model is that it

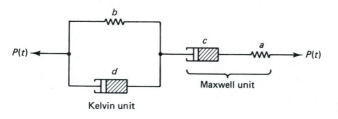

| Figure 3.12   Burgers model.

continues to deform at a uniform rate as long as the load is sustained by the Maxwell dashpot, a behavior not similar to concrete, where creep reaches a limiting value with time, as shown in Fig. 3.8.

A modification in the form of the Ross rheological model in Fig. 3.13 can eliminate this deficiency. *A* in this model represents the Hookian direct proportionality of stress-to-strain element, *D* represents the Newtonian element, and *B* and *C* are the elastic springs that can transmit the applied load *P(t)* to the enclosing cylinder walls by direct friction. Since each coil has a defined frictional resistance, only those coils whose resistances equal the applied load *P(t)* are displaced; the others remain unstressed, symbolizing the irrecoverable deformation in concrete. As the load continues to increase, it overcomes the spring resistance of unit *B*, pulling out the spring from the dashpot and signifying failure in a concrete element. More rigorous models have been used, such as Roll's model to assist in predicting the creep strains. Mathematical expressions for such predictions can be very rigorous. One convenient expression due to Ross defines creep *C* under load after a time interval *t* as follows:

$$C = \frac{t}{a + bt} \tag{3.5}$$

where *a* and *b* are constants determinable from tests.

Work by Branson (Refs. 3.10 and 3.11) has simplified creep evaluation. The additional strain ϵ_{cu} due to creep can be defined as

$$\epsilon_{cu} = \rho_u f_{ci} \tag{3.6a}$$

where ρ_u = unit creep coefficient, generally called *specific creep*
$\quad\quad f_{ci}$ = stress intensity in the structural member corresponding to unit strain
$\quad\quad \epsilon_{ci}$

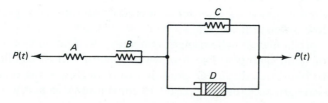

Figure 3.13   Ross model.

If C_u is the ultimate creep coefficient,

$$C_u = \rho_u E_c \qquad (3.6b)$$

An average value of $C_u \simeq 2.35$.

Branson's model, verified by extensive tests, relates the creep coefficient C_t at any time to the ultimate creep coefficient as follows:

$$C_t = \frac{t^{0.6}}{10 + t^{0.6}} C_u \qquad (3.7)$$

or, alternatively,

$$\rho_t = \frac{t^{0.6}}{10 + t^{0.6}} \qquad (3.8)$$

where t is the time in days.

The selected references at the end of the chapter give detailed information on the creep coefficients and constants to be used to evaluate creep effect. The brief discussion in this section is intended to provide exposure to the procedures considered in any fundamental study of creep and shrinkage behavior.

3.10 HIGH-STRENGTH CONCRETE

3.10.1 General Principles

Concretes with compressive strength f'_c of at least 6000 psi (44.4 MPa) can be classified as high-strength concrete, with the possibility today of achieving 20,000-psi (137.9-MPa) concrete under field conditions. To produce such concrete, chemical and mineral admixtures as well as air-entraining agents have to be used. Chemical retarders are used to retard the setting time for the cement-rich, high-strength concrete. Mineral admixtures such as fly ash, slag cement, and silica fume are also frequently used.

It is found that silica-fume admixtures in the range from 5% to 30% by weight of cement are an ideal additive for drastic increase in the compressive strength and considerable reduction in permeability. The increase in concrete density and strength is due to the dispersion of ultrafine particles of silica fume between the cement grains. This in turn results in a reduction of workability, which is enhanced further by the reduced water/cement ratios of the high-strength concrete mix. Consequently, high-range, water-reducing admixtures, called plasticizers, would have to be added in the required proportions in order to increase workability appreciably while maintaining a low water/cement ratio.

As with other ingredients of high-strength concrete, the fine and coarse aggregates should be of good quality. For low water/cement ratios, smaller-size coarse aggregates give better results. The grading of the aggregates is relatively unimportant in high-strength concrete compared to conventional concrete due to the high content of fine cementitious materials.   However, it is sometimes helpful to increase the fineness modulus to make the concrete consistency less viscous. Gap grading provides better results than continuous grading.   For compressive strength above 8000 psi, it is advisable to use a maximum size of aggregate less than $\frac{3}{8}$ to $\frac{1}{2}$ in. Cleanliness of both the fine and coarse aggregates deserves particular attention.   In general, three characteristics of the coarse aggregate—compressive strength, bonding potential with cement paste, and low water absorption capacity— are important in the production of high-strength concrete.

In addition to stringent quality control of materials, high-strength concrete requires proper proportioning to attain the desired mix along with careful mixing, handling, placing, and curing.   Available mix proportions data could be used as guidelines for trial mix designs.   However, to attain the desired strength and characteristics, extensive trial mix designs are required.   In addition, the importance of curing increases due to the use of low water/cement ratios, as one must not only avoid moisture escape but also provide extra water for hydration.   Similarly, proper mixing, handling, and placing are important to prevent moisture loss and produce workable concrete.

As to water content, it is important to consider the *total* water content, including that from the coarse and fine aggregate and all admixtures.   Whereas in conventional practice a range of 0.40 to 0.45w/c is used, the following are the recommended values for higher-strength concretes:

f_c' (psi)	w/c
6,000–10,000	0.35–0.40
10,000–12,000	0.30–0.35
12,000–20,000	0.30–0.22

The very low w/c ratio ranges are achieved by utilizing large amounts of super-plasticizer and high cement content.

In summary, four basic principles have to be considered in the production of high-strength concrete: (1) improved aggregate–matrix bond, (2) reduced porosity, (3) improved compaction, and (4) application of internal agents such as silica fumes and plasticizers and external agents such as lateral confinement through internal steel hoops, heat or steam curing, proper handling, and strict quality control.

3.10.2 Design Criteria

Available expressions defining concrete properties are based primarily on experimental data of concrete with compressive strength below 6000 psi.   Such expressions do not necessarily define the relevant parameters when high-strength concrete is being used.   When the concrete compressive strength exceeds 6000 psi for high-strength concrete, particularly in the range from 8000 to 12,000 psi, engineering properties of the concrete, such as elasticity, flexural strength, tensile resistance, and bond strength, may be affected.

The principal mechanical properties of concrete are compressive and tensile strength, creep and shrinkage, and modulus of elasticity. The actual values of tensile strength, modulus of elasticity, creep, and shrinkage are a function of compressive strength for most low- and moderate-strength concretes. But such correlation is not always the case for high compressive strengths.

Modulus of elasticity E_c. The modulus of elasticity is strongly influenced by the concrete materials and proportions used. An increase in the modulus E_c is expected with the increase in compressive strength since the slope of the ascending branch of the stress–strain diagram becomes steeper for higher-strength concretes, but at a *lower* rate than the compressive strength. The value of the secant modulus E_c for normal-strength concretes at 28 days is usually approximately 4×10^6 psi, whereas for higher-strength concretes values in the range from 7 to 8×10^6 psi have been recorded. These higher values can be used to reduce short- and long-term deflection of flexural members and eccentricity of columns and other biaxially loaded members.

For concretes in the strength range up to 6000 psi, the ACI Code empirical equation for the secant modulus of concrete E_c given in Section 3.9.5 is reasonably applicable. However, as the strength of concrete increases, in the range from 12,000 to 20,000 psi, the value of E_c increases at a faster rate than that generated by the ACI expression ($E_c = 33w_c^{1.5} \sqrt{f_c'}$), thereby underestimating the true E_c value.

Available expressions for E_c applicable to concrete strength up to 12,000 psi are inconclusive. The expression due to Carrasquillo et al. (Ref. 3.14) for normal-weight concrete of strengths up to 12,000 psi and lightweight concrete up to 9000 psi is

$$E_c = (40,000 \sqrt{f_c'} + 1 \times 10^6) \left(\frac{w_c}{145} \right)^{1.5} \tag{3.9}$$

where w_c is the unit weight of the hardened concrete in pcf. Other investigations report that as f_c' approaches 12,000 psi for normal-weight concrete and less for lightweight concrete, Eq. 3.9 can underestimate the true value of E_c. At the present state of the art, it is advisable in cases of very high strength use in major structures where f_c' is in the range of 20,000 psi or higher that adequate stress–strain cylinder compression tests be performed with stress–strain readings. In this manner, the deduced secant modulus value of E_c at an $f_c = 0.45f_c'$ intercept could predict more accurately the true value of the E_c for the particular mix and aggregate size and properties until an acceptable expression is available to the designer. The long-term stiffness and deflection computations would thereby be more representative.

Recent work at Rutgers (Ref. 3.18) on high-strength composite construction has resulted in considerable enhancement of the ductility of high-strength reinforced concrete beams. Prestressed concrete prisms of high-strength concrete were used in place of the normal mild steel bar reinforcement. The mix proportions in lb/yd³ were as shown in Table 3.9. The mix was designed for 7-day compressive strength

TABLE 3.9   MIX PROPORTIONS (lb/yd³) FOR COMPOSITE BEAMS = $f'_c > 13,000$ PSI

Coarse aggregate, $\frac{3}{8}$ in.	Fine aggregate (natural sand)	Portland cement, type II	Water	Powder silica fume, Force 10,000	Liquid super plasticizer (W. R. Grace)
(1)	(2)	(3)	(4)	(5)	(6)
1851	1100	720	288	180	54

1 lb/yd³ = 0.59 Kg/m.³

of 12,000 psi (84 MPa). The ratio of the cementations/fine/coarse aggregate was 1:1.22:2.06, and the slump varied between 4 and 6 in. (100 and 150 mm). The prestressing strands were stress relieved 270-kips (1900-MPa), 7-wire, 3/8-in.- (9.5-mm)-diameter strands.

Figure 3.14 shows the cross section of the composite beams, and Fig. 3.15 gives a typical stress–strain relationship of the concrete, which achieved in some of the mixes a 7-day strength of 13,250 psi (91.4 MPa). The tested specimens were instrumented with a fiber-optic sensor system developed by the author using Bragg grating sensors both internally and externally.

Modulus of rupture f'_r.   An evaluation of the tensile behavior of concrete can be made by the modulus of rupture or bending test.   A range of $7.5\sqrt{f'_c}$ to $12\sqrt{f'_c}$ has been reported for high-strength concrete modulus of rupture values with a reasonable expression in terms of compressive strength as follows for plain normal-weight concrete (Refs. 3.13 and 3.16):

$$f'_r \text{ (psi)} = 11.7\sqrt{f'_c} \tag{3.10}$$

Tensile splitting strength f'_t.   Tensile strength of concrete is an important parameter in determining when the first flexural crack may develop.   A general expression for high-strength concrete for an f'_c range of 3000 to 12,000 psi is given as follows:

$$f'_t = 7.4\sqrt{f'_c} \tag{3.11}$$

However, with deliberate selection of materials and proportions, including the use of silica fume and smaller coarse aggregates, the tensile strength may be increased to almost twice that predicted by this expression.

Creep and shrinkage.   With high-strength concrete, greater stress may be applied with little or no increase in long-term deformation above the level expected in moderate-strength concrete.   Since high-strength concrete has low water/cement ratios that could be as low as 0.22 for $f'_c \simeq 20,000$ psi, shrinkage can be very limited, with a range of shrinkage strain of 250×10^{-6} to 500×10^{-6} in./in.

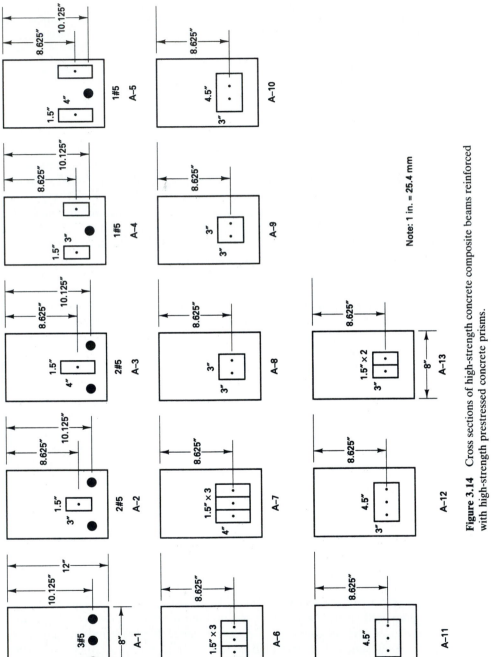

Figure 3.14 Cross sections of high-strength concrete composite beams reinforced with high-strength prestressed concrete prisms.

58

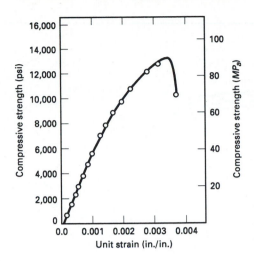

Figure 3.15 Stress–strain diagram of high-strength concrete (13,250 psi) with mix proportions given on Table 3.9.

3.10.3 Confining Effect on High-strength Concrete

Use of high-strength concrete in compression members, such as in tall structures, leads to considerable reduction in the size of the concrete sections. Widely accepted properties of such concretes are their higher modulus E_c values, less ductile mode of failure, and larger strain at maximum stress. The use of confining circular or rectangular spiral reinforcement leads to increased strength and ductility of the confined concrete. Published experimental results on the effects of rectilinear confinement in very high strength concrete (in excess of 12,000 psi) are scarce. Results of tests in Ref. 3.15 for concretes of up to 13,560-psi compressive strength indicate general improvement of the behavior of the concrete when confined. Instead of collapsing in a very brittle fashion, the concrete failed in a more ductile and gradual manner.

The peak stress f_0 in Figure 3.16b, the strain ϵ_0, and especially the ductility increased with the increase in the volumetric ratio, but not proportionately. If the peak stress $f_0 = Kf_c'$, where K is the effective confinement, then K can be expressed as

$$K = 1 + 0.0091\left(1 - \frac{0.245s}{h''}\right)\left(\rho'' + \frac{nd''}{8sd}\rho\right)\frac{f_y''}{\sqrt{f_c'}} \qquad (3.12)$$

where s = center-to-center spacing of the lateral ties, in.
h'' = length of one side of the rectangular ties, in.
$n \leq$ number of longitudinal steel bars
d'' = nominal diameter of lateral ties, in.
d = nominal diameter of longitudinal steel bars, in.
ρ'' = volumetric ratio of lateral reinforcement
ρ = volumetric ratio of longitudinal reinforcement
f_y'' = yielding stress of the lateral steel, psi

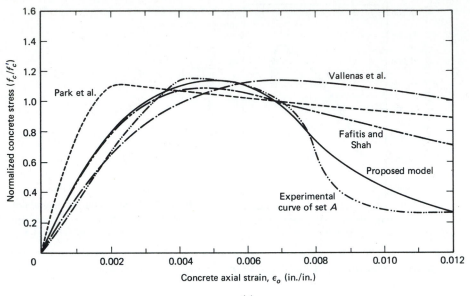

(a)

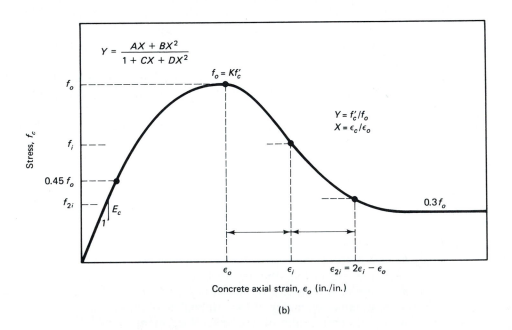

(b)

Figure 3.16 (a) Normalized complete stress–strain curve by various authors; (b) stress and strain parameters. (From Ref. 3.15.)

The peak strain ϵ_0 can be predicted by the following expression (Ref. 3.15):

$$\epsilon_0 = 0.00265 + \frac{0.035\left(1 - \frac{0.734s}{h''}\right)(\rho''f''_y)^{2/3}}{\sqrt{f'_c}} \tag{3.13}$$

3.10.4 Mix Proportions for 20,000-psi Concrete

For very high compressive strength in excess of 12,000 psi (82.74 MPa), it is essential to make a large number of trial mixes (five or more) and take extra care in the selection of aggregate size and source. Steel cylinder molds are preferred for uniformity of test results, using 4 in. × 8 in. molds and applying the appropriate dimensional correction. It is also necessary to grind the cylinder ends and then either cap them with high-strength capping compound prior to loading or apply the load directly to the ground ends or through a removable steel cap with a hard neoprene pad bearing directly on the ground specimen ends. Preparation of the cylinders should resemble as closely as possible the field conditions of concrete placement. Mock-up placement of the high-strength concrete is advisable in order to evaluate the construction procedures and performance of the concrete in field conditions and to identify potential problems with batching, placement, and testing of the concrete at early ages, with corrective measures taken immediately.

A good example of the use of high-strength concrete in the 20,000-psi range (137.9 MPa) at 56 days and a concrete modulus $E_c = 7.8 \times 10^6$ psi (53.8×10^3 MPa) is the Two Union Square Building, Seattle, Washington (Ref. 3.17). Actual typical mix obtained is listed in Table 3.10, with the design mix values in parentheses.

A slump of 8 in. with $w/c \simeq 0.22$ resulted from the mix proportions indicated. A typical compressive age plot for the indicated mix based on 4 in. × 8 in. cylinder tests is shown in Fig. 3.17.

TABLE 3.10   MIX PROPORTIONS FOR $f'_c > 18{,}000$ PSI

					Superplasticizer	
Coarse aggregate ($\frac{3}{8}$ in.) (lb)	Fine aggregate (paving sand) (lb)	Cement (lb)	Water (lb)	Silica fume (gal)	W. R. Grace Dartard 40 (oz/100)	Mighty 150 (lb cement)
1872	1165	957	217	13	2.1	9.8
1894	1165	956	217	13	2.1	16.4
(1805)	(1100)	(950)	($w/c = 0.22$)	(70 lb)[a]	(6.0)	(up to 24)

[a]Weight of solid silica fume only. Water contained as part of the emulsion must be subtracted from the total water allowed.

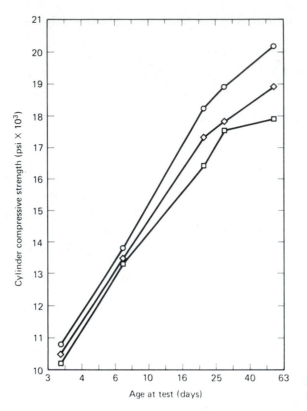

Figure 3.17 Compressive strength versus age of high-strength concrete.

SELECTED REFERENCES

3.1. ACI Committee 211, *Standard Practice for Selecting Proportions for Normal, Heavyweight, and Mass Concrete*, ACI 211.1-91, American Concrete Institute, Detroit, 38 pp.

3.2. ACI Committee 211, *Standard Practice for Selecting Proportions for Structural Lightweight Concrete*, ACI 211.2-91, American Concrete Institute, Detroit, 14 pp.

3.3. Portland Cement Association, *Design and Control of Concrete Mixtures*, 12th ed., PCA, Skokie, Ill., 1979, 140 pp.

3.4. ACI Committee 318, *Building Code Requirements for Reinforced Concrete*, ACI Standard 318–95; and the *Commentary on Building Code Requirements for Reinforced Concrete*, American Concrete Institute, Detroit, 1995.

3.5. American Society for Testing and Materials, *Significance of Tests and Properties of Concrete and Concrete Making Materials*, Special Technical Publication 169B, ASTM, Philadelphia, 1978, 882 pp.

3.6. Ross, A. D., "The Elasticity, Creep and Shrinkage of Concrete," in *Proceedings of the Conference on Non-metallic Brittle Materials*, Interscience Publishers, London, 1958, pp. 157–174.

3.7. Neville, A. M., *Properties of Concrete*, 3rd ed., Pitman Books, London, 1981, 779 pp.

3.8. Freudenthal, A. M., and Roll, F., "Creep and Creep Recovery of Concrete under High Compressive Stress," *Journal of the American Concrete Institute*, Proc. Vol. 54, June 1958, pp. 1111–1142.

3.9. Ross, A. D., "Creep Concrete Data," *Proceedings, Institution of Structural Engineers*, London, Vol. 15, 1937, pp. 314–326.

3.10. Branson, D. E., *Deformation of Concrete Structures*, McGraw-Hill, New York, 1977, 546 pp.

3.11. Branson, D. E., "Compression Steel Effects on Long Term Deflections," *Journal of the American Concrete Institute*, Proc. Vol. 68, August 1971, pp. 555–559.

3.12. Mindess, S., and Young, J. F., *Concrete*, Prentice Hall, Englewood Cliffs, N.J., 1981, 671 pp.

3.13. Nawy, E. G., and Balaguru, P. N., "High Strength Concrete," Chapter 5 in *Handbook of Structural Concrete*, Pitman Books, London/McGraw-Hill, New York, 1983, 1968 pp.

3.14. Carrasquillo, R. L., Nilson, A. H., and Slate, F. O., "Properties of High Strength Concrete Subjected to Short Term Loads," *Journal of the American Concrete Institute*, Proc. Vol. 78, No. 3, May–June 1981, pp. 171–178.

3.15. Yong, Y. K., Nour, M. G., and Nawy, E. G., "Behavior of Laterally Confined High-Strength Concrete under Axial Loads," *Journal of the Structural Division, ASCE*, Vol. 114, No. 2, February 1988, pp. 332–351.

3.16. Shah, S. P., and Ahmad, S. H., "Structural Properties of High Strength Concrete and Its Implications for Precast Prestressed Concrete," *Journal of the Prestressed Concrete Institute*, November–December 1985, pp. 92–119.

3.17. Hester, W. T., *Two Union Square Summary of Mock-Up Concrete Placement*, Report, University of California, Berkeley, 1987, 23 pp.

3.18. Chen, B., and Nawy, E. G., "Structural Behavior Evaluation of High-Strength Concrete Reinforced with Prestressed Prisms Using Fiber Optic Sensors," *Proceedings, ACI Structural Journal*, American Concrete Institute, Detroit, Nov.-Dec. 1994, pp. 708–718.

3.19. Chen, B., Maher, M. H., and Nawy, E. G., "Fiber Optic Bragg Grating Sensor for Non-Destructive Evaluation of Composite Beams," *Proceedings, ASCE Journal of the Structural Division*, Vol. 120 No. 12, American Society of Civil Engineers, New York, 1995, pp. 3456–3470.

PROBLEMS FOR SOLUTION

3.1. Design a concrete mix using the following data:

> Required strength $f_c' = 5000$ psi (34.5 MPa)
> Type of structure: beam
> Maximum size of aggregate $= \frac{3}{4}$ in. (18 mm)
> Fineness modulus of sand $= 2.6$
> Dry-rodded weight of coarse aggregate $= 100$ lb/ft³
> Moisture absorption: 2% for coarse aggregate and 2% for fine aggregate

3.2. Using the data of Ex. 3.1, design a 6% air-entrained mix.

3.3. Repeat Exs. 3.1 and 3.2 for a mix design strength $f_c' = 3000$ psi (20.7 MPa).

3.4. Estimate the strength of the trial mix, f_{cr}', for the following cases:

(a) $f_c' = 3500$ psi (24.15 MPa); s (using 40 consecutive tests) = 300 psi (2.07 MPa).

(b) $f_c' = 3000$ psi (20.7 MPa); s (using 20 consecutive tests) = 250 psi (1.73 MPa).

(c) $f_c' = 3000$ psi (20.7 MPa); test results are not available.

(d) $f_c' = 4000$ psi (27.6 MPa); s (using 15 tests) = 375 psi (2.59 MPa).

REINFORCED CONCRETE

4.1 INTRODUCTION

Concrete is strong in compression but weak in tension. Therefore, reinforcement is needed to resist the tensile stresses resulting from the induced loads. Additional reinforcement is occasionally used to reinforce the compression zone of concrete beam sections. Such steel is necessary for heavy loads in order to reduce long-term deflections. Whereas Chapters 2 and 3 dealt with plain concrete and its constituent materials, this chapter discusses composite reinforced concrete, which can withstand high tensile as well as compressive forces. A discussion of the types of reinforcing material, the variety of structural systems, and their components is presented.

Additionally, concrete structures have to perform adequately under service-load conditions in addition to having the necessary reserve strength to resist ultimate load. The subjects of reliability, safety, and load factors are also presented.

Photo 22 University of Illinois Assembly Hall at Urbana. (Courtesy of Ammann & Whitney.)

4.2 TYPES AND PROPERTIES OF STEEL REINFORCEMENT

Steel reinforcement for concrete consists of bars, wires, and welded wire fabric, all of which are manufactured in accordance with ASTM standards. The most important properties of reinforcing steel are:

1. Young's modulus, E_s
2. Yield strength, f_y
3. Ultimate strength, f_u
4. Steel grade designation
5. Size or diameter of the bar or wire

To increase the bond between concrete and steel, projections called *deformations* are rolled on the bar surface as shown in Fig. 4.1 in accordance with ASTM specifications. The deformations shown must satisfy ASTM Specification A616-76 to be accepted as deformed bars. The deformed wire has indentations pressed into the wire or bar to serve as deformations. Except for wire used in spiral reinforcement in columns, only deformed bars, deformed wires, or wire fabric made from smooth or deformed wire may be used in reinforced concrete under approved practice.

Figure 4.2 shows typical stress–strain curves for grade 40, 60, and 80 steels. They have corresponding yield strengths of 40,000, 60,000, and 80,000 psi (276, 414, and 552 N/mm², respectively) and generally have well-defined yield points. For steels that lack a well-defined yield point, the yield-strength value is taken as

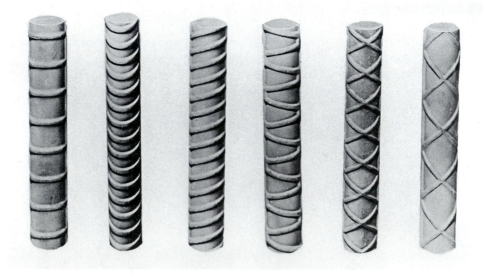

Figure 4.1   Various forms of ASTM-approved deformed bars.

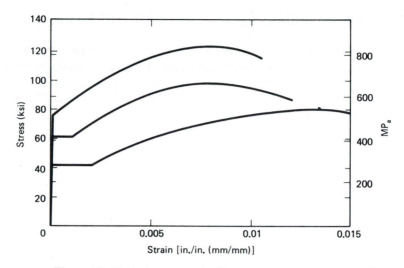

Figure 4.2   Typical stress–strain diagrams for various steels.

the strength corresponding to a unit strain of 0.005 for grades 40 and 60 steels and 0.0035 for grade 80 steel.   The ultimate tensile strengths corresponding to the 40, 60, and 80 grade steels are 70,000, 90,000, and 100,000 psi (483, 621, and 690 N/mm^2), and some steel types are given in Table 4.1.   The percent elongation at fracture, which varies with the grade, bar diameter, and manufacturing source, ranges from 4.5% to 12% over an 8-in. (203.2-mm) gage length.

For most steels, the behavior is assumed to be elastoplastic, and Young's modulus is taken as 29×10^6 psi (200×10^6 MPa).   Table 4.1 presents the rein-

TABLE 4.1   REINFORCEMENT GRADES AND STRENGTHS

1982 Standard type	Minimum yield point or yield strength, f_y (psi)	Ultimate strength, f_u (psi)
Billet steel (A615)		
Grade 40	40,000	70,000
Grade 60	60,000	90,000
Axle steel (A617)		
Grade 40	40,000	70,000
Grade 60	60,000	90,000
Low-alloy steel		
(A706): Grade 60	60,000	80,000
Deformed wire		
Reinforced	75,000	85,000
Fabric	70,000	80,000
Smooth wire		
Reinforced	70,000	80,000
Fabric	65,000, 56,000	75,000, 70,000

forcement-grade strengths, and Table 4.2(a) and (b) presents geometrical properties of the various sizes of bars.

Welded wire fabric is increasingly used for slabs because of the ease of placing the fabric sheets, control of reinforcement spacing, and better bond. The fabric reinforcement is made of smooth or deformed wires that run in perpendicular directions and are welded together at intersections. Table 4.3 presents geometrical properties of some standard wire reinforcement.

TABLE 4.2(a)   WEIGHT, AREA, AND PERIMETER OF INDIVIDUAL BARS

Bar designation number	Weight per foot (lb)	1982 Standard nominal dimensions		
		Diameter, d_b [in. (mm)]	Cross-sectional area, A_b (in.²)	Perimeter (in.)
3	0.376	0.375 (9)	0.11	1.178
4	0.668	0.500 (13)	0.20	1.571
5	1.043	0.625 (16)	0.31	1.963
6	1.502	0.750 (19)	0.44	2.356
7	2.044	0.875 (22)	0.60	2.749
8	2.670	1.000 (25)	0.79	3.142
9	3.400	1.128 (28)	1.00	3.544
10	4.303	1.270 (31)	1.27	3.990
11	5.313	1.410 (33)	1.56	4.430
14	7.65	1.693 (43)	2.25	5.32
18	13.60	2.257 (56)	4.00	7.09

TABLE 4.2(b)   ASTM STANDARD METRIC REINFORCING BARS

Bar size designation (No.)	Nominal Dimensions		
	Mass (kg/m)	Diameter (mm)	Area (mm²)
10 M	0.785	11.3	100
15 M	1.570	16.0	200
20 M	2.355	19.5	300
25 M	3.925	25.2	500
30 M	5.495	29.9	700
35 M	7.850	35.7	1000
45 M	11.775	43.7	1500
55 M	19.625	56.4	2500

ASTM A615M Grade 300 is limited to size No. 5, 10 M through No. 20 M, otherwise grades 400 or 500 MPa for all the sizes. Check availability with local suppliers for No. 45 M and 55 M.

TABLE 4.3 STANDARD WIRE REINFORCEMENT

W&D size Smooth	W&D size Deformed	Nominal diameter (in.)	Nominal area (in.²)	Nominal weight (lb/ft)	Area (in.²/ft of width for various spacings) Center-to-center spacing (in.)						
					2	3	4	6	8	10	12
W31	D31	0.628	0.310	1.054	1.86	1.24	0.93	0.62	0.465	0.372	0.31
W30	D30	0.618	0.300	1.020	1.80	1.20	0.90	0.60	0.45	0.366	0.30
W28	D28	0.597	0.280	0.952	1.68	1.12	0.84	0.56	0.42	0.336	0.28
W26	D26	0.575	0.260	0.934	1.56	1.04	0.78	0.52	0.39	0.312	0.26
W24	D24	0.553	0.240	0.816	1.44	0.96	0.72	0.48	0.36	0.288	0.24
W22	D22	0.529	0.220	0.748	1.32	0.88	0.66	0.44	0.33	0.264	0.22
W20	D20	0.504	0.200	0.680	1.20	0.80	0.60	0.40	0.30	0.24	0.20
W18	D18	0.478	0.180	0.612	1.08	0.72	0.54	0.36	0.27	0.216	0.18
W16	D16	0.451	0.160	0.544	0.96	0.64	0.48	0.32	0.24	0.192	0.16
W14	D14	0.422	0.140	0.476	0.84	0.56	0.42	0.28	0.21	0.168	0.14
W12	D12	0.390	0.120	0.408	0.72	0.48	0.36	0.24	0.18	0.144	0.12
W11	D11	0.374	0.110	0.374	0.66	0.44	0.33	0.22	0.165	0.132	0.11
W10.5		0.366	0.105	0.357	0.63	0.42	0.315	0.21	0.157	0.126	0.105
W10	D10	0.356	0.100	0.340	0.60	0.40	0.30	0.20	0.15	0.12	0.10
W9.5		0.348	0.095	0.323	0.57	0.38	0.285	0.19	0.142	0.114	0.095
W9	D9	0.338	0.090	0.306	0.54	0.36	0.27	0.18	0.135	0.108	0.09
W8.5		0.329	0.085	0.289	0.51	0.34	0.255	0.17	0.127	0.102	0.085
W8	D8	0.319	0.080	0.272	0.48	0.32	0.24	0.16	0.12	0.096	0.08
W7.5		0.309	0.075	0.255	0.45	0.30	0.225	0.15	0.112	0.09	0.075
W7	D7	0.298	0.070	0.238	0.42	0.28	0.21	0.14	0.105	0.084	0.07
W6.5		0.288	0.065	0.221	0.39	0.26	0.195	0.13	0.097	0.078	0.065
W6	D6	0.276	0.060	0.204	0.36	0.24	0.18	0.12	0.09	0.072	0.06
W5.5		0.264	0.055	0.187	0.33	0.22	0.165	0.11	0.082	0.066	0.055
W5	D5	0.252	0.050	0.170	0.30	0.20	0.15	0.10	0.075	0.06	0.05
W4.5		0.240	0.045	0.153	0.27	0.18	0.135	0.09	0.067	0.054	0.045
W4	D4	0.225	0.040	0.136	0.24	0.16	0.12	0.08	0.06	0.048	0.04
W3.5		0.211	0.035	0.119	0.21	0.14	0.105	0.07	0.052	0.042	0.035
W3		0.195	0.030	0.102	0.18	0.12	0.09	0.06	0.045	0.036	0.03
W2.9		0.192	0.029	0.098	0.174	0.116	0.087	0.058	0.043	0.035	0.029
W2.5		0.178	0.025	0.085	0.15	0.10	0.075	0.05	0.037	0.03	0.025
W2		0.159	0.020	0.068	0.12	0.08	0.06	0.04	0.03	0.024	0.02
W1.4		0.135	0.014	0.049	0.084	0.056	0.042	0.028	0.021	0.017	0.014

4.3 BAR SPACING AND CONCRETE COVER FOR STEEL REINFORCEMENT

It is necessary to guard against honeycombing and ensure that the wet concrete mix passes through the reinforcing steel without separation. Since the graded aggregate size in structural concrete often contains $\frac{3}{4}$-in. (19-mm diameter) coarse aggregate, minimum allowable bar spacing and minimum required cover are needed. Additionally, to protect the reinforcement from corrosion and loss of strength in case of fire, codes specify a minimum required concrete cover. Some of the major requirements of ACI Code 318 are:

1. Clear distance between parallel bars in a layer must not be less than the bar diameter d_b or 1 in. (25.4 mm).
2. Clear distance between longitudinal bars in columns must not be less than $1.5d_b$ or 1.5 in. (38.1 mm).
3. Minimum clear cover in cast-in-place concrete beams and columns should not be less than 1.5 in. (38.1 mm) when there is no exposure to weather or contact with the ground; this same cover requirement also applies to stirrups, ties, and spirals.

In the case of slabs, plates, shells, and folded plates, where concrete is not exposed to a severe environment and where the reinforcement size does not exceed a No. 11 bar diameter (85.8 mm), the clear cover should not be less than $\frac{3}{4}$ in. (19 mm). Detailed requirements as to thickness of cover for various conditions can be found in various codes of practice, such as the Underwriters' National Building Code and the ACI Code.

4.4 CONCRETE STRUCTURAL SYSTEMS

Every structure is proportioned as to both architecture and engineering to serve a particular function. Form and function go hand in hand, and the best structural system is the one that fulfills most of the needs of the user while being serviceable, attractive, and economically cost efficient. Although most structures are designed for a life span of 50 years, the durability performance record indicates that properly proportioned concrete structures have generally had longer useful lives.

Numerous concrete landmarks can be cited where major credit is due to the art and science of structural design applied with ingenuity, logic, and imagination. Such concrete structural systems as the TWA Terminal, New York; the Newark Terminal, New Jersey; Symphony Hall, Melbourne, Australia; Chicago's Marina Towers and Water Tower Place; the Dallas Super Dome; Two Union Square Towers, Seattle; Trump Tower, New York; and many others are a testimony to the marriage of form and function with superior engineering judgment. Photographs of several such landmarks appear throughout the book.

Such concrete systems are composed of a variety of concrete structural elements that, when synthesized, produce a total system. The components can be broadly classified into (1) floor slabs, (2) beams, (3) columns, (4) walls, and (5) foundations.

4.4.1 Floor Slabs

Floor slabs are the main horizontal elements that transmit the moving live loads as well as the stationary dead loads to the vertical framing supports of a structure. They can be slabs on beams, as in Fig. 4.3, or waffle slabs, slabs without beams (flat plates) resting directly on columns, or composite slabs on joists. They can be proportioned such that they act in one direction (one-way slabs) or proportioned so that they act in two perpendicular directions (two-way slabs and flat plates). A detailed discussion of the analysis and design of such floor systems is given in subsequent chapters.

4.4.2 Beams

Beams are the structural elements that transmit the tributory loads from floor slabs to vertical supporting columns. They are normally cast monolithically with the slabs and are structurally reinforced on one face, the lower tension side, or both the top and bottom faces. As they are cast monolithically with the slab, they form a T-beam section for interior beams or an L beam at the building exterior, as seen

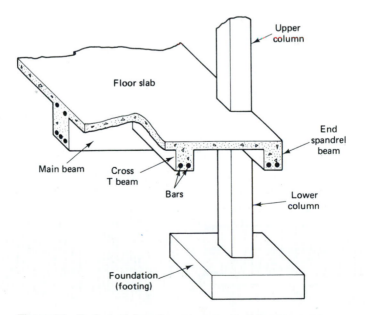

Figure 4.3 Typical reinforced concrete structural framing system.

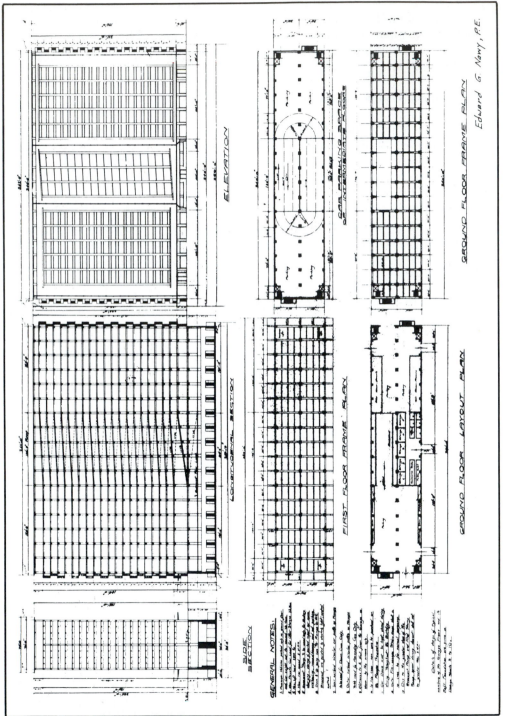

Figure 4.4 Typical working drawing for a reinforced concrete parking structure. (Design by E. G. Nawy.)

in Fig. 4.3.   The plan dimensions of a slab panel determine whether the floor slab behaves essentially as a one-way or two-way slab.

4.4.3 Columns

The vertical elements support the structural floor system.   They are compression members subjected in most cases to both bending and axial load and are of major importance in the safety considerations of any structure.   If a structural system is also composed of horizontal compression members, such members would be considered as beam-columns.

4.4.4 Walls

Walls are the vertical enclosures for building frames.   They are not usually or necessarily made of concrete but of any material that esthetically fulfills the form and functional needs of the structural system.   Additionally, structural concrete walls are often necessary as foundation walls, stairwell walls, and shear walls that resist horizontal wind loads and earthquake-induced loads.

4.4.5 Foundations

Foundations are the structural concrete elements that transmit the weight of the superstructure to the supporting soil.   They could be in many forms, the simplest being the isolated footing shown in Fig. 4.3.   It can be viewed as an inverted slab transmitting a distributed load from the soil to the column.   Other forms of foundations are piles driven to rock, combined footings supporting more than one column, mat foundations, and rafts, which are basically inverted slab and beam construction.

   The results of the analysis and design process of a structure have to be presented in concise and standardized form, which the constructor can use for building the entire system.   Hence knowledge and easy reading of working drawings is important.   A typical layout drawing of a multilevel parking garage structure is shown in Fig. 4.4.   The *ACI Manual of Detailing* gives an adequate coverage of typical working drawings for various structural systems and of the layout and detailing of reinforcement.

4.5 RELIABILITY AND STRUCTURAL SAFETY OF CONCRETE COMPONENTS

Three developments in recent decades have had a major influence on present and future design procedures.   They are the vast increase in experimental and analytical evaluation of concrete elements, the probabilistic approach to the interpretation of behavior, and the digital computational tools available for rapid analysis of the safety and reliability of systems.   Until recently, most safety factors in design have had an empirical background based on local experience over an extended period of time. As additional experience is accumulated and more knowledge is gained from failures as well as familiarity with the properties of concrete, factors of safety are adjusted and in most cases lowered by the codifying bodies.

A. L. L. Baker in 1956 proposed a simplified method of safety factor determination, as shown in Table 4.4, based on probabilistic evaluation. This method expects the design engineer to make critical choices regarding the magnitudes of safety margins in a design. The method takes into consideration that different weights should be assigned to the various factors affecting a design. The weighted failure effects W_t for the various factors of workmanship, loading conditions, results of failure, and resistance capacity are tabulated in Table 4.4.

The safety factor against failure is

$$SF = 1.0 + \frac{\Sigma W_t}{10} \tag{4.1}$$

where the maximum total weighted value ΣW_t of all parameters affecting performance equals 10. In other words, for the worst combination of conditions affecting structural performance, the safety factor SF = 2.0.

This method assumes adequate information on prior performance data similar to a design in progress. Such data in many instances are not readily available for determining safe weighted values W_t in Eq. 4.1. Additionally, if the weighted

TABLE 4.4 BAKER'S WEIGHTED SAFETY FACTOR

Weighted failure effect		Maximum W_t
1. Results of failure: 1.0 to 4.0		
Serious, either human or economic		4.0
Less serious, only the exposure of	1.0	
nondamageable material		
2. Workmanship: 0.5 to 2.0		
Cast in place		2.0
Precast "factory manufactured"	0.5	
3. Load conditions: 1.0 to 2.0		2.0
(high for simple spans and overload		
possibilities; low for load combina-		
tions such as live loads and wind)		
4. Importance of member in structure		0.5
(beams may use lower value than		
columns)		
5. Warning of failure		1.0
6. Depreciation of strength		0.5
	Total = ΣW_t =	10.0

$$SF = 1.0 + \frac{\Sigma W_t}{10}$$

factors are numerous, a probabilistic determination of them is more difficult to codify. Hence an undue value-judgment burden is probably placed on the design engineer if the full economic benefit of the approach is to be achieved.

Another method with a smaller number of probabilistic parameters deals primarily with loads and resistances. Its approach for both steel and concrete structures is generally similar; both the load and resistance factor design methods (LRFD) and first-order second-moment method (FOSM) propose general reliability procedures for evaluating probability-based factored load design criteria, as in Refs. 4.7, 4.8, and 4.10. They are intended for use in proportioning structural members on the basis of load types such that the resisting strength levels are *greater* than the factored load or moment distributions. As these approaches are basically load oriented, they reduce the number of individual variables that have to be considered, such as those listed in Table 4.4.

Assume that ϕ_i represents the resistance factors of a concrete element and that γ_i represents the load factors for the various types of load. If R_n is the nominal resistance of the concrete element and W_i represents the load effect of various types of superimposed load,

$$\phi_i R_n \geq \gamma_i W_i \tag{4.2}$$

where i represents the various types of load, such as dead, live, wind, earthquake, or time-dependent effects.

Figure 4.5a and b shows a plot of the separate frequency distributions of the actual load W and the resistance R with mean values $\overline{R}$ and $\overline{W}$. Figure 4.5c gives the two distributions superimposed and intersecting at point C.

It can be recognized that safety and reliable integrity of the structure can be expected to exist if the load effect W falls at a point to the left of intersection C on the W curve and to the right of intersection C on the resistance curve R. Failure, on the other hand, would be expected to occur if the load effect or the resistance falls within the shaded area in Fig. 4.5c. If β is a safety index, then

$$\beta = \frac{\overline{R} - \overline{W}}{\sqrt{\sigma_R^2 + \sigma_W^2}} \tag{4.3}$$

where σ_R and σ_W are the standard deviations of the resistance and the load, respectively.

A plot of the safety index β for a hypothetical structural system is shown in Fig. 4.6 against the probability of failure of the system. One can observe that such a probability is reduced as the difference between the mean resistance $\overline{R}$ and load effect $\overline{W}$ is increased, or the variability of resistance and load effect as measured by their standard deviations σ_R and σ_W is decreased, thereby reducing the shaded area under intersection C in Fig. 4.5c.

The extent of increasing the $\overline{R} - \overline{W}$ difference or decreasing the degree of scatter of σ_R or σ_W is naturally dictated by economic considerations. It is economically unreasonable to design a structure for zero failure, particularly since types of risk other than load are an accepted matter, such as the risks of severe

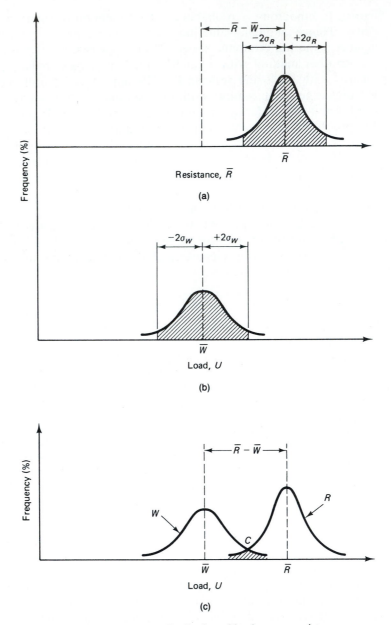

Figure 4.5 Frequency distribution of loads versus resistance.

earthquake, hurricane, volcanic eruption, or fire. Safety factors and corresponding load factors would thus have to disregard those types or levels of load, stress, and overstress whose probability of occurrence is very low. In spite of this, it is still possible to achieve reliable safety conditions by choosing such a safety index value

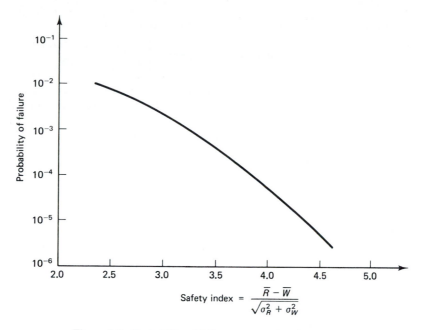

Figure 4.6   Probability of failure versus safety index β.

β through a proper choice of R_n and W_i values using the appropriate resistance factors ϕ_i and load factors γ_i in Eq. 4.2.   A safety index β having the value 1.75 to 3.2 for concrete structures is suggested in Ref. 4.8, where the lower value accounts for load contributions from wind and earthquake.

If the factored external load is expressed as U_i, then $\Sigma\gamma_i W = U_i$ for the different loading combinations.   The following are U_i values recommended in Ref. 4.8 to choose a maximum U for use in Eq. 4.2, that is, $\phi_i R_n \geq \gamma_i W_i \geq U_{i(\text{max})}$:

$$\gamma_i U_W = \text{maximum} \begin{cases} 1.4D_n \\ 1.2D_n + 1.6L_n \\ 1.2D_n + 1.6S_n + (0.5L_n \text{ or } 0.8W_n) \\ 1.2D_n + 1.3W_n + 0.5L_n \\ 1.2D_n + 1.5E_n + (0.5L_n \text{ or } 0.2S_n) \\ 0.9D_n - (1.3W_n \text{ or } 1.5E_n) \end{cases} \tag{4.4}$$

where the subscript n stands for the nominal value of the variable working load:

D_n = dead load L_n = live load S_n = snow load
W_n = wind load E_n = earthquake load

ϕ_i and γ_i are considered to have optimal values; a resistance factor ϕ value of 0.7 to 0.85 is recommended.   As discussed in Section 4.6, in this probabilistic approach the present ACI Code would require that

$$\phi_i R_n = \text{maximum} \begin{cases} 1.4D_n + 1.7L_n \\ 0.9D_n - 1.3W_n \end{cases} \tag{4.5}$$

As more substantive records of performance are compiled with time, the details of the foregoing approach to reliability, safety, and reserve strength evaluation of structural components can be more universally accepted and extended beyond treatment of the component elements to the treatment of the total structural system.

4.6 ACI LOAD FACTORS AND SAFETY MARGINS

The general concept of safety and reliability of performance presented in the preceding sections is inherent in a more simplified but less accurate fashion in the ACI Code. The γ load factors and the ϕ strength reduction factors give an overall safety factor based on load types such that

$$\text{SF} = \frac{\gamma_1 D + \gamma_2 L}{D + L} \times \frac{1}{\phi} \tag{4.6}$$

where ϕ is the strength reduction factor and γ_1 and γ_2 are the respective load factors for the dead load D and the live load L. Basically, a single common factor is used for dead load and another for live load. Variation in resistance capacity is considered in the ϕ reduction factor. Hence the method is a simplified empirical approach to the safety and reliability of structural performance that is not economically efficient for every case and not fully adequate in other instances, such as combinations of dead and wind loads.

The ACI factors are termed *load factors*, because they restrict the estimation of reserve strength to the loads only as compared to the other parameters listed in Table 4.4. The estimated service or working loads are magnified by the coefficients, such as a coefficient of 1.4 for dead loads and 1.7 for live load. The types of normally occurring loads can be identified as (1) dead load, D; (2) live load, L; (3) wind load, W; (4) loads due to lateral pressure such as from soil in a retaining wall, H; (5) lateral fluid pressure loads, F; (6) loads due to earthquake, E; and (7) loads due to time-dependent effects, such as creep or shrinkage.

The basic combination of vertical loads is dead load plus live load. The *dead load*, which constitutes the weight of the structure and other relatively permanent features, can be estimated more accurately than the live load. The *live load* is estimated using the weight of nonpermanent loads, such as people and furniture. The transient nature of live loads makes them difficult to estimate more accurately. Therefore, a higher load factor is normally used for live loads than for dead loads. If the combination of loads consists only of live and dead loads, the ultimate load can be taken as

$$U = 1.4D + 1.7L \tag{4.7a}$$

Structures are seldom subjected to dead and live loads alone; wind load is often present. For structures in which wind load should be considered, the recommended combination is

$$U = 0.75(1.4D + 1.7L + 1.7W) \qquad (4.7b)$$

Maximum dead, live, and wind loads rarely, if ever, occur simultaneously. Hence the total factored load has to be reduced using a reduction factor of 0.75. Since wind load is applied laterally, it is possible that the absence of vertical live load while wind load is present can produce maximum stress. The following load combination should also be used to arrive at the maximum value of the factored load U:

$$U = 0.9D + 1.3W \qquad (4.7c)$$

Structures that have to resist lateral pressure due to earth fill or fluid pressure should be designed for the worst of the following combinations of factored loads:

$$U = 1.4D + 1.7L + 1.7H \qquad (4.8a)$$
$$U = 0.9D + 1.7H \qquad (4.8b)$$
$$U = 1.4D + 1.7L \qquad (4.8c)$$
$$U = 1.4D + 1.7L + 1.4F \qquad (4.8d)$$
$$U = 0.9D + 1.4F \qquad (4.8e)$$
$$U = 1.4D + 1.7L \qquad (4.8f)$$

The following combinations must be considered for earthquake loading:

$$U = 0.75(1.4D + 1.7L + 1.87E) \qquad (4.9a)$$
$$U = 0.9D + 1.43E \qquad (4.9b)$$

or

$$U \geq 1.4D + 1.7L \qquad (4.9c)$$

whichever is largest. The philosophy used for combining the various load components for earthquake loading is essentially the same as that used for wind loading.

4.7 DESIGN STRENGTH VERSUS NOMINAL STRENGTH: STRENGTH REDUCTION FACTOR φ

The strength of a particular structural unit calculated using the current established procedures is termed *nominal strength*. For example, in the case of a beam, the resisting moment capacity of the section calculated using the equations of equilibrium and the properties of concrete and steel is called the *nominal resisting moment capacity M_n* of the section. This nominal strength is reduced using a strength reduction factor, φ, to account for inaccuracies in construction, such as in the dimensions or position of reinforcement or variations in properties. The reduced strength of the member is defined as the design strength of the member.

TABLE 4.5   RESISTANCE OR STRENGTH REDUCTION
FACTOR ϕ

Structural element	Factor ϕ
Beam or slab: bending or flexure	0.9
Columns with ties	0.7
Columns with spirals	0.75
Columns carrying very small axial loads (refer to Chapter 9 for more details)	0.7–0.9 or 0.75–0.9
Beam: shear and torsion	0.85

For a beam, the design moment strength ϕM_n should be at least equal to or slightly greater than the external factored moment M_u for the worst condition of factored load U. The factor ϕ varies for the different types of behavior and for the different types of structural elements. For beams in flexure, for instance, the reduction factor is 0.9.

For tied columns that carry dominant compressive loads, the ϕ factor equals 0.7. The smaller strength reduction factor used for columns is due to the structural importance of the columns in supporting the total structure compared to other members and to guard against progressive collapse and brittle failure with no advance warning of collapse. Beams, on the other hand, are designed to undergo excessive deflections before failure. Hence the inherent capability of the beam for advanced warning of failure permits the use of a higher strength reduction factor or resistance factor.

Table 4.5 summarizes the resistance factors ϕ for various structural elements as given in the ACI Code. A comparison of these values to those given in Ref. 4.8 indicates that the ϕ values in this table, as well as the load factors of Eq. 4.8, are in some cases more conservative than they should be. In cases of earthquakes, wind, and shear forces, the probability of load magnitude and reliability of performance is subject to higher randomness and hence a higher coefficient of variation than the other types of loading.

4.8 QUALITY CONTROL AND QUALITY ASSURANCE

Quality control assures the reliability of performance of the designed system in accordance with assumed and expected reserve strengths in the design. To exercise "quality control" and achieve "quality assurance" encompasses monitoring the roles and performance of all participants: the client or owner, the designer, the concrete producer, the laboratory tester, the constructor, and the user.

Most of the different phases of the total process of construction are affected by complex standards and regulations of the various codifying agencies. Also, in contrast to mechanized production such as in the case of machines, building construction does not follow the moving-belt or chain-production process, where the products move but the workers are relatively stationary; the contrary is true.   Consequently, complications are more profound in constructed systems such as those made with concrete.   This is due partially to the fact that concrete is a nonhomogeneous material with properties dependent on many variables, requiring extra effort in quality control due to the greater effect of the human factor on the quality of the finished product.

The reliability of the performance of personnel involved in the various stages of creating a concrete structural system from conception through design, construction, and use depends on knowledge, training, and communication at all levels. A smooth flow of correct information among all participants and a shared systematic understanding of the developing problems lead to increased motivation toward improved solutions and hence improved quality control and a resulting high level of quality assurance.   In summary, a quality assurance system needs to be provided based on the exercise of quality control at the various phases and interacting parameters of a total system, as shown in Fig. 4.7.

4.8.1 The User

Construction of a designed system is governed basically by five primary tasks: planning, design, materials selection, construction, and use (including maintenance).   Figure 4.8 presents schematically the sequence of these enumerated tasks and the respective divisions of responsibility.   As seen from this diagram, the process starts with the user, since the principal aim of a project is to satisfy the user's needs, and it culminates with the user as the primary beneficiary of the final product.

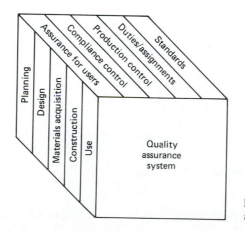

Figure 4.7 Components of a quality assurance system. (From Ref. 4.12.)

Photo 23   Flexural behavior of a reinforced concrete beam.   (Test by Nawy et al.)

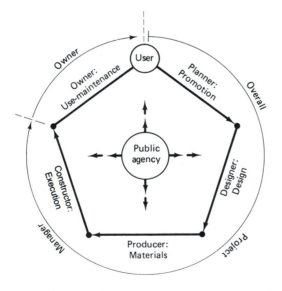

Quality control:

●   Compliance control at these points

――――   Production control at these bars

◀― ◀―   Agency overall inspection

Figure 4.8   Quality control schematic.

Quality assurance is necessary to satisfy the user's needs and rights. It ensures that the activities influencing the final quality of a concrete structure are:

1. Based on clearly defined fundamental requirements that satisfy the operational, environmental, and boundary conditions set for the project at the outset
2. Properly presented in accurate, well-dimensioned engineering working drawings based on optimal design procedures
3. Correctly and efficiently carried out by competent personnel in accordance with predetermined plans and working drawings well supervised during the design stage
4. Systematically executed in accordance with detailed specifications that comform to the applicable codes and local regulations

To achieve these aims, the expertise denoted by the other components of the polygon in Fig. 4.8 are called upon, starting with the planner and designer and culminating with the constructor.

4.8.2 Planning

In order to plan the successful execution of a proposed constructed system, all main and subactivities have to be clearly defined. This is accomplished through dividing the total project into a network plan of separately defined activities, relating each activity with time, analyzing the input control and the resulting output control, and expressing these conditions in the form of a checklist. In such a manner, the successful decision-making process concerning performance requirements becomes easier to accomplish. Such a process usually entails decisions on *what* function needs to be accomplished in a construction project, *where* and *when* that function will be executed, *how* the system will be constructed, and *who* the user will be. Correct determination of these factors leads to a decision as to the level of quality control needed and the degree of quality assurance that is expected to result.

4.8.3 Design

Quality control in design aims at verifying that the designed system has the safety, serviceability, and durability required for the use to which the system is intended as required by the applicable codes, and that such a design is correctly presented in the working drawings and the accompanying specifications. The degree of quality control depends on the type of system to be constructed: the more important the system, the more control that is required.

As a minimum, a design must always be checked by an engineer other than the originating design engineer. Usually, one of three types of verifications is used, depending on the practice of the designing agency: (1) total direct checking, in which all computations are verified; (2) total parallel checking, in which calculations are *independently* made and the two sets of calculations compared; and (3) partial checking, in which selected parts are checked in both direct and parallel checking.

Quality control of the design calculations can generally be achieved through assuring that:

1. A clear understanding exists of the structural concept that applies to the particular system.
2. There is knowledge and compliance with the relevant fundamental requirements of the design and the environmental, operational, and boundary conditions.
3. Where possible, applicable calculation models utilizing available computer programs are used for checks.
4. No discrepancies exist between the different phases or parts of the total design computations.
5. All expected load cases and load combinations as described in Section 4.6 are considered.
6. The appropriate safety factors are adopted and the required reliability levels verified.
7. Verifiable computer programs are used in the design, and the experienced designer is well acquainted with the programming steps and background of the programs, particularly when total computer-aided designs are used.

Since engineering working drawings are the primary link between the design process and the construction process, they should be a major object of design quality assurance. Consequently, the student has to be well acquainted with reading and interpreting working drawings and must be able to produce clear sketches that accurately express the design details if the constructed system is to reflect the actual design. Figure 4.4, as well as Figs. 10.12 to 10.21, are intended to give general guidance on the systematic detailing necessary for composing sets of logical engineering working drawings.

Quality control of working drawings normally encompasses a verification of whether the following parameters are included in the project set of drawings:

1. General definition of the structure
2. Consistency among the working drawings
3. Compliance with the site boundary conditions, including soil test boring requirements
4. Listing of the type, grade, quality, and structural strength of the various construction materials involved, such as cement, concrete mix proportion and strength, reinforcing steel, and formwork
5. No ambiguity and risk of misunderstanding of the details in the drawings

6. Compliance with the design calculation results and correct dimensioning
7. Adequate cross-sectional and construction details, as well as explicit dimensional tolerances
8. Sequence of formwork placing and removal

4.8.4 Materials Selection

It has to be emphasized at the outset that the quality of materials such as reinforced concrete is not determined only by compressive or tensile strength tests. As seen from previous sections of this book, many other factors affect the quality of the finished product, such as water/cement ratio, cement content, creep and shrinkage characteristics, freeze and thaw properties, and other durability aspects and conditions.

Two types of quality are involved: (1) *required quality*, which is the specified contractural requirement for the material, and (2) *usage quality*, which is the ability of the material to satisfy the needs of the user (Ref. 4.12).

Required quality of the material, such as ready-mix concrete, is assured by production control. Such a process involves:

1. General organization of the production staffing and operation.
2. Production sequence and supply line of the constituent materials, such as stone, fine aggregate, cement, and additives.
3. Internal control, involving frequency of verifications and tests, analysis of test results, the recording and observation methods used, and the procedures applied in dealing with discrepancies and deviations.
4. Use of statistical control charts to classify the specified requirements of quality levels into measurable main variables and nonmeasurable variables, selection of the main variables to be controlled by the control charts, and the preparation of a mean chart and a range chart for each variable selected. The measurable variables are to be controlled using $\overline{X}$ and $\overline{R}$ charts, while the nonmeasurable variables are controlled by $\overline{p}$ and $\overline{c}$ charts, denoting the averages of means and variations. An action limit and a warning limit with an upper and a lower boundary need to be specified.
5. Classification of defects as a measure of the nondefinable variables.

Usage quality is determined by the compliance control set in the specifications. These are prescribed by the client for the mix proportioning and design and the constituent materials of the concrete, as well as the quality of the reinforcement, whether normal or prestressing reinforcement. A projection of the expected quality control record and hence quality assurance of the concrete, for instance, can normally be made if the concrete producer has maintained good statistical quality control of strength test results over a lengthy period of operation. Hence compliance control levels can vary depending on the reliability and confidence in the effectiveness of quality management of the material producer to conform with the specifications of the delivered lots.

4.8.5 Construction

Construction is the execution stage of a project, which can be used to satisfy all design and specification requirements within prescribed time limits at minimum cost.   To achieve the desired quality assurance, the construction phase has to be preceded by an elaborate and correct preparation stage, which can be part of the design phase.   The preparation or planning phase is very critical since it gives an overall clear view of the various activities involved and the possible problems that could arise at the various phases of execution.   The use of computers in the planning phase is essential today for large projects in order that relevant input as to product quality, output, time scheduling, and costs can be charted and monitored.

The human factor is of major significance at the construction phase.   In most instances, the major site activities involve labor-force use and path scheduling of its utilization.   An improved information flow system, clear delineation of the chain of command, and reward for superior performance increase motivation and lead to overall improvement in the entire quality assurance system and an optimization of the efficiency/cost ratio of a project.   The steps to be carried out to achieve quality assurance, as proposed in Ref. 4.12, are summarized briefly next and are represented graphically in Fig. 4.9.

1. Organizational preparation covering planning, time scheduling, contract details, and definition and assignment of duties
2. On-site preparation, involving access roads, site trailers and offices, energy provisions, amenities, and so on
3. Formwork acquisition, type, and preparation for installation
4. Reinforcement procurement, fabrication, and planning
5. Concrete mix proportioning, laboratory mix designing, and coordination with design engineers
6. Concrete delivery, placing in the forms, and field slump testing
7. Curing and surface treatment of the hardened concrete

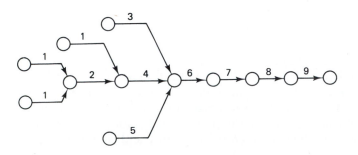

Figure 4.9 Operations sequence network for concrete structures.

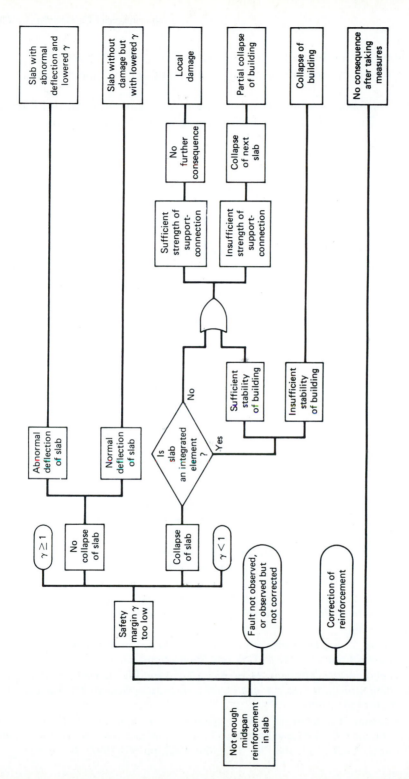

Figure 4.10 Cause–effect flow chart. (From Ref. 4.12.)

8. Quality control tests of the concrete at 7- and 28-day intervals

9. Removal of formwork, sequential removal of shoring supports, then reshoring

The probability of errors in execution for quality control can normally be expected. The extent and importance of such errors depend on a variety of factors described in previous sections. To apply corrective measures, a logical sequence of steps has to be followed for the detection and analysis of the undesired occurrence. A quantitative analysis of the error impact can often be made provided that the probabilities of occurrence of all the basic events are known to the investigators.

The flow chart in Fig. 4.10 depicts the cause–effect sequence that can be followed in identifying an undesired event in a quality assurance program. γ in the chart is the safety margin factor available in the original design.

In summary, the brief discussion in Section 4.8 should provide the reader with an introduction to a continuously evolving topic that has a profound impact on the strength and durability of constructed systems, that is, quality control and quality assurance. It should complement Section 4.5 on reliability and structural safety and Sections 4.6 and 4.7 on load factors and design strengths.

SELECTED REFERENCES

4.1. American Society for Testing and Materials, *Standard Specification for Deformed and Plain Billet-Steel Bars for Concrete Reinforcement*, A6 15-79, ASTM, Philadelphia, 1980, pp. 588–599.

4.2. American Society for Testing and Materials, *Standard Specification for Rail-Steel Deformed and Plain Bars for Concrete Reinforcement*, A6 16-79, ASTM, Philadelphia, 1980, pp. 600–605.

4.3. American Society for Testing and Materials, *Standard Specification for Axle Steel Deformed and Plain Bars for Concrete Reinforcement*, A6 17-79, ASTM, Philadelphia, 1980, pp. 606–611.

4.4. American Society for Testing and Materials, *Standard Specification for Cold-Drawn Steel Wire for Concrete Reinforcement*, A8 2-79, ASTM, Philadelphia, 1980, pp. 154–157.

4.5. American Society for Testing and Materials, *Standard Specification for Low-Alloy Steel Deformed Bars for Concrete Reinforcement*, A706-79, ASTM, Philadelphia, 1980, pp. 755–760.

4.6. Baker, A. L. L., *The Ultimate Load Theory Applied to the Design of Reinforced and Prestressed Concrete Frames*, Concrete Publications, London, 1956, 91 pp.

4.7. Galambos, T. V., "Proposed Criteria for Load and Resistance Factor Design of Steel Building Structures," *Steel Research for Construction Bulletin 27*, American Iron and Steel Institute, January 1978.

4.8. Ellingwood, B., McGregor, J. G., Galambos, T. V., and Cornell, C. A., "Probability Based Load Criteria: Load Factors and Load Combinations," *Journal of the Structural Division, ASCE*, Vol. 108, No. ST5, May 1982, pp. 978–997.

4.9. National Bureau of Standards, *Development of a Probability Based Load Criterion for American National Standard A58—Building Code Requirements for Minimum Design Loads in Buildings and Other Structures*, Special Publication 577, NBS, Washington, D.C., June 1980, 221 pp.

4.10. American National Standards Institute, *Minimum Design Loads for Buildings and Other Structures*, ANSI A58, ANSI, New York, 1982, 100 pp.

4.11. ACI Committee 318, *Building Code Requirements for Reinforced Concrete*, ACI Standard 318-95; and the *Commentary on Building Code Requirements for Reinforced Concrete*, American Concrete Institute, Detroit, 1995.

4.12. Comité Euro-International du Béton, *Quality Control and Quality Assurance for Concrete Structures*, CEB Commission I (Dr. A. G. Meseguer, Chairman), Bulletin d'Information 157, Paris, March 1983, 98 pp.

4.13. Riggs, J. L., *Production Systems: Planning, Analysis, and Control*, Wiley, New York, 1981, 649 pp.

FLEXURE IN BEAMS

5

5.1 INTRODUCTION

Loads acting on a structure, be they live gravity loads or other types, such as horizontal wind loads or those due to shrinkage and temperature, result in bending and deformation of the constituent structural elements. The bending of the beam element is the result of the deformational strain caused by the flexural stresses due to the external load.

As the load is increased, the beam sustains additional strain and deflection, leading to development of flexural cracks along the span of the beam. Continuous increases in the level of the load lead to failure of the structural element when the external load reaches the capacity of the element. Such a load level is termed the *limit state of failure in flexure*. Consequently, the designer has to design the cross section of the element or beam such that it would not develop excessive cracking at service load levels and have adequate safety and reserve strength to withstand the applied loads or stresses without failure.

Flexural stresses are a result of the external bending moments. They control in most cases the selection of the geometrical dimensions of a reinforced concrete

Photo 24 Priory Church, St. Louis, Missouri. (Courtesy Portland Cement Association.)

section.   The design process through the selection and analysis of a section is usually started by satisfying the flexural (bending) requirements, except for special components such as footings.   Thereafter, other factors, such as shear capacity, deflection, cracking, and bond development of the reinforcement, are analyzed and satisfied.

While the input data for the analysis of sections differ from the data needed for design, every design is essentially an analysis.   One assumes the geometrical properties of a section in a design and proceeds to analyze such a section to determine if it can safely carry the required external loads.   Hence a good understanding of the fundamental principles in the analysis procedure significantly simplifies the task of designing sections.   The basic mechanics of materials principles of equilibrium of internal couples have to be adhered to at all stages of loading.

Photo 25 Empire State Performing Arts Center (Albany, New York) during construction.

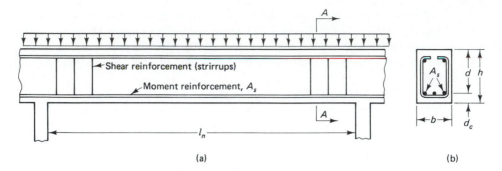

Figure 5.1 Typical reinforced concrete beam: (a) elevation; (b) section $A-A$.

If a beam is made up of homogeneous, isotropic, and linearly elastic material, the maximum bending stress can be obtained using the well-known beam flexure formula $f = Mc/I$. At ultimate load, the reinforced concrete beam is neither homogeneous nor elastic, thereby making that expression inapplicable for evaluating the stresses. However, the basic principles of the theory of bending can still be used to analyze reinforced concrete beam cross sections. Figure 5.1 shows a typical continuous reinforced concrete beam. If the beam is so proportioned that all its constituent materials attain their capacity prior to failure, both the concrete and the steel fail simultaneously at midspan when the ultimate strength of the beam is reached. The corresponding strain and stress diagrams are shown in Fig. 5.2.

The following assumptions are made in defining the behavior of the section:

1. Strain distribution is assumed to be linear. This assumption is based on Bernoulli's hypothesis that plane sections before bending remain plane and perpendicular to the neutral axis after bending.
2. Strain in the steel and the surrounding concrete is the same prior to cracking of the concrete or yielding of the steel.
3. Concrete is weak in tension. It cracks at an early stage of loading at about 10% of its limit compressive strength. Consequently, concrete in the tension zone of the section is neglected in the flexural analysis and design computations, and the tension reinforcement is assumed to take the total tensile force.

To satisfy the equilibrium of the horizontal forces, the compressive force C in the concrete and the tensile force T in the steel should balance each other, that is,

$$C = T \tag{5.1}$$

The terms in Fig. 5.2 are defined as follows:

b = width of the beam at the compression side

d = depth of the beam measured from the extreme compression fiber to the centroid of steel area

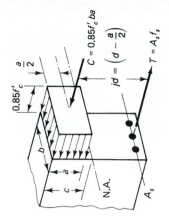

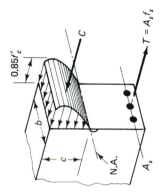

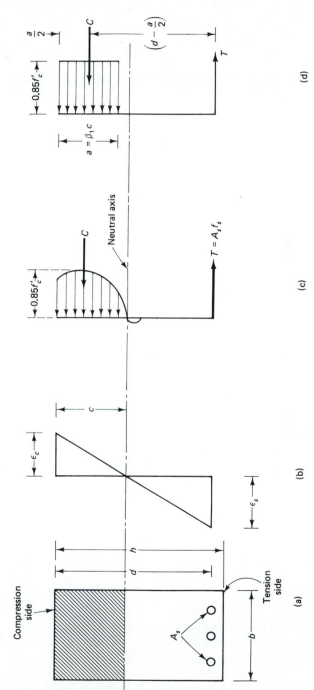

Figure 5.2 Stress and strain distribution across beam depth: (a) beam cross section; (b) strains; (c) actual stress block; (d) assumed equivalent stress block.

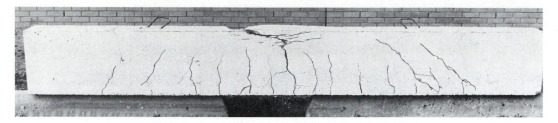

Photo 26   Simply supported beam in flexural failure.   (Tests by Nawy.)

h = total depth of the beam

A_s = area of the tension steel

ϵ_c = strain in extreme compression fiber

ϵ_s = strain at the level of tension steel

f'_c = compressive strength of the concrete

f_s = stress in the tension steel

f_y = yield strength of the tension reinforcement

c = depth of the neutral axis measured from extreme compression fibers

5.2 THE EQUIVALENT RECTANGULAR BLOCK

The actual distribution of the compressive stress in a section has the form of a rising parabola, as shown in Fig. 5.2c. It is time consuming to evaluate the volume of the compressive stress block if it has a parabolic shape. An equivalent rectangular stress block due to Whitney can be used with ease and without loss of accuracy to calculate the compressive force and hence the flexural moment strength of the section. This equivalent stress block has a depth a and an average compressive strength $0.85f'_c$. As seen from Fig. 5.2d, the value of $a = \beta_1 c$ is determined using a coefficient β_1 such that the area of the equivalent rectangular block is approximately the same as that of the parabolic compressive block, resulting in a compressive force C of essentially the same value in both cases.

The $0.85f'_c$ value for the average stress of the equivalent compressive block is based on the core test results of concrete in the structure at a minimum age of 28 days. Based on exhaustive experimental tests, a maximum allowable strain of 0.003 in./in. was adopted by the ACI as a safe limiting value. Even though several forms of stress blocks including trapezoidal have been proposed to date, the simplified equivalent rectangular block is accepted as the standard in the analysis and design of reinforced concrete. The behavior of the steel is assumed to be elastoplastic, as shown in Fig. 5.3a.

Using all the preceding assumptions, the stress distribution diagram shown in Fig 5.2c can be redrawn as shown in Fig. 5.2d. One can easily deduce that the

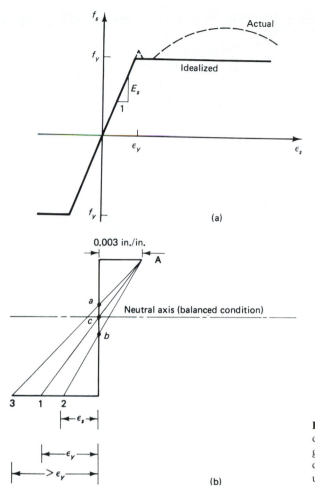

Figure 5.3 Strain distribution across depth: (a) idealized stress–strain diagram of the reinforcement; (b) strain distribution for various modes of flexural failure.

compression force C can be written as $0.85f'_c ba$, that is, the *volume* of the compressive block at or near the ultimate when the tension steel has yielded, $\epsilon_s > \epsilon_y$. The tensile force T can be written as $A_s f_y$. Thus equilibrium Eq. 5.1 can be rewritten as

$$0.85f'_c ba = A_s f_y \tag{5.2}$$

or

$$a = \frac{A_s f_y}{0.85f'_c b} \tag{5.3}$$

The moment of resistance of the section, that is, the nominal strength M_n, can be expressed as

$$M_n = (A_s f_y)jd \quad \text{or} \quad M_n = (0.85f'_c ba)jd \tag{5.4a}$$

Photo 27   Close-up of flexural cracks in Photo 26.

where jd is the lever arm, denoting the distance between the compression and tensile forces of the internal resisting couple.   Using the simplified equivalent rectangular stress block from Fig. 5.2d, the lever arm is

$$jd = d - \frac{a}{2}$$

Hence the nominal moment of resistance becomes

$$M_n = A_s f_y \left(d - \frac{a}{2}\right) \tag{5.4b}$$

Since $C = T$, the moment equation can also be written as

$$M_n = 0.85 f'_c ba \left(d - \frac{a}{2}\right) \tag{5.4c}$$

If the reinforcement ratio $\rho = A_s/bd$, Eq. 5.3 can be rewritten as

$$a = \frac{\rho d f_y}{0.85 f'_c}$$

If $r = b/d$, Eq. 5.4c becomes

$$M_n = \rho r d^2 f_y \left(d - \frac{\rho d f_y}{1.7 f'_c}\right) \tag{5.5a}$$

or $$M_n = [\omega r f'_c (1 - 0.59\omega)] d^3 \tag{5.5b}$$

where $\omega = \rho f_y / f'_c$.   Equation 5.5b is sometimes expressed as

$$M_n = R b d^2 \tag{5.6a}$$

where

$$R = \omega f'_c(1 - 0.59\omega) \tag{5.6b}$$

Equations 5.5 and 5.6 are useful for the development of charts.   A plot of the R value for singly reinforced beams is shown in Fig. 5.4.

If f'_c, f_y, b, d, and A_s are given for a rectangular section and the beam is so proportioned and reinforced that failure occurs by simultaneous yielding of the tension steel and crushing of the concrete at the compression side, the resisting moment strength can be obtained using Eq. 5.4 or 5.5, but using the balanced steel area A_{sb} and the balanced rectangular block depth a_b instead of A_s and a.   However, beams have to be designed to fail in tension by initial yielding of the reinforcement, for reasons explained in subsequent sections.

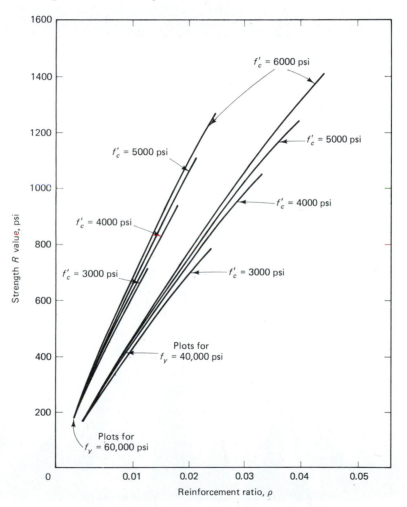

Figure 5.4   Strength–R curves for singly reinforced beams.

Photo 28 Beam at failure subjected to combined compression and bending.   (Tests by Nawy et al.)

Depending on the type of failure, that is, yielding of the steel or crushing of the concrete, three types of beams can be identified.

1. *Balanced section.*   The steel starts yielding when the concrete just reaches its ultimate strain capacity and commences to crush.   At the start of failure, the permissible extreme fiber compressive strain is 0.003 in./in., while the tensile strain in the steel equals the yield strain $\epsilon_y = f_y/E_s$.   The "balanced" strain distribution follows line $Ac1$ in Fig. 5.3b across the depth of the beam.

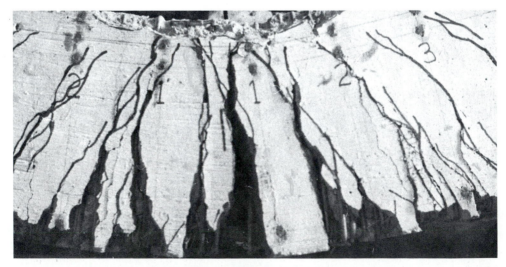

Photo 29 Cracking level at rupture.

2. *Overreinforced section.* Failure occurs by initial crushing of the concrete. At the initiation of failure, the steel strain ϵ_s will be lower than the yield strain ϵ_y, as in line $Ab2$, Fig. 5.3b; hence the steel stress f_s will be lower than its yield strength f_y. Such a condition is accomplished by using more reinforcement at the tension side than that required for the balanced condition.

3. *Underreinforced section.* Failure occurs by initial yielding of the steel, as in line $Aa3$, Fig. 5.3b. The steel continues to stretch as the steel strain increases beyond ϵ_y. This condition is accomplished when the area of the tension reinforcement used in the beam is less than that required for the balanced strain condition.

Note from positions c, b, and a of the neutral axis that the axis rises toward the compressive fibers in the underreinforced beam as the limit state of failure is reached. This behavior is easily identified in tests as the flexural cracks propagate toward the compression fibers until the concrete crushes. It should also be recognized that the vertical distances between points c, b, and a of the neutral axis from the extreme compression fibers for the three types of failure depend largely on the percentage ratio $\rho = A_s/bd$, but do not differ significantly since low values of strain are involved.

Concrete failure is sudden since it is a brittle material. Therefore, almost all codes of practice recommend designing underreinforced beams to provide sufficient warning, such as excessive deflection before failure. In the case of statically indeterminate structures, ductile failure is essential for proper moment redistribution. Hence for beams the ACI Code limits the maximum amount of steel to 75% of that required for a balanced section. For practical purposes, however, the reinforcement ratio A_s/bd should not normally exceed 50%, to avoid congestion of the reinforcement and facilitate proper placing of the concrete. If the actual reinforcement ratio and the balanced reinforcement ratio are denoted as ρ and $\bar{\rho}_b$, respectively, then

$$\rho \leq 0.75\bar{\rho}_b \tag{5.7a}$$

The code also stipulates the minimum steel requirement as

$$A_{s,min} = \frac{3\sqrt{f_c'}}{f_y} b_w d \geq \frac{200 b_w d}{f_y} \tag{5.7b}$$

and for statically determinate T section with the flange in tension, or for cantilevers,

$$A_{s,min} = \frac{6\sqrt{f_c'}}{f_y} b_w d \geq \frac{200 b_w d}{f_y} \tag{5.7c}$$

but not greater than that calculated by Eq. 5.7b with b_w set equal to the width of the flange. Both Eqs. 5.7b and c need not be applied to each section, provided that A_s is at least *one third greater* than required by analysis.

5.3 BALANCED REINFORCEMENT RATIO $\bar{\rho}_b$

To analyze a given beam, one has to determine first the maximum allowable reinforcement ratio $0.75\bar{\rho}_b$. For rectangular sections reinforced only at the tension side, $\bar{\rho}_b$ is a function of only concrete strength and properties of steel, that is, modulus of elasticity E_s and yield strength f_y, irrespective of the section geometry. Using the strain distribution diagram in Fig. 5.2 for the balanced strain condition and from similar triangles, the relationship between the depth c (c_b for the balanced condition) of the neutral axis and the effective depth d can be written as

$$\frac{c_b}{d} = \frac{0.003}{0.003 + f_y/E_s}$$

If E_s is taken as 29×10^6 psi,

$$\frac{c_b}{d} = \frac{87,000}{87,000 + f_y} \tag{5.8a}$$

The relationship between the depth a of the equivalent rectangular stress block and the depth c of neutral axis is

$$a = \beta_1 c \tag{5.8b}$$

The value of the stress block depth factor β_1 is

$$\beta_1 = \begin{cases} 0.85 & \text{for } 0 < f_c' \leq 4000 \text{ psi} \\ 0.85 - 0.05\left(\dfrac{f_c' - 4000}{1000}\right) & \text{for } 4000 \text{ psi} < f_c' \leq 8000 \text{ psi} \\ 0.65 & \text{for } f_c' > 8000 \text{ psi} \end{cases}$$

Hence, for the balanced strain condition, the depth of the rectangular stress block is

$$a_b = \beta_1 c_b$$

For equilibrium of the horizontal forces

$$A_{sb}f_y = 0.85f_c'ba_b$$

or

$$\bar{\rho}_b = \frac{A_{sb}}{bd} = \frac{0.85f_c'}{f_y}\frac{a_b}{d}$$

From Eq. 5.8, the balanced steel ratio becomes

$$\bar{\rho}_b = \beta_1 \frac{0.85f_c'}{f_y}\frac{87,000}{87,000 + f_y} \tag{5.9}$$

where f_c' and f_y are expressed in psi. Thus, if f_c' and f_y are known, $\bar{\rho}_b$ and thus $0.75\bar{\rho}_b$ can be readily obtained regardless of the geometry of the concrete section.

TABLE 5.1   MAXIMUM PERMISSIBLE REINFORCEMENT RATIO $(0.75\,\bar{\rho}_b \times 10^{-4})$ FOR BEAMS WITH TENSION REINFORCEMENT ONLY (SINGLY REINFORCED BEAMS)[a]

f_y (psi)	$f'_c = 4000$ $\beta_1 = 0.85$	$f'_c = 5000$ $\beta_1 = 0.80$	$f'_c = 6000$ $\beta_1 = 0.75$	$f'_c = 9000$ $\beta_1 = 0.65$
40,000	371	437	491	638
60,000	214	252	283	367
80,000	141	166	187	243

[a]The values for the reinforcement ratio ρ in both the standard and SI units are almost identical since the ρ is a dimensionless quantity. ACI 318 M–89, for example, rounds up the conversion values such that, for $f_y = 60,000$ psi ≈ 400 MPa and for $f'_c = 4000$ psi ≈ 27.6 MPa, $\rho = 214$ as above.

Representative values of the maximum permissible reinforcement ratio ρ for singly reinforced beams are given in Table 5.1 both in pounds and in SI units. These values are 75% of the balanced reinforcement ratio $\bar{\rho}_b$ and should aid the student in eliminating tedious computations of these frequently used values.

5.4 ANALYSIS OF SINGLY REINFORCED RECTANGULAR BEAMS FOR FLEXURE

The sequence of calculations presented in the flow chart of Fig. 5.5 can be used for the analysis of a given beam for both longhand and computers.   The flow chart was developed using the method of analysis presented in Section 5.2.   The following examples illustrate typical analysis calculations following the flow-chart logic in Fig. 5.5.

5.4.1 Example 5.1: Flexural Analysis of a Singly Reinforced Beam (Tension Reinforcement Only)

A singly reinforced concrete beam ($f'_c = 4000$ psi or 27.6 MPa) has the cross section shown in Fig. 5.6.   Determine if the beam is overreinforced or underreinforced and if it satisfies the ACI Code requirements for maximum and minimum reinforcement ratios for (a) $f_y = 60,000$ psi (413.4 MPa); (b) $f_y = 40,000$ psi (275.6 MPa).

Solution   (a) From Eq. 5.9 for the balanced condition,

$$\bar{\rho}_b = \beta_1 \frac{0.85 f'_c}{f_y} \frac{87,000}{87,000 + f_y}$$

$$f'_c = 4000 \text{ psi}$$

$$f_y = 60,000 \text{ psi}$$

$$\beta_1 = 0.85 \quad \text{for } f'_c = 4000 \text{ psi}$$

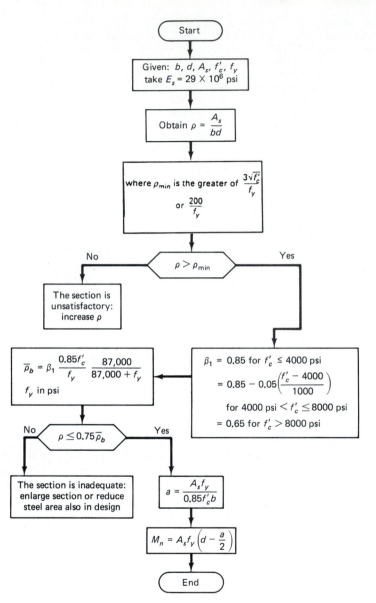

Figure 5.5 Flow chart for analysis of singly reinforced rectangular beams in bending.

Therefore,

$$\bar{\rho}_b = 0.85\left(\frac{0.85 \times 4000}{60,000}\right)\frac{87,000}{87,000 + 60,000} = 0.029$$

$$A_{sb} = \bar{\rho}_b bd = 0.029 \times 10 \times 18$$

$$= 5.22 \text{ in.}^2 \text{ (3367 mm}^2) < A_s = 6 \text{ in.}^2 \text{ (3870 mm}^2)$$

$$\rho = \frac{6.0}{10 \times 18} = 0.033$$

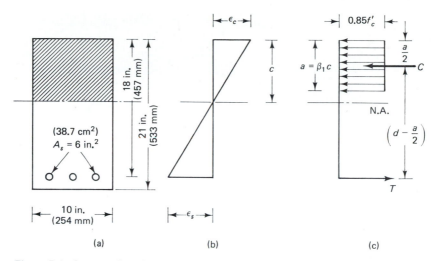

Figure 5.6 Stress and strain distribution in a typical singly reinforced rectangular section: (a) cross section; (b) strains; (c) stresses.

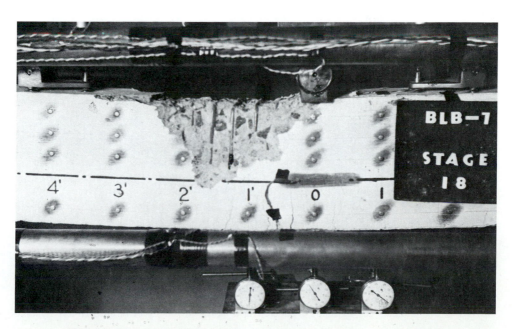

Photo 30 Beam subjected to combined axial load and bending. The neutral axis is at 70% of depth. (Tests by Nawy et al.)

Photo 31 Flexural cracking and deflection of beam subjected to flexure only prior to failure. (Tests by Nawy et al.)

Thus this section is overreinforced because $A_s > A_{sb}$ or $\rho > \bar{\rho}_b$ and does not satisfy ACI Code requirements for ductility and maximum allowable reinforcement.

(b) In a new trial with $f_y = 40,000$ psi, one finds that

$$\bar{\rho}_b = 0.85 \left(\frac{0.85 \times 4000}{40,000} \right) \frac{87,000}{87,000 + 40,000} = 0.0495$$

$$A_{sb} = 8.91 \text{ in.}^2 \ (5746.95 \text{ mm}^2) > 6 \text{ in.}^2 \ (3367 \text{ mm}^2)$$

Therefore, the section is underreinforced. From Eq. 5.7b,

$$\text{minimum allowable reinforcement ratio } \rho_{\min} = 3\sqrt{f_c'} f_y = \frac{3\sqrt{4000}}{40,000} = 0.0047$$

Photo 32 Crushing of concrete at compression side of beam subjected to flexure.

or

$$\rho_{min} = \frac{200}{f_y} = \frac{200}{40,000} = 0.005 \ll \text{actual } \rho$$

maximum allowable steel area $= 0.75 \times 8.91 = 6.68$ in.$^2 > 6$ in.^{2}

Therefore, the cross section satisfies ACI Code requirements for maximum and minimum reinforcement. Note that the actual steel area in case (b) is only slightly less than the maximum allowable 75% of the balanced steel area. Hence congestion of steel is likely and the design can be improved by increasing the beam section size and reducing A_s.

5.4.2 Example 5.2: Nominal Resisting Moment in a Singly Reinforced Beam

For the beam cross section shown in Fig. 5.7, calculate the nominal moment strength if f_y is 60,000 psi (413.4 MPa) and f'_c is (a) 3000 psi (20.68 MPa); (b) 5000 psi (34.47 MPa); (c) 9000 psi (62.10 MPa).

Solution
$$b = 10 \text{ in. (254.0 mm)}$$
$$d = 18 \text{ in. (457.2 mm)}$$
$$A_s = 4 \text{ in.}^2 \text{ (2580 mm}^2\text{)}$$
$$f_y = 60,000 \text{ psi}$$

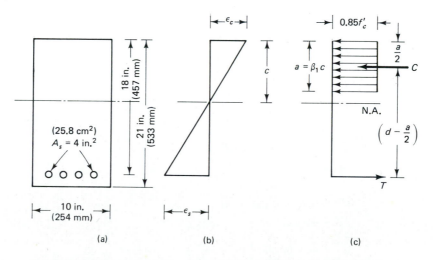

Figure 5.7   Beam cross-section strain and stress diagrams, Ex. 5.2: (a) cross section; (b) strains; (c) stresses.

Note that f_y should be in psi units in the ρ_{min} expression.

(a) $f_c' = 3000$ psi (20.68 MPa).

$$\rho_{min} = \frac{3\sqrt{f_c'}}{f_y} = \frac{3\sqrt{4000}}{60,000} = 0.0032 \quad \text{(controls)}$$

or

$$\rho_{min} = \frac{200}{f_y} = \frac{200}{60,000} = 0.0033$$

$$\rho = \frac{A_s}{bd} = \frac{4}{10 \times 18} = 0.0222 > 0.0032 \quad \text{O.K.}$$

$$\beta_1 = 0.85$$

Using Eq. 5.9,

$$\bar{\rho}_b = \beta_1 \frac{0.85 f_c'}{f_y} \frac{87,000}{87,000 + f_y}$$

$$= 0.85 \left(\frac{0.85 \times 3,000}{60,000} \right) \frac{87,000}{87,000 + 60,000} = 0.021$$

$$0.75\bar{\rho}_b = 0.016$$

$$\rho > 0.75\bar{\rho}_b$$

Hence the beam is considered overreinforced and does not satisfy the ACI requirements for ductility and maximum allowable reinforcement ratio.

(b) $f_c' = 5000$ psi (34.47 MPa).

$$\rho_{min} = \frac{3\sqrt{5000}}{60,000} = 0.0035$$

It is less than available, hence O.K.

$$\beta_1 = 0.85 - 0.05 \left(\frac{5000 - 4000}{1000} \right)$$

$$= 0.8$$

$$\bar{\rho}_b = 0.8 \left(\frac{0.85 \times 5000}{60,000} \right) \frac{87,000}{87,000 + 60,000}$$

$$= 0.034$$

$$0.75\bar{\rho}_b = 0.025 > \rho = 0.0222 \quad \text{O.K.}$$

$$A_s = 4 \text{ in.}^2$$

$$a = \frac{4 \times 60,000}{0.85 \times 5000 \times 10}$$

$$= 5.65 \text{ in.}$$

$$M_n = 4 \times 60,000 \left(18 - \frac{5.65}{2} \right) = 3,642,000 \text{ in.-lb}$$

$$= 303,500 \text{ ft-lb} \text{ (411.52 kN-m)}$$

Or, using Eq. 5.5 gives

$$\omega = \frac{0.0222 \times 60,000}{5000} = 0.267$$

$$M_n = \left[0.267 \times \frac{10}{18} \times 5000 \left(1 - 0.59 \times 0.267 \right) \right] 18^3$$

$$= 3,644,019 \text{ in.-lb } (411.77 \text{ kN-m})$$

(c) $f_c' = 9000$ psi (62.10 MPa).

ρ_{min} from parts (a) and (b) less than available ρ, hence O.K.

$\beta_1 = 0.65$ for $f_c' \geq 8000$ psi

$$\bar{\rho}_b = 0.65 \left(\frac{0.85 \times 9000}{60,000} \right) \frac{87,000}{87,000 + 60,000}$$

$$= 0.049$$

$$0.75\bar{\rho}_b = 0.037 > \rho = 0.022 \qquad \text{O.K.}$$

$$a = \frac{4.0 \times 60,000}{0.85 \times 9000 \times 10} = 3.14 \text{ in. } (79.8 \text{ mm})$$

$$M_n = 4.0 \times 60,000 \left(18 - \frac{3.14}{2} \right) = 3,943,200 \text{ in.-lb}$$

$$= 328,600 \text{ ft-lb } (445.6 \text{ kN-m})$$

5.5 TRIAL-AND-ADJUSTMENT PROCEDURES FOR THE DESIGN OF SINGLY REINFORCED BEAMS

In Ex. 5.2, the geometrical properties of the beam, that is, b, d, and A_s, were given.   In a design example, an assumption of width b (or the ratio b to d) and the level of reinforcement ratio ρ have to be made.   The ratio b/d varies between 0.25 and 0.6 in usual practice.   Although the ACI Code permits a reinforcement ratio ρ up to $0.75\bar{\rho}_b$, a ratio of $0.5\bar{\rho}_b$ is advisable to prevent congestion of steel, secure a good bond between the reinforcement and the adjacent concrete, and provide good deflection control.

Studies on cost optimum design indicate that cost-effective sections can be obtained using a minimum practical b/d ratio and a maximum practical reinforcement ratio ρ within the above-stated limitations.   Hence one could use the following steps to design the beam cross section following the flow-chart logic of Fig. 5.8.

1. Calculate the external factored moment.   To obtain the beam self-weight, an assumption has to be made for the value of d.   The minimum thickness for deflection specified in the ACI Code can be used as a guide.   Assume a b/d ratio r between 0.25 and 0.6 and calculate $b = rd$.   A first trial assumption $b \simeq d/2$ is recommended.

2. Choose a reinforcement ratio of approximately $0.5\bar{\rho}_b$.

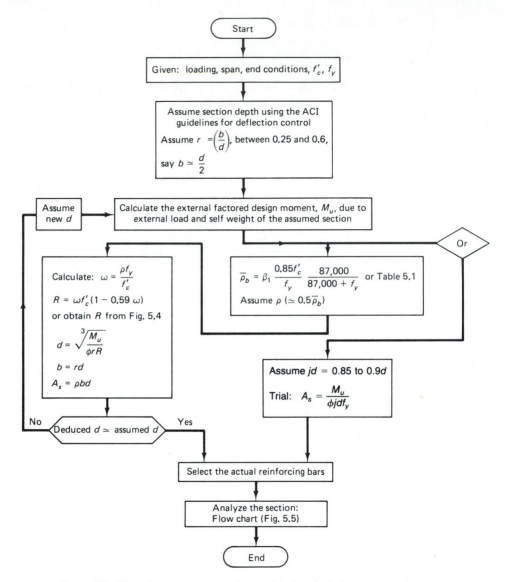

Figure 5.8 Flow chart for sequence of operations for the design of singly reinforced rectangular sections.

3. (a) Select a value of moment factor R based on an assumed ρ value $\simeq 0.5\bar{\rho}_b$. Assuming that $b \simeq d/2$, calculate d for $M_n = Rbd^2$ and proceed to analyze the section.

 (b) Alternatively, choose d on the basis of minimum deflection requirement. Choose a width b as in 3(a). Assume a moment arm $jd \simeq 0.85d$ to $0.90d$. Calculate A_s as a first trial, then analyze the section using $b = d/2$.

The process of arriving at the final section is highly convergent even by longhand computations in that it should not require more than three trial cycles. The use of hand-held or desktop personal computers (see Appendix A) enormously simplifies the design–analysis process and permits the student or engineer to proportion sections at a fraction of the time needed when using handbooks, charts, or longhand computations, easy as these other means can be.

For designers who prefer charts, Eq. 5.6 ($M_n = Rbd^2$) can be used for the first trial in design.   The value of R can be obtained from charts (see Fig. 5.4) for various values of ρ, f'_c, and f_y available in handbooks.

5.5.1 Example 5.3: Design of a Singly Reinforced Simply Supported Beam for Flexure

A reinforced concrete simply supported beam has a span of 30 ft (9.14 m) and is subjected to a service uniform live load $w_w = 1500$ lb/ft (21.9 kN/m), as shown in Fig. 5.9.   Design a beam section to resist the factored external bending load.   Given:

$$f'_c = 4000 \text{ psi (27.6 MPa)}$$
$$f_y = 60{,}000 \text{ psi (414 MPa)}$$

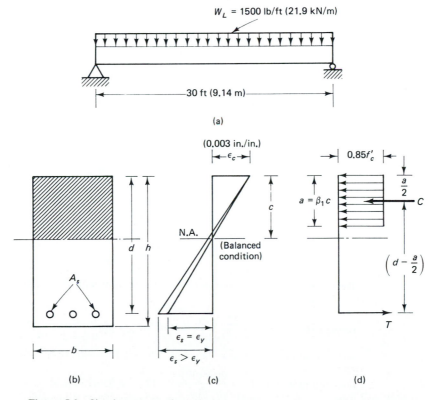

Figure 5.9 Simply supported reinforced concrete uniformly loaded beam: (a) elevation; (b) cross section; (c) strains; (d) stress.

Solution   Assume a minimum thickness from the ACI Code deflection table:

$$\frac{l_n}{16} = \frac{30 \times 12}{16} = 22.5 \text{ in.}$$

For the purpose of estimating the preliminary self-weight, assume total thickness $h = 24.0$ in., effective depth $d = 20$ in., and width of the beam $b = 10$ in. ($r = b/d = 0.5$).

$$\text{beam self-weight} = \frac{24 \times 10}{144} \times 150 = 250 \text{ lb/ft}$$

$$\text{factored load } w_u = 1.4D + 1.7L = 1.4 \times 250 + 1.7 \times 1500 = 2900 \text{ lb/ft}$$

$$\text{required factored moment } M_u = \frac{w_u l_n^2}{8} = \frac{2900 \times 30^2}{8} \times 12 = 3,915,000 \text{ in.-lb}$$

$$\text{required nominal resisting moment } M_n = \frac{M_u}{\phi} = \frac{3,915,000}{0.9} = 4,350,000 \text{ in.-lb}$$

Get $\bar{\rho}_b$ from Table 5.1, which gives $0.75\bar{\rho}_b$, or calculate:

$$\bar{\rho}_b = \beta_1 \frac{0.85 f_c'}{f_y} \frac{87,000}{87,000 + f_y} = 0.85 \left(\frac{0.85 \times 4000}{60,000} \right) \frac{87,000}{87,000 + 60,000} = 0.0285$$

Assume a reinforcement ratio $\rho = 0.5\bar{\rho}_b = 0.0143$.

$$\omega = \frac{\rho f_y}{f_c'} = \frac{0.0143 \times 60,000}{4000} = 0.215$$

Using Eq. 5.6b yields

$$R = \omega f_c'(1 - 0.59\omega) = 0.215 \times 4000(1 - 0.59 \times 0.215)$$

$$\approx 750$$

The value of R can also be obtained from the chart in Fig. 5.4 using the chosen ρ and the given values for f_c' and f_y.

Using Eq. 5.6a, one has $M_n = Rbd^2$ and, assuming that $b = 0.5d$,

$$d = \sqrt[3]{\frac{M_n}{0.5R}} = \sqrt[3]{\frac{4,350,000}{0.5 \times 750}} = 22.64 \text{ in.}$$

$$b = 0.5 \times 22.64 = 11.32 \text{ in.}$$

Based on practical considerations, try a section with $b = 12$ in., $d = 23$ in., and $h = 26$ in.

$$\text{revised self-weight} = \frac{12 \times 26}{144} \times 150 = 325 \text{ lb/ft}$$

$$\text{factored load } w_u = 1.4 \times 325 + 1.7 \times 1500 = 3005 \text{ lb/ft}$$

$$\text{factored moment } M_u = \frac{3005(30)^2}{8} \times 12 = 4,056,750 \text{ in.-lb}$$

$$\text{required resisting moment, } M_n = \frac{M_u}{\phi} = \frac{4,056,750}{0.9} = 4,507,500 \text{ in.-lb}$$

$$\rho_{min} = \frac{3\sqrt{f_c}}{f_y} \quad \text{or} \quad \frac{200}{f_y} \quad \frac{3\sqrt{f_c}}{f_y} = \frac{3\sqrt{4000}}{60,000} = 0.00316, \quad \frac{200}{f_y} = \frac{200}{60,000} = 0.0033$$

$$A_s = \rho bd = 0.0143 \times 12 \times 23 = 3.95 \text{ in.}^2$$

Try three No. 10 bars (32.3-mm diameter) with $A_s = 3.81$ in.^{2}

$$\rho = \frac{3.81}{12 \times 23} = 0.0138 < 0.75\overline{\rho}_b > \rho_{min} \quad \text{O.K.}$$

Check the nominal moment strength of the assumed section:

$$a = \frac{A_s f_y}{0.85 f'_c b} = \frac{3.81 \times 60,000}{0.85 \times 4000 \times 12} = 5.60 \text{ in.}$$

Available $M_n = 3.81 \times 60,000\left(23 - \dfrac{5.60}{2}\right) = 4,617,720$ in.-lb (521.8 kN-m)

$$> \text{required } M_n = 4,507,500 \text{ in.-lb}$$

Adopt the section. Note that the designed section resists a slightly larger moment than the required moment:

$$\text{percent overdesign} = \frac{4,617,720 - 4,507,500}{4,507,500} = 2.45\%$$

This is a reasonable level expected in proportioning concrete elements. It is always necessary to check that the web width can accommodate the number of bars in each layer based on concrete cover and minimum spacing requirements. In this example, the minimum web width to accommodate three No. 10 bars = 10.5 in. < b = 12.0 in., which is O.K.

Alternative Solution by Trial and Adjustment

$$\text{minimum thickness } h = \frac{l_n}{16} = \frac{30 \times 12}{16} = 22.5 \text{ in.}$$

Try $h = 26$ in. (660.4 mm), $d = 23.0$ in (584.2 mm), and $b \approx \frac{1}{2}d \approx 12$ in. (304.8 mm).

$$\text{self-weight} = \frac{12 \times 26}{144} \times 150 = 325.0 \text{ lb/ft}$$

$$\text{factored load } U = 1.4 \times 325.0 + 1.7 \times 1500 = 3005 \text{ lb/ft}$$

$$\text{factored moment } M_u = \frac{3005(30.0)^2}{8} \times 12$$

$$= 4,056,750 \text{ in.-lb (458.1 kN-m)}$$

$$\text{required nominal resisting moment } M_n = \frac{M_u}{\phi} = \frac{4,056,750}{0.9}$$

$$= 4,507,500 \text{ in.-lb (509.3 kN-m)}$$

Assume that moment arm $jd \approx 0.85d$:

$$\text{moment arm } jd \approx 0.85 \times d \approx 0.85 \times 23.0 = 19.55 \text{ in.}$$

$$M_n = A_s f_y\left(d - \frac{a}{2}\right) = A_s f_y jd \quad \text{or} \quad 4,507,500 = A_s \times 60,000 \times 19.55$$

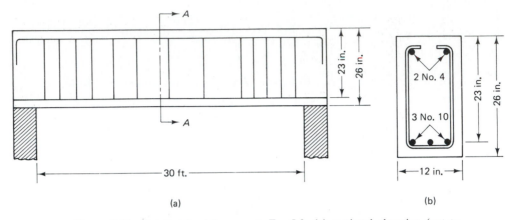

Figure 5.10 Details of reinforcement, Ex. 5.3: (a) sectional elevation (not to scale); (b) midspan section $A-A$.

Hence

$$A_s = \frac{4,507,500}{60,000 \times 19.55} = 3.84 \text{ in.}^2$$

Try three No. 10 bars (32.3-mm diameter = 3.81 in.² (2457.5 mm²). Continue the design following the flow chart in Fig. 5.8.

5.5.2 Arrangement of Reinforcement

Figure 5.10 shows the cross section of the beam at midspan. In arranging the reinforcing bars, one should satisfy the minimum cover requirements explained in Section 4.3. The required clear cover for beams is 1.5 in. (38 mm).

The stirrups shown in Fig. 5.10 should be designed to satisfy the shear requirements of the beam explained in Chapter 6. Two bars called *hangers* are placed on the compression side to support the stirrups. Reinforcement detailing provisions and bar development length requirements are discussed in Chapter 10.

5.6 ONE-WAY SLABS

One-way slabs are concrete structural floor panels for which the ratio of the long span to the short span equals or exceeds a value of 2.0. When this ratio is less than 2.0, the floor panel becomes a two-way slab or plate, as discussed in Chapter 11. A one-way slab is designed as a singly reinforced 12-in. (304.8-mm) wide beam strip using the same design and analysis procedure discussed earlier for singly reinforced beams. Figure 5.11 shows a one-way slab floor system.

Loading for slabs is normally specified in pounds per square foot (psf). One has to distribute the reinforcement over the 12-in. strip and specify the center-to-center spacing of the reinforcing bars. In slab design, a thickness is normally

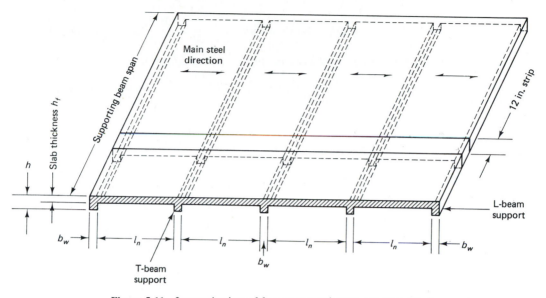

Figure 5.11 Isometric view of four-span continuous one-way-slab floor system.

assumed and the reinforcement is calculated using a trial lever arm $(d - a/2)$ or $0.9d$.

Supported slabs, that is, slabs not on grade, do not normally require shear reinforcement for typical loads. Transverse reinforcement has to be provided perpendicular to the direction of bending in order to resist shrinkage and temperature stresses. Shrinkage and temperature reinforcement should not be less than 0.002 times the gross area for grade 40 bars and 0.0018 for grade 60 steel and welded fire fabric. For structural slabs and footings of uniform thickness, the maximum spacing of the tension reinforcement should not exceed three times the thickness or 18 in.

5.6.1 Example 5.4: Design of a One-way Slab for Flexure

A one-way single-span reinforced concrete slab has a simple span of 10 ft (3.05 m) and carries a live load of 120 psf (5.75 kPa) and a dead load of 20 psf (0.96 kPa) in addition to its self-weight. Design the slab and the size and spacing of the reinforcement at midspan assuming a simple support moment. Given:

$f'_c = 4000$ psi (27.5 MPa), normal-weight concrete

$f_y = 60,000$ psi (413.4 MPa)

Minimum thickness for deflection $= \dfrac{l}{20}$

Solution Minimum depth for deflection, $h = \dfrac{l}{20} = \dfrac{10 \times 12}{20} = 6$ in. (152.4 mm)

Assume for flexure an effective depth $d = 5$ in. (127 mm).

$$\text{self-weight of a 12-in. strip} = \frac{6 \times 12}{144} \times 150 = 75 \text{ lb/ft (3.59 kN/m)}$$

Therefore,

$$\text{factored external load } w_u = 1.7 \times 120 + 1.4(20 + 75) = 337 \text{ lb/ft}$$

$$\text{factored external moment } M_u = \frac{337 \times 10^2}{8} \times 12 \text{ in.-lb}$$

$$= 50{,}550 \text{ in.-lb (5.7 N-m)}$$

Assume that the arm $(d - a/2) = 0.9d = 0.9 \times 5 = 4.50$ in.

$$M_u = \phi A_s f_y \left(d - \frac{a}{2} \right)$$

Therefore,

$$50{,}550 = 0.9 \times A_s \times 60{,}000(4.50)$$

or $A_s = 0.21$ in.^{2} per 12 in. of slab.

Trial-and-adjustment check for assumed moment arm:

$$a = \frac{A_s f_y}{0.85 f_c' b} = \frac{0.21 \times 60{,}000}{0.85 \times 4000 \times 12} = 0.31 \text{ in.} \quad (8.6 \text{ mm})$$

$$50{,}550 = 0.9 \times A_s \times 60{,}000 \left(5 - \frac{0.31}{2} \right)$$

$$A_s = 0.193 \text{ in.}^2 \text{ per 12-in. slab strip}$$

Use No. 4 bars at 12-in. center-to-center spacing (13-mm-diameter bars at 304.8 mm center to center) with an area of 0.20 in.^{2} or No. 3 bars at $6\frac{1}{2}$-in. center-to-center spacing.

$$\rho = \frac{0.20}{5.0 \times 12} = 0.0033, \qquad \rho_{min} = \frac{200}{60{,}000} = 0.0033$$

or

$$\frac{3\sqrt{f_c'}}{f_y} = \frac{3\sqrt{4000}}{60{,}000} = 0.0031 \qquad \text{O.K.}$$

$$\rho_{max} = 0.75\bar{\rho}_b = 0.75 \left(\frac{0.85 \times 4{,}000}{60{,}000} \times 0.85 \times \frac{87{,}000}{87{,}000 + 60{,}000} \right)$$

$$= 0.0214 > \rho_{min} \qquad \text{O.K.}$$

Shrinkage and temperature reinforcement:

$$\rho = 0.0018$$

$$\text{area of steel} = 0.0018 \times 6 \times 12 = 0.13 \text{ in.}^2 = \text{No. 4 bars at 18 in. c-c}$$

Provide No. 4 bars at 18 in. center to center (maximum allowable spacing = $3h$ = $3 \times 6 = 18$ in.).

Hence this design can be adopted with slab thickness $h = 6$ in. (152.4 mm) and effective depth $d = 6.0 - (0.75 + 0.25) = 5$ in. (127.0 mm) to satisfy the $\frac{3}{4}$-in. minimum concrete cover requirement. Use for main reinforcement No. 4 bars at 12 in. center to center and for temperature reinforcement No. 4 bars at 18 in. center to center, as shown in Fig. 5.12.

5.7 DOUBLY REINFORCED SECTIONS

Doubly reinforced sections contain reinforcement both at the tension and at the compression face, usually at the support section only. They become necessary when either architectural limitations restrict the beam web depth at midspan, or the midspan section dimensions are not adequate to carry the support negative moment even when the tensile steel at the support is sufficiently increased. In such cases, most of the bottom bars at midspan are extended and well anchored at the supports to act as compression reinforcement. The bar development length has to be well established and the compressive and tensile steel at the support section well tied with closed stirrups to prevent buckling of the compressive bars.

In analysis or design of beams with compression reinforcement A'_s, the analysis is so divided that the section is theoretically split into two parts, as shown in

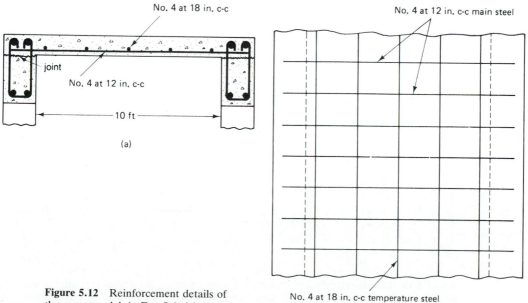

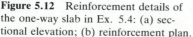

Figure 5.12 Reinforcement details of the one-way slab in Ex. 5.4: (a) sectional elevation; (b) reinforcement plan.

Fig. 5.13. The two parts of the solution comprise (1) the singly reinforced part involving the equivalent rectangular block, as discussed in Section 5.2, with the area of tension reinforcement being $(A_s - A'_s)$; and (2) the two areas of equivalent steel A'_s at both the tension and compression sides to form the couple T_2 and C_2 as the second part of the solution.

It can be seen from Fig. 5.13 that the total nominal resisting moment $M_n = M_{n1} + M_{n2}$, that is, the summation of the moments for parts 1 and 2 of the solution.

Part 1.   The tension force $T_1 = A_{s1}f_y = C_1$. But $A_{s1} = A_s - A'_s$ since equilibrium requires that A_{s2} at the tension side be balanced by an equivalent A'_s at the compression side.   Hence the nominal resisting moment

$$M_{n1} = A_{s1}f_y \left(d - \frac{a}{2} \right) \qquad \text{or} \qquad M_{n1} = (A_s - A'_s)f_y \left(d - \frac{a}{2} \right) \qquad (5.10a)$$

where

$$a = \frac{A_{s1}f_y}{0.85f'_c b} = \frac{(A_s - A'_s)f_y}{0.85f'_c b}$$

Part 2.

$$A'_s = A_{s2} = (A_s - A_{s1})$$
$$T_2 = C_2 = A_{s2}f_y$$

Taking the moment about the tension steel, we have

$$M_{n2} = A_{s2}f_y(d - d') \qquad (5.10b)$$

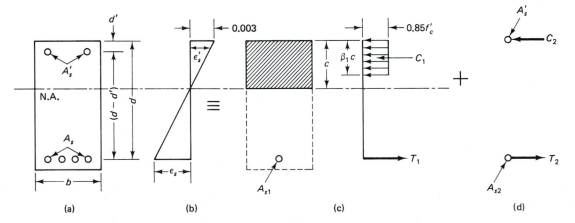

Figure 5.13 Doubly reinforced beam design: (a) cross section; (b) strains; (c) part 1 of the solution, singly reinforced part; (d) part 2 of the solution, contribution of compression reinforcement.

Adding the moments for parts 1 and 2 yields

$$M_n = M_{n1} + M_{n2} = (A_s - A_s')f_y\left(d - \frac{a}{2}\right) + A_s'f_y(d - d') \qquad (5.11a)$$

The design moment strength ϕM_n must be equal to or greater than the external factored moment M_u such that

$$M_u = \phi\left[(A_s - A_s')f_y\left(d - \frac{a}{2}\right) + A_s'f_y(d - d')\right] \qquad (5.11b)$$

This equation is valid *only* if A_s' yields. Otherwise, the beam has to be treated as a singly reinforced beam neglecting the compression steel, or one has to find the actual stress f_s' in the compression reinforcement A_s' and use the actual force in the moment equilibrium equation.

Strain-compatibility check. It is always necessary to verify that the strains across the depth of the section follow the linear distribution indicated in Fig. 5.13. In other words, a check is necessary to ensure that strains are compatible across the depth at the strength design levels. Such a verification is called a *strain-compatibility check*.

For A_s' to yield, the strain ϵ_s' in the compression steel should be greater than or equal to the yield strain of reinforcing steel, which is f_y/E_s. The strain ϵ_s' can be calculated from similar triangles. Referring to Fig. 5.13b, one has

$$\epsilon_s' = \frac{0.003(c - d')}{c}$$

or

$$\epsilon_s' = 0.003\left(1 - \frac{d'}{c}\right)$$

Since

$$c = \frac{a}{\beta_1} = \frac{(A_s - A_s')f_y}{\beta_1 \times 0.85f_c'b} = \frac{(\rho - \rho')f_y d}{\beta_1 \times 0.85f_c'}$$

$$\epsilon_s' = 0.003\left[1 - \frac{0.85\beta_1 f_c' d'}{(\rho - \rho')df_y}\right] \qquad (5.12)$$

As mentioned earlier, for compression steel to yield, the following condition must be satisfied:

$$\epsilon_s' \geq \frac{f_y}{E_s} \quad \text{or} \quad \epsilon_s' \geq \frac{f_y}{29 \times 10^6}$$

The compression steel yields if

$$0.003\left[1 - \frac{0.85\beta_1 f_c'}{(\rho - \rho')f_y}\frac{d'}{d}\right] \geq \frac{f_y}{29 \times 10^6}$$

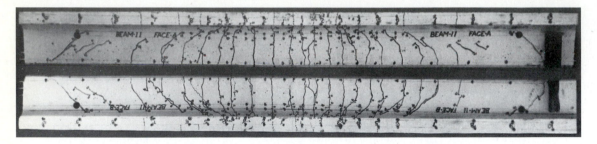

Photo 33 Flexural cracking in heavily reinforced beam. (Tests by Nawy, Potyondy, et al.)

or   (5.13)

$$-\frac{0.85\beta_1 f_c' d'}{(\rho - \rho')f_y d} \geq \frac{f_y - 87,000}{87,000}$$

or

$$\rho - \rho' \geq \frac{0.85\beta_1 f_c' d'}{f_y d} \frac{87,000}{87,000 - f_y}$$ (5.14)

If ϵ_s' is less than ϵ_y the stress in the compression steel, f_s', can be calculated as

$$f_s' = E_s \epsilon_s' = 29 \times 10^6 \epsilon_s'$$ (5.15)

Using Eqs. 5.12 and 5.15 yields

$$f_s' = 29 \times 10^6 \times 0.003 \left[1 - \frac{0.85\beta_1 f_c' d'}{(\rho - \rho')f_y d} \right]$$ (5.16)

This value of f_s' can be used as a first approximation in the strain compatibility check in cases where the compression reinforcement did *not* yield. The reinforcement ratio for the balanced section can be written as

$$\rho_b = \bar{\rho}_b + \rho' \frac{f_s'}{f_y}$$ (5.17a)

where $\bar{\rho}_b$ corresponds to the balanced steel ratio for a singly reinforced beam that has a tension steel area A_{s1}.

The singly reinforced part of the solution in a doubly reinforced section normally utilizes the maximum allowable reinforcement ratio, $0.75\rho_b$. Consequently, the maximum allowable reinforcement ratio for a doubly reinforced beam can be expressed as

$$\rho \leq 0.75\bar{\rho}_b + \rho' \frac{f_s'}{f_y}$$ (5.17b)

In this discussion, adjustment for the concrete area replaced by the compression reinforcement is disregarded as being insignificant for practical design purposes.

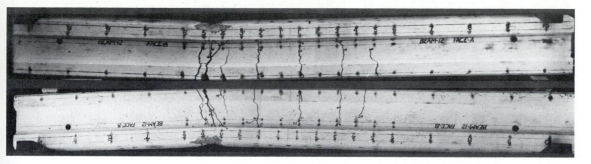

Photo 34 Flexural cracking in lightly reinforced beam. (Tests by Nawy et al.)

Note that in cases where the compression reinforcement A'_s did not yield the depth of the rectangular compressive block should be calculated using the actual stress in the compression steel from the calculated strain value ϵ'_s at the compression reinforcement level so that

$$a = \frac{A_s f_y - A'_s f'_s}{0.85 f'_c b} \tag{5.18}$$

Equation 5.16 can be used for the f'_s value in the first trial in order to obtain an "a" value and hence the first trial neutral axis depth value c. Once c is known, ϵ'_s can be evaluated from similar triangles in Fig. 5.13b, thereby obtaining the first approximation of f'_s to be used in recalculating a more refined value. More than one or two additional trials for calculating f'_s are not justified since undue refinement has negligible practical effect on the true value of the nominal moment strength M_n.

The nominal moment strength in Eq. 5.11 becomes in this case

$$M_n = (A_s f_y - A'_s f'_s)\left(d - \frac{a}{2}\right) + A'_s f'_s (d - d') \tag{5.19}$$

The flow chart in Fig. 5.14 can be used for the sequence of calculations in the analysis of doubly reinforced beams. Examples 5.5 and 5.6 illustrate the analysis and design of doubly reinforced sections.

5.7.1 Example 5.5: Analysis of a Doubly Reinforced Beam for Flexure

Calculate the nominal moment strength M_n of the doubly reinforced section shown in Fig. 5.15. Given:

$$f'_c = 5000 \text{ psi (34.46 MPa), normal-weight concrete}$$
$$f_y = 60{,}000 \text{ psi (413.4 MPa)}$$
$$d' = 2.5 \text{ in. (63.5 mm)}$$

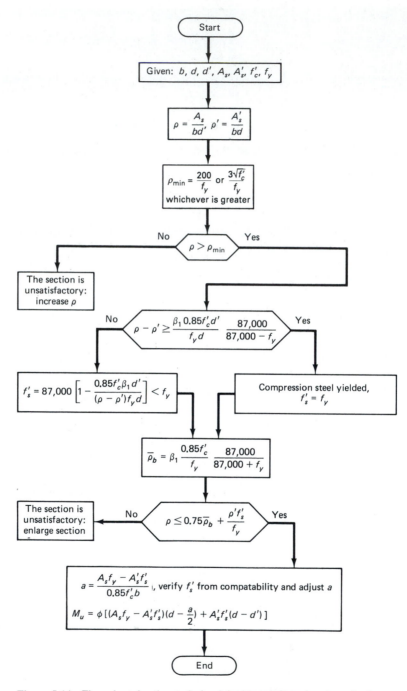

Figure 5.14 Flow chart for the analysis of doubly reinforced rectangular beam.

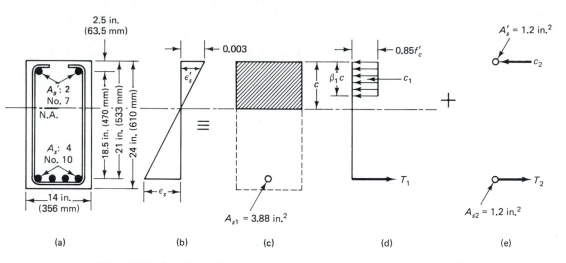

Figure 5.15 Doubly reinforced cross-section geometry and stress and strain distribution: (a) cross section; (b) strains; (c) part 1 section; (d) part 1 forces; (e) part 2 forces.

Solution

$$A_s = 5.08 \text{ in.}^2, \qquad \rho = \frac{A_s}{db} = \frac{5.08}{14 \times 21} = 0.0173$$

$$A_s' = 1.2 \text{ in.}^2, \qquad \rho' = \frac{A_s'}{bd} = \frac{1.2}{14 \times 21} = 0.0041$$

$$A_s - A_s' = A_{s1} = 5.08 - 1.2 = 3.88 \text{ in.}^2$$

$$\rho - \rho' = 0.0173 - 0.0041 = 0.0132$$

To check whether the compression steel has yielded, use Eq. 5.14:

$$\rho - \rho' \geq \frac{0.85\beta_1 f_c'd'}{f_y d} \frac{87,000}{87,000 - f_y}$$

$$\geq \frac{0.85 \times 0.80 \times 5000 \times 2.5}{60,000 \times 21} \frac{87,000}{87,000 - 60,000}$$

$$\geq 0.0217$$

The actual $(\rho - \rho') = 0.0132 < 0.0217$. Therefore, the compression steel did not yield and f_s' is less than f_y. For the first trial in cases where the compression steel did not yield

$$f_s' = 87,000\left[1 - \frac{0.85\beta_1 f_c'}{(\rho - \rho')f_y}\frac{d'}{d}\right]$$

$$= 87,000\left(1 - \frac{0.85 \times 0.80 \times 5000}{0.0132 \times 60,000} \times \frac{2.5}{21}\right) = 42,538 \text{ psi}$$

$$a = \frac{A_s f_y - A_s' f_s'}{0.85 f_c'b} = \frac{5.08 \times 60,000 - 1.2 \times 42,538}{0.85 \times 5000 \times 14} = 4.26 \text{ in. (108.33 mm)}$$

$$\text{neutral-axis depth } c = \frac{4.26}{0.80} = 5.325 \text{ in.}$$

From similar triangles in Fig. 5.15b, the strain ϵ'_s at the compression steel level $=$ 0.00159 in./in., giving $f'_s = 0.00159 \times 29 \times 10^6 = 46,155$ psi. An additional trial cycle for a more refined value of $a = 4.21$ in.; hence $c = 5.26$ in. gives $f'_s = 45,650$ psi (314.76 kN).

$$\bar{\rho}_b = \beta_1 \frac{0.85 f'_c}{f_y} \frac{87,000}{87,000 + f_y}$$

$$= 0.8 \frac{0.85 \times 5000}{60,000} \frac{87,000}{87,000 + 60,000} = 0.0335$$

Hence $0.75\bar{\rho}_b = 0.0252$. From Eq. 5.17a, the maximum allowable reinforcement ratio

$$\rho \le 0.75\bar{\rho}_b + \rho' \frac{f'_s}{f_y}$$

$$0.75\bar{\rho}_b + \rho' \frac{f'_s}{f_y} = 0.0252 + 0.0041 \times \frac{45,650}{60,000}$$

$$= 0.0283 > \rho = 0.0173 \qquad \text{O.K.}$$

$$a = \frac{5.08 \times 60,000 - 1.2 \times 45,650}{0.85 \times 5000 \times 14} = 4.20 \text{ in. (106.73 mm)}$$

$$M_n = (A_s f_y - A'_s f'_s)\left(d - \frac{a}{2}\right) + A'_s f'_s(d - d')$$

$$= (5.08 \times 60,000 - 1.2 \times 45,650)\left(21.0 - \frac{4.20}{2}\right)$$

$$+ 1.2 \times 45,650(21.0 - 2.5) = 5,738,808 \text{ in.-lb (648 kN-m)}$$

Note that, if the first trial value of $f'_s = 42,538$ psi from Eq. 5.16 is used, $M_n = 5,732,689$ in.-lb, which differs by less than 1% from the final M_n value. Such a low percent difference can justify using the value of f'_s obtained from Eq. 5.16 for all practical purposes without resort to additional trials.

$$M_u = \phi M_n = 0.9 \times 5,738,808 = 5,164,927 \text{ in.-lb (583.64 kN-m)}$$

5.7.2 Trial-and-Adjustment Procedure for the Design of Doubly Reinforced Sections for Flexure

1. *Midspan section.* The trial-and-adjustment procedure described in Section 5.5 is followed in order to design the section at midspan if it is a rectangular section; otherwise, follow the same procedure as that for the design of T beams and L beams (Section 5.10).

2. *Support section.* The width b and the effective depth d are already known from part 1 together with the value of the external negative factored moment M_u.
    (a) Find the strength M_{n1} of a singly reinforced section using the already established b and d dimensions of the section at midspan and a reinforcement ratio $\rho \le 0.75\bar{\rho}_b$.

(b) From step (a), find $M_{n2} = M_n - M_{n1}$ and determine the resulting $A_{s2} = A_s'$. The total steel area at the tension side would be $A_s = A_{s1} + A_s'$.

(c) *Alternatively*, determine how many bars are extended from the midspan to the support to give the A_s' to be used in calculating M_{n2}.

(d) From step (c), find the value of $M_{n1} = M_n - M_{n2}$. Calculate A_{s1} for a singly reinforced beam as the first part of the solution. Then determine total $A_s = A_{s1} + A_s'$. Verify that A_{s1} does not exceed $0.75\overline{\rho}_b$ if it is revised in the solution.

(e) Check for the compatibility of strain in both alternatives to verify whether the compression steel yielded or not and use the corresponding stress in the steel for calculating the forces and moments.

(f) Check for satisfactory minimum reinforcement requirements.

(g) Select the appropriate bar sizes.

   If it is necessary to design a doubly reinforced rectangular precast continuous beam, alternative method 3(a) or 3(b) of Section 5.5 for singly reinforced beams can be followed. An assumption is made of an R value higher than the R value that is used for singly reinforced beams for selection of the first trial section. Since it is not advisable to use an A_s' value larger than $\frac{1}{3}A_s$ to $\frac{1}{2}A_s$, assume that $R' \simeq 1.3R$ to $1.5R$.

5.7.3 Example 5.6: Design of a Doubly Reinforced Beam for Flexure

A doubly reinforced concrete beam section has a maximum effective depth $d = 25$ in. (635 mm) and is subjected to a total factored moment $M_u = 9.4 \times 10^6$ in.-lb (1062 kN-m), including its self-weight. Design the section and select the appropriate reinforcement at the tension and the compression faces to carry the required load. Given:

$$f_c' = 4000 \text{ psi (27.58 MPa)}$$
$$f_y = 60,000 \text{ psi (413.4 MPa)}$$
Minimum effective cover $d' = 2.5$ in. (63.5 mm)

Solution   Assume that $b = 14$ in. $\simeq 0.55d$.

$$\overline{\rho}_b = \beta_1 \frac{0.85f_c'}{f_y} \frac{87,000}{87,000 + 60,000} = 0.85 \left(\frac{0.85 \times 4000}{60,000}\right) \frac{87,000}{87,000 + 60,000}$$

$$= 0.0285$$

Or obtain $0.75\overline{\rho}_b$ from Table 5.1. Assume a tension reinforcement ratio 0.016 ($\simeq 0.5\overline{\rho}_b$) for the singly reinforced part of the solution

tension reinforcement area $A_{s1} = (A_s - A_s') = 0.016 \times 14 \times 25 = 5.6$ in.^{2}

The resisting strength of a singly reinforced section of dimensions 14 in. $\times$ 25 in. and a tensile steel area $A_{s1} = 5.6$ in.^{2} is

$$M_{n1} = 5.6 \times 60,000\left(25.0 - \frac{5.6 \times 60,000}{2 \times 0.85 \times 4000 \times 14}\right)$$

$$= 7,214,118 \text{ in.-lb } (815.2 \text{ kN-m})$$

ϕM_{n1} is less than the total factored $M_u = 9.4 \times 10^6$ in.-lb; that is, the section is too small to support the required factored moment M_u. Therefore, the section should be designed as doubly reinforced. The resisting moment corresponding to the singly reinforced part is

$$M_{n1} = 7,214,118 \text{ in.-lb}$$

The moment to be resisted by the doubly reinforced part is

$$\frac{9,400,000}{0.9} - 7,214,118 = 3,230,326 \text{ in.-lb } (365.0 \text{ kN-m})$$

To verify if the compression steel A_s' has yielded, check

$$\rho - \rho' \geq \frac{0.85 f_c' \beta_1 d'}{f_y d} \frac{87,000}{87,000 - f_y} \tag{5.14}$$

$$\geq \frac{0.85 \times 4000 \times 0.85 \times 2.5}{60,000 \times 25} \frac{87,000}{87,000 - 60,000}$$

$$\geq 0.0155$$

The actual $\rho - \rho' = 0.016 > 0.0155$; hence compression steel yielded

$$f_s' = f_y$$

Since $M_{n2} = A_s' f_y (d - d')$, $3,230,326 = A_s' \times 60,000(25.0 - 2.5)$, to give $A_s' = 2.39$ in.^{2}, corresponding to A_{s2}.

$$A_s = A_{s1} + A_s' = 5.6 + 2.39 = 7.99 \text{ in.}^2$$

Use eight No. 9 (28.6-mm-diameter) bars in *two layers* at the tension side ($A_s = 8.0$ in.^{2}) and four No. 7 (22.2-mm-diameter) bars in one layer at the compression side ($A_s' = 2.4$ in.^{2}), as shown in Fig. 5.16. Check if compression steel yielded in the final design:

$$\rho = \frac{8.0}{14 \times 25} = 0.02286, \qquad \rho' = \frac{2.4}{14 \times 25} = 0.00686$$

$$\rho - \rho' = 0.0160 < 0.75\bar{\rho}_b \qquad \text{O.K.}$$

$$\rho - \rho' > 0.0155, \quad \text{hence } f_s' = f_y \qquad \text{O.K.}$$

$$\text{minimum reinforcement ratio} = \frac{200}{f_y} = \frac{200}{60,000} = 0.0033 < \rho$$

Also, $\rho_{\min} = \dfrac{3\sqrt{f_c'}}{f_y} = \dfrac{3\sqrt{4000}}{60,000} = 0.0031$, both $< \rho$ O.K.

$$A_s - A_s' = 8.00 - 2.4 = 5.6 \text{ in.}^2$$

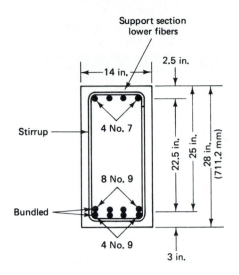

Figure 5.16 Reinforcing details of the doubly reinforced beam in Ex. 5.6.

$$\text{design moment } M_u = 0.9\left[5.6 \times 60,000\left(25.0 - \frac{5.6 \times 60,000}{2 \times 0.85 \times 4,000 \times 14}\right)\right.$$

$$\left. + 2.4 \times 60,000(25.0 - 2.5)\right]$$

$$= 9,408,706 \text{ in.-lb} > 9,400,000 \text{ in.-lb} \,(1063 \text{ kN-m} > 1062.0 \text{ kN-m})$$

Adopt the design.

Alternative Solution   Assume an R' value $\simeq 1.5R \simeq 1350$ (larger than $R = 900$ for singly reinforced beams of the same material properties).

$$M_n = \frac{9.4 \times 10^6}{0.9} = 10.44 \times 10^6 \text{ in.-lb}$$

$$M_n = Rbd^2 \quad \text{or} \quad 10.44 \times 10^6 = 1350bd^2$$

$$bd^2 = \frac{10.44 \times 10^6}{1350} = 7737 \text{ in.}^3$$

Assume that $b \simeq \frac{1}{2}d$; hence $d^3 = 15,474$ in.^{3} and $d = 24.92$ in. Assume a trial section with $b = 14$ in., $d = 25$ in., and proceed to analyze the section in the usual manner, first choosing $\rho - \rho' \leq \bar{\rho}_b$, as given in the preceding alternative solution.

5.8 NONRECTANGULAR SECTIONS

T beams and L beams are the most commonly used flanged sections. Because slabs are cast monolithically with the beams as shown in Fig. 5.17, additional stiffness or strength is added to the rectangular beam section from participation of the slab. Based on extensive tests and longstanding engineering practice, a segment

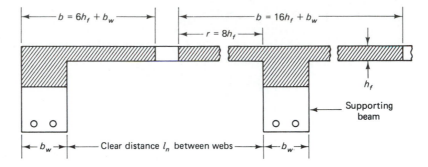

Figure 5.17   T and L beams as part of a slab beam floor system (cross section at beam midspan).

of the slab can be considered to act as a monolithic part of the beam across the beam flange.   In the case of composite sections, if the beam and slab are continuously shored during construction (supported continuously), the slab and beam can be assumed to act together in supporting all loads, including their self-weight. However, if the beam is not shored, the beam must carry its weight plus the weight of the slab while it hardens.   After the slab has hardened, the two together will support the additional loads.

Photo 35   Structural behavior of simply supported prestressed flanged beam.   (Tests by Nawy et al.)

The flange width accepted for inclusion with the beam in forming the flanged section has to satisfy the following requirements:

T beams:

Effective overhang $\not> 8h_f$
Overhang width on each side $\not> \frac{1}{2}$ the clear distance to the face of the next web ($\not> \frac{1}{2}l_n$)
Flange width $b \not> \frac{1}{4}L$ of supporting beam span $= \frac{1}{4}L$

Spandrel or edge beams (beams with a slab on one side only):

The effective overhang $\not> 6h_f$ nor $\not> \frac{1}{2}$ the clear distance to the next web ($\not> \frac{1}{2}l_n$) nor $\frac{1}{12}$ the span length of the beam.
Beams with overhang on one side are called *L beams*.

5.9 ANALYSIS OF T AND L BEAMS

Flanged beams are considered primarily for use as sections at midspans, as shown in Fig. 5.17. This is because the flange is in compression at midspan and can contribute to the moment strength of the midspan section. At the support, the flange is in tension; consequently, it is disregarded in the flexural strength computations of the support section. In other words, the support section would be an inverted doubly reinforced section having the compressive steel A_s' at the bottom fibers and tensile steel A_s at the top fibers. Figure 5.18 shows an elevation of a continuous beam with sections taken at midspan and at the supports to illustrate this discussion.

The basic principles used for the design of rectangular beams are also valid for the flanged beams. The major difference between the rectangular and flanged sections is in the calculation of compressive force C_c. Depending on the depth of the neutral axis, c, the following cases can be identified.

Case 1: Depth of neutral axis c less than flange thickness h_f (Fig. 5.19).   This case can be treated similarly to the standard rectangular section provided that the depth a of the equivalent rectangular block is less than the flange thickness. The flange width b of the compression side *should be used* as the beam width in the analysis.

Referring to Fig. 5.19 for force equilibrium, where C is equal to T, gives

$$0.85f_c'ba = A_sf_y \quad \text{or} \quad a = \frac{A_sf_y}{0.85f_c'b}$$

The nominal moment strength would thus be $M_n = A_sf_y(d - a/2)$. This expression is the same as that of Eq. 5.4 for the rectangular section. Since the force contribution of concrete in the tension zone is neglected, it does not matter whether part of the flange is in the tension zone.

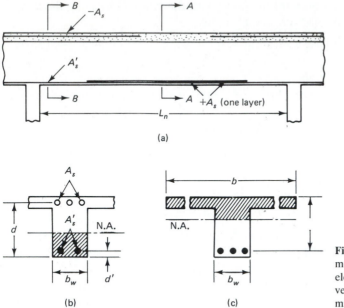

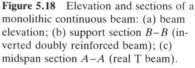

Figure 5.18   Elevation and sections of a monolithic continuous beam: (a) beam elevation; (b) support section B–B (inverted doubly reinforced beam); (c) midspan section A–A (real T beam).

Case 2: Depth of neutral axis c larger than flange thickness h_f (Fig. 5.20).

In this case, $c > h_f$, the depth of the equivalent rectangular stress block a could be smaller or larger than the flange thickness h_f. If c is greater than h_f and a is less than h_f, the beam could still be considered as a rectangular beam for design purposes. Hence the design procedure explained for case 1 is applicable to this case.

If both c and a are greater than h_f, the section has to be considered as a T section. This type of T beam ($a > h_f$) can be treated in a manner similar to that for a doubly reinforced rectangular cross section (Fig. 5.20). The contribution of the flange overhang compressive force is considered analogous to the contribution

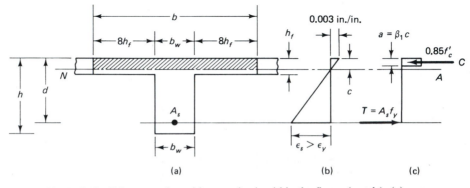

Figure 5.19   T-beam section with neutral axis within the flange ($c < h_f$): (a) cross section; (b) strains; (c) stresses.

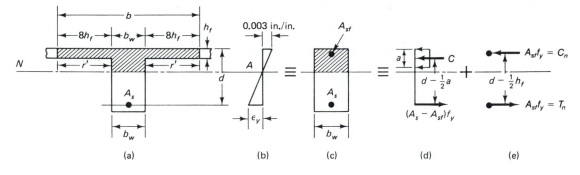

Figure 5.20   Stress and strain distribution in flanged sections design (T-beam transfer): (a) cross section; (b) strains; (c) transformed section; (d) part 1 forces; (e) part 2 forces.

of imaginary compressive reinforcement.   In Fig. 5.20, the compressive force C_n is equal to the average concrete strength f'_c multiplied by the cross-sectional area of the flange overhangs.

Thus $C_n = 2r'h_f \times 0.85f'_c = 0.85f'_c (b - b_w)h_f$, where r' is the overhang length on each side of the web.   The compressive force C_n is *equated* to a tensile force T_n for equilibrium such that $T_n = (A_{sf} \times f_y)$, where A_{sf} is an imaginary compressive steel area whose force capacity is equivalent to the force capacity of the compression flange overhang.   Consequently, an equivalent area A_{sf} of compression reinforcement to develop the overhang flanges would have a value of

$$A_{sf} = \frac{0.85f'_c(b - b_w)h_f}{f_y} \tag{5.20}$$

For a beam to be considered as a *real* T beam, the tension force $A_s f_y$ generated by the steel should be greater than the compression force capacity of the total flange area $0.85 f'_c bh_f$.   Hence

$$a = \frac{A_s f_y}{0.85f'_c b} > h_f \tag{5.21a}$$

or $$h_f < (1.18\bar{\omega}d = a) \tag{5.21b}$$

where $\bar{\omega} = (A_s/bd)(f_y/f'_c)$.

The concrete stress block is, in reality parabolic and extends to the neutral-axis depth c.   Consequently, from a theoretical viewpoint, if one were using a parabolic stress block, Eq. 5.21b for a T beam can also be written as

$$h_f < \frac{1.18\bar{\omega}d}{\beta_1} \tag{5.21c}$$

The percentage for the balanced condition in a T beam can be written as

$$\rho_b = \frac{b_w}{b}(\bar{\rho}_b + \rho_f) \tag{5.22}$$

where $\bar{\rho}_b = \dfrac{0.85\beta_1 f'_c}{f_y} \dfrac{87{,}000}{87{,}000 + f_y}$

ρ_f = reinforcement ratio for tension steel area necessary to develop the compressive strength of the overhanging flanges

$= 0.85f'_c(b - b_w)\dfrac{h_f}{f_y b_w d}$

As in the case of singly and doubly reinforced beams, the maximum allowable percentage of the steel ρ at the tension side should not exceed 75% of the balanced steel percentage ρ_b to ensure ductile failure. Hence, in the case of a T beam,

$$\rho = \frac{A_s}{bd} \leq 0.75\rho_b \qquad (5.23)$$

A strain-compatibility check is not needed since the imaginary steel area A_{sf} is assumed to yield in all cases. To satisfy the requirement of minimum reinforcement so that the beam does not behave as nonreinforced, for positive reinforcement,

$$\rho_w = \frac{A_s}{b_w d} \geq \frac{200}{f_y} \geq \frac{3\sqrt{f'_c}}{f_y} \qquad (5.24)$$

For negative reinforcement and T sections with flanges in tension,

$$\rho_{\min} \geq \frac{6\sqrt{f'_c}}{f_y} \geq \frac{200}{f_y}$$

It is to be noted that b_w is used in Eq. 5.24 instead of width b, which is used in the case of singly or doubly reinforced beams.

As in the case of design and analysis of doubly reinforced sections, the reinforcement at the tension side is considered to be composed of two areas: A_{s1} to balance the rectangular block compressive force on area $b_w a$, and A_{s2} to balance the imaginary steel area A_{sf}. Consequently, the total nominal moment strength for parts 1 and 2 of the solution is

$$M_n = M_{n1} + M_{n2} \qquad (5.25a)$$

$$M_{n1} = A_{s1}f_y\left(d - \frac{a}{2}\right) = (A_s - A_{sf})f_y\left(d - \frac{a}{2}\right) \qquad (5.25b)$$

$$M_{n2} = A_{s2}f_y\left(d - \frac{h_f}{2}\right) = A_{sf}f_y\left(d - \frac{h_f}{2}\right) \qquad (5.25c)$$

The design moment strength ϕM_n, which has to be at least equal to the external factored moment M_u, becomes

$$M_u = \phi M_n = \phi\left[(A_s - A_{sf})f_y\left(d - \frac{a}{2}\right) + A_{sf}f_y\left(d - \frac{h_f}{2}\right)\right] \qquad (5.26)$$

The flow chart in Fig. 5.21 presents the sequence of calculations for the analysis of the T beam.   The following analysis example illustrates the nominal moment strength calculations for a typical precast T beam.

5.9.1 Example 5.7: Analysis of a T Beam for Moment Capacity

Calculate the nominal moment strength and the design ultimate moment of the precast T beam shown in Fig. 5.22 if the beam span is 30 ft (9.14 m).   Given:

$f'_c = 4000$ psi (27.58 MPa), normal-weight concrete
$f_y = 60,000$ psi (413.4 MPa)

Photo 36   The Trump Towers under construction, Fifth Avenue, New York City.   (Courtesy of Concrete Industry Board.)

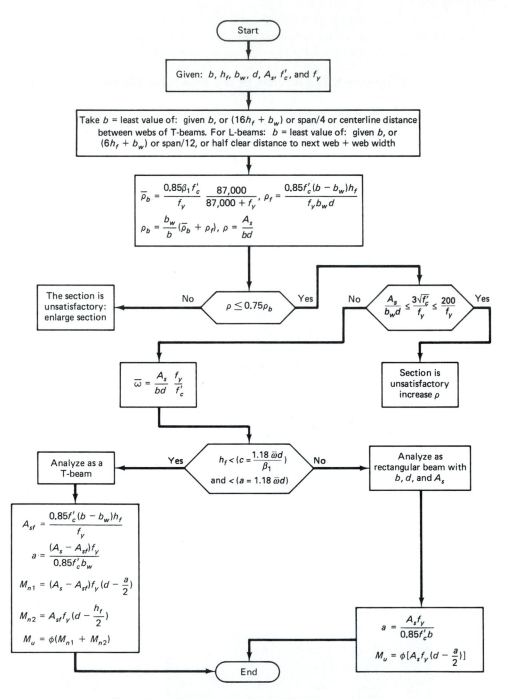

Figure 5.21   Flow chart for the analysis of T and L beams.

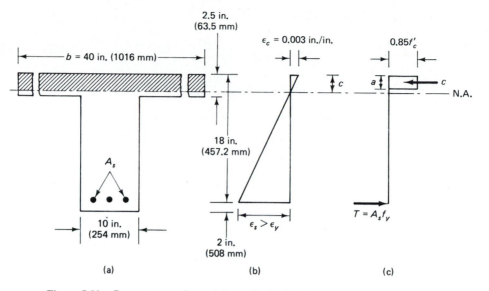

Figure 5.22 Geometry, strain, and force distributions in the T beam of Ex. 5.7: (a) cross section; (b) strains; (c) stresses.

Reinforcement area at the tension side:

(a) A_s = 4.0 in.² (2580 mm²)
(b) A_s = 6.0 in.² (3870 mm²)

Solution Flange-width check is not necessary for a precast beam since the precast section can act independently depending on the construction system. Check for ρ_{max}:

$$\rho_{max} \le 0.75\rho_b$$

$$\bar{\rho}_b = \frac{0.85\beta_1 f'_c}{f_y} \frac{87,000}{87,000 + f_y}$$

$$= \frac{0.85 \times 0.85 \times 4000}{60,000} \frac{87,000}{87,000 + 60,000} = 0.029$$

$$\rho_f = \frac{0.85f'_c(b - b_w)h_f}{f_y b_w d}$$

$$= \frac{0.85 \times 4000(40 - 10) \times 2.5}{60,000 \times 10 \times 18} = 0.024$$

$$\rho_b = \frac{b_w}{b}(\bar{\rho}_b + \rho_f) = \frac{10}{40}(0.029 + 0.024) = 0.013$$

$$\rho_{max} = 0.75\rho_b = 0.010$$

$$\rho_{min} = \frac{200}{60,000} = 0.0033 \quad \text{and} \quad \frac{3\sqrt{4000}}{60,000} = 0.0032$$

$$\rho_{min} = 0.0033 \text{ controls.}$$

(a) $A_s = 4$ in.²:

$$\rho_w = \frac{A_s}{b_w d} = \frac{4.0}{10 \times 18} = 0.0222 > \rho_{min} = 0.0033 \qquad \text{O.K.}$$

$$\rho = \frac{A_s}{bd} = \frac{4.0}{40 \times 18} = 0.006 < \rho_{max} \qquad \text{O.K.}$$

Check whether the section will act as a T beam.

$$\overline{\omega} = \frac{A_s f_y}{bd f_c'} = \frac{4.0}{40 \times 18}\left(\frac{60}{4}\right) = 0.083$$

$$c = \frac{1.18\overline{\omega}d}{\beta_1} = \frac{1.18 \times 0.083 \times 18}{0.85} = 2.10 \text{ in.} < (h_f = 2.5 \text{ in.})$$

Therefore, the beam can be analyzed as a rectangular beam using b, d, and A_s.

$$M_n = A_s f_y\left(d - \frac{a}{2}\right)$$

$$a = \frac{4.0 \times 60,000}{0.85 \times 4000 \times 40} = 1.765 \text{ in.}$$

$$M_n = 4.0 \times 60,000\left(18 - \frac{1.765}{2}\right) = 4,108,200 \text{ in.-lb}$$

$$M_u = \phi M_n = 0.9 \times 4,108,200 \text{ in.} = 3,697,380 \text{ in.-lb (417.8 kN-m)}$$

(b) $A_s = 6.0$ in.²:

$$\rho_w = \frac{A_s}{b_w d} = \frac{6.0}{10 \times 18} = 0.033 > \rho_{min} = 0.0033 \qquad \text{O.K.}$$

$$\rho = \frac{A_s}{bd} = \frac{6.0}{40 \times 18} = 0.0083 < \rho_{max} = 0.010 \qquad \text{O.K.}$$

$$\overline{\omega} = \frac{6.0}{40 \times 18}\left(\frac{60}{4}\right) = 0.125$$

$$\frac{1.18\overline{\omega}d}{\beta_1} = \frac{1.18 \times 0.125 \times 18}{0.85} = 3.124 > (h_f = 2.5)$$

Therefore, the neutral axis is *below* the flange. The beam has to be treated as a T beam or equivalent doubly reinforced rectangular beam with imaginary compression steel area A_{sf}.

$$A_{sf} = \frac{0.85 f_c'(b - b_w)h_f}{f_y}$$

$$= \frac{0.85 \times 4000(40 - 10) \times 2.5}{60,000} = 4.25 \text{ in.}^2$$

$$a = \frac{(A_s - A_{sf})f_y}{0.85 f_c' b_w}$$

$$= \frac{(6.00 - 4.25) \times 60,000}{0.85 \times 4000 \times 10}$$

$$= 3.09 \text{ in.}$$

$$M_{n1} = (A_s - A_{sf})f_y \left(d - \frac{a}{2} \right)$$

$$= (6.00 - 4.25) \times 60,000 \left(18 - \frac{3.09}{2} \right)$$

$$= 1,727,780 \text{ in.-lb}$$

$$M_{n2} = A_{sf}f_y \left(d - \frac{h_f}{2} \right) = 4.25 \times 60,000 \left(18 - \frac{2.5}{2} \right) = 4,271,220 \text{ in.-lb}$$

$$M_n = M_{n1} + M_{n2} = (1,727,780 + 4,271,220) = 6,000,000 \text{ in.-lb}$$

$$= 6000.00 \text{ in.-kips}$$

$$M_u = \phi M_n = 0.9 \times 6,000,000 = 5,400,000 \text{ in.-lb (610.2 kN-m)}$$

5.10 TRIAL-AND-ADJUSTMENT PROCEDURE FOR THE DESIGN OF FLANGED SECTIONS

The slab thickness h_f of the flange overhang is known at the outset since the slab is designed first. Also available is the external factored moment M_u at midspan. The trial-and-adjustment steps for proportioning the web of the beam section can be summarized as follows.

1. Choose a singly reinforced beam section that can resist the external factored moment M_u and the moment due to self-weight. Remember that a T section or an L section would have a smaller size or depth than a singly reinforced section.

2. Check whether the span/depth ratio is reasonable, between 12 and 18. If not, adjust the preliminary section.

3. Calculate the flange width on the basis of the criteria in Section 5.8.

4. Determine if the neutral axis is within or outside the flange, where the neutral-axis depth $c = 1.18\overline{\omega}d/\beta_1$ for rectangular singly reinforced sections.
   (a) If $c < h_f$, the beam has to be treated as a singly reinforced beam with a width b equivalent to the flange width determined in step 3.
   (b) If $c > h_f$ and the equivalent block depth a is also $> h_f$, design as a T beam or an L beam, as the case may be.

5. Find the equivalent compressive steel area A_{sf} for the flange overhang and analyze the assumed section as in Ex. 5.7(b). Calculate the nominal resisting capacities M_{n1} and M_{n2}.

6. Repeat steps 4 and 5 until the calculated $\phi M_n = \phi(M_{n1} + M_{n2})$ is close in value to the factored moment M_u and verify that the assumed self-weight of the web is correct.

7. *Alternatively*, the first trial section can be chosen using a moment factor R'' $> R$ in step 3(a) of Section 5.5 for singly reinforced beams such that $R'' \simeq$ $1.35 - 1.50R$. Select a trial section depth from $M_n = R''bd^2$ and proceed to analyze the section.

5.10.1 Example 5.8: Design of an End-span L Beam

A roof-garden floor is composed of a monolithic one-way slab system on beams as in Fig. 5.23. The effective beam span is 35 ft (10.67 m) and all beams are spaced at 7 ft 6 in. (2.29 m) clear span. The floor supports a 6-ft 4-in. (1.52-m) depth of soil in addition to its self-weight. Assume also that the slab edges support a 12-in.-wide, 7-ft-high wall weighing 840 lb per linear foot. Design the midspan section of the edge spandrel L beam AB assuming that the moist soil weighs 125 lb/ft³ (2.56 tons/m³). Given:

$$f'_c = 3000 \text{ psi } (20.68 \text{ MPa}), \text{ normal-weight concrete}$$

$$f_y = 60,000 \text{ psi } (413.7 \text{ MPa})$$

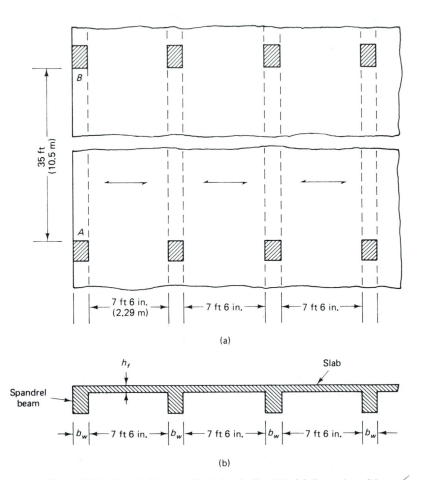

Figure 5.23 Spandrel beam AB design in Ex. 5.8: (a) floor plan; (b) transverse section.

Solution *Slab design*

The weight of soil $= 6.33 \times 125 = 791$, say 800 psf (38.32 kPa). Assume a slab thickness $h = 4$ in. (101.6 mm) $= \frac{4}{12} \times 150 = 50$ psf. $d = h - (\frac{3}{4}$ in. cover $+ \frac{1}{2}$ diameter of No. 4 bars) $= 4.0 - 1.0 = 3.0$ in.

factored load intensity $w_u = 1.4(800 + 50) = 1190$ lb/ft² (57.0 kPa)

From the ACI Code, the negative moment for the interior support of a continuous slab is

$$-M_u = \frac{w_u l_n^2}{12} = \frac{1190(7.5)^2}{12} \times 12 = 66{,}938 \text{ in.-lb}$$

The required slab negative moment strength

$$-M_n = \frac{66{,}938}{0.9} = 74{,}376 \text{ in.-lb}$$

$$M_n = A_s f_y \left(d - \frac{a}{2} \right)$$

Assume that $(d - a/2) \simeq 0.9d = 0.9 \times 3.0 = 2.7$ in.

$$A_s = \frac{74{,}376}{60{,}000 \times 2.7} = 0.459 \text{ in.}^2 \text{ on a 12-in. strip}$$

Try $b = 12$ in., $d = 3$ in., and No. 5 bars at 7.5 in. center to center. $A_s = 0.50$ in.²

$$a = \frac{A_s f_y}{0.85 f'_c b} = \frac{0.50 \times 60{,}000}{0.85 \times 3{,}000 \times 12} = 0.97 \text{ in.}$$

$$\text{nominal resisting moment } M_n = 0.50 \times 60{,}000 \left(3.0 - \frac{0.97}{2} \right) = 74{,}846 \text{ in.-lb}$$

$$> \text{required } M_n = 74{,}376 \text{ in.-lb} \qquad \text{O.K.}$$

$$\rho = \frac{0.50}{12 \times 3} = 0.0138$$

$$\bar{\rho}_b = \frac{0.85 \beta_1 f'_c}{f_y} \left(\frac{87{,}000}{87{,}000 + f_y} \right) = \frac{0.85 \times 0.85 \times 3000}{60{,}000} \left(\frac{87{,}000}{87{,}000 + f_y} \right) = 0.0214$$

maximum allowable $\rho = 0.75 \bar{\rho}_b = 0.016 > 0.0138$ \qquad O.K.

Similarly, for the positive moment, $+M_u = w_u l_n^2/16$, requiring No. 5 bars at 10 in. c-c. Use No. 5 bars at 7½-in. center-to-center main negative reinforcement (15.9-mm diameter at 190.5-mm spacing) and No. 5 bars at 10 in. center to center for main positive reinforcement

temperature steel $= 0.0018bh = 0.0018 \times 12 \times 4.0 = 0.0864$ in.²

maximum allowable spacing $= 3h = 3 \times 4 = 12$ in. (or 18 in. max. mm)

Use No. 3 bars at 12 in. $= 0.11$ in.² for temperature (9.53-mm diameter at 305-mm spacing).

Beam web design

In order to choose a trial web section, assume that $h = l_n/18$ for deflection or

$$h = \frac{35.0 \times 12}{18} = 23.33 \text{ in.}$$

Assume that $h = 26$ in. (660.4 mm), $d = 22.5$ in. (571.6 mm), and $b_w = 14$ in. (355.6 mm).

$$\text{load area on L beam } AB = \frac{7.5}{2} + \frac{14}{12} = 4.92 \text{ ft}$$

$$\text{superimposed working load } w_w = (4.92 - 1.0) \times 800 = 3136 \text{ lb/ft}$$

$$\text{slab weight} = \frac{4.0}{12} \times 150 \times 4.92 = 246 \text{ lb/ft}$$

$$\text{weight of beam web} = \frac{14(26 - 4)}{144} \times 150 = 321 \text{ lb/ft}$$

$$\text{7-ft wall weight} = 840 \text{ lb/ft}$$

$$\text{total service load} = 3136 + 246 + 321 + 840 = 4543 \text{ lb/ft}$$

$$\text{factored load } w_u = 1.4 \times 4543 = 6360 \text{ lb/ft}$$

Try a factored external moment:

$$M_u = \frac{w_u l_n^2}{11} = \frac{6360(35.0)^2}{11} \times 12 = 8,499,273 \text{ in.-lb}$$

$$\text{required } M_n = \frac{M_u}{\phi} = \frac{8,499,273}{0.9} = 9,443,637 \text{ in.-lb (1067.06 kN-m)}$$

To determine whether the beam is an actual L beam or not, it is necessary to find if the neutral axis falls outside the flange. Consequently, the area of the tension steel A_s at midspan has to be assumed. If a rectangular section is initially assumed with an appropriate moment arm $jd \approx 0.85d = 0.85 \times 22.5 = 19.3$ in.,

$$M_n = A_s f_y jd \quad \text{or} \quad 9,443,637 = A_s \times 60,000 \times 19.3$$

$$A_s = \frac{9,443,637}{60,000 \times 19.3} = 8.16 \text{ in.}^2$$

Assume eight No. 9 bars in two layers = 8.0 in.² (5160 mm²).

$$\rho = \frac{A_s}{bd}, \quad \text{where } b = b_w + 6h_f = 14 + 6 \times 4.0 = 38 \text{ in. (965.2 mm)}$$

$$\rho = \frac{8.0}{38 \times 22.5} = 0.00936$$

$$\overline{\omega} = \rho \frac{f_y}{f_c'} = 0.00936 \times \frac{60,000}{3000} = 0.187$$

$$\text{depth of neutral axis } c = \frac{1.18 \overline{\omega} d}{\beta_1} = \frac{1.18 \times 0.187 \times 22.5}{0.85}$$

or

$$c = 5.84 \text{ in.} > 4.0 \text{ in.}$$

$$a = \beta_1 c = 0.85 \times 5.84 = 4.96 > 4.0 \text{ in.}$$

Hence, the section is an L beam since the neutral axis is below the flange, as shown in Fig. 5.24.

For $f'_c = 3000$ psi and $f_y = 60,000$ psi, $\bar{\rho}_b = 0.0214$ and

$$A_{sf} = \frac{h_f(b - b_w)0.85f'_c}{f_y} = \frac{4.0(38 - 14) \times 0.85 \times 3000}{60,000}$$

$$= 4.08 \text{ in.}^2 \ (2631.6 \text{ mm}^2)$$

$$\rho_f = \frac{A_{sf}}{b_w d} = \frac{4.08}{14 \times 22} = 0.01295$$

$$\rho_b = (\bar{\rho}_b + \rho_f)\frac{b_w}{b} = (0.0214 + 0.01295)\frac{14}{38} = 0.01266$$

$$0.75\rho_b = 0.00949 > \text{actual } \rho = 0.00936$$

Hence, the section is underreinforced and satisfies the ACI Code requirements.

$$a = \frac{(A_s - A_{sf})f_y}{0.85f'_c b_w} = \frac{(8.0 - 4.08)60,000}{0.85 \times 3000 \times 14} = 6.59 \text{ in.} \ (167.4 \text{ mm})$$

$$M_n = (A_s - A_{sf})f_y\left(d - \frac{a}{2}\right) + A_{sf}f_y\left(d - \frac{1}{2}h_f\right)$$

$$= (8.0 - 4.08)60,000\left(22.5 - \frac{6.59}{2}\right) + 4.08 \times 60,000\left(22.5 - \frac{4.0}{2}\right)$$

$$= 9,535,416 \text{ in.-lb}$$

design moment $M_u = 0.9 \times 9,535,416$ in.-lb $= 8,581,874$ in.-lb

actual factored $M_u = 8,499,273$ in.-lb $< 8,581,874$ in.-lb

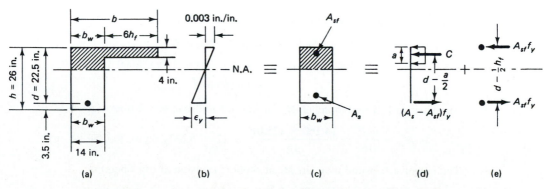

Figure 5.24   Forces and stresses in L beams: (a) cross section; (b) strain diagram; (c) transformed section; (d) part 1 forces; (e) part 2 forces.

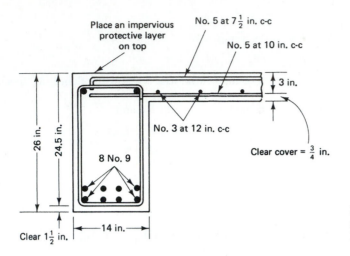

Place an impervious
protective layer
on top

No. 5 at $7\frac{1}{2}$ in. c-c

No. 5 at 10 in. c-c

3 in.

No. 3 at 12 in. c-c

Clear cover = $\frac{3}{4}$ in.

26 in.

24.5 in.

8 No. 9

14 in.

Clear $1\frac{1}{2}$ in.

Figure 5.25 Midspan section flexural reinforcement details for beam *AB* of Ex. 5.8.

Adopt the design. Flexural reinforcement details for the L beam *AB* are shown in Fig. 5.25.

5.10.2 Example 5.9: Design of an Interior Continuous Floor Beam for Flexure

Design a rectangular interior beam having a clear span of 25 ft (7.62 m) and carrying a working live load of 8000 lb per linear foot (35.58 kN) in addition to its self-weight and slab weight, as shown in Fig. 5.26. Assume the beam to have a 4-in. (101.6-mm) slab cast monolithically with it. Given:

$f'_c = 4000$ psi (27.58 MPa), normal-weight concrete
$f_y = 60,000$ psi (413.4 MPa)

Assume no wind or earthquake.

Solution Assume that the web self-weight = 400 lb/ft.

$$\text{factored load } w_u = 1.4 \times 400 + 1.7 \times 8000$$
$$= 14,160 \text{ lb/ft}$$

The positive factored moment M_u for interior midspan lower fibers (ACI) is

$$+M_u = \frac{w_u l_n^2}{16} = \frac{14,160 \times (25.0)^2}{16} \times 12 = 6,637,500 \text{ in.-lb (750 kN-m)}$$

The negative factored moment M_u at support (tension at top fibers) is

$$-M_u = \frac{14,160 \ (25.0)^2}{11} \times 12 = 9,654,546 \text{ in.-lb (1091 kN-m)}$$

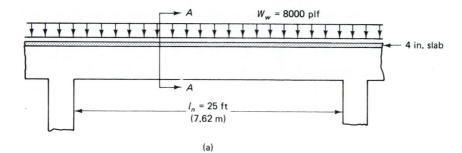

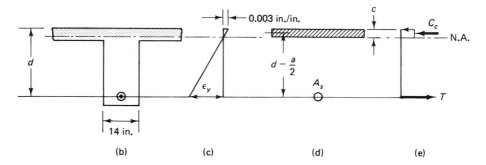

Figure 5.26 Continuous beam midspan section in Ex. 5.9: (a) beam elevation;
(b) section A–A; (c) strain distribution; (d) N.A. inside flange; (e) force and stress.

Section at midspan (T beam)

Assume that b_w = 14 in. (0.3556 m).

$$b \ngtr 16 \times 4 + 14 = 78 \text{ in.}$$

$$\ngtr \frac{25 \times 12}{4} = 75 \text{ in.}$$

$$\ngtr \text{ half clear distance to next web not known}$$

Therefore, b = 75 in. (1.905 m) controls.

If the depth a of the stress block is assumed equivalent to the flange thickness
h_f = 4 in., the compressive force C_n (volume of the compressive block) is

$$C_n = 0.85 f'_c b h_f = 0.85 \times 4000 \times 75 \times 4.0 = 1,020,000 \text{ lb}$$

$$\text{required positive} + M_n = \frac{6,637,500}{\phi = 0.90} = 7,375,000 \text{ in.-lb}$$

In order to obtain on first trial a reasonable area A_s of steel at the tension side
and also to determine if the beam section is flanged, find for a first trial an A_s for a
rectangular section that can resist M_n = 7,375,000 in.-lb.

For deflection purposes assume that interior partitions would be damaged by excessive deflection; hence use

$$d = \frac{l}{12} = \frac{25.0 \times 12}{12} = 25 \text{ in. (635 mm)}$$

For a self-weight of 400 lb/ft, try $b_w = 14$ in. (354 mm) and $h = 28.0$ in. (711 mm).

$$\text{self-weight} = \frac{14 \times 28}{144} \times 150 = 408 \text{ lb/ft}$$

$$\text{revised factored } U = 1.4 \times 408 + 1.7 \times 8000 = 14,171 \text{ lb/ft}$$

$$\text{revised} + M_n = 7,375,500 \times \frac{14,171}{14,160} = 7,381,230 \text{ in.-lb}$$

$$\text{revised} - M_n = \frac{9,654,546}{0.9} \times \frac{14,171}{14,160} = 10,736,607 \text{ in.-lb}$$

Assume that moment arm $jd \simeq 0.9d = 0.9 \times 25 = 22.5$ in.

$$A_s = \frac{M_n}{f_y \times 0.9d} = \frac{7,381,230}{60,000 \times 22.5} = 5.47 \text{ in.}^2$$

For a T beam, a smaller A_s is needed. Try four No. 10 bars = 5.08 in.^{2}.

$$T_n = 5.08 \times 60,000 = 304,800 \text{ lb} \ll 1,020,000 \text{ lb}$$

The neutral axis has to be well within the flange in order to balance $T_n = C_n$. Hence treat the T beam as a singly reinforced section having a compression flange width $b = 75$ in. and

$$\rho = \frac{5.08}{b \times d} = \frac{5.08}{75 \times 25} = 0.0027$$

$$< 0.75 \, \rho_b \text{ for T section, hence O.K.}$$

Another check for the neutral-axis position can be accomplished using the following expression, where

$$\overline{\omega} = \rho \frac{f_y}{f'_c} = 0.0027 \times \frac{60,000}{4000} = 0.0406$$

$$c = \frac{1.18\overline{\omega}d}{\beta_1} = 1.18 \times 0.0406 \times \frac{25}{0.85} = 1.41 \text{ in.} \ll h_f = 4.0 \text{ in.}$$

The nominal resisting moment $M_n = A_s f_y (d - a/2)$ or

$$M_n = 5.08 \times 60,000(25 - \tfrac{1}{2} \times 0.85 \times 1.41) = 7,437,349 \text{ in.-lb (840.4 kN-m)}$$

The actual M_n is larger than the required $M_n = 7,381,230$ in.-lb.

$$\frac{A_s}{b_w d} = \frac{5.08}{14 \times 25} = 0.0145 > \rho_{\min} = \frac{200}{f_y} = 0.0033 \qquad \text{and}$$

$$\rho_{\min} = \frac{3\sqrt{f'_c}}{f_y}, \qquad \text{hence O.K.}$$

Adopt a midspan section with $b_w = 14$ in. (355.6 mm), $h = 28$ in. (685.8 mm), $d = 25.0$ in. (635.0 mm), and $A_s =$ four No. 10 bars (diameter 32.25 mm).

Section at support (doubly reinforced rectangular section)

This section is subjected to moments similar to the moments acting on the section in Ex. 5.6 and has the same cross-sectional dimensions. The required nominal moment of resistance $M_n = 10,736,607$ in.-lb (1212.1 kN-m). Assume that two No. 10 bars extend from the midspan to the support $= 2.54$ in.² $M_{n2} = A'_s f_y(d - d')$, assuming that A'_s has yielded since the area is so close to 2.4 in.² in Ex. 5.6, to be subsequently verified, or

$$M_{n2} = 2.54 \times 60,000(25 - 2.5) = 3,429,000 \text{ in.-lb}$$

$$M_{n1} = 10,736,607 - 3,429,000 = 7,307,607 \text{ in.-lb}$$

Assume that moment arm $jd \simeq 0.85d = 25 \times 0.85 = 21.5$ in.

$$\text{trial } A_{s1} = \frac{7,307,607}{60,000 \times 21.5} = 5.73 \text{ in.}^2$$

$$\text{total } A_s = A_{s1} + A_{s2} = 5.73 + 2.54 = 8.27 \text{ in.}^2$$

Try seven No. 10 bars (50.0 mm) in *two* layers, $A_s = 8.89$ in.².

$$A_{s1} = 8.89 - 2.54 = 6.35 \text{ in.}^2$$

$$\rho - \rho' = \frac{6.35}{14 \times 25} = 0.0181 > 0.0155$$

from Ex. 5.6; hence the assumption that A'_s yielded is valid.

$$a = \frac{(A_s - A'_s)f_y}{0.85f'_c b_w} = \frac{6.35 \times 60,000}{0.85 \times 4000 \times 14} = 8.00 \text{ in.}$$

$$M_{n1} = (A_s - A'_s)f_y\left(d - \frac{a}{2}\right) = 6.35 \times 60,000\left(25.0 - \frac{8.00}{2}\right)$$

$$= 8,000,200 \text{ in.-lb}$$

$$\text{available } M_n = M_{n1} + M_{n2} = 8,000,200 + 3,429,000$$

$$= 11,429,200 \text{ in.-lb (1291.5 kN-m)}$$

The available nominal moment strength M_n is larger than the required $M_n = 10,736,607$ in.-lb (1215.4 kN-m). Hence the design is adequate. Therefore, use seven No. 10 bars on top at the support in *two* layers and two No. 10 bars at the bottom fibers of the section at the support. Provide closed stirrups to tie the tension and the compression steel at the support. It is to be noted that bar sizes larger than No. 11 should be avoided where possible in superstructure beams, because they are difficult to cut and less efficient for crack control.

For the design to be complete, diagonal tension capacity and serviceability and bar development checks have to be made, as discussed in Chapters 6, 8, and 10. Details of the reinforcement over the span are shown in Fig. 5.27.

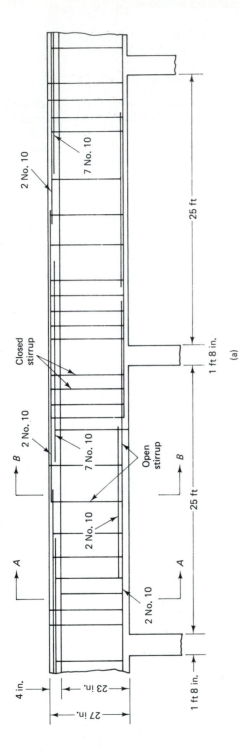

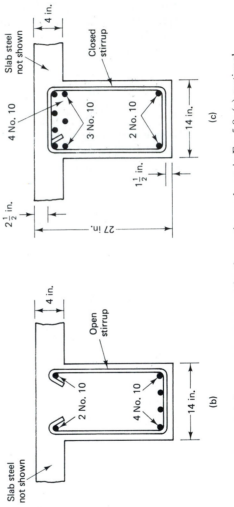

Figure 5.27 Reinforcement arrangement for the continuous beam in Ex. 5.9: (a) sectional elevation (not to scale); (b) midspan section B–B; (c) support section A–A.

5.11 SI EXPRESSIONS AND EXAMPLE FOR FLEXURAL DESIGN OF BEAMS

$$E_c = w_c^{1.5} \, 0.043 \, \sqrt{f_c'} \; \text{MPa}$$

where $w_c = 1500$ to $2500 \; \text{kg/m}^3$ (90 to 155 lb/ft³). For standard, normal-weight concrete, $w_c = 2400 \; \text{kg/m}^3$ to give $E_c = 29{,}700 \; \text{MPa}$.

$$E_s = 200{,}000 \; \text{MPa}$$

modulus of rapture $f_r = 0.7\sqrt{f_c'}$

$$c_b = \left(\frac{600}{600 + f_y}\right)d \quad \text{and} \quad a_b = \frac{\beta d}{1}$$

$$A_{s,\min} = \frac{\sqrt{f_c'}}{4f_y} b_w d \geq \frac{1.4}{f_y} bd$$

where f_y is in MPa units.

For cantilevers and negative moment zone,

$$A_{s,\min} = \frac{\sqrt{f_c'}}{2f_y} b_w d$$

$$\beta_1 = 0.85 - 0.008 \quad (f_c' = 30)$$

The value of β_1 for strengths above 30 MPa should be reduced continuously at the rate of 0.008 for each 1 MPa of strength in excess of 30 MPa, but β_1 cannot be less than 0.65.

Spacing of reinforcement for structural slabs and footings in the direction of the span should not exceed $3 \times$ slab thickness or 450 mm.

For singly reinforced beams, from Eq. 5.4 or 5.5,

$$M_n = A_s f_y \left(d - \frac{a}{2}\right) \quad \text{or} \quad \omega r f_c'(1 - 0.59\omega)d^3$$

$$\bar{\rho}_b = \beta_1 \frac{0.85 f_c'}{f_y} \frac{0.003 E_s}{0.003 E_s + f_y}$$

where f_c', f_y and E_s are in MPa units and $r = b/d$

For doubly reinforced beams, from Eq. 5.19,

$$M_n = (A_s f_y - A_s' f_s') \left(d - \frac{a}{2}\right) + A_a' f_s'(d - d')$$

or

$$M_n = 0.85f_c'ba\left(d - \frac{a}{2}\right) + A_s'f_s'(d - d')$$

$$\rho \leq 0.75\bar{\rho}_b + \frac{\rho'f_s'}{f_y}$$

where $\bar{\rho}_b$ = reinforcement percentage for singly reinforced section.

$$f_s' = 0.003E_s\left[1 - \frac{0.85\beta_1 f_c'd'}{(\rho - \rho')f_y d}\right]$$

For flanged sections, from Eq. 5.26,

$$M_n = (A_s - A_{sf})f_y\left(d - \frac{a}{2}\right) + A_{sf}f_y\left(d - \frac{a}{2}\right)$$

where

$$A_{sf} = \frac{0.85f_c'(b - b_w)h_f}{f_y}$$

and

$$a = \frac{(A_s - A_{sf})f_y}{0.85f_c'b_w}$$

5.11.1 Example 5.10: SI Flexural Design

Solve Ex. 5.3 using SI units.

Solution:

$$f_c' = 27.6 \text{ MPa}$$
$$f_y = 414 \text{ MPa}$$
$$w_w = 21.9 \text{ kN/m}$$
$$\ell_n = 9.14 \text{ m} = 9140 \text{ mm}$$
$$\text{concrete unit weight} = 23.6 \text{ kN/m}^3$$
$$= 23.6 \times 10^{-3} \text{ kN/mm}^3$$
$$\text{Pa} = \text{N/m}^2$$
$$\text{MPa} = \text{N/mm}^2$$

Assume a minimum thickness from the ACI deflection table:

$$\frac{\ell_n}{16} = \frac{9.14}{16} = 0.571 \text{ m} = 571 \text{ mm}$$

For the purpose of estimating the preliminary self-weight, assume a total thickness $h = 600$ mm, effective depth $d = 500$ mm, and width of beam $b = 250$ mm.

beam self-weight $= 250 \times 600(23.6 \times 10^{-3})$ kN/mm $= 3540$ kN/mm $= 3.54$ kN/m

factored load $w_u = 1.4D + 1.7L = 1.4 \times 3.54 + 1.7 \times 21.9$ kN/m $= 42.2$ kN/m

Required factored moment:

$$M_u = \frac{w_u \ell_n^2}{8} = \frac{42.2(9.14)^2}{8} = 440 \text{ kN-m}$$

Required nominal resisting moment:

$$\text{(moment strength)} = \frac{M_u}{\phi} = \frac{440}{0.9} = 489 \text{ kN-m}$$

From Table 5.1, $\bar{\rho}_b = 0.0285$. Assume a reinforcement ratio $\rho = 0.5\bar{\rho}_b = 0.0143$.

$$\text{reinforced index } \omega = \frac{\rho f_y}{f'_c} = \frac{0.0143 \times 414}{27.6} = 0.214$$

Using Eq. 5.6b yields

$$R = \omega f'_c (1 - 0.59\omega) = 0.214 \times 27.6(1 - 0.59 \times 0.0143) = 5.86 \text{ MPa}$$

From Eq. 5.6a, $M_n = Rbd^2$ and assuming

$$d = \sqrt[3]{\frac{M_n}{0.5R}} = \sqrt[3]{\frac{489 \times 10^3 \text{ N-m}}{0.5 \times 5.86 \times 10^6 \text{ N/m}^2}}$$

$$= 0.55 \text{ m} = 550 \text{ mm}$$

Try $h = 700$ mm ($d = 600$ mm, $b = 300$ mm).

revised self-weight $= 300 \times 600(23.6 \times 10^{-3})$ kN/mm $= 4.25$ kN/m

revised factored load $w_u = 1.4 \times 4.25 + 1.7 \times 21.9 = 43.2$ kN/m

$$M_u = \frac{43.2(9.14)^2}{8} = 451 \text{ kN-m}$$

$$\text{required } M_n = \frac{M_u}{\phi} = \frac{451}{0.9} = 501 \text{ kN-m}$$

Either use three No. 10 bars (32.3-mm diameter), giving $A_s = 2460$ mm² or, from Fig. B.2b in the appendix, two No. 30 M and one No. 35 M metric bars: $2 \times 700 + 1000 = 2400$ mm.

$$\rho = \frac{2400}{300 \times 600} = 0.0133 < 0.75 \, \bar{\rho}_b < 0.021$$

$$\rho_{min} = \frac{\sqrt{f'_c}}{4f_y} = \frac{\sqrt{27.6}}{4 \times 414} = 0.003 < 0.133, \quad \text{hence O.K.}$$

$$a = \frac{A_s f_y}{0.85 f'_c b} = \frac{2400 \times 413}{0.85 \times 27.6 \times 300} = 141 \text{ mm}$$

$$\text{available } M_n = A_s f_y \left(d - \frac{a}{2} \right) = 2400 \times 414 \left(600 - \frac{141}{2} \right)$$

$$= 525 \times 10^{-6} \text{ N-mm} = 525 \text{ kN-m} > \text{required } M_n = 501 \text{ kN-m}$$

Adopt the section.

Note that the designed section resists a slightly larger moment than the required moment:

$$\text{percent overdesign} = \frac{525 - 501}{501} = 4.8\%$$

which is a reasonable level expected in proportioning concrete elements. It is always necessary to check that the web width can accommodate the number of bars in each layer based on the minimum spacing requirements. In this example, the minimum web width to accommodate the two No. 30 M and one No. 35 M bars = 270 mm < 300 mm available.

SELECTED REFERENCES

5.1. Whitney, C. S., "Plastic Theory of Reinforced Concrete Design," *Transactions of the ASCE*, Vol. 107, 1942, pp. 251–326.

5.2. Hognestad, E. N., Hanson, N. W., and McHenry, D., "Concrete Stress Distribution in Ultimate Strength Design," *Journal of the American Concrete Institute*, Proc. Vol. 52, December 1955, pp. 455–479.

5.3. Whitney, C. S., and Cohn, E., "Guide for Ultimate Strength Design of Reinforced Concrete," *Journal of the American Concrete Institute*, Proc. Vol. 53, November 1956, pp. 455–475.

5.4. Mattock, A. H., Kriz, L. B., and Hognestad, E. N., "Rectangular Stress Distribution in Ultimate Strength Design," *Journal of the American Concrete Institute*, Proc. Vol. 58, February 1961, pp. 825–928.

5.5. ACI–ASCE Joint Committee: "Report of ASCE–ACI Joint Committee on Ultimate Strength Design," *ASCE, Proceedings*, Vol. 81, October 1955, 68 pp. See also *Journal of the American Concrete Institute*, Proc. Vol. 52, January 1956, pp. 550–524.

5.6. Balaguru, P., "Cost Optimum Design of Singly Reinforced Sections," *Journal of Civil Engineering Design*, Vol. 2., No. 2, 1981, pp. 149–169.

5.7. Concrete Reinforcing Steel Institute, *CRSI Handbook*, CRSI, Chicago, 1988, 928 pp.

5.8. ACI Committee 340, *Design Handbook*, Vol. 1, Special Publication 17, American Concrete Institute, Detroit, 1991, 508 pp.

5.9. Nawy, E. G., "Strength, Serviceability and Ductility," Chapter 12, in *Handbook of Structural Concrete*, Pitman Books, London/McGraw-Hill, New York, 1983, 1968 pp.

5.10. Chen, B., and Nawy, E. G., "Structural Behavior Evaluation of High Strength Concrete Reinforced with Prestressed Prisms Using Fiber Optic Sensors," *Proceedings, ACI Structural Journal*, American Concrete Institute, Detroit, Dec. 1994, pp. 708–718.

5.11. Chen, B., Maher, M. H. and Nawy, E. G., "Fiber Optic Bragg Grating Sensor for Non-Destructive Evaluation of Composite Beams," *Proceedings, ASCE Journal of the Structural Division*, Vol. 120, No. 12, American Society of Civil Engineers, New York, Dec. 1994, pp. 3456–3470.

PROBLEMS FOR SOLUTION

5.1. For the beam cross section shown in Fig. 5.28, determine whether the failure of the beam will be initiated by crushing of concrete or yielding of steel.   Given:

f'_c = 4000 psi (27.58 MPa) for case (a), A_s = 10 in.²
f'_c = 7000 psi (48.26 MPa) for case (b), A_s = 5 in.²
f_y = 60,000 psi (413.7 MPa)

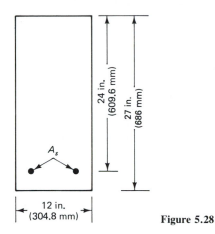

Figure 5.28

Also determine whether the section satisfies ACI Code requirements.

5.2. Calculate the nominal moment strength of the beam sections shown in Fig. 5.29.
Given:

f'_c = 3000 psi (20.68 MPa) for case (a)
f'_c = 6000 psi (41.36 MPa) for case (b)
f'_c = 9000 psi (62.10 MPa) for case (a)
f_y = 60,000 psi (413.7 MPa)

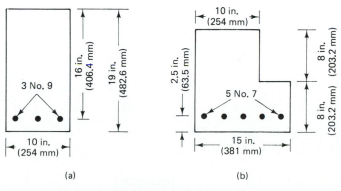

(a)   (b)

Figure 5.29

5.3. Calculate the safe distributed load intensity that the beam shown in Fig. 5.30 can carry.   Given:

f'_c = 4000 psi (27.58 MPa), normal-weight concrete
f_y = 60,000 psi (413.7 MPa)

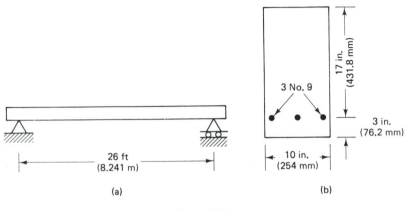

26 ft
(8.241 m)

(a)

17 in.
(431.8 mm)

3 No. 9

3 in.
(76.2 mm)

10 in.
(254 mm)

(b)

Figure 5.30

5.4. Design a one-way slab to carry a live load of 100 psf and an external dead load of 50 psf.   The slab is simply supported over a span of 12 ft.   Given:

f'_c = 4000 psi (27.58 MPa), normal-weight concrete
f_y = 60,000 psi (413.7 MPa)

5.5. Design the simply supported beams shown in Fig. 5.31.   Given:

f'_c = 5000 psi (34.47 MPa), normal-weight concrete
f_y = 60,000 psi (413.7 MPa)
ρ = 0.5ρ_b

Assume that $b/h \simeq \frac{1}{2}$.

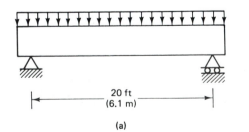

W_L = 1000 lb/ft (14.6 kN/m)
W_D = 500 lb/ft (including self-weight of the beam) (7.3 kN/m)

20 ft
(6.1 m)

(a)

Figure 5.31

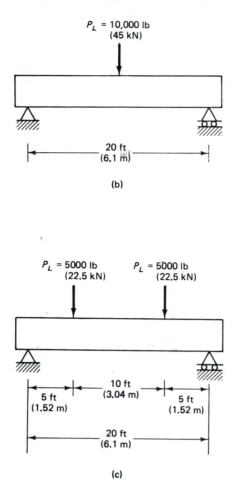

No external dead load

(b)

(c)

Figure 5.31   Continued

5.6. Check whether the sections shown in Fig. 5.32 satisfy ACI 318 Code requirements for maximum and minimum reinforcement.   Given:

$$f'_c = 5000 \text{ psi (34.48 MPa), normal-weight concrete}$$
$$f_y = 60,000 \text{ psi (413.7 MPa)}$$

The compression fibers in all the figures are the top fibers of the sections.

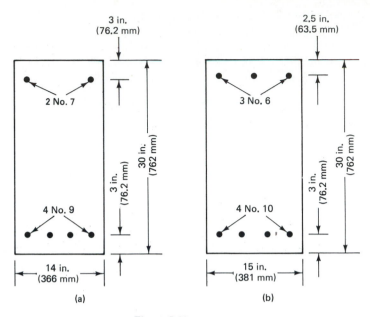

Figure 5.32

5.7. Calculate the stresses in the compression steel, f'_s, for the cross sections shown in Fig. 5.33. Also compute the nominal moment strength for the section in part (b). Given:

$f'_c = 6000$ psi (41.37 MPa), normal-weight concrete
$f_y = 60,000$ psi (413.7 MPa)

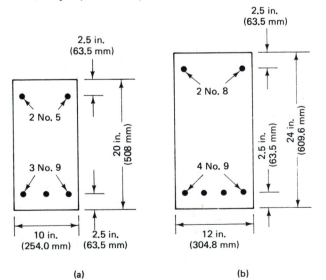

Figure 5.33

5.8. Calculate the ultimate moment capacity of the beam sections of Problem 5.2.   Assume two No. 6 bars for compression reinforcement.

5.9. Solve Problem 5.3 if two No. 6 bars are added as compression reinforcement.

5.10. At failure, determine whether the precast sections shown in Fig. 5.34 will act similarly to rectangular sections or as flanged sections.   Given:

$$f'_c = 4000 \text{ psi } (27.58 \text{ MPa}), \text{ normal-weight concrete}$$
$$f_y = 60,000 \text{ psi } (413.7 \text{ MPa})$$

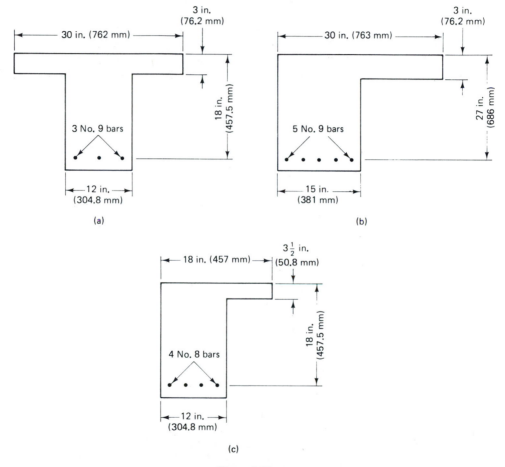

Figure 5.34

5.11. Check whether the sections of Problem 5.10 satisfy ACI Code requirements.

5.12. Calculate the nominal moment strength of the sections shown for Problem 5.10.

5.13. Repeat Problem 5.5 using a T section instead of a rectangular section.   Use a flange thickness of 3 in. (76.2 mm) and a flange width of 30 in. (762 mm).

5.14. Using the details of Problem 5.4, design a reinforced concrete T beam for the slab floor system shown in Fig. 5.35. The floor area is 30 ft × 60 ft (9.14 m × 18.29 m) with an effective T-beam span of 30 ft (9.14 m).

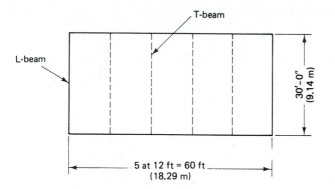

Figure 5.35 Plan of one-way-slab floor system.

SHEAR AND DIAGONAL TENSION IN BEAMS

6

6.1 INTRODUCTION

This chapter presents procedures for the analysis and design of reinforced concrete sections to resist the shear forces resulting from externally applied loads. Since the strength of concrete in tension is considerably lower than its strength in compression, design for shear becomes of major importance in concrete structures.

The behavior of reinforced concrete beams at failure in shear is distinctly different from their behavior in flexure. They fail abruptly without sufficient advanced warning, and the diagonal cracks that develop are considerably wider than the flexural cracks. The accompanying photographs show typical beam shear failure in diagonal tension as discussed in the subsequent sections. Because of the brittle nature of such failures, the designer has to design sections that are adequately strong to resist the external factored shear loads without reaching their shear strength capacity. Shear is also a significant parameter in the behavior of brackets, corbels, and deep beams. Consequently, the design of these elements is also discussed in detail.

Photo 37   Water Tower Place, Chicago.   (Courtesy of Portland Cement Association.)

Photo 38   Typical diagonal tension failure at rupture load level.   (Test by Nawy et al.)

Photo 39   Simply supported beam prior to developing diagonal tension crack (load stage 11).   (Test by Nawy et al.)

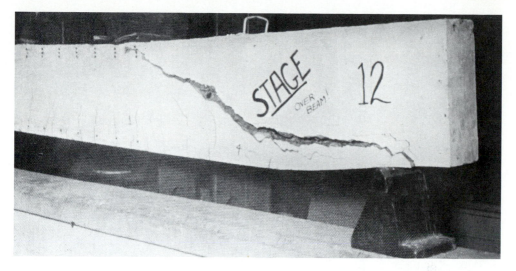

Photo 40   Principal diagonal tension crack at failure of beam in Photo 39 (load stage 12).

6.2 BEHAVIOR OF HOMOGENEOUS BEAMS

Consider the two infinitesimal elements A_1 and A_2 of a rectangular beam in Fig. 6.1a made of homogeneous, isotropic, and linearly elastic material. Figure 6.1b shows the bending stress and shear stress distributions across the depth of the section. The tensile normal stress f_t and the shear stress v are the values in element A_1 across plane a_1-a_1 at a distance y from the neutral axis. From the principles of classical mechanics, the normal stress f and the shear stress v for element A_1 can be written as

$$f = \frac{My}{I} \tag{6.1}$$

and

$$v = \frac{VA\bar{y}}{Ib} \tag{6.2}$$

where M and V = bending moment and shear force at section a_1-a_1

A = cross-sectional area of the section at the plane passing through the centroid of element A_1

y = distance from the element to the neutral axis

$\bar{y}$ = distance from the centroid of A to the neutral axis

I = moment of inertia of the cross section

b = width of the beam

Figure 6.2 shows the internal stresses acting on the infinitesimal elements A_1 and A_2. Using Mohr's circle in Fig. 6.2b, the principal stresses for element A_1 in

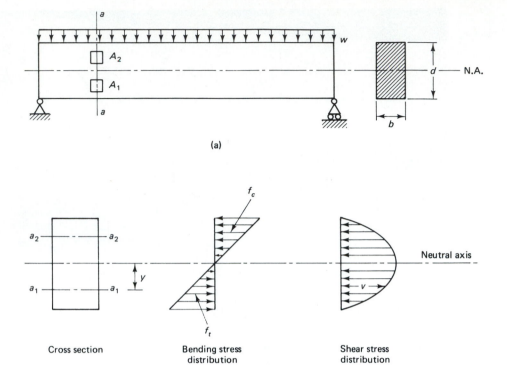

Figure 6.1 Stress distribution for a typical homogeneous rectangular beam.

the tensile zone below the neutral axis become

$$f_{t(\max)} = \frac{f_t}{2} + \sqrt{\left(\frac{f_t}{2}\right)^2 + v^2} \quad \text{principal tension} \tag{6.3a}$$

$$f_{c(\max)} = \frac{f_t}{2} - \sqrt{\left(\frac{f_t}{2}\right)^2 + v^2} \quad \text{principal compression} \tag{6.3b}$$

and

$$\tan 2\theta_{\max} = \frac{v}{f_t/2} \tag{6.3c}$$

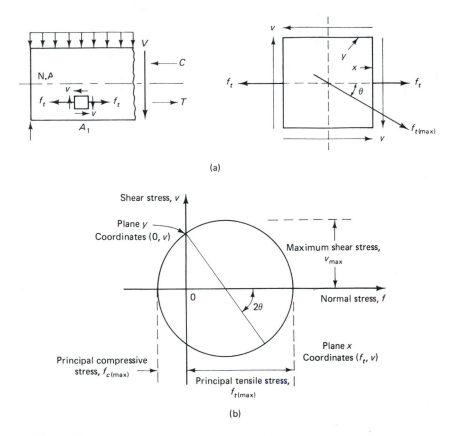

(a)

(b)

Figure 6.2   Stress state in elements A_1 and A_2: (a) stress state in element A_1; (b) Mohr's circle representation, element A_1; (c) stress state in element A_2; (d) Mohr's circle representation, element A_2.

6.3 BEHAVIOR OF REINFORCED CONCRETE BEAMS AS NONHOMOGENEOUS SECTIONS

The behavior of reinforced concrete beams differs in that the tensile strength of concrete is about one-tenth of its strength in compression. The compressive stress f_c in element A_2 of Fig. 6.2b above the neutral axis prevents cracking because the maximum principal stress in the element is in compression. For element A_1 below the neutral axis, the maximum principal stress is in tension; hence cracking ensues. As one moves toward the support, the bending moment and hence f_t decrease, accompanied by corresponding increase in the shear stress. The principal stress $f_{t(max)}$ in tension acts at an approximately 45° plane to the normal at sections close to the support, as seen in Fig. 6.3. Because of the low tensile strength of concrete, diagonal cracking develops along planes perpendicular to the planes of principal tensile stress; hence the term *diagonal tension cracks*. To prevent such cracks from opening, special "diagonal tension" reinforcement has to be provided.

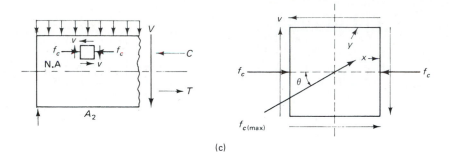

(c)

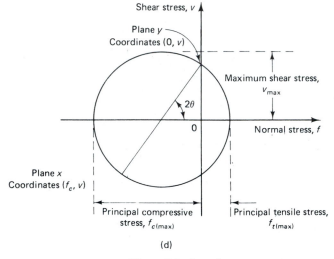

(d)

Figure 6.2 (*cont.*)

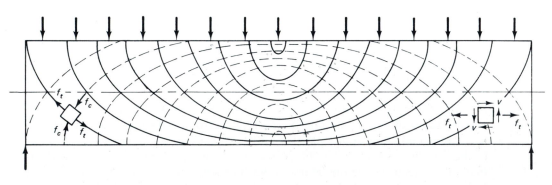

Figure 6.3 Trajectories of principal stresses in a homogeneous isotropic beam. Solid lines: tensile trajectories; dashed lines: compressive trajectories.

If f_t close to the support in Fig. 6.3 is assumed equal to zero, the element becomes nearly in a state of pure shear and the principal tensile stress, using Eq. 6.3b, would be equal to the shear stress v on a 45° plane.   It is this diagonal tension stress that causes the inclined cracks.

Definitive understanding of the correct shear mechanism in reinforced concrete is still incomplete.   However, the approach of the ACI–ASCE Joint Committee 426 gives a systematic empirical correlation of the basic concepts developed from extensive test results.

6.4 REINFORCED CONCRETE BEAMS WITHOUT DIAGONAL TENSION REINFORCEMENT

In regions of large bending moments, cracks develop almost perpendicular to the axis of the beam.   These cracks are called *flexural cracks*.   In regions of high shear due to the diagonal tension, the inclined cracks develop as an extension of the flexural crack and are termed *flexure shear cracks*.   Figure 6.4 portrays the types of cracks expected in a reinforced concrete beam with or without adequate diagonal tension reinforcement.

6.4.1 Modes of Failure of Beams without Diagonal Tension Reinforcement

The slenderness of the beam, that is, its shear span/depth ratio, determines the failure mode of the beam.   Figure 6.5 demonstrates schematically the failure patterns.   The shear span a for concentrated load is the distance between the point of application of the load and the face of support.   For distributed loads, the shear span l_c is the clear beam span.   Fundamentally, three modes of failure or their combinations occur: (1) flexural failure, (2) diagonal tension failure, and (3) shear compression failure.   The more slender the beam, the stronger the tendency toward flexural behavior, as seen from the following discussion.

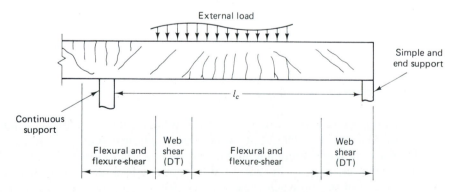

Figure 6.4   Crack categories.

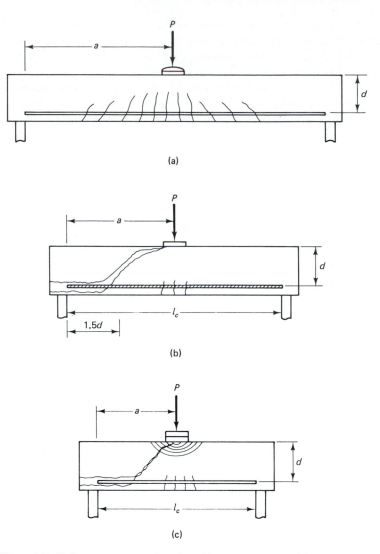

Figure 6.5 Failure patterns as a function of beam slenderness: (a) flexural failure; (b) diagonal tension failure; (c) shear compression failure.

6.4.2 Flexural Failure (F)

In this region, cracks are mainly vertical in the middle third of the beam span and perpendicular to the lines of principal stress. These cracks result from a very small shear stress v and a dominant flexural stress f of a value close to an almost horizontal principal stress $f_{t(\max)}$. In such a failure mode, a few very fine vertical cracks start to develop in the midspan area at about 50% of the failure load in flexure. As the external load increases, additional cracks develop in the central region of the span, and the initial cracks widen and extend deeper toward the neutral axis and beyond, with a marked increase in the deflection of the beam. If the beam is underreinforced, failure occurs in a ductile manner by initial yielding of the main longitudinal flexural reinforcement. This type of behavior gives ample warning of the imminence of collapse of the beam. The shear span/depth ratio for this behavior exceeds a value of 5.5 in the case of concentrated loading and is in excess of 16 for distributed loading.

6.4.3 Diagonal Tension Failure (DT)

This failure precipitates if the strength of the beam in diagonal tension is lower than its strength in flexure. The shear span/depth ratio is of *intermediate* magnitude, with the ratio a/d varying between 2.5 and 5.5 for the case of concentrated loading. Such beams can be considered of intermediate slenderness. Cracking starts with the development of a few fine vertical flexural cracks at midspan, followed by the destruction of the bond between the reinforcing steel and the surrounding concrete at the support. Thereafter, without ample warning of impending failure, two or three diagonal cracks develop at about $1\frac{1}{2}d$ to $2d$ distance from the face of the support. As they stabilize, one of the diagonal cracks widens into a principal diagonal tension crack and extends to the top compression fibers of the beam, as seen in Fig. 6.5b or 6.7. Notice that the flexural cracks do not propagate to the neutral axis in this essentially brittle failure mode, which has relatively small deflection at failure.

6.4.4 Shear Compression Failure (SC)

These beams have a small shear span/depth ratio, a/d, of magnitude 1 to 2.5 for the case of concentrated loading and less than 5.0 for distributed loading. As in the diagonal tension case, few fine flexural cracks start to develop at midspan and stop propagating as destruction of the bond occurs between the longitudinal bars and the surrounding concrete at the support region. Thereafter, an inclined crack steeper than in the diagonal tension case suddenly develops and proceeds to propagate toward the neutral axis. The rate of its progress is reduced with crushing of the concrete in the top compression fibers and a redistribution of stresses within the top region. Sudden failure takes place as the principal inclined crack dynamically joins the crushed concrete zone, as illustrated in Fig. 6.5c. This type of failure can be considered relatively less brittle than the diagonal tension failure due to the stress redistribution. Yet it is, in fact, a brittle type of failure with limited warning, and such a design should be avoided completely.

    A reinforced concrete beam or element is not homogeneous, and the strength of the concrete throughout the span is subject to a normally distributed variation. Hence one cannot expect that a stabilized failure diagonal crack occurs at both ends of the beam. Also, because of these properties, overlapping combinations of flexure–diagonal tension failure and diagonal tension–shear compression failure

TABLE 6.1 BEAM SLENDERNESS EFFECT ON MODE OF FAILURE

Beam category	Failure mode	Shear span/depth ratio as a measure of slenderness[a]	
		Concentrated load, a/d	Distributed load, l_c/d
Slender	Flexure (F)	Exceeds 5.5	Exceeds 16
Intermediate	Diagonal tension (DT)	2.5–5.5	11–16[b]
Deep	Shear compression (SC)	1–2.5	1–5[b]

[a]a = shear span for concentrated loads
l_c = shear span for distributed loads
d = effective depth of beam
[b]For a uniformly distributed load, a transition develops from deep beam to intermediate beam effect.

can occur at overlapping shear span/depth ratios. If the appropriate amount of shear reinforcement is provided, brittle failure of horizontal members can be eliminated with little additional cost to the structure. Table 6.1 summarizes the effect of the slenderness values on the mode of failure.

6.5 DIAGONAL TENSION ANALYSIS OF SLENDER AND INTERMEDIATE BEAMS

The occurrence of the first inclined crack determines the shear strength of a beam without web reinforcement. Because crack development is a function of the tensile strength of the concrete in the beam web, a knowledge of the principal stress in the critical sections is necessary, as discussed in Sections 6.2 and 6.3. The controlling principal stress in concrete is the result of the shearing stress v_u due to the external factored shear V_u and the horizontal flexural stress f_t due to the external factored bending moment M_u. The ACI Code provides an empirical model based on results of extensive tests of failure of a large number of beams without web reinforcement. The model is a regression solution to the basic equation for two-dimensional principal stress at a point.

$$f_{t(\max)} = f_t' = \frac{f_t}{2} + \sqrt{\left(\frac{f_t}{2}\right)^2 + v^2} \qquad (6.3a)$$

where $f_{t(\max)}$ is the principal stress in tension and can be assumed to be equal to a constant multiplied by the tensile splitting strength f_t' of plain concrete. Since f_t'

has been proven to be a function of $\sqrt{f'_c}$, Eq. 6.3a becomes

$$\sqrt{f'_c} = K_1 \left[\frac{f_t}{2} + \sqrt{\left(\frac{f_t}{2}\right)^2 + v^2} \right] \qquad (6.4)$$

where K_1 is a constant.

The flexural stress f_t in the concrete is a function of the steel stress in the longitinal reinforcement or the moment of resistance of the section, or

$$f_t \propto \frac{E_c}{E_s} f_s \propto \frac{E_c M_n}{E_s A_s d}$$

But the reinforcement ratio $\rho_w = A_s/bd$ at the tension side and E_c/E_s have a constant value. Hence

$$f_t = F_1 \frac{M_n}{\rho_w bd^2} \qquad (6.5)$$

where F_1 is a constant to be determined by test and M_n is the nominal moment strength of a given section. The shear stress v at the specific cross section bd due to the vertical external factored shear force V_u is

$$v = F_2 \frac{V_n}{bd} \qquad (6.6)$$

where V_n is the nominal shear resistance at the section under consideration and F_2 is the other constant, to be determined from the beam tests. Coefficients F_1 and F_2 both depend on several variables, including the geometry of the beam, type of loading, amount and arrangement of reinforcement, and the interaction between the steel reinforcement and the concrete.

Substituting f_t of Eq. 6.5 and v of Eq. 6.6, rearranging terms, and evaluating the constants K_1, F_1, and F_2 of the experimental model yields the following regression expression:

$$\frac{V_n}{bd\sqrt{f'_c}} = 1.9 + 2500\rho_w \frac{V_n d}{M_n \sqrt{f'_c}} \leq 3.5 \qquad (6.7)$$

A plot of Eq. 6.7 is shown in Fig. 6.6. Note that $M_u/V_u d = a/d$ (see Fig. 6.5); consequently, Eq. 6.7 accounts indirectly for the shear span/depth ratio, hence it accounts for the slenderness of the member. If the nominal shear resistance of the *plain* concrete web is termed V_c, V_n in the left-hand side of Eq. 6.7 has to be expressed as V_c. Transforming Eq. 6.7 into a force format for evaluation of the nominal shear resistance of the web of a beam of normal concrete and having no diagonal tension steel gives

$$V_c = 1.9 b_w d\sqrt{f'_c} + 2500\rho_w \frac{V_n d}{M_n} b_w d \leq 3.5 b_w d\sqrt{f'_c} \qquad (6.8)$$

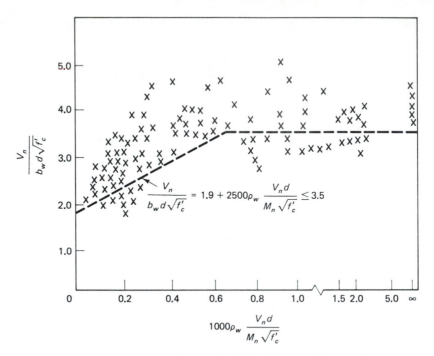

Figure 6.6   Shear resistance of reinforced concrete beam webs.

It is to be emphasized that the ratio $V_u d/M_u$ or $V_n d/M_n$ cannot exceed 1.0, where $V_n = V_u/\phi$ and $M_n = M_u/\phi$ as the values of shear and moment at the section for which V_c is being evaluated.   Also note that using nominal values (subscript n) rather than factored values (subscript u) results in a minor inaccuracy since ϕ is 0.9 for moment while it is 0.85 for shear.   However, such inaccuracy can be justifiable in maintaining consistency by presenting nominal strength values in all the expressions.

The first critical values of V_n and M_n are taken at a distance d from the face of the support since the stabilized (principal) diagonal tension cracks develop in that zone, as seen from Fig. 6.5b.   As one moves toward the midspan of the beam, the values of M_n and V_n will change.   The appropriate moments M_n and shears V_n have to be calculated for the particular section that is being analyzed for web steel reinforcement.

For simplicity of calculations, a more conservative ACI expression can be applied, particularly if the same beam section is not repetitively used in the structure:

$$V_c = \lambda \times 2.0\sqrt{f_c'}b_w d \qquad (6.9)$$

where λ is a factor dependent on the type of concrete, with values of 1.0 for normal-weight concrete, 0.85 for sand-lightweight concrete, and 0.75 for all lightweight concrete.   When axial compression also exists, V_c in Eq. 6.9 becomes

$$V_c = 2\lambda\left(1 + \frac{N_u}{2000A_g}\right)\sqrt{f'_c}b_w d \qquad (6.10a)$$

When significant axial tension exists,

$$V_c = 2\lambda\left(1 + \frac{N_u}{500A_g}\right)\sqrt{f'_c}b_w d \qquad (6.10b)$$

N_u/A_g is expressed in psi, where N_u is the axial load on the member and A_g is the gross area of the section; N_u is negative in tension.

6.6 WEB STEEL PLANAR TRUSS ANALOGY

As discussed previously, web reinforcement has to be provided to prevent failure due to diagonal tension.   Theoretically, if the necessary steel bars in the form of the tensile stress trajectories shown in Fig. 6.3 are placed in the beam, no shear failure can occur.   However, practical considerations eliminate such a solution, and other forms of reinforcing are improvised to neutralize the principal tensile stresses at the critical shear failure planes.   The mode of failure in shear reduces the beam to a simulated arched section in compression at the top and tied at the bottom by the longitudinal beam tension bars, as seen in Fig. 6.7A(a).   If one isolates the main concrete compression element shown in Fig. 6.7A(b), it can be considered as the compression member of a triangular truss, as shown in Fig. 6.7A(c) with the polygon of forces C_c, T_b, and T_s representing the forces acting on the truss members—hence the expression *truss analogy*.   Force C_c is the compression in the simulated concrete strut, force T_b is the tensile force increment of the main longitudinal tension bar, and T_s is the force in the bent bar.   Figure 6.7B(a) shows the analogy truss for the case of using vertical stirrups instead of inclined bars, with the forces polygon having a vertical tensile force T_s instead of the inclined one in Fig. 6.7A(c).

As can be seen from the previous discussion, the shear reinforcement basically performs four main functions:

1. Carries a portion of the external factored shear force V_u
2. Restricts the growth of the diagonal cracks
3. Holds the longitudinal main reinforcing bars in place so that they can provide the dowel capacity needed to carry the flexural load
4. Provides some confinement to the concrete in the compression zone if the stirrups are in the form of closed ties

This discussion, however, does not adequately account for the equilibrium role of additional longitudinal tensile reinforcement in enhancing the shear strength

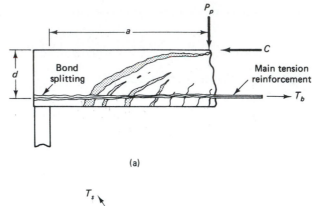

(a)

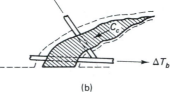

(b)

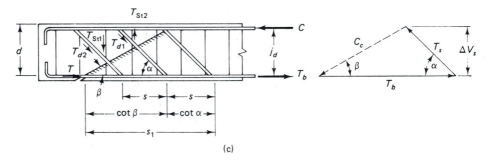

(c)

Figure 6.7A Diagonal tension failure mechanism: (a) failure pattern; (b) concrete simulated strut; (c) planar truss analogy.

of a beam (Ref. 6.16 Sec. 3.2.1). To do so and hence maintain total equilibrium in beam shear caused by shear-bending interaction, one has to consider the horizontal tensile component $V_n \cot \theta$ of the vertical external nominal shear force V_n. This component is considered as equally shared by the top compression bars (truss compression chords) and the bottom longitudinal tensile bars (truss ties), as shown in Fig. 6.8. While neglecting this tensile component is not significant when only shear is present, it has to be accounted for when torsion is also acting, as is discussed in Sec. 7.2. In such a case, the shear flow concept in a membrane element model should be applied in which both the longitudinal and transverse reinforcement have to be considered, as in Chapter 7, Fig. 7.10.

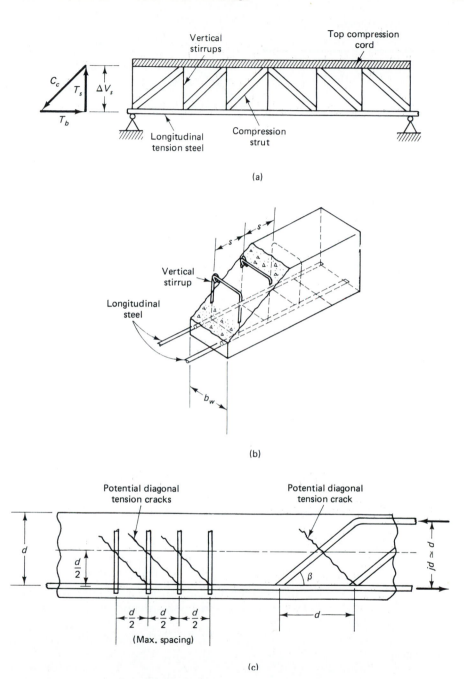

Figure 6.7B   Web steel arrangement: (a) truss analogy for vertical stirrups; (b) three-dimensional view of vertical stirrups; (c) spacing of web steel.

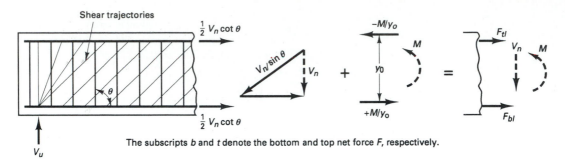

The subscripts b and t denote the bottom and top net force F, respectively.

Figure 6.8   Shear–flexure interaction equilibrium.

6.6.1 Web Steel Resistance

If V_c, the nominal shear resistance of the plain web concrete, is less than the nominal total vertical shearing force $V_u/\phi = V_n$, web reinforcement has to be provided to carry the difference in the two values; hence

$$V_s = V_n - V_c \tag{6.11}$$

The nominal resisting shear V_c can be calculated from Eq. 6.8 or 6.9, and V_s can be determined from equilibrium analysis of the bar forces in the analogous triangular truss cell.   From Fig. 6.7A(c),

$$V_s = T_s \sin \alpha = C_c \sin \beta \tag{6.12a}$$

where T_s is the force resultant of all web stirrups across the diagonal crack plane and n is the number of spacings s.   If $s_1 = ns$ in the bottom tension chord of the analogous truss cell, then

$$s_1 = jd(\cot \alpha + \cot \beta) \tag{6.12b}$$

Assuming that moment arm $jd \simeq d$, the stirrup force per unit length from Eqs. 6.12a and 6.12b, where $s_1 = ns$, becomes

$$\frac{T_s}{s_1} = \frac{T_s}{ns} = \frac{V_s}{\sin \alpha} \frac{1}{d(\cot \beta + \cot \alpha)} \tag{6.12c}$$

If there are n inclined stirrups within the s_1 length of the analogous truss chord, and if A_v is the area of one inclined stirrup,

$$T_s = nA_v f_y \tag{6.13a}$$

Hence

$$nA_v = \frac{V_s ns}{d \sin \alpha(\cot \beta + \cot \alpha)f_y} \tag{6.13b}$$

But it can be assumed that in the case of diagonal tension failure the compression diagonal makes an angle $\beta = 45°$ with the horizontal; Eq. 6.13b becomes

$$V_s = \frac{A_v f_y d}{s} [\sin \alpha (1 + \cot \alpha)]$$

to get

$$V_s = \frac{A_v f_y d}{s} (\sin \alpha + \cos \alpha) \tag{6.14a}$$

or

$$s = \frac{A_v f_y d}{V_n - V_c} (\sin \alpha + \cos \alpha) \tag{6.14b}$$

If the inclined web steel consists of a single bar or a single group of bars all bent at the same distance from the face of the support,

$$V_s = A_v f_y \sin \alpha \leq 3.0\sqrt{f_c'} b_w d$$

If vertical stirrups are used, angle α becomes 90°, giving

$$V_s = \frac{A_v f_y d}{s} \tag{6.15a}$$

or

$$s = \frac{A_v f_y d}{(V_u/\phi) - V_c} = \frac{A_v \phi f_y d}{V_u - \phi V_c} \tag{6.15b}$$

6.6.2 Limitations on Size and Spacing of Stirrups

Equations 6.14 and 6.15 give inverse relationships between the spacing of the stirrups and the shear force or shear stress they resist, with the spacing s decreasing with the increase $(V_n - V_c)$. In order for every *potential* diagonal crack to be resisted by a vertical stirrup as seen in Fig. 6.7A(c), maximum spacing limitations are to be applied as follows for vertical stirrups:

1. $V_n - V_c > 4\sqrt{f_c'} b_w d$: $s_{max} = d/4 \leq 24$ in.

2. $V_n - V_c \leq 4\sqrt{f_c'} b_w d$: $s_{max} = d/2 \leq 24$ in.

3. $V_n - V_c > 8\sqrt{f_c'} b_w d$: enlarge section

The minimum web steel area A_v has to be provided if the factored shear force V_u exceeds one-half the shear strength ϕV_c of the plain concrete web. This precaution is necessary to prevent brittle failure, thus enabling both the stirrups and the beam compression zone to continue carrying the increasing shear after the formulation of the first inclined crack.

$$\text{minimum } A_v = \frac{50 b_w s}{f_y} \tag{6.16}$$

where A_v is the area of all the vertical stirrup legs in the cross section.

6.7 WEB REINFORCEMENT DESIGN PROCEDURE FOR SHEAR

The following is a summary of the recommended sequence of design steps.

1. Determine the critical section and calculate the factored shear force V_u. When the reaction, in the direction of applied shear, introduces compression into the end regions of a member, the critical section can be assumed at a distance of d from the support, provided that no concentrated load acts between the support face and distance d thereafter.

2. Check whether

$$V_u \leq \phi(V_c + 8\sqrt{f_c'}b_w d)$$

where b_w is the web width or diameter of the circular section. If this condition is not satisfied, the cross section has to be enlarged.

3. Use minimum shear reinforcement A_v if V_u is larger than one-half ϕV_c, with the following exceptions:
   (a) Concrete joist construction
   (b) Slabs and footings
   (c) Small shallow beams of depth not exceeding 10 in. (254 mm) or $2\frac{1}{2}$ times the flange thickness:

$$\text{minimum } A_v = \frac{50 b_w s}{f_y}$$

   Good construction practice dictates that some stirrups always be used to facilitate proper handling of the reinforcement cage.

4. If $V_u > \phi V_c$, shear reinforcement must be provided such that $V_u \leq \phi(V_c + V_s)$, where

$$V_s = \begin{cases} \dfrac{A_v f_y d}{s} & \text{for vertical stirrups} \\[2ex] \dfrac{A_v f_y d}{s}(\sin \alpha + \cos \alpha) & \text{for inclined stirrups} \end{cases}$$

5. Maximum spacing s must be $s = d/2 \leq 24$ in., except that in cases where $V_s > 4\sqrt{f'_c}b_w d$ the spacing then becomes $s \leq d/4 \leq 24$ in.

Figure 6.9 presents a flow chart for the performance of the sequence of calculations necessary for the design of vertical stirrups. Simple corresponding modifications of this chart can be made so that the chart can be used in the design of inclined web steel.

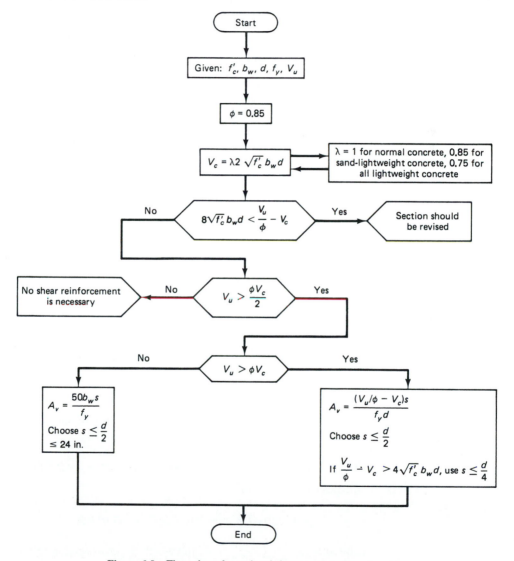

Figure 6.9 Flow chart for web reinforcement design procedure.

6.8 EXAMPLES OF THE DESIGN OF WEB STEEL FOR SHEAR

6.8.1 Example 6.1: Design of Web Stirrups

A rectangular isolated beam has an effective span of 25 ft (7.62 m) and carries a working live load of 8000 lb per linear foot (116.8 kN/m) and no external dead load except its self-weight. Design the necessary shear reinforcement. Use the simplified term of Eq. 6.9 for calculating the capacity V_c of the plain concrete web. Given:

$f'_c = 4000$ psi (27.6 MPa), normal-weight concrete

$f_y = 60,000$ psi (414 MPa)

$b_w = 14$ in. (356 mm)

$d = 28$ in. (712 mm)

$h = 30$ in. (762 mm)

longitudinal tension steel: six No. 9 bars (diameter 28.6 mm)

no axial force acts on the beam

Solution   *Factored shear force (Step 1)*

$$\text{beam self-weight} = \frac{14 \times 30}{144} \times 150 = 437.5 \text{ lb/ft}$$

$$\text{total factored load} = 1.7 \times 8000 + 1.4 \times 437.5 = 14{,}212.5 \text{ lb/ft}$$

The factored shear force at the face of the support is

$$V_u = \frac{25}{2} \times 14{,}212.5 = 177{,}656 \text{ lb}$$

The first critical section is at a distance $d = 28$ in. from the face of the support of this beam (half-span $= 150$ in.).

$$V_u \text{ at } d = \frac{150 - 28}{150} \times 177{,}656 = 144{,}494 \text{ lb}$$

Shear capacity (Step 2)

The shear capacity of the plain concrete in the web from the simplified equation for normal-weight concrete ($\lambda = 1.0$) is

$$V_c = 2.0\lambda\sqrt{f'_c}b_w d = 2 \times 1.0\sqrt{4000} \times 14 \times 28 = 49{,}585 \text{ lb}$$

Check for adequacy of section for shear:

$$(8 + 2.0)\sqrt{f'_c}b_w d = 10\sqrt{f'_c}b_w d = 247{,}923 \text{ lb}$$

$$\text{required } V_n = \frac{V_u}{\phi} = \frac{144{,}494}{0.85} = 169{,}993 \text{ lb} \qquad \text{cross section O.K.}$$

$$V_n > \tfrac{1}{2} V_c \qquad \text{hence stirrups are necessary}$$

Shear reinforcement (Steps 3 to 5)

Try No. 4 two-legged stirrups (area per leg $= 0.20$ in.^{2}).

$$A_v = 2 \times 0.2 = 0.40 \text{ in.}^2$$

From Eq. 6.15b,

$$s = \frac{A_v f_y d}{(V_u/\phi) - V_c} = \frac{0.4 \times 60,000 \times 28}{169,993 - 49,584}$$

$$= 5.58 \text{ in. (141.7 mm)}$$

Since $V_n - V_c > 4\sqrt{f_c'}b_w d$, the maximum allowable spacing $s = d/4 = 28/4 = 7$ in. At the critical section, $d = 28$ in. from the face of the support, the maximum allowable spacing would in this case be 5.58 in.

The shear force for distributed load decreases linearly from the support to midspan of the beam. Hence the web reinforcement can be reduced accordingly after determining the zone where minimum reinforcement is necessary and the zone where no web reinforcement is needed. The same size and spacing of stirrups needed at the critical section d from face of support should be continued to the support. Figure 6.10 illustrates the various values being calculated:

Critical phase x_d (consider the midspan as the origin): $V_n = 169,993$ lb and from before, $s = 5.58$ in. x_d from the midspan point $= 150 - 28 = 122$ in.

Plane x_1 at $s = d/4$ maximum spacing:

$$V_{s1} = 4\sqrt{f_c'}b_w d = 4\sqrt{4000} \times 14 \times 28 = 99,169 \text{ lb}$$

$$V_{n1} = 99,169 + 49,585 = 148,754 \text{ lb}$$

$$x_1 \text{ from midspan point} = (150 - 28) \times \frac{148,754}{169,993} = 106.76 \text{ in.}$$

Plane x_2 at $s = d/2$ maximum spacing:

$$s = \frac{A_v f_y d}{V_n - V_c} \quad \text{or} \quad \frac{28}{2} = \frac{0.4 \times 60,000 \times 28}{V_s}$$

or

$$V_{s2} = 48,000 \text{ lb}$$

$$V_{n2} = 48,000 + 49,585 = 97,585 \text{ lb}$$

$$x_2 \text{ from midspan point} = 122 \times \frac{97,585}{169,993} = 70.03 \text{ in.}$$

From Fig. 6.10a, the distance 36.73 in. is the transition zone from $s = 7$ in. to $s = 14$ in.; hence a stirrup spacing of 8 in. center to center is shown in Fig. 6.10b.

Plane x_3 at shear force V_c:

$$V_c = 2\sqrt{f_c'}b_w d = 49,585 \text{ lb}$$

$$x_3 \text{ from midspan point} = 122 \times \frac{49,585}{169,993} = 35.59 \text{ in.}$$

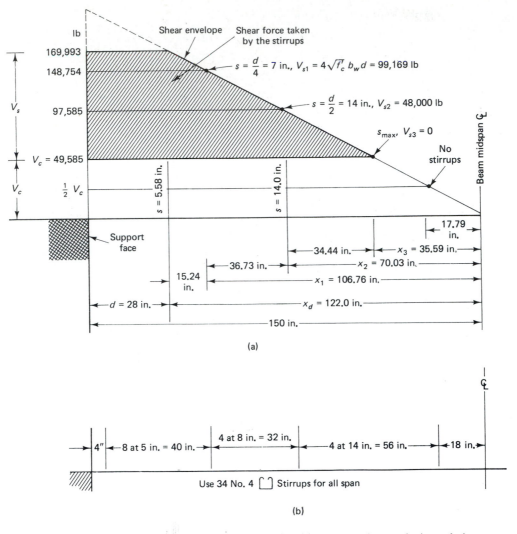

Figure 6.10 Stirrups arrangements for Ex. 6.1: (a) shear envelope and stirrup design segments; (b) vertical stirrups spacing.

Discontinue the stirrups at plane where $V_n \leq \frac{1}{2}V_c$.

Minimum web steel: Test when $V_u > \frac{1}{2}\phi V_c$ or $V_n > \frac{1}{2}V_c$

$$V_u = 144,494$$

$$\tfrac{1}{2}V_c = \tfrac{1}{2} \times 49,585 = 24,793 \text{ lb}$$

$$\text{minimum } A_v = \frac{50b_w s}{f_y} = \frac{50 \times 14 \times 14}{60,000} = 0.16 \text{ in.}^2$$

$$< \text{ actual } A_v = 0.40 \text{ in.}^2 \qquad \text{O.K.}$$

or

$$\text{maximum allowed } s = \frac{A_v f_y}{50 b_w} = \frac{0.40 \times 60,000}{50 \times 14} = 34.3 \text{ in.}$$

$$\text{versus maximum used } s = \frac{d}{2} = 14 \text{ in.} \qquad \text{O.K.}$$

$$x_y = 122.0 \times \frac{24,793}{169,993} = 17.79 \text{ in. from midspan}$$

Proportion the spacing of the vertical stirrups accordingly.

The shaded area in Fig. 6.10a is the shear force area for which stirrups must be provided.   The spacing of the stirrups in Fig. 6.10b is based on the practical consideration of the desirability of using whole spacing dimensions and varying the spacing as little as possible.

6.8.2 Example 6.2: Alternative Solution to Example 6.1

Find the force V_c and the change in stirrup spacing for the beam in Ex. 6.1 if the more refined Eq. 6.8 is used where the separate contribution of the main longitudinal steel at the tension side is more accurately reflected.

Photo 41   Two Union Square, Seattle, Washington, 20,000-psi high-strength concrete used for this high-rise building; design by the NBBJ Group, Architects. (Courtesy Turner Construction and the RBBJ Group, Dr. Weston Hester of the University of California, materials consultant.)

Solution   The shear capacity of the plain concrete in the web is

$$V_c = 1.9\lambda\sqrt{f_c'}\,b_w d + 2500\rho_w \frac{V_u d}{M_u}\,b_w d \le 3.5\sqrt{f_c'}\,b_w d$$

where ρ_w is the longitudinal steel ratio in the web at the tension side only.

$$\rho_w = \frac{6.0}{14 \times 28} = 0.0153$$

V_u at d from support = 144,494 lb   (Ex. 6.1)

$V_u d = 144,494 \times 28 = 4,045,832$ in.-lb

$$M_u \text{ at } d \text{ from support} = V_u d - \frac{w_u d^2}{2} = \left(14,212.5 \times \frac{25}{2}\right) \times 28 - \frac{14,212.5(28)^2}{12 \times 2}$$

$$= 4,510,100 \text{ in.-lb}$$

$$\frac{V_u d}{M_u} \quad \text{or} \quad \frac{V_n d}{M_n} = \frac{4,045,832}{4,510,100} = 0.9 < 1.0 \qquad \text{use 0.9}$$

$$V_c = 1.9\sqrt{4000} \times 14 \times 28 + 2500 \times 0.0153 \times 0.9 \times 14 \times 28$$
$$= 47,105.4 \text{ lb} + 13,494.6 = 60,600 \text{ lb}$$

Use a two-legged No. 4 size vertical stirrup, as in Ex. 6.1.

$$s = \frac{A_v f_y d}{V_u/\phi - V_c} = \frac{0.4 \times 60,000 \times 28}{169,993 - 60,600} = 6.14 \text{ in.} \quad (156.0 \text{ mm})$$

For $s = d/4 = 7$ in.: By trial and adjustment and applying the expression in Eq. 6.8 in the trials,

$$V_{c1} = 57,738 \text{ lb,} \qquad V_{n1} = 152,941 \text{ lb}$$

$$x_1 = (150 - 28)\frac{152,941}{169,993} = 109.86 \text{ in.}$$

For $s = d/2 = 14$ in.:

$$V_{c2} = 50,718 \text{ lb,} \qquad V_{n2} = 98,824 \text{ lb,}$$

$$x_2 = 122 \times \frac{98,824}{169,993} = 70.92 \text{ in.}$$

At the point in the shear envelope where $V_s = 0$, the value of $V_u d/M_u$ in Eq. 6.8 is close to zero for uniformly distributed loads.   Hence assume that

$$V_c \simeq 2.0\lambda\sqrt{f_c'}\,b_w d \quad \text{instead of} \quad V_c = 1.9\lambda\sqrt{f_c'}\,b_w d + 2500\rho_w \frac{V_u d}{M_u}\,b_w d$$

$\lambda = 1.0$ for normal-weight concrete.   Therefore, use $x_3 = 35.59$ in. in Fig. 6.11 (as in Fig. 6.10 of Ex. 6.1) as being accurate enough for all practical purposes.   Proportion the spacing of the stirrups as in Fig. 6.11b.

The shear diagram showing all these details is given in Fig. 6.11.   It can be seen that this refined solution reduced the shearing force taken by the stirrups at the d critical section by the difference between the 49,585 lb of Ex. 6.1 and the 60,600 lb

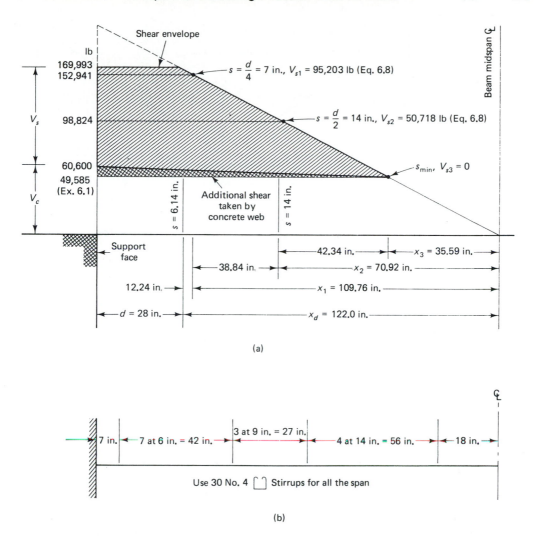

(a)

(b)

Figure 6.11 Stirrups arrangement for Ex. 6.2: (a) shear envelope and stirrup design segments; (b) vertical stirrups spacing.

of Ex. 6.2, as shown in the darkly shaded portion. This difference is taken by the plain concrete in the web. There is a small saving in the number of stirrups that can be justified only if the beam section designed by the refined method is extensively and repetitively used in a multifloor multispan building.

Note in these shear problems that if concentrated loads act on the beam close to the midspan or, in the case of reversible loads, Fig. 6.12, almost constant stirrup spacing throughout the span becomes necessary. The spacing to be used would be that required at the critical section at distance d from the face of the support. Superposition of the shear diagram for a concentrated live load over that of the distributed load due to self-weight or otherwise gives the total shear force for stirrup spacing determination.

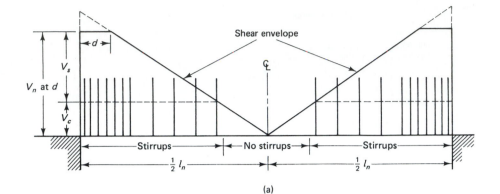

(a)

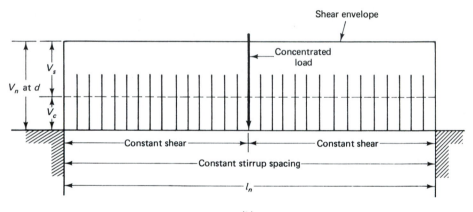

(b)

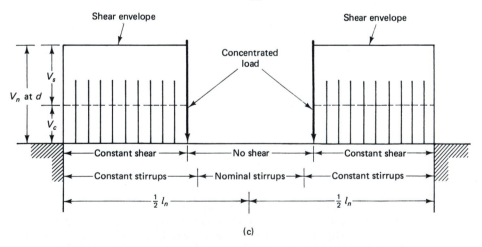

(c)

Figure 6.12 Schematic stirrups distribution: (a) stirrups spacing for uniformly distributed load on beam; (b) stirrups spacing for centrally loaded beam; (c) stirrups spacing for third-point loaded beam.

6.9 DEEP BEAMS

Deep beams are structural elements loaded as beams but having a large depth/thickness ratio and a shear span/depth ratio not exceeding 2 to 2.5, where the shear span is the clear span of the beam for distributed load.   Floor slabs under horizontal loads, wall slabs under vertical loads, short-span beams carrying heavy loads, and some shear walls are examples of this type of structural element.

Because of the geometry of deep beams, they behave as two-dimensional rather than one-dimensional members and are subjected to a two-dimensional state of stress.   As a result, plane sections before bending do not necessarily remain plane after bending.   The resulting strain distribution is no longer considered as linear, and shear deformations that are neglected in normal beams become significant compared to pure flexure.   Consequently, the stress block becomes nonlinear even at the elastic stage.   At the limit state of ultimate load, the compressive stress distribution in the concrete would no longer follow the same parabolic shape or intensity as that shown in Fig. 5.2c for a normal beam.

Figure 6.13 illustrates the linearity of the stress distribution at midspan prior to cracking in a normal beam where the effective span/depth ratio exceeds a value of $3\frac{1}{2}$ to 4.   In contrast, Fig. 6.14a shows the nonlinearity of stress at midspan corresponding to the nonlinear strain under discussion.   It can also be recognized that the magnitude of the maximum tensile stress at the bottom fiber far exceeds the magnitude of the maximum compressive stress.   The stress trajectories in Fig. 6.14b and c confirm this observation.   Note the steepness and concentration of the principal tensile stress trajectories at midspan and the concentration of the compressive stress trajectories at the support for both cases of loading of the beam at top or bottom.

The concrete cracks in a direction perpendicular to the tensile principal stress trajectories.   As the load increases, the cracks widen and propagate, and more cracks open.   Hence less and less concrete remains to resist the indeterminate state of stress.   Because the shear span is small, the compressive stresses in the support region affect the magnitude and direction of the principal tensile stresses such that they become less inclined and lower in value.

In many cases the cracks would almost be vertical or follow the direction of the compression trajectories, with the beam almost shearing off from the support in a total shear failure.   Hence, in the case of deep beams, horizontal reinforcement is needed throughout the height of the beams, in addition to the vertical shear reinforcement along the span.   From Fig. 6.14b and c and the steep gradient of

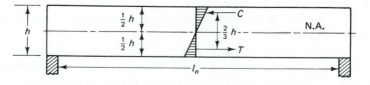

Figure 6.13   Elastic distribution in normal beams ($l_n/h \geq 3\frac{1}{2}$ to 5).

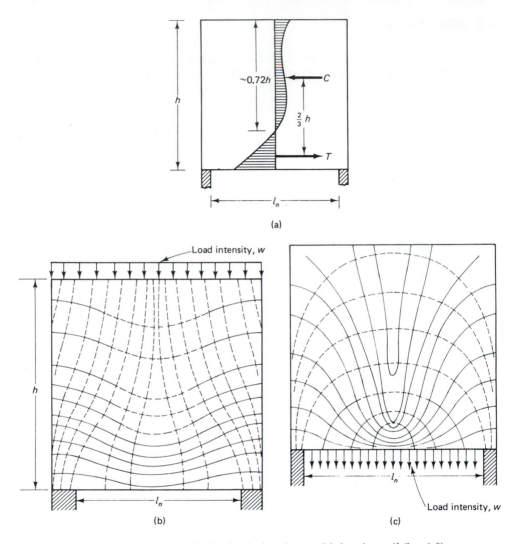

Figure 6.14 Elastic stress distribution in deep beams: (a) deep beam ($l_n/h \leq 1.0$); (b) principal stress trajectories in deep beams loaded on top; (c) principal stress trajectories in deep beams loaded at bottom.

the tensile stress trajectories at the lower fibers, a concentration of horizontal reinforcing bars is required to resist the high tensile stresses at the lower regions of the deep beam.

Additionally, the high depth/span ratio of the beam should provide an increased resistance to the external shear load due to a higher compressive arch action. Consequently, it should be expected that the nominal resisting shear force V_c for deep beams will considerably exceed the V_c value for normal beams.

In summary, shear in deep beams is a major consideration in their design. The magnitude and spacing of both the vertical and horizontal shear reinforcement differ considerably from those used in normal beams, as well as the expressions that have to be used for their design.

6.9.1 Design Criteria for Shear in Deep Beams Loaded at the Top

From the discussion in Section 6.9, it can be inferred that deep beams ($a/d < 2.5$ and $l_n/d < 5.0$) have a higher nominal shear resistance V_c than do normal beams. While the critical section for calculating the factored shear force V_u is taken at distance d from the face of the support in normal beams, the shear plane in the deep beam is considerably steeper in inclination and closer to the support. If x is the distance of the failure plane from the face of the support, l_n the clear span for uniformly distributed load, and a the shear arm or span for concentrated loads, the expression for distance is

$$\text{uniform load:} \qquad x = 0.15l_n \tag{6.17a}$$

$$\text{concentrated load:} \quad x = 0.50a \tag{6.17b}$$

In either case, the distance x should not exceed the effective depth d.

The factored shear force V_u has to satisfy the condition

$$V_u \leq \phi(8\sqrt{f_c'}b_w d) \qquad \text{for } \frac{l_n}{d} < 2.0 \tag{6.18a}$$

or

$$V_u \leq \phi\left[\frac{2}{3}\left(10 + \frac{l_n}{d}\right)\sqrt{f_c'}b_w d\right] \qquad \text{for } 2 \leq \frac{l_n}{d} \leq 5 \tag{6.18b}$$

If not, the section has to be enlarged. The strength reduction factor $\phi = 0.85$.

The nominal shear resisting force V_c of the plain concrete can be taken as

$$V_c = \left(3.5 - 2.5\frac{M_u}{V_u d}\right)\left(1.9\sqrt{f_c'} + 2500\rho_w \frac{V_u d}{M_u}\right)b_w d \leq 6\sqrt{f_c'}b_w d \tag{6.19a}$$

where $1.0 < 3.5 - 2.5(M_u/V_u d) \leq 2.5$. This factor is a multiplier of the basic equation for V_c in normal beams to account for the higher resisting capacity of deep beams. The ACI Code allows this higher resisting capacity provided that some minor unsightly cracking is tolerated if V_u exceeds the first shear cracking load; otherwise, the designer can use

$$V_c = 2\sqrt{f_c'}b_w d \tag{6.19b}$$

When the factored shear V_u exceeds ϕV_c, shear reinforcement has to be provided such that $V_u \leq \phi(V_c + V_s)$, where V_s is the force resisted by the shear reinforcement:

$$V_s = \left(\frac{A_v}{s_v}\frac{1 + l_n/d}{12} + \frac{A_{vh}}{s_h}\frac{11 - l_n/d}{12}\right)f_y d \tag{6.20}$$

where A_v = total area of vertical reinforcement spaced at s_v in the horizontal direction at both faces of the beam

A_{vh} = total area of horizontal reinforcement spaced at s_h in the vertical direction at both faces of the beam

$$\text{maximum } s_v \leq \frac{d}{5} \quad \text{or} \quad 18 \text{ in.}$$

$$\text{maximum } s_h \leq \frac{d}{3} \quad \text{or} \quad 18 \text{ in.} \qquad \text{whichever is smaller} \qquad (6.21a)$$

and

$$\text{minimum } A_v = 0.0015bs_v \qquad (6.21b)$$

$$\text{minimum } A_{vh} = 0.0025bs_h$$

The shear reinforcement required at the critical section must be provided throughout the deep beams.

In the case of continuous deep beams, because of the large stiffness and negligible rotation of the beam section at the supports, the continuity factor at the first interior support has a value close to 1.0. Consequently, the same reinforcement for shear can be used in all spans for all practical purposes if all the spans are equal and similarly loaded.

6.9.2 Design Criteria for Flexure in Deep Beams

6.9.2.1 Simply supported beams.   The ACI Code does not specify a design procedure but requires a rigorous nonlinear analysis for the flexural analysis and design of deep beams. The simplified provisions presented in this section are based on the recommendations of the Euro-International Concrete Committee (CEB).

Figure 6.14a shows a schematic stress distribution in a homogeneous deep beam having a span/depth ratio $l_n/h \simeq 1.0$. It was experimentally observed that the moment lever arm does not change significantly even after initial cracking. Since the nominal resisting moment is

$$M_n = A_s f_y (\text{moment arm } jd) \qquad (6.22a)$$

the reinforcement area A_s for flexure is

$$A_s = \frac{M_u}{\phi f_y jd} \geq \frac{3\sqrt{f_c'}}{f_y} bd \geq \frac{200bd}{fy} \qquad (6.22b)$$

The lever arm as recommended by CEB is

$$jd = 0.2(l + 2h) \qquad \text{for } 1 \leq \frac{l}{h} < 2 \qquad (6.23a)$$

and $$jd = 0.6l \qquad \text{for } \frac{l}{h} < 1 \qquad (6.23b)$$

where l is the effective span measured center to center of supports or 1.15 clear span l_n, whichever is smaller.   The tension reinforcement has to be placed in the lower segment of beam height such that the segment height is

$$y = 0.25h - 0.05l < 0.20h \tag{6.24}$$

It should consist of closely spaced small-diameter bars well anchored into the supports.

6.9.2.2. Continuous beams.

Continuous deep beams can be treated in the same manner as simply supported deep beams, except that additional reinforcement has to be provided for the negative moment at the support.   Figure 6.15 presents stress trajectories of the principal tensile and compressive stresses in a continuous deep beam.   Comparing this diagram to Fig. 6.14b for the simply supported case, one can observe the similarity of the steepness of the tensile stress trajectories at midspan.   At the continuous supports, the total section is in tension.

The concentration of the tensile stress trajectories at the support regions of the continuous deep beam necessitates a concentration of well-anchored horizontal shear reinforcement.   The required total flexural reinforcement area

$$A_s = \frac{M_u}{\phi f_y jd} \geq \frac{200bd}{f_y} \geq \frac{3\sqrt{P_c'}}{f_y} bd$$

as in Eq. 6.22b for the simply supported beam.   The lever arm jd is, however, different and has a value

$$jd = 0.2(l + 1.5h) \qquad \text{for } 1 \leq \frac{l}{h} \leq 2.5 \tag{6.25a}$$

$$jd = 0.5l \qquad\qquad \text{for } \frac{l}{h} < 1.0 \tag{6.25b}$$

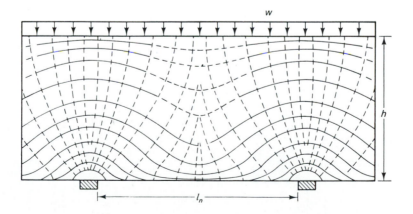

Figure 6.15 Tensile and compression trajectories in a continuous deep beam. Solid line: tension trajectories; dashed line: compression trajectories.

The distribution of the negative flexural reinforcement A_s in continuous beams should be such that the steel area A_{s1} should be placed in the top 20% of the beam depth, and the balance steel area A_{s2} at the next 60% of the beam depth, as shown in Fig. 6.16. The value of

$$A_{s1} = 0.5 \left(\frac{l}{h} - 1 \right) A_s \tag{6.26a}$$

$$A_{s2} = A_s - A_{s1} \tag{6.26b}$$

For cases where the ratio l/h has a value equal to or less than 1.0, use nominal steel for A_{s1} in the top 20% of the beam depth and provide the total A_s in the next 60% of the depth. In the lower h_3 zone the positive reinforcement coming from the beam span should pass through the support for anchorage and continuity.

6.9.3 Sequence of Deep Beams Design Steps for Shear

The following is a recommended procedure for the design of shear reinforcement in deep beams based on ACI requirements. The sequence of steps should essentially be similar to that in Section 6.7 for web reinforcement design in normal beams. Additionally, flexural reinforcement has to be provided to resist the stresses due to bending.

1. Check whether the beam can be classified as a deep beam, that is, $a/d < 2.5$ or $l_n/d < 5.0$ for a concentrated or a uniform load, respectively.
2. Determine the critical section distances x from the face of support: $x = 0.5a$ for concentrated load and $x = 0.15l_n$ for distributed load. Calculate the factored V_u at the critical section, and check whether it is less than the maximum $\phi V_n = V_u$ permitted by Eqs. 6.18a and 6.18b; if not, enlarge the beam section.
3. Calculate the shear resisting capacity V_c of the plain concrete from Eq. 6.19.

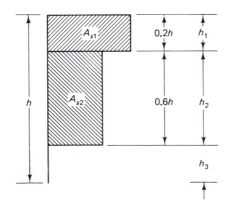

Figure 6.16 Distribution of horizontal flexural steel in continuous deep beams.

4. Calculate V_s if $V_u > \phi V_c$ and choose s_v and s_h by assuming the size of shear reinforcement in both the horizontal and vertical directions.

5. Verify if the size and maximum spacing from step 4 satisfy Eqs. 6.21a and 6.21b; if not, revise and recheck using Eq. 6.20.

6. Select reasonable size and spacing of the shear reinforcement in both horizontal and vertical directions. Where possible, use welded wire fabric mats since they provide superior anchorage of the reinforcement to tied bar mats and are easier to handle and keep in position at both faces of the deep beam.

7. Design the flexural reinforcement as in Section 6.9.2 after determining the moment lever arm jd for the particular case of simply supported or continuous deep beams.

8. Distribute the flexural reinforcement in accordance with Eqs. 6.26a and 6.26b and Fig. 6.16 if the beam is continuous. If the beam is simply supported, concentrate the flexural horizontal longitudinal bars in the lower $(0.25h - 0.05l) \leq 0.20h$ part of the beam depth.

9. Sketch a detailed schematic of the distribution of both the shear and the flexural reinforcement.

6.9.4 Example 6.3: Design of Shear Reinforcement in Deep Beams

A simply supported beam having a clear span $l_n = 10$ ft (3.05 m) is subjected to a uniformly distributed live load of 86,000 lb/ft (1255.6 kN/m) on the top. The height h of the beam is 6 ft (1.83 m) and its thickness b is 20 in. (508 mm). The area of its horizontal tension steel is 8.0 in.^{2} (5161 mm^2), determined in Ex. 6.4. Given:

$$f_c' = 4000 \text{ psi (27.6 MPa)}$$
$$f_y = 60,000 \text{ psi (414 MPa)}$$

Design the shear reinforcement for this beam.

Solution   *Check l_n/d and evaluate factored shear force V_u (Step 1)*

Assume that $d \simeq 0.9h = 0.9 \times 6 \times 12 \simeq 65$ in. (1651 mm).

$$\frac{l_n}{d} = \frac{10.0 \times 12}{65} = 1.85 < 5$$

Hence treat as a deep beam.

$$\text{beam self-weight} = \frac{20 \times 72}{144} \times 150 = 1500 \text{ lb/ft (21.9 kN/m)}$$

$$\text{total factored load} = 1.7 \times 86,000 + 1.4 \times 1500$$

$$= 148,300 \text{ lb/ft (2165.2 kN/m)}$$

$$\text{distance of the critical section} = 0.15l_n = 0.15 \times 10.0$$

$$= 1.5 \text{ ft} = 18 \text{ in. (457.2 mm)}$$

The factored shear force V_u at the critical section is

$$V_u = \frac{148{,}300 \times 10}{2} - 148{,}300 \times \frac{18}{12} = 519{,}050 \text{ lb (2308.7 kN)}$$

Nominal shear strength V_n and resisting capacity V_c (Steps 2 and 3)

$$\phi V_n = \phi(8\sqrt{f'_c}\, b_w d) = 0.85(8\sqrt{4000} \times 20 \times 65)$$

$$= 559{,}091 \text{ lb (2486.8 kN)} > 519{,}050 \qquad \text{O.K.}$$

$$M_u = \frac{148{,}300 \times 10 \text{ ft}}{2} \times 1.5 - \frac{148{,}300 \times (1.5)^2}{2}$$

$$= 945{,}412.5 \text{ ft-lb} = 11{,}344{,}950 \text{ in.-lb}$$

$$\frac{M_u}{V_u d} = \frac{11{,}344{,}950}{519{,}050 \times 65} = 0.3363$$

$$3.5 - 2.5\frac{M_u}{V_u d} = 3.5 - 2.5 \times 0.3363 = 2.66 > 2.5 \qquad \text{use 2.5}$$

$$\text{with } \frac{M_u}{V_u d} = 0.4$$

$$\rho_w = \frac{8.0}{20 \times 65} = 0.0062$$

$$\frac{V_u d}{M_u} = 2.97$$

From Eq. 6.19,

$$V_c = 2.5\left(1.9\sqrt{f'_c} + 2500\,\rho_w\,\frac{V_u d}{M_u}\right)b_w d$$

$$= 2.5(1.9\sqrt{4000} + 2500 \times 0.0062 \times 2.5) \times 20 \times 65 = 516{,}479 \text{ lb}$$

$$6\sqrt{f'_c}\, b_w d = 493{,}315 \text{ lb} < 516{,}479 \text{ lb}$$

Hence $V_c = 493{,}315$ lb (2144.3 kN) controls.

Shear reinforcement (Steps 4 and 5)

Assume No. 3 (9.52-mm diameter) bars placed both horizontally and vertically on both faces of the beam.

$$A_v = 2 \times 0.11 = 0.22 \text{ in.}^2 (141.9 \text{ mm}^2) = A_{vh}$$

$$\phi V_s = V_u - \phi V_c$$

or

$$V_s = \frac{V_u}{\phi} - V_c = \frac{519{,}050}{0.85} - 493{,}315 = 117{,}332 \text{ lb (521.9 kN)}$$

$$V_s = \left(\frac{A_v}{s_v}\frac{1 + l_n/d}{12} + \frac{A_{vh}}{s_h}\frac{11 - l_n/d}{12}\right)f_y d$$

Assume that $s_v = s_h = s$ (similar spacing in both the vertical and horizontal directions).

Hence

$$117,332 = \left(\frac{0.22}{s}\frac{1\ +\ 120/65}{12} + \frac{0.22}{s}\frac{11\ -\ 120/65}{12}\right)60,000 \times 65$$

$$s = 7.31 \text{ in. (186 mm)}$$

The maximum permissible spacing of vertical bars $s_v = d/5$ or 18 in., whichever is smaller.

$$s_v = \frac{65}{5} = 13 \text{ in.} \qquad \text{hence } s_v = 7.31 \text{ in. controls}$$

The maximum permissible spacing of horizontal bar $s_h = d/3$ or 18 in., whichever is smaller.

$$s_h = \frac{65}{3} = 21.7 \text{ in.} \qquad \text{hence } s_h = 18 \text{ in. controls}$$

Since similar spacing assumed in both directions, $s_h = 7.31$ in.
Use spacing $s_v = s_h = 7$ in. (178 mm).
   Check for minimum steel:

minimum $A_v = 0.0015bs_v = 0.0015 \times 20 \times 7 = 0.21$ in.$^2 < 0.22$ in.^{2} \qquad O.K.

minimum $A_{vh} = 0.0025bs_h = 0.0025 \times 20 \times 7 = 0.35$ in.$^2 > 0.22$ in.^{2}

Hence No. 3 bars are not adequate for horizontal steel. No. 4 bars on both faces = 2 × 0.20 in.^{2} = 0.40 in.^{2}. Use vertical No. 3 bars at 7 in. center to center (9.53-mm diameter at 177.8 mm center to center) and horizontal No. 4 bars at 7 in. center to center (12.70 mm diameter at 177.8 mm center to center), Fig. 6.17. Use of No. 4 bars instead of No. 3 bars in Eq. 6.20 for V_s would give a higher value of the force V_s that the shear reinforcement is resisting.

   A better reinforcing system would be to use welded wire fabric in deep beams. For a comparable reinforcing area needed, use size D20 welded wire fabric (0.5 in. = 12.7 mm diameter) spaced at $s_h = 6$ in. (152 mm) center to center in the vertical direction and $s_v = 8$ in. (203 mm) center to center in the horizontal direction.

6.9.5 Example 6.4: Flexural Steel in Deep Beams

Design the flexural reinforcement for the beam in Ex. 6.3.

Solution $l_n = 120$ in. (3048 mm) and $h = 72$ in. (1828 mm). Since the width of the supports is not given, assume that $l = 1.15l_n = 138$ in. (3505 mm). The external factored load $U = 148,300$ lb/ft.

$$\text{external factored moment } M_u = \frac{w_u l_n^2}{8} = \frac{148,300(10.0)^2}{8} = 1,853,750 \text{ ft-lb}$$

$$= 22,245,000 \text{ in.-lb (2513.7 kNm)}$$

$$\frac{l}{h} = \frac{138}{72} = 1.92 > 1 < 2$$

$$jd = 0.2(138 + 2 \times 72) = 56.4 \text{ in.}$$

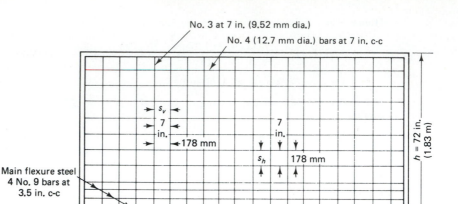

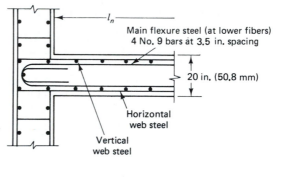

Figure 6.17 Reinforcement for a simply supported deep beam (Ex. 6.4): (a) sectional elevation of beam; (b) cross-sectional plan of beam at support.

$$A_s = \frac{22{,}245{,}000}{0.9 \times 56.4 \times 60{,}000} = 7.30 \text{ in.}^2 \, (4708.5 \text{ mm}^2)$$

$$> \left(\frac{3\sqrt{f_c'}}{f_y} bd = 4.1 \text{ in.}^2 \right) \qquad \text{O.K.}$$

Use four No. 9 horizontal bars on each face, area = 8.00 in.². The height over which A_s is to be distributed above the lower beam face is

$$0.25h - 0.05l = 0.25 \times 72 - 0.05 \times 138 = 11.1 \text{ in.}$$

$$\text{Spacing of flexural steel} = \frac{11.1}{3} = 3.7 \text{ in.}$$

Space four No. 9 bars at 3.5-in. center-to-center vertical spacing on each face of the deep beam to be well anchored into the supports (28.6-mm-diameter bars at 76.2-mm spacing). Figures 6.17a and b give, respectively, a sectional elevation and a horizontal cross section showing the details of the vertical and horizontal shear reinforcement as well as the flexural reinforcement concentrated at the lower 10.5 in. of the deep beam.

6.9.6 Example 6.5: Reinforcement Design for Continuous Deep Beams

Design the reinforcement necessary for an interior span of a continuous beam over several supports if the loading and the properties of the beam are the same as those of Ex. 6.3.

Solution   *Shear reinforcement*

Since the deep beam has a large stiffness, the shear continuity factor for the first interior support is assumed to equal 1.0. Hence use the same vertical and horizontal shear reinforcement as in Ex. 6.3. Use size D20 welded wire fabric (0.5 in. = 12.7-mm diameter) spaced at 6 in. (152 mm) center to center in the horizontal direction and 8 in. (203 mm) center to center in the vertical direction (Fig. 6.18).

Flexural reinforcement

Assume that $d = 65$ in. from Ex. 6.3. The approximate positive factored moment at midspan is

$$+M_u = \frac{w_u l_n^2}{16} = \frac{148,300(10)^2}{16} = 926,875 \text{ ft-lb}$$

$$= 11,122,500 \text{ in.-lb (1257 kNm)}$$

$$+M_n = \frac{M_u}{\phi} = 12,358,333 \text{ in.-lb (1397 kNm)}$$

For continuous deep beams,

$$\text{lever arm } jd = 0.20(l + 1.5h)$$

$$= 0.20(138 + 1.5 \times 72) = 49.2 \text{ in.}$$

$$+A_s = \frac{M_n}{f_y jd} = \frac{12,358,333}{60,000 \times 49.2} = 4.19 \text{ in.}^2$$

$$< +A_s = \frac{200bd}{f_y} = \frac{200(20 \text{ in.})(0.9 \times 72 \text{ in.})}{60,000} = 4.32 \text{ in.}^2$$

$$< \frac{3\sqrt{f_c'}}{f_y}bd = \frac{3\sqrt{4000}}{60,000} 20(0.9 \times 72) = 4.10 \text{ in.}^2$$

Hence $+A_s = 4.32$ in.^{2} controls.

Use three No. 8 bars on each face (three 25.4-mm diameter on each face), area = 4.74 in.^{2} (3057 mm^2). Continue the reinforcement over all the beam span into the support as in Fig. 6.18.

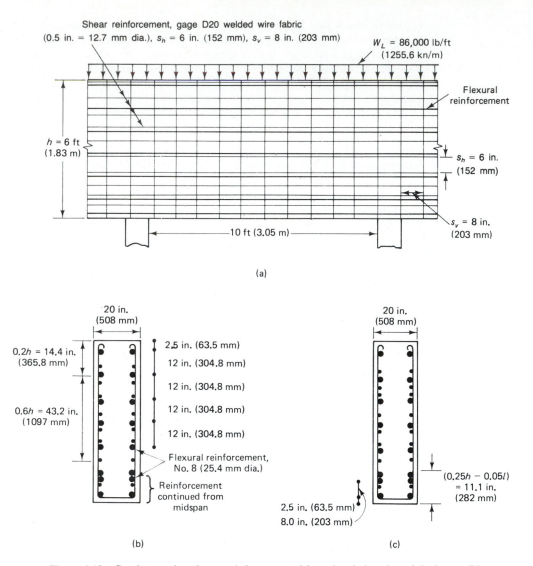

Figure 6.18 Continuous deep beam reinforcement: (a) sectional elevation of the beam; (b) section over support; (c) section at midspan.

The maximum negative factored moment at an interior span is

$$-M_u = \frac{w_u l_n^2}{12} = \frac{148,300(10)^2}{12} = 14,830,000 \text{ in.-lb (1675.8 kN-m)}$$

lever arm $jd = 0.2(l + 1.5h) = 0.2(138 + 1.5 \times 72) = 49.2$ in.

The negative nominal moment of resistance is

$$-M_n = \frac{M_u}{\phi} = \frac{14,830,000}{0.9} = 16,477,778 \text{ in.-lb}$$

$$\text{total negative steel } A_s = \frac{16,477,778}{60,000 \times 49.2} = 5.58 \text{ in.}^2 \text{ (3599 mm}^2)$$

The reinforcement area A_{s1} to be provided for the upper zone is

$$0.5\left(\frac{l}{h} - 1\right)A_s = 0.5\left(\frac{138}{72} - 1\right)5.58 = 2.56 \text{ in.}^2$$

$$h_1 = 0.2 \times 72 = 14.4 \text{ in.}$$

$$A_{s2} = 5.58 - 2.56 = 3.02 \text{ in.}^2 \text{ over } h_2 = 72.0 - 14.4 - 14.4 = 43.2 \text{ in.}$$

Using No. 8 bars (25.4-mm diameter):

Zone h_1:   two No. 8 bars on each face (3.14 in.² > 2.56 in.²)

Zone h_2:   three No. 8 bars on each face (4.74 in.² > 3.02 in.²)

Figure 6.18 shows in elevation and cross section the arrangement of reinforcement for this beam.

6.10 BRACKETS OR CORBELS

Brackets or corbels are short-haunched cantilevers that project from the inner face of columns to support heavy concentrated loads or beam reactions. They are very important structural elements for supporting precast beams, gantry girders, and any other forms of precast structural systems. Precast and prestressed concrete is becoming increasingly dominant, and larger spans are being built, resulting in heavier shear loads at supports. Hence the design of brackets and corbels has become increasingly important. The safety of the total structure could depend on the sound design and construction of the supporting element, in this case the corbel, necessitating a detailed discussion of this subject.

In brackets or corbels, the ratio of the shear arm or span to the corbel depth is often less than 1.0. Such a small ratio changes the state of stress of a member into a two-dimensional one, as discussed in the case of deep beams. Shear deformations would hence affect their nonlinear stress behavior in the elastic state and beyond, and the shear strength becomes a major factor. They differ from deep beams in the existence of potentially large horizontal forces transmitted from the supported beam to the corbel or bracket. These horizontal forces result from long-term shrinkage and creep deformation of the supported beam, which in most cases is anchored to the bracket.

The cracks are usually mostly vertical or steeply inclined pure shear cracks. They often start from the point of application of the concentrated load and propagate toward the bottom reentrant corner junction of the bracket to the column face as in Fig. 6.19a, or start at the upper reentrant corner of the bracket or corbel and proceed almost vertically through the corbel toward its lower fibers, as shown in Fig. 6.19b. Other failure patterns in such elements are shown in Fig. 6.19c and d. They can also develop through a combination of the ones illustrated. Bearing failure can also occur by crushing of the concrete under the concentrated load-bearing plate, if the bearing area is not adequately proportioned.

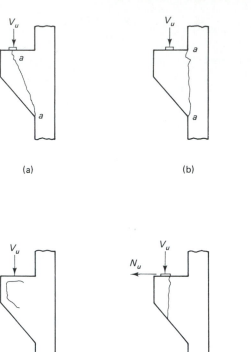

Figure 6.19 Failure patterns: (a) diagonal shear; (b) shear friction; (c) anchorage splitting; (d) vertical splitting.

As will be noticed in the subsequent discussion, detailing of the corbel or bracket reinforcement is of major importance. Failure of the element can be attributed in many cases to incorrect detailing that does not realize full anchorage development of the reinforcing bars.

6.10.1 Shear Friction Hypothesis for Shear Transfer in Corbels

Corbels cast at different times than the main supporting columns can have a potential shear crack at the interface between the two concretes through which shear transfer has to develop. As discussed in the case of deep beams, the smaller the ratio a/d is, the larger the tendency for pure shear to occur through essentially vertical planes. This behavior is more accentuated in the case of corbels with a potential interface crack between two dissimilar concretes.

The shear friction approach in this case is recommended by the ACI, as shown in Fig. 6.19b. An assumption is made of an already cracked vertical plane ($a-a$ in Fig. 6.20) along which the corbel is considered to slide as it reaches its limit state of failure. A coefficient of friction μ is used to transform the horizontal resisting forces of the well-anchored closed ties into a vertical nominal resisting

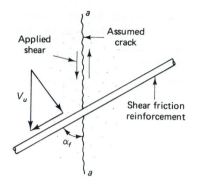

Figure 6.20   Shear friction reinforcement at crack.

force larger than the external factored shear load.   Hence the nominal vertical resisting shear force

$$V_n = A_{vf} f_y \mu \tag{6.27a}$$

to give

$$A_{vf} = \frac{V_n}{f_y \mu} \tag{6.27b}$$

where A_{vf} is the total area of the horizontal, anchored closed shear ties.

The external factored vertical shear has to be $V_u \le \phi V_n$, where for normal concrete,

$$V_n \le 0.2 f'_c b_w d \tag{6.28a}$$

or
$$V_n \le 800 b_w d \tag{6.28b}$$

whichever is smaller.   The required effective depth d of the corbel can be determined from Eq. 6.28a or 6.28b, whichever gives a larger value.

For all-lightweight or sand-lightweight concretes, the shear strength V_n should not be taken greater than $(0.2 - 0.07 \, a/d) f'_c b_w d$ nor $(800 - 280 \, a/d) \, b_w d$ in pounds.

If the shear friction reinforcement is inclined to the shear plane such that the shear force produces some tension in the shear friction steel,

$$V_n = A_{vf} f_y (\mu \sin \alpha_f + \cos \alpha_f) \tag{6.28c}$$

where α_f is the angle between the shear friction reinforcement and the shear plane. The reinforcement area becomes

$$A_{vf} = \frac{V_n}{f_y (\mu \sin \alpha_f + \cos \alpha_f)} \tag{6.28d}$$

The assumption is made that all the shear resistance is due to the resistance at the crack interface between the corbel and the column. The ACI coefficient of friction μ has the following values:

Concrete cast monolithically	1.4λ
Concrete placed against hardened roughened concrete	1.0λ
Concrete placed against unroughened hardened concrete	0.6λ
Concrete anchored to structural steel	0.7λ

where $\lambda = 1.0$ for normal-weight concrete, 0.85 for sand-lightweight concrete, and 0.75 for all-lightweight concrete.

High values of the friction coefficient μ are used so that the calculated shear strength values are in agreement with experiments. If considerably higher strength concretes are used in the corbels, such as polymer-modified concretes, to interface with the normal concrete of the supporting columns, higher μ values could logically be used for such cases than those listed above. Work by the author in Ref. 6.13 substantiates the use of higher values.

Part of the horizontal steel A_{vf} is incorporated in the top tension tie, and the remainder of A_{vf} is distributed along the depth of the corbel as in Fig. 6.21. Evaluation of the top horizontal primary reinforcement layer A_s will be discussed in the next section.

6.10.2 Horizontal External Force Effect

When the corbel or bracket is cast monolithically with the supporting column or wall and is subjected to a large horizontal tensile force N_{uc} produced by the beam supported by the corbel, a modified approach is used, often termed the *strut theory approach*. In all cases, the horizontal factored force N_{uc} cannot exceed the vertical

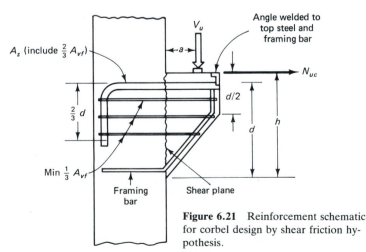

Figure 6.21 Reinforcement schematic for corbel design by shear friction hypothesis.

factored shear V_u.   As seen in Fig. 6.22, reinforcing steel A_n has to be provided
to resist the force N_{uc}.

$$A_n = \frac{N_{uc}}{\phi f_y} \tag{6.29}$$

and

$$A_f = \frac{V_u a + N_{uc}(h - d)}{\phi f_y jd} \tag{6.30}$$

Reinforcement A_f also has to be provided to resist the bending moments caused
by V_u and N_{uc}.

The value of N_{uc} considered in the design should not be less than $0.20V_n$.
The flexural steel area A_f can be approximately obtained by the usual expression
for the limit state at failure of beams, that is,

$$A_f = \frac{M_u}{\phi f_y jd} \tag{6.31}$$

where $M_u = V_u a + N_{uc} (h - d)$.   The axis of such an assumed section lies along
a compression strut inclined at an angle β to the tension tie A_s, as shown in Fig.
6.22.   The volume C_c of the compressive block is

$$C_c = 0.85 f_c' \beta_1 cb = \frac{T_s}{\cos \beta} = \frac{A_s f_y}{\cos \beta} = \frac{V_u}{\sin \beta} \tag{6.32a}$$

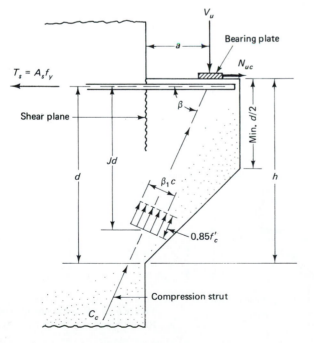

Figure 6.22   Compression strut in corbel.

for which the depth $\beta_1 c$ of the block is obtained perpendicular to the *direction* of the compressive strut,

$$\beta_1 c = \frac{A_s f_y}{0.85 f_c' b \cos \beta} \tag{6.32b}$$

The effective depth d minus the $\beta_1 c/2 \cos \beta$ in the vertical direction gives the lever arm jd between the force T_s and the horizontal component of C_c in Fig. 6.22. Therefore,

$$jd = d - \frac{\beta_1 c}{2 \cos \beta} \tag{6.32c}$$

If jd is substituted in Eq. 6.31,

$$A_f = \frac{M_u}{\phi f_y (d - \beta_1 c/2 \cos \beta)} \tag{6.33}$$

To eliminate several trials and adjustments, the lever arm jd from Eq. 6.32c can be approximated for all practical purposes in most cases as

$$jd \simeq 0.85d \tag{6.34a}$$

so that

$$A_f = \frac{M_u}{0.85 \phi f_y d} \tag{6.34b}$$

The area A_s of the primary tension reinforcement (tension tie) can now be calculated and placed as shown in Fig. 6.23.

$$A_s \geq \tfrac{2}{3} A_{vf} + A_n \tag{6.35}$$

or
$$A_s \geq A_f + A_n \tag{6.36}$$

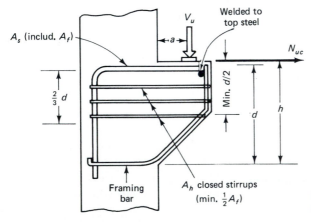

Figure 6.23   Reinforcement schematic for corbel design by strut theory.

whichever is larger:

$$\rho = \frac{A_s}{bd} \geq 0.04\frac{f_c'}{f_y}$$

If A_h is assumed to be the total area of the closed stirrups or ties parallel to A_s,

$$A_h \geq 0.5(A_s - A_n) \tag{6.37}$$

The bearing area under the external load V_u on the bracket should not project beyond the straight portion of the primary tension bars A_s, nor should it project beyond the interior face of the transverse welded anchor bar shown in Fig. 6.23.

6.10.3 Sequence of Corbel Design Steps

As discussed in the preceding section, a horizontal factored force N_{uc}, a vertical factored force V_u, and a bending moment $[V_u a + N_{uc} (h - d)]$ basically act on the corbel.   To prevent failure, the corbel has to be designed to resist these three parameters simultaneously by one of the following two methods, depending on the type of corbel construction sequence, that is, whether the corbel is cast monolithically with the column or not:

1. For monolithically cast corbel with the supporting column, by evaluating the steel area A_h of the closed stirrups that are placed below the primary steel ties A_s.   Part of A_h is due to the steel area A_n from Eq. 6.29 resisting the horizontal force N_{uc}.
2. Calculating the steel area A_{vf} by the shear friction hypothesis if the corbel and the column are *not* cast simultaneously, using part of A_{vf} along the depth of the corbel stem and incorporating the balance in the area A_s of the primary top steel reinforcing layer.

The primary tension steel area A_s is the major component of both methods 1 and 2.   Calculations of A_s depend on whether Eq. 6.35 or 6.36 governs.   If Eq. 6.35 controls, $A_s = \frac{2}{3}A_{vf} + A_n$ is used and the remaining $\frac{1}{3}A_{vf}$ is distributed over a depth $\frac{2}{3}d$ adjacent to A_s.

If Eq. 6.36 controls, $A_s = A_f + A_n$ with the addition of $\frac{1}{2}A_f$ provided as closed stirrups parallel to A_s and distributed within $\frac{2}{3}d$ vertical distance adjacent to A_s.

In both cases, the primary tension reinforcement and the closed stirrups automatically yield the total amount of reinforcement needed for either type of corbel.   Since the mechanism of failure is highly indeterminate and randomness can be expected in the propagation action of the shear crack, it is sometimes advisable to choose the larger calculated value of the primary top steel area A_s in the corbel regardless of whether the corbel element is cast simultaneously with the supporting column.

The horizontal closed stirrups are also a major element in reinforcing the corbel, as seen from the foregoing discussions. Occasionally, additional inclined closed stirrups are also used.

The following sequence of steps is proposed for the design of the corbel:

1. Calculate the factored vertical force V_u and the nominal resisting force V_n of the section such that $V_n \geq V_u/\phi$, where $\phi = 0.85$ for all calculations. V_u/ϕ should be $\leq 0.20f'_c b_w d$ or $\leq 800b_w d$. If not, the concrete section at the support should be enlarged.

2. Calculate $A_{vf} = V_n/f_y\mu$ for resisting the shear friction force and use in the subsequent calculation of the primary tension top steel A_s.

3. Calculate the flexural steel area A_f and the direct tension steel area A_n, where

$$A_f = \frac{V_u a + N_{uc}(h - d)}{\phi f_y jd} \quad \text{and} \quad A_n = \frac{N_{uc}}{\phi f_y}$$

4. Calculate the primary steel area:
   (a) $A_s = \frac{2}{3}A_{vf} + A_n$
   (b) $A_s = A_f + A_n$

   and select whichever is larger. If case (a) controls, the remaining $\frac{1}{3}A_{vf}$ has to be provided as closed stirrups parallel to A_s and distributed within a $\frac{2}{3}d$ distance adjacent to A_s, as in Fig. 6.21. If case (b) controls, use in addition $\frac{1}{2}A_f$ as closed stirrups distributed within a distance $\frac{2}{3}d$ adjacent to A_s as in Fig. 6.23.

$$A_h \geq 0.5(A_s - A_n)$$

and
$$\rho = \frac{A_s}{bd} \geq 0.04\frac{f'_c}{f_y}$$

or
$$\text{minimum } A_s = 0.04\frac{f'_c}{f_y}bd$$

5. Select the size and spacing of the corbel reinforcement with special attention to the detailing arrangements, as many corbel failures are due to incorrect detailing.

A flow chart for proportioning corbels is given in Fig. 6.24.

6.10.4 Example 6.6: Design of a Bracket or Corbel

Design a corbel to support a factored vertical load $V_u = 90,000$ lb (180 kN) acting at a distance $a = 5$ in. (127 mm) from the face of the column. It has a width $b = 10$ in. (254 mm), a total thickness $h = 18$ in. (457 mm), and an effective depth $d = 14$ in. (356 mm). Given:

$f'_c = 5000$ psi (34.5 MPa), normal-weight concrete
$f_y = 60,000$ psi (414 MPa)

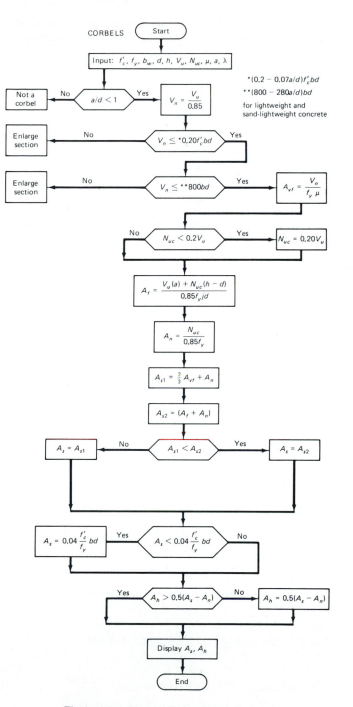

Figure 6.24 Flow chart for design of corbels.

Assume the corbel to be either cast after the supporting column was constructed or both cast simultaneously. Neglect the weight of the corbel.

Solution

Step 1

$$V_n \geq \frac{V_u}{\phi} = \frac{90{,}000}{0.85} = 105{,}882 \text{ lb}$$

$$0.2f'_c bd = 0.2 \times 5000 \times 10 \times 14 = 140{,}000 \text{ lb} > V_n$$

$$800bd = 800 \times 10 \times 14 = 112{,}000 \text{ lb} > V_n \qquad \text{O.K.}$$

Step 2

(a) Monolithic construction; normal-weight concrete $\mu = 1.4\lambda$:

$$A_{vf} = \frac{V_u}{\phi f_y \mu} = \frac{105{,}882}{60{,}000 \times 1.4} = 1.261 \text{ in.}^2 \ (813.3 \text{ mm}^2)$$

(b) Nonmonolithic construction; $\mu = 1.0\lambda$:

$$A_{vf} = \frac{105{,}882}{60{,}000 \times 1.0} = 1.765 \text{ in.}^2 \ (1138.2 \text{ mm}^2)$$

Choose the larger $A_{vf} = 1.765 \text{ in.}^2$ as controlling.

Step 3

Since no value of the horizontal external force N_{uc} transmitted from the super-imposed beam is given, use

$$\text{minimum } N_{uc} = 0.20V_u = 0.2 \times 90{,}000 = 18{,}000 \text{ lb}$$

$$A_f = \frac{M_u}{\phi f_y jd} = \frac{V_u a + N_{uc}(h - d)}{\phi f_y jd} \qquad \text{where } jd \simeq 0.85d$$

$$= \frac{90{,}000 \times 5 + 18{,}000(18 - 14)}{0.90 \times 60{,}000(0.85 \times 14)} = 0.812 \text{ in.}^2 \ (523.9 \text{ mm}^2)$$

$$A_n = \frac{N_{uc}}{\phi f_y} = \frac{18{,}000}{0.85 \times 60{,}000} = 0.353 \text{ in.}^2 \ (277.6 \text{ mm}^2)$$

Step 4

Check the controlling area of primary steel A'_s:

(a) $A_s = (\frac{2}{3}A_{vf} + A_n) = \frac{2}{3} \times 1.765 + 0.353 = 1.529 \text{ in.}^2$

(b) $A_s = A_f + A_n = 0.812 + 0.353 = 1.165 \text{ in.}^2$

$$\text{minimum } A_s = 0.04\frac{f'_c}{f_y}bd = 0.04 \times \frac{5{,}000}{60{,}000} \times 10 \times 14 = 0.47 \text{ in.}^2$$

$$< 1.529 \qquad \text{O.K.}$$

Provide A_s = 1.529 in.² (986.2 mm²). Horizontal closed stirrups: Since case (a) controls,

$$A_h = 0.5(A_s - A_n) = 0.5(1.529 - 0.353) = 0.588 \text{ in.}^2$$

Step 5

Select bar sizes:

(a) Required A_s = 1.529 in.²; use three No. 7 bars = 1.80 in.² (three bars of diameter 22.2 mm = 1161 mm²).

(b) Required A_h = 0.588 in.²; use three No. 3 closed stirrups = 2 × 3 × 0.11 = 0.66 in.² spread over $\frac{2}{3}d$ = 9.33 in. vertical distance. Hence use three No. 3 closed stirrups at 3 in. center to center. Also use three framing size No. 3 bars and one welded No. 3 anchor bar.

Details of the bracket reinforcement are shown in Fig. 6.25. The bearing area under the load has to be checked and the bearing pad designed such that the bearing

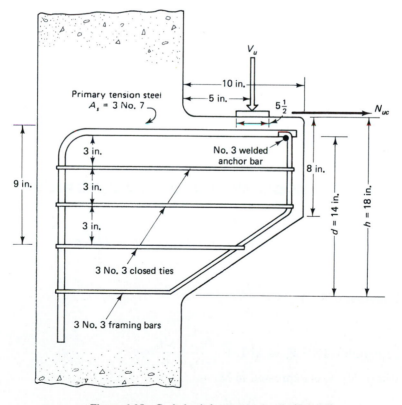

Figure 6.25 Corbel reinforcement details (Ex. 6.5).

Photo 42 High-strength concrete corbel at failure. (Nawy et al.)

stress at the factored load V_u should not exceed $\phi(0.85f'_c A_1)$, where A_1 is the pad area.

$$V_u = 90{,}000 \text{ lb} = 0.70(0.85 \times 5000)A_1$$

$$A_1 = \frac{90{,}000}{0.70 \times 0.85 \times 5000} = 30.25 \text{ in.}^2 \ (19{,}516 \text{ mm}^2)$$

Use a plate $5\frac{1}{2}$ in. $\times$ $5\frac{1}{2}$ in. Its thickness has to be designed based on the manner in which V_u is applied.

6.11 SI DESIGN EXPRESSIONS AND EXAMPLE FOR SHEAR DESIGN

Equation 6.8: $V_c = \lambda\left[\left(\sqrt{f'_c} + 120\rho_w\,\frac{V_u d}{M_u}\right) \div 7\right]b_w d$

Equation 6.9: $V_c = \lambda\dfrac{\sqrt{f'_c}}{6}b_w d$

Equation 6.10: $V_c = \lambda\left(1 + \dfrac{N_u}{14A_g}\right)\dfrac{\sqrt{f'_c}}{6}b_w d$
where N_u/A_g is expressed in MPa.

Equation 6.15a: $V_s = \dfrac{A_v f_y d}{s}$

Equation 6.15b: $V_s = \dfrac{A_v f_y d}{(V_u/\phi) - V_c}$

Min $A_v = \dfrac{b_w s}{3 f_y}$ where b_w and s are expressed in mm and f_y in MPa.

Limitations on Spacing Stirrups

1. $V_n - V_c > \dfrac{\sqrt{f_c'}}{3} b_w d$: $s_{max} = \dfrac{d}{4} \le 610$ mm

2. $V_n - V_c \le \dfrac{\sqrt{f_c'}}{3} b_w d$: $s_{max} = \dfrac{d}{2} \le 610$ mm

3. $V_n - V_c > \dfrac{2}{3} \sqrt{f_c'} b_w d$:   enlarge section

6.11.1 Example 6.7: SI Shear Design

Solve Ex. 6.1 using SI units (see Fig. 6.26).

$f_c' = 27.6$ MPA $\lambda = 1.0$ for normal-weight concrete

$f_y = 414$ MPa $l = 7.62$ m

$b_w = 356$ mm $w_L = 117$ kN/m

$d = 710$ mm   No axial load

$h = 765$ mm   No wind or earthquake

$w_w = 117$ kN/m

$A_s = 6$ No. 9 bars (diameter 28.6 mm.) $= 3850$ mm.^{2}

Closest area using metric bars from Fig. B.2b in the appendix:

2 No. 25 M + 4 No. 30 M $= 2 \times 500 + 4 \times 7000 = 3800$ mm.^{2}

Solution:
beam self-weight $= 356 \times 765 (23.6 \times 10^{-3})$ kN/mm $= 6430$ kN/mm $= 6.4$ kN/m
total factored load $w_u = 1.4 \times 6.4 + 1.7 \times 117 = 208$ kN/m
factored shear force at face of support,

$$V_u = \frac{208 \times 7.62}{2} = 792 \text{ kN}$$

Half-span $= \dfrac{7620}{2} = 3810$ mm

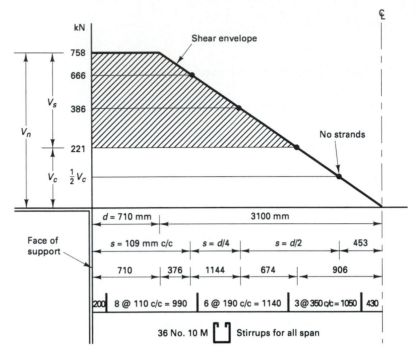

Figure 6.26 Shear envelope and stirrups arrangement (SI units) for Ex. 6.7.

$$V_u \text{ at } d \text{ from support} = \frac{3810 - 710}{3810} \times 792 = 644 \text{ kN}$$

$$\text{required } V_n = \frac{V_u}{\phi} = \frac{644}{0.85} = 758 \text{ kN}$$

$$V_c = \lambda \frac{\sqrt{f_c'}}{6} b_w d = 1.0 \frac{\sqrt{27.6}}{6} 356 \times 710$$

$$= 221 \text{ kN}$$

Check for adequacy of section:

$$V_c + \left(\frac{2}{3} \sqrt{f_c'}\right) b_w d = 212 + \left(\frac{2}{3} \sqrt{27.6}\right) 356 \times 710 \times \frac{1}{1000}$$

$$= 1106 \text{ kN} > 758 \text{ kN}$$

Hence, the section is adequate. Since $V_n > \frac{1}{2}V_c$, stirrups are needed.

Web Steel Reinforcement: Try No. 4 stirrups or, from Fig. B.2b, try No. 10 M metric bar, $A_v = 2 \times 100 = 200 \text{ mm.}^2$.

$$s = \frac{A_v f_y d}{V_u/\phi - V_c} = \frac{200 \times 414 \times 10^{-6} \times 710 \text{ N-mm}}{(758 - 221)10^3 \text{N}} = 109 \text{ mm}$$

Plane x_1 at s = d/4 maximum spacing:

$$V_n - V_c = 758 - 221 = 537 \text{ kN}$$

$$\frac{1}{3}\sqrt{f_c'}b_w d = \frac{\sqrt{27.6}}{3} \times 358 \times 710 \times \frac{1}{1000} = 445 \text{ kN}$$

$$< V_n - V_c = 537 \text{ kN}$$

Find plane for s = d/4 at a distance x_1 from midspan.

$$V_{n1} = V_c + 445 = 221 + 445 = 666 \text{ kN}$$

$$x_1 \text{ from midspan} = \frac{(3810 - 710) \times 666}{758} = 2724 \text{ mm}$$

Plane x_2 at s = d/2 maximum spacing:

$$s = \frac{d}{2} = \frac{A_v f_y d}{V_{n2} - V_c}$$

or

$$V_{n2} = V_c + \frac{A_v f_y d}{s = 710/2}$$

$$= 221 + \frac{200 \times 414 \times 710}{710/2} \times \frac{1}{1000} = 386 \text{ kN}$$

$$x_2 \text{ from midspan} = \frac{(3810 - 710) \times 386}{758} = 1580 \text{ mm}$$

Plane x_3 at shear force V_c:

$$V_c = 221 \text{ kN}$$

$$x_3 \text{ from midspan} = \frac{(3810 - 710)221}{756} = 90$$

Discontinue stirrups at plane where $V_n \leq \frac{1}{2}V_c$.

$$\text{minimum } A_v = \frac{b_w s}{3f_y} = \frac{356 \times 710/2}{3 \times 414} = 102 \text{ mm}^2$$

$$< \text{actual } A_v = 200 \text{ mm}^2, \quad \text{O.K.}$$

x_4 for $\frac{1}{2}V_c$ = 906/2 = 453 mm from midspan.

SELECTED REFERENCES

6.1. ACI–ASCE Committee 426, "The Shear Strength of Reinforced Concrete Members," *Journal of the Structural Division, ASCE*, Vol. 99, No. ST6, June 1973, pp. 1091–1187.

6.2. Taylor, H. P. J., *The Fundamental Behavior of Reinforced Concrete Beams in Bending and Shear*, Special Publication SP-42, Vol. 1, American Concrete Institute, Detroit, 1974, pp. 43–77.

6.3. Zsutty, T. C., "Beam Shear Strength Prediction by Analysis of Existing Data," *Journal of the American Concrete Institute*, Proc. Vol. 65, November 1968, pp. 943–951.

6.4. Mattock, A. H., "Diagonal Tension Cracking in Concrete Beams with Axial Forces," *Journal of the Structural Division*, ASCE, Vol. 95, No. ST9, September 1969, pp. 1887–1900.

6.5. Laupa, A., Seiss, C. P., and Newmark, N. M., *Strength in Shear of Reinforced Concrete Beams*, University of Illinois Bulletin 428, Vol. 52, March 1955, 73 pp.

6.6. Moody, K. G., Viest, I. M., Elstner, R. C. and Hognestad, E., "Shear Strength of Reinforced Concrete Beams," *Journal of the American Concrete Institute*, Proc. Vols. 51-15, 51-21, 51-28, and 51-34, February 1954, pp. 317–332, 417–434, 525–539, 697–732, respectively.

6.7. Comité Euro-International du Béton, *International Recommendations for the Design and Construction of Concrete Structures*, June 1970, 80 pp.; and "*CEB-FIP*" *Model Code for Concrete Structures*, Vol. 2, CEB, Paris, April 1990.

6.8. Park, R., and Paulay, T., *Reinforced Concrete Structures*, Wiley, New York, 1975, 768 pp.

6.9. Raths, C. H., and Kriz, L. B., "Connections in Precast Concrete Structures—Strength of Corbels," *Journal of the Prestressed Concrete Institute*, Proc. Vol. 10, No. 1, February 1965, pp. 16–47.

6.10. Leonhardt, F., "Uber die Kunst des Bewehrens von Stahlbetontragwerken," *Beton- und Stahlbetonbau*, Vol. 60, No. 8, pp. 181–192; No. 9, pp. 212–220.

6.11. ACI–ASCE Committee 426, *Suggested Revisions to Shear Provisions for Building Codes*, ACI 426 IR-77, American Concrete Institute, Detroit, 1979, 84 pp.

6.12. Mattock, A. H., Chen, K. C., and Soongswang, K., "The Behavior of Reinforced Concrete Corbels," *Journal of the Prestressed Concrete Institute*, Vol. 21, No. 2, April 1976, pp. 52–77.

6.13. Nawy, E. G., and Ukadike, M. M., "Shear Transfer in Concrete and Polymer Modified Concrete Members Subjected to Shear Load," *Journal of the American Society for Testing and Materials*, *Proceedings*, Philadelphia, March 1983, pp. 83–97.

6.14. ACI Committee 340, *Strength Design Handbook*, Vol. 1, *Beams, Slabs, Brackets, Footings and Pile Caps*, Special Publication SP-17(91), American Concrete Institute, Detroit, 1991, 508 pp.

6.15. Yong, Y. K., McCloskey, D. H., and Nawy, E. G., *Reinforced Corbels of High Strength*, Special Publication SP-87, American Concrete Institute, Detroit, 1985, pp. 197–212.

6.16. Hsu, T. T. C., *Unified Theory of Reinforced Concrete*, CRC Press, Boca Raton, Fla., 1993, 313 pp.

PROBLEMS FOR SOLUTION

6.1. A simply supported beam has a clear span $l_n = 22$ ft (6.70 m) and is subjected to an external uniform service dead load $w_D = 1200$ lb per ft (17.5 kN/m) and live load $w_L = 900$ lb per ft (13.1 kN/m). Determine the maximum factored vertical shear

V_u at the critical section.   Also determine the nominal shear resistance V_c by both the short method and by the more refined method of taking the contribution of the flexural steel into account.   Design the size and spacing of the diagonal tension reinforcement.   Given:

$$b_w = 12 \text{ in. } (305 \text{ mm})$$

$$d = 17 \text{ in. } (432 \text{ mm})$$

$$h = 20 \text{ in. } (508 \text{ mm})$$

$$A_s = 6.0 \text{ in.}^2 (3780 \text{ mm}^2)$$

$$f'_c = 4000 \text{ psi } (27.6 \text{ MPa}), \text{ normal-weight concrete}$$

$$f_y = 60,000 \text{ psi } (413.7 \text{ MPa})$$

Assume that no torsion exists.

6.2. Solve Problem 6.1 assuming that the beam is made of sand-lightweight concrete and that it is subjected to an axial service compressive load of 2500 lb acting at its plastic centroid.

6.3. A cantilever beam is subjected to a concentrated service live load of 25,000 lb   (111.2 kN) acting at a distance of 3 ft 6 in. (1.07 m) from the wall support.   Its cross section is 10 in. × 20 in. with an effective depth $d = 17$ in. (432 mm).   Design the stirrups needed.   Given:

$$f'_c = 3000 \text{ psi } (20.7 \text{ MPa}), \text{ normal-weight concrete}$$

$$f_y = 40,000 \text{ psi } (275.8 \text{ MPa})$$

6.4. The first interior span of a continuous beam has a clear span $l_n = 18$ ft (5.49 m) and is subjected to an intensity of external uniform service live load $w_L = 1800$ lb per linear foot (26.3 kN/m) and a service dead load $w_D = 2200$ lb per linear foot (32.1 kN/m) not including its self-weight.   Design the section for flexure and diagonal tension, including the size and spacing of the stirrups, assuming that the beam width $b_w = 15$ in. (381.0 mm).   Assume that the beam is not subjected to torsion and that all spans are equal.   Given:

$$f'_c = 5000 \text{ psi } (34.47 \text{ MPa}), \text{ normal-weight concrete}$$

$$f_y = 60,000 \text{ psi } (413.7 \text{ MPa})$$

6.5. A continuous beam has two equal spans $l_n = 18$ ft (5.49 m) and is subjected to an external service dead load w_D of 350 lb per ft. (5.1 kN/m) and a service live load w_L of 900 lb per ft. (13.2 kN/m).   In addition, an external service concentrated dead load P_D of 20,000 lb and an external service concentrated live load P_L of 28,500 lb (12.8 kN) are applied to one midspan only.   Design the diagonal tension reinforcement necessary.   Given:

$$f'_c = 5000 \text{ psi } (34.47 \text{ MPa}), \text{ normal-weight concrete}$$

$$f_y = 60,000 \text{ psi } (413.7 \text{ MPa})$$

6.6. Design the vertical stirrups for a beam having the shear diagram shown in Fig. 6.27 assuming that $V_c = 2\sqrt{f_c'}b_w d$.   Given:

$$b_w = 14 \text{ in. (356 mm)}$$
$$d_w = 20 \text{ in. (508 mm)}$$
$$V_{u1} = 75,000 \text{ lb (333.6 kN)}$$
$$V_{u2} = 60,000 \text{ lb (266.9 kN)}$$
$$V_{u3} = 45,000 \text{ lb (200.2 kN)}$$
$$f_c' = 4000 \text{ psi (27.6 MPa), normal-weight concrete}$$
$$f_y = 60,000 \text{ psi (414 MPa)}$$

6.8. Calculate the nominal shear strength V_c of the plain concrete in the web of the continuous normal-weight concrete beam shown in Fig. 6.28 using the more refined expression for evaluating the shear.   Given:

$$\rho_w = 0.025$$
$$\frac{l_n}{d} = 16$$
$$\frac{x}{l_n} = 0.45$$
$$M_0 = -\frac{w_u l_n^2}{8} = 120,000 \text{ ft-lb (162.7 kNm)}$$
$$M_x = 55,000 \text{ ft-lb (74.6 kNm)}$$
$$f_c' = 5000 \text{ psi (34.5 MPa), lightweight concrete}$$
$$f_y = 60,000 \text{ psi (414 MPa)}$$

Also compute the intensity of factored load w_u per foot to which this span is subjected.

6.8. A simply supported deep beam has a clear span $l_n = 10$ ft (3.1 m) and an effective center-to-center span $l = 11$ ft 6 in. (3.5 m).   The total depth of the beam is $h = 8$ ft 10 in. (2.7 m).   It is subjected to a uniform factored load on the top fibers of

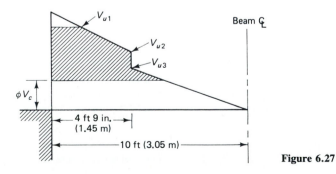

Figure 6.27

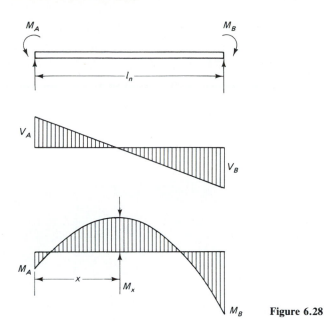

Figure 6.28

intensity w = 120,000 lb/ft (1601.8 kN/m), including its self-weight.   Design the beam for flexure and shear.   Given:

$$f'_c = 4500 \text{ psi (31.03 MPa), normal-weight concrete}$$
$$f_y = 60,000 \text{ psi (413.7 MPa)}$$

Assume the beam to be loaded only in its plane and that wind and earthquakes are not a consideration.

6.9. Solve Problem 6.8 if the same beam was continuous over three spans and was subjected to the same intensity of load.

6.10. Design a bracket to support a concentrated factored load V_u = 125,000 lb (556.0 kN) acting at a lever arm a = 4 in. (101.6 mm) from the column face; horizontal factored force N_{uc} = 40,000 lb (177.9 kN).   Given:

$$b = 14 \text{ in. (356 mm)}$$
$$f'_c = 5000 \text{ psi (34.5 MPa), normal-weight concrete}$$
$$f_y = 60,000 \text{ psi (414 MPa)}$$

Assume that the bracket was cast after the supporting column cured and that the column surface at the bracket location was not roughened before casting the bracket. Detail the reinforcing arrangements for the bracket.

6.11. Solve Problem 6.10 if the structural system was made from monolithic sand-light-weight concrete in which the corbel or bracket is cast simultaneously with the supporting column.

TORSION

7

7.1 INTRODUCTION

Torsion occurs in monolithic concrete construction primarily where the load acts at a distance from the longitudinal axis of the structural member. An end beam in a floor panel, a spandrel beam receiving load from one side, a canopy or a bus-stand roof projecting from a monolithic beam on columns, peripheral beams surrounding a floor opening, or a helical staircase are all examples of structural elements subjected to twisting moments. These moments occasionally cause excessive shearing stresses. As a result, severe cracking can develop well beyond the allowable serviceability limits unless special torsional reinforcement is provided. Photos 44 and 45 illustrate the extent of cracking at failure of a beam in torsion. They show the curvilinear plane of twist caused by the imposed torsional moments. In actual spandrel beams of a structural system, the extent of damage due to torsion is usually not as severe, as seen in Photos 46 and 47. This is due to the redistribution of stresses in the structure. However, loss of integrity due to torsional distress should always be avoided by proper design of the necessary torsional reinforcement.

Photo 43 Newark International Airport terminal, New Jersey. (Courtesy of Port of New York–New Jersey Authority.)

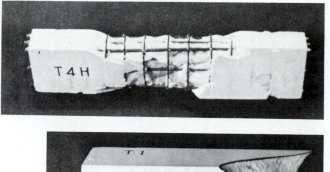

Photo 44   Reinforced plaster beam at failure in pure torsion. .(Rutgers tests: Law, Nawy, et al.)

(a)

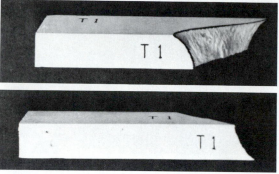

(b)

Photo 45   Plain mortar beam in pure torsion: (a) top view; (b) bottom view. (Rutgers tests: Law, Nawy, et al.)

An introduction to the subject of torsional stress distribution has to start with the basic elastic behavior of simple sections, such as circular or rectangular sections. Most concrete beams subjected to twist are components of rectangles. They are usually flanged sections such as T beams and L beams. Although circular sections are rarely a consideration in normal concrete construction, a brief discussion of torsion in circular sections serves as a good introduction to the torsional behavior of other types of sections.

Shear stress is equal to shear strain times the shear modulus at the elastic level in circular sections. As in the case of flexure, the stress is proportional to its distance from the neutral axis (i.e., the center of the circular section) and is maximum at the extreme fibers. If r is the radius of the element, $J = \pi r^4/2$, its polar moment of inertia, and v_{te} the elastic shearing stress due to an elastic twisting moment T_e,

$$v_{te} = \frac{T_e r}{J} \qquad\qquad (a)$$

When deformation takes place in the circular shaft, the axis of the circular cylinder is assumed to remain straight. All radii in a cross section also remain straight (i.e., without warping) and rotate through the same angle about the axis. As the circular element starts to behave plastically, the stress in the plastic outer ring becomes constant while the stress in the inner core remains elastic, as shown in Fig. 7.1. As the whole cross section becomes plastic, $b = 0$ and the shear stress

$$v_{tf} = \frac{3}{4}\frac{T_p r}{J} \qquad\qquad (b)$$

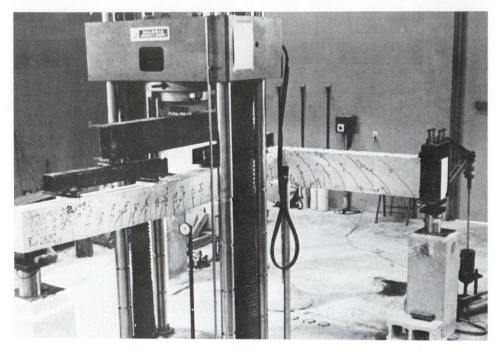

Photo 46   Reinforced concrete beams in torsion: testing setup.   (Courtesy of Thomas T. C. Hsu.)

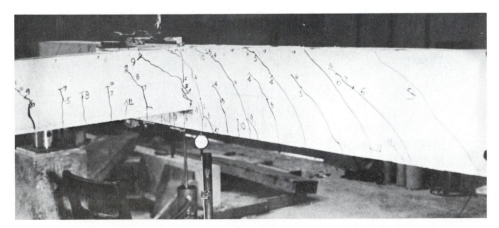

Photo 47   Close-up of torsional cracking of beams in Photo 44.   (Courtesy of Thomas T. C. Hsu.)

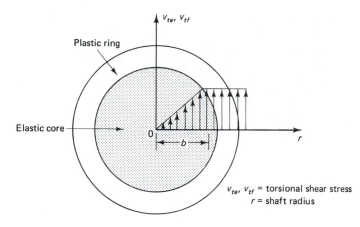

Figure 7.1   Torsional stress distribution through circular section.

where v_{tf} is the nonlinear shear stress due to an ultimate twisting moment T_p, where the subscript f denotes failure.

In rectangular sections, the torsional problem is considerably more complicated. The originally plane cross sections undergo warping due to the applied torsional moment. This moment produces axial as well as circumferential shear stresses with zero values at the corners of the section and the centroid of the rectangle and maximum values on the periphery at the middle of the sides, as seen in Fig. 7.2. The maximum torsional shearing stress would occur at midpoints A and B of the larger dimension of the cross section. These complications plus the fact that reinforced concrete sections are neither homogeneous nor isotropic make it difficult to develop exact mathematical formulations based on physical models such as Eqs. (a) and (b) for circular sections.

For over 60 years, the torsional analysis of concrete members has been based on either (1) the classical theory of elasticity developed through mathematical

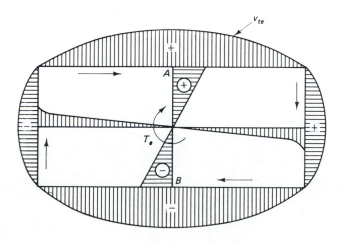

Figure 7.2   Pure torsion stress distribution in a rectangular section.

formulations coupled with membrane analogy verifications (St. Venant's) or (2) the theory of plasticity represented by the sand-heap analogy (Nadai's). Both theories were applied essentially to the state of pure torsion. But it was found experimentally that the elastic theory is not entirely satisfactory for the accurate prediction of the state of stress in concrete in pure torsion. The behavior of concrete was found to be better represented by the plastic approach. Consequently, almost all developments in torsion as applied to concrete and reinforced concrete have been in the latter direction.

7.2 PURE TORSION IN PLAIN CONCRETE ELEMENTS

7.2.1 Torsion in Elastic Materials

St. Venant presented in 1853 his solution to the elastic torsional problem with warping due to pure torsion that develops in noncircular sections. Prandtl in 1903 demonstrated the physical significance of the mathematical formulations by his membrane analogy model. The model establishes particular relationships between the deflected surface of the loaded membrane and the distribution of torsional stresses in a bar subjected to twisting moments. Figure 7.3 shows the membrane analogy behavior for rectangular as well as L-shaped forms.

For small deformations, it can be proved that the differential equation of the deflected membrane surface has the same form as the equation that determines the stress distribution over the cross section of the bar subjected to twisting moments. Similarly, it can be demonstrated that (1) the tangent to a contour line at any point of a deflected membrane gives the direction of the shearing stress at the corresponding cross section of the actual membrane subjected to twist; (2) the maximum slope of the membrane at any point is proportional to the magnitude of shear stress τ at the corresponding point in the actual member; (3) the twisting moment to which the actual member is subjected is proportional to *twice* the volume under the deflected membrane.

It can be seen from Figs. 7.2 and 7.3b that the torsional shearing stress is inversely proportional to the distance between the contour lines. The closer the lines are, the higher the stress, leading to the previously stated conclusion that the maximum torsional shearing stress occurs at the middle of the longer side of the rectangle. From the membrane analogy, this maximum stress has to be proportional to the steepest slope of the tangents at points A and B.

If δ = maximum displacement of the membrane from the tangent at point A, then from basic principles of mechanics and St. Venant's theory,

$$\delta = b^2 G\theta \tag{7.1a}$$

where G is the shear modulus and θ is the angle of twist. But $v_{t(\max)}$ is proportional to the slope of tangent; hence

$$v_{t(\max)} = k_1 b G\theta \tag{7.1b}$$

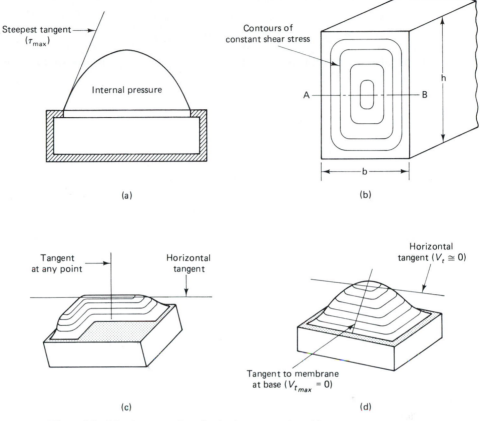

Figure 7.3 Membrane analogy in elastic pure torsion: (a) membrane under pressure; (b) contours in a real beam or in a membrane; (c) L section; (d) rectangular section.

where the k's are constants. The corresponding torsional moment T_e is proportional to *twice* the volume under the membrane, or

$$T_e \propto 2(\tfrac{2}{3}\delta bh) = k_2\,\delta bh$$

or

$$T_e = k_3 b^3 h G\theta \tag{7.1c}$$

From Eqs. 7.1b and 7.1c,

$$v_{t(\max)} = \frac{T_e b}{kb^3 h} \simeq \frac{T_e b}{J_1} \tag{7.1d}$$

The denominator kb^3h in Eq. 7.1d represents the polar moment of inertia J_1 of the section. Comparing Eq. 7.1d to Eq. (a) for the circular section shows the similarity of the two expressions except that the factor k in the equation for the rectangular section takes into account the shear strains due to warping. Equation

7.1d can be further simplified to give

$$v_{t(\max)} = \frac{T_e}{kb^2h} \tag{7.2}$$

It can also be written to give the stress at planes inside the section, such as an inner concentric rectangle of dimensions x and y, where x is the shorter side, so that

$$v_{t(\max)} = \frac{T_e}{kx^2y} \tag{7.3}$$

It is important to note in using the membrane analogy approach that the torsional shear stress changes from one point to another along the same axis as AB in Fig. 7.3, because of the changing slope of the analogous membrane, rendering the torsional shear stress calculations lengthy.

7.2.2 Torsion in Plastic Materials

As indicated earlier, the plastic sand-heap analogy provides a better representation of the behavior of brittle elements such as concrete beams subjected to pure torsion. The torsional moment is also proportional to *twice* the volume under the heap, and the maximum torsional shearing stress is proportional to the slope of the sand heap. Figure 7.4 is a two- and three-dimensional illustration of the sand heap. The torsional moment T_p in Fig. 7.4d is proportional to twice the volume of the rectangular heap shown in parts (b) and (c). It can also be recognized that the slope of the sand-heap sides as a measure of the torsional shearing stress is *constant* in the sand-heap analogy approach, whereas it is continuously variable in the membrane analogy approach. This characteristic of the sand heap considerably simplifies the solutions.

7.2.3 Sand-heap Analogy Applied to L Beams

Most concrete elements subjected to torsion are flanged sections, most commonly L beams comprising the external wall beams of a structural floor. The L beam in Fig. 7.5 is chosen in applying the plastic sand-heap approach to evaluate its torsional moment capacity and shear stresses to which it is subjected.

The sand heap is broken into three volumes:

V_1 = pyramid representing a square cross-sectional shape = $y_1b_w^2/3$

V_2 = tent portion of the web representing a rectangular cross-sectional shape = $y_1b_w(h - b_w)/2$

V_3 = tent representing the flange of the beam, transferring part *PDI* to *NQM* = $y_2h_f(b - b_w)/2$

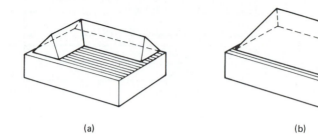

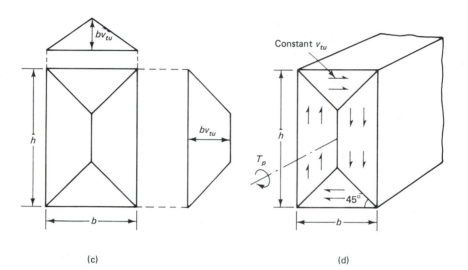

Figure 7.4 Sand-heap analogy in plastic pure torsion: (a) sand-heap L section; (b) sand-heap rectangular section; (c) plan of rectangular section; (d) torsional shear stress.

Torsional moment is proportional to twice the volume of the sand heaps; hence

$$T_p \simeq 2\left[\frac{y_1 b_w^2}{3} + \frac{y_1 b_w(h - b_w)}{2} + \frac{y_2 h_f(b - b_w)}{2}\right] \tag{7.4}$$

Also, torsional shear stress is proportional to the slope of the sand heaps; hence

$$y_1 = \frac{v_t b_w}{2} \tag{7.5a}$$

$$y_2 = \frac{v_t h_f}{2} \tag{7.5b}$$

Substituting y_1 and y_2 from Eqs. 7.5a and 7.5b into Eq. 7.4 gives us

$$v_{t(\text{max})} = \frac{T_p}{(b_w^2/6)(3h - b_w) + (h_f^2/2)(b - b_w)} \tag{7.6}$$

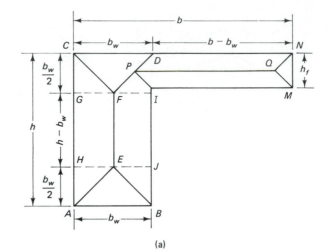

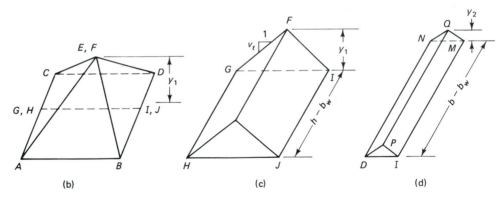

Figure 7.5 Sand-heap analogy of flanged section: (a) sand heap on L-shaped cross section; (b) composite pyramid from web (V_1); (c) tent segment from web (V_2); (d) transformed tent of beam flange (V_3).

If both the numerator and denominator of Eq. 7.6 are divided by $(b_w h)^2$ and the terms rearranged, we have

$$v_{t(\max)} = \frac{T_p h / (b_w h)^2}{[\frac{1}{6}(3 - b_w/h)] + \frac{1}{2}(h_f/b_w)2(b/h - b_w/h)]} \tag{7.7a}$$

If one assumes that C_t is the denominator in Eq. 7.7a and $J_E = C_t/(b_w h)^2$, Eq. 7.7a becomes

$$v_{t(\max)} = \frac{T_p h}{J_E} \tag{7.7b}$$

where J_E is the equivalent polar moment of inertia and a function of the shape of the beam cross section.   Note that Eq. 7.7b is similar in format to Eq. 7.1d from the membrane analogy except for the different values of the denominators J and J_E. Equation 7.7a can be readily applied to rectangular sections by setting $h_f = 0$.

It must also be recognized that concrete is not a perfectly plastic material; hence the actual torsional strength of the plain concrete section has a value lying between the membrane analogy and the sand-heap analogy values.

Equation 7.7b can be rewritten designating $T_p = T_c$ as the nominal torsional resistance of the plain concrete and $v_{t(max)} = v_{tc}$ using ACI terminology, so that

$$T_c = k_2 b^2 h v_{tc} \tag{7.8a}$$

$$T_c = k_2 x^2 y v_{tc} \tag{7.8b}$$

where x is the smaller dimension of the rectangular section.

Extensive work by Hsu confirmed by others, has established that k_2 can be taken as $\frac{1}{3}$. This value originated from research in the skew-bending theory of plain concrete. It was also established that $6\sqrt{f'_c}$ can be considered as a limiting value of the pure torsional strength of a member without torsional reinforcement. Using a reduction factor of 2.5 for the first cracking torsional load $v_{tc} = 2.4\sqrt{f'_c}$ and using $k_2 = \frac{1}{3}$ in Eq. 7.8 results in

$$T_c = 0.8\sqrt{f'_c}\, x^2 y \tag{7.9a}$$

where x is the shorter side of the rectangular section. The high reduction factor of 2.5 is used to offset any effect of shear and bending moments that might be present.

If the cross section is a T or L section, the area can be broken into component rectangles as in Fig. 7.6, such that

$$T_c = 0.8\sqrt{f'_c} \sum x^2 y \tag{7.9b}$$

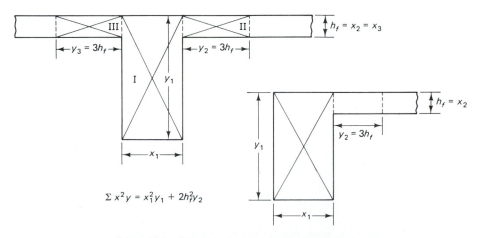

Figure 7.6   Component rectangles for T_c calculation.

7.2.4 Skew-bending Theory

This theory considers in detail the internal deformational behavior of the series of
transverse warped surfaces along the beam. Initially proposed by Lessig, it had
subsequent contributions from Collins, Hsu, Zia, Gesund, Mattock, and Elfgren
among the several researchers in this field. T. C. C. Hsu made a major contribution
experimentally to the development of the skew-bending theory as it presently
stands. In his book (Ref. 7.13), Hsu details the development of the theory of
torsion as applied to concrete structures and how the skew-bending theory formed
the basis of the 1989 ACI Code provisions on torsion. The complexity of the
torsional problem can thus permit in this textbook only the brief discussion that
follows.

The failure surface of the normal beam cross section subjected to bending
moment M_u remains plane after bending, as in Fig. 7.7a. If a twisting moment
T_u is also applied exceeding the capacity of the section, cracks develop on three
sides of the beam cross section and compressive stresses on portions of the fourth
side along the beam. As torsional loading proceeds to the limit state at failure,
a skewed failure surface results due to the combined torsional moment T_u and
bending moment M_u. The neutral axis of the skewed surface and the shaded area
in Fig. 7.7b denoting the compression zone would no longer be straight but subtend
a varying angle θ with the original plane cross sections.

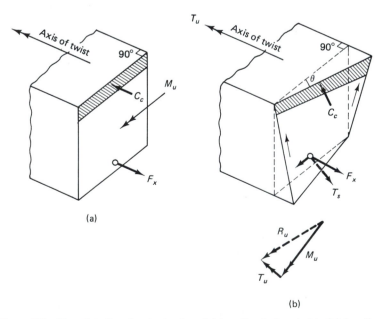

Figure 7.7 Skew bending due to torsion: (a) bending before twist; (b) bending
and torsion.

Prior to cracking, neither the longitudinal bars nor the closed stirrups make any appreciable contribution to the torsional stiffness of the section. At the post-cracking stage of loading, the stiffness of the section is reduced, but its torsional resistance is considerably increased, depending on the amount and distribution of *both* the longitudinal bars and the transverse *closed* ties. It has to be emphasized that little additional torsional strength can be achieved beyond the capacity of the plain concrete in the beam unless both longitudinal torsion bars and transverse ties are used.

The skew-bending theory idealizes the compression zone by considering it to be of uniform depth. It assumes the cracks on the remaining three faces of the cross section to be uniformly spread, with the steel ties (stirrups) at those faces carrying the tensile forces at the cracks and the longitudinal bars resisting shear through dowel action with the concrete. Figure 7.8a shows the forces acting on

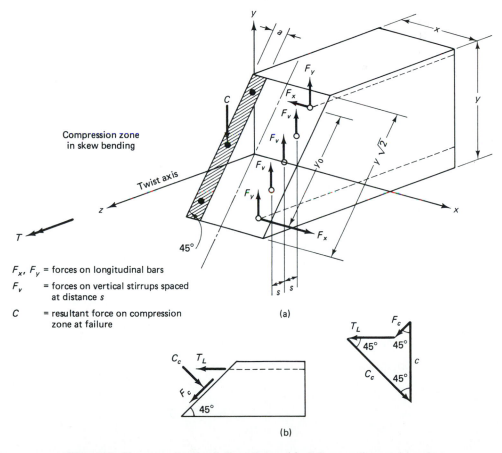

F_x, F_y = forces on longitudinal bars

F_v = forces on vertical stirrups spaced at distance s

C = resultant force on compression zone at failure

(a)

(b)

Figure 7.8   Forces on the skewly bent planes: (a) all forces acting on skew plane at failure; (b) vector forces on compression zone.

the skewly bent plane. The polygon in Fig. 7.8b gives the shear resistance F_c of the concrete, the force T_l of the active longitudinal steel bars in the compression zone, and the normal compressive block force C_c.

The torsional moment T_c of the resisting shearing force F_c generated by the shaded compressive block area in Fig. 7.8a is thus

$$T_c = \frac{F_c}{\cos 45°} \times \text{its arm about forces } F_v \text{ in Fig. 7.8a}$$

or

$$T_c = \sqrt{2} \, F_c(0.8x) \tag{7.10a}$$

where x is the shorter side of the beam. Extensive tests (Refs. 7.9 and 7.13) to evaluate F_c in terms of internal stress in concrete, $k_1\sqrt{f_c'}$, and the geometrical torsional constants of the section, k_2x^2y, led to the expression

$$T_c = \frac{2.4}{\sqrt{x}} x^2y \sqrt{f_c'} \tag{7.10b}$$

7.3 TORSION IN REINFORCED CONCRETE ELEMENTS

Torsion rarely occurs in concrete structures without being accompanied by bending and shear. The foregoing should give a sufficient background on the contribution of the plain concrete in the section toward resisting *part* of the combined stresses resulting from torsional, axial, shear, or flexural forces. The capacity of the plain concrete to resist torsion when in combination with other loads could, in many cases, be lower than when it resists the same factored external twisting moments alone. Consequently, torsional reinforcement has to be provided to resist the excess torque.

Inclusion of longitudinal and transverse reinforcement to resist part of the torsional moments introduces a new element in the set of forces and moments in the section. If

T_n = required total nominal torsional resistance of the section including the re-
    inforcement

T_c = nominal torsional resistance of the plain concrete

T_s = torsional resistance of the reinforcement

then

$$T_n = T_c + T_s \tag{7.11}$$

T_c is assumed equal to zero for design simplification, and all the torsion is assumed to be borne by the longitudinal steel bars and the closed transverse stirrups.

To study the contribution of the longitudinal steel bars and the closed stirrups, one has to analyze the system of forces acting on the warped cross sections of the structural element at the limit state of failure.

A modified space truss analogy is presented comparable to the plane truss analogy used for the design of shear stirrups. In this theory, both the longitudinal reinforcement and the transverse stirrups (ties) are utilized as components of the space truss (see Sec. 7.3.2).

7.3.1 Space Truss Analogy Theory

This theory was originally developed by Rausch and extended later by Lampert and Collins, with additional work by Hsu, Thurliman, Elfgren, and others. Further refinement was introduced by Collins and Mitchell (Ref. 7.12) as a compression field theory.

Hsu (Refs. 7.14, 7.15) proposed combining the equilibrium, compatibility, and the softened constitutive laws of concrete in a unified theory that can predict with reasonable accuracy the shear and torsional behavior of beams (the softened truss model). The shear flow concept is utilized in deriving the relevant expressions for shear equilibrium.

The space truss analogy is an extension of the model used in the design of the shear-resisting stirrups, in which the diagonal tension cracks, once they start to develop, are resisted by the stirrups. Because of the nonplanar shape of the cross sections due to the twisting moment, a space truss composed of the stirrups is used as the diagonal tension members, and the idealized concrete strips at a variable angle between the cracks are used as the compression members, as shown in Fig. 7.9.

It is assumed in this theory that the concrete beam behaves in torsion similar to a thin-walled box with a constant shear flow in the wall cross section, producing a constant torsional moment. The use of hollow-walled sections rather than solid sections proved to give essentially the same ultimate torsional moment, provided that the walls are not too thin. Such a conclusion is borne out by tests, which have shown that the torsional strength of the solid sections is composed of the resistance of the closed stirrup cage, consisting of the longitudinal bars and transverse stirrups, and the idealized concrete inclined compression struts in the plane of the cage wall. The compression struts are the inclined concrete strips between the cracks in Fig. 7.9.

The CEB–FIP code is based on the space truss model. In this code, the effective wall thickness of the hollow beam is taken as $\frac{1}{6}D_0$, where D_0 is the diameter of the circle inscribed in the rectangle connecting the corner longitudinal bars; that is, $D_0 = x_0$ in Fig. 7.9. A rational method to derive the effective wall thickness was given by Hsu (Ref. 7.15). This nonlinear analysis takes into account the warping compatibility condition of the wall. In summary, the absence of the core does not affect the strength of such members in torsion; hence the acceptability of the space truss analogy approach based on hollow sections.

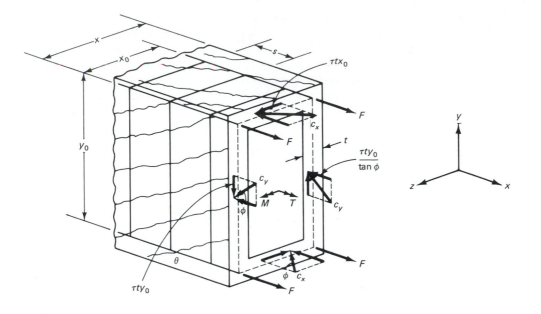

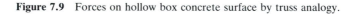

F = tensile force in each longitudinal bar

c_x = inclined compressive force on horizontal side

c_y = inclined compressive force on vertical side

τt = shear flow force per unit length of wall = q

Figure 7.9   Forces on hollow box concrete surface by truss analogy.

7.3.2 Equilibrium in Element Shear

A unit square membrane element of thickness t is subjected to shear flow q due to pure shear as in Fig. 7.10 (Ref. 7.15).   Reinforcement in both the longitudinal (E–W) direction ℓ and transverse (N–S) direction t is subjected to a unit stress f_ℓ/s_ℓ and f_v/s, respectively, such that the shear flow q can be defined by the equilibrium equations

$$q = (F_\ell) \tan \theta \qquad (7.12a)$$

where unit $F_\ell = A_\ell f_\ell/s_\ell$ and

$$q = (F_t) \cot \theta \qquad (7.12b)$$

where unit $F_t = A_t f_v/s$. A_ℓ and A_t are the cross-sectional areas of the reinforcement, and s_ℓ and s are the spacings in the ℓ and t directions, respectively.

From the geometry of the triangles in Fig. 7.10, the shear flow can also be defined as

$$q = (f_D t) \sin \theta \cos \theta \qquad (7.12c)$$

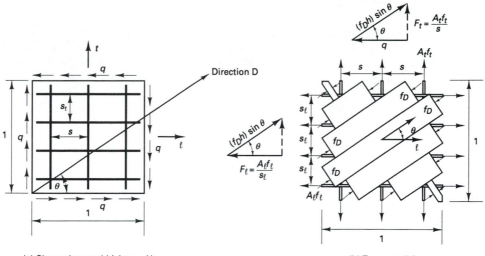

(a) Shear element (thickness *h*) (b) Truss model

Figure 7.10 Equilibrium forces in element shear (Ref. 7.15).

If the reinforcement in both directions is assumed to have yielded, Eqs. 7.12a and b give.

$$\tan \theta = \sqrt{\frac{F_{ty}}{F_{\ell y}}} \tag{7.13a}$$

and

$$q_y = \sqrt{F_{\ell y} F_{ty}} \tag{7.13b}$$

where the subscript *y* denotes the yielding of reinforcement.

7.3.3 Equilibrium in Element Torsion

The case of a hollow tube of any shape and variable thicknenss is considered (Fig. 7.11). It is subjected to pure torsion. St. Venant's theory stipulates that the cross-sectional shape remains unchanged in elastic small deformations, and the warping deformation perpendicular to the cross section would be the same along the member's axis. Hence it can be assumed that only shear stresses develop in the tube wall in the form of shear flow *q* in Fig. 7.11a and that the in-plane normal stresses in the wall vanish. If an infinitesimal wall element *ABCD* is isolated as in Fig. 7.11b, the shear flow in the ℓ direction has to be equal to the shear flow in the *t* direction or

$$\tau_\ell t_1 = \tau_t t_2 \tag{7.14}$$

On this basis, the shear flow *q* is considered constant throughout the cross section (Ref. 7.15). The torsional force over an infinitesimal distance *dt* along the shear

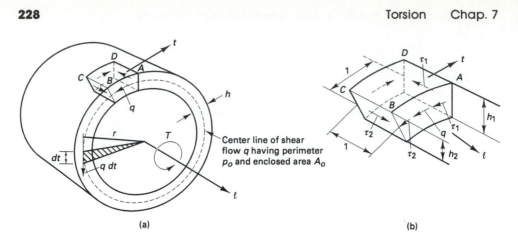

Center line of shear
flow q having perimeter
p_0 and enclosed area A_0

(a)   (b)

Figure 7.11   Hollow tube equilibrium torsion forces: (a) section of tube subjected
to torsion T; (b) unit shear element from tube wall of varying thickness h.
Note: l and t denote the longitudinal and transverse directions, respectively.

flow path is qdt so that the torsional resistance to the external torsional moment
T in Fig. 7.11a becomes

$$T = q \oint r \, dt \qquad (7.15)$$

It can be seen from Fig. 7.11a that $r \, dt$ in the integral is equal to *twice* the area of
the shaded triangle formed by r and dt. A summation of the total area around the
cross section gives

$$\oint r \, dt = 2A_0 \qquad (7.16)$$

where A_0 = cross-sectional area bounded by the shear flow center line. Substituting
$2A_0$ into Eq. 7.15 gives

$$q = \frac{T}{2A_0} \qquad (7.17)$$

By neglecting warping, the shear element subjected to pure torsion in the
tube wall of Fig. 7.11a becomes identical to the membrane shear element in Fig.
7.10a. Hence, substituting for the shear flow q from Eq. 7.17 into Eqs. 7.12a, b,
and c, three equations of equilibrium for torsion result,

$$T = \frac{\overline{F_\ell}}{p_0} (2A_0) \tan \theta \qquad (7.18a)$$

where $\overline{F_\ell} = F_\ell p_0$ and p_0 = perimeter of the shear flow path. $\overline{F_\ell}$ is the *total* longitudinal force due to torsion.

$$T = F_t(2A_0)\cot \theta \qquad (7.18b)$$

$$T = (f_D t)(2A_0) \sin \theta \cos \theta \qquad (7.18c)$$

Equation 7.18b can be written at yield as

$$T_n = \frac{2A_0 A_t f_{yv}}{s} \cot \theta \tag{7.19}$$

where T_n is the maximum torsional moment strength.

The required torsional reinforcements in the transverse and longitudinal directions become

$$A_t = \frac{T_n s}{2A_0 f_{yv} \cot \theta} \tag{7.20}$$

$$A_{\ell 1} = \frac{A_t}{s}\left(\frac{f_{yv}}{f_{y\ell}}\right)(s_\ell \cot^2 \theta) \tag{7.21a}$$

where $A_{\ell 1}$ is the area of one longitudinal bar.

If s_ℓ as the longitudinal reinforcement spacing represents the perimeter p_h of the center line of the outermost closed transverse torsional reinforcement, then

$$A_\ell = \frac{A_t}{s} p_h \frac{f_{yv}}{f_{y\ell}} \cot^2 \theta \tag{7.21b}$$

where A_ℓ = *total* area of all longitudinal torsional steel in the section.

The factored torsional moment strength, ϕT_n, must equal or exceed the external torsion, T_u, due to the factored loads. In the calculation of T_n (ACI 318–95, Ref. 7.16), all the torque is assumed to be resisted by the closed stirrups and longitudinal steel, with the torsional moment T_c resisted by the concrete compression struts assumed as zero. At the same time, the shear resisted by concrete, V_c, is assumed to be unchanged by the presence of torsion. This simplification eliminates the need for the rigor of the lengthy interaction expressions for V, T, and M used in the previous codes. In summary, the web reinforcement for shear is determined by the value of $V_s = V_n - V_c$, whereas the web reinforcement for torsion uses the T_n value alone.

7.4 SHEAR–TENSION–BENDING INTERACTION

Consider the rectangular box in Figs. 7.9 and 7.12. The shear flow q will not be the same on the four walls of the box when subjected to combined shear and torsion, as shown in Fig. 7.12. Failure can precipitate in two distinct modes:

(a) Yielding of the longitudinal bottom tension steel and the transverse stirrups
(b) Yielding of the longitudinal top compression steel and the transverse stirrups

(a) *Bottom tension steel yielding.*   If the failure mode is caused by yielding of the longitudinal bottom stringer (tensile steel) and the transverse stirrups

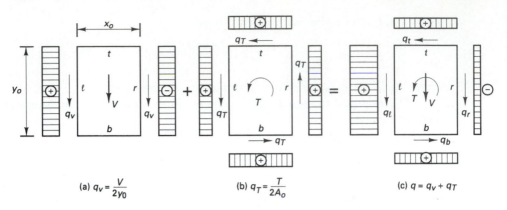

(a) $q_v = \dfrac{V}{2y_0}$ (b) $q_T = \dfrac{T}{2A_0}$ (c) $q = q_v + q_T$

Figure 7.12 Hollow section shear flow q due to combined shear and torsion.

due to combined shear and torsion, the following expression can be derived from equilibrium (Ref. 7.15):

$$\frac{M}{F_B y_0} + \left(\frac{V}{2y_0}\right)^2 \frac{y_0}{F_B} \frac{s}{A_t f_v} + \left(\frac{T}{2A_0}\right)^2 \frac{y_0 + x_0}{F_B} \frac{s}{A_t f_v} = 1 \qquad (7.22)$$

If M_0, V_0, and T_0 are the moments and forces acting *alone*, they can be defined as follows:

$$M_0 = F_B y_0 \qquad\qquad\qquad\qquad\qquad\qquad (7.23a)$$

$$V_0 = 2y_0 \sqrt{\left(\frac{F_T}{y_0}\right)\frac{A_t f_v}{s}} \qquad \text{for a two-web box} \qquad (7.23b)$$

$$T_0 = 2A_0 \sqrt{\left(\frac{2F_T}{p_0}\right)\frac{A_t f_v}{s}} \qquad\qquad\qquad (7.23c)$$

where $p_0 = 2(y_0 + x_0)$.

$$R = \frac{F_T}{F_B} \qquad\qquad\qquad\qquad\qquad\qquad (7.23d)$$

A nondimensional interaction surface relationship can be obtained by introducing Eq. 7.23 into Eq. 7.22 such that

$$\frac{M}{M_0} + \left(\frac{V}{V_0}\right)^2 R + \left(\frac{T}{T_0}\right)^2 R = 1 \qquad (7.24a)$$

(*b*) *Top compression steel yielding.*   If the failure mode is caused by yielding of the longitudinal top chord (compression steel) and the transverse stirrups,

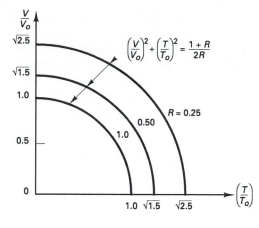

Figure 7.13 Shear–torsion interaction diagram.

Eq. 7.24a becomes

$$-\left(\frac{M}{M_0}\right)\frac{1}{R} + \left(\frac{V}{V_0}\right)^2 + \left(\frac{T}{T_0}\right)^2 = 1 \tag{7.24b}$$

From both Eqs. 7.24a and b, the interaction of V and T is *circular* for a constant bending moment M for both failure surfaces. The intersection of the two failure surfaces for these two failure modes forms a peak interaction curve between V and T such that Eqs. 7.24a and b give

$$\left(\frac{V}{V_0}\right)^2 + \left(\frac{T}{T_0}\right)^2 = \frac{1 + R}{2R} \tag{7.25a}$$

Equation 7.25a for $R = 0.25$, 0.5, and 1.0 on the peak planes gives the circular plots shown in Fig. 7.13.

A third mode of failure is caused by yielding in the top bar, in the bottom bar, and in the transverse reinforcement, all on the side where shear flows due to shear and torsion are additive, that is, the left wall (Ref. 7.15). A modified form of Eq. 7.25a results as follows:

$$\left(\frac{V}{V_0}\right)^2 + \left(\frac{T}{T_0}\right)^2 + \sqrt{2}\left(\frac{VT}{V_0T_0}\right) = \frac{1 + R}{2R} \tag{7.25b}$$

7.5 ACI DESIGN OF REINFORCED CONCRETE BEAMS SUBJECTED TO COMBINED TORSION, BENDING, AND SHEAR

7.5.1 Torsional Behavior of Structures

The torsional moment acting on a particular structural component such as a spandrel beam can be calculated using normal structural analysis procedures. Design of the particular component needs to be based on the limit state at failure. Therefore, the nonlinear behavior of a structural system after torsional cracking must be identified in one of the following two conditions: (1) no redistribution of torsional stresses to other members after cracking and (2) redistribution of torsional stresses and moments after cracking to effect deformation compatibility between intersecting members.

Stress resultants due to torsion in statically determinate beams can be evaluated from equilibrium conditions alone. Such conditions require a design for the full factored external torsional moment, because no redistribution of torsional stresses is possible. This state is often termed *equilibrium torsion*. An edge beam supporting a cantilever canopy as in Fig. 7.14 is such an example.

The edge beam has to be designed to resist the *total* external factored twisting moment due to the cantilever slab; otherwise, the structure will collapse. Failure would be caused by the beam not satisfying conditions of equilibrium of forces and moments resulting from the large external torque.

In statically indeterminate systems, stiffness assumptions, compatibility of strains at the joints, and redistribution of stresses may affect the stress resultants, leading to a reduction of the resulting torsional shearing stresses. A reduction is permitted in the value of the factored moment used in the design of the member if part of this moment can be redistributed to the intersecting members. The ACI Code permits a maximum factored torsional moment at the critical section d from the face of the supports for reinforced concrete members as follows:

$$T_u = \phi 4 \sqrt{f_c'} \frac{A_{cp}^2}{p_{cp}} \tag{7.26}$$

where A_{cp} = area enclosed by outside perimeter of concrete cross section

$\qquad\quad = x_0 y_0$

$\quad p_{cp}$ = outside perimeter of concrete cross section A_{cp} in.

$\qquad\quad = 2(x_0 + y_0)$

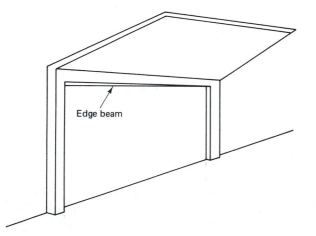

Figure 7.14 No redistribution torsion (equilibrium torsion).

For prestressed concrete members at $\frac{1}{2}h$ from the face of the support

$$T_u = \phi 4 \sqrt{f_c'} \frac{A_{cp}^2}{p_{cp}} \sqrt{1 + \frac{\bar{f_c}}{4\sqrt{f_c'}}} \tag{7.27}$$

where $\bar{f_c}$ = average compressive stress in the concrete at the centroidal axis due to effective prestress only after allowing for all losses ($\bar{f_c}$ in the ACI Code is denoted as f_{pc}).

Neglect of the full effect of the total value external torsional moment in this case does not, in effect, lead to failure of the structure but may result in excessive cracking if $\phi 4 \sqrt{f_c'} \left(\dfrac{A_{cp}^2}{p_{cp}}\right)$ is considerably smaller in value than the actual factored torque. An example of compatibility torsion can be seen in Fig. 7.15.

Beams B_2 apply twisting moments T_u at sections 1 and 2 of spandrel beam AB in Fig. 7.15b. The magnitudes of relative stiffnesses of beam AB and transverse beams B_2 determine the magnitudes of rotation at intersecting joints 1 and 2. Because of the development of torsional plastic hinges near joints A and B, the end moments for beams B_2 at their intersections with spandrel beam AB will not be fully transferred as twisting moments to the column supports at A and B. They would be greatly reduced, because moment redistribution results in transfer for most of the end bending moments from ends 1 and 2 to ends 3 and 4, as well as the midspan of beams B_2. T_u at each spandrel beam supports A and B and at the critical section at distance d from these supports is determined from Eq. 7.26 for reinforced concrete and Eq. 7.27 for prestressed concrete.

If the actual factored torque due to beams B_2 is less than that given by Eqs. 7.26 or 7.27, the beam could be designed for the lesser torsional value. Torsional moments are neglected, however, if for reinforced concrete

$$T_u < \phi \sqrt{f_c'} \frac{A_{cp}^2}{p_{cp}} \tag{7.28}$$

and for prestressed concrete

$$T_u < \phi \sqrt{f_c'} \frac{A_{cp}^2}{p_{cp}} \sqrt{1 + \frac{\bar{f_c}}{4\sqrt{f_c'}}} \tag{7.29}$$

7.5.2 Torsional Moment Strength

The size of a cross section is chosen on the basis of reducing unsightly cracking and preventing the crushing of the surface concrete caused by the inclined compressive stresses due to shear and torsion defined by the left-hand side of the expressions in Eqs. 7.30a and b. The geometrical dimensions for torsional moment

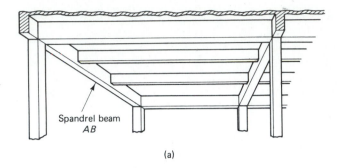

(a)

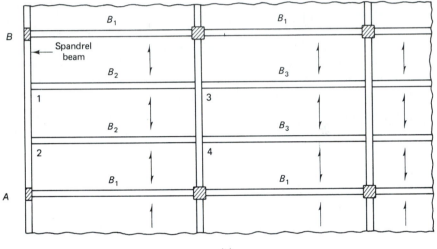

(b)

Figure 7.15 Torsion redistribution (compatibility): (a) isometric view of one end panel; (b) plan of a typical one-way floor system.

strength in both reinforced and prestressed members are limited by the following expressions

(a) Solid sections

$$\sqrt{\left(\frac{V_u}{b_w d}\right)^2 + \left(\frac{T_u p_h}{1.7 A_{oh}^2}\right)^2} \leq \phi \left(\frac{V_c}{b_w d} + 8 \sqrt{f_c'}\right) \qquad (7.30a)$$

Photo 48 River City Apartment Complex Atrium and Galleries, Chicago, Illinois (Courtesy Bertrand Goldberg Associates, Architects and Engineers, Chicago, Illinois.)

(b) Hollow sections

$$\sqrt{\left(\frac{V_u}{b_w d}\right)} + \left(\frac{T_u p_h}{1.7A_{oh}^2}\right) \leq \phi \left(\frac{V_c}{b_w d} + 8\sqrt{f_c'}\right) \tag{7.30b}$$

Reinforced concrete:

$$V_c = 2\lambda \sqrt{f_c'}\, b_w d \tag{7.30c}$$

Prestressed concrete for $f_{pe} > 0.4f_{pu}$:

$$V_c = \left(0.6\lambda \sqrt{f_c'} + 700\frac{V_u d}{M_u}\right)b_w d, \qquad \frac{V_u d}{M_u} \leq 1.0 \tag{7.30d}$$

$$\geq 1.7\lambda b_w d \leq 5.0\lambda \sqrt{f_c'}\, b_w d$$

where A_{oh} = area enclosed by the center line of the outermost closed transverse torsional reinforcement, in.^{2}

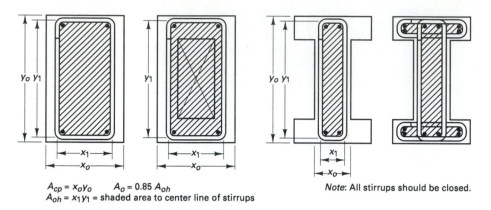

$$A_{cp} = x_o y_o \qquad A_o = 0.85 A_{oh}$$
$$A_{oh} = x_1 y_1 = \text{shaded area to center line of stirrups}$$

Note: All stirrups should be closed.

Figure 7.16   Torsional geometric parameters.

p_h = perimeter of center line of outermost closed transverse torsional reinforcement, in.

The areas A_{oh} for different sections are given in Fig. 7.16.

The sum of the stresses at the left-hand side of Eqs. 7.30a and b should not exceed the stresses causing shear cracking plus $8\sqrt{f'_c}$. This is similar to the limiting strength $V_s \leq 8\sqrt{f'_c}$ for shear without torsion.   As stipulated in the ACI 318-95 commentary, the upper limit of stress in terms of the nominal shear strength V_c of the plain concrete in the web permits applying the two expressions in both reinforced and prestressed concrete elements.

7.5.2.1 Hollow sections wall thickness.   The shear stresses due to shear and to torsion both develop in the walls of the hollow section, as seen in Fig. 7.17a.

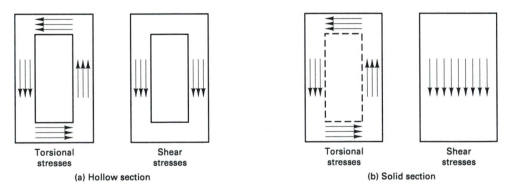

Torsional stresses	Shear stresses		Torsional stresses	Shear stresses
(a) Hollow section			(b) Solid section	

Figure 7.17   Superposition of torsional and shear stresses. Case (a): directly additive occurring in the left wall of the box (Eq. 7.30b). Case (b): torsion acts on "tubular" outer wall section while shear stress acts on the full width of solid section; stresses combined using square root of sum of squares (Eq. 7.30a).

Note that in a solid section the shear stresses due to torsion still concentrate in the outer zones of the section as in Fig. 7.17b and as discussed in Section 7.3.1.

    If the wall thickness in the hollow section varies around its perimeter, the section geometry has to be evaluated at such a location where the left-hand side of Eq. 7.30b has a maximum value. Also, if the wall thickness $t < A_{oh}/p_h$, the left-hand side of Eq. 7.30b should be taken as

$$\frac{V_u}{b_w d} + \frac{T_u}{1.7 A_{oh} t}$$

The wall thickness t is the thickness where stresses are being checked.

7.5.3 Torsional Web Reinforcement

As indicated in Section 7.3.1, meaningful additional torsional strength due to the addition of torsional reinforcement can be achieved only by using both stirrups and longitudinal bars. Ideally, equal volumes of steel in both the closed stirrups and the longitudinal bars should be used so that both participate equally in resisting the twisting moments. This principle is the basis of the ACI expressions for proportioning the torsional web steel. If s is the spacing of the stirrups, A_ℓ is the total cross-sectional area of the longitudinal bars, and A_t is the cross section of one stirrup leg, the transverse reinforcement for torsion has to be based on the full external torsional moment strength value T_n (T_u/ϕ), where

$$T_n = \frac{2 A_0 A_t f_{yv}}{s} \cot \theta \tag{7.31a}$$

(see the derivation of Eq. 7.19).

where A_0 = gross area enclosed by the shear flow path, in.^{2}

        A_t = cross-sectional area of one leg of the transverse closed stirrups, in.^{2}

       f_{yv} = yield strength of closed transverse torsional reinforcement not to exceed 60,000 psi

       θ = angle of the compression diagonals (struts) in the space truss analogy for torsion (see Fig. 7.9)

Transposing terms in Eq. 7.31a, the transverse reinforcement area becomes

$$\frac{A_t}{s} = \frac{T_n}{2 A_0 f_{yv}} \cot \phi \tag{7.31b}$$

    The area A_0 has to be determined by analysis (Refs. 7.14 and 7.15) except that the ACI 318 Code permits taking $A_0 = 0.85 A_{oh}$ in lieu of the analysis.

    As discussed in Sec. 7.3, the factored torsional resistance ϕT_n must equal or exceed the factored external torsional moment T_u. All the torsional moment is assumed in the ACI 318-95 code to be resisted by the closed stirrups and the longitudinal steel with the torsional resistance, T_c, of the concrete disregarded;

that is, $T_c = 0$. The shear V_c resisted by the concrete is assumed to be unchanged by the presence of torsion.

The angle θ subtended by the concrete compression diagonals (struts) should not be taken smaller than 30° nor larger than 60°. It can be obtained by analysis as detailed in Refs. 7.13 and 7.15. According to Eq. 7.2lb, the additional longitudinal reinforcement for torsion should not be less than

$$A_l = \frac{A_t}{s} p_h \frac{f_{yv}}{f_{yl}} \cot^2 \theta \tag{7.32}$$

where f_{yl} = yield strength of the longitudinal torsional reinforcement, not to exceed 60,000 psi, and A_l = total area of longitudinal torsional steel in the cross section.

The same angle θ should be used in both equations 7.31 and 7.32. It should be noted that as θ gets smaller the amount of stirrups required by Eq. 7.31 decreases. At the same time the amount of longitudinal steel required by Eq. 7.32 increases.

In lieu of determining the angle θ by analysis (Ref. 7.15), the ACI Code allows a value of θ equal to:

 (i) 45° for nonprestressed members or members with less prestress than in (ii)
 (ii) 37.5° for prestressed members with an effective prestressing force larger than 40% of the tensile strength of the Longitudinal reinforcement.

7.5.3.1 Minimum torsional reinforcement.

It is necessary to provide a minimum area of torsional reinforcement in all regions where the factored torsional moment T_u exceeds the value given by Eqs. 7.28 and 7.29. In such a case, the minimum area of the transverse closed stirrups required should be

$$A_v + 2A_t \geq \frac{50 b_w s}{f_{yv}} \tag{7.33}$$

The maximum spacing should not exceed the smaller of $p_h/8$ or 12 in.

The minimum total area of the additional longitudinal torsional reinforcement should be determined by

$$A_{l,min} = \frac{5\sqrt{f'_c}}{f_{yl}} A_{cp} - \frac{A_t t}{s} p_h \frac{f_{yu}}{f_{yl}} \tag{7.34}$$

where A_t/s should not be taken less than $25 b_w/f_{yu}$.

The additional longitudinal reinforcement required for torsion should be distributed around the perimeter of the closed stirrups with a maximum spacing of 12 in. The longitudinal bars or tendons should be placed inside the closed stirrups, with at least one longitudinal bar or tendon in each corner of the stirrup. The bar diameter should be at least *one-sixteenth* of the stirrup spacing, but not less than a No. 3 bar. Also, the torsional reinforcement should extend for a minimum distance of $b_t + d$ beyond the point theoretically required for torsion, because torsional diagonal cracks develop in a helical form extending beyond the cracks caused by shear and flexure. b_t is the width of that part of the cross section containing the stirrups resisting torsion. The critical section in beams is at a distance d from the face of the support for reinforced concrete elements and at $h/2$ for prestressed concrete elements, d being the effective depth and h the total depth of the section.

7.5.4 Design Procedure for Combined Torsion and Shear

The following is a summary of the recommended sequence of design steps. A flow chart describing the sequence of operations in graphical form is shown in Fig. 7.18.

1. Classify whether the applied torsion is equilibrium compatibility torsion. Determine the critical section and compute the factored torsional moment T_u. The critical section is taken as d from the face of the support in reinforced concrete beams and $h/2$ in prestressed concrete beams. If T_u is less than $\phi \sqrt{f'_c} \dfrac{A_{cp}^2}{P_{cp}}$ for nonprestressed members or less than $\phi \sqrt{f'_c} \dfrac{A_{cp}^2}{P_{cp}} \sqrt{1 + \dfrac{\bar{f}_c}{4\sqrt{f'_c}}}$ for prestressed members, torsional effects are neglected ($\bar{f}_c$ in the ACI Code is denoted as f_{pc}).

2. Check whether the factored torsional moment T_u causes equilibrium or compatibility torsion. For compatibility torsion, limit the design torsional moment to the lesser of the actual moment T_u or $T_u = \phi 4 \sqrt{f'_c} \dfrac{A_{cp}^2}{P_{cp}}$ for reinforced concrete members and $T_u = \phi 4 \sqrt{f'_c} \dfrac{A_{cp}^2}{P_{cp}} \sqrt{\dfrac{1 + \bar{f}_c}{4 \sqrt{f'_c}}}$ for prestressed concrete members. The value of the design nominal strength T_n has to be at least equivalent to the factored T_u/ϕ, proportioning the section such that:

(a) for solid sections:

$$\sqrt{\left(\dfrac{V_u}{b_w d}\right)^2 + \left(\dfrac{T_u p_h}{1.7 A_{oh}^2}\right)^2} \le \phi \left(\dfrac{V_c}{b_w d} + 8 \sqrt{f'_c}\right)$$

(b) For hollow sections:

$$\sqrt{\dfrac{V_u}{b_w d}} + \left(\dfrac{T_u p_h}{1.7 A_{oh}^2}\right) \le \phi \left(\dfrac{V_c}{b_w d} + 8 \sqrt{f'_c}\right)$$

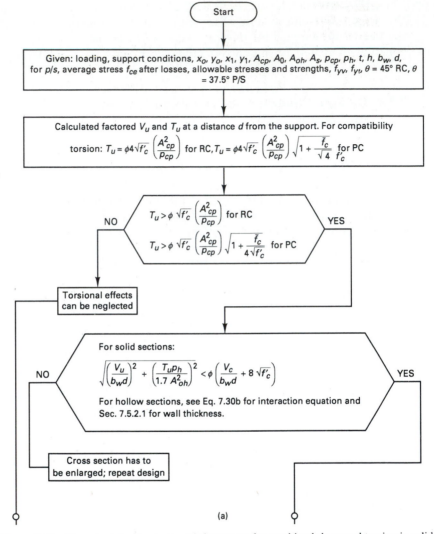

Figure 7.18 Flow chart for the design reinforcement for combined shear and torsion in solid sections: (a) torsional web steel; (b) shear web steel.

If the wall thickness is less than A_{oh}/p_h, the second term should be taken as $T_u/1.7A_{oh}t$.

3. Select the required *torsional* closed stirrups to be used as transverse reinforcement, using a maximum yield strength of 60,000 psi, such that

$$\frac{A_t}{s} = \frac{T_n}{2A_0 f_{yv}} \cot \theta$$

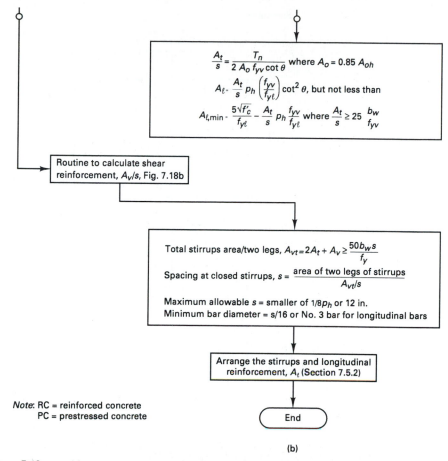

Routine to calculate shear reinforcement, A_v/s, Fig. 7.18b

Figure 7.18   cont'd

Unless using A_0 and θ values obtained from analysis (Ref. 7.14), use $A_0 = 0.85A_{oh}$ and $\theta = 45°$ for nonprestressed members with an effective prestress not less than the tensile strength of the longitudinal reinforcement. The additional longitudinal reinforcement should be

$$A_l = \frac{A_t}{s} \, p_h \, \frac{f_{yv}}{f_{yl}} \, \cot^2 \theta$$

but not less than

$$A_{l,\min} = \frac{5\sqrt{f'_c}\,A_{cp}}{f_{yl}} - \frac{A_t}{s}\,p_h\,\frac{f_{yv}}{f_{yl}}$$

where A_t/s shall not be less than $25b_w/f_{yv}$. Maximum allowable spacing of transverse stirrups is the smaller of $\frac{1}{8}p_h$ or 12 in., and bars should have a diameter of at least one-sixteenth of the stirrup spacing, but not less than a No. 3 bar size.

Subroutine

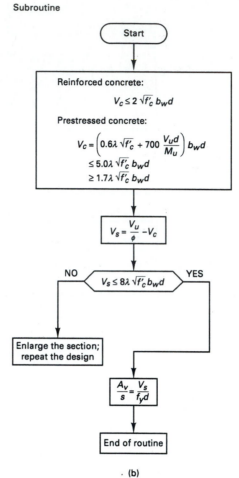

· (b)

Figure 7.18 cont'd

4. Calculate the required *shear* reinforcement A_v per unit spacing in a transverse section. V_u is the factored external shear force at the critical section, V_c is the nominal shear resistance of the concrete in the web, and V_s is the shearing force to be resisted by the stirrups:

$$\frac{A_v}{s} = \frac{V_s}{f_y d}$$

where $V_s = V_n - V_c$ and

$$V_c = 2\lambda \sqrt{f_c'} b_w d$$

for reinforced concrete.

$$V_c = \left(0.6\lambda \sqrt{f_c'} + \frac{700 V_u d}{M_u}\right) b_w d$$

for prestressed concrete if $f_{pe} > 0.4 f_{pu}$. Limits of V_c for prestressed beams are

$$V_c \geq 1.7\lambda \sqrt{f_c'}\, b_w d \leq 5.0\lambda \sqrt{f_c'}\, b_w d; \qquad \frac{V_u d}{M_u} \leq 1.0$$

where $\lambda = 1.0$ for normal-weight concrete
    $= 0.85$ for sand-lightweight concrete
    $= 0.75$ for all-lightweight concrete

The value of V_n has to be at least equal to the factored V_u/ϕ.

   5.   Obtain the total A_{vt}, the area of the closed stirrups for torsion and shear, and design the stirrups such that

$$A_{vt} = 2A_t + A_v \geq \frac{50 b_w s}{f_{yv}}$$

Extend the stirrups a distance $b_t + d$ beyond the point theoretically no longer required, where b_t = width of the cross section containing the closed stirrup resisting torsion.

7.5.5 Example 7.1: Design of Web Reinforcement for Combined Torsion and Shear in a T-beam Section

A T-beam cross section has the geometrical dimensions shown in Fig. 7.19. A factored external shear force acts at the critical section, having a value $V_u = 45,000$ lb (203 kN). It is subjected to the following torques: (a) equilibrium factored external torsional moment $T_u = 500,000$ in.-lb (57.2 kN-m); (b) compatibility factored $T_u = 75,000$ in.-lb (8.47 kN-m); (c) compatibility factored $T_u = 300,000$ in.-lb. Given:

   bending reinforcement $A_s = 3.4$ in.^{2} (2190 mm^2)
   $f_c' = 4000$ (27.6 MPa), normal-weight concrete
   $f_y = 60,000$ (414 MPa)

Design the web reinforcement needed for this section.

Solution   (a) Equilibrium torsion:

Factored torsional moment (Step 1)

Assume that the flanges are not confined with ties.

       given equilibrium torsional moment $= 500,000$ in.-lb. (57.2 kN-m)

The total torsional moment must be provided for in the design.

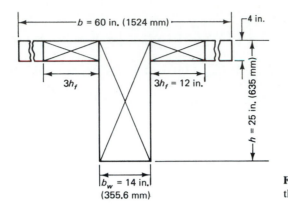

Figure 7.19 Component rectangles of the T beam.

$$\text{required } T_n = \frac{T_u}{\phi} = \frac{500{,}000}{0.85} = 588{,}235 \text{ in.-lb (66.5 kN-m)}$$

$$A_{cp} = 14 \times 25 = 350 \text{ in.}^2$$

$$p_{cp} = 2(x_0 + y_0) = 2(14 + 25) = 78 \text{ in.}$$

If the flanges were confined with closed ties,

$$A_{cp} = 14 \times 25 + 2(4 \times 12) = 446 \text{ in.}^2$$

$$p_{cp} = 2[(14 + 25) + 2(4 + 3 \times 4)] = 142 \text{ in.}^2$$

From Eq. 7.28, torsional moment for which torsion can be neglected is

$$T_u = \phi \sqrt{f_c'} \frac{A_{cp}^2}{p_{cp}} = 0.85 \sqrt{4000} \frac{350^2}{78}$$

$$= 84{,}429 \text{ in.-lb.} < 588{,}235$$

Hence design for full torsion.

Sectional properties (Step 2)

$A_0 = 0.85 A_{oh}$, where A_{oh} is the area enclosed by the center line of the outermost closed stirrups. Assuming 1.5-in. clear cover and No. 4 stirrups, from Fig. 7.19,

$$x_1 = 14 - 2(1.5 + 0.25) = 10.5 \text{ in.}$$

$$y_1 = 25 - 2(1.5 + 0.25) = 21.5 \text{ in.}$$

$$A_{oh} = 10.5 \times 21.5 = 226 \text{ in.}^2$$

$$A_0 = 0.85 (x_1 \times y_1) = 0.85(10.5 \times 21.5) = 192 \text{ in.}^2$$

$$d = 25 - (1.5 + 0.5 + 0.25) = 22.75, \text{ say } 22.5 \text{ in.}$$

$$p_h = 2(x_1 + y_1) = 2(10.5 + 21.5) = 64 \text{ in.}$$

Use $\theta = 45°$, $\cot \theta = 1.0$.

Check adequacy of section (Step 3)

For the section to be adequate, it should satisfy Eq. 7.30a:

$$\sqrt{\left(\frac{V_u}{b_w d}\right)^2 + \left(\frac{T_u p_h}{1.7 A_{oh}^2}\right)^2} \leq \phi \left(\frac{V_c}{b_w d} + 8\sqrt{f_c'}\right)$$

$$V_c = 2\sqrt{f_c'}b_w d = 2\sqrt{4000} \times 14 \times 22.5 = 39,845 \text{ lb.}$$

$$\sqrt{\left(\frac{V_u}{b_w d}\right)^2 + \left(\frac{T_u p_h}{1.7 A_{oh}^2}\right)^2} = \sqrt{\left(\frac{45,000}{14 \times 22.5}\right)^2 + \left(\frac{500,000 \times 64}{1.7(226)^2}\right)^2}$$

$$= \sqrt{20,408 + 135,821} = 396 \text{ psi (2.73 MPa)}$$

$$\phi\left(\frac{V_c}{b_w d} + 8\sqrt{f_c'}\right) = 0.85\left(\frac{39,845}{14 \times 225} + 8\sqrt{4000}\right)$$

$$= 0.85(126.5 + 506.0) = 538 \text{ psi (3.71 MPa)} > 396 \text{ psi}$$

Hence the section is adequate.

Torsional reinforcement (Step 4)

From Eq. 7.31,

$$\frac{A_t}{s} = \frac{T_n}{2 A_o f_{yv} \cot \theta} = \frac{588,235}{2 \times 192 \times 60,000 \times 1.0}$$

$$= 0.026 \text{ in.}^2/\text{in./one leg}$$

Shear reinforcement

$$V_c = 2\sqrt{f_c'}b_w d = 39,845$$

$$V_n = \frac{45,000}{0.85} = 52,940 \text{ lb} > V_c; \text{ also} > \frac{1}{2}V_c$$

for minimum shear web reinforcement. Hence, provide shear stirrups.

$$V_s = V_n - V_c = 52,940 - 39,845 = 13,095 \text{ lb}$$

$$\frac{A_v}{s} = \frac{V_s}{f_y d} = \frac{13,095}{60,000 \times 22.5} = 0.01 \text{ in.}^2/\text{in./two legs}$$

$$\frac{A_{vt}}{s} = \frac{2A_t}{s} + \frac{A_v}{s} = 2 \times 0.026 + 0.01 = 0.062 \text{ in.}^2/\text{in./two legs}$$

Try No. 3 (9.5-mm diameter) closed stirrups. Area of two legs = 0.22 in.².

$$s = \frac{\text{area of stirrup cross section}}{\text{required } A_{vt}/s} = \frac{0.22}{0.062} = 3.5 \text{ in.}$$

Maximum allowable spacing s_{max} = smaller of $\frac{1}{8}p_h$ or 12 in., where $p_h = 2(x_1 + y_1)$ = 64 in. From before $\frac{1}{8}p_h = \frac{64}{8}$ = 8 in. > 3.5 in.

$$\text{minimum } A_{vt} = \frac{50b_w s}{f_{yv}} = \frac{50 \times 14 \times 3.5}{60,000} = 0.04 \text{ in.}^2$$

less than 0.22 in.²; does not control. Hence use No. 3 closed stirrups at 3.5 in. center to center.

$$A_\ell = \frac{A_t}{s} p_h \frac{f_{yv}}{f_{y\ell}} \cot^2 \theta$$

$$= 0.026 \times 64 \frac{60,000}{60,000} \times 1.0 = 1.66 \text{ in.}^2$$

$$\text{minimum } A_\ell = \frac{5 \sqrt{f_c'} A_{cp}}{f_{y\ell}} - \frac{A_t}{s} p_h \frac{f_{yv}}{f_{y\ell}}$$

$$= \frac{5 \sqrt{4000} \times 350}{60,000} - 0.026 \times 64 \times \frac{60,000}{60,000}$$

$$= 1.84 - 1.66 = 0.18 \text{ in.}^2 < 1.66 \text{ in.}^2$$

Hence $A_\ell = 1.66$ in.² controls.

Distribution of torsion longitudinal steel

Torsional $A_\ell = 1.66$ in.² Assume that $\frac{1}{4}A_\ell$ goes to the top corners and $\frac{1}{4}A_\ell$ goes to the bottom corners of the stirrups, to be added to the flexural bars. The balance, $\frac{1}{2}A_\ell$, would thus be distributed equally to the vertical faces of the beam web cross section at a spacing not to exceed 12 in. center to center.

$$\text{midspan } \Sigma A_s = \frac{A_\ell}{4} + A_s = \frac{1.66}{4} + 3.4 = 3.81 \text{ in.}^2$$

Provide five No. 8 (25.4-mm-diameter) bars at the bottom. Provide two No. 4 (12.7-mm-diameter) bars with an area of 0.40 in.² at the top. Provide two No. 4 (12.7-mm-diameter) bars on each vertical face. Figure 7.20 shows the geometry of the cross section.

(b) Compatibility Torsion

Factored torsional moment (Step 1)

Given $T_u = 75,000$ in.-lb. (8.47 kN-m) $< T_u = 84,429$ in.-lb. from part (a). Hence disregard torsion and provide stirrups for shear only.

From part (a),

$$\frac{A_v}{s} = 0.01 \quad \text{and} \quad \text{min.} \frac{A_v}{s} = \frac{0.04}{3.5} = 0.011 \text{ in./in./two legs}$$

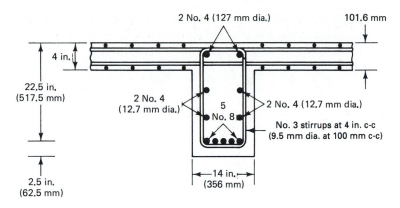

Figure 7.20   Web reinforcement details, Ex. 7.1(a).

For No. 3 stirrups, $s = 0.22/0.011 = 20$ in. center to center.

$$\text{maximum } s = \frac{d}{2} = \frac{22.5}{2} = 11.25 \text{ in.}$$

Use No. 3 closed stirrups at 10 in. c-c at the critical section.

(c) Compatibility Torsion

Factored torsional moment (Step 1)

Since $T_u = 300,000$ in.-lb. (33.9 kN-m) is greater than 177,861 in.-lb from case (a); hence stirrups have to be provided. Because this is a compatibility torsion, the section can be designed by Eq. 7.27 for

$$T_u = \phi 4 \sqrt{f_c'} \frac{A_{cp}^2}{P_{cp}} \geq 0.85 \times 4 \sqrt{4000} \frac{350^2}{78}$$

$$= 4 \times 84,429 \text{ from case (a)} = 337,716 \text{ in.-lb}$$

This is $>300,000$; hence use $T_u = 300,000$ in.-lb for the torsional design of the section.

$$\text{required } T_n = \frac{T_u}{\phi} = \frac{300,000}{0.85} = 352,941 \text{ in.-lb}$$

Torsional reinforcement (Step 2)

From case (a) $A_0 = 192$ in.^{2}, $p_h = 64$ in.

$$\frac{A_t}{s} = \frac{T_n}{2A_0 f_{yv} \cot \theta} = \frac{352,941}{2 \times 192 \times 60,000 \times 1.0}$$

$$= 0.015 \text{ in.}^2/\text{in./one leg}$$

From case (a),

$$\frac{A_v}{s} = 0.011 \text{ in.}^2/\text{in./two legs}$$

$$\frac{A_{vt}}{s} = 2\frac{A_t}{s} + \frac{A_v}{s} = 2 \times 0.015 + 0.011 = 0.041 \text{ in.}^2/\text{in./two legs}$$

Using No. 3 stirrups, $s = 0.22/0.041 = 5.37$ in.   This is less than $\frac{1}{8}p_h = 8$ in. or 12 in.   Hence, use No. 3 closed stirrups at 5 in. c-c (9.5-mm diameter at 127 mm c-c) at the critical section.

$$A_\ell = \frac{A_t}{s}\, p_h\, \frac{f_{yv}}{f_{y\ell}}\, \cot^2\theta = 0.015 \times 64 \times \frac{60{,}000}{60{,}000} \times 1.0 = 0.96 \text{ in.}^2$$

$$\text{min } A_\ell = \frac{5\,\sqrt{f_c'}\, A_{cp}}{f_{y\ell}} - \frac{A_t}{s}\, p_h\, \frac{f_{yv}}{f_{y\ell}}$$

$$= \frac{5\,\sqrt{4000} \times 350}{60{,}000} - 0.015 \times 64 \times \frac{60{,}000}{60{,}000}$$

$$= 1.84 - 0.96 = 0.88 \text{ in.}^2 < 0.96 \text{ in.}^2$$

$A_\ell = 0.96$ in.² controls.

Distribution of torsion longitudinal bars

Torsional $A_\ell = 0.96$ in.², so $A_\ell/4 = 0.24$ in.².   Using the same logic as that followed in case (a), provide five No. 8 (25.4-mm-diameter) bars at the bottom face. The area required, $A_s + A_\ell/4 = 3.64$ in.²; the area provided $= 3.95$ in.².   The required area at top corners and at each vertical face $= 0.24$ in.².   Provide two No. 4 bars (12.7-mm diameter) at the top two corners and at each of the vertical sides, giving 0.40 in.² in each area.   Figures 7.20 and 7.21 show the geometry of the section reinforcement.

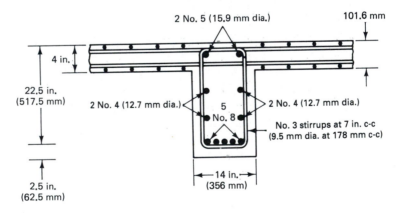

Figure 7.21   Web reinforcement details, Ex. 7.1(c).

7.5.6 Example 7.2: Equilibrium Torsion Web Steel Design

A normal-weight 7-ft cantilever concrete canopy slab on continuous beams spans 24 ft (7.32 m) on several supports, as shown in Fig. 7.22.   It carries a uniform service live load of 30 psf (1.44 kPa) on the cantilever.   Design the interior span spandrel

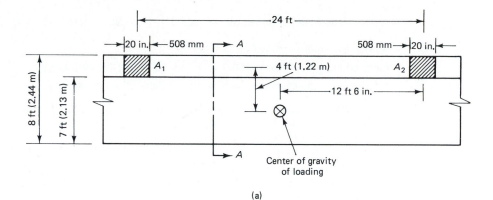

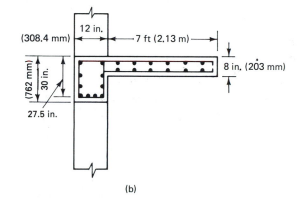

Figure 7.22 Plan and sectional elevation, Ex. 7.2: (a) plan; (b) section $A–A$.

beam $A1–A2$ for diagonal tension and torsion.   Assume no wind or earthquake and neglect creep and shrinkage effects.   Given:

$f_c' = 4000$ psi (27.6 MPa)

$f_y = 60,000$ psi (413.7 MPa)

exterior columns $= 12$ in. $\times 20$ in. (304.8 $\times$ 508 mm)

midspan $A_s = 1.50$ in.^{2} (967.74 mm^2)

support $A_s = 2.4$ in.^{2} (1548 mm^2)

support $A_s' = 0.8$ in.^{2} (516.13 mm^2)

Solution

Factored torsional moment (Step 1)

Beam $A1–A2$ is a case of nonredistribution torsion because the torsional resistance of the beam is required to maintain equilibrium.   Hence the section has to be designed to resist the total external factored torsional moment.

service dead load of the cantilever slab $= \dfrac{8.0}{12} \times 150 = 100.0$ psf (5.08 kPa)

service live load $= 30$ psf (1.44 kPa)

factored load $U = 1.4 \times 100.0 + 1.7 \times 30 = 191$ psf (9.1 kPa)

total load on the cantilever slab $= 191 \times 24 \times 7 = 32,088$ lb (144.4 kN)

This load acts at center of gravity of loading shown in Fig. 7.22a, having a moment arm $= 4.0$ ft (1.22 m). Hence the maximum factored moment at the center line of the support $= \frac{1}{2}(32,088 \times 4) = 64,176$ ft-lb.

Note that the reaction at the supports is half of the total torque acting on the slab, as shown in Fig. 7.23, because the center of gravity of the twisting moment is midway between the supports. Since the load is uniformly distributed, the torsional moment variation will be linear along the span. Figure 7.24 shows the torsional envelope for this beam. The factored torsional moment at the critical section d (27.5 in.) from the face of the support is

$$T_u = 64,176.0 \left(\frac{12 - \dfrac{10 + 27.5}{12}}{12} \right) = 47,464 \text{ ft-lb}$$

$$= 569,570 \text{ in.-lb } (70.4 \text{ kN-m})$$

$$\text{required } T_n = \frac{47,464}{0.85} = 55,840 \text{ ft-lb } (64 \text{ kN-m})$$

$$= 670,080 \text{ in.-lb}$$

Shear force distribution (Step 2)

Since the beam is to be designed for combined shear and torsion, the distribution of the shear force along the span needs to be determined.

$$\text{stem load} = \left(\frac{22 \times 12}{144} \times 150 \right) 1.4 = 385 \text{ lb/ft}$$

Total factored shear at face of support is

$$V_u = \frac{1}{2} (385 \times 24 + 32,088) = 20,644 \text{ lb}$$

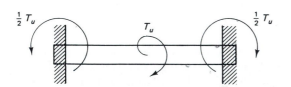

Figure 7.23   Distribution of torsional moment.

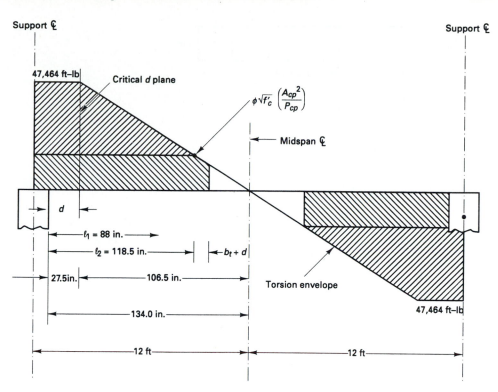

Figure 7.24   Torsion envelope for beam A_1–A_2, Ex. 7.2.

V_u at distance of d from face of support

$$= 20{,}644 \left[\frac{12 - \dfrac{10 + 27.5}{12}}{12} \right] = 15{,}268 \text{ lb}$$

$$V_n = \frac{15{,}268}{0.85} = 17{,}962 \text{ lb}$$

$$T_u = 569{,}570 \text{ in.-lb}$$

Section properties (Step 3)

From Fig. 7.25, assuming 1.5-in. clear cover and No. 4 stirrups and that the flange is not confined with closed ties,

$$A_{cp} = 12 \times 30 = 360 \text{ in.}^2$$
$$p_{cp} = 2(x + y) = 2(12 + 30) = 84 \text{ in.}$$

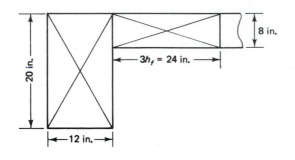

Figure 7.25 Component rectangles.

If the flange was confined, a flange width = 3 × slab thickness would have been taken. In such a case,

$$A_{cp} = 12 \times 30 + 24 \times 8 = 552 \text{ in.}^2$$

$$p_{cp} = 2(12 + 30) + 2(24 + 8) = 148 \text{ in.}$$

$$x_1 = 12 - 2(1.5 + 0.25) = 8.5 \text{ in.}$$

$$y_1 = 30 - 2(1.5 + 0.25) = 26.5 \text{ in.}$$

$$p_h = 2(x_1 + y_1) = 2(8.5 + 26.5) = 70 \text{ in.}$$

$$d = 30 - (1.5 + 0.5 + 0.25) = 27.75, \text{ say } 27.5 \text{ in.}$$

$$A_{oh} = 8.5 \times 26.5 = 225 \text{ in.}^2$$

$$A_0 = 0.85 A_{oh} = 191 \text{ in.}^2$$

$$\theta = 45°, \quad \cot \theta = 1.0$$

Check if torsion has to be considered

From Eq. 7.28,

$$T_u = \phi \sqrt{f_c'} \frac{A_{cp}^2}{p_{cp}} = 0.85 \sqrt{4000} \frac{360^2}{84}$$

$$= 82,942 \text{ in.-lb} < 569,570 \text{ in.-lb.}$$

Hence, torsional moment has to be considered.

Section adequacy check (Step 3)

$$V_c = 2 \sqrt{f_c'} b_w d = 2 \sqrt{4000} \times 12 \times 27.5$$

$$= 41,740 \text{ lbs}$$

$$\sqrt{\left(\frac{V_u}{b_w d}\right)^2 + \left(\frac{T_u p_h}{1.7 A_{oh}^2}\right)^2} = \sqrt{\left(\frac{15,268}{12 \times 27.5}\right)^2 + \left(\frac{569,570 \times 70}{1.7 (225)^2}\right)^2}$$

$$= \sqrt{2140 + 214,616} = 465 \text{ psi}$$

$$\phi\left(\frac{V_c}{b_w d} + 8\sqrt{f_c'} +\right) = 0.85\left(\frac{41,740}{12 \times 27.5} + 8\sqrt{4000}\right) = 0.85(126 + 505)$$

$$= 536 \text{ psi} > 465 \text{ psi}; \text{ hence section is adequate}$$

Since this is an equilibrium torsion, there is no need to evaluate the value of T_n that the section can sustain using Eq. 7.27.

Torsional reinforcement (Step 3)

From Eq. 7.31b,

$$\frac{A_t}{s} = \frac{T_n}{2A_0 f_{yv} \cot \theta} = \frac{670,080}{2 \times 191 \times 60,000 \times 1.0}$$

$$= 0.03 \text{ in.}^2/\text{in./one leg}$$

Shear reinforcement (Step 4)

$V_c = 41,740$ lb from before. Required $V_n = 17,962$ lb. from before $< 41,740$ lb. Also $< \frac{1}{2} V_c$; hence no minimum shear reinforcement needed.

$$\frac{A_{vt}}{s} = \frac{2A_t}{s} + \frac{A_v}{s} = 2 \times 0.03 + 0 = 0.06 \text{ in.}^2/\text{in./two legs}$$

$$\min \frac{A_{vt}}{s} = \frac{50 b_w}{y} = \frac{50 \times 12}{60,000} = 0.01 < 0.06, \qquad \text{O.K.}$$

Try No. 3 closed stirrups $= 2 \times 0.11 = 0.22$ in.^{2} (bar size has the larger of at least No. 3 bar or $s/16$).

$$s = \frac{\text{area of the cross section}}{\text{required } A_{vt}/s} = \frac{0.22}{0.06} = 3.67 \text{ in.c-c}$$

Maximum allowable $s =$ lesser of $p_h/8$ or 12 in.

$$\frac{p_h}{8} = \frac{70}{8} = 8.75 \text{ in.}$$

Therefore, provide No. 3 (9.5-mm diameter) closed stirrups at 3.5 in. center to center (89 mm c-c) at the critical section up to the face of the support. Since the maximum spacing is 8.75 in. and V_c is larger than the factored V_u/ϕ, the increase in spacing along the span toward midspan is determined only with respect to the decrease in T_n along the span. Assume that the stirrups start being spaced at $s = 8.5$ in. at a plane x_1 distance from face of the support, having a torsional moment T_{n1}.
For s = 8.5,

$$\frac{A_{vy}}{s} = \frac{0.22}{8.5} = 0.026$$

$$T_{n1} = \frac{0.026}{2 \times 0.03} \times 55,840 = 24,197 \text{ ft-lb}$$

From similar triangles in Fig. 7.24,

$$x_1 = 27.5 + \left(106.5 - \frac{24,197}{55,840} \times 106.5 \right) = 88 \text{ in.}$$

Torsion is disregarded at T_{n2} if

$$T_u < \phi \sqrt{f_c'} \frac{A_{cp}^2}{p_{cp}}$$

$$T_{n2} = \sqrt{4000} \frac{360^2}{84} = 97,579 \text{ in.-lb} = 8132 \text{ ft-lb}$$

$$x_2 = 27.5 + \left(106.5 - \frac{8132}{55,840} \times 106.5\right) = 118.5 \text{ in.}$$

Extend closed stirrups a distance $b_t + d$ beyond x_2, that is, $118.5 + 12 + 27.5 = 158$ in.; thus use closed stirrups throughout the span. Figure 7.26 shows schematically the spacing of the closed No. 3 stirrups.

Longitudinal torsional reinforcement

From Eq. 7.32,

$$A_\ell = \frac{A_t}{s} p_n \frac{f_{yv}}{f_{y\ell}} \cot^2 \theta = 0.03 \times 70 \times \frac{60,000}{60,000} \times 1.0 = 2.1 \text{ in.}^2$$

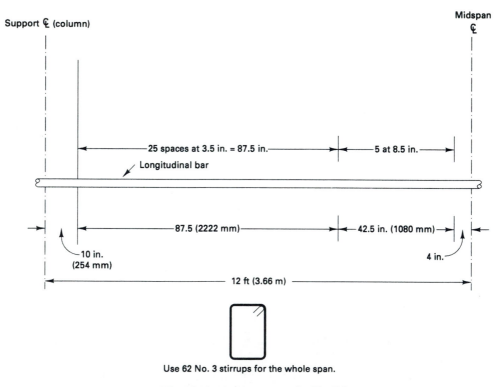

Use 62 No. 3 stirrups for the whole span.

Figure 7.26   Closed stirrup arrangement for Ex. 7.2.

From Eq. 7.34,

$$A_{\ell, \, min} = \frac{5\sqrt{f_c'}A_{cp}}{f_{y\ell}} - \frac{A_t}{s} \, p_h \, \frac{f_{yv}}{f_{y\ell}}$$

$$= \frac{5\sqrt{4000}\;360}{60,000} - 0.03 \times 70 \times \frac{60,000}{60,000} \simeq 0$$

Use $A_\ell = 2.1$ in.^{2} (1355 mm^2). To distribute A_ℓ evenly on all four faces of the beam, use ¼ A_ℓ at each vertical faced with ¼A_ℓ at the top two corners and ¼A_ℓ at the bottom two corners or tension side to be added to the flexural reinforcement. $A_\ell/4 = 2.1/4 = 0.53$. Use two No. 5 bars $= 0.62$ in.^{2} (12.7-mm diameter) on each vertical side for both the support and midspan sections.

Support section:

$$\sum A_s = \frac{A_\ell}{4} + A_s = 0.53 + 2.4 = 2.93 \text{ in.}^2$$

Use four No. 8 bars $= 3.16$ in.^{2} (25.4-mm diameter).

$$A_s' = \frac{A_\ell}{4} + A_s' = 0.53 + 0.8 = 1.33 \text{ in.}^2$$

Use two No. 8 bars $= 1.58$ in.^{2}.

Midspan section:

$$\sum A_s = \frac{A_\ell}{4} + A_s = 0.53 + 1.50 = 2.03 \text{ in.}^2$$

Use three No. 8 bars $= 2.37$ in.^{2} at bottom.

Since the torque decreases as the midspan is approached, two of the top No. 8 longitudinal bars can be cut off prior to reaching the midspan section. Figures 7.27a and b give the reinforcing details of the beam at the support and midspan sections, respectively.

7.5.7 Example 7.3: Compatibility Torsion Web Steel Design

A parking-garage floor system of one-way slabs on beams is shown in Fig. 7.28. Typical panel dimensions are 12 ft 6 in. × 50 ft (3.81 m × 15.24 m) on centers. Design the exterior spandrel beam A_1–B_1 for combined torsion and shear, assuming that the sections are adequately designed for bending. Given:

    service live load $= 50$ psf (2.4 kPa)
    slab thickness $= 5$ in. (127 mm)
    $f_c' = 4000$ psi (27.58 MPa), normal-weight concrete
    $f_y = 60,000$ psi (413.7 MPa)
    height floor to floor $= 10$ ft

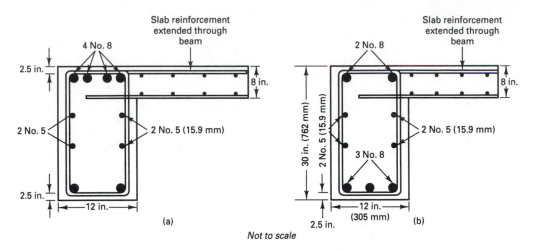

Figure 7.27   Web reinforcement details: (a) support section; (b) midspan section.

exterior columns = 14 in. × 24 in. (356 mm × 610 mm)

interior columns = 24 in. × 24 in. (610 mm × 610 mm)

all beams = 14 in. × 30 in. (356 mm × 762 mm)

required flexural reinforcement for beam A_1-B_1:

  midspan A_s = 1.69 in.²

  support A_s = 2.16 in.²

  support A_s' = 0.90 in.²

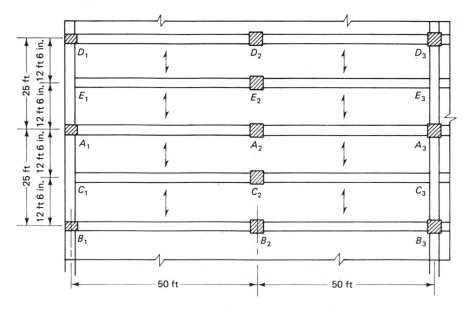

Figure 7.28   Plan of floor systems.

Solution

Beam A_1-B_1 is a case of compatibility torsion because it is part of a continuous floor system where redistribution of moments takes place. The torsional moment due to C_2-C_1 at intersection C_1 is redistributed in directions C_1-C_2 due to the flexibility and rotation of the beam section at C_1 compared to its rigidity at A_1 and B_1. Hence the maximum factored torsional value to be applied to the section at each of the two ends (Fig. 7.29a) is to be the lesser value of the actual T_u or that obtained from Eq. 7.26.

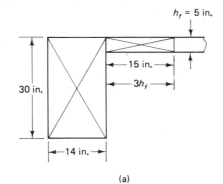

h_f = 5 in.

—15 in.—

—$3h_f$—

30 in.

14 in.

(a)

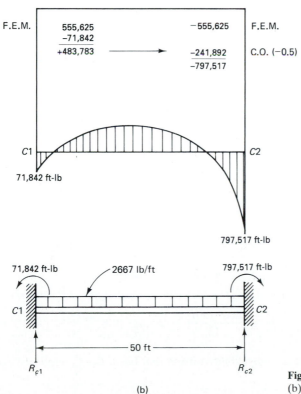

F.E.M.   555,625   −555,625   F.E.M.
           −71,842
           +483,783   −241,892   C.O. (−0.5)
                          −797,517

C1

71,842 ft-lb

C2

797,517 ft-lb

71,842 ft-lb   2667 lb/ft   797,517 ft-lb

C1   C2

—— 50 ft ——

R_{c1} R_{c2}

(b)

Figure 7.29 (a) Component rectangles; (b) bending moments for beam C_1-C_2.

Section properties (Step 1)

From Fig. 7.29, assuming 1.5-in. cover and No. 4 closed stirrups, x and y are the smaller and larger dimensions, respectively, of the section, and x_1 and y_1 are the inner dimensions to the center of the stirrups. The flange is not confined with torsional closed ties.

$$A_{cp} = 14 \times 30 = 420 \text{ in.}^2$$
$$p_{cp} = 2(x + y) = 2(14 + 30) = 88 \text{ in.}$$
$$x_1 = 14 - 2(1.5 + 0.25) = 10.5$$
$$y_1 = 30 - 2(1.5 + 0.25) = 26.5 \text{ in.}$$
$$p_h = 2(x_1 + y_1) = 2(10.5 + 26.5) = 74 \text{ in.}$$
$$d = 30 - (1.5 + 0.5 + 0.25) \approx 27.5 \text{ in.}$$
$$A_{oh} = 10.5 \times 26.5 = 278 \text{ in.}^2$$
$$A_0 = 0.85 A_{oh} = 236 \text{ in.}^2$$
$$\theta = 45°, \qquad \cot \theta = 1.0$$

Factored torsional moments (Steps 2, 3)

The maximum factored torsional value to be applied to the section at each of the two ends (Fig. 7.29a), a compatibility torsional moment, is from Eq. 7.26 for compatibility torsion

$$T_u = \phi 4 \sqrt{f_c'} \frac{A_{cp}^2}{p_{cp}}$$

$$= 0.85 \times 4 \sqrt{4000} \ \frac{420^2}{88} = 431{,}047 \text{ in.-lb}$$

$$= 35{,}921 \text{ ft-lb (48.7 kN-m)}$$

$$T_n = \frac{35{,}921 \times 12}{0.85} = 507{,}120 \text{ in.-lb (57.3 kN-m)}$$

2. *Fixed end moments in beam C_1–C_2:*

service dead load $= \left[\dfrac{5.0}{12} \times 12.5 + \dfrac{(30 - 5) \times 14}{144} \right] 150 = 1146 \text{ lb/ft (16.7 kN/m)}$

service live load $= 50 \times 12.5 = 625 \text{ lb/ft (9.1 kN/m)}$

factored load $U = 1.4 \times 1146 + 1.7 \times 625 = 2667 \text{ lb/ft (38.9 kN/m)}$

fixed end moment $= \dfrac{w_u \ell^2}{12} = \dfrac{2667 \ (50)^2}{12} = 555{,}625 \text{ ft-lb}$

The factored torque in compatibility torsion that beam C_2–C_1 applies at connection C_1 is

$$T_u = 2 \times 35{,}921 = 71{,}842$$

This value is less than the factored end moment $w_u \ell^2 / 12$ at end C_1. Hence the torsional moment to be used at midspan of $A_1 - B_1$ is $T_u = 71{,}842$ ft-lb. Perform the moment distribution shown in Fig. 7.29b to determine the reaction R_{c1} and R_{c2}.

3. *Beam reaction at C_1 and the resulting shear in beam $A_1 - B_1$:*

$$\sum M_{c2} = 0 \quad \text{or} \quad 50 R_{c1} + 797{,}517 - 71{,}842 - \frac{2667(50)^2}{2} = 0$$

$$R_{c1} = \frac{-797{,}517 + 71{,}842 + 3{,}333{,}750}{50} = 52{,}162 \text{ lb}$$

$$\text{factored self-weight of } A_1 - B_1 = 1.4 \frac{14 \times 30}{144} \, 150 = 613 \text{ lb/ft}$$

Distance of critical section in $A_1 - B_1$ from column center line $= d + 14/2 = (30 - 2.5) + 14/2 = 34.5$ in.

$$V_u = \frac{52{,}162}{2} + 613 \left(12.5 - \frac{34.5}{12} \right) = 31{,}981 \text{ lb}$$

Beams $A_1 - B_1$ would be subjected to the torsion and shear envelopes shown in Fig. 7.30.

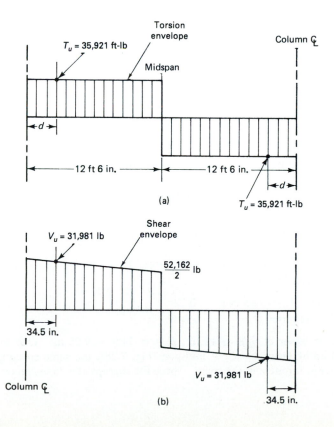

Figure 7.30   (a) Torsion and (b) shear factored force envelopes for beam $A_1 - B_1$, Ex. 7.3.

Section adequacy check (Step 4)

$$V_c = 2\sqrt{f_c'}\,b_w d = 2\,\sqrt{4000} \times 14 \times 27.5$$
$$= 48{,}700 \text{ lb}$$

From Eq. 7.30a,

$$\sqrt{\left(\frac{V_u}{b_w d}\right)^2 + \left(\frac{T_u p_h}{1.7 A_{oh}^2}\right)^2} = \sqrt{\left(\frac{31{,}981}{14 \times 27.5}\right)^2 + \left(\frac{431{,}047 \times 74}{1.7\,(278)^2}\right)^2}$$

$$= \sqrt{6900 + 58{,}943} = 257 \text{ psi}$$

$$\phi\left(\frac{V_c}{b_w d} + 8\sqrt{f_c'}\right) = 0.85\left(\frac{48{,}700}{14 \times 27.5} + 8\,\sqrt{4000}\right)$$

$$= 538 \text{ psi} > 257 \text{ psi; hence section is adequate.}$$

Torsional reinforcement (Step 5)

From Eq. 7.31b,

$$\frac{A_t}{s} = \frac{T_n}{2 A_0 f_{yv} \cot\theta} = \frac{507{,}120}{2 \times 236 \times 60{,}000 \times 1.0}$$

$$= 0.017 \text{ in.}^2/\text{in.}/\text{one leg}$$

Shear reinforcement (Step 6)

$V_c = 48{,}700$ lb and $V_u = 31{,}981$ lb from before.

$$\text{required } V_n = \frac{31{,}981}{0.85} = 37{,}625 \text{ lb} < V_c$$

but larger than $\tfrac{1}{2}V_c = 18{,}812$ lb; hence minimum shear reinforcement needed.

$$\frac{A_v}{s} = \frac{50 b_w}{f_y} = \frac{50 \times 14}{60{,}000} = 0.012 \text{ in.}^2/\text{in.}/\text{two legs}$$

$$\frac{A_v t}{s} = \frac{2 A_t}{s} + \frac{A_v}{s} = 2 \times 0.017 + 0.012 = 0.046 \text{ in.}^2/\text{in.}/\text{two legs}$$

Try No. 3 closed stirrups $= 2 \times 0.11 = 0.22$ in.² (bar size has to be the larger of at least No. 3 or $s/16$ for longitudinal bars).

$$s = \frac{\text{area of cross section}}{\text{required } A_{vt}/s} = \frac{0.22}{0.046} = 4.78 \text{ in.c-c}$$

Maximum allowable s = lesser of $p_h/8$ or 12 in.; $p_h/8 = 74/8 = 9.25$ in. Due to constant torsion imposed by beam C_2-C_1 at midspan (Fig. 7.28), use same spacing of the closed No. 3 stirrups throughout the span. Space the stirrups at 4.75 in. center to center.

Longitudinal torsional reinforcement

From Eq. 7.32,

$$A_\ell = \frac{A_t}{s} \, p_h \frac{f_{yv}}{f_{y\ell}} \cot^2 \theta = 0.017 \times 74 \times \frac{60,000}{60,000} \times 1.0 = 1.26 \text{ in.}^2$$

From Eq. 7.34,

$$A_{\ell,\min} = \frac{5\sqrt{f'_c}\, A_{cp}}{f_{y\ell}} - \frac{A_t}{s} \, p_h \frac{f_{yv}}{f_{y\ell}}$$

$$= \frac{5\sqrt{4000} \times 420}{60,000} - 0.017 \times 74 \times \frac{60,000}{60,000} = 0.95 \text{ in.}^2$$

Use $A_\ell = 1.26$ (813 mm²).

To distribute A_ℓ evenly on all four faces of the beam, use $\frac{1}{4}A_\ell$ at each vertical face with $\frac{1}{4}A_\ell$ at the top two corners and $\frac{1}{4}A_\ell$ at the bottom two corners or tension side to be added to the flexural reinforcement. $A_\ell/4 = 1.26/4 = 0.32$ in.². Use three No. 4 bars = 0.60 in.² (12.7-mm diameter) on each vertical face for both the support and midspan sections. (Three No. 3 bars could be used, but are less rigid in handling.)

Support section:

$$\sum A_s = \frac{A_\ell}{4} + A_s = 0.46 + 2.16 = 2.62 \text{ in.}^2$$

Use six No. 6 bars = 2.64 in.² (six bars, 19.1-mm diameter) at top.

$$\sum A'_s = 0.46 + 0.90 = 1.34 \text{ in.}^2$$

Use three No. 6 bars = 1.32 in.² at bottom.

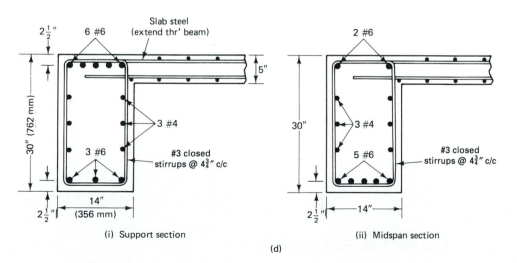

(d)

Figure 7.31   Web reinforcement details: (i) support section; (ii) midspan section.

Midspan section:

$$\sum A_s = 0.46 + 1.69 = 2.15 \text{ in.}^2$$

Use five No. 6 bars = 2.30 in.² at bottom, with three of these bars to continue up to the support (five bars, 19.1-mm diameter).

Figures 7.31a and b give details of the combined torsion–shear reinforcement in the spandrel beam.

7.6 SI METRIC TORSION EXPRESSIONS AND EXAMPLE FOR TORSION DESIGN

In order to design for combined torsion and shear using the SI (System International) method, the following equations replace the corresponding expressions in the PI (pound-inch) method:

Equation 7.26: $T_u \leq \dfrac{\phi \sqrt{f'_c}}{3} \dfrac{A^2_{cp}}{P_{cp}}$

Equation 7.27: $T_u \leq \dfrac{\phi \sqrt{f'_c}}{3} \dfrac{A^2_{cp}}{P_{cp}} \sqrt{1 + \dfrac{3 \bar{f}_c}{\sqrt{f'_c}}}$

Equation 7.28: $T_u \leq \dfrac{\phi \sqrt{f'_c}}{12} \dfrac{A^2_{cp}}{P_{cp}}$

Equation 7.29: $T_u \leq \dfrac{\phi \sqrt{f'_c}}{12} \dfrac{A^2_{cp}}{P_{cp}} \sqrt{1 + \dfrac{3 \bar{f}_c}{\sqrt{f'_c}}}$

Equation 7.30a: $\sqrt{\left(\dfrac{V_u}{b_w d}\right)^2 + \left(\dfrac{T_u p_h}{1.7 A^2_{oh}}\right)^2} \leq \phi \left(\dfrac{V_c}{b_w d} + \dfrac{8 \sqrt{f'_c}}{12}\right)$

Equation 7.30b: $\dfrac{V_u}{b_w d} + \dfrac{T_u p_n}{1.7 A^2_{oh}} \leq \phi \left(\dfrac{V_c}{b_w d} + \dfrac{8 \sqrt{f'_c}}{12}\right)$

Equation 7.30c (reinforced): $V_c = \lambda \dfrac{\sqrt{f'_c}}{6} b_w d$

Equation 7.30d, (prestressed): $V_c = \left(\dfrac{\lambda \sqrt{f'_c}}{20} + \dfrac{5V_u d}{M_u} \right) b_w d$

$$\geq (\lambda \sqrt{f'_c})\, b_w d$$

$$\leq (0.4\lambda \sqrt{f'_c})\, b_w d \quad \text{and} \quad \dfrac{V_u d}{M_u} \leq 1.0$$

Equation 7.31a: $T_n = \dfrac{2A_0 A_t f_{yv}}{s} \cot \theta$

where f_{yv} is in MPa, s in mm, A_0 and A_t in mm^2, and T_n in kN-m.

Equation 7.31b: $A_t = \dfrac{T_n}{2A_0 f_{yv} \cot \theta}$

Equation 7.32: $A_\ell = \dfrac{A_t}{s} p_h \dfrac{f_{yv}}{f_{y\ell}} \cot^2 \theta$

where f_{yv} and $f_{y\ell}$ are in MPa, p_h and s in mm, and A_ℓ and, A_t in mm^2.

Equation 7.33: $A_v \dfrac{0.35 b_w s}{f_y}$

Equation 7.34: $A_{,min} = \dfrac{5\sqrt{f'_c}\, A_{cp}}{12 f_{y\ell}} - \dfrac{A_t}{s} p_h \dfrac{f_{yv}}{f_{y\ell}}$

where A_t/s should not be taken less than $0.175 b_w/f_{yv}$.

Maximum allowable spacing of transverse stirrups is the smaller of $\frac{1}{8}p_h$ or 300 mm, and bars should have a diameter of at least one-sixteenth of the stirrups spacing, but not less than No. 10 M bar size. Maximum f_{yv} or $f_{y\ell}$ should not exceed 400 MPa.

7.6.1 Example 7.4: SI Torsion Design

Solve Ex. 7.1 using SI units.

Data

$f'_c = 27.6$ MPa   (MPa = N/mm^2)

$f_y = 414$ MPa

$V_u = 203$ kN
(a) equilibrium $T_u = 57.2$ kN-m
(b) compatibility $T_u = 8.47$ kN-m
(c) compatibility $T_u = 34.3$ kN-m

(c) compatibility T_u = 34.3 kN-m

$$b_w = 356 \text{ mm}, \qquad A_s = 2190 \text{ mm}^2$$
$$d = 570 \text{ mm}$$
$$h = 635 \text{ mm}$$
$$h_f = 101 \text{ mm}$$

Solution

(a) Equilibrium torsion, T_u = 57.2 kN-m

(No confining ties in the flanges; hence disregarded when computing A_{cp}. Same applies to p_{cp}).

$$A_{cp} = 356 \times 635 = 226{,}060 \text{ mm}^2$$
$$p_{cp} = 2(x + y) = 2(356 + 635) = 1982 \text{ mm}$$

From Eq. 7.28, torsional moment for which torsion can be neglected is

$$T_u = \frac{\phi \sqrt{f_c'}}{12} \frac{A_{cp}^2}{p_{cp}} = \frac{0.85 \sqrt{27.6}}{12} \frac{226{,}060^2}{1982}$$
$$= 9.6 \times 10^6 \text{ N-mm} = 9.6 \text{ kN-m} < 57.2 \text{ kN-m}$$

in case (a); hence design for torsion for in case.

$$T_n = \frac{T_u \, \phi = 57.2}{0.85} = 67.3 \text{ kN-m}$$

Sectional properties

$A_0 = 0.85 A_{oh}$
where A_{oh} is the area enclosed by the center line of the outermost closed stirrups. Assume 40-mm clear cover and No. 10 M bars (diameter = 11.3 mm, A_s = 100 mm²).

$$x_1 = 356 - 2\left(40 + \frac{11.3}{2}\right) = 264 \text{ mm}$$
$$y_1 = 635 - 2\left(40 + \frac{11.3}{2}\right) = 543 \text{ mm}$$
$$A_{oh} = x_1 y_1 = 264 \times 543 = 143{,}400 \text{ mm}^2$$
$$A_0 = 0.85 A_{oh} \simeq 122{,}000 \text{ mm}^2$$
$$d = 635 - \left(40 + 11.3 + \frac{11.3}{2}\right) = 578; \quad \text{use } d = 570 \text{ mm}$$
$$p_h = 2(x_1 + y_1) = 2(264 + 544) = 1616 \text{ mm}$$

Use $\theta = 45°$; cot $\theta = 1.0$.

Check adequacy of section

For the section to be adequate, it should satisfy Eq. 7.30a:

$$\sqrt{\left(\frac{V_u}{b_w d}\right)^2 + \left(\frac{T_u p_h}{1.7 A_{oh}^2}\right)^2} \leq \phi \left(\frac{V_c}{b_w d} + \frac{8\sqrt{f_c'}}{12}\right)$$

$$V_c = \lambda \frac{\sqrt{f_c'}}{6} b_w d = \frac{1.0\sqrt{27.6}}{6} \times 356 \times 570$$

$$= 177,800 \text{ N} = 177.8 \text{ kN}$$

$$\sqrt{\left(\frac{V_u}{b_w d}\right)^2 + \left(\frac{T_u p_h}{1.7 A_{oh}^2}\right)^2} = \sqrt{\left(\frac{203 \times 10^3}{356 \times 570}\right)^2 + \left(\frac{47.2 \times 10^6 \times 1616}{1.7(143,400)^2}\right)^2}$$

$$= \sqrt{(1)^2 + (2.6)^2}$$

$$= 2.76 \text{ N/mm}^2, \text{ say } 2.8 \text{ MPa}$$

$$\phi\left(\frac{V_c}{b_w d} + \frac{8\sqrt{f_c'}}{12}\right) = 0.85\left(\frac{177.8 \times 10^3}{356 \times 570} + \frac{8\sqrt{27.6}}{12}\right)$$

$$= 0.85(0.88 + 3.50)$$

$$= 3.71 \text{ MPa} > 2.8 \text{ MPa}$$

Hence, the section is adequate.

Torsional reinforcement (Step 3)

$$T_n = 67.3 \text{ kN-m} = 67.3 \times 10^6 \text{ N-mm}$$

From Eq. 7.31 b,

$$\frac{A_t}{s} = \frac{T_n}{2A_0 f_{yv} \cot \theta} = \frac{67.3 \times 10^6}{2 \times 122,000 \times 414 \times 1.0}$$

$$= 0.666 \text{ mm}^2/\text{mm/one leg}$$

Shear reinforcement

$$V_c = \lambda \frac{\sqrt{f_c'}}{6} b_w d = 177.8 \text{ kN}$$

From before, required $V_n = 203/0.85 = 239 \text{ kN} > V_c > \frac{1}{2}V_c$ for minimum shear web reinforcement. Provide closed stirrups.

$$V_s = V_n - V_c = 239 - 177.8 = 61.2 \text{ kN}$$

$$\frac{A_v}{s} = \frac{V_s}{f_y d} = \frac{61.2 \times 10^3}{414 \times 570} = 0.26 \text{ mm}^2/\text{mm/two legs}$$

$$\frac{A_{vt}}{s} = \frac{2A_t}{s} + \frac{A_v}{s} = 2 \times 0.666 + 0.26 = 1.6 \text{ mm}^2/\text{mm/two legs}$$

Try No. 10 M closed stirrups (11.3-mm diameter, $A_s = 100$ mm²).   Area of two legs $= 2 \times 100 = 200$ mm².

$$s = \frac{\text{area stirrups cross section}}{\text{required } A_{vt}/s} = \frac{200}{1.6} = 125 \text{ mm}$$

Maximum allowable spacing, s_{max} = smaller or $\frac{1}{8}p_h$, where $p_h = 2(x_i + y_1) = 1616$ mm from before; $\frac{1}{8}p_h = 1616/8 = 202$ mm > 125 mm. From Eq. 7.33,

$$A_{vt} = \frac{0.35b_w s}{f_{yv}} = \frac{0.35 \times 356 \times 125}{414} = 37 \text{ mm}^2$$

Hence, use No. 10 M closed stirrups at 125 mm center to center.
      From Eq. 7.32,

$$A_\ell = \frac{A_t}{s} p_h \frac{f_{yv}}{f_{y\ell}} \cot^2 \theta$$

$$= 0.666 \times 1616 = 1076 \text{ mm}^2$$

From Eq. 7.34,

$$A_{\ell min} = \frac{5\sqrt{f_c'}\, A_{cp}}{12 f_{y\ell}} - \frac{A_t}{s} p_h \frac{f_{yv}}{f_{y\ell}}$$

$$= \frac{5\sqrt{27.6} \times 226{,}060}{12 \times 414} - 0.666 \times 1616$$

$$= 1195 - 1076 = 119 \text{ mm}^2$$

where

$$\frac{A_t}{s} \geq \frac{0.175 b_w}{f_{yv}} = \frac{0.175 \times 356}{414} = 0.15$$

$$< 0.666 \quad \text{O.K.}$$

Hence, $A_t = 1076$ mm² controls.
      Assume that $\frac{1}{4}A_\ell$ goes to the top corners and $\frac{1}{4}A_\ell$ goes to the bottom of the stirrups to be added to the flexural bars The balance, $\frac{1}{2}A_\ell$, would thus be distributed equally on the vertical faces of the beam web cross section at a spacing not to exceed 300 mm c-c.

$$\text{midspan } \sum A_s = \frac{A_\ell}{4} + A_s = \frac{1076}{4} + 2190 = 2460 \text{ mm}^2$$

From Fig. B.2b, provide five No. 25 M longitudinal bars, $A_s = 2500$ mm² at the bottom.   Provide two No. 15 M bars at the top corners of the stirrups (400 mm²) and two No. 15 M bars at each vertical face of the web.

(b) Compatibility torsion, $T_u = 8.47$ kN-m (Step 4)

From part (a), T_u value for torsion to be neglected $= 9.6$ kN-m > 8.47 kN-m. Hence disregard torsion and provide stirrups for shear only.

From part (a), $A_v/s = 0.26$ mm²/mm/two legs.

For No. 10 m stirrups, $s = 200/0.26 = 770$ mm.

Maximum $s = d/2 = 570/2 = 285$ mm.

Use No. 10 M closed stirrups at 220 mm center to center at critical section.

(c) Compatibility Torsion, $T_u = 34.3$ kN-m

$T_u = 34.3 > 9.6$ kN-m from part (a); hence, closed stirrups have to be provided. Since this is a compatibility torsion, the section can be designed from Eq. 7.26 for

$$T_u = \frac{\phi \sqrt{f'_c}}{3} \frac{A_{cp}^2}{P_{cp}} = 4 \times 9.6 \text{ from part (a)}$$

$$= 38.4 \text{ kN-m} > 34.3 \text{ kN-m}$$

Hence, use $T_u = 34.3$ kN-m for the torsional design of the section.

required $T_n = \dfrac{T_u}{\phi} = \dfrac{34.3}{0.85} = 40.4$ kN-m

Torsional reinforcement (Step 3)

From part (a), $A_0 = 122{,}000$ mm², $p_h = 1616$ mm.

$$\frac{A_t}{s} = \frac{T_n}{2A_0 f_{yv} \cot \theta} = \frac{40.4 \times 10^6 \text{ N-mm}}{2 \times 122{,}000 \times 414} = 0.40 \text{ mm}^2/\text{mm/one leg}$$

$A_v/s = 0.26$ mm²/mm/two legs

$$\frac{A_{vt}}{s} = 2\frac{A_t}{s} + \frac{A_v}{s} = 2 \times 0.40 + 0.26 = 1.06 \text{ mm}^2/\text{mm/two legs}$$

Using No. 10 M closed stirrups,

$$s = \frac{2 \times 100}{1.06} = 189 \text{ mm, say } 180 \text{ mm}$$

This is less than $\frac{1}{8}p_h$ or 300 mm. Hence, use No. 10 M closed stirrups at 180 mm c-c (diameter of 11.3 mm) at the critical section.

$$A_\ell = \frac{A_t}{s} p_h \frac{f_{yv}}{f_{y\ell}} \cot^2 \theta = 0.40 \times 1616 = 646 \text{ mm}^2$$

$$A_{\ell,min} = \frac{5\sqrt{f'_c}A_{cp}}{f_{y\ell}} - \frac{A_t}{s} p_h \frac{f_{yv}}{f_{y\ell}}$$

$$= 1195 \text{ (from before)} - (0.40 \times 1616) = 549 \text{ mm}^2, \text{ controls}$$

Use $A_\ell = 876$ mm².

Distribution of torsion longitudinal bars

$$\text{torsional } A_\ell = 876 \text{ mm}^2, \qquad \frac{A_\ell}{4} = 220 \text{ mm}^2$$

Using the same logic as that followed in part (a), provide at bottom an area $A_s = 2190 + 220 = 2410$ mm², that is, five No. 25 M bars ($A_s = 2500$ mm²) and two No. 15 M (400 mm²) bars at top corners and each of the two vertical faces of the web.

SELECTED REFERENCES

7.1. Timoshenko, S., *Strength of Materials,* Part II, *Advanced Theory,* Van Nostrand Reinhold, New York, 1952, 501 pp.

7.2. Nadai, A., *Plasticity: A Mechanics of the Plastic State of Matter,* McGraw-Hill, New York, 1931, 349 pp.

7.3. Cowan, H. J., "Design of Beams Subject to Torsion Related to the New Australian Code," *Journal of the American Concrete Institute,* Proc. Vol. 56, January 1960, pp. 591–618.

7.4. Gesund, H., Schnette, F. J., Buchanan, G. R., and Gray, G. A., "Ultimate Strength in Combined Bending and Torsion of Concrete Beams Containing Both Longitudinal and Transverse Reinforcement," *Journal of the American Concrete Institute,* Proc. Vol. 61, December 1964, pp. 1509–1521.

7.5. Lessig, N. N., "Determination of Carrying Capacity of Reinforced Concrete Elements with Rectangular Cross-Section Subjected to Flexure with Torsion," *Zhelezonbeton,* 1959, pp. 5–28.

7.6. Zia, P., *Tension Theories for Concrete Members,* Special Publication SP 18-4, American Concrete institute, Detroit, 1968, pp. 103–132.

7.7. Hsu, T. T. C., "Ultimate Torque of Reinforced Concrete Members," *Journal of the Structural Division, ASCE,* Vol. 94, No. ST2, February 1968, pp. 485–510.

7.8. Rangan, B. V., and Hall, A. J., "Strength of Rectangular Prestressed Concrete Beams in Combined Torsion, Bending and Shear," *Journal of the American Concrete Institute,* Proc. Vol. 70, April 1973, pp. 270–279.

7.9. Wang, C. K., and Salmon, C. G., *Reinforced Concrete Design,* 4th ed., Harper & Row, New York, 1991, 918 pp.

7.10. Thurliman, B., *Torsional Strength of Reinforced and Prestressed Concrete Beams— CEB Approach, U.S. and European Practices,* Special Publication, American Concrete Institute, Detroit, 1979, pp. 117–143.

7.11. Rabbat, B. G., and Collins, M. P., "*A Variable Angle Space Truss Model for Structural Concrete Members Subjected to Complex Loading*, Special Publication SP 55-22, American Concrete Institute, Detroit, 1978, pp. 547–587.

7.12. Collins, M. P., and Mitchell, D., "Shear and Torsion Design of Prestressed and Nonprestressed Concrete Beams," *Journal of the Prestressed Concrete Institute,* Proc. Vol. 25, No. 5, September–October 1980, pp. 32–100.

7.13. Hsu, T. T. C., *Torsion of Reinforced Concrete,* Van Nostrand Reinhold, New York, 1983, 510 pp.

7.14. Hsu, T. T. C., "Shear Flow Zone in Torsion of Reinforced Concrete," *Journal of the Structural Division, ASCE*, Vol. 116, No. 11, New York, November 1990, pp. 3206–3225.

7.15. Hsu, T. T. C., *Unified Theory of Reinforced Concrete*, CRC Press, Boca Raton, Fla., 1993, 313 pp.

7.16. American Concrete Institute, *Building Code Requirements for Concrete (ACI 318-95) and Commentary (ACI 318R-95)*, ACI, Detroit, 1996.

PROBLEMS FOR SOLUTION

7.1. Calculate the torsional capacity T_n for the sections shown in Fig. 7.32. For compatibility torsion. Given:

$$f'_c = 4000 \text{ psi } (27.6 \text{ MPa}), \text{ normal-weight concrete}$$
$$f_y = 60,000 \text{ psi } (413.7 \text{ MPa})$$

7.2. A cantilever beam is subjected to a concentrated service live load of 20,000 lb (90 kN) acting at a distance of 3 ft 6 in. (1.07 m) from the wall support. In addition, the beam has to resist an equilibrium factored torsion T_u = 300,000 in.-lb (33.89 kN/m). The beam cross section is 12 in. × 25 in. (305 mm × 635 mm) with an effective depth of 22.5 in. (571.5 mm). Design the stirrups and the additional longitudinal steel needed.

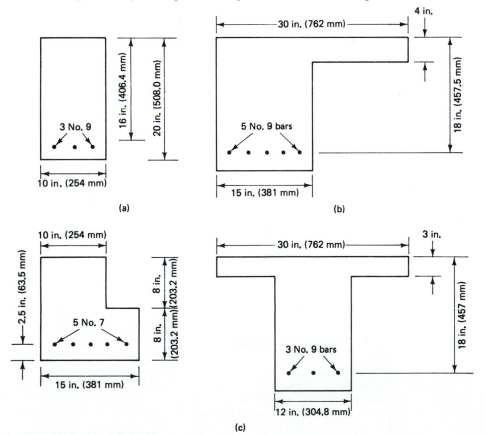

Figure 7.32 Cross sections for Problem 7.1.

Given:

$$f_c' = 3500 \text{ psi}$$
$$f_y = 60,000 \text{ psi}$$
$$A_s = 4.0 \text{ in.}^2 \ (2580.64 \text{ mm}^2)$$

7.3. The first interior span of a four-span continuous beam has a clear span l_n = 18 ft (5.49 m). The beam is subjected to a uniform external service dead load w_D = 1700 plf (24 pkN/m) and a service live load w_L = 2200 plf (32.1 kN/m). Design the section for flexure, diagonal tension, and torsion. Select the size and spacing of the closed stirrups and extra longitudinal steel that might be needed for torsion. Assume that the beam width b_w = 15 in. (381.0 mm) and that redistribution of torsional stresses is possible such that the external torque T_u can be assumed as $\phi(4\sqrt{f_c'}\, A_{cp}^2/p_{cp})$. Given:

$$f_c' = 5000 \text{ psi (34.47 MPa), normal-weight concrete}$$
$$f_y = 60,000 \text{ psi (413.7 MPa)}$$

7.4. A continuous beam has the shear and torsion envelopes shown in Fig. 7.33. The beam dimensions are b_w = 14 in. (355.6 mm) and d = 25 in. (635 mm). It is subjected to factored shear forces V_{u1} = 75,000 lb (333.6 kN), V_{u2} = 60,000 lb, and V_{u3} = 45,000 lb. Design the beam for torsion and shear and detail the web reinforcement.

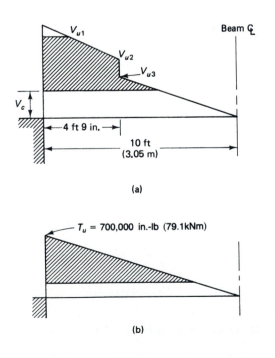

(a)

(b)

Figure 7.33   (a) Shear and (b) torsion envelopes.

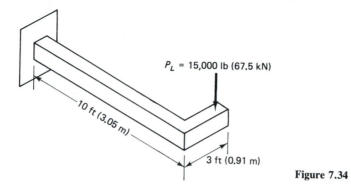

P_L = 15,000 lb (67.5 kN)

10 ft (3.05 m)

3 ft (0.91 m)

Figure 7.34

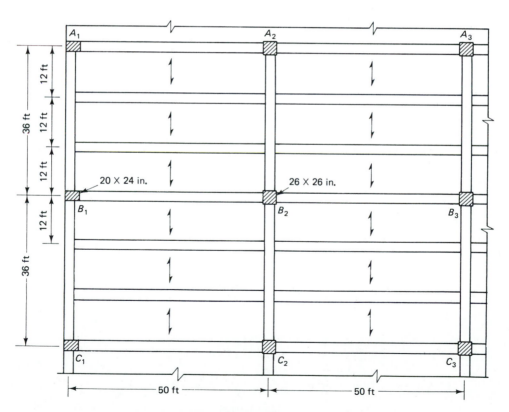

Figure 7.35

Given:

f'_c = 4000 psi (27.58 MPa), lightweight concrete
f_y = 60,000 psi (413.7 MPa)

The required reinforcement is as follows:

midspan A_s = 3.0 in.²
support A_s = 3.6 in.², A'_s = 0.7 in.²

7.5. Design the rectangular beam shown in Fig. 7.34 for bending, shear, and torsion. Assume that the beam width b = 12 in. (305 mm). Given:

f'_c = 4000 psi (27.58 MPa)
f_y = 60,000 psi (413.8 MPa)

7.6. An exterior spandrel beam A_1–B_1, part of the monolithic floor system shown in Fig. 7.35, has a center-to-center span of 36 ft and a slab thickness h_f = 6 in. (153 mm) on beams 15 in. × 36 in. in cross section. It is subject to a service live load = 50 psf (2.4 kPa). Design the shear and torsion reinforcement necessary to resist the external factored loads. Given:

f'_c = 4000 psi (27.58 MPa), normal-weight concrete
f_y = 60,000 psi (413.7 MPa)

Assume that the required flexural reinforcement for beam A_1–B_1 is

midspan A_s = 2.09 in.²
support A_s = 3.0 in.², A'_s = 1.6 in.²

SERVICEABILITY OF BEAMS AND ONE-WAY SLABS

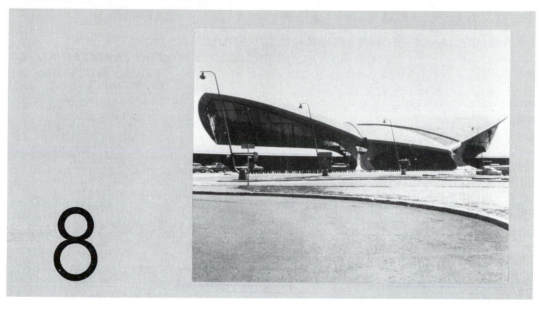

8

8.1 INTRODUCTION

Serviceability of a structure is determined by its deflection, cracking, extent of corrosion of its reinforcement, and surface deterioration of its concrete. Surface deterioration can be minimized by proper control of mixing, placing, and curing of the concrete. If the surface is exposed to potentially damaging chemicals, such as in a chemical factory or a sewage plant, a special type of cement with appropriate additives should be used in the concrete mix. Use of adequate cover as recommended in Chapters 4 and 5, proper quality control of the materials, and the application of proper crack control and deflection control criteria to the design can minimize and in most cases eliminate these problems.

This chapter deals with the evaluation of deflection and cracking behavior of beams and one-way slabs in some detail. It is intended to give the designer adequate basic background on the effect of cracking on the stiffness of the member, the short- and long-term deflection performance, and the manner in which the

Photo 49   Kennedy International airport TWA terminal, New York. (Courtesy of Ammann & Whitney.)

cracked concrete beam element can still perform adequately and esthetically without loss of reliability in its performance. Deflection of two-way action slabs and plates is given in Chapter 11 with numerical examples of deflection calculations for both short- and long-term loading.

8.2 SIGNIFICANCE OF DEFLECTION OBSERVATION

The working stress method of design and analysis used prior to the 1970s limited the stress in concrete to about 45% of its compressive strength and the stress in the steel to less than 50% of its yield strength. Elastic analysis was applied to the design of structural frames as well as reinforced concrete sections. The structural elements were proportioned to carry the highest service-level moment along the span of the member, with redistribution of moment effect often largely neglected. As a result, heavier sections with higher reserve strength resulted as compared to those obtained by the current ultimate strength approach.

Higher-strength concretes having f_c' values in excess of 12,000 psi (82.74 MPa) and higher-strength steels are being used in strength design, and expanding knowledge of the properties of the materials has resulted in lower values of load factors and reduced reserve strength. Hence more slender and efficient members are specified, with deflection becoming a more pronounced controlling criteria.

Beams and slabs are rarely built as isolated members, but a monolithic part of an integrated system. Excessive deflection of a floor slab may cause dislocations in the partitions it supports. Excessive deflection of a beam can damage a partition below, and excessive deflection of a lintel beam above a window opening could crack the glass panels. In the case of open floors or roofs such as top garage floors, ponding of water can result. For these reasons, deflection control criteria are necessary, such as those given in Table 11.3.

8.3 DEFLECTION BEHAVIOR OF BEAMS

The load–deflection relationship of a reinforced concrete beam is basically trilinear, as idealized in Fig. 8.1. It is composed of three regions prior to rupture:

Region I: precracking stage, where a structural member is crack-free (Fig. 8.2)

Region II: postcracking stage, where the structural member develops acceptable controlled cracking both in distribution and width

Region III: postserviceability cracking stage, where the stress in the tension reinforcement reaches the limit state of yielding

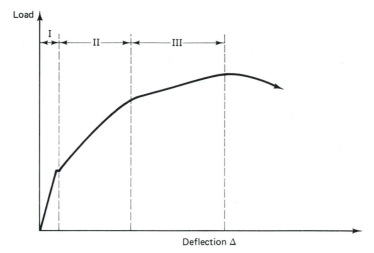

Figure 8.1 Beam load–deflection relationship. Region I, precracking stage; region II, postcracking stage; region III, postserviceability stage (steel yields).

8.3.1 Precracking Stage: Region I

The precracking segment of the load–deflection curve is essentially a straight line defining full elastic behavior. The maximum tensile stress in the beam in this region is less than its tensile strength in flexure, that is, less than the modulus of rupture f_r of concrete. The flexural stiffness EI of the beam can be estimated using Young's modulus E_c of concrete and the moment of inertia of the uncracked reinforced concrete cross section. The load–deflection behavior depends on the stress–strain relationship of the concrete as a significant factor. A typical stress–strain diagram for concrete is shown in Fig. 8.3.

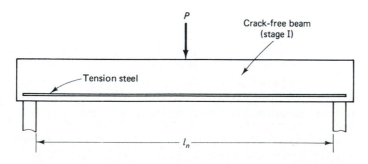

Figure 8.2 Centrally loaded beam at the precracking stage.

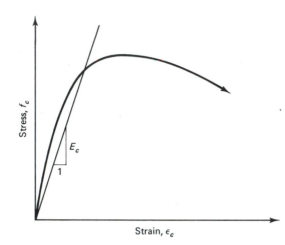

Figure 8.3 Stress–strain diagram of concrete.

The value of E_c can be estimated using the ACI empirical expression given in Chapter 3:

$$E_c = 33w_c^{1.5}\sqrt{f_c'}$$

or $\qquad E_c = 57,000\ \sqrt{f_c'}\qquad$ for normal-weight concrete

An accurate estimation of the moment of inertia I necessitates consideration of the contribution of the steel reinforcement A_s. This can be done by replacing the steel area by an equivalent concrete area $(E_s/E_c)A_s$ since the value of Young's modulus E_s of the reinforcement is higher than E_c. One can transform the steel area to an equivalent concrete area, calculate the center of gravity of the transformed section, and obtain the transformed moment of inertia I_{gt}.

Example 8.1 presents a typical calculation of I_{gt} for a transformed rectangular section. Most designers, however, use a gross moment of inertia I_g based on the uncracked concrete section, disregarding the additional stiffness contributed by the steel reinforcement as insignificant.

The precracking region stops at the initiation of the first flexural crack when the concrete stress reaches its modulus of rupture strength f_r. Similarly to the direct tensile splitting strength, the modulus of rupture of concrete is proportional to the square root of its compressive strength. For design purposes, the value of the modulus for normal-weight concrete may be taken as

$$f_r = 7.5\sqrt{f_c'} \tag{8.1}$$

If lightweight concrete is used, the value of f_r from Eq. 8.1 is multiplied by 0.75 for all lightweight concrete and by 0.85 for sand-lightweight concrete.

If the distance of the extreme tension fiber from the center of gravity of the section is y_t and the cracking moment is M_{cr},

$$M_{cr} = \frac{I_g f_r}{y_t} \tag{8.2}$$

For a rectangular section

$$y_t = \frac{h}{2} \tag{8.3}$$

where h is the total thickness of the beam. Equation 8.2 is derived from the classical bending equation $\sigma = Mc/I$ for elastic and homogeneous materials.

Calculations of deflection for this region are not important since very few reinforced concrete beams remain uncracked under actual loading. However, mathematical knowledge of the variation in stiffness properties is important since segments of the beam along the span in the actual structure can remain uncracked.

8.3.1.1 Example 8.1: Alternative methods of cracking moment evaluation

Calculate the cracking moment M_{cr} for the beam cross section shown in Fig. 8.4, using both (a) transformed and (b) gross cross-section alternatives in the solution. Given:

$f'_c = 4000$ psi (27.6 MPa)

$f_y = 60,000$ psi (414 MPa)

$E_s = 29 \times 10^6$ psi (200,000 MPa), normal-weight concrete

Reinforcement: four No. 9 bars (four bars, 28.6-mm diameter) placed in two bundles.

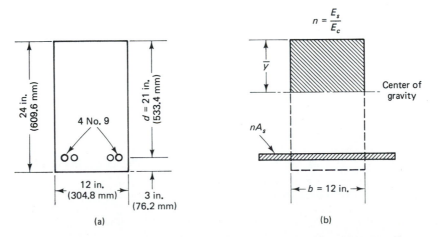

Figure 8.4 Cross-section transformation in Ex. 8.1: (a) midspan section; (b) transformed section.

Solution   (a) *Transformed section solution:* Depth of center-of-gravity axis, $\bar{y}$, can be obtained using the first moment of area:

$$\left[bh + \left(\frac{E_s}{E_c} - 1 \right) A_s \right] \bar{y} = bh\frac{h}{2} + \left(\frac{E_s}{E_c} - 1 \right) A_s d$$

Note that $(E_s/E_c) - 1$ is used instead of E_s/E_c to account for the concrete displaced by the reinforcing bars.

It is customary to denote $n = E_s/E_c$ as the modular ratio. Taking moments about the top extreme fibers of the section,

$$\bar{y} = \frac{(bh^2/2) + (n - 1)A_s d}{bh + (n - 1)A_s}$$

For normal-weight 4000-psi concrete,

$$E_c = 57,000\sqrt{4000}$$
$$= 3.6 \times 10^6 \text{ psi } (24.8 \times 10^6 \text{ MPa})$$
$$n = \frac{29 \times 10^6}{3.6 \times 10^6} = 8.1$$
$$\bar{y} = \frac{\dfrac{12 \times (24)^2}{2} + (8.1 - 1)4.0 \times 21}{12 \times 24 + (8.1 - 1)4.0} = 12.8 \text{ in. } (325.1 \text{ mm})$$

If the moment of inertia of steel reinforcement about its own axis is neglected as insignificant,

$$\text{transformed section } I_{gt} = \frac{bh^3}{12} + bh(12.8 - 12.0)^2 + (n - 1)A_s(d - \bar{y})^2$$

or

$$I_{gt} = \frac{12 \times 24^3}{12} + 12 \times 24 \times 0.8^2 + 7.1 \times 4.0(21 - 12.8)^2$$
$$= 15,918 \text{ in.}^4 \ (66.22 \times 10^8 \text{ mm}^4)$$

The distance of the center of gravity of the transformed section from the lower extreme fibers is

$$y_t = 24 - 12.8 = 11.2 \text{ in. } (284.5 \text{ mm})$$

$$f_r = 7.5\sqrt{4000} = 474.3 \text{ psi } (3.27 \text{ MPa})$$
$$M_{cr} = \frac{I_g f_r}{y_t} = \frac{15,918 \times 474.3}{11.2} = 674,100 \text{ in.-lb } (76.17 \text{ kN-m})$$

(b) *Gross section solution*

$$\bar{y} = \frac{h}{2} = 12 \text{ in.}$$

$$\text{gross section } I_g = \frac{bh^3}{12} = \frac{12 \times 24^3}{12} = 13,824 \text{ in.}^4$$

$$y_t = 12 \text{ in. (304.8 mm)}$$

$$f_r = 474.3 \text{ psi}$$

$$M_{cr} = \frac{13{,}824 \times 474.3}{12} = 546{,}394 \text{ in.-lb (61.74 kN-m)}$$

There is a difference of about 15% in the value of I_g and 19% in the value of M_{cr}. Even though this percentage difference in the values of the I_g and M_{cr} obtained by the two methods seems somewhat high, such a difference in the deflection calculation values is not of real significance and in most cases does not justify using the transformed-section method for evaluating M_{cr}.

8.3.2 Postcracking Service Load Stage: Region II

The precracking region ends at the initiation of the first crack and moves into region II of the load–deflection diagram in Fig. 8.1. Most beams lie in this region at service loads. A beam undergoes varying degrees of cracking along the span corresponding to the stress and deflection levels at each section. Hence cracks are wider and deeper at midspan, whereas only narrow minor cracks develop near the supports in a simple beam.

When flexural cracking develops, the contribution of the concrete in the tension zone reduces substantially. Hence the flexural rigidity of the section is reduced, making the load–deflection curve less steep in this region than in the precracking stage segment. As the magnitude of cracking increases, stiffness continues to decrease, reaching a lower-bound value corresponding to the reduced moment of inertia of the cracked section. At this limit state of service load cracking, the contribution of tension-zone concrete to the stiffness is neglected. The moment of inertia of the cracked section designated as I_{cr} can be calculated from the basic principles of mechanics.

Strain and stress distributions across the depth of a typical cracked rectangular concrete section are shown in Fig. 8.5. The following assumptions are made with respect to deflection computation based on extensive testing verification: (1) the strain distribution across the depth is assumed to be linear; (2) concrete does not resist any tension; (3) both concrete and steel are within the elastic limit; and (4) strain distribution is similar to that assumed for strength design, but the magnitudes of strains, stresses, and stress distribution are different.

To calculate the moment of inertia, the value of the neutral axis depth, c, should be determined from horizontal force equilibrium.

$$A_s f_s = bc\frac{f_c}{2} \tag{8.4a}$$

Since the steel stress $f_s = E_s \epsilon_s$ and concrete stress $f_c = E_c \epsilon_c$, Eq. 8.4a can be rewritten as

$$A_s E_s \epsilon_s = \frac{bc}{2} E_c \epsilon_c \tag{8.4b}$$

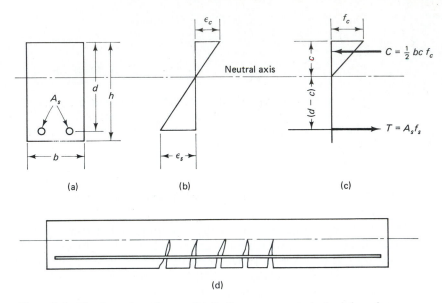

Figure 8.5 Elastic strain and stress distributions across a cracked reinforced concrete section: (a) cross section; (b) strain; (c) elastic stress and force; (d) cracked beam prior to failure in flexure.

From similar triangles in Fig. 8.5b,

$$\frac{\epsilon_c}{c} = \frac{\epsilon_s}{d - c} \tag{8.5a}$$

or
$$\epsilon_s = \epsilon_c \left(\frac{d}{c} - 1 \right) \tag{8.5b}$$

From Eqs. 8.4b and 8.5b,

$$A_s E_s \epsilon_c \left(\frac{d}{c} - 1 \right) = \frac{bc}{2} E_c \epsilon_c \tag{8.6a}$$

or
$$\frac{A_s E_s}{E_c} \left(\frac{d}{c} - 1 \right) = \frac{bc}{2} \tag{8.6b}$$

Replacing the modular ratio E_s/E_c by n, Eq. 8.6b can be rewritten as

$$\frac{bc^2}{2} + nA_s c - nA_s d = 0 \tag{8.6c}$$

The value of c can be obtained by solving the quadratic equation, 8.6c. The moment of inertia I_{cr} can be obtained from

$$I_{cr} = \frac{bc^3}{3} + nA_s(d - c)^2 \tag{8.7}$$

where the term $bc^3/3$ in Eq. 8.7 denotes the moment of inertia of the *compressive* area bc about the neutral axis, that is, the base of the compression rectangle, neglecting the section area in tension below the neutral axis.   The reinforcing area is multiplied by n to transform it to its equivalent in concrete for contribution to the section stiffness.   The moment of inertia of the steel about its own axis is disregarded as negligible.

Only part of the beam cross section is cracked in the case under discussion. As seen from Fig. 8.5d, the uncracked segments below the neutral axis along the beam span possess some degree of stiffness, which contributes to the overall beam rigidity.   The actual stiffness of the beam lies between $E_c I_g$ and $E_c I_{cr}$, depending on such other factors as (1) extent of cracking, (2) distribution of loading, and (3) contribution of the concrete, as seen in Fig. 8.5d between the cracks.   Generally, as the load approaches the steel yield load level, the stiffness value approaches $E_c I_{cr}$.

Branson developed simplified expressions for calculating the effective stiffness $E_c I_e$ for design.   The Branson equation, verified as applicable to most cases of reinforced and prestressed beams and universally adopted for deflection calculations, defines the effective moment of inertia as

$$I_e = \left(\frac{M_{cr}}{M_a}\right)^3 I_g + \left[1 - \left(\frac{M_{cr}}{M_a}\right)^3\right]I_{cr} \le I_g \qquad (8.8a)$$

Equation 8.8a is also written in the form

$$I_e = I_{cr} + \left(\frac{M_{cr}}{M_a}\right)^3 (I_g - I_{cr}) \le I_g \qquad (8.8b)$$

The effective moment of inertia I_e as shown in Eq. 8.8b depends on the maximum moment M_a along the span in relation to the cracking moment capacity M_{cr} of the section.

8.3.2.1 Example 8.2: Effective moment of inertia of cracked beam sections

Calculate the moment of inertia I_{cr} and the effective moment of inertia I_e of the beam cross section in Ex. 8.1 if the external maximum service load moment is 2,000,000 in.-lb (226 kN-m).   Given (Ex. 8.1):

$$b = 12 \text{ in. (305 mm)}$$
$$d = 21 \text{ in. (533 mm)}$$
$$h = 24 \text{ in. (610 mm)}$$
$$A_s = 4.0 \text{ in.}^2 \text{ (2580 mm}^2\text{)}$$
$$f'_c = 4000 \text{ psi (27.6 MPa)}$$
$$f_y = 60,000 \text{ psi (413.7 MPa)}$$
$$E_s = 29 \times 10^6 \text{ psi (200} \times 10^3 \text{ MPa)}$$
$$E_c = 3.6 \times 10^6 \text{ psi (24.8} \times 10^3 \text{ MPa)}$$
$$n = 8.1$$

Solution   From Eq. 8.6c,

$$\frac{12c^2}{2} + 8.1 \times 4.0c - 8.1 \times 4.0 \times 21 = 0$$

Hence neutral axis depth $c = 8.3$ in. (210.8 mm). From Eq. 8.7,

$$I_{cr} = \frac{12.0 \times 8.3^3}{3} + 8.1 \times 4.0(21.0 - 8.3)^2 = 7513 \text{ in.}^4 \ (31.25 \times 10^8 \text{ mm}^4)$$

Using the I_{gt} and M_{cr} values of Ex. 8.1, which include the effect of the transformed steel area,

$$I_e = 7513 + \left(\frac{674{,}100}{2{,}000{,}000}\right)^3 (15{,}918 - 7513)$$

$$= 7835 \text{ in.}^4 \ (32.59 \times 10^8 \text{ mm}^4) < I_g \quad \text{as expected}$$

If the gross cross-section values for I_g and M_{cr} are used without including the effect of transformed A_s, the effective moment of inertia becomes

$$I_e = 7513 + \left(\frac{546{,}394}{2{,}000{,}000}\right)^3 (13{,}824 - 7513)$$

$$= 7642 \text{ in.}^4 (31.79 \times 10^8 \text{ mm}^4) < I_g$$

Comparison of the two values of effective I_e calculated by the two methods (7835 in.^{4} versus 7642 in.^{4}) shows an insignificant difference. Hence, use of the cross-section properties in Eq. 8.8 is, in most cases, adequate, particularly when one considers the variability in the loads and the randomness in the properties of concrete.

8.3.3 Postserviceability Cracking Stage and Limit State of Deflection Behavior at Failure: Region III

The load–deflection diagram of Fig. 8.1 is considerably flatter in region III than in the preceding regions. This is due to substantial loss in stiffness of the section because of extensive cracking and considerable widening of the stabilized cracks throughout the span. As the load continues to increase, the strain ϵ_s in the steel bars at the tension side continues to increase beyond the yield strain ϵ_y with no additional stress. The beam is considered at this stage to have structurally failed by initial yielding of the tension steel. It continues to deflect without additional loading, the cracks continue to open, and the neutral axis continues to rise toward the outer compression fibers. Finally, a secondary compression failure develops, leading to total crushing of the concrete in the maximum moment region, followed by rupture.

The increase in the beam load level between first yielding of the tension reinforcement in a simple beam and the rupture load level varies between 4% and 10%. The deflection value before rupture, however, can be several times that at the steel yield level, depending on the beam span/depth ratio, the steel percentage, the type of loading, and the degree of confinement of the beam section. An ultimate deflection value 8 to 12 times the first yield deflection has frequently been observed in tests.

Postyield deflection and limit deflection at failure are not of major significance in design and hence are not being discussed in any detail in this text. It is important, however, to recognize the reserve deflection capacity as a measure of ductility in structures in earthquake zones and in other areas where the probability of overload is high.

8.4 LONG-TERM DEFLECTION

Time-dependent factors magnify the magnitude of deflection with time. Consequently, the design engineer has to evaluate immediate as well as long-term deflection in order to ensure that their values satisfy the maximum permissible criteria for the particular structure and its particular use.

Time-dependent effects are caused by the superimposed creep, shrinkage, and temperature strains. These additional strains induce a change in the distribution of stresses in the concrete and the steel, resulting in an increase in the curvature of the structural element for the same external load.

The calculation of creep and shrinkage strains at a given time is a complex process, as discussed in Chapter 3. One has to consider how these time-dependent concrete strains affect the stress in the steel and the curvature of the concrete element. In addition, consideration has to be given to the effect of progressive cracking on the change in stiffness factors, considerably complicating the analysis

Photo 50 Deflected simply supported beam prior to failure. (Tests by Nawy et al.)

Photo 51 Deflected continuous prestressed beam prior to failure. (Tests by Nawy, Potyondy, et al.)

and design process. Consequently, an empirical approach to evaluate deflection under sustained loading is, in many cases, more practical.

The additional deflection under sustained loading and long-term shrinkage in accordance with the ACI procedure can be calculated using a multiplying factor:

$$\lambda = \frac{T}{1 + 50\rho'} \tag{8.9}$$

where ρ' is the compression reinforcement ratio calculated at midspan for simple and continuous beams and T is a factor that is taken as 1.0 for loading time duration of 3 months, 1.2 for 6 months, 1.4 for 12 months, and 2.0 for 5 years or more.

If the instantaneous deflection is Δ_i, the additional time-dependent deflection becomes $\lambda\Delta_i$, and the total long-term deflection would be $(1 + \lambda)\Delta_i$. Since live loads are not present at all times, only part of the live load in addition to the more permanent dead load is considered as the sustained load. Figure 8.6 gives the relationship between the load duration in months and the multiplier T in Eq. 8.6. It is seen from this plot that the maximum multiplier value $T = 2.0$ represents a nominal limiting time-dependent factor for 5 years' duration of loading. In effect, the expression for the long-term factor λ in Eq. 8.9 has similar characteristics as

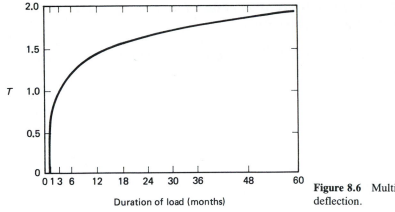

Figure 8.6 Multipliers for long-term deflection.

the stiffness EI of a section in that it is a function of the material property T and the section property $(1 + 50\rho')$.

The total long-term deflection is

$$\Delta_{LT} = \Delta_L + \lambda_\infty \Delta_D + \lambda_t \Delta_{LS} \qquad (8.10)$$

where Δ_L = initial live-load deflection
   Δ_D = initial dead-load deflection
   Δ_{LS} = initial sustained live-load deflection (a percentage of the immediate Δ_L determined by expected duration of sustained load)
   λ_∞ = time-dependent multiplier for infinite duration of sustained load
   λ_t = time-dependent multiplier for limited load duration

The value of the multiplier λ is the same for normal-weight or lightweight concrete.

8.5 PERMISSIBLE DEFLECTIONS IN BEAMS AND ONE-WAY SLABS

Permissible deflections in a structural system are governed primarily by the amount that can be sustained by the interacting components of a structure without loss of esthetic appearance and without detriment to the deflecting member. The level of acceptability of deflection values is a function of such factors as the type of building, the use or nonuse of partitions, the presence of plastered ceilings, or the sensitivity of equipment or vehicular systems that are being supported by the floor. Since deflection limitations have to be placed at service load levels, structures designed conservatively for low concrete and steel stresses would normally have no deflection problems. Present-day structures, however, are designed by ultimate load procedures efficiently utilizing high-strength concretes and steels. More slender members resulting from such designs would have to be better controlled for serviceability deflection performance, both immediate and long term.

8.5.1 Empirical Method of Minimum Thickness Evaluation for Deflection Control

The ACI Code recommends in Table 8.1 minimum thickness for beams as a function of the span length, where no deflection computations are necessary if the member is not supporting or attached to construction likely to be damaged by large deflections. Other deflections would have to be calculated and controlled as in Table 8.2. If the total beam thickness is less than required by the table, the designer should verify the deflection serviceability performance of the beam through detailed computations of the immediate and long-term deflections.

8.5.2 Permissible Limits of Calculated Deflection

The ACI Code requires that the calculated deflection for a beam or one-way slab has to satisfy the serviceability requirement of minimum permissible deflection for the various structural conditions listed in Table 8.2 of Section 8.5.1 if Table 8.1 is *not* used. However, long-term effects cause measurable increases in deflection with time and result sometimes in excessive overstress in the steel and concrete. Hence it is always advisable to calculate the total time-dependent deflection Δ_{LT} in Eq. 8.10 and design the beam size based on the permissible span/deflection ratios of Table 8.2.

TABLE 8.1 MINIMUM THICKNESS OF BEAMS AND ONE-WAY SLABS UNLESS DEFLECTIONS ARE COMPUTED[a]

| | Minimum thickness, h | | | |
Member[b]	Simply supported	One end continuous	Both ends continuous	Cantilever
Solid one-way slabs	$l/20$	$l/24$	$l/28$	$l/10$
Beams or ribbed one-way slabs	$l/16$	$l/18.5$	$l/21$	$l/8$

[a]Clear span length l is in inches. Values given should be used directly for members with normal-weight concrete ($w_c = 145$ pcf) and grade 60 reinforcement. For other conditions, the values should be modified as follows: (1) For structural lightweight concrete having unit weights in the range from 90 to 120 lb/ft^3, the values should be multiplied by $(1.65 - 0.005w_c)$, but not less than 1.09, where w_c is the unit weight in lb/ft^3. (2) For f_y other than 60,000 psi, the values should be multiplied by $(0.4 + f_y/100,000)$.

[b]Members not supporting or attached to partitions or other construction likely to be damaged by large deflections.

TABLE 8.2 MINIMUM PERMISSIBLE RATIOS OF SPAN (l) TO DEFLECTION (Δ)
(l = longer span)

Type of member	Deflection, Δ, to be considered	$(l/\Delta)_{min}$
Flat roofs not supporting and not attached to nonstructural elements likely to be damaged by large deflections	Immediate deflection due to live load L	180[a]
Floors not supporting and not attached to nonstructural elements likely to be damaged by large deflections	Immediate deflection due to live load L	360
Roof or floor construction supporting or attached to nonstructural elements likely to be damaged by large deflections	That part of total deflection occurring after attachment of nonstructural elements: sum of long-term deflection due to all sustained loads (dead load plus any sustained portion of live load) and immediate deflection due to any additional live load[b]	480[c]
Roof or floor construction supporting or attached to nonstructural elements not likely to be damaged by large deflections		240[c]

[a]Limit not intended to safeguard against ponding. Ponding should be checked by suitable calculations of deflection, including added deflections due to ponded water, and considering long-term effects of all sustained loads, camber, construction tolerances, and reliability of provisions for drainage.

[b]Long-term deflection has to be determined, but may be reduced by the amount of deflection calculated to occur before attachment of nonstructural elements. This reduction is made on the basis of accepted engineering data relating to time–deflection characteristics of members similar to those being considered.

[c]Ratio limit may be lower if adequate measures are taken to prevent damage to supported or attached elements, but should not be lower than tolerance of nonstructural elements.

8.6 COMPUTATION OF DEFLECTIONS

Deflection of structural members is a function of the span length, support, or end conditions, such as simple support or restraint due to continuity, the type of loading, such as concentrated or distributed load, and the flexural stiffness EI of the member.

The general expression for the maximum deflection Δ_{max} in an elastic member can be expressed from basic principles of mechanics as

$$\Delta_{max} = K\frac{Wl_n^3}{48EI_c} \tag{8.11}$$

where W = total load on the span
l_n = clear span length
E = modulus of concrete

I_c = moment of inertia of the section

K = a factor depending on the degree of fixity of the support

Equation 8.11 can also be written in terms of moment such that the deflection at any point in a beam is

$$\Delta = k\frac{ML^2}{E_c I_e} \qquad (8.12)$$

where k = a factor depending on support fixity and load conditions

M = moment acting on the section

I_e = effective moment of inertia

Table 8.3 gives the maximum elastic deflection values in terms of the gravity load for typical beams loaded with uniform or concentrated load.

8.6.1 Example 8.3: Deflection Behavior of a Uniformly Loaded Simple Span Beam

A simply supported uniformly loaded beam has a clear span $l_n = 27$ ft (8.23 m), a width $b = 10$ in. (254 mm), and a total depth $h = 16$ in. (406 mm), $d = 13.0$ in. (330 mm), and $A_s = 1.32$ in.^{2} (852 mm^2). It is subjected to a service dead-load moment $M_D = 215,000$ in.-lb (24.3 kN-m), and a service live-load moment $M_L = 250,000$ in.-lb (28.3 kN-m). Determine if the beam satisfies the various deflection criteria for short- and long-term loading. Assume that 60% of the live load is continuously applied for 24 months. Given:

$f_c' = 5000$ psi (34.5 MPa), normal-weight concrete

$f_y = 60,000$ psi (413.7 MPa)

Solution

$$E_c = 33w^{1.5}\sqrt{5000} = 4.29 \times 10^6 \text{ psi (29,700 MPa)}$$

$$E_s = 29 \times 10^6 \text{ psi (200,000 MPa)}$$

$$\text{modular ratio } n = \frac{E_s}{E_c} = \frac{29.0 \times 10^6}{4.29 \times 10^6} = 6.76$$

$$f_r = 7.5\sqrt{f_c'} = 7.5\sqrt{5000} = 530 \text{ psi (3.66 MPa)}$$

Minimum required depth

From Table 8.1,

$$h_{\min} = \frac{l_n}{16} = \frac{27.0 \times 12}{16} \cong 17 \text{ in. (432 mm)} > \text{actual } h = 16.0 \text{ in.}$$

Hence deflection calculations have to be made.

TABLE 8.3 MAXIMUM DEFLECTION EXPRESSIONS FOR MOST COMMON LOAD
AND SUPPORT CONDITIONS

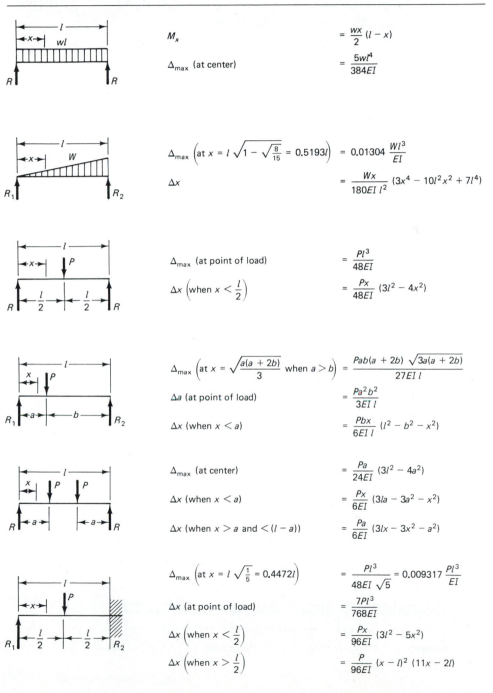

$$M_x = \frac{wx}{2}(l - x)$$

$$\Delta_{max} \text{ (at center)} = \frac{5wl^4}{384EI}$$

$$\Delta_{max}\left(\text{at } x = l\sqrt{1 - \sqrt{\tfrac{8}{15}}} = 0.5193l\right) = 0.01304\,\frac{Wl^3}{EI}$$

$$\Delta x = \frac{Wx}{180EI\,l^2}(3x^4 - 10l^2x^2 + 7l^4)$$

$$\Delta_{max} \text{ (at point of load)} = \frac{Pl^3}{48EI}$$

$$\Delta x \left(\text{when } x < \frac{l}{2}\right) = \frac{Px}{48EI}(3l^2 - 4x^2)$$

$$\Delta_{max}\left(\text{at } x = \sqrt{\frac{a(a + 2b)}{3}} \text{ when } a > b\right) = \frac{Pab(a + 2b)\,\sqrt{3a(a + 2b)}}{27EI\,l}$$

$$\Delta a \text{ (at point of load)} = \frac{Pa^2b^2}{3EI\,l}$$

$$\Delta x \text{ (when } x < a) = \frac{Pbx}{6EI\,l}(l^2 - b^2 - x^2)$$

$$\Delta_{max} \text{ (at center)} = \frac{Pa}{24EI}(3l^2 - 4a^2)$$

$$\Delta x \text{ (when } x < a) = \frac{Px}{6EI}(3la - 3a^2 - x^2)$$

$$\Delta x \text{ (when } x > a \text{ and } < (l - a)) = \frac{Pa}{6EI}(3lx - 3x^2 - a^2)$$

$$\Delta_{max}\left(\text{at } x = l\sqrt{\tfrac{1}{5}} = 0.4472l\right) = \frac{Pl^3}{48EI\,\sqrt{5}} = 0.009317\,\frac{Pl^3}{EI}$$

$$\Delta x \text{ (at point of load)} = \frac{7Pl^3}{768EI}$$

$$\Delta x \left(\text{when } x < \frac{l}{2}\right) = \frac{Px}{96EI}(3l^2 - 5x^2)$$

$$\Delta x \left(\text{when } x > \frac{l}{2}\right) = \frac{P}{96EI}(x - l)^2(11x - 2l)$$

TABLE 8.3 (*Cont.*)

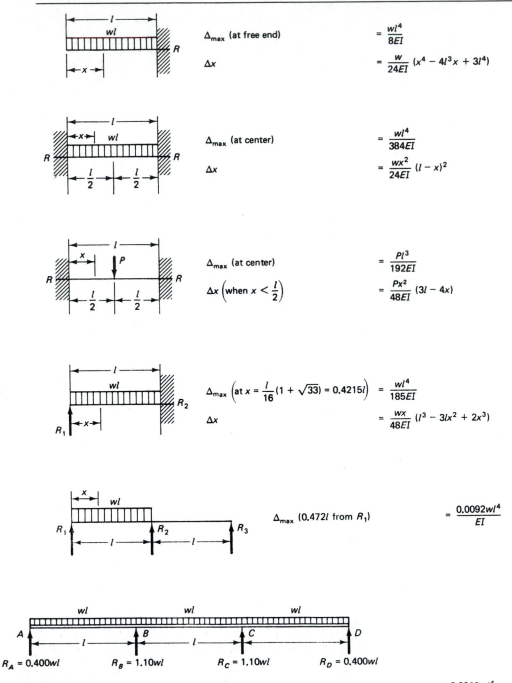

Δ_{max} (at free end) $= \dfrac{wl^4}{8EI}$

Δx $= \dfrac{w}{24EI}(x^4 - 4l^3x + 3l^4)$

Δ_{max} (at center) $= \dfrac{wl^4}{384EI}$

Δx $= \dfrac{wx^2}{24EI}(l - x)^2$

Δ_{max} (at center) $= \dfrac{Pl^3}{192EI}$

$\Delta x \left(\text{when } x < \dfrac{l}{2} \right)$ $= \dfrac{Px^2}{48EI}(3l - 4x)$

$\Delta_{max} \left(\text{at } x = \dfrac{l}{16}(1 + \sqrt{33}) = 0.4215l \right)$ $= \dfrac{wl^4}{185EI}$

Δx $= \dfrac{wx}{48EI}(l^3 - 3lx^2 + 2x^3)$

Δ_{max} (0.472*l* from R_1) $= \dfrac{0.0092wl^4}{EI}$

Δ_{max} (0.446*l* from *A* or *D*) $= \dfrac{0.0069wl^4}{EI}$

TABLE 8.3   (*Cont.*)

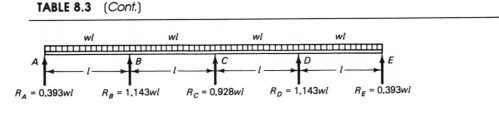

$$\Delta_{max} \text{ (0.440}l \text{ from } A \text{ or } E) = \frac{0.0065wl^4}{EI}$$

Effective moment of inertia I_e

$$I_g = \frac{bh^3}{12} = \frac{10(16)^3}{12} = 3410 \text{ in.}^4$$

$$y_t = \frac{16.0}{2} = 8.0 \text{ in.}$$

$$M_{cr} = \frac{f_r I_g}{y_t} = \frac{530 \times 3410}{8.0} = 225,900 \text{ in.-lb}$$

Depth of neutral axis c

$$d = 16.0 - 3.0 = 13.0 \text{ in.} \qquad A_s = 1.32 \text{ in.}^2$$

$$\frac{10c^2}{2} = nA_s(d - c)$$

or $5c^2 = 6.76 \times 1.32(13.0 - c)$, to get $c = 4.03$ in.

$$I_{cr} = \frac{10c^3}{3} + 6.76 \times 1.32(13.0 - c)^2 = \frac{10(4.03)^3}{3} + 8.923(13.0 - 4.03)^2$$

$$= 940 \text{ in.}^4$$

Dead load

$$\frac{M_{cr}}{M_a} = \frac{225,900}{215,000} = 1.05 > 1.0$$

Use $M_{cr} = M_a$ and $I_e = I_g$ since the dead-load moment is smaller than the cracking moment (the beam will not crack at the dead-load level).

Dead load + 60% live load

$$\frac{M_{cr}}{M_a} = \frac{225,900}{215,000 + 0.6 \times 250,000} = 0.62$$

Dead load + live load

$$\frac{M_{cr}}{M_a} = \frac{225,900}{215,000 + 250,000} = 0.49$$

$$I_e = \left(\frac{M_{cr}}{M_a}\right)^3 I_g + \left[1 - \left(\frac{M_{cr}}{M_a}\right)^3\right] I_{cr}$$

Dead load

$$I_e = 3410 \text{ in.}^4$$

Dead load + 0.6 live load

$$I_e = 0.24 \times 3410 + 0.76 \times 940 = 1530 \text{ in.}^4$$

Dead load + live load

$$I_e = 0.12 \times 3410 + 0.88 \times 940 = 1230 \text{ in.}^4$$

Short-term deflection

$$\Delta = \frac{5wl^4}{384EI} = \frac{5Ml_n^2}{48EI} = \frac{5(27.0 \times 12)^2 M}{48 \times 4.29 \times 10^6 I} = 0.0025\frac{M}{I} \text{ in.}$$

Initial live-load deflection

$$\Delta_L = \frac{0.0025(215,000 + 250,000)}{1230} - \frac{0.0025(215,000)}{3410}$$
$$= 0.943 - 0.158 \simeq 0.8 \text{ in.}$$

Initial dead-load deflection

$$\Delta_D = \frac{0.0025 \times 215,000}{3410} = 0.16 \text{ in.}$$

Initial 60% sustained live-load deflection

$$\Delta_{LS} = 0.0025\left(\frac{215,000 + 250,000 \times 0.6}{1530} - \frac{215,000}{3410}\right)$$
$$= 0.60 - 0.16 = 0.44 \text{ in.}$$

Long-term deflection

From Eq. 8.10,

$$\Delta_{LT} = \Delta_L + \lambda_\infty\Delta_D + \lambda_t\Delta_{LS}$$

$$\lambda = \frac{T}{1 + 50\,\rho'}, \qquad \text{where } \rho' = 0 \text{ for singly reinforced beam}$$

$$T \text{ for 5 years or more} = 2.0 \qquad \lambda_\infty = \frac{2.0}{1 + 0} = 2.0$$

$$T \text{ for 24 months} = 1.65 \qquad \lambda_t = \frac{1.65}{1} = 1.65$$

$$\Delta_{LT} = 0.8 + 2.0 \times 0.16 + 1.65 \times 0.44 = 1.9 \text{ in.}$$

Deflection requirements (Table 8.2)

$$\frac{l_n}{180} = \frac{27 \times 12}{180} = 1.80 \text{ in.} > \Delta_L$$

$$\frac{l_n}{360} = 0.90 \text{ in.} \qquad\qquad > \Delta_L$$

$$\frac{l_n}{480} = 0.68 \text{ in.} \qquad\qquad < \Delta_{LT}$$

$$\frac{l_n}{240} = 1.35 \text{ in.} \qquad\qquad < \Delta_{LT}$$

Hence the use of this beam is limited to floors or roofs not supporting or attached to nonstructural elements such as partitions.

8.7 DEFLECTION OF CONTINUOUS BEAMS

As discussed in Chapter 5, a continuous reinforced concrete beam would have a flanged section at midspan, and sometimes a doubly reinforced section at the support if the reinforcement at the bottom fibers of the support section are adequately tied and anchored. Consequently, it is necessary to be able to find the effective moment of inertia I_e of T sections and of doubly reinforced sections. A simple procedure is to use the weighted-average section properties as required by the ACI code:

1. Beams with both ends continuous:

$$\text{average } I_e = 0.70I_m + 0.15(I_{e1} + I_{e2}) \tag{8.13}$$

2. Beams with one end continuous

$$\text{average } I_e = 0.85I_m + 0.15(I_{ec}) \tag{8.14}$$

where I_m = midspan section I_e
I_{e1}, I_{e2} = I_e for the respective beam ends
I_{ec} = I_e of continuous end

It is seen from Eqs. 8.13 and 8.14 that the controlling moment of inertia for deflection evaluation is the midspan-section effective moment of inertia.

Moment envelopes have to be used to calculate the positive and negative values of I_e. If the continuous beam is subjected to a single heavy concentrated load, only the midspan effective moment of inertia I_e is to be used.

8.7.1 Deflection of T Beams

The most common nonrectangular sections are the flanged T and L beams. The same principles used for deflection computations of rectangular sections can be applied to the nonrectangular ones. The contribution of the compressive resisting force can be obtained using the appropriate concrete area, as explained below.

As in the case of rectangular beams, the contribution of steel to the moment of inertia of the uncracked section is disregarded. The cross section of the beam in Fig. 8.7a is divided into two areas for the purpose of calculating I_g.

$$\text{depth of center of gravity } \bar{y} = \frac{A_1 y_1 + A_2 y_2}{A_1 + A_2} \tag{8.15a}$$

$$y_t = h - \bar{y} \tag{8.15b}$$

The gross moment of inertia, I_g, for the two rectangles is

$$I_g = \frac{bh_f^3}{12} + bh_f\left(\bar{y} - \frac{h_f}{2}\right)^2 + \frac{b_w(h - h_f)^3}{12} + b_w(h - h_f)\left(y_t - \frac{h - h_f}{2}\right)^2 \tag{8.16}$$

For the cracked section, the depth c of the neutral axis is calculated from the horizontal force equilibrium, as in Fig. 8.7b and c. If the depth of neutral axis falls within the flange thickness, the beam behaves as a rectangular section having a width b of the flange and an effective depth d.

When the depth c of neutral axis falls below the flange thickness h_f, the appropriate areas of concrete in the flange and the web of the section and corresponding stresses are applied in the calculation of the compression force. The average stress in the flange area bh_f would be $(f_c + f_{c1})/2$, where f_{c1} is the stress at the bottom of the flange. Using similar triangles yields

$$f_{c1} = f_c \frac{c - h_f}{c} \tag{8.17}$$

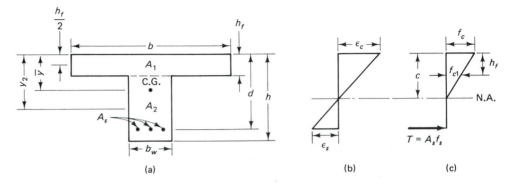

(a)   (b)   (c)

Figure 8.7 Stress and strain distribution across depth of flanged sections: (a) geometry; (b) strains; (c) stresses.

Photo 52 Deflected simply supported beam at failure. (Tests by Nawy et al.)

The average stress in compression for the web area, $b_w(c - h_f)$, would be $f_{c1}/2$ based on the triangular distribution of stress. Hence the force equilibrium equation can be written as

$$A_s f_y = bh_f \frac{f_c + f_{c1}}{2} + b_w(c - h_f)\frac{f_{c1}}{2} \qquad (8.18a)$$

Using Eqs. 8.17 and 8.18a,

$$2A_s E_s \epsilon_s = bh_f E_c \epsilon_c \left(1 + \frac{c - h_f}{c}\right) + b_w(c - h_f)E_c \epsilon_c \frac{c - h_f}{c} \qquad (8.18b)$$

Expressing ϵ_s in terms of ϵ_c and using modular ratio n gives us

$$2nA_s \frac{d - c}{c} = bh_f \frac{2c - h_f}{c} + b_w(c - h_f)\frac{c - h_f}{c} \qquad (8.18c)$$

or

$$b_w(c - h_f)^2 - 2nA_s(d - c) + bh_f(2c - h_f) = 0 \qquad (8.18d)$$

The quadratic equation 8.18d has to be solved to obtain c. Once c is known, the moment of inertia I_{cr} of the cracked section can be calculated using the following expression:

$$I_{cr} = \frac{1}{3} b_w(c - h_f)^3 + \frac{1}{12} bh_f^3 + bh_f \left(c - \frac{h_f}{2}\right)^2 + nA_s(d - c)^2 \qquad (8.19)$$

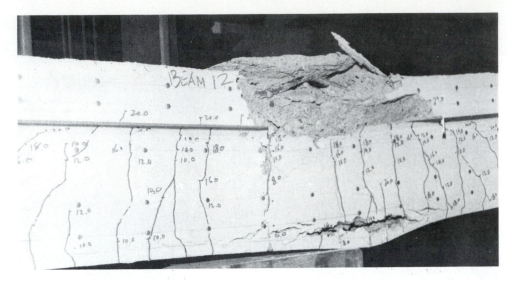

Photo 53 Flexural stabilized cracks at failure.   (Tests by Nawy et al.)

The effective moment of inertia I_e and deflection Δ can be computed as in the case of rectangular sections using Eqs. 8.8a and 8.8b.   In the case of L sections, expressions for I_{cr} such as those of Eq. 8.19 can be developed in a similar manner as for T sections.

8.7.2 Deflection of Beams with Compression Steel

Beams with compression reinforcement can be treated similarly to singly reinforced sections except that the contribution of the compression reinforcement to the stiffness of the beam should be considered because of its high stiffening effect.   For the moment of inertia of the uncracked section, I_g can be used with sufficient accuracy.   The contribution of the compression steel A'_s to the cracked moment of inertia I_{cr} has to be included.   Also, Eq. 8.6c has to be modified for calculating the neutral-axis depth c of the beam.   If the compressive force $A'_s f'_s$ of the steel is added to the compressive force of the concrete, Eq. 8.4a as seen from Fig. 8.8 becomes

$$A_s f_s = bc\frac{f_c}{2} - A'_s f_c \frac{c - d'}{c} + A'_s f'_s \qquad (8.20a)$$

where d' is the effective cover of compression reinforcement.

As in the case of singly reinforced concrete beams (Eqs. 8.4 to 8.6), Eq. 8.20a can be written in the form

$$\frac{bc^2}{2} + [nA_s + (n - 1)A'_s]c - nA_s d - (n - 1)A'_s d' = 0 \qquad (8.20b)$$

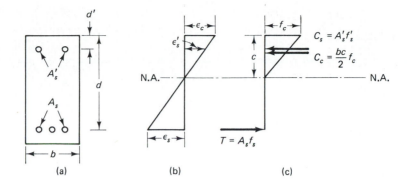

Figure 8.8 Stress and strain distribution at service load in doubly reinforced beam: (a) geometry; (b) strains; (c) stresses.

The moment of inertia I_{cr} of the cracked section can therefore be expressed as

$$I_{cr} = \frac{bc^3}{3} + nA_s(d - c)^2 + (n - 1)A_s'(c - d')^2 \qquad (8.21)$$

The procedure for calculating the effective moment of inertia I_e, and the deflection Δ is the same as in the case of singly reinforced beams.

8.7.3 Deflection Bending Moments in Continuous Beams

The flexural moment envelope has to be constructed for the total continuous beam span in order to evaluate the effective moment of inertia I_e. The usual methods of structural analysis are followed in finding the continuity moments at supports and the positive midspan moments for the various spans. Once these moments are determined, the immediate central postelastic (i.e., postcracking) deflection can be evaluated.

As in the case of simply supported beams, the deflection Δ can be written either in terms of load as in Eq. 8.11 or in terms of moment as in Eq. 8.12. If an interior span AB subjected to a uniform load is isolated as in Fig. 8.9, the midspan deflection Δ_c is

$$\Delta_c = \frac{5l^2}{48EI}[M_m + 0.1(M_a + M_b)] \qquad (8.22)$$

where M_a, M_b = negative service load bending moments
$\quad M_0$ = simple span service load static moment
$\quad M_m$ = midspan moment

Use the correct algebraic sign for the moments in Eq. 8.22, with M_a and M_b due to the *same* loading generally negative. As the exterior span is subjected to the largest positive and negative moments, deflection calculations control for this span in most cases.

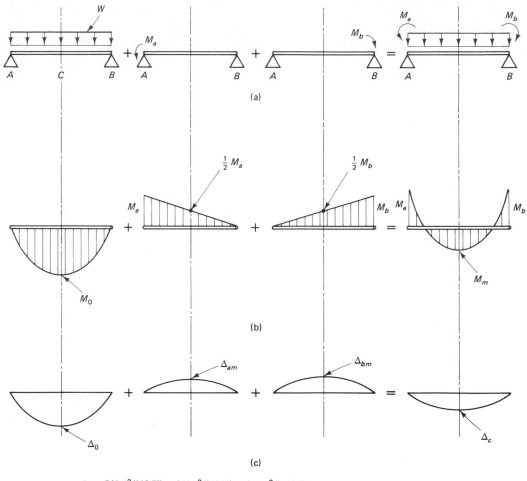

$$\Delta_c = 5M_0 l^2/(48EI) + 3M_a l^2/(48EI) + 3M_b l^2/(48EI)$$

$$= \frac{5l^2}{48EI} [M_m + 0.1 (M_a + M_b)] \text{ where } M_m \text{ is positive and } M_a \text{ and } M_b \text{ generally negative}$$

Figure 8.9 Deflection bending moments in continuous beams: (a) loads; (b) moments; (c) deflections, using superposition.

8.7.4 Example 8.4: Deflection of a Continuous Four-span Beam

A reinforced concrete beam supporting a 4-in. (102-mm) slab is continuous over four equal spans $l = 36$ ft (11 m) as shown in Fig. 8.10. It is subjected to a uniformly distributed load $w_D = 700$ lb/ft (10.22 kN/m), including its self-weight and a service live load $w_L = 1200$ lb/ft (17.52 kN/m). The beam has the dimensions $b = 14$ in. (356 mm), $d = 18.25$ in. (464 mm) at midspan, and a total thickness $h = 21.0$ in. (533 mm). The first interior span is reinforced with four No. 9 bars at

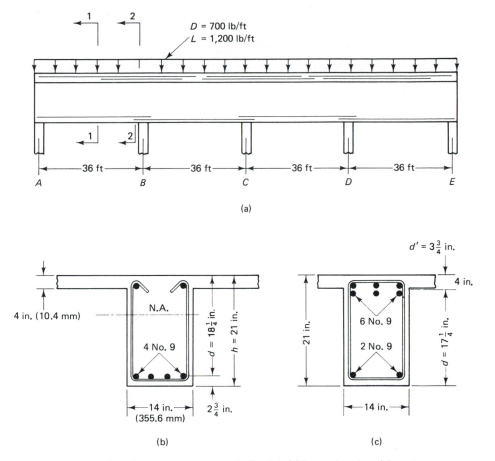

Figure 8.10   Details of continuous beam in Ex. 8.4: (a) beam elevation; (b) section 1–1; (c) section 2–2.

midspan (28.6 mm diameter) at the bottom fibers and six No. 9 bars at the top fibers of the support section.

Calculate the maximum deflection of the continuous beam and determine what code deflection criteria it meets and what limitations, if any, have to be placed on its use.   Given:

$f'_c = 4000$ psi (27.8 MPa), normal-weight concrete

$f_y = 60,000$ psi (414 MPa)

50% of the live load is sustained 36 months on the structure

Solution

Minimum depth requirement

From Table 8.1,

$$\text{minimum } h = \frac{l}{18.5} = \frac{36.0 \times 12}{18.5} = 23.35 \text{ in.}$$

$$\text{actual } h = 21.0 \text{ in.} < 23.35 \text{ in.}$$

Deflection calculations have to be made.

Material properties and bending moment envelope

$$E_c = 57,000 \sqrt{f'_c} = 57,000 \sqrt{4000} = 3.6 \times 10^6 \text{ psi (24,822 MPa)}$$

$$E_s = 29 \times 10^6 \text{ psi (200,000 MPa)}$$

$$\text{modular ratio } n = \frac{E_s}{E_c} = \frac{29 \times 10^6}{3.6 \times 10^6} = 8.1$$

$$\text{modulus of rupture } f_r = 7.5\sqrt{f'_c} = 7.5\sqrt{4000} = 474.3 \text{ psi (3.27 MPa)}$$

From bending moment analysis, the bending moment diagram for the beam is shown in Fig. 8.11.   For deflection, the largest moments are in end spans AB and ED.

$$\text{positive moment} = 0.0772wl^2$$

$$+M_D = 0.0772 \times 700(36.0)^2 \times 12 = 840,430 \text{ in.-lb}$$

$$+M_L = 0.0772 \times 1200(36.0)^2 \times 12 = 1,440,737 \text{ in.-lb}$$

$$+(M_D + M_L) = 0.0772 \times 1900(36.0)^2 \times 12 = 2,281,167 \text{ in.-lb}$$

$$\text{negative moment} = 0.1071wl^2$$

$$-M_D = 0.1071 \times 700(36.0)^2 \times 12 = 1,165,933 \text{ in.-lb}$$

$$-M_L = 0.1071 \times 1200(36.0)^2 \times 12 = 1,998,743 \text{ in.-lb}$$

$$-(M_D + M_L) = 0.1071 \times 1900(36.0)^2 \times 12 = 3,164,676 \text{ in.-lb}$$

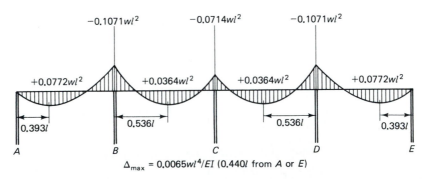

Figure 8.11   Bending moment envelope.

Effective moment of inertia I_e

Figure 8.12 shows the theoretical midspan and support cross sections to be used for calculations of the gross moment of inertia I_g.

1. *Midspan section*

width of T-beam flange $= b_w + 16h_f = 14.0 + 16 \times 4.0 = 78$ in. (1981 mm)

Depth from compression flange to the elastic centroid from Eq. 8.15a:

$$\bar{y} = \frac{A_1 y_1 + A_2 y_2}{A_1 + A_2}$$

$$= \frac{78(4 \times 2) + 14 \times (21 - 4) \times 12.5}{78 \times 4 + 14 \times 17} = 6.54 \text{ in.}$$

$$y_t = h - \bar{y} = 21.0 - 6.54 = 14.46 \text{ in.}$$

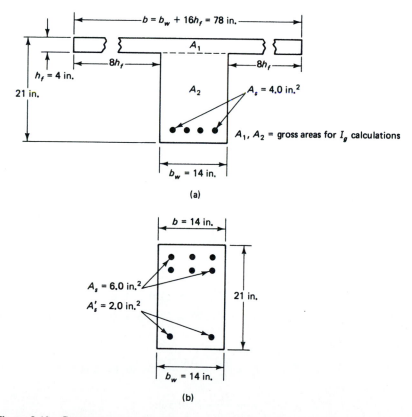

Figure 8.12 Gross moment of inertia I_g cross sections in Ex. 8.4: (a) midspan section; (b) support section.

From Eq. 8.16,

$$I_g = \frac{78(4)^3}{12} + 78 \times 4\left(6.54 - \frac{4}{2}\right)^2 + \frac{14(21 - 4)^3}{12} + 14(21 - 4)\left(14.46 - \frac{21 - 4}{2}\right)^2$$

$$= 21{,}033 \text{ in.}^4$$

$$M_{cr} = \frac{f_r I_g}{y_t} = \frac{474.3 \times 21{,}033}{14.46} = 689{,}900 \text{ in.-lb}$$

Depth of neutral axis

$$A_s = \text{four No. 9 bars} = 4.0 \text{ in.}^2$$

From Eq. 8.18d,

$$14(c - 4.0)^2 - 2 \times 8.1 \times 4.0(18.25 - c) + 78 \times 4(2c - 4.0) = 0$$

or $c^2 + 41.17c - 157.0$ to give $c = 3.5$ in.   Hence the neutral axis is inside the flange and the section is analyzed as a rectangular section.

From Eq. 8.6c, for rectangular sections

$$\frac{78c^2}{2} + 8.1 \times 4 \times c - 8.1 \times 4 \times 18.25 = 0$$

Therefore, $c = 3.5$ in.

$$I_{cr} = \frac{78.0(3.5)^3}{3} + 8.1 \times 4(18.25 - 3.5)^2 = 8163.8 \text{ in.}^4$$

Ratio M_{cr}/M_a

$$D \text{ ratio} = \frac{689{,}900}{840{,}430} = 0.821$$

$$D + 50\% \, L \text{ ratio} = \frac{689{,}900}{840{,}430 + 0.5 \times 1{,}440{,}737} = 0.442$$

$$D + L \text{ ratio} = \frac{689{,}900}{2{,}281{,}167} = 0.302$$

Effective moment of inertia for midspan section

$$I_e = \left(\frac{M_{cr}}{M_a}\right)^3 I_g + \left[1 - \left(\frac{M_{cr}}{M_a}\right)^3\right] I_{cr}$$

I_e for dead load $= 0.5534 \times 21{,}033 + 0.4466 \times 8163.8 = 15{,}286 \text{ in.}^4$

I_e for $D + 0.5L = 0.0864 \times 21{,}033 + 0.9136 \times 8163.8 = 9276 \text{ in.}^4$

I_e for $D + L = 0.0275 \times 21{,}033 + 0.9725 \times 8163.8 = 8518 \text{ in.}^4$

2. *Support section*

$$I_g = \frac{bh^3}{12} = \frac{14(21)^3}{12} = 10{,}804.5 \text{ in.}^4$$

$$y_t = \frac{21.0}{2} = 10.5 \text{ in.}$$

$$M_{cr} = \frac{f_r I_g}{y_t} = \frac{474.3 \times 10{,}804.5}{10.5} = 488{,}055 \text{ in.-lb}$$

Depth of neutral axis

$$A_s = \text{six No. 9} = 6.0 \text{ in.}^2 \text{ (3870 mm}^2\text{)}$$
$$A_s' = \text{two No. 9} = 2.0 \text{ in.}^2 \text{ (1290 mm}^2\text{)}$$
$$d = 21.0 - 3.75 = 17.25 \text{ in. (438.2 mm)}$$

From Eq. 8.20b,

$$\frac{14c^2}{2} + [8.1 \times 6.0 + (8.1 - 1)2.0]c - 8.1 \times 6.0$$
$$\times 17.25 - (8.1 - 1) \times 2.0 \times 3.75 = 0$$

or $c^2 + 8.97c - 125.34 = 0$, to give $c = 7.58$ in.

From Eq. 8.21, the cracking moment of inertia is

$$I_{cr} = \frac{bc^3}{3} + nA_s(d - c)^2 + (n - 1)A_s'(c - d')^2$$
$$= \frac{14(7.58)^3}{3} + 8.1 \times 6.0(17.25 - 7.58)^2 + (8.1 - 1)2.0(7.58 - 3.75)^2$$
$$= 6908.2 \text{ in.}^4$$

Ratio M_{cr}/M_a

$$D \text{ ratio} = \frac{488,055}{1,165,933} = 0.42$$
$$D + 50\% L \text{ ratio} = \frac{488,055}{1,165,933 + 0.5 \times 1,998,743} = 0.225$$
$$D + L = \frac{488,055}{3,164,676} = 0.15$$

Effective moment of inertia for support section

I_e for dead load $= 0.0741 \times 10,804.5 + 0.9259 \times 6908.2 = 7196.9$ in.^{4}

I_e for $D + 0.5L = 0.0122 \times 10,804.5 + 0.9878 \times 6908.2 = 6955.7$ in.^{4}

I_e for $D + L = 0.0034 \times 10,845.5 + 0.9966 \times 6908.2$ in.$^4 = 6921.6$ in.^{4}

Average effective I_e for continuous span

From Eq. 8.14,

$$\text{average } I_e = 0.85I_m + 0.15I_{ec}$$

dead load: $I_e = 0.85 \times 15,286 + 0.15 \times 7196.9 = 14,073$ in.^{4}

$D + 0.5L:$ $I_e = 0.85 \times 9276 + 0.15 \times 6955.7 = 8928$ in.^{4}

$D + L:$ $I_e = 0.85 \times 8518 + 0.15 \times 6921.6 = 8278$ in.^{4}

Short-term deflection

From Table 8.3, the maximum deflection for span AB or DE is

$$\Delta = \frac{0.0065wl^4}{EI}$$

l assumed $\simeq l_n$ for all practical purposes

$$\Delta = \frac{0.0065(36.0 \times 12)^4}{36 \times 10^6} \times \frac{w}{I_e} \times \frac{1}{12} = 5.240 \frac{w}{I_e} \text{ in.}$$

(A more accurate result can be obtained from Eq. 8.22.)

Initial live-load deflection

$$\Delta_L = \Delta_{i,L+D} - \Delta_{i,D}$$

$$\Delta_L = \frac{5.240(1900)}{8278} - \frac{5.240(700)}{14,073} = 1.20 - 0.26 = 0.94 \text{ in.}$$

Initial dead-load deflection

$$\Delta_D = \frac{5.240(700)}{14,073} = 0.26 \text{ in.}$$

Initial 50% sustained live-load deflection

$$\Delta_{LS} = \frac{5.240(1300)}{8928} - \frac{5.240(700)}{14,073} = 0.76 - 0.26 = 0.50 \text{ in.}$$

Long-term deflection

$$\rho' = \frac{A_s'}{bd} = 0 \qquad \text{at midspan in this case}$$

From Eq. 8.9,

$$\text{multiplier } \lambda = \frac{T}{1 + 50\rho'}$$

From Fig. 8.6,

$$T = 1.75 \text{ for 36-month sustained load}$$
$$T = 2.0 \text{ for 5-year loading}$$

Therefore,

$$\lambda_\infty = 2.0 \quad \text{and} \quad \lambda_t = 1.75$$

From Eq. 8.10, the total sustained load deflection is

$$\Delta_{LT} = \Delta_L + \lambda_\infty \Delta_D + \lambda_t \Delta_{LS}$$

or

$$\Delta_{LT} = 0.94 + 2.0 \times 0.26 + 1.75 \times 0.50 = 2.35 \text{ in. (59.6 mm)}$$

Deflection requirements (Table 8.2)

$$\frac{l}{180} = \frac{36 \times 12}{180} = 2.4 \text{ in.} > \Delta_L$$

$$\frac{l}{360} = 1.2 \text{ in.} > \Delta_L$$

$$\frac{l}{480} = 0.9 \text{ in.} < \Delta_{LT}$$

$$\frac{l}{240} = 1.8 \text{ in.} < \Delta_{LT}$$

Hence the continuous beam is limited to floors or roofs not supporting or attached to nonstructural elements such as partitions.

8.8 OPERATIONAL DEFLECTION CALCULATION PROCEDURE AND FLOW CHART

Deflection of structures affects their esthetic appearance as well as their long-term serviceability. The following step-by-step procedure should be followed after the structural member is designed for flexure.

1. Compare the total design depth of the member with the minimum allowable value obtained from Table 8.1. If it is less than the allowable, proceed to perform a detailed calculation of short- and long-term deflection. It is, however, always advisable to perform the detailed calculations regardless of the comparison with Table 8.1.
2. The detailed calculations should establish as a first step:
   (a) The gross moment of inertia I_g
   (b) The cracking moment M_{cr}, which is a function of the modulus of rupture of concrete
3. Calculate the depth c of the neutral axis of the *transformed* section. Find the cracking moment of inertia I_{cr}.
4. Find the effective moment of inertia I_e as follows:

$$I_e = \left(\frac{M_{cr}}{M_a}\right)^3 I_g + \left[1 - \left(\frac{M_{cr}}{M_a}\right)^3\right] I_{cr} \leq I_g$$

or

$$I_e = I_{cr} + \left(\frac{M_{cr}}{M_a}\right)^3 (I_g - I_{cr}) \leq I_g$$

The effective I_e has to be calculated for the following service load-level combinations:
   (a) Dead load (D)
   (b) Dead load + sustained portion of live load ($D + \alpha L$, where α is less than 1.0)
   (c) Dead load + live load ($D + L$)
5. Calculate the immediate deflection based on I_e of the three combinations in step 4, using the elastic deflection expression in Table 8.3. If the beam is continuous over more than two supports, find the average I_e as follows:

   Both ends continuous: average $I_e = 0.70I_m + 0.15(I_{e1} + I_{e2})$

   One end continuous: average $I_e = 0.85I_m + 0.15I_{ec}$

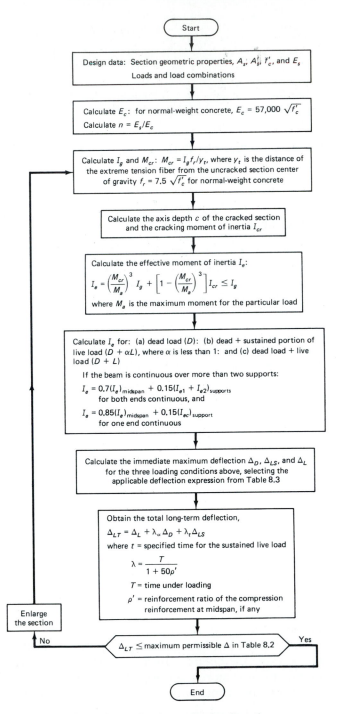

Figure 8.13 Deflection evaluation flow chart.

6. Calculate the long-term deflection, finding first the time-dependent multiplier $\lambda = T/(1 + 50\rho')$ from values in Fig. 8.6.   The total long-term deflection is

$$\Delta_{LT} = \Delta_L = \lambda_\infty \Delta_D + \lambda_t \Delta_{LS}$$

7. If $\Delta_{LT} <$ maximum permissible Δ in Table 8.2, limit the use of the structure to particular loading types or conditions, or enlarge the section.   Figure 8.13 gives a flow chart of the operational sequence of deflection control checks that the designer engineer should use when deflection computations are necessary.

8.9 DEFLECTION CONTROL IN ONE-WAY SLABS

One-way slabs can be treated as rectangular beams of 12-in. (304.8-mm) width. Because floor loads are specified as load intensity per unit area, such intensity on a one-way slab over a 1-ft width becomes pounds per linear foot.   Reinforcement is chosen in terms of bar spacing instead of number of bars, and the area of steel for a 12-in. width of slab can be easily calculated for the total number of bars in a 12-in.-wide strip.

8.9.1 Example 8.5: Deflection Calculations for a Simply Supported One-way Slab

A 5-in.-thick ($h = 127$ mm) one-way slab has a span of 12 ft (3.66 m.).   It is subjected to a live load of 60 psf (2.88 kPa) in addition to its self-weight.   Calculate the immediate and long-term deflections of this slab, assuming that 45% of the live load is sustained over a 24-month period.   Determine what type of elements it should support. Given:

$$f'_c = 3500 \text{ psi (24.1 MPa)}$$
$$f_y = 60,000 \text{ psi (414 MPa)}$$
$$E_s = 29 \times 10^6 \text{ psi (200,000 MPa)}$$

Steel reinforcement: No. 4 bars at 6 in. center-to-center spacing (12.7-mm diameter at 152.4 mm center to center)

Solution

Minimum depth requirement

From Table 8.1,

$$\text{minimum } h = \frac{l}{20} = \frac{12 \times 12}{20} = 7.20 \text{ in.}$$
$$\text{actual } h = 5 \text{ in.} < 7.20 \text{ in.}$$

Deflection calculations have to be made.

Material properties and bending moments

$$E_c = 57,000 \sqrt{f_c'} = 57,000 \sqrt{3500} = 3.37 \times 10^6 \text{ psi (23,256 MPa)}$$

$$E_s = 29 \times 10^6 \text{ psi (200,000 MPa)}$$

$$\text{modular ratio } n = \frac{E_s}{E_c} = \frac{29 \times 10^6}{3.37 \times 10^6} = 8.61$$

$$\text{modulus of rupture } f_r = 7.5 \sqrt{f_c'} = 7.5 \sqrt{3500} = 443.7 \text{ psi}$$

$$\text{gross moment of inertia } I_g = \frac{bh^3}{12} = \frac{12(5.0)^3}{12} = 125.0 \text{ in.}^4$$

$$\text{cracking moment } M_{cr} = \frac{f_r I_g}{y_t} = \frac{443.7 \times 125.0}{2.5} = 22,185 \text{ in.-lb}$$

$$\text{service load bending moment} = \frac{wl_n^2}{8} = \frac{w(12.0)^2}{8} \times 12 \text{ in.-lb} = 216w \text{ in.-lb}$$

Neutral-axis depth of transformed section

If c is the depth from the compression fibers to the neutral axis of the transformed section,

$$A_s = \text{No. 4 at 6 in.} = 0.40 \text{ in.}^2 \text{ per 12-in.-wide strip}$$

$$d = h - 0.75 - \frac{\phi}{2} = 5.0 - 0.75 - 0.25 = 4.0 \text{ in.}$$

From Eq. 8.6(c) for rectangular sections,

$$\frac{bc^2}{2} + nA_s c - nA_s d = 0$$

$$\frac{12c^2}{2} + 8.61 \times 0.40c - 8.61 \times 0.40 \times 4.0 = 0$$

or $c^2 + 0.574c - 2.296 = 0$, giving $c = 1.255$ in.

Effective moment of inertia

Dead load

$$w_D = \text{self-weight of slab} = \frac{5}{12} \times 150 \text{ pcf} = 62.5 \text{ psf}$$

$$M_a = 216w = 216 \times 62.5 = 13,500 \text{ in.-lb} < M_{cr}$$

Hence the slab will not crack under dead load and $I_e = I_g = 125.0$ in.^{4}.
Dead load + 45% live load:

$$M_a = 216(62.5 + 0.45 \times 60) = 19,332 \text{ in.-lb} < M_{cr}$$

Hence the slab will not crack under dead load and 45% sustained live load and $I_e = I_g = 125.0$ in.^{4}.

Dead + live load

$$M_a = 216(62.5 + 60.0) = 26,460 \text{ in.-lb} > M_{cr}$$

This section is cracked.

$$I_{cr} = \frac{bc^3}{3} + nA_s(d - c)^2 \qquad \text{from Eq. 8.7}$$

or

$$I_{cr} = \frac{12(1.255)^3}{3} + 8.61 \times 0.40(4.0 - 1.255)^2 = 33.86 \text{ in.}^4$$

$$\frac{M_{cr}}{M_a} = \frac{22,185}{26,460} = 0.838$$

$$I_e = \left(\frac{M_{cr}}{M_a}\right)^3 I_g + \left[1 - \left(\frac{M_{cr}}{M_a}\right)^3\right]I_{cr} = 0.59 \times 125.0 + 0.41 \times 33.86 = 87.63 \text{ in.}^4$$

Short-term deflection

From Table 8.3,

$$\Delta = \frac{5wl_n^4}{384E_cI_e} = \frac{5w(12.0 \times 12)^4}{384 \times 3.37 \times 10^6 I_e} \times \frac{1}{12} = \frac{0.1384}{I_e} w \text{ in.}$$

Initial live-load deflection

$$\Delta_L = \frac{0.1384(62.5 + 60.0)}{87.63} - \frac{0.1384(62.5)}{125.0} = 0.194 - 0.069 = 0.125 \text{ in. (3.2 mm)}$$

Initial dead-load deflection

$$\Delta_D = \frac{0.1384(62.5)}{125.0} = 0.069 \text{ in. (1.8 mm)}$$

Initial 45% sustained LL deflection

$$\Delta_{LS} = \frac{0.1384(62.5 + 0.45 \times 60)}{125.0} - \frac{0.1384(62.5)}{125.0}$$

$$= 0.099 - 0.069 = 0.030 \text{ in. (0.8 mm)}$$

Long-term deflection

From Eq. 8.9, multiplier $\lambda = T/(1 + 50 \rho')$. From Fig. 8.6, $T = 1.65$ for 24-month sustained load. Therefore,

$$\lambda_\infty = 2.0 \qquad \text{and} \qquad \lambda_t = 1.65$$

From Eq. 8.10, the total sustained load deflection is

$$\Delta_{LT} = \Delta_L + \lambda_\infty\Delta_D + \lambda_t\Delta_{LS}$$

or

$$\Delta_{LT} = 0.125 + 2.0 \times 0.069 + 1.65 \times 0.030 = 0.313 \text{ in. (7.9 mm)}$$

Deflection requirements (Table 8.2)

$$\frac{l}{180} = \frac{12 \times 12}{180} = 0.80 \text{ in.} > \Delta_L$$

$$\frac{l}{360} = 0.40 \text{ in.} > \Delta_L$$

$$\frac{l}{480} = 0.30 \text{ in.} \simeq \Delta_{LT}$$

$$\frac{l}{240} = 0.60 \text{ in.} > \Delta_{LT}$$

Therefore, the slab can support sensitive attached nonstructural elements that are otherwise damaged by large deflections.

It should be noted that actual deflections can vary by as much as 20%–30% depending on several factors, such as concrete constituents and environmental effects. Hence all calculated values should be rounded to the nearest quarter-inch.

8.10 FLEXURAL CRACKING IN BEAMS AND ONE-WAY SLABS

8.10.1 Fundamental Behavior

Concrete cracks at an early stage of its loading history because it is weak in tension. Consequently, it is necessary to study its cracking behavior and control the width of the flexural cracks. Cracking contributes to the corrosion of the reinforcement, surface deterioration, and its long-term detrimental effects.

Increased use of high-strength reinforcing steels having 60,000- to 100,000-psi (413.7- to 551.6-MPa) yield strength and with high stresses occurring at low load levels is becoming prevalent. Also, higher-strength concrete in excess of 9000- to 20,000-psi strength in compression (62.1 to 138 MPa) and optimal utilization of the material in the strength theories of analysis and design are possible today. Hence prediction and control of cracking and crack widths are essential for reliable serviceability performance under long-term loading.

Two types of stresses act on the tensile stretched zones of the concrete surrounding the tension reinforcement shown in Fig. 8.14a. They are longitudinal and lateral sets of stresses. As the longitudinal bending stress acts, the tensile zone undergoes a lateral contraction before cracking, resulting in lateral compression between the concrete and the reinforcing bars or wires. At the moment that a flexural crack starts to develop, this biaxial lateral compression has to disappear at the crack because the longitudinal tension in the concrete becomes zero at the crack location.

The longitudinal bond stress gradually reaches its peak at the crack. This causes the tensile stress f_t in the concrete at that location suddenly to reach its maximum value. The concrete can no longer withstand any tension because of

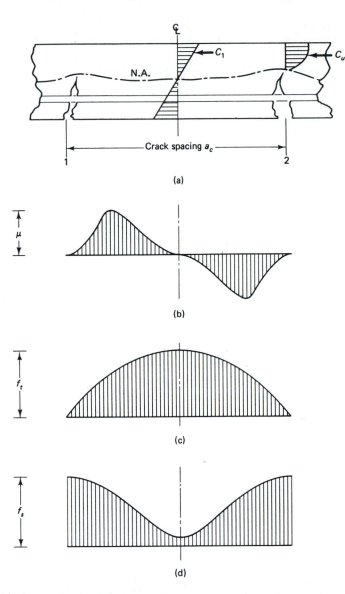

Figure 8.14 Longitudinal stress distribution between two adjacent cracks when cracks are fully developed: (a) crack development geometry; (b) ultimate bond stress μ; (c) longitudinal tensile stress f_t in the concrete; (d) longitudinal tensile stress f_s in the steel.

the high stress concentration at the moment of incipient fracture, and it splits, as seen in Fig. 8.14a.

The stress in the concrete is dynamically transferred to the reinforcing steel (Fig. 8.14d). At the transfer of stress, the tensile stress in the concrete at the cracked section is relieved, becoming zero at the crack (Fig. 8.14c). Laterally, the neutral-axis position rises at the cracked section in order to maintain equilibrium at that section.

The distance a_c between two adjacent cracks is the stabilized crack spacing, that is, the distance between two cracks when they continue to widen under load as principal cracks while other previously formed cracks between them close due to redistribution of stress. In other words, cracks stabilize when no new cracks form in the structural member. A schematic plot of the crack width versus crack spacing is given in Fig. 8.15. It illustrates in the almost horizontal plateau of the diagram the load at which the crack spacing becomes stabilized.

The width of each of the two cracks would be a function of the difference in elongation between the reinforcing bars and the surrounding stretched concrete over a length a_c. From a practical viewpoint, the elongation of the concrete and the shrinkage strain can be neglected as insignificant. Hence

$$\text{crack width } w = \alpha a_c^\beta \epsilon_s^\gamma \tag{8.23}$$

The value of γ partly depends on whether the reinforced concrete member is one- or two-dimensional, while α and β are experimental nonlinearity constants.

It has been proved that a_c varies with $(1/k_1\mu')$, $k_2 f_t'$, and d_b/k_3, ρ_t, where μ is the bond stress, f_t' is the tensile strength of the concrete, d_b is the diameter of the steel bar, $\rho_t = A_s/A_t$ is the ratio of the steel area at the tension side of the section, and A_t is the area of concrete in tension; k_1, k_2, and k_3 are constants.

8.10.2 Crack-width Evaluation

While Eq. 8.23 is the basic mathematical model for the evaluation of the maximum crack width, the large number of variables involved, the randomness of cracking behavior, and the large degree of scatter require extensive idealization and sim-

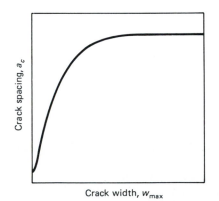

Figure 8.15 Schematic variation of crack width versus crack spacing.

plification.   One simplification based on a statistical study of test data of several investigators is the Gergely–Lutz expression,

$$w_{max} = 0.076\beta f_s \sqrt[3]{d_c A} \tag{8.24}$$

where w_{max} = crack width in units of 0.001 in. (0.0254 mm)

  $\beta = (h - c)/(d - c)$ = depth factor; average value = 1.20

  d_c = thickness of cover to the center of the first layer of bars (in.)

  f_s = maximum stress (ksi) in the steel at service load level with $0.6f_y$ to be used if no computations are available

  A = area of concrete in tension divided by the number of bars (in.²)
  = bt/γ_{bc}, where γ_{bc} is defined as the number of bars at the tension side

It is to be noted that allowance of $f_s = 0.6f_y$ in lieu of actual steel stress computations is applicable only to normal structures.   Special precautions have to be taken for structures exposed to very aggressive climates, such as chemical factories or offshore structures.   Additionally, the depth of the concrete area in tension in reinforced concrete is determined by having the center of gravity of the bars as the centroid of the concrete area in tension.   Hence, for a single layer of bars, the depth t of the concrete area in tension equals $2d_c$.   The shaded area in Fig. 8.16 gives the total concrete area in tension.

8.10.3 Example 8.6: Maximum Crack Width in a Reinforced Concrete Beam

Calculate the maximum crack width for a rectangular simply supported beam that has the cross section shown in Fig. 8.16.   The beam span is 30 ft (9.14 m).   It carries a working uniform load of 1000 lb/ft, including its own weight (14.6 kN/m).   Given:

  f'_c = 5000 psi, normal-weight concrete (34.47 MPa)

  f_y = 60,000 psi (413.7 MPa)

  E_s = 29 × 10⁶ psi (200,000 MPa)

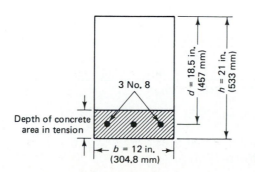

Figure 8.16   Beam geometry.

Solution

Alternative using the actual steel stress

$$\text{gross moment of inertia } I_g = \frac{bh^3}{12} = \frac{12(21)^3}{12} = 9261.0 \text{ in.}^4$$

$$\text{modulus of rupture } f_r = 7.5\sqrt{f_c'} = 7.5\sqrt{5000} = 530.3 \text{ psi (2.66 MPa)}$$

$$\text{cracking moment } M_{cr} = \frac{I_g}{y_t} f_r = \frac{9261.0 \times 530.3}{10.5} = 467,725 \text{ in.-lb}$$

$$\text{maximum beam moment } M_a = \frac{wl_n^2}{8} = \frac{1000(30.0)^2}{8} = 112,500 \text{ lb-ft}$$

$$= 1,350,000 \text{ in.-lb}$$

$$\frac{bc^2}{2} + nA_sc - nA_sd = 0$$

$$A_s = 2.37 \text{ in.}^2 \text{ (1528.7 mm}^2\text{)}$$

$$E_c = 57,000\sqrt{5000} = 4.03 \times 10^6 \text{ psi (27,797 MPa)}$$

$$n = \frac{E_s}{E_c} = \frac{29 \times 10^6}{4.03 \times 10^6} = 7.20$$

$$6c^2 + 7.2 \times 2.37c - 7.2 \times 2.37 \times 18.5 = 0 \qquad \text{to give } c = 5.97 \text{ in. (149.2 mm)}$$

From Eq. 8.7, the cracked moment of inertia is

$$I_{cr} = \frac{bc^3}{3} + nA_s(d - c)^2$$

$$= \frac{12(5.97)^3}{3} + 7.2 \times 2.37(18.5 - 5.97)^2 = 3530.2 \text{ in.}^4$$

$$\text{steel stress } f_s = \frac{M_a}{I_{cr}}(d - c)n$$

$$= \frac{1,350,000}{3530.2}(18.5 - 5.97) \times 7.2$$

$$= 34,500 \text{ psi (238.05 MPa)} < 36,000 \text{ psi} \qquad \text{O.K.}$$

$$\text{steel stress } f_s = 34.5 \text{ ksi} \qquad \text{to be used in Eq. 8.24}$$

$$\beta = \frac{h - c}{d - c} = \frac{21.0 - 5.97}{18.5 - 5.97} = 1.20$$

$$A = \frac{bt}{\text{no. of bars}} = \frac{b(2d_c)}{\gamma_{bc}} = \frac{12.0(2 \times 2.5)}{3} = 20 \text{ in.}^2$$

$$w_{max} = 0.076\beta f_s\sqrt[3]{d_c A} \times 10^{-3}$$

$$= 0.076 \times 1.20 \times 34.5 \sqrt[3]{2.5 \times 20.0} \times 10^{-3}$$

$$= 0.0116 \text{ in. (0.29 mm)}$$

Alternative using $f_s = 0.6f_y$

$$\beta = 1.20 \quad \text{for beams}$$
$$f_s = 0.6f_y = 0.6 \times 60.0 = 36.0 \text{ ksi}$$

$$w_{max} = 0.076 \times 1.20 \times 36.0\sqrt[3]{3.0 \times 2.4} \times 10^{-3} = 0.0137 \text{ in. } (\approx 0.35 \text{ mm})$$

The previous alternative solution is usually unnecessary due to its length and rigor. It is presented to illustrate the computation of the actual value of the stress in the main longitudinal steel at service load levels. Such computations might be necessary for crack-width evaluation where low service stress levels have to be used in such flexural designs as in the case of water-retaining and sanitary engineering structures. The stress $f_s \leq 0.6f_y$ gives a load factor of 1.67 for the limit state at failure.

8.10.4 Crack-width Evaluation for Beams Reinforced with Bundled Bars

The bond stress between the reinforcing bars and the surrounding concrete is a major parameter affecting flexural crack spacing and hence crack width. The contact area of bundled bars is less than that of the isolated bars if they act independently. Using the perimetric reduction factor deduced from Fig. 8.17, the

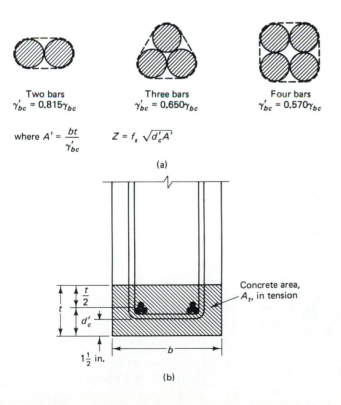

Two bars
$\gamma'_{bc} = 0.815\gamma_{bc}$

Three bars
$\gamma'_{bc} = 0.650\gamma_{bc}$

Four bars
$\gamma'_{bc} = 0.570\gamma_{bc}$

where $A' = \dfrac{bt}{\gamma'_{bc}}$ $Z = f_s \sqrt{d'_c A'}$

(a)

Concrete area, A_t, in tension

(b)

Figure 8.17 Perimetric reduction factors for beams with bundled bars: (a) perimetric factors; (b) section geometry of the concrete area in tension.

cracking equation becomes

$$w_{max} = 0.076\beta f_s \sqrt[3]{d_c' A'} \tag{8.25}$$

where w_{max} is the crack width in units of 0.001 in., and $A' = bt/\gamma_{bc}'$ with the factor for γ_{bc} shown in Fig. 8.17a. d_c' is the depth of cover to the center of gravity of the bundle. The steps for calculation of w_{max} are identical to those for beams reinforced with nonbundled bars.

8.10.5 Example 8.7: Maximum Crack Width in a Beam Reinforced with Bundled Bars

Find the maximum flexural crack width for a reinforced concrete beam that has the cross-sectional geometry shown in Fig. 8.18. Given:

$$f_y = 60,000 \text{ psi}$$
$$f_s = 0.6f_y = 36,000 \text{ psi}$$
$$A_s = \text{two bundles of three No. 8 bars each (25.4-mm diameter)}$$
$$\text{size of stirrups} = \text{No. 4 (12.7-mm diameter)}$$

Solution

$d_c' = $ center of gravity of the three bars from the outer tension fibers

$$= (1.5 + 0.5) + \frac{2 \times 0.5 + 1 \times 1.5}{3} = 2.83 \text{ in.}$$

$t = $ depth of the concrete area in tension

$$= 2 \times 2.83 = 5.66 \text{ in.}$$

$\gamma_{bc} = $ number of bars if all are of the same diameter, or the total steel area divided by the area of the largest bar if more than one size is used

$$= 6 \text{ in this case}$$

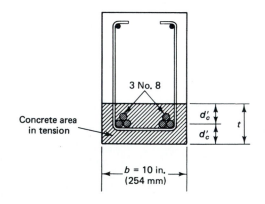

Figure 8.18 Beam geometry.

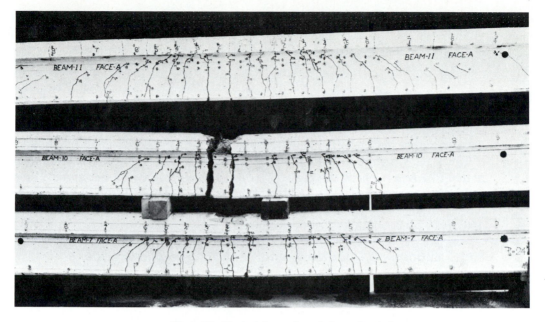

Photo 54 Typical flexural crack formation in beams.

$$\gamma'_{bc} = 0.650\gamma_{bc} = 0.650 \times 6 = 3.9$$

$$A' = \frac{bt}{\gamma'_{bc}} = \frac{10 \times 5.66}{3.9} = 14.51 \text{ in.}^2$$

$$w_{\max} = 0.076 \times 1.20 \times 36.0\sqrt[3]{2.83 \times 14.51} \times 10^{-3} = 0.011 \text{ in. } (0.3 \text{ mm})$$

8.10.6 Z Factor for Crack-control Check in Beams

Crack-control checks are necessary only when tension steel is used with yield strength f_y exceeding 40,000 psi (275.8 MPa).   The ACI Code, in order to reduce the length of calculations, recommends a Z factor where

$$Z = f_s\sqrt[3]{d_c A} \quad \text{kips per inch} \tag{8.26a}$$

for beams reinforced with nonbundled bars.   For beams reinforced with bundled bars:

$$Z = f_s\sqrt[3]{d'_c A'} \quad \text{kips per inch} \tag{8.26b}$$

A Z value not exceeding 175 for interior exposure corresponds to $w_{\max} = 0.016$ in. (40 mm) and 145 for exterior exposure corresponds to $w_{\max} = 0.013$ in. (0.33 mm).   It is to be emphasized that smaller reinforcing bars at closer spacing within the tensile zone give more even distribution of cracks and hence are preferable for crack control.   Also, use of the Z factor should be discouraged since it gives the designer *no* physical indication of the crack-width magnitude.

8.11 TOLERABLE CRACK WIDTHS

The maximum crack width that a structural element should be permitted to develop depends on the particular function of the element and the environmental conditions to which the structure is liable to be subjected to.   Table 8.4 from the ACI Committee 224 report on cracking serves as a reasonable guide on permissible crack widths in concrete structures under the various environmental conditions encountered.   Engineering judgment has to be exercised in the determination of the maximum crack width that can be tolerated.

8.11.1 Example 8.8: Crack Control in a Rectangular Beam

Check whether the beam in Ex. 8.6 satisfies serviceability criteria for crack control for the following three environmental conditions using $f_s = 0.6f_y$: (a) interior exposure using Eq. 8.24; (b) interior exposure using the Z-factor approach of Section 8.10.6; (c) exposure to deicing chemicals.

Solution   (a) The permissible crack width = 0.016 in.   The expected crack width = 0.0137 in. from the solution in Ex. 8.6.   Hence the beam satisfies the crack-control criteria.

(b) $z = f_s \sqrt[3]{d_c A} = 36.0 \sqrt[3]{2.5 \times 20} = 132.63$ kips/in. < 175; O.K.

(c) The maximum tolerable crack width from Table 8.4 is $w_{max} = 0.007$ in. The expected crack width = 0.0137 in. > 0.007 in.   Hence the beam does not satisfy the crack-control criteria, and the reinforcement content at the tension side has to be redesigned and increased.

TABLE 8.4   TOLERABLE CRACK WIDTHS

	Tolerable crack width	
Exposure condition	in.	mm
Dry air or protective membrane	0.016	0.41
Humidity, moist air, soil	0.012	0.30
Deicing chemicals	0.007	0.18
Seawater and seawater spray; wetting and drying	0.006	0.15
Water-retaining structures (excluding nonpressure pipes)	0.004	0.10

8.11.2 Example 8.9: Crack Control in a Beam Reinforced with Bundled Bars

Verify if the beam in Ex. 8.7 satisfies the crack-control criteria for (a) aggressive chemical environment; (b) water-retaining structures; and (c) exterior exposure. Also find the maximum width of the beam based on a Z-factor value $= 175$ kips/in. for interior exposure.

Solution   The expected crack width $= 0.011$ in. (from Ex. 8.7).
(a) The maximum tolerable crack width from Table 8.4 is $w_{max} = 0.007$ in. $<$ 0.011 in.   Hence the beam does not satisfy the crack-control criteria.
(b) The tolerable $w_{max} = 0.004$ in. < 0.011 in.   The beam does not satisfy the crack-control criteria.
(c) The tolerable $w_{max} = 0.013$ in. (ACI code) $>$ expected $w_{max} = 0.011$ in. The beam satisfies the crack-control criteria.
(d) $t = 5.66$ in. from Ex. 8.7:

$$Z = 175 = f_s \sqrt[3]{d_c' A} = 36.0 \sqrt[3]{2.83A}$$

$$A = \left(\frac{175}{36.0}\right)^3 \times \frac{1}{2.83} = 40.6 \text{ in.}^2$$

But $A = bt/\gamma_{bc}'$, where $\gamma_{bc}' = 0.650\gamma_{bc} = 3.9$ from Ex. 8.7, or $40.6 = bt/3.90$.

$$b = \frac{40.6 \times 3.90}{5.66} = 28 \text{ in. (711 mm)}$$

Hence the maximum allowable width of the beam web is 28 in. to satisfy the crack-control criteria for interior exposure.

8.12 SI CONVERSION EXPRESSIONS AND EXAMPLE OF DEFLECTION EVALUATION

1. $E_c = w_c^{1.5} \, 0.043 \sqrt{f_c'}$ MPa, where $w_c = 1500$ to 2500 kg/m³ (90 to 155 lb/ft³). For standard, normal-weight concrete, $w_c = 2400$ kg/m³ to give $E_c = 29,700$ MPa.

2. $E_s = 200,000$ MPa

3. Modulus of rupture $f_r = 0.7 \sqrt{f_c'}$

4. For rectangular sections, $I_g = \dfrac{bh^3}{12}$, and $I_{cr} = \dfrac{bc^3}{3} + nA_s(d - c)^2$, where $n = E_s/E_c$.

5. $M_{cr} = \dfrac{f_r I_g}{y_t}$, where y_t is the distance from the neutral axis to the tensile extreme fibers $= \frac{1}{2}h$ for rectangular sections

6. $I_e = \dfrac{M_{cr}}{M_a} I_g + \left[1 - \left(\dfrac{M_{cr}}{M_a} \right)^3 \right] I_{cr}$

7. Long-term deflection multiplier $\lambda = \dfrac{T}{1 + 50 \, \rho'}$

8.12.1 SI Example on Deflection

Solve Ex. 8.3 using SI units.

Solution

$$
\begin{array}{lll}
f'_c & = 34.5 \text{ MPa, normal weight} & A_s = 852 \text{ mm}^2 \\
f_y & = 414 \text{ MPa} \quad (\text{MPa} = \text{N/mm}^2) & \\
\ell_n & = 8.23 \text{ m} & \\
b & = 254 \text{ mm} & \\
h & = 406 \text{ mm} & \\
d & = 330 \text{ mm} & \\
\text{service } M_D & = 24.3 \text{ kN/m} & \\
M_L & = 28.3 \text{ kN/m} &
\end{array}
$$

Assume 60% live load sustained for 24 months.

$$ E_c = w_c^{1.5} \, 0.043 \, \sqrt{f'_c} \quad (\text{MPa}) $$

where $w_c = 1500$ to 2500 kg/m³ (90 to 155 lb/ft³).   For standard normal-weight concrete, $w_c = 2400$ kg/m³.

$$ E_c = 2400^{1.5} \times 0.043 \, \sqrt{34.5} = 29{,}700 \text{ MPa} $$

$$ E_s = 200{,}000 \text{ MPa} $$

$$ \text{modular ratio } n = \dfrac{E_s}{E_c} = \dfrac{200{,}000}{29{,}700} = 6.7 $$

$$ f_r = 0.7\sqrt{f'_c} = 0.7\sqrt{34.5} = 4.1 \text{ MPa} $$

From Table 8.1,

$$ h_{\min} = \dfrac{\ell_n}{15} = \dfrac{8230}{16} = 520 \text{ mm} > \text{actual } h = 406 \text{ mm} $$

Hence, deflection calculations have to be made.

Effective moment of inertia

$$I_g = \frac{bh^3}{12} = \frac{254(406)^3}{12} = 14.2 \times 10^8 \text{ mm}^4$$

$$y_t = \frac{h}{2} = \frac{406}{2} = 203 \text{ mm}$$

$$M_{cr} = \frac{f_r I_g}{y_t} = \frac{4.1 \times 14.2 \times 10^8}{203} = 28.7 \times 10^6 \text{ N-mm}$$

$$= 28.7 \text{ kN-m}$$

Depth of neutral axis c

$$d = 30 \text{ mm} \qquad A_s = 852 \text{ mm}^2$$

$$\frac{254c^2}{2} = nA_s(d-c)$$

or

$$127c^2 = 6.7 \times 852(330 - c) \qquad \text{to get } c = 102 \text{ mm}$$

$$I_{cr} = \frac{bc^3}{3} + nA_s(d - c)^2$$

$$= \frac{254(102)^3}{3} + 6.7 \times 852(330 - 102)^2$$

$$= 89.8 \times 10^6 + 296.7 \times 10^6 = 3.86 \times 10^8 \text{ mm}^4$$

Dead load

$$M_a = 24.3 \text{ kN-m} \quad \text{(given)}$$

$$\frac{M_{cr}}{M_a} = \frac{28.7}{24.3} = 1.18 > 1.0$$

Use $M_{cr} = M_a$ and $I_e = I_g$ since the dead-load moment is smaller than the cracking moment (the beam will not crack at dead-load level).

Dead load + 60% live load

$$\left(\frac{M_{cr}}{M_a}\right)^3 = \left(\frac{28.7}{24.3 + 0.6 \times 28.3}\right)^3 = 0.30$$

Dead load + live load

$$\left(\frac{M_{cr}}{M_a}\right)^3 = \left(\frac{28.7}{24.3 + 28.3}\right)^3 = 0.16$$

$$I_c = \left(\frac{M_{cr}}{M_a}\right)^3 I_g + \left[1 - \left(\frac{M_{cr}}{M_a}\right)^3\right] I_{cr}$$

Dead load

$$I_e = 14.2 \times 10^8 \text{ mm}^4$$

Dead load + 0.6 live load

$$I_e = 0.3 \times 14.2 \times 10^8 + 0.7 \times 3.86 \times 10^8 = 7.0 \times 10^8 \text{ mm}^4$$

Dead load + live load

$$I_e = 0.16 \times 14.2 \times 10^8 + 0.84 \times 3.86 \times 10^8 = 5.5 \times 10^8 \text{ mm}^4$$

Short-term deflection

$$\Delta = \frac{5\omega\ell^4}{384EI} = \frac{5M\ell_n^2}{48EI}$$

$$= \frac{5(8230)^2M}{48 \times 29{,}700\,I} = 238\frac{M}{I} \text{ mm}$$

Initial live-load deflection

$$\Delta_L = \frac{238(24.3 + 28.3) \times 10^6}{5.5 \times 10^8} - \frac{238(24.3) \times 10^6}{14.2 \times 10^8}$$

$$= 23 - 4 = 19 \text{ mm}, \quad \text{say } 20 \text{ mm } (0.8 \text{ in.})$$

Initial dead-load deflection

$$\Delta_D = \frac{238(24.3) \times 10^6}{14.2 \times 10^8} = 4 \text{ mm}$$

Initial 60% sustained live-load deflection

$$\Delta_{LS} = 238\left[\frac{(24.3 + 0.6 \times 28.3) \times 10^6}{7.0 \times 10^8} - \frac{24.3 \times 10^6}{14.2 \times 10^8}\right]$$

$$= 14 - 4 = 10 \text{ mm}$$

Long-term deflection

From Eq. 8.10,

$$\Delta_{LT} = \Delta_L + \lambda_\infty\Delta_D + \lambda_t\Delta_{LS}$$

$$\lambda = \frac{T}{1 + 50\rho'}$$

where $\rho' = 0$ for singly reinforced beam.

T for 5 years or more $= 2.0$, $\lambda_\infty = 2.0$

T for 24 months $= 1.65$, $\lambda_t = 1.65$

$\Delta_{LT} = 20 + 2.0 \times 4 + 1.65 \times 10 = 45 \text{ mm}$

Deflection requirements (Table 8.2)

$$\frac{\ell_n}{180} = \frac{8230}{180} = 46 \text{ mm} > \Delta_L$$

$$\frac{\ell_n}{360} = \frac{8230}{360} = 23 \text{ mm} > \Delta_L$$

$$\frac{\ell_n}{480} = \frac{8230}{480} = 17 \text{ mm} < \Delta_{LT}$$

$$\frac{\ell_n}{240} = \frac{8230}{240} = 34 \text{ mm} < \Delta_{LT}$$

Hence, the use of the beam is limited to floors or roofs not supporting or attached to nonstructural elements such as partitions.

SELECTED REFERENCES

8.1. ACI Committee 435, "Control of Deflection in Concrete Structures," ACI Committee Report, 1995, 7 pp.

8.2. Branson, D. E., "Design Procedures for Computing Deflections," *Journal of the American Concrete Institute*, Proc. Vol. 65, September 1968, pp. 730–742.

8.3. ACI Committee 435, "Variability of Deflections of Simply Supported Reinforced Concrete Beams," *Journal of the American Concrete Institute*, Proc. Vol. 69, January 1972, pp. 29–35.

8.4. Branson, D. E., *Deformation of Concrete Structures*, McGraw-Hill, New York, 1977.

8.5. Nawy, E. G., "Crack Control in Reinforced Concrete Structures," *Journal of the American Concrete Institute*, Proc. Vol. 65, October 1968, pp. 825–838.

8.6. Gergely, P., and Lutz, L. A., *Maximum Crack Width in Reinforced Concrete Flexural Members*, Special Publication SP-20, American Concrete Institute, Detroit, 1968.

8.7. Nawy, E. G., "Crack Control in Beams Reinforced with Bundled Bars," *Journal of the American Concrete Institute*, Proc. Vol. 69, October 1972, pp. 637–640.

8.8. ACI Committee 224, "Control of Cracking in Concrete Structures," *Journal of the American Concrete Institute*, Proc. Vol. 77, October 1980, pp. 35–76; Proc. Vol. 69, December 1972, pp. 717–753.

8.9. Nawy, E. G., and Blair, K. W., "Further Studies on Flexural Crack Control in Structural Slab Systems"; "Discussion" by ACI Code Committee 318; and "Authors' Closure," *Journal of the American Concrete Institute*, Proc. Vol. 70, January 1973, pp. 61–63.

PROBLEMS FOR SOLUTION

8.1. Calculate I_g and I_{cr} for cross sections (a) through (f) in Fig. 8.19. Given:

$$f_c' = 4000 \text{ psi (27.6 MPa), normal-weight concrete}$$

$$f_y = 60,000 \text{ psi (414 MPa)}$$

$$E_s = 29 \times 10^6 \text{ psi (200,000 MPa)}$$

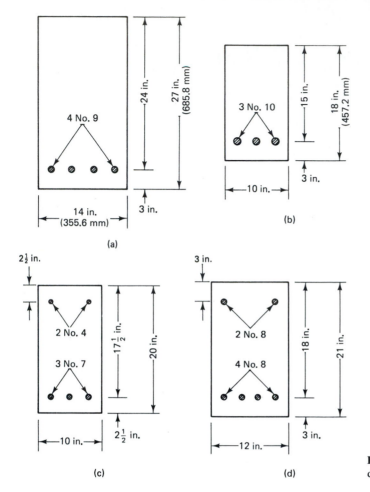

(a)

(b)

(c)

(d)

Figure 8.19 Beam cross sections for deflection calculations.

8.2. Calculate the maximum immediate and long-term deflection for a 6-in.-thick slab on simple supports spanning over 13 ft. The service dead and live loads are 70 psf (33.5 kPa) and 120 psf (57.46 kPa), respectively. The reinforcement consists of No. 5 bars (16-mm diameter) at 6 in. center to center (154.2 mm center to center). Also check which limitations, if any, need to be placed on its usage. Assume that 60% of the live load is sustained over a 30-month period. Given:

$$f'_c = 4500 \text{ psi (31 MPa)}$$

$$f_y = 60{,}000 \text{ psi (414 MPa)}$$

$$E_s = 29 \times 10^6 \text{ psi (200,000 MPa)}$$

8.3. Calculate the deflection due to dead load and dead load plus live load for the following cases in Problem 8.1: cross sections (a), (d), and (e). Use for service-load levels $0.2M_u$ as maximum dead-load moment and $0.35M_u$ as maximum live-load moment. Assume that all beams are simply supported and have a span of 22 ft (6.71 m).

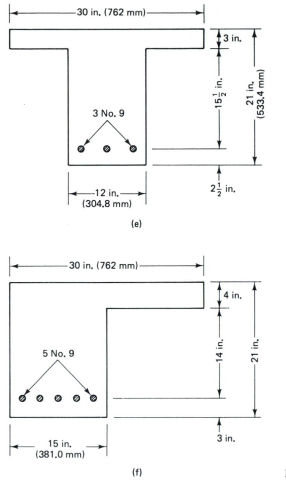

(e)

(f)

Figure 8.19   (*cont.*)

8.4. Repeat Problem 8.2 assuming the slab to be continuous over four supports. The top tension reinforcement at the support consists of No. 5 bars at 4 in. center to center, and the compression reinforcement consists of No. 5 bars at 12-in. centers.

8.5. A beam supporting a 4-in. slab is continuous over four supports. The center-to-center spans are 26 ft with the end span resting on an outer wall. It has a web width $b_w =$ 12 in. and a total thickness $h = 18$ in. and carries a service live load $W_L = 6000$ lb/ft and a service dead load $W_D = 1800$ lb/ft, including its self-weight. The midspan tension reinforcement consists of $A_s =$ four No. 8 bars (28.6 mm), and the support reinforcement is comprised of $A_s =$ six No. 10 bars (32.3 mm) and $A_s' =$ three No. 8 bars. Calculate the maximum immediate and long-term deflections of this beam assuming that 55% of the sustained live load acts over a 24-month period. Also check what deflection serviceability criteria this beam satisfies and whether it can support attached partitions and other elements that can be damaged by large deflections.

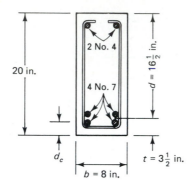

$b = 8$ in.

d_c

$t = 3\frac{1}{2}$ in.

$d = 16\frac{1}{2}$ in.

2 No. 4

4 No. 7

20 in.

Figure 8.20 Beam geometry.

Given:

$$f'_c = 5000 \text{ psi (34.5 MPa)}$$
$$f_y = 60{,}000 \text{ psi (413.7 MPa)}$$
$$E_s = 29 \times 10^6 \text{ psi (200{,}000 MPa)}$$

8.6. Calculate the maximum expected flexural crack width in the beam of Problem 8.5 and verify if it satisfies the serviceability criteria for crack control if it is subjected to (a) interior exposure and (b) freeze–thaw and deicing cycles.

8.7. A rectangular beam under simple bending has the dimensions shown in Fig. 8.20. It is subjected to an aggressive chemical environment. Calculate the maximum expected flexural crack width and whether the beam satisfies the serviceability criteria for crack control. Given:

$$f'_c = 5000 \text{ psi (34.5 MPa)}$$
$$f_y = 60{,}000 \text{ psi (414 MPa)}$$
minimum clear cover $= 1\frac{1}{2}$ in. (38.1 mm)

8.8. Find the Z factor in Problem 8.7 and determine if it satisfies the serviceability criteria for crack control if it is subjected to exterior exposure conditions.

8.9. Find the maximum web of a beam reinforced with bundled bars to satisfy the crack-control criteria for interior exposure conditions. Given:

$$f'_c = 4000 \text{ psi (27.6 MPa)}$$
$$f_y = 60{,}000 \text{ psi (414 MPa)}$$
$$A_s = \text{two bundles of three No. 9 bars each (three bars of 28.6-mm diameter each in a bundle)}$$
No. 4 stirrups used (13-mm diameter)

COMBINED COMPRESSION AND BENDING: COLUMNS

9.1 INTRODUCTION

Columns are vertical compression members of a structural frame intended to support the load-carrying beams. They transmit loads from the upper floors to the lower levels and then to the soil through the foundations. Since columns are compression elements, failure of one column in a critical location can cause the progressive collapse of the adjoining floors and the ultimate total collapse of the entire structure.

Structural column failure is of major significance in terms of economic as well as human loss. Thus extreme care needs to be taken in column design, with a higher reserve strength than in the case of beams and other horizontal structural elements, particularly since compression failure provides little visual warning.

As will be seen in subsequent sections, the ACI Code requires a considerably lower strength reduction factor ϕ in the design of compression members than the ϕ factors in flexure, shear, or torsion. The discussion presented in Chapter 4 on the probability of failure and reliability of performance explains and justifies in

Photo 55 High-strength concrete high-rise building at 535 Madison Avenue, New York. (Courtesy of Construction Industry Board, New York.)

more detail the reasons for the additional reserve strength needed in proportioning compression members.

The principles of stress and strain compatibility used in the analysis (design) of beams discussed in Chapter 5 are equally applicable to columns. A new factor is introduced, however: the addition of an external axial force to the bending moments acting on the critical section. Consequently, an adjustment becomes necessary to the force and moment equilibrium equations developed for beams to account for combined compression and bending.

The amount of reinforcement in the case of beams was controlled so as to have ductile failure behavior. In the case of columns, the axial load will occasionally dominate; hence compression failure behavior in cases of a large axial load/bending moment ratio cannot be avoided.

As the load on a column continues to increase, cracking becomes more intense along the height of the column at the transverse tie locations. At the limit state of failure the concrete cover in tied columns or the shell of concrete outside the spirals of spirally confined columns spalls and the longitudinal bars become exposed. Additional load leads to failure and local buckling of the individual longitudinal bars at the unsupported length between the ties. It is noted that at the limit state of failure the concrete cover to the reinforcement spalls first after the bond is destroyed.

As in the case of beams, the strength of columns is evaluated on the basis of the following principles:

1. A linear strain distribution exists across the thickness of the column.
2. There is no slippage between the concrete and the steel (i.e., the strain in steel and in the adjoining concrete is the same).
3. The maximum allowable concrete strain at failure for the purpose of strength calculations = 0.003 in./in.
4. The tensile resistance of the concrete is negligible and is disregarded in computations.

9.2 TYPES OF COLUMNS

Columns can be classified on the basis of the form and arrangement of reinforcement, the position of the load on the cross section, and the length of the column in relation to its lateral dimensions.

The form and arrangement of the reinforcement identify three types of columns, as shown in Fig. 9.1:

1. Rectangular or square columns reinforced with longitudinal bars and lateral ties (Fig. 9.1a).
2. Circular columns reinforced with longitudinal reinforcement and spiral reinforcement, or lateral ties (Fig. 9.1b).

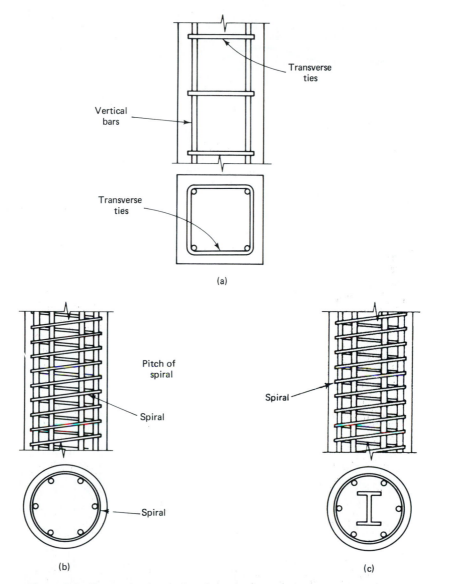

Figure 9.1   Types of columns based on the form and type of reinforcement: (a) tied column; (b) spiral column; (c) composite column.

3. Composite columns where steel structural shapes are encased in concrete. The structural shapes could be placed inside the reinforcement cage, as shown in Fig. 9.1c.

Although tied columns are the most commonly used because of lower construction costs, spirally bound rectangular or circular columns are also used where

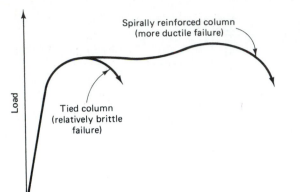

Load

Spirally reinforced column
(more ductile failure)

Tied column
(relatively brittle
failure)

Midheight displacement or deformation

Figure 9.2 Comparison of load–deformation behavior of tied and spirally bound columns.

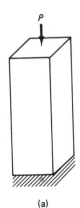

(a)

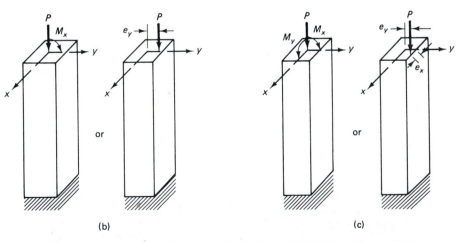

(b)   (c)

Figure 9.3 Types of columns based on the position of the load on the cross section: (a) concentrically loaded column; (b) axial load plus uniaxial moment; (c) axial load plus biaxial moment.

increased ductility is needed, such as in earthquake zones. The ability of the spiral column to sustain the maximum load at excessive deformations prevents the complete collapse of the structure before total redistribution of moments and stresses is complete. Figure 9.2 shows the large increase in ductility (toughness) due to the effect of spiral binding.

Based on the position of the load on the cross section, columns can be classified as concentrically or eccentrically loaded, as shown in Fig. 9.3. Concentrically loaded columns (Fig. 9.3a) carry no moment. In practice, however, all columns have to be designed for some unforeseen or accidental eccentricity due to such causes as imperfections in the vertical alignment of formwork.

Eccentrically loaded columns (Fig. 9.3b and c) are subjected to moment in addition to the axial force. The moment can be converted to a load P and an eccentricity e, as shown in Fig. 9.3b and c. The moment can be uniaxial, as in the case of an exterior column in a multistory building frame or when two adjacent panels are not similarly loaded, such as columns A and B in Fig. 9.4. A column is considered biaxially loaded when bending occurs about both the X and Y axes, such as in the case of corner column C of Fig. 9.4b.

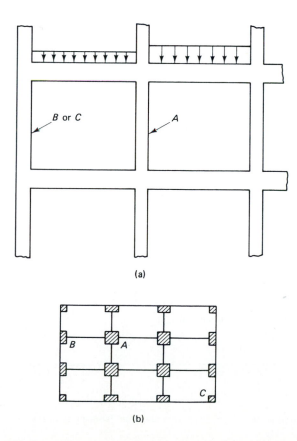

(a)

(b)

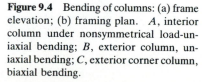

Figure 9.4 Bending of columns: (a) frame elevation; (b) framing plan. A, interior column under nonsymmetrical load-uniaxial bending; B, exterior column, uniaxial bending; C, exterior corner column, biaxial bending.

Failure of columns could occur as a result of material failure by initial yielding of the steel at the tension face or initial crushing of the concrete at the compression face, or by loss of lateral structural stability (i.e., through buckling).

If a column fails due to initial material failure, it is classified as a *short column*. As the length of the column increases, the probability that failure will occur by buckling also increases. Therefore, the transition from the short column (material failure) to the long column (failure due to buckling) is defined by using the ratio of the effective length kl_u to the radius of gyration r. The height, l_u, is the unsupported length of the column, and k is a factor that depends on end conditions of the column and whether it is braced or unbraced. For example, in the case of unbraced columns, if kl_u/r is less than or equal to 22, such a column is classified as a short column, in accordance with the ACI load criteria. Otherwise, it is defined as a long or a slender column. The ratio kl_u/r is called the *slenderness ratio*.

9.3 STRENGTH OF SHORT CONCENTRICALLY LOADED COLUMNS

Consider a column of gross cross-sectional area A_g with width b and total depth h, reinforced with a total area of steel A_{st} on all faces of the column. The net cross-sectional area of the concrete is $A_g - A_{st}$.

Figure 9.5 presents the stress history in the concrete and the steel as the column load is increased. Both the steel and the concrete behave elastically at first. At a strain of approximately 0.002 in./in. to 0.003 in./in., the concrete reaches its maximum strength f_c'. Theoretically, the maximum load that the column can take occurs when the stress in the concrete reaches f_c'. Further increase is possible if strain hardening occurs in the steel at about 0.003-in./in. strain levels.

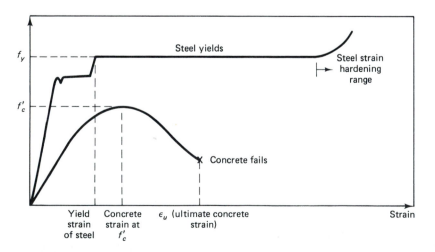

Figure 9.5 Stress–strain behavior of concrete and steel (concentric load).

Therefore, the maximum concentric load capacity of the column can be obtained by adding the contribution of the concrete, which is $(A_g - A_{st})0.85f'_c$, and the contribution of the steel, which is $A_{st}f_y$, where A_g is the total gross area of the concrete section and A_{st} is the total steel area $= A_s + A'_s$. The value of $0.85f'_c$ instead of f'_c is used in the calculation since it is found that the maximum attainable strength in the actual structure approximates $0.85f'_c$. Thus the nominal concentric load capacity, P_0, can be expressed as

$$P_0 = 0.85f'_c(A_g - A_{st}) + A_{st}f_y \qquad (9.1)$$

It should be noted that concentric load causes uniform compression throughout the cross section. Consequently, at failure the strain and stress will be uniform across the cross section, as shown in Fig. 9.6.

It is highly improbable to attain zero eccentricity in actual structures. Eccentricities could easily develop because of factors such as slight inaccuracies in the layout of columns and unsymmetric loading due to the difference in thickness of the slabs in adjacent spans or imperfections in the alignment, as indicated earlier. Hence a minimum eccentricity of 10% of the thickness of the column in the direction perpendicular to its axis of bending is considered as an acceptable assumption for columns with ties and 5% for spirally reinforced columns.

To reduce the calculations necessary for analysis and design for minimum eccentricity, the ACI Code specifies a reduction of 20% in the axial load for tied columns and a 15% reduction for spiral columns. Using these factors, the maximum nominal axial load capacity of columns cannot be taken greater than

$$P_{n(max)} = 0.8[0.85f'_c(A_g - A_{st}) + A_{st}f_y] \qquad (9.2)$$

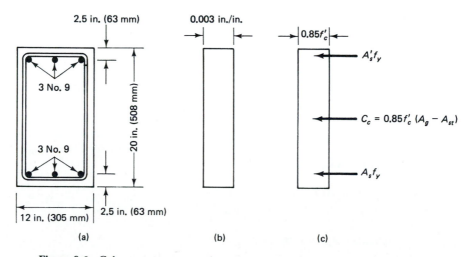

Figure 9.6 Column geometry: strain and stress diagrams (concentric load): (a) cross section; (b) concrete strain; (c) stress (forces).

for tied reinforced columns and

$$P_{n(max)} = 0.85[0.85f_c'(A_g - A_{st}) + A_{st}f_y] \qquad (9.3)$$

for spirally reinforced columns.

These nominal loads should be reduced further using strength reduction factors ϕ, as explained in later sections. Normally, for design purposes, $A_g - A_{st}$ can be assumed to be equal to A_g without great loss in accuracy.

9.3.1 Example 9.1: Analysis of an Axially Loaded Short Rectangular Tied Column

A short tied column is subjected to axial load only. It has the geometry shown in Fig. 9.6a and is reinforced with three No. 9 bars (28.6-mm diameter) on each of the two faces parallel to the x axis of bending. Calculate the maximum nominal axial load strength $P_{n(max)}$. Given:

$$f_c' = 4000 \text{ psi (27.6 MPa)}$$
$$f_y = 60,000 \text{ psi (414 MPa)}$$

Solution $A_s = A_s' = 3$ in.². Therefore, $A_{st} = 6$ in.². Using Eq. 9.2 yields

$$P_{n(max)} = 0.8\{0.85 \times 4000[(12 \times 20) - 6] + 6 \times 60,000\}$$
$$= 924,480 \text{ lb (4110 kN)}$$

If $A_g - A_{st}$ is taken as equal to A_g, it results in

$$P_{n(max)} = 0.8(0.85 \times 4000 \times 12 \times 20 + 6 \times 60,000)$$
$$= 940,800 \text{ lb (4180 kN)}$$

Note from Fig. 9.6b and c that the entire concrete cross section is subjected to a uniform stress of $0.85f_c'$ and a uniform strain of 0.003 in./in.

9.3.2 Example 9.2: Analysis of an Axially Loaded Short Circular Column

A 20-in.-diameter, short, spirally reinforced circular column is symmetrically reinforced with six No. 8 bars, as shown in Fig. 9.7. Calculate the strength $P_{n(max)}$ of this column if subjected to axial load only. Given:

$$f_c' = 4000 \text{ psi (27.6 MPa)}$$
$$f_y = 60,000 \text{ psi (414 MPa)}$$

Solution $A_{st} = 4.74$ in.²

$$A_g = \frac{\pi}{4}(20)^2 = 314 \text{ in.}^2$$

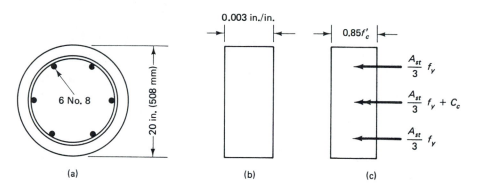

Figure 9.7 Column geometry: strain and stress diagrams (concentric load): (a) cross section; (b) concrete strain; (c) stress (forces).

Using Eq. 9.3 yields

$$P_{n(\max)} = 0.85[0.85 \times 4000(314 - 4.74) + 4.74 \times 60,000]$$
$$= 1,135,501 \text{ lb } (5065 \text{ kN})$$

or, assuming that $A_g - A_{st} \simeq A_g$,

$$P_{n(\max)} = 0.85[0.85 \times 4000 \times 314 + 4.74 \times 60,000]$$
$$= 1,149,200 \text{ lb } (5062 \text{ kN})$$

9.4 STRENGTH OF ECCENTRICALLY LOADED COLUMNS: AXIAL LOAD AND BENDING

9.4.1 Behavior of Eccentrically Loaded Short Columns

The same principles concerning the stress distribution and the equivalent rectangular stress block applied to beams are equally applicable to columns. Figure 9.8 shows a typical rectangular column cross section with strain, stress, and force distribution diagrams. The diagram differs from Fig. 5.13 in the introduction of an additional longitudinal nominal force P_n at the limit failure state acting at an eccentricity e from the plastic (geometric) centroid of the section. The depth of neutral axis primarily determines the strength of the column.

The equilibrium expressions for forces and moments from Fig. 9.8 can be expressed as follows for nonslender columns.

$$\text{nominal axial resisting force } P_n \text{ at failure } = C_c + C_s - T_s \qquad (9.4)$$

Nominal resisting moment M_n, which is equal to $P_n e$, can be obtained by writing the moment equilibrium equation about the plastic centroid. For columns with symmetrical reinforcement, the plastic centroid is the same as the geometric centroid.

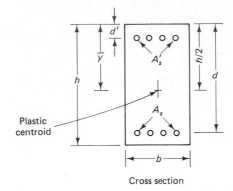

Cross section

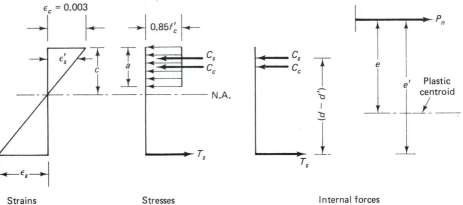

Strains

$$\epsilon_s = 0.003 \frac{d - c}{c}$$

$$\epsilon'_s = 0.003 \frac{c - d'}{c}$$

Stresses

$$f_s = E_s \epsilon_s \leq f_y$$

$$f'_s = E_s \epsilon'_s \leq f_y$$

Internal forces

$$C_c = 0.85 f'_c \, ba$$

$$C_s = A'_s f'_s$$

$$T_s = A_s f_s$$

c = distance to neutral axis

$\bar{y}$ = distance of plastic centroid

e = eccentricity of load to plastic centroid

e' = eccentricity of load to tension steel

d' = effective cover of compression steel

Figure 9.8 Stresses and forces in columns.

$$M_n = P_n e = C_c \left(\bar{y} - \frac{a}{2} \right) + C_s(\bar{y} - d') + T_s(d - \bar{y}) \qquad (9.5)$$

Since

$$C_c = 0.85 f'_c ba$$
$$C_s = A'_s f'_s$$
$$T_s = A_s f_s$$

Eqs. 9.4 and 9.5 can be rewritten as

$$P_n = 0.85f_c'ba + A_s'f_s' - A_sf_s \tag{9.6}$$

$$M_n = P_n e = 0.85f_c'ba\left(\bar{y} - \frac{a}{2}\right) + A_s'f_s'(\bar{y} - d') + A_sf_s(d - \bar{y}) \tag{9.7}$$

In Eqs. 9.6 and 9.7, the depth of the neutral axis c is assumed to be less than the effective depth d of the section, and the steel at the tension face is in actual tension. Such a condition changes if the eccentricity e of the axial force P_n is very small. For such small eccentricities, where the total cross section is in compression, contribution of the tension steel should be added to the contribution of concrete and compression steel. The term A_sf_s in Eqs. 9.6 and 9.7 in such a case would have a positive sign since all the steel is in compression. It is also assumed that $ba - A_s' \approx ba$; that is, the volume of concrete displaced by compression steel is negligible.

If a computer is used for the analysis (design), more refined solutions can be obtained with minimum effort. Hence the area of concrete displaced with the steel is considered in the computer-aided solutions presented in Appendix A. The examples in this chapter include for comparison purposes those more refined results from the computer solutions. It can be observed that the error in neglecting the contribution of the displaced concrete is not significant.

It should be noted that the axial force P_n cannot exceed the maximum axial load strength $P_{n(max)}$, calculated using Eq. 9.2. Depending on the magnitude of the eccentricity e, the compression steel A_s' or the tension steel A_s will reach the yield strength f_y. Stress f_s' reaches f_y when failure occurs by crushing of the concrete. If failure develops by yielding of the tension steel, f_s should be replaced by f_y. When the magnitude of f_s' or f_s is less than f_y, the actual stresses can be calculated using the following equations obtained from similar triangles in the strain distribution across the depth of the section (Fig. 9.8).

$$f_s' = E_s\epsilon_s' = E_s\frac{0.003(c - d')}{c} \leq f_y \tag{9.8}$$

$$f_s = E_s\epsilon_s = E_s\frac{0.003(d - c)}{c} \leq f_y \tag{9.9}$$

9.4.2 Basic Column Equations 9.6 and 9.7 and Trial-and-Adjustment Procedure for Analysis (Design) of Columns

Equations 9.6 and 9.7 determine the nominal axial load P_n that can be safely applied at an eccentricity e for any eccentrically loaded column. If we examine these two expressions, the following unknowns can be identified:

1. Depth of the equivalent stress block, a
2. Stress in compression steel, f_s'
3. Stress in tension steel, f_s
4. P_n for the given e, or vice versa

The stresses f'_s and f_s can be expressed in terms of the depth of neutral axis c as in Eqs. 9.8 and 9.9 and thus in terms of a. The two remaining unknowns, a and P_n, can be solved using Eqs. 9.6 and 9.7. However, combining Eqs. 9.6 to 9.9 leads to a cubical equation in terms of the neutral-axis depth c. We also must check whether the steel stresses are less than the yield strength f_y. Hence the following trial-and-adjustment procedure is suggested for a general case of analysis (design).

For a given section geometry and eccentricity e, assume a value for the distance c down to the neutral axis. This value is a measure of the compression block depth a since $a = \beta_1 c$. Using the assumed value of c, calculate the axial load P_n using Eq. 9.6 and $a = \beta_1 c$. Calculate the stresses f'_s and f_s in compression and tension steel, respectively, using Eqs. 9.8 and 9.9. Also, calculate the eccentricity corresponding to the calculated load P_n using Eq. 9.7. This calculated eccentricity should match the given eccentricity e. If not, repeat the steps until a convergence is accomplished. If the calculated eccentricity is larger than the given eccentricity, this indicates that the assumed value for c and the corresponding depth a of the compression block are less than the actual depth. In such a case, try another cycle, assuming a larger value of c.

This trial-and-adjustment process converges rapidly and becomes exceedingly simpler if a computer program is used, as explained in Appendix A. This discussion pertains to a general case. Simplifying assumptions can be made in most cases to shorten the iteration process.

9.5 MODES OF MATERIAL FAILURE IN COLUMNS

Based on the magnitude of strain in the steel reinforcement at the tension side (Fig. 9.8), the section is subjected to one of two initial conditions of failure as follows:

1. Tension failure by initial yielding of steel at the tension side
2. Compression failure by initial crushing of the concrete at the compression side

The balanced condition occurs when failure develops simultaneously in tension and in compression.

If P_n is the axial load and P_{nb} is the axial load corresponding to the balanced condition, then

$$P_n < P_{nb} \quad \text{tension failure}$$
$$P_n = P_{nb} \quad \text{balanced failure}$$
$$P_n > P_{nb} \quad \text{compression failure}$$

In all these cases the strain–compatibility relationship must be maintained.

9.5.1 Balanced Failure in Rectangular Column Sections

As the eccentricity decreases, a gradual transition takes place from a primary tension failure to a primary compression failure. The balanced failure condition is reached when the tension steel reaches its yield strain ϵ_y at precisely the same load level as the concrete reaches its ultimate strain ϵ_c (0.003 in./in.) and starts crushing.

From similar triangles, an expression for the depth of neutral axis c_b at balanced condition can be written as (Fig. 9.8)

$$\frac{c_b}{d} = \frac{0.003}{0.003 + f_y/E_s} \tag{9.10a}$$

or, using $E_s = 29 \times 10^6$ psi,

$$c_b = d\frac{87,000}{87,000 + f_y} \tag{9.10b}$$

$$a_b = \beta_1 c_b = \beta_1 d\frac{87,000}{87,000 + f_y} \tag{9.11}$$

The axial load corresponding to balanced condition P_{nb} and the corresponding eccentricity e_b can be determined by using this a_b in Eqs. 9.6 and 9.7.

$$P_{nb} = 0.85f_c'ba_b + A_s'f_s' - A_sf_y \tag{9.12}$$

$$M_{nb} = P_{nb}e_b = 0.85f_c'ba_b\left(\bar{y} - \frac{a_b}{2}\right) + A_s'f_s'(\bar{y} - d') + A_sf_y(d - \bar{y}) \tag{9.13}$$

where

$$f_s' = 0.003E_s\frac{c_b - d'}{c_b} \leq f_y \tag{9.14}$$

and $\bar{y}$ is the distance from the compression fibers to the plastic or geometric centroid. Note that, since a_b and f_s' are known, both P_{nb} and e_b can be calculated without going through any trial runs. If $A_s' = A_s$, then $\bar{y} = 0.5h$.

9.5.2 Example 9.3: Analysis of a Column Subjected to Balanced Failure

Calculate the nominal balanced load, P_{nb}, in Ex. 9.1 and the corresponding eccentricity, e_b, for the balanced failure condition if the column shown in Fig. 9.9 is subjected to combined bending and axial load. Given:

$b = 12$ in.

$d = 17.5$ in.

$h = 20$ in.

$d' = 2.5$ in.

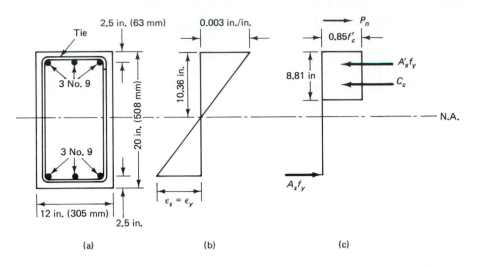

Figure 9.9 Column geometry: strain and stress diagrams (balanced failure): (a) cross section; (b) balanced strains; (c) stress.

$A_s = A'_s = 3.0$ in.^{2}

$f'_c = 4000$ psi

$f_y = 60,000$ psi

Solution   Using Eq. 9.10b gives

$$c_b = 17.5\left(\frac{87,000}{87,000 + 60,000}\right) = 10.36 \text{ in.}$$

$$a_b = \beta_1 c_b = 0.85 \times 10.36 = 8.81 \text{ in.}$$

$$f'_s = 0.003E_s\frac{c_b - d'}{c_b} \leq f_y$$

$$= 0.003(29 \times 10^6)\left(\frac{10.36 - 2.5}{10.36}\right) = 66,006 \text{ psi} > f_y$$

Therefore,

$$f'_s = f_y = 60,000 \text{ psi}$$

Using Eq. 9.12, we have

$$P_{nb} = 0.85 \times 4000 \times 12 \times 8.81 + 3 \times 60,000 - 3 \times 60,000$$

$$= 359,448 \text{ lb}$$

Using Eq. 9.13 and $\bar{y} = h/2 = 10$ in. yields

$$M_{nb} = 0.85 \times 4000 \times 12 \times 8.81\left(10 - \frac{8.81}{2}\right) + 3 \times 60,000(10 - 2.5) + 3$$

$$\times 60,000(17.5 - 10)$$

$$= 4{,}711{,}111 \text{ in.-lb (532 kN-m)}$$

$$e_b = \frac{M_{nb}}{P_{nb}} = \frac{4{,}711{,}111}{359{,}448} = 13.1 \text{ in. (333 mm)}$$

(If displaced concrete is taken into account in the calculation, $P_{nb} = 348{,}986$ lb and $e_b = 13.3$ in.)

9.5.3 Tension Failure in Rectangular Column Sections

The initial limit state of failure in cases of large eccentricity occurs by yielding of steel at the tension side. The transition from compression failure to tension failure takes place at $e = e_b$. If e is larger than e_b or $P_n < P_{nb}$, the failure will be in tension through initial yielding of the tensile reinforcement. Equations 9.6 and 9.7 are applicable in the analysis (design) by substituting the yield strength f_y for the stress f_s in the tension reinforcement. The stress f_s' in the compression reinforcement may or may not be the yield strength, and the actual stress f_s' should be calculated using Eq. 9.8.

Symmetrical reinforcement is usually used such that $A_s' = A_s$ in order to prevent the possible interchange of the compression reinforcement with the tension reinforcement during bar cage placement. Symmetry of reinforcement is also often necessary where the possibility exists of stress reversal due to change in wind direction.

If the compression steel is assumed to have yielded and $A_s = A_s'$, Eqs. 9.6 and 9.7 can be rewritten as

$$P_n = 0.85 f_c' ba \tag{9.15}$$

$$M_n = P_n e = 0.85 f_c' ba \left(\bar{y} - \frac{a}{2} \right) + A_s' f_y \left(\bar{y} - d' \right) + A_s f_y (d - \bar{y}) \tag{9.16a}$$

or

$$M_n = P_n e = 0.85 f_c' ba \left(\frac{h}{2} - \frac{a}{2} \right) + A_s f_y (d - d') \tag{9.16b}$$

In Eq. 9.16b, the geometric centroid is replaced by $h/2$ for symmetrical reinforcement and A_s' is replaced by A_s.

Additionally, Eqs. 9.15 and 9.16b can be combined to obtain a single equation for P_n. Replacing $0.85 f_c' ba$ in Eq. 9.16b by Eq. 9.15 gives

$$P_n e = P_n \left(\frac{h}{2} - \frac{a}{2} \right) + A_s f_y (d - d') \tag{9.16c}$$

Since $a = P_n / 0.85 f_c' b$ from Eq. 9.15,

$$P_n e = P_n \left(\frac{h}{2} - \frac{P_n}{1.7 f_c' b} \right) + A_s f_y (d - d') \tag{9.16d}$$

$$\frac{P_n^2}{1.7 f_c' b} - P_n \left(\frac{h}{2} - e \right) - A_s f_y (d - d') = 0 \tag{9.16e}$$

Photo 56 Eccentrically loaded column at limit state of failure. (Tests by Nawy et al.)

If

$$\rho = \rho' = \frac{A_s}{bd} \tag{9.16f}$$

$$P_n = 0.85f_c'b\left[\left(\frac{h}{2} - e\right) + \sqrt{\left(\frac{h}{2} - e\right)^2 + \frac{2A_sf_y(d - d')}{0.85f_c'b}}\right] \tag{9.17}$$

and if

$$m = \frac{f_y}{0.85f_c'} \tag{9.18}$$

then Eq. 9.16e can be rewritten as

$$P_n = 0.85f_c'bd\left[\frac{h - 2e}{2d} + \sqrt{\left(\frac{h - 2e}{2d}\right)^2 + 2m\rho\left(1 - \frac{d'}{d}\right)}\right] \tag{9.19}$$

Replacing the eccentricity e (distance between the plastic centroid and the load) with e' (distance between the tension steel and the load), Eq. 9.19 can also be rewritten as

$$P_n = 0.85f_c'bd\left[\left(1 - \frac{e'}{d}\right) + \sqrt{\left(1 - \frac{e'}{d}\right)^2 + 2m\rho\left(1 - \frac{d'}{d}\right)}\right] \qquad (9.20)$$

Note that $e' = [e + (d - h/2)]$ in Fig. 9.8 and

$$\frac{h - 2e}{2d} = 1 - \frac{e'}{d}$$

For nonstandard cases where the reinforcement is not symmetrical (i.e., if ρ is not equal to ρ') and if the concrete displaced by compression steel is taken into consideration (i.e., in Eqs. 9.15 and 9.16a), the compressive force contribution of concrete, C_c, is changed from $0.85f_c'ba$ to $0.85f_c'(ba - A_s')$, then Eq. 9.19 changes to

$$P_n = 0.85f_c'bd\left\{\rho'(m - 1) - \rho m + \left(1 - \frac{e'}{d}\right)\right]$$

$$+ \sqrt{\left(1 - \frac{e'}{d}\right)^2 + 2\left[\frac{e'}{d}(\rho m - \rho'm + \rho') + \rho'(m - 1)\left(1 - \frac{d'}{d}\right)\right]}\right\} \qquad (9.21)$$

where e' is the distance between the axial force P_n and the tension steel (or eccentricity to tension steel).

$$\rho = \frac{A_s}{bd} \qquad (9.22a)$$

$$\rho' = \frac{A_s'}{bd} \qquad (9.22b)$$

Equations 9.20 and 9.21 are valid only if the compression steel yields. Otherwise, Eqs. 9.6, 9.7, and 9.8 should be used for obtaining P_n. Examples 9.4 and 9.5 illustrate the design process of a column controlled by initial tension failure.

9.5.4 Example 9.4: Analysis of a Column Controlled by Tension Failure; Stress in Compression Steel Equals Yield Strength

Calculate the nominal axial load strength P_n of the section in Ex. 9.1 (see Fig. 9.10) if the load acts at an eccentricity $e = 14$ in. (356 mm). Given:

$b = 12$ in.
$d = 17.5$ in.

$h = 20$ in.

$d' = 2.5$ in.

$A_s = A_s' = 3$ in.^{2}

$f_c' = 4000$ psi

$f_y = 60,000$ psi

Solution   Using the results of Ex. 9.3, $e_b = 13.1$ in. $< e = 14$ in.   Therefore, failure will occur by initial yielding of tension steel.

$$\rho = \rho' = \frac{A_s}{bd} = \frac{3}{12 \times 17.5} = 0.0143$$

$$m = \frac{60,000}{0.85 \times 4000} = 17.65$$

$$\frac{h - 2e}{2d} \quad \text{or} \quad 1 - \frac{e'}{d} = \frac{20 - 2 \times 14}{2 \times 17.5} = -0.2286$$

$$1 - \frac{d'}{d} = 1 - \frac{2.5}{17.5} = 0.8571$$

Using Eq. 9.19 or 9.20, we have

$$P_n = 0.85 \times 4000 \times 12 \times 17.5$$

$$[-0.2286 + \sqrt{(-0.2286)^2 + 2 \times 17.65 \times 0.0143 \times 0.8571}] = 333,979 \text{ lb}$$

$$a = \frac{P_n}{0.85 f_c' b} = \frac{333,979}{0.85 \times 4000 \times 12} = 8.19 \text{ in.}$$

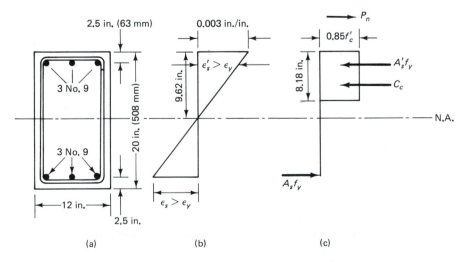

Figure 9.10   Column geometry: strain and stress diagrams (tension failure): (a) cross section; (b) strains; (c) stresses.

$$c = \frac{8.19}{0.85} = 9.63 \text{ in.}$$

$$f_s' = 0.003 \times 29 \times 10^6 \left(\frac{9.63 - 2.5}{9.63} \right)$$

$$= 64,414 \text{ psi} > f_y \qquad \text{therefore,} \ f_s' = f_y \qquad \text{O.K.}$$

$$P_n = 333,979 \text{ lb (1500.47 kN) at } e = 14 \text{ in. (356.6 mm)}$$

(If the area of displaced concrete is taken into account, $P_n = 328,970$ lb.)   If f_s' is less than f_y, a trial-and-adjustment procedure has to be used for the analysis.

It must be emphasized that in each analysis (design) problem the balanced P_{nb}, M_{nb}, and hence e_b have to be evaluated to verify whether the appropriate expressions for tension failure or compression failure are applied in the solution.

9.5.5 Example 9.5: Analysis of a Column Controlled by Tension Failure; Stress in Compression Steel Less Than Yield Strength

A short, rectangular, reinforced concrete column is 12 in. × 15 in. (305 mm × 381 mm), as shown in Fig. 9.11, and is subjected to a load eccentricity $e = 12$ in. (305 mm). Calculate the safe nominal load strength P_n and the nominal moment strength M_n of the column section.   Given:

$f_c' = 4000$ psi (27.6 MPa), normal-weight concrete
$f_y = 60,000$ psi (414 MPa)

eccentricity $e = 12$ in. (305 mm), and three No. 9 bars (28.7-mm diameter) for each of the compression and tension reinforcements

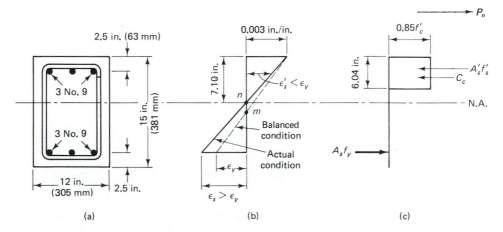

Figure 9.11 Column geometry: strain and stress diagrams (tension failure $f_s' < f_y$): (a) cross section; (b) strains; (c) stresses.

Solution To determine whether the failure is by crushing of concrete or yielding of steel, P_{nb} and e_b have to be calculated first.

$$A_s = A_s' = 3 \text{ in.}^2 \ (19.4 \text{ cm}^2)$$

$$d = 12.5 \text{ in.}$$

$$c_b = \left(\frac{87,000}{87,000 + 60,000}\right) 12.5 = 7.4 \text{ in.}$$

$$a_b = 0.85 \times 7.4 = 6.3 \text{ in.}$$

$$f_s' = 29 \times 10^6 \times 0.003\left(\frac{7.4 - 2.5}{7.4}\right) = 57,609 \text{ psi} < f_y$$

Hence the compression steel did not yield.

$$P_{nb} = 0.85 \times 4000 \times 12 \times 6.3 + 3 \times 57,609 - 3 \times 60,000 = 249,867 \text{ lb}$$

$$M_{nb} = 0.85 \times 4000 \times 12 \times 6.3\left(\frac{15}{2} - \frac{6.3}{2}\right) + 3 \times 57,609\left(\frac{15}{2} - 2.5\right)$$

$$+ 3 \times 60,000\left(12.5 - \frac{15}{2}\right) = 2,882,259 \text{ in.-lb}$$

$$e_b = \frac{M_{nb}}{P_{nb}} = \frac{2,882,259}{249,867} = 11.5 \text{ in.}$$

The specified eccentricity $e = 12$ in. is $> e_b$. Hence failure will occur by initial yielding of steel. Point m on the vertical axis of the strain diagram in Fig. 9.11b denotes the position of the neutral axis at the balanced condition. For this condition, calculations showed that the strain in the compression steel is less than its yield strain. As the neutral-axis position rises to point n in the case of initial failure by yielding of the tension steel, the strain ϵ_s' in the compression steel will be less than that of the balanced condition and hence less than ϵ_y. Therefore, the trial-and-adjustment method should be used for the calculation of P_n. Since $c_b = 7.4$ in., assume a slightly smaller depth to the neutral axis for initial tension failure. Try $c = 7.10$ in.

$$a = 0.85 \times 7.10 = 6.04 \text{ in.}$$

$$f_s' = 29 \times 10^6 \times 0.003\left(\frac{7.10 - 2.5}{7.10}\right) = 56,366 \text{ psi}$$

Since failure is by yielding of tension steel, the stress in the tension steel is equal to the yield strength, or

$$f_s = f_y = 60,000 \text{ psi}$$

$$P_n = 0.85 \times 4000 \times 12 \times 6.04 + 3 \times 56,366 - 3 \times 60,000$$

$$= 235,531 \text{ lb}$$

$$M_n = 0.85 \times 4000 \times 12 \times 6.04(7.5 - 3.02) + 3 \times 56,366 \times 5 + 3$$

$$\times 60,000 \times 5$$

$$= 2,849,508 \text{ in.-lb} \ (322 \text{ kN-m})$$

$$e = \frac{M_n}{P_n} = 12.10 \text{ in.} \ (305 \text{ mm})$$

Therefore,

$$P_n \simeq 236,000 \text{ lb } (1050 \text{ kN})$$

(If displaced concrete is taken into account in the calculations, $P_n = 235,050$ lb.)

9.5.6 Compression Failure in Rectangular Column Sections

For initial crushing of the concrete, the eccentricity e has to be less than the balanced eccentricity e_b and the stress in the tensile reinforcement below yield, that is, $f_s < f_y$.

The analysis (design) process necessitates applying the basic equilibrium equations 9.6 and 9.7, using the trial-and-adjustment procedure and ensuring strain-compatibility checks at all stages. The procedure is summarized in Section 9.4.2, and the following example illustrates its use in the analysis and design of reinforced concrete columns.

9.5.7 Example 9.6: Analysis of a Column Controlled by Compression Failure; Trial-and-Adjustment Procedure

Calculate the nominal load P_n of the section in Ex. 9.1 (see Fig. 9.12) if the column is subjected to a load eccentricity $e = 10$ in. (254 mm). Given:

$b = 12$ in. (305 mm)
$d = 17.5$ in. (445 mm)
$h = 20$ in. (508 mm)
$d' = 2.5$ in.
$A_s = A'_s = 3.0$ in.^{2} (1940 mm^2)
$f'_c = 4000$ psi (27.6 MPa)
$f_y = 60,000$ psi (414 MPa)

Solution Using the results of Ex. 9.3, eccentricity for balanced failure $e_b = 13.1$ in., which is larger than the given eccentricity of 10 in. Therefore, failure will occur by initial crushing of concrete at the compression face.

Trial 1

Assume that

$$c = 11.0 \text{ in. } (279 \text{ mm})$$
$$a = \beta_1 c = 0.85 \times 11.0 = 9.35 \text{ in.}$$

Using Eq. 9.8,

$$f'_s = 29 \times 10^6 \times 0.003 \frac{11.0 - 2.5}{11.0} = 67,227 \text{ psi} > f_y$$

Therefore,

$$f'_s = f_y = 60,000 \text{ psi}$$

Photo 57 Compression side of eccentrically loaded column at failure. (Tests by Nawy et al.)

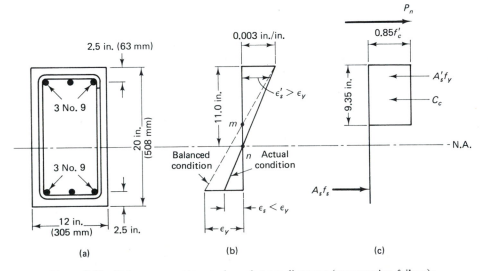

Figure 9.12 Column geometry: strain and stress diagrams (compression failure): (a) cross section; (b) strains; (c) stresses.

Using Eq. 9.9,

$$f_s = 29 \times 10^6 \times 0.003 \left(\frac{17.5 - 11.0}{11.0} \right) = 51,409 \text{ psi}$$

Using Eq. 9.6,

$$P_n = 0.85 \times 4000 \times 12 \times 9.35 + 3 \times 60,000 - 3 \times 51,409$$
$$= 407,253 \text{ lb}$$

Using Eq. 9.7,

$$M_n = 0.85 \times 4000 \times 12 \times 9.35 \left(10 - \frac{9.35}{2} \right) + 3 \times 60,000(10 - 2.5)$$
$$+ 3 \times 51,409(17.5 - 10) = 4,538,083 \text{ in.-lb}$$
$$e = \frac{M_n}{P_n} = 11.14 \text{ in.} > 10 \text{ in.}$$

Therefore, an axial load of 407,253 lb can be applied at an eccentricity of 11.14 in.

Trial 2

Assume that $c = 11.5$ in.; hence $a = 0.85 \times 11.5 = 9.77$ in.

$$f'_s = 60,000 \text{ psi}$$
$$f_s = 45,391 \text{ psi}$$
$$P_n = 442,647 \text{ lb}$$
$$M_n = 4,410,218 \text{ in.-lb}$$
$$e = 9.96 \text{ in.} \approx \text{given eccentricity of 10 in.}$$

Therefore, for $e = 10$ in. (254 mm), P_n can be assumed to be 442,647 lb. (If displaced concrete is taken into account in the calculations, $P_n = 432,445$ lb.)

9.5.8 General Case of Columns Reinforced on All Faces: Exact Solution

When columns are reinforced with bars on all faces and those where the reinforcement in the parallel faces is nonsymmetrical, solutions have to be based on using first principles. Equations 9.6 and 9.7 have to be adjusted for this purpose and the trial-and-adjustment procedure adhered to. Strain-compatibility checks for strain in each reinforcing bar layer have to be performed at all load levels.

Figure 9.13 illustrates the case of a column reinforced on all four faces. Assume that

$$G_{sc} = \text{center of gravity of the steel compressing force}$$
$$G_{st} = \text{center of gravity of the steel tensile force}$$
$$F_{sc} = \text{resultant steel compressive force} = \Sigma A'_s f_{sc}$$
$$F_{st} = \text{resultant steel tensile force} = \Sigma A_s f_{st}$$

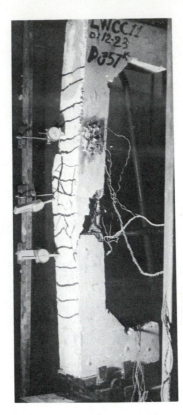

Photo 58   Tension side of eccentrically loaded column at rupture and cover spalling.   (Tests by Nawy et al.)

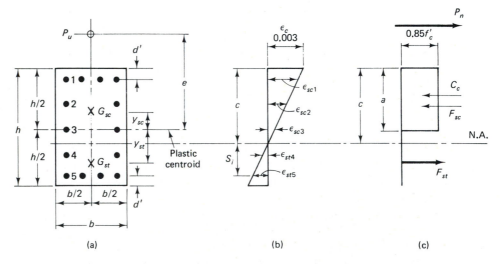

Figure 9.13   Column reinforced with steel on all faces: (a) cross section; (b) strain; (c) forces.

Equilibrium of the internal and external forces and moments requires that

$$P_n = 0.85f_c'b \ \beta_1 c + F_{sc} - F_{st} \tag{9.23}$$

$$P_n e = 0.85f_c'b \ \beta_1 c \left(\frac{h}{2} - \frac{1}{2} \beta_1 c \right) + F_{sc}y_{sc} + F_{st}y_{st} \tag{9.24}$$

Trial and adjustment is applied assuming a neutral-axis depth c and consequently a depth a of the equivalent rectangular block. The strain values in each bar layer are determined by the linear strain distribution in Fig. 9.13b to ensure strain compatibility. The stress in each reinforcing bar is obtained using the expression

$$f_{si} = E_s \epsilon_{si} = E_s \epsilon_c \frac{s_i}{c} = 87,000 \frac{s_i}{c} \tag{9.25}$$

where f_{si} has to be $\le f_y$.

Find P_n corresponding to the assumed c in Eq. 9.23. Substitute into Eq. 9.24 the P_n value thus obtained with the parameter c as the unknown. If the resulting c is not close to the assumed value, proceed to another trial. The nominal resisting load P_n of the section would be the one corresponding to the trial depth c of the last trial cycle.

It is advisable in many instances also to add steel to the column faces that are perpendicular to the axis of bending such that their area does not exceed 25% of the area of the main steel.

9.6 WHITNEY'S APPROXIMATE SOLUTION IN LIEU OF EXACT SOLUTIONS

Empirical expressions proposed by Whitney can be used for rapidity in lieu of the trial-and-adjustment method, although with some loss in accuracy.

9.6.1 Rectangular Concrete Columns

These expressions are presented particularly for circular columns, since longhand trial-and-adjustment procedures for their analysis or design can be time consuming. Strain-compatibility checks for the reinforcement require that the strain in the bars be evaluated at each level across the depth of the section; hence the use of an empirical one-step expression becomes useful for a quick analysis. The use of hand-held computers and personal computers reduces or preempts the need for the use of the Whitney empirical approach in the case of designers familiar with computer use. Appendix A applies the exact method of analysis and design of circular and rectangular columns with strain-compatibility checks at all stages through the use of personal computer programs developed for this purpose.

Whitney's solution is based on the following assumptions.

1. Reinforcement is symmetrically placed in single layers parallel to the axis of bending in rectangular sections.
2. Compression steel has yielded.
3. Concrete displaced by the compression steel is negligible compared to the total concrete area in compression; hence no correction is made for the concrete displaced by the compression steel.
4. For the purpose of calculating the contribution C_c of the concrete, the depth of the stress block is assumed to be $0.54d$, corresponding to an average value of a for balanced conditions in rectangular sections.
5. The interaction curve in the compression zone is a straight line, as shown later in Fig. 9.17.

For most cases, Whitney's method leads to a conservative solution except when the factored load P_u has a value higher than the balanced load P_{ub}, as in Ex. 9.7(b), and the external eccentricity e is very small. Otherwise, the method leads to a nonconservative solution as illustrated in Ex. 9.7(b) and as seen in the shaded area *BST* of Fig. 9.17.

If compression controls, the equation can be written as

$$P_n = \frac{A'_s f_y}{[e/(d - d')] + 0.5} + \frac{bhf'_c}{(3he/d^2) + 1.18} \tag{9.26}$$

The following example illustrates the use of this equation.

9.6.2 Example 9.7: Analysis of a Column Controlled by Compression Failure; Whitney's Equation

Calculate the nominal strength load P_n for the section in Ex. 9.6 using Whitney's equation if the load eccentricity is (a) $e = 6$ in. (152.4 mm); (b) $e = 10$ in. (254 mm).

Solution   (a) $e = 6$ in.

$$P_n = \frac{3 \times 60{,}000}{[6/(17.5 - 2.5)] + 0.5} + \frac{12 \times 20 \times 4000}{[(3 \times 20 \times 6)/17.5^2] + 1.18}$$

$$= 607{,}555 \text{ lb } (2734.0 \text{ kN})$$

Exact solution, using trial and adjustment and including the displaced concrete, gives $P_n = 608{,}458$ lb (2738.0 kN). The approximate solution is conservative.

(b) $e = 10$ in. Using Eq. 9.26,

$$P_n = \frac{3 \times 60{,}000}{[10/(17.5 - 2.5)] + 0.5} + \frac{12 \times 20 \times 4000}{[(3 \times 20 \times 10)/17.5^2] + 1.18}$$

$$= 460{,}098 \text{ lb } (2070.4 \text{ kN})$$

The exact solution, using the trial-and-adjustment procedure and including the effect of the displaced concrete, gives $P_n = 433{,}138$ lb (1960 kN), showing that the approximate solution is not always conservative, as discussed above.

9.6.3 Circular Concrete Columns

As in the case of rectangular columns, force and moment equilibrium equations can be used to solve for the unknown nominal axial load P_n for any given eccentricity. The equilibrium equations are similar to Eqs. 9.6 and 9.7 except that (1) the shape of the area under compressive stress will be a segment of a circle, and (2) reinforcing bars are not grouped together parallel to the compression and tension sides. Therefore, the force and stress in each bar should be considered separately. The area and the center of gravity of the segment of a circle in compression should be calculated using the appropriate mathematical expressions. This accurate approach can be easily adopted if we choose to use hand-held or desktop computers. The following simplified empirical Whitney's approach can be used for longhand calculations.

9.6.4 Empirical Method of Analysis of Circular Columns

Transform the circular column to an idealized equivalent rectangular column as shown in Ex. 9.8 and Fig. 9.14. For compression failure, the equivalent rectangular column would have (1) the thickness in the direction of bending equal to $0.8h$, where h is the outside diameter of the circular column (Fig. 9.14b); (2) the width of the idealized rectangular column to be obtained from the same gross area A_g of the circular column such that $b = A_g/0.8h$; and (3) the total area of reinforcement A_{st} to be equally divided in two parallel layers and placed at a distance of $2D_s/3$ in the direction of bending, where D_s is the diameter of the cage measured center to center of the outer vertical bars. For tension failure, use the actual column for evaluating C_c, but place 40% of the steel A_{st} in parallel at a distance of $0.75D_s$, as shown in Fig. 9.14. The equivalent column method provides satisfactory results for most cases.

Once the dimensions of the equivalent rectangular column are established, the analysis (design) can be made as for rectangular columns. The equations for tension and compression failure can also be expressed in terms of the dimensions of the circular column, as follows:

For tension failure,

$$P_n = 0.85f'_ch^2\left[\sqrt{\left(\frac{0.85e}{h} - 0.38\right)^2 + \frac{\rho_g m D_s}{2.5h}} - \left(\frac{0.85e}{h} - 0.38\right)\right] \qquad (9.27)$$

For compression failure,

$$P_n = \frac{A_{st}f_y}{(3e/D_s) + 1.00} + \frac{A_gf'_c}{[9.6he/(0.8h + 0.67D_s)^2] + 1.18} \qquad (9.28)$$

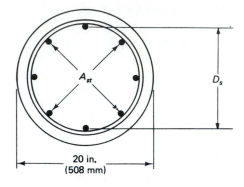

(a)

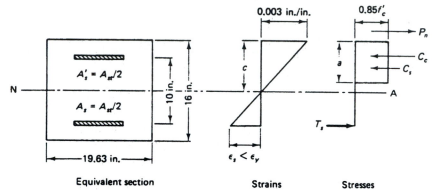

(b)

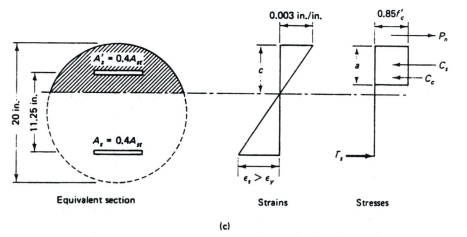

(c)

Figure 9.14 Equivalent column section: (a) given circular section (A_{st}, total reinforcement area); (b) equivalent rectangular section (compression failure); (c) equivalent column (tension failure).

where h = diameter of section

D_s = diameter of the reinforcement cage center to center of the outer vertical bars

e = eccentricity to the plastic centroid of section

$$\rho_g = \frac{A_{st}}{A_g} = \frac{\text{gross steel area}}{\text{gross concrete area}}$$

$$m = \frac{f_y}{0.85f'_c}$$

9.6.5 Example 9.8: Calculation of Equivalent Rectangular Cross Section for a Circular Column

Obtain an equivalent rectangular cross section for the circular column shown in Fig. 9.14a.   Assume that D_s = 15 in.

Solution   thickness of the rectangular section = $0.8 \times 20 = 16$ in.

$$\text{width of the rectangular section} = \frac{\pi}{4}\frac{(20)^2}{16} = 19.63 \text{ in.}$$

$$d - d' = \frac{2}{3} \times 15 = 10 \text{ in.}$$

$$A_s = A'_s = \frac{A_{st}}{2}$$

9.6.6 Example 9.9: Analysis of a Circular Column

A circular column 20 in. (508 mm) in diameter is reinforced with six No. 8 equally spaced bars.   Calculate (a) the load and eccentricity for the balanced failure condition; (b) the load P_n for e = 16.0 in. (406 mm); (c) the load P_n for e = 5.0 in. (127 mm). Assume the column to be nonslender (short) and spirally reinforced.   Given:

f'_c = 4000 psi (27.6 MPa)
f_y = 60,000 psi (414 MPa)

Solution   For case (a), an equivalent rectangular column shown in Fig. 9.15 is used for the analysis.   Using the results of Ex. 9.8 for the equivalent column, h = 16 in., d = 13 in., d' = 3 in., b = 19.63 in., and $A'_s = A_s = 3 \times 0.79 = 2.37$ in.^{2}.

(a) *Balanced failure*

$$c_b = 13\frac{87,000}{87,000 + 60,000} = 7.69 \text{ in.}$$

$$a_b = 0.85 \times 7.69 = 6.54 \text{ in.}$$

$$f'_s = 0.003 \times 29 \times 10^6\frac{7.69 - 3}{7.69} = 53,060 \text{ psi}$$

$$f_y = 60,000 \text{ psi}$$

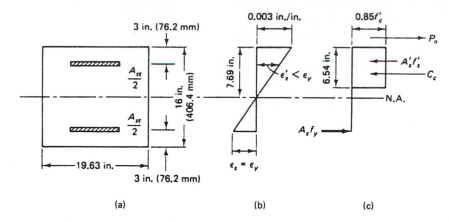

Figure 9.15 Column geometry: strain and stress diagrams (balanced failure): (a) equivalent section; (b) strains; (c) stresses.

$$P_{nb} = 0.85 \times 4000 \times 19.63 \times 6.54 + 2.37 \times 53,060 - 2.37 \times 60,000$$
$$= 420,045 \text{ lb}$$

If the actual circular cross section is used in the analysis instead of the equivalent section and an exact strain-compatibility analysis is made, $P_{nb} = 454,330$ lb and $e_b = 7.18$ in.

$$M_{nb} = 0.85 \times 4000 \times 19.63 \times 6.54 \left(8.0 - \frac{6.54}{2}\right) + 2.37 \times 53,060$$
$$\times 5 + 2.37 \times 60,000 \times 5$$
$$= 3,404,371 \text{ in.-lb}$$
$$e_b = \frac{M_{nb}}{P_{nb}} = \frac{3,404,371}{420,045} = 8.10 \text{ in. (205.9 mm)}$$

For cases (b) and (c), Whitney's formulas involving the actual dimensions of the circular column can be used directly.

(b) *Large eccentricity:* For $e = 16$ in. $> e_b$, tension failure controls. Assuming an effective cover of 2.5 in. to the center of the longitudinal reinforcement,

$$D_s = 20 - 2 \times 2.5 = 15 \text{ in.}$$

$$\rho_g = \frac{2 \times 2.37}{314} = 0.015$$

$$m = \frac{60,000}{0.85 \times 4000} = 17.65$$

Using Eq. 9.27 gives

$$P_n = 0.85 \times 4000 \times 400 \left[\sqrt{\left(\frac{0.85 \times 16}{20} - 0.38\right)^2 + \frac{0.015 \times 17.65 \times 15}{2.5 \times 20}} \right.$$
$$\left. - \left(\frac{0.85 \times 16}{20} - 0.38\right)\right]$$

$$= 151{,}793 \text{ lb } (675 \text{ kN})$$
$$\phi P_n = 0.75 \times 151{,}793 = 113{,}845 \text{ lb}$$

[Using strain compatibility, $P_n = 173{,}940$ lb.]

   (c) *Small eccentricity:* For $e = 5.0$ in. $< e_b$, compression failure controls.   Using Eq. 9.28, we have

$$\text{total steel area } A_{st} = A_s + A_s' = 2 \times 2.37 = 4.74 \text{ in.}^2 \, (3057.3 \text{ mm}^2)$$

$$\text{gross concrete area } A_g = \frac{\pi(20.0)^2}{4} = 314.2 \text{ in.}^2 \, (2025.3 \text{ mm}^2)$$

$$P_n = \frac{4.74 \times 60{,}000}{\dfrac{3 \times 5.0}{15} + 1} + \frac{314.2 \times 4000}{\dfrac{9.6 \times 20 \times 5}{(0.8 \times 20 + 0.67 \times 15)^2} + 1.18}$$

$$= 626{,}577 \text{ lb } (2780 \text{ kN})$$

(Using strain compatibility $P_n = 621{,}653$ lb, indicating that the Whitney solution is in this case not conservative.)

9.7 COLUMN STRENGTH REDUCTION FACTOR ϕ

For members subject to flexure and relatively small axial loads, failure is initiated by yielding of the tension reinforcement and takes place in an increasingly ductile manner.   Hence for small axial loads it is reasonable to permit an increase in the ϕ factor from that required for pure compression members.   When the axial load vanishes, the member is subjected to pure flexure, and the strength reduction factor ϕ becomes 0.90.   Figure 9.16a and b shows the zone in which the value of ϕ can be increased from 0.7 to 0.9 for tied columns and 0.75 to 0.9 for spiral columns. As the factored design compression load ϕP_n decreases beyond $0.1A_g f_c'$, in Fig. 9.16a, the ϕ factor is increased from 0.7 to 0.9 for tied columns and 0.75 to 0.9 for spiral columns.   For those cases where the value of P_{nb} is less than $0.1A_g f_c'$, ϕ values are increased when the load $P_u < P_{ub}$ or $\phi P_n < \phi P_{nb}$, as in Fig. 9.16b.

   The value of $0.10 f_c' A_g$ is chosen by the ACI Code as the design axial load value ϕP_n below which the ϕ factor could safely be increased for most compression members.   In summary, if initial failure is in compression, the strength reduction factor ϕ is always 0.70 for tied columns and 0.75 for spirally reinforced columns.

   The following expressions give variations in the value of ϕ for symmetrically reinforced compression members.   The columns should have an effective depth not less than 70% of the total depth and the steel reinforcement a yield strength not exceeding 60,000 psi if the calculations resulted in a higher value.   For tied columns,

$$\phi = 0.90 - \frac{0.20\phi P_n}{0.1 f_c' A_g} \geq 0.70 \qquad (9.29)$$

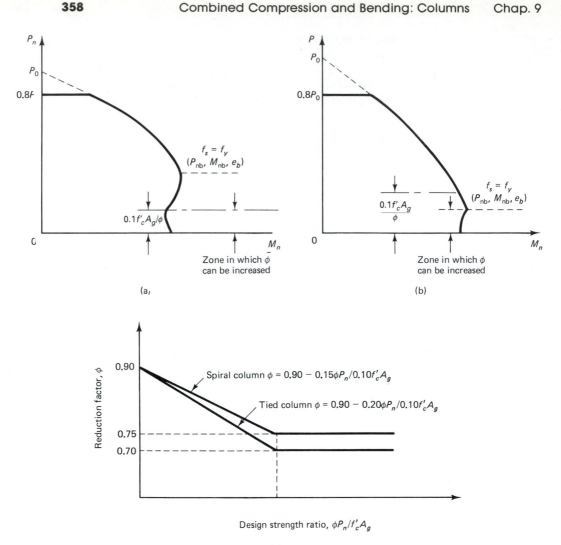

Figure 9.16   Controlling zones for modification of reduction factor ϕ in columns: (a) $0.1f'_c A_g$ $< \phi P_{nb}$; (b) $0.1f'_c A_g > \phi P_{nb}$; (c) variation of ϕ for symmetrically reinforced compression members (P_n is the nominal axial strength at the given eccentricity).

For spirally reinforced columns,

$$\phi = 0.90 - \frac{0.15\phi P_n}{0.1f'_c A_g} \geq 0.75 \tag{9.30}$$

where

$$P_u = \phi P_n \tag{9.31}$$

In both Eqs. 9.29 and 9.30, if ϕP_{nb} is less than $0.1f'_c A_g$, then ϕP_{nb} should be substituted for $0.1f'_c A_g$ in the denominator, using $0.7P_{nb}$ for tied and $0.75P_{nb}$ for spirally reinforced columns. Figure 9.16c presents the graphical representation of Eqs. 9.29 and 9.30.

9.7.1 Example 9.10: Calculation of Design Load Strength ϕP_n from Nominal Resisting Load P_n

Calculate the design loads P_u in Exs. 9.1 to 9.7 and 9.9 using the appropriate ϕ reduction factors.

Solution   *Example 9.1*

$$P_{n(\text{max})} = 924,480 \text{ lb}, \quad \text{tied column}$$

Therefore,

$$\phi = 0.7$$
$$\phi P_{0(\text{max})} = 0.7 \times 924,480 = 647,136 \text{ lb}$$

Example 9.2

$$P_{n(\text{max})} = 1,135,501 \text{ lb}, \quad \text{spiral column}$$

Therefore,

$$\phi = 0.75$$
$$\phi P_{0(\text{max})} = 0.75 \times 1,135,501 = 851,626 \text{ lb}$$

Example 9.3

$$P_{nb} = 359,448 \text{ lb}, \quad \text{tied column}$$

Therefore,

$$\phi = 0.7$$
$$\phi P_{nb} = 251,614 \text{ lb}$$
$$e_b = 13.1 \text{ in.}$$

Example 9.4

$P_n = 333,979$ lb, tied column. Failure by yielding of steel at the tension side. There-fore, it has to be checked if the ϕ value is larger than 0.7.

$$0.1A_g f'_c = 0.1 \times 12 \times 20 \times 4000 = 96,000 \text{ lb} < 0.7P_n = 233,785 \text{ lb}$$

Therefore,

$$\phi = 0.7$$
$$\phi P_n = 0.7 \times 333,979 = 233,785 \text{ lb}$$

Example 9.5

$$\phi = 0.7 \quad \text{(similar to previous case)}$$
$$\phi P_n = 0.7 \times 236,200 = 165,340 \text{ lb}$$

Example 9.6

$P_n = 442{,}647$ lb, tied column.   Failure is by crushing of concrete; axial force is greater than balanced load.   Therefore,

$$\phi = 0.7$$
$$\phi P_n = 0.7 \times 442{,}647 = 309{,}853 \text{ lb}$$

Example 9.7

(a) $P_n = 607{,}555$ lb, tied column.   Failure is by crushing of concrete.   Therefore,

$$\phi = 0.7$$
$$\phi P_n = 0.7 \times 607{,}555 = 425{,}288 \text{ lb}$$

(b) $P_n = 460{,}098$ lb.

$$\phi P_n = 0.7 \times 460{,}098 = 322{,}070 \text{ lb}$$

Example 9.9

Spiral column: (a) Balanced $P_{ub} = 0.75 \times P_{nb} = 0.75 \times 420{,}045 = 315{,}034$ lb.
(b) Tension failure, $P_n = 151{,}793$ lb

$$0.1 f'_c A_g = 0.1 \times 4000 \times 314 = 125{,}600 \text{ lb} < \phi P_{nb} = 315{,}034 \text{ lb}$$

$$\phi P_n = 0.75 \times 151{,}793 = 113{,}845 \text{ lb}$$

$$f'_c A_g = 4000 \times 314 = 1{,}256{,}000 \text{ lb}$$

Therefore, using Eq. 9.30 and assuming that $\phi = 0.75$, we have

$$\phi = 0.9 - \frac{1.5 \times 0.75 \times 151{,}793}{1{,}256{,}000}$$

$$= 0.764 > 0.75$$

Note: To obtain the exact answer, we should reiterate the calculation for ϕ until the answer converges.   Therefore,

$$\phi P_n = 0.764 \times 151{,}793 = 115{,}970 \text{ lb}$$

(c) Compression failure:

$$\phi P_n = 0.75 P_n = 0.75 \times 626{,}577 = 469{,}933 \text{ lb}$$

9.8 LOAD–MOMENT STRENGTH INTERACTION DIAGRAMS (*P–M* DIAGRAMS) FOR COLUMNS CONTROLLED BY MATERIAL FAILURE

From the discussion in Sections 9.3 and 9.4 and the numerical examples presented, we can postulate that the capacity of reinforced concrete sections to resist combined axial and bending loads can be expressed by *P–M* interaction diagrams to relate the axial load to the bending moment in compression members.   Figure 9.17 presents one such diagram.   Notice that the approximation made in using the

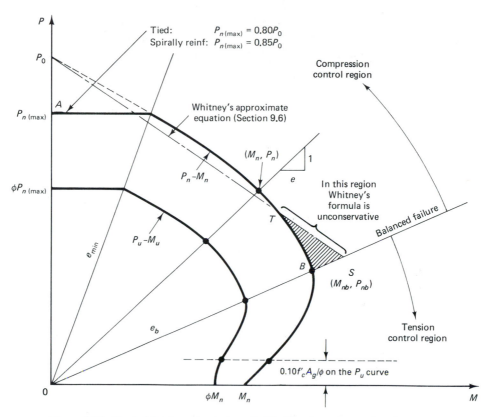

Figure 9.17   Typical load-moment strength $(P-M)$ column interaction diagram.

Whitney empirical approach is not always conservative, particularly when the factored P_u is close to the balanced case.

Each point on the curve represents one combination of nominal load strength P_n and nominal moment strength M_n corresponding to a particular neutral-axis location. The interaction diagram is separated into the tension control region and the compression control region by the balanced condition at point B. The following example illustrates the construction of the $P-M$ diagram for a typical rectangular section.

9.8.1 Example 9.11: Construction of a Load-Moment Interaction Diagram

Construct a $P-M$ diagram for a rectangular column (see Fig. 9.18) having the following geometry: width $b = 12$ in. (305 mm), thickness $h = 14$ in. (356 mm), reinforcement is four No. 11 bars (35.8-mm diameter). Given:

$$f'_c = 6000 \text{ psi (41.4 MPa)}$$

$$f_y = 60,000 \text{ psi (414 MPa)}$$

$$d' = 3.0 \text{ in. (76.2 mm)}$$

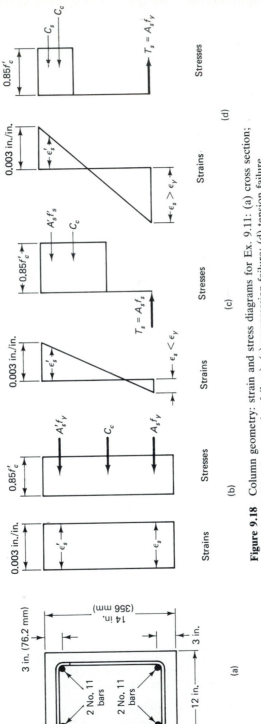

Figure 9.18 Column geometry: strain and stress diagrams for Ex. 9.11: (a) cross section; (b) concentric load (compression failure); (c) compression failure; (d) tension failure.

Solution   *Concentric load*

$$A_s = A_s' = 3.12 \text{ in.}^2$$

$$P_{n(max)} = 0.80(0.85f_c'A_g + A_{st}f_y)$$

$$= 0.80(0.85 \times 6000 \times 14 \times 12 + 2 \times 3.12 \times 60,000)$$

$$= 984,960 \text{ lb}$$

$$\phi P_{n(max)} = 0.7P_{n(max)} = 689,472 \text{ lb}$$

(If displaced concrete is taken into account in the calculations, $P_{n(max)} = 959,500$ lb and $\phi P_{n(max)} = 671,650$ lb.)

Balanced condition

$$d = 14.0 - 3.0 = 11.0 \text{ in.}$$

$$c_b = \frac{87,000}{87,000 + 60,000} \times 11.0 = 6.51 \text{ in.}$$

$$\epsilon_s' = 0.003 \times \frac{6.51 - 3.0}{6.51} = 0.0016 \text{ in./in.}$$

$$< \frac{f_y}{E_s} = \frac{60}{29,000} = 0.00207$$

$$f_s' = 29 \times 10^6 \times 0.0016 = 46,400 \text{ psi}$$

$$\beta_1 = 0.85 - \frac{0.05(6000 - 4000)}{1000} = 0.75$$

$$a_b = \beta_1 c_b = 0.75 \times 6.51 = 4.88 \text{ in.}$$

$$P_{nb} = 0.85f_c'ba_b + A_s'f_s' - A_sf_y$$

or

$$P_{nb} = 0.85 \times 6000 \times 12 \times 4.88 + 3.12 \times 46,400 - 3.12 \times 60,000$$

$$= 298,656 + 144,768 - 187,200 = 256,224 \text{ lb}$$

$$M_{nb} = 0.85f_c'ba_b\left(\bar{y} - \frac{a}{2}\right) + A_s'f_s'(\bar{y} - d') + A_sf_y(d - \bar{y})$$

$$\bar{y} = \text{distance from extreme compression fibers to the plastic centroid}$$

$$= \frac{h}{2} = \frac{14.0}{2} = 7 \text{ in.}$$

or

$$M_{nb} = 298,656\left(7.0 - \frac{4.88}{2}\right) + 144,768(7.0 - 3.0) + 187,200(11.0 - 7.0)$$

$$= 1,361,871 + 579,072 + 748,800 = 2,689,743 \text{ in.-lb}$$

$$e_b = \frac{M_{nb}}{P_{nb}} = \frac{2,689,743}{256,224} = 10.50 \text{ in.}$$

$$\phi P_{nb} = 0.7P_{nb} = 179,357 \text{ lb} \qquad \phi M_{nb} = 1,882,820 \text{ in.-lb}$$

$$e_b = 10.5 \text{ in.}$$

(If displaced concrete is considered in the calculations, $\phi P_{nb} = 169{,}444$ lb and $e_b = 10.9$ in.)

Pure bending M_{no}

Neglect the contribution of A_s' to the moment strength as insignificant when $P_u = 0$. Hence

$$a = \frac{A_s f_y}{0.85 f_c' b} = \frac{3.12 \times 60{,}000}{0.85(6000)(12)} = \frac{187{,}200}{61{,}200} = 3.06 \text{ in.}$$

$$c = \frac{a}{\beta_1} = \frac{3.06}{0.75} = 4.08 \text{ in.} \qquad \epsilon_s' = 0.00079 \qquad f_s' = 22{,}910 \text{ psi}$$

$$M_{no} = A_s f_y \left(d - \frac{a}{2} \right) = 187{,}200(11.0 - 1.53) = 1{,}772{,}784 \text{ in.-lb}$$

$$\phi M_{no} = 0.9 \times 1{,}772{,}784 = 1{,}595{,}506 \text{ in.-lb}$$

For $c = 10$ in. $> c_b$: compression controls

$$\epsilon_s' = 0.003 \times \frac{10 - 3.0}{10} = 0.0021 \text{ in./in.}$$

$$\epsilon_y = \frac{f_y}{E_s} = \frac{60{,}000}{29 \times 10^6} = 0.0021 \simeq \epsilon_s' \qquad \text{therefore, } f_s' = f_y$$

$$\epsilon_s = 0.003 \times \frac{11.0 - 10.0}{10} = 0.0003$$

$$f_s = 0.0003 \times 29 \times 10^6 = 8700 \text{ psi}$$

$$a = \beta_1 c = 0.75 \times 10 = 7.5 \text{ in.}$$

$$C_c = 0.85(6000)(12)(7.5) = 459{,}000 \text{ lb}$$

$$C_s = 3.12(60{,}000) = 187{,}200 \text{ lb}$$

$$T_s = 3.12(8700) = 27{,}144 \text{ lb}$$

$$P_n = C_c + C_s - T_s$$

$$P_n = 459{,}000 + (187{,}200 - 27{,}144) = 619{,}056 \text{ lb}$$

$$M_n = C_c \left(\bar{y} - \frac{a}{2} \right) + C_s(\bar{y} - d') + T_s(d - \bar{y})$$

$$= 459{,}000(7.0 - 3.75) + 187{,}200(7.0 - 3.0) + 27{,}144(11.0 - 7.0)$$

$$= 1{,}491{,}750 + 748{,}800 + 108{,}576 = 2{,}349{,}126 \text{ in.-lb}$$

$$\phi P_n = 0.7 P_n = 0.7 \times 619{,}056 = 433{,}339 \text{ lb}$$

(With displaced concrete taken into account, $\phi P_n = 422{,}200$ lb and $e = 3.79$ in.)

For $c = 4.2$ in. $< c_b$: tension control

$$a = \beta_1 c = 0.75 \times 4.2 = 3.15 \text{ in.}$$

$$\epsilon_s' = 0.003 \times \frac{4.2 - 3.0}{4.2} = 0.0009 < \epsilon_y$$

$$f_s' = 0.0009 \times 29 \times 10^6 = 24{,}857 \text{ psi}$$

$$f_s = f_y = 60{,}000 \text{ psi}$$

$$C_c = 0.85(6000)(12)(3.15) = 192{,}780 \text{ lb}$$

$$C_s = 3.12(24{,}857) = 77{,}554 \text{ lb}$$

$$T_s = 3.12(60{,}000) = 187{,}200 \text{ lb}$$

$$P_n = C_c + C_s - T_s = 192{,}780 + 77{,}554 - 187{,}200 = 83{,}134 \text{ lb}$$

$$M_n = C_c\left(\bar{y} - \frac{a}{2}\right) + C_s(\bar{y} - d') + T_s(d - \bar{y})$$

or

$$M_n = 192{,}780\left(7.0 - \frac{3.15}{2}\right) + 77{,}554(7.0 - 3.0) + 187{,}200(11.0 - 7.0)$$

$$= 1{,}044{,}868 + 310{,}216 + 748{,}800 = 2{,}103{,}884 \text{ in.-lb}$$

$$e = \frac{M_n}{P_n} = \frac{2{,}103{,}884}{83{,}134} = 25.31 \text{ in.}$$

Assuming that $\phi = 0.70$,

$$\phi P_n = 0.70 \times 83{,}134 = 58{,}194 \text{ lb}$$

$$0.1A_g f_c' = 0.10(12.0 \times 14.0) \times 6000 = 100{,}800 \text{ lb}$$

$$\phi P_n < 0.1A_g f_c' \qquad \text{hence } \phi > 0.70$$

$$A_g f_c' = 1{,}008{,}000 \text{ lb}$$

Therefore,

$$\phi = 0.90 - \frac{0.2 \times 58{,}194}{0.1 \times 1{,}008{,}000} = 0.792$$

$$\phi P_n = 0.792 \times 83{,}134 = 65{,}842 \text{ lb}$$

Verify the first trial ϕ value:

$$\phi = 0.90 - \frac{0.2 \times 65{,}842}{0.1 \times 1{,}008{,}000} = 0.78$$

$$\phi P_n = 0.78 \times 83{,}134 = 64{,}845 \text{ lb}$$

(If displaced concrete is considered, $\phi P_n = 47{,}056$ lb and $e = 30.4$ in.)

For $0.10f_c'A_g = \phi P_n$

Assume by trial and adjustment a value of $c = 4.85$ in.

$$a = \beta_1 c = 0.75 \times 4.85 = 3.64 \text{ in.}$$

$$\epsilon_s' = 0.003 \times \frac{4.85 - 3.0}{4.85} = 0.00114 < \epsilon_y$$

$$f_s' = 0.00114 \times 29 \times 10^6 = 33{,}060 \text{ psi}$$

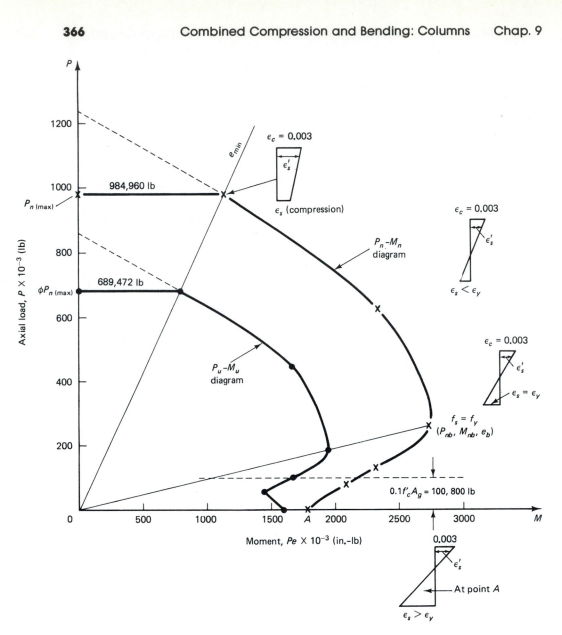

Figure 9.19   Interaction $P-M$ diagram for Ex. 9.11.

$$C_c = 0.85 \times 6000 \times 12 \times 3.64 = 227,768 \text{ lb}$$

$$C_s = 3.12 \times 33,060 = 103,147 \text{ lb}$$

$$T_s = 3.12 \times 60,000 = 187,200 \text{ lb}$$

$$P_n = C_c + C_s - T_s = 227,768 + 103,147 - 187,200 = 143,715 \text{ lb}$$

$$\phi P_n = 0.70 \times 143,715 = 100,619 \text{ lb}$$

$$0.1 f'_c A_g = 100,800 \text{ lb} \cong \phi P_n; \text{ assumed } c \text{ value is O.K.}$$

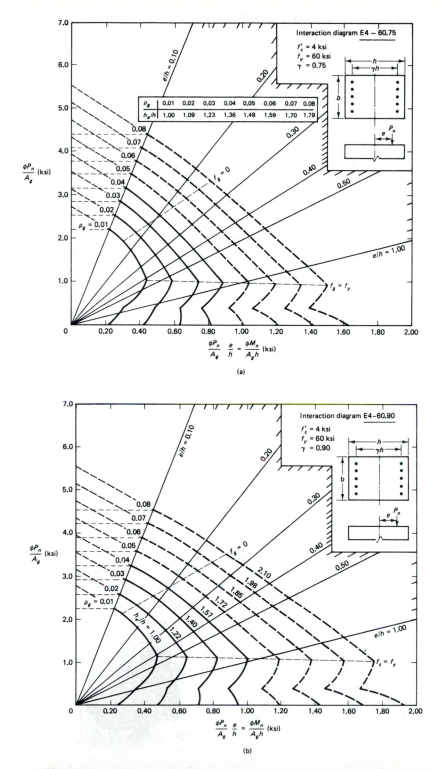

Figure 9.20 Typical nondimensional column interaction charts: (a) chart for small column sizes; (b) chart for larger column sizes.

$$M_n = C_c\left(\bar{y} - \frac{a}{2}\right) + C_s(\bar{y} - d') + T_s(d - \bar{y}) = 227,768\left(7.0 - \frac{3.64}{2}\right)$$
$$+ 103,147(7.0 - 3.0) + 187,200(11.0 - 7.0) = 2,341,226 \text{ lb}$$
$$e = \frac{M_n}{P_n} = \frac{2,341,226}{143,715} = 16.29 \text{ in.}$$

Additional points on the diagram are calculated by assigning other values for the neutral-axis depth c. The interaction diagram is presented in Fig. 9.19. Sets of charts of nondimensional interaction diagrams are available for various codes and units to facilitate speedy analysis and design in engineering offices. A typical chart from the ACI 340 SP-17 *Handbook* is shown in Fig. 9.20.

9.9 PRACTICAL DESIGN CONSIDERATIONS

The following guidelines should be followed in the design and arrangement of reinforcement to arrive at a practical design.

9.9.1 Longitudinal or Main Reinforcement

Most columns are subjected to bending moment in addition to axial force. For this reason and to ensure some ductility, a minimum of 1% reinforcement should be provided in the columns. A reasonable reinforcement ratio is between 1.5% and 3.0%. Occasionally, in high-rise buildings where column loads are very large, 4% reinforcement is not unreasonable. Even though the code allows a maximum of 8% for longitudinal reinforcement in columns, it is not advisable to use more than 4% in order to avoid reinforcement congestion, especially at beam–column junctions.

A minimum of four longitudinal bars should be used in the case of tied columns. For spiral columns, at least six longitudinal bars should be used to provide hoop action in the spirals; see the ACI Code for further discussion.

9.9.2 Lateral Reinforcement for Columns

Lateral ties. Lateral reinforcement is required to prevent spalling of the concrete cover or local buckling of the longitudinal bars. The lateral reinforcement could be in the form of ties evenly distributed along the height of the column at specified intervals. Longitudinal bars spaced more than 6 in. apart should be supported by lateral ties, as shown in Fig. 9.21.

The following guidelines are to be followed for the selection of the size and spacing of ties.

1. The size of the tie should not be less than a No. 3 (9.5-mm) bar. If the longitudinal bar size is larger than No. 10 (32 mm), then No. 4 (12-mm) bars at least should be used as ties.

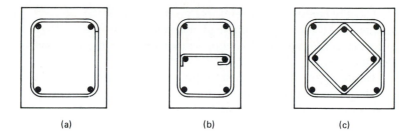

(a)   (b)   (c)

Figure 9.21   Typical ties arrangement for four, six, and eight longitudinal bars in a column: (a) one tie; (b) two ties; (c) two ties.

2. The vertical spacing of the ties must not exceed:
    (a) Forty-eight times the diameter of the tie
    (b) Sixteen times the diameter of the longitudinal bar
    (c) Least lateral dimension of the column

Figure 9.21 shows a typical arrangement of ties for four, six, and eight longitudinal bars in a column cross section.

Spirals.   The other type of lateral reinforcement is spirals or helical lateral reinforcement, as shown in Fig. 9.22.   They are particularly useful in increasing ductility or member toughness and hence are mandatory in high-earthquake-risk regions.   Normally, concrete outside the confined core of the spirally reinforced column can totally spall under unusual and sudden lateral forces such as earthquake-induced forces.   The columns have to be able to sustain most of the load even after the spalling of the cover in order to prevent the collapse of the building. Hence the spacing and size of spirals are designed to maintain most of the load-carrying capacity of the column, even under such severe load conditions.

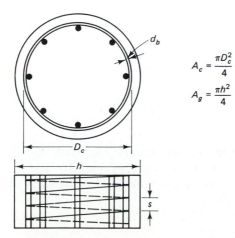

$$A_c = \frac{\pi D_c^2}{4}$$

$$A_g = \frac{\pi h^2}{4}$$

Figure 9.22   Helical or spiral reinforcement for columns.

Closely spaced spiral reinforcement increases the ultimate load capacity of columns. The spacing or pitch of the spiral is so chosen that the load capacity due to the confining spiral action compensates for the loss due to spalling of the concrete cover.

Equating the increase in strength due to confinement and the loss of capacity in spalling and incorporating a safety factor of 1.2, the following minimum spiral reinforcement ratio ρ_s is obtained:

$$\rho_s = 0.45\left(\frac{A_g}{A_c} - 1\right)\frac{f'_c}{f_{sy}} \tag{9.32}$$

where $\rho_s = \dfrac{\text{volume of the spiral steel per one revolution}}{\text{volume of concrete core contained in one revolution}}$

$$A_c = \frac{\pi D_c^2}{4}$$

$$A_g = \frac{\pi h^2}{4} \tag{9.33a}$$

h = diameter of the column $\tag{9.33b}$

a_s = cross-sectional area of the spiral

d_b = nominal diameter of the spiral wire

D_c = diameter of the concrete core out-to-out of the spiral

f_{sy} = yield strength of the spiral reinforcement

To determine the pitch s of the spiral, calculate ρ_s using Eq. 9.32, choose a bar diameter d_b for the spiral, and calculate a_s; then obtain pitch s using Eq. 9.35b. The spiral reinforcement ratio ρ_s can be written as

$$\rho_s = \frac{a_s\pi(D_c - d_b)}{(\pi/4)D_c^2 s} \tag{9.34}$$

Therefore,

$$\text{pitch } s = \frac{a_s\pi(D_c - d_b)}{(\pi/4)D_c^2\rho_s} \tag{9.35a}$$

or

$$s = \frac{4a_s(D_c - d_b)}{D_c^2\rho_s} \tag{9.35b}$$

The spacing or pitch of spirals is limited to a range of 1 to 3 in. (25.4 to 76.2 mm), and the diameter should be at least $\frac{3}{8}$ in. (9.53 mm). The spirals should be well anchored by providing at least one and one-half extra turns when splicing of spirals rather than welding is used.

9.10 OPERATIONAL PROCEDURE FOR THE DESIGN OF NONSLENDER COLUMNS

The following steps can be used for the design of nonslender (short) columns where the behavior is controlled by material failure.

1. Evaluate the factored external axial load P_u and factored moment M_u. Calculate the eccentricity $e = M_u/P_u$.
2. Assume a cross section and the type of vertical reinforcement to be used. Fractional dimensions are to be avoided in selecting column sizes.
3. Assume a reinforcement ratio ρ between 1% and 4% and obtain the reinforcement area.
4. Calculate P_{nb} for the assumed section and determine the type of failure, whether by initial yielding of the steel or initial crushing of the concrete.
5. Check for the adequacy of the assumed section.   If the section cannot support the factored load or it is oversized, hence uneconomical, revise the cross section and (or) the reinforcement and repeat steps 4 and 5.
6. Design the lateral reinforcement.

Figure 9.23 presents a flow chart for the sequence of calculations.

9.11 NUMERICAL EXAMPLES FOR ANALYSIS AND DESIGN OF NONSLENDER COLUMNS

9.11.1 Example 9.12: Design of a Column with Large Eccentricity; Initial Tension Failure

The tied reinforced concrete column in Fig. 9.24 is subjected to a service axial force due to dead load = 65,000 lb (289 kN) and a service axial force due to live load = 125,000 lb (556 kN).   Eccentricity to the plastic centroid is $e = 16$ in. (406 mm).

Design the longitudinal and lateral reinforcement for this column, assuming a nonslender column with a total reinforcement ratio between 2% and 3%. Given:

$$f'_c = 4000 \text{ psi (27.6 MPa), normal-weight concrete}$$
$$f_y = 60,000 \text{ psi (414 MPa)}$$

Solution   *Calculate the factored external load and moment (step 1)*

$$P_u = 1.4D + 1.7L = 1.4 \times 65,000 + 1.7 \times 125,000 = 303,500 \text{ lb (1350 kN)}$$
$$P_ue = 303,500 \times 16 = 4,856,000 \text{ in.-lb (549 kN-m)}$$

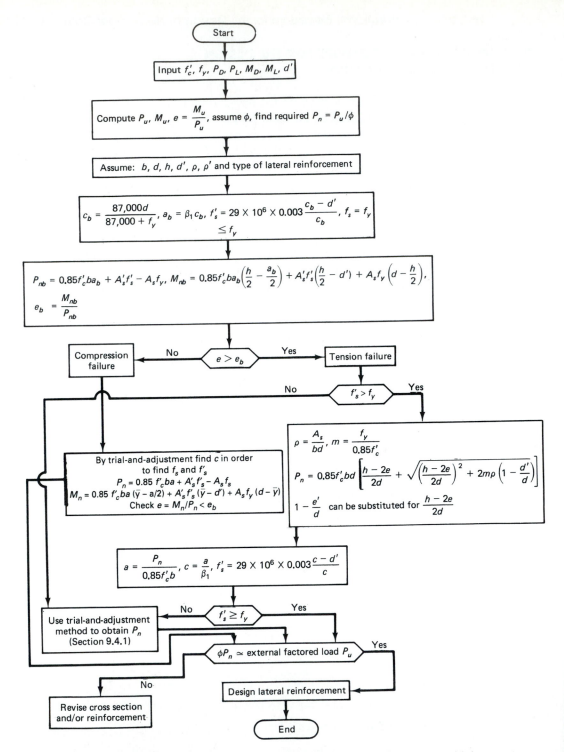

Figure 9.23 Flow chart for design of nonslender rectangular columns with bars on two faces only.

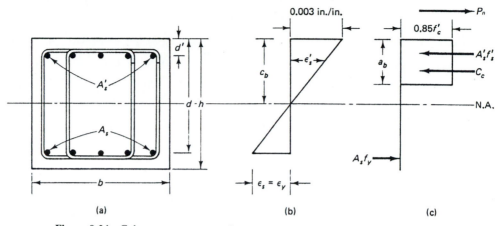

Figure 9.24 Column geometry: strain and stress diagrams in Ex. 9.12 (balanced failure): (a) cross section; (b) strains; (c) stresses.

Assume a section 20 in. × 20 in. and a total reinforcement ratio of 3% (steps 2 and 3)

Assume that $\rho = \rho' = A_s/bd = 0.015$ and $d' = 2.5$ in.

$$A_s = A_s' = 0.015 \times 20(20 - 2.5) = 5.25 \text{ in.}^2$$

Try five No. 9 bars, 5.00 in.² on each face (3225 mm²).

$$\rho = \frac{5.00}{20 \times 17.5} = 0.0143$$

Check whether the given factored axial load P_u is greater than the balanced load, ϕP_{nb} (step 4)

$$c_b = d\,\frac{87,000}{87,000 + f_y} = 17.5\left(\frac{87,000}{87,000 + 60,000}\right) = 10.4 \text{ in.}$$

$$a_b = \beta_1 c_b = 0.85 \times 10.4 = 8.82 \text{ in.}$$

$$f_s' = 0.003 \times 29 \times 10^6\left(\frac{10.4 - 2.5}{10.4}\right)$$

$$= 66,086 \text{ psi} > f_y$$

Therefore, use $f_s' = f_y$. Using Eq. 9.6 gives

$$P_{nb} = 0.85 f_c' b a_b + A_s' f_y - A_s f_y$$

$$= 0.85 \times 4000 \times 20 \times 8.82 = 599,760 \text{ lb (2670 kN)}$$

$$\phi P_{nb} = 0.7 \times P_{nb} = 419,832 \text{ lb}$$

Since the given load $P_u = 303,500$ lb is less than ϕP_{nb}, the column will fail by initial yielding of the tension reinforcement.

Check the adequacy of the section (step 5)

$$\rho = 0.0143 \qquad m = \frac{60,000}{0.85 \times 4000} = 17.65$$

$$\frac{h - 2e}{2d} \qquad \text{or} \qquad 1 - \frac{e'}{d} = \frac{20 - 32}{2 \times 17.5} = -0.34$$

$$1 - \frac{d'}{d} = 1 - \frac{2.5}{17.5} = 0.857$$

Using Eq. 9.19 or 9.20 yields

$$P_n = 0.85 \times 4000 \times 20 \times 17.5(-0.34 + \sqrt{0.12 + 2 \times 0.0143 \times 17.65 \times 0.857})$$

$$= 480,015 \, \text{lb} \quad \text{(applies when compression steel yields)}$$

$$\phi P_n = 0.7 \times 480,015 = 336,011 \, \text{lb} \, (1512 \, \text{kN})$$

$$\phi P_n > 0.1 A_g f'_c \qquad \text{therefore,} \, \phi = 0.7 \qquad \text{O.K.}$$

Check if the compression steel stress $f'_s = f_y$.

$$a = \frac{480,015}{0.85 \times 4000 \times 20} = 7.06 \, \text{in.}$$

$$c = \frac{a}{\beta_1} = 8.3 \, \text{in.}$$

$$f'_s = 0.003 \times 29 \times 10^6 \left(\frac{8.3 - 2.5}{8.3}\right) = 60,795 \, \text{psi} > f_y$$

$$\text{therefore,} \, f'_s = f_y \qquad \text{O.K.}$$

An external load of 303,500 lb is less than 336,011 lb. Hence the design is satisfactory. (If displaced concrete is considered, $\phi P_n = 327,964$ lb, which shows that using the results above can sometimes be unconservative.) Therefore, adopt a section 20 in. × 20 in. (508 mm × 508 mm) with five No. 9 bars on each side (5 bars of 28.6-mm diameter) having $d = 17.5$ in.(445 mm).

Design of ties (step 6)

Using No. 3 ties, spacing will be the minimum of:

(1) $16 \times \frac{9}{8} = 18$ in.
(2) $48 \times \frac{3}{8} = 18$ in.
(3) Least dimension of 20 in.

Therefore, provide No. 3 ties at 18 in. center to center (9.53-mm diameter at 457 mm center to center).

9.11.2 Example 9.13: Design of a Column with Small Eccentricity; Initial Compression Failure

The nonslender column shown in Fig. 9.25 is subjected to a factored P_u = 365,000 lb (1620 kN) and a factored M_u = 1,640,000 in.-lb (185 kNm). Assume that the gross reinforcement ratio ρ_g = 1.5% to 2% and that the effective cover to the center of the longitudinal steel is d' = $2\frac{1}{2}$ in. (63.5 mm). Design the column section and the necessary longitudinal and transverse reinforcement. Given:

f_c' = 4500 psi (31.03 MPa), normal-weight concrete
f_y = 60,000 psi (414 MPa)

Solution   *Calculation of factored design loads (step 1)*

$$P_u = 365,000 \text{ lb}$$

$$e = \frac{1,640,000}{365,000} = 4.5 \text{ in. (114 mm)}$$

Assume a 15 in. × 15 in. (d = 12.5 in.) section (steps 2 and 3)

Assume that the reinforcement ratio $\rho = \rho' = 0.01$.

$$A_s = A_s' \approx 0.01 \times 15 \times 12.5 = 1.875 \text{ in.}^2$$

Provide two No. 9 bars on each side.

$$A_s = A_s' = 2.0 \text{ in.}^2 \text{ (1290 mm}^2)$$

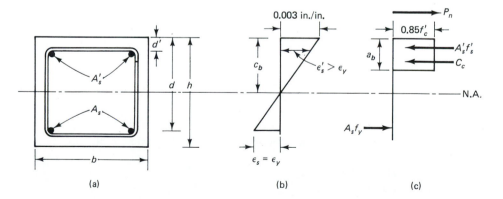

Figure 9.25   Column geometry: strain and stress diagrams in Ex. 9.13: (a) cross section; (b) strains (balanced case); (c) stresses.

Check whether the given load is less or greater than P_{ub} (step 4)

$$d = 15.0 - 2.5 = 12.5 \text{ in. } (317.5 \text{ mm})$$

$$c_b = \frac{87,000}{87,000 + 60,000} \times 12.5 = 7.4 \text{ in.}$$

$$\beta_1 = 0.85 - 0.05 \frac{4500 - 4000}{1000} = 0.825$$

$$a_b = \beta_1 c_b = 0.825 \times 7.4 = 6.11 \text{ in.}$$

$$\epsilon'_s = 0.003 \times \frac{7.4 - 2.5}{7.4}$$

$$= 0.00199 \text{ in./in. } < \frac{f_y}{E_s}$$

$$f'_S = E_A \epsilon'_S = 29,000 \times 10^3 \times 0.00199 = 57,610 \text{ psi}$$

$$\phi P_{nb} = 0.7(0.85 f'_c b a_b + A'_s f'_s - A_s f_y)$$

$$= 0.7(0.85 \times 4500 \times 15 \times 6.11 + 2.0 \times 57,610 - 2 \times 60,000)$$

$$= 242,050 \text{ lb } (1080 \text{ kN})$$

$$\phi P_{nb} < P_u \qquad \text{compression failure controls}$$

Check the adequacy of the section (step 5)

Using Eq. 9.26,

$$P_n = \frac{A'_s f_y}{\dfrac{e}{d - d'} + 0.5} + \frac{bhf'_c}{\dfrac{3he}{d^2} + 1.18}$$

$$= \frac{2 \times 60,000}{\dfrac{4.5}{12.5 - 2.5} + 0.5} + \frac{15 \times 15 \times 4500}{\dfrac{3 \times 15 \times 4.5}{12.5^2} + 1.18}$$

$$= 535,200 \text{ lb } (2380 \text{ kN})$$

$$= \phi P_n = 0.7 \times 535,200 = 374,000 \text{ lb } > 365,000 \text{ lb}$$

Therefore, the section is adequate to carry the load. Results from more exact computer solutions in Appendix A give $\phi P_n = 372,247$ lb. Use a column 15 in. × 15 in. ($d = 12.5$ in.) with two No. 9 bars on each face (381 mm × 381 mm size with two bars of 28.6-mm diameter on each face).

Design of ties (step 6)

Using No. 3 ties, the spacing will be the minimum of:

(1) $16 \times \frac{9}{8} = 18$ in.
(2) $48 \times \frac{3}{8} = 18$ in.
(3) 15 in.

Therefore, provide No. 3 ties at 15 in. (9.53-mm diameter at 381 mm center to center).

9.11.3 Example 9.14: Design of a Circular Spirally Reinforced Column

A spirally reinforced circular column is subjected to an external factored load $P_u = 110,000$ lb (489 kN) acting at an eccentricity to the plastic centroid of magnitude $e = 16$ in. (406 mm).   Design the column cross section and the longitudinal and spiral reinforcement necessary, assuming a nonslender column with a total reinforcement ratio of about 2%.   Given:

$$f'_c = 4000 \text{ psi (27.58 MPa), normal-weight concrete}$$
$$f_y = 60,000 \text{ psi (414 MPa)}$$
$$f_{sy} = 60,000 \text{ psi (414 MPa)}$$

Solution   *Calculation of factored external loads (step 1)*
        Given:

$$P_u = 110,000 \text{ lb}$$
$$e = 16 \text{ in.}$$
$$\text{required } P_n = \frac{P_u}{\phi} = \frac{110,000}{0.75} = 146,670 \text{ lb}$$

Try a 20-in. circular column with six No. 8 bars (area = 4.74 in.²) (steps 2 and 3)

        Provide a clear cover of 1.5 in. and effective cover to the center of the bar of 2.5 in.

Check the adequacy of the section (steps 4 and 5)

        Using the results of Ex. 9.9(b), the available $P_n = 151,793$ lb with eccentricity $e = 16$ in. > required $P_n = 146,670$ or $\phi P_n = 113,845$ lb > $P_u = 110,000$ lb.   Hence adopt the section.   Use six No. 8 longitudinal bars.

Design the spiral reinforcement (Step 6)

        Using Eq. 9.32,

$$\text{required } \rho_s = 0.45 \left(\frac{A_g}{A_c} - 1 \right) \frac{f'_c}{f_{sy}}$$

Using No. 3 spirals with a yield strength $f_y = 60,000$ psi:

    clear concrete cover $d_c = 1.5$ in. (38.1 mm)

$$f_{sy} = 60,000 \text{ psi}$$
$$D_c = h - 2d_c = 20.0 - 2 \times 1.5 = 17.0 \text{ in. (431.8 mm)}$$
$$A_c = \frac{\pi (17.0)^2}{4} = 226.98 \text{ in.}^2$$
$$A_g = 314.0 \text{ in.}^2$$
$$\rho_s = 0.45 \left(\frac{314.0}{226.98} - 1 \right) \frac{4000}{60,000} = 0.0115$$

For No. 3 spirals, $a_s = 0.11$ in.^{2}.   Using Eq. 9.35b,

$$\text{pitch } s = \frac{4a_s(D_c - d_b)}{D_c^2 \rho_s} = \frac{4 \times 0.11(17.0 - 0.375)}{(17.0)^2 \times 0.0115} = 2.20 \text{ in. } (55.9 \text{ mm})$$

Provide No. 3 spirals at $2\frac{1}{4}$-in. pitch (9.53-mm-diameter spiral at 54.0-mm pitch).

9.12 LIMIT STATE AT BUCKLING FAILURE (SLENDER OR LONG COLUMNS)

Considerable literature exists on the behavior of columns subjected to stability considerations.   If the column slenderness ratio exceeds the limits for short columns, the compression member will buckle prior to reaching its limit state of material failure.   The strain in the compression face of the concrete at buckling load will be less than the 0.003 in./in. shown in Fig. 9.8.   Such a column would be a slender member subjected to combined axial load and bending, deforming laterally and developing additional moment due to the $P\Delta$ effect, where P is the axial load and Δ is the deflection of the column's buckled shape at the section being considered, as seen in Fig. 9.26.

k is the column length factor, as shown in Fig. 9.27. M_1 and M_2 are the moments at the opposite ends of the compression member. M_2 is always larger than M_1, and the ratio M_1/M_2 is taken as positive for single curvature and negative for double curvature, as shown in Fig. 9.28a.

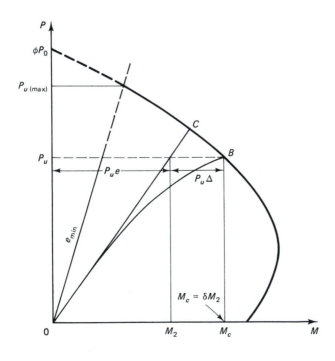

Figure 9.26 Loading moment $(P-M)$ magnification interaction diagram.

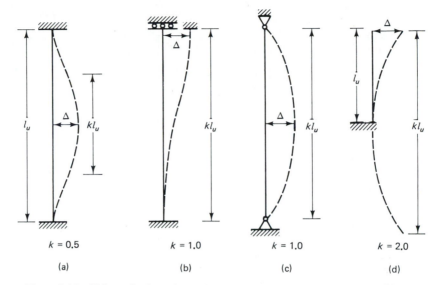

Figure 9.27   Values of column length factor k for typical end conditions: (a) fixed–fixed; (b) fixed–fixed with lateral motion; (c) pinned; (d) fixed–free.

The effective length kl_u is used as the modified length of the column to account for end restraints other than pinned. kl_u represents the length of an auxiliary pin-ended column, which has an Euler buckling load equal to that of the column under consideration. Alternatively, it is the distance between the points of contraflexure of the member in its buckled form.

The value of the end restraint effective length factor k varies between 0.5 and 2.0.

Both column ends fixed	$k = 0.5$
Both column ends fixed, lateral motion exists	$k = 1.0$
Both column ends pinned, no lateral motion	$k = 1.0$
One end fixed, other end free	$k = 2.0$

Typical cases illustrating the buckled shape of the column for several end conditions and the corresponding length factors k are shown in Fig. 9.27.

For members in a structural frame, the end restraint lies between the hinged and fixed conditions. The actual k value can be determined from the Jackson and Moreland alignment charts in Fig. 9.28. In lieu of these charts, the following equations suggested in the ACI Code commentary can also be used for calculating k.

1. *Braced compression members.* An upper bound to the effective length factor may be taken as the smaller of the following two expressions:

$$k = 0.7 + 0.05(\psi_A + \psi_B) \le 1.0 \qquad (9.36a)$$

$$k = 0.85 + 0.05\psi_{\min} \le 1.0 \qquad (9.36b)$$

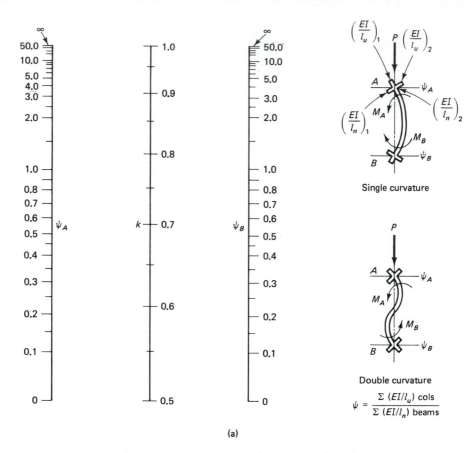

(a)

Figure 9.28 Effective length factor k for (a) braced and (b) unbraced frames.

where ψ_A and ψ_B are the values of ψ at the two ends of the column and ψ_{min} is the smaller of the two values. ψ is the ratio of the stiffness of all compression members to the stiffness of all flexural members in a plane at one end of the column.   Or

$$\psi = \frac{\Sigma\ EI/l_u\ \text{columns}}{\Sigma\ EI/l_n\ \text{beams}} \tag{9.37}$$

where l_u is the unsupported length of the column; l_n is the clear beam span.

2. *Unbraced compression members restrained at both ends.*   The effective length may be taken as follows:

$$\text{For } \psi_m < 2: \quad k = \frac{20 - \psi_m}{20}\sqrt{1 + \psi_m} \tag{9.38a}$$

$$\text{For } \psi_m \geq 2: \quad k = 0.9\sqrt{1 + \psi_m} \tag{9.38b}$$

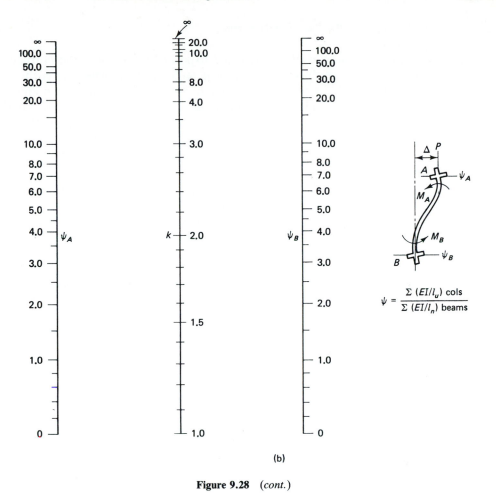

(b)

Figure 9.28   (*cont.*)

where ψ_m is the average of the ψ values at the two ends of the compression member.

3. *Unbraced compression members hinged at one end.*   The effective length factor may be taken as

$$k = 2.0 + 0.3\psi \qquad (9.39)$$

where ψ is the value at the restrained end.

The radius of gyration $r = \sqrt{I_g/A_g}$ can be taken as $r = 0.3h$ for rectangular sections, where h is the column dimension perpendicular to the axis of bending. For circular sections, the radius of gyration r is taken as $0.25h$.

9.12.1 Buckling Considerations

Consider a slender column subjected to axial load P_u at an eccentricity e. The buckling effect produces an additional moment $P_u\Delta$, where Δ is the maximum lateral displacement of the compression member between its two ends from the vertical plumb position. This additional moment reduces the load capacity from point C to point B in the interaction diagram (Fig. 9.26). The total moment $P_u e + P_u\Delta$ is represented by point B in the diagram, and the column should be designed for a larger magnified moment M_c as a nonslender column by the usual *first-order analysis*.

In such an analysis, the moments and axial forces in a frame are obtained by the classical elastic procedures. These procedures do not consider the effects of the lateral displacement Δ on the axial force P_u and the bending moment M_c. Consequently, the resulting load–deflection and load–moment relationships are linear. If the P–Δ effect is taken into account, a second-order analysis becomes necessary with a resulting nonlinear relationship of the load to the lateral displacement (deflection) and the moment. The ACI 318-95 Code permits using either a first- or second-order analysis for columns of intermediate slenderness and requires a second-order analysis for long columns having a slenderness ratio of 100 or more. One ACI Code method where the P–Δ effect is ignored, the moment magnification method, is presented in Section 9.13.

9.13 MOMENT MAGNIFICATION: FIRST-ORDER ANALYSIS

The factored axial forces P_u, the factored moments M_1 and M_2 at the column ends, and, where required, the relative story deflections are computed in this method using an elastic first-order analysis with the section properties determined taking into account the influence of axial loads, the presence of cracked regions along the length of the member, and the effects of duration of the load.

As discussed later in conjunction with Fig. 9.26, the moment M_2 is magnified by a magnification factor δ. The column is subjected to moments M_1 and M_2 at its ends, where M_2 is considered larger than M_1. The factored axial force P_u and the factored moments M_1 and M_2 are resisted by analytically chosen sectional properties, taking into account the cracked regions along the compression member's length or height and the load duration. In lieu of these computations, the ACI 318-95 Code allows using the following average values for properties of members in a structure:

1. Modulus of elasticity $E_c = 33w_c^{1.5}\sqrt{f_c'}$, and for concrete strength $f_c' > 5000$ psi $< 12{,}000$ psi.

$$E_c = (40{,}000\sqrt{f_c'} + 1 \times 10^6)\left(\frac{w_c}{145}\right)^{1.5}$$

2. Moment of inertia

Beams	$0.35I_g$
Columns	$0.70I_g$
Walls: uncracked	$0.70I_g$
cracked	$0.35I_g$
Flat plates and flat slabs	$0.25I_g$

3. Area: $1.0A_s$

4. Radius of gyration $r = 0.30h$ for rectangular members, where h is in the direction stability is being considered, or $r = 0.25D$ for circular members, where D is the diameter of the compression member.

The moments of inertia should be divided by $1 + \beta_d$ when sustained lateral loads act or for stability checks when β_d is a creep factor, where

$$\beta_d = \frac{\text{maximum factored sustained axial}}{\text{total factored axial load}}$$

The column load is assumed to act at an eccentricity $e + \Delta$ in Fig. 9.26 to produce a moment M_c. The ratio M_c/M_2 is termed the magnification factor δ. The degree of magnification is dependent on the slenderness ratio $k\ell_u/r$, where k is the effective length factor for compression members, a function of the relative stiffnesses at the joint of each end of the member.

The magnification factor is controlled by the type of the magnified moments δM_2 and δM_1 acting at the respective ends 2 and 1 of a column, that is, whether sidesway of the structural frame occurs or not. Note in the case of compression members subjected to bending about both principal axes that the moment about each axis should be separately considered based on the restraint condition corresponding to that axis.

9.13.1 Moment Magnification in Nonsway Frames

In the case of compression members in nonsway frames, that is, braced frames, the effective length factor k can be taken as 1.0 unless analysis gives a lower value. In such a case, k values are calculated on the basis of the EI tabulated values and the monograms in Fig. 9.28.

The slenderness effects can be disregarded if

$$\frac{k\ell_u}{r} \leq 34 - 12\frac{M_1}{M_2} \tag{9.40}$$

$k\ell_u$ = effective length between points of inflection, and M_1/M_2 is not taken less than -0.5. The term M_1/M_2 is positive if the column is bent in a single curvature so that the two terms subtract in Eq. 9.40 and is negative in double curvature so that the two terms add (see Fig. 9.28a). If the nonsway magnification factor is δ_{ns} and the sway factor $\delta_s = 0$, the magnified moment becomes

$$M_c = \delta_{ns}M_2 \tag{9.41}$$

where

$$\delta_{ns} = \frac{C_m}{1 - (P_u/0.75P_c)} \geq 1.0 \tag{9.42a}$$

$$P_c = \frac{\pi^2 EI}{(k\ell_u)^2} \tag{9.42b}$$

where P_c is the Euler buckling load for pin-ended columns.

Stiffness EI is to be taken as

$$EI = \frac{0.2E_c I_g + E_s I_{se}}{1 + \beta_d} \tag{9.42c}$$

or conservatively as

$$EI = \frac{0.4 E_c I_g}{1 + \beta_d} \tag{9.42d}$$

C_m = a factor relating the actual moment diagram to an equivalent uniform moment diagram. For members without transverse loads, that is, subjected to end loads only,

$$C_m = 0.6 + \frac{M_1}{M_2} \geq 0.4 \tag{9.43}$$

where $M_2 \leq M_1$ and $M_1/M_2 > 0$ if no inflection point exists between the column ends [Fig. 9.28a (single curvature)]. For other conditions, such as members with transverse loads between supports, $C_m = 1.0$.

The minimum allowed value of M_2 is

$$M_{2,\min} = P_u (0.6 + 0.03h) \tag{9.44}$$

where h is in inches.

In SI units $M_{2,\min} = P_u(15 + 0.03h)$, where h is in millimeters. In other words, the minimum eccentricity in the slender columns is $e_{\min} = 0.6 + 0.03h$. If $M_{2,\min}$ exceeded the applied moment M_2, the value of C_m in Eq. 9.43 should either be taken as 1.0 or be based on the actual computed end moments M_1 and M_2.

Frames braced against sidesway or braced with shear walls would normally have a lateral deflection less than total height $h_s/1500$. Once this ratio is exceeded, appropriate measures have to be taken to minimize the additional moments caused by sidesway and hence to reduce lateral drift of the frame and its constituent columns.

9.13.2 Moment Magnification in Sway Frames

For compression members not braced against sidesway, the effective length factor k can also be determined from the EI values presented in Section 9.12, but its value should not exceed 1.0.   The slenderness effects can be disregarded if

$$\frac{k\ell_u}{r} < 22 \tag{9.45}$$

The end moments M_1 and M_2 should be magnified as follows with the nonsway moments unmagnified, provided that ℓ_u/r is less than $35/\sqrt{\{p_u/f_c'A_g\}}$:

$$
\begin{aligned}
M_1 &= M_{1ns} + \delta_s M_{1s} \\
M_2 &= M_{2ns} + \delta_s M_{2s}
\end{aligned}
\tag{9.46}
$$

On the assumption that $M_2 > M_1$, the design moment should be

$$M_c = M_{2ns} + \delta_s M_{2s} \tag{9.47}$$

where M_{2ns} = factored end moment at the end of the compression member due to loads that cause no appreciable sidesway, calculated using a first-order elastic frame analysis. M_{2s} = factored end moment at the end of the compression members due to loads that cause appreciable sidesway, calculated using a first-order elastic frame analysis.

$$\delta_s M_s = \frac{M_s}{1 - (\Sigma\, P_u/0.75\Sigma\, P_c)} \geq M_s, \quad \text{where } \delta_s \leq 2.5 \tag{9.48}$$

where $\Sigma\, P_u$ is the summation for all the vertical loads in a story and $\Sigma\, P_c$ is the summation of the Euler buckling loads, P_c, for pin-ended columns for all sway resisting columns in a story [$P_c = \pi^2 EI/(k\ell_u)^2$ from Eq. 9.42b] with the EI values obtained from Eq. 9.42c or d.

In the case of an individual compression member having

$$\frac{\ell_u}{r} > \frac{35}{\sqrt{P_u/f_c'A_g}} \tag{9.49}$$

the member has to be designed for a factored axial load P_u and magnified moment $M_c = \delta_{ns} M_2$, where in this case $M_c = \delta_{ns} M_{2ns} + \delta_s M_{2s}$.   This condition can develop in slender columns with high axial loads when the maximum moment may develop between the ends of the column, so the end moments might not necessarily be the maximum moments.

It is important to summarize that the moment magnification method, originally developed for prismatic columns, should work well for columns of slenderness ratio $k\ell_u/r$ less than 100, particularly if the frame is braced.   In the case of unbraced frames of comparable slenderness ratios, taking into account the $P-\Delta$ effect on the moments and deflections through a second-order analysis can give more accurate results.   Such an analysis can be either of the following:

1. Execute several applications of the first-order analysis where the lateral load (h_i in Fig. 9.29) is incremented by $\Sigma P_u \Delta_i$ in each cycle, and consider the final result a second-order result.
2. Use a real second-order analysis computer program in which the reduction in the relative sidesway resistance is used in a global stiffness matrix for the elements involved.

The ACI 318-95 Code allows a stability index, Q, to calculate δ_s in lieu of Eq. 9.48 as an alternative method, provided that δ_s does not exceed 1.5.

9.14 SECOND-ORDER FRAMES ANALYSIS AND THE P–Δ EFFECT

A second-order analysis is a frame analysis that includes the internal force effects resulting from lateral displacment (deflection) of a column.   When such an analysis is performed in order to evaluate $\delta_s M_s$ in a nonbraced frame, the deflections must be computed on the basis of fully cracked sections with reduced EI stiffness values. Approximations such as the use of several first-order analysis cycles and idealizations of nonprismatic sections can be made in the analysis.   But the analysis should verify that the predicted strength of the compression members of a structural frame are in good agreement within a 15% range with results of frame analysis for columns in indeterminate reinforced concrete structures.   The structure being analyzed should result in geometry of members similar to the geometry of the sections to be built.   If the members in the final structure have cross-sectional dimensions differing by more than 10% from those assumed in the analysis, a new computation cycle has to be performed.

A second-order analysis is an iterative procedure of the P–Δ effects on the slender column, including shear deformations. Hence it is reasonable to expect that canned computer programs have to be used rather than long-hand computations in the design of the slender columns of a frame structure. An attempt will be made here to illustrate the iteration procedure involved in the use of several cycles of lateral load increments to the P–Δ values.   It must be stated, however, that the large majority of columns in concrete building frames do not necessitate such an analysis since the $K\ell_u/r$ ratio is in most cases below 100.

Consider the column between the two floors $i-1$ and i in the frame shown in Fig. 9.29.   Assume that the maximum lateral displacement or drift at the upper end of the top column in the frame is x_{max} and that the total height of the building is h_s.   A large drift or lateral displacement of the building upper floors results in cracking of the masonry and interior finishes.   Unless precautions are taken to permit movement of interior partitions without damage, the maximum lateral deflection limitation should be $h_s/500$.   Hence a good assumption is to choose x_{max} in the range of $h_s/350$ to $h_s/500$, considering that a *fully braced* frame has normally a ratio of maximum drift x_{max} to frame height h_s less than 1/1500.

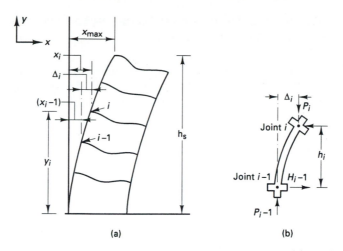

Figure 9.29   Second-order frame parameters: (a) Δ–P drift of frame; (b) idealized column between i and $i - 1$ floors.

If x_i is the drift at floor level i, and h_i is the height of the column between floors $i - 1$ and i in Fig. 9.29a, it can be assumed that the proportional horizontal drift for a particular floor is directly proportional to the square of the ratio of the height h_i of the floor and the total height h_s of the entire frame.   Hence

$$x_i = x_{max} \frac{h_i}{h_s^{\,2}} \tag{9.50}$$

The procedure can be summarized as follows:

1. Choose geometrical sections of the frame and its columns and their stiffness EI by approximate procedures.
2. Calculate the drifts, that is, the lateral deflections Δ_i, and the corresponding ultimate loads $P_{u,i}$ at joints $i = 1, \ldots, n$ (Fig. 9.29).
3. Find the equivalent horizontal forces H_i from $H_i = P_i \Delta_i / h_i$ (Fig. 9.29b).
4. Add the values obtained in step 3 to the actual lateral loads acting on the frame.
5. Perform a frame analysis using the appropriate computer program.
6. The iterative computer program, using the stiffnesses, EI, chosen for the input data gives Δ_i results that have to be compared with the x_i values allowed.
7. If all Δ_i values are $\leq$ all the x_i values, accept the solution and the design as a second-order solution.   If not, run additional computer cycles with modified stiffnesses until the desired results are achieved.

Any of several computer programs can be utilized to account for the P–Δ effects in frame sideways.   Strudel and PCA Frame are an example of such general-purpose programs.

9.15 OPERATIONAL PROCEDURE AND FLOW CHART FOR THE DESIGN OF SLENDER COLUMNS

1. Determine whether the frame has an appreciable sidesway. If it does, use the magnification factors δ_{ns} and δ_s. If the sidesway is negligible, assume that $\delta_s = 0$. Assume a cross section. Calculate the eccentricity using the greater of the end moments and check whether it is more than the minimum allowable eccentricity; that is,

$$\frac{M_2}{P_u} \geq (0.6 + 0.03h) \text{ in.}$$

If the given eccentricity is less than the specified minimum, use the minimum value.

2. Calculate ψ_A and ψ_B using Eq. 9.37. Obtain k using Fig. 9.28 or Eqs. 9.38a and b. Calculate kl_u/r and determine whether the column is a short or long column. If the column is slender and kl_u/r is less than 100, calculate the magnified moment M_c. Using the M_c value obtained, calculate the equivalent eccentricity to be used if the column is to be designed as a short column. If kl_u/r is greater than 100, perform a second-order analysis.

3. Design the equivalent nonslender column. The flow chart (Fig. 9.30) presents the sequence of calculations. The necessary equations are provided in Section 9.11 and in the flow chart.

9.15.1 Example 9.15: Design of a Slender (Long) Column

A rectangular tied column is part of a 5 × 3 bays frame building subjected to uniaxial bending. Its clear height is l_u = 18 ft (5.55 m) and it is not braced against sidesway. The factored external load P_u = 726,000 lb (3229 kN). The factored end moments are M_1 = 550,000 in.-lb (203.88 kN-m) and M_2 = 1,525,000 in.-lb (172.25 kN-m). Design the column section and the reinforcement necessary for the following two conditions:

1. Consider gravity loads only, assuming lateral sidesway due to wind as negligible.
2. Consider sidesway wind effects to cause a factored P_u = 90,000 lb (400.3 kN) and a factored M_u = 1,220,000 in.-lb. (137.9 kN-m).

Loads per floor of all columns at that level are

$$\Sigma P_u = 15.5 \times 10^6 \text{ lb (68,944 kN)}$$

$$\Sigma P_c = 32.0 \times 10^6 \text{ lb (142,336 kN)}$$

Given:

$$\beta_d = 0.5, \qquad \psi_A = 2.0, \qquad \psi_B = 3.0$$
$$f'_c = 5000 \text{ psi (34.5 MPa)}$$
$$f_y = 60,000 \text{ psi (413.7 MPa)}$$
$$d' = 2.5 \text{ in. (64 mm)}$$

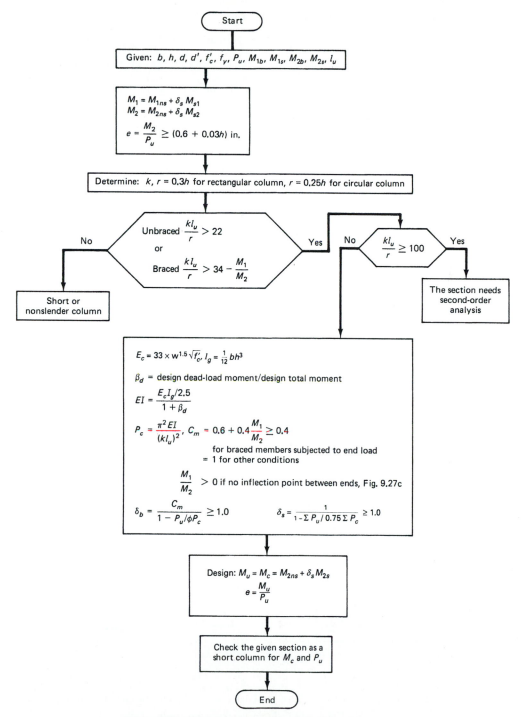

Figure 9.30 Flowchart for design of slender columns.

Solution for Case 1: *Gravity Loading Only*

Check for sidesway and minimum eccentricity (step 1)

Since the frame has no appreciable sidesway, the entire moment M_2 is taken as M_{2b}, and magnification factor δ_s is taken as equal to zero. By trial and adjustment, a column section is assumed and analyzed. Try a section 21 in. × 21 in. (533 mm × 533 mm) as shown in Fig. 9.31a.

$$\text{actual eccentricity} = \frac{M_{2ns}}{P_u} = \frac{1,525,000}{726,000} = 2.1 \text{ in. (52 mm)}$$

$$\text{minimum allowable eccentricity} = 0.6 + 0.03 \times 21 = 1.23 \text{ in.} < 2.1 \text{ in.}$$

Use $M_{2ns} = 1,525,000$ in.-lb (172.26 kN-m).

Calculate the eccentricity to be used for equivalent short column (step 2)

From the chart in Fig. 9.28b, $k = 1.7$.

$$\text{actual slenderness ratio} \; \frac{kl_u}{r} = \frac{1.7 \times 18 \times 12}{0.3 \times 21} = 58.29$$

$$\text{allowable slenderness ratio} \; \frac{kl_u}{r} \text{ for unbraced column} = 22$$

As 58.29 is > 22 and < 100, use the moment magnification method.

$$E_c = 33w^{1.5}\sqrt{f'_c} = 33 \times 150^{1.5}\sqrt{5000}$$

$$= 4.29 \times 10^6 \text{ psi } (29.6 \times 10^6 \text{ kPa})$$

$$I_g = \frac{21(21)^3}{12} = 16,206.8 \text{ in.}^4$$

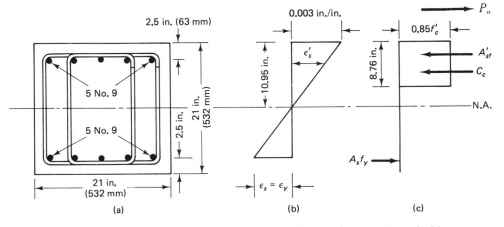

Figure 9.31 Column geometry: strain and stress diagrams (balanced failure): (a) cross section; (b) strains; (c) stresses.

$$EI = \frac{0.4 \, E_c I_g}{1 + \beta_d} = \frac{0.40 \, (4.29) \times 10^6 \times 16{,}206.8}{1 + 0.5}$$

$$= 18.54 \times 10^9 \text{ lb-in.}^2$$

$$(kl_u)^2 = (1.7 \times 18 \times 12)^2 = 134.8 \times 10^3 \text{ in.}^2$$

Hence

$$P_c = \text{Euler buckling load} = \frac{\pi^2 EI}{(kl_u)^2}$$

$$= \frac{\pi^2 \times 18.54 \times 10^9}{134.8 \times 10^3} = 1.356 \times 10^6 \text{ lb} = 1356 \text{ kips (6032 kN)}$$

$C_m = 1.0$ for nonbraced column

$$\text{moment magnifier } \delta_{ns} = \frac{C_m}{1 - P_u/0.75 P_c} = \frac{1.0}{1 - \dfrac{726}{0.75 \times 1356}} = 3.495$$

$$\text{design moment } M_c = \delta_{ns} M_{2ns} = 3.495 \times 1{,}525{,}000$$

$$= 5{,}329{,}875 \text{ in.-lb (561 kN-m)}$$

Assume that the reduction factor $\phi = 0.70$.

$$\text{required } P_n = \frac{P_u}{\phi} = \frac{726{,}000}{0.7} = 1{,}037{,}143 \text{ lb (4612.8 kN)}$$

$$\text{required } M_n = \frac{5{,}329{,}875}{0.7} = 7{,}614{,}107 \text{ in.-lb (1608 kN-m)}$$

Hence design a nonslender column section for an axial load strength $P_n = 1{,}037{,}143$ lb and a moment strength $M_n = 7{,}614{,}107$ in.-lb.

$$e = \frac{7{,}614{,}107}{1{,}037{,}143} = 7.34 \text{ in. (186 mm)}$$

Design of an equivalent nonslender column (step 3)

Analyze the assumed 21 in. $\times$ 21 in. square section. Assume that $\rho = p' \approx 1.25\%$.

$$A_s = A_s' = 0.0125(21 \times 18.5) = 4.86 \text{ in.}^2$$

Provide five No. 9 bars (five of 28-mm diameter) on each face; $A_s = A_s' = 5.0 \text{ in.}^2$ (3226 mm²).

$$c_b = d \times \frac{87{,}000}{87{,}000 + f_y} = 18.5 \times \frac{87{,}000}{87{,}000 + 60{,}000} = 10.95 \text{ in. (278 mm)}$$

$$a_b = \beta_1 c_b = (0.85 - 0.05) \times 10.95 = 8.76 \text{ in. (222.5 mm)}$$

$$f_s' = 0.003 \times 29 \times 10^6 \, \frac{10.95 - 2.5}{10.94} = 67{,}137 \text{ psi} > f_y$$

therefore, $f_s' = f_y$

$$P_{nb} = 0.85f'_cba = 0.85 \times 5000 \times 21 \times 8.76$$
$$= 781,830\,\text{lb}\,(3473.8\,\text{kN})$$
$$M_{nb} = 9,584,800\,\text{in.-lb}\,(2024.7\,\text{kN-m})$$
$$e_b = 12.25\,\text{in.}$$
$$>e = 7.34\,\text{in.}$$

The failure will be in compression.

Solving for P_n in the same manner as in Ex. 9.13, $c = 13.53$ in., $\phi = 0.7$, $P_n = 1,101,650$ lb $>$ required, $P_n = 1,037,143$ lb, and $M_n = 8,081,983$ in.-lb with eccentricity $e = 8,081,983/1,037,143 \simeq 7.34$ in.   For the required equivalent column, $e = 7.34$ in.; hence column O.K. for the nonsway case.

Design of ties (step 4)

Try No. 3 ties (9.52-mm diameter).   The spacing must be at least $h = 21$ in. (533.4 mm).

$$16 \text{ diameter No. 9 bar} = 18 \text{ in. (457.2 mm)}$$
$$48 \text{ diameter No. 3 tie} = 18 \text{ in. (457.2 mm)}$$

Hence use No. 3 (9.52-mm diameter) closed ties spread at 18 in. (455 mm) center to center.

Solution for Case 2: Gravity and Wind Loading (Sidesway)

$$\frac{\ell_u}{r} = \frac{18 \times 12}{0.3 \times 21} = 34.3$$

$$\frac{35}{\sqrt{P_u/f'_c A_g}} = \frac{35}{\sqrt{726,000/5000\,(21 \times 21)}} = 60 > 34.3$$

Hence, nonsway moment M_{2ns} need not be magnified.

$$U = 0.75(1.4D + 1.7L + 1.7W)$$

Also

$$U = 0.9D + 1.3W \quad (\text{did not control})$$

Therefore,

$$P_u = 0.75(726,000 + 90,000) = 612,000\,\text{lb}$$
$$M_{2ns} = 0.75 \times 1,525,000 = 1,143,750\,\text{in.-lb} \quad (\text{unmagnified})$$
$$M_{2s} = 0.75 \times 1,220,000 = 915,000\,\text{in.-lb}$$

$$\delta_s = \frac{1.0}{1 - \dfrac{\Sigma P_u}{0.75\Sigma P_c}} = \frac{1.0}{1 - \dfrac{15.5 \times 10^6}{0.75 \times 32.0 \times 10^6}} = 2.82 > 1.0$$

$$\text{hence } \delta_s M_s > M_s \quad \text{O.K.}$$

$$M_c = 1,143,750 + 2.82 \times 915,000 = 3,724,050\,\text{in.-lb}$$

$$\text{required } P_n = \frac{612,000}{0.7} = 874,286\,\text{lb}$$

$$\text{required } M_n = \frac{3{,}724{,}050}{0.7} = 5{,}320{,}070 \text{ in.-lb}$$

$$\text{eccentricity } e = \frac{5{,}320{,}070}{874{,}286} = 6.09 \text{ in.} < e_b = 12.25 \text{ in.}$$

Hence failure will be in compression.

Conditions for case 2 do not control since failure is still in compression and the required P_n is less than that for case 1. Adopt the same section 21 in. × 21 in. with five No. 9 reinforcing bars on each of the two faces parallel to the neutral axis.

9.16 COMPRESSION MEMBERS IN BIAXIAL BENDING

9.16.1 Exact Method of Analysis

Columns in corners of buildings are compression members subjected to biaxial bending about both the x and the y axes, as shown in Fig. 9.32. Also, biaxial bending occurs due to imbalance of loads in adjacent spans and almost always in bridge piers. Such columns are subjected to moments M_{xx} about the x axis, creating a load eccentricity e_y, and a moment M_{yy} about the y axis, creating a load eccentricity e_x. Thus the neutral axis is inclined at an angle θ to the horizontal.

The angle θ depends on the interaction of the bending moments about both axes and the magnitude of the load P_u. The compressive area in the column section can have one of the alternative shapes shown in Fig. 9.32c. Since such a column has to be designed from first principles, the trial-and-adjustment procedure has to be followed where compatibility of strain has to be maintained at all levels of the reinforcing bars. The process is similar to the one briefly outlined in Section 9.5.8 for columns with reinforcing bars on all faces. Additional computational effort is needed because of the position of the inclined neutral-axis plane and the four different possible forms of the concrete compression area.

Figure 9.33 shows the strain distribution and forces on a biaxially loaded rectangular column cross section. G_c is the center of gravity of the concrete compression area having coordinates x_c and y_c from the neutral axis in the directions x and y, respectively. G_{sc} is the resultant position of steel forces in the compression area having coordinates x_{sc} and y_{sc} from the neutral axis in the directions x and y, respectively. G_{st} is the resultant position of steel forces in the tension area having coordinates x_{st} and y_{st} from the neutral axis in directions x and y, respectively. From equilibrium of internal and external forces,

$$P_n = 0.85 f_s' A_c + F_{sc} - F_{st} \tag{9.51}$$

where A_c = area of the compression zone covered by the rectangular stress block
F_{sc} = resultant steel compressive forces ($\Sigma\, A_s' f_{sc}$)
F_{st} = resultant steel tensile force ($\Sigma\, A_s f_{st}$)

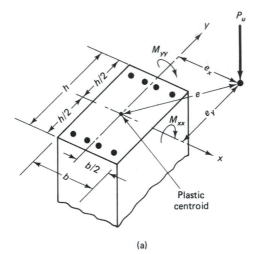

(a)

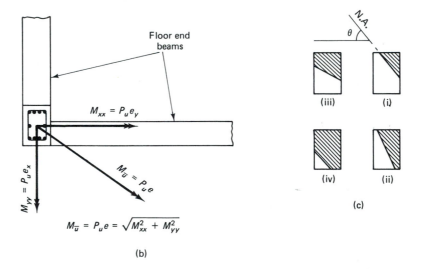

(b)

(c)

Figure 9.32 Corner column subjected to axial load: (a) biaxially stressed column cross section; (b) vector moments M_{xx} and M_{yy} in column plan.

From equilibrium of internal and external moments,

$$P_n e_x = 0.85 f_c' A_c x_c + F_{sc} x_{sc} + F_{st} x_{st} \qquad (9.52a)$$

$$P_n e_y = 0.85 f_c' A_c y_c + F_{sc} y_{sc} + F_{st} y_{st} \qquad (9.52b)$$

The position of the neutral axis has to be assumed in each trial and the stress calculated in *each* bar using

$$f_{si} = E_s \epsilon_{si} = E_s \epsilon_c \frac{s_i}{c} < f_y \qquad (9.53)$$

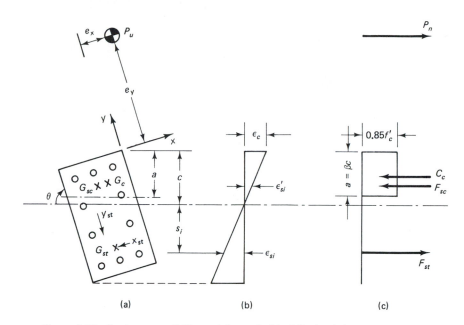

Figure 9.33 Strain-compatibility and forces in biaxially loaded rectangular columns: (a) cross section; (b) strain; (c) forces.

9.16.2 Load Contour Method

One method that gives a rapid solution is to design the column for the vector sum of M_{xx} and M_{yy} and use a circular reinforcing cage in a square section for the corner column.   However, such a procedure cannot be economically justified in most cases.   Another design approach well proven by experimental verification is to transform the biaxial moments into an equivalent uniaxial moment and an equivalent uniaxial eccentricity.   The section can then be designed for uniaxial bending, as previously discussed in this chapter, to resist the actual factored biaxial bending moments.

Such a method considers a failure surface instead of failure planes and is generally termed the *Bresler–Parme contour method.* This method involves cutting the three-dimensional failure surfaces in Fig. 9.34 at a constant value P_n to give an interaction plane relating M_{nx} and M_{ny}.   In other words, the contour surface S can be viewed as a curvilinear surface that includes a family of curves, termed the *load contours.*

The general nondimensional equation for the load contour at a constant load P_n may be expressed as follows:

$$\left(\frac{M_{nx}}{M_{ox}}\right)^{\alpha_1} + \left(\frac{M_{ny}}{M_{oy}}\right)^{\alpha_2} = 1.0 \qquad (9.54)$$

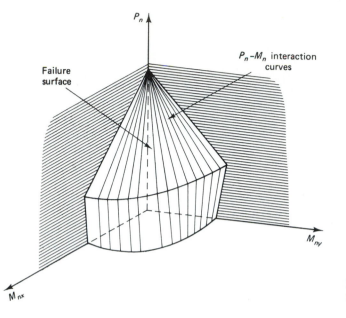

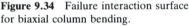

Figure 9.34 Failure interaction surface for biaxial column bending.

where $M_{nx} = P_n e_y$ and $M_{ny} = P_n e_x$

$M_{ox} = M_{nx}$ at such an axial load P_n where M_{ny} or $e_x = 0$

$M_{oy} = M_{ny}$ at such an axial load P_n when M_{nx} or $e_y = 0$

The moments M_{ox} and M_{oy} are the *required* equivalent resisting moment strengths about the x and y axes, respectively.

α_1, α_2 = exponents depending on the cross-section geometry, steel percentage, and its location and material stresses f_c' and f_y

Equation 9.54 can be simplified using a common exponent and introducing a factor β for one particular axial load value P_n such that the M_{nx}/M_{ny} ratio would have the same value as the M_{ox}/M_{oy} as detailed by Parme and associates. Such simplification leads to

$$\left(\frac{M_{nx}}{M_{ox}}\right)^\alpha + \left(\frac{M_{ny}}{M_{oy}}\right)^\alpha = 1.0 \tag{9.55}$$

where α would have a value of (log 0.5/log β). Figure 9.35 gives a contour plot *ABC* from Eq. 9.55.

For design purposes, the contour is approximated by two straight lines *BA* and *BC*, and Eq. 9.55 can be simplified to two conditions:

1. For *AB* when $M_{ny}/M_{oy} < M_{nx}/M_{ox}$,

$$\frac{M_{nx}}{M_{ox}} + \frac{M_{ny}}{M_{oy}}\left[\frac{1-\beta}{\beta}\right] = 1.0 \tag{9.56a}$$

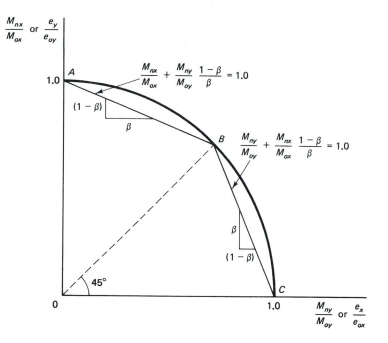

Figure 9.35   Modified interaction contour plot of constant P_n for biaxially loaded column.

2. For BC when $M_{ny}/M_{oy} > M_{nx}/M_{ox}$

$$\frac{M_{ny}}{M_{oy}} + \frac{M_{nx}}{M_{ox}}\left[\frac{1-\beta}{\beta}\right] = 1.0 \qquad (9.56b)$$

In both Eqs. 9.56 a and b, the *actual* controlling equivalent uniaxial moment strength M_{oxn} or M_{oyn} should be at least equivalent to the *required* controlling moment strength M_{ox} or M_{oy} of the chosen column section.

For rectangular sections where the reinforcement is evenly distributed along all the column faces, the ratio M_{oy}/M_{ox} can be approximately taken as equal to b/h. Hence Eqs. 9.56a and b can be modified as follows:

1. For $\dfrac{M_{ny}}{M_{nx}} > b/h$,

$$M_{ny} + M_{nx}\frac{b}{h}\frac{1-\beta}{\beta} \simeq M_{oy} \qquad (9.57a)$$

2. For $\dfrac{M_{ny}}{M_{nx}} \le b/h$,

$$M_{nx} + M_{ny}\frac{h}{b}\frac{1-\beta}{\beta} \simeq M_{ox} \qquad (9.57b)$$

The controlling required moment strength M_{ox} or M_{oy} for designing the section is the larger of the two values as determined from Eq. 9.57a or b.

Plots of Fig. 9.36 are used in the selection of β in the analysis and design of such columns. In effect, the modified load–contour method can be summarized in Eq. 9.55 as a method for finding an equivalent required moment strength M_{ox} and M_{oy} for designing the columns as if they were uniaxially loaded.

9.16.3 Step-by-Step Operational Procedure for the Design of Biaxially Loaded Columns

The following steps can be used as a guideline for the design of columns subjected to bending in both the x and y directions. The procedure assumes an equal area of reinforcement on all four faces.

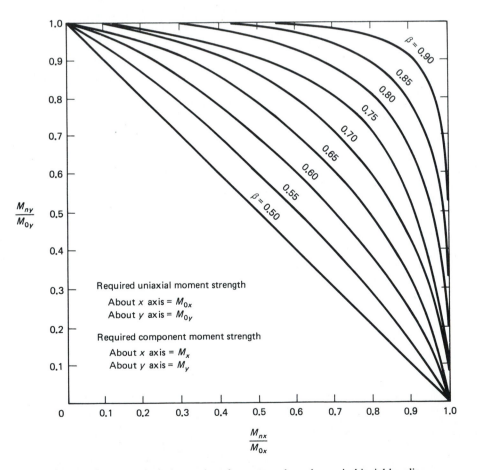

Figure 9.36   Contour β-factor chart for rectangular columns in biaxial bending.

1. Calculate the uniaxial bending moments assuming an equal number of bars on each column face. Assume a value of an interaction contour β factor between 0.50 and 0.70. Assume a ratio of h/b. This ratio can be approximated to M_{nx}/M_{ny}. Using Eqs. 9.57a and b, determine the equivalent required uniaxial moment M_{ox} or M_{oy}. If M_{nx} is larger than M_{ny}, use M_{ox} for the design, and vice versa.

2. Assume a cross section for the column and a reinforcement ratio $\rho = \rho' \simeq$ 0.01 to 0.02 on each of the two faces parallel to the axis of bending of the larger equivalent moment. Make a preliminary selection of the steel bars. Verify the capacity P_n of the assumed column cross section. In the completed design, the same amount of longitudinal steel should be used on all four faces.

3. Calculate the *actual* nominal moment strength M_{oxn} for equivalent uniaxial bending about the x axis when $M_{oy} = 0$. Its value has to be at least equivalent to the *required* moment strength M_{ox}.

4. Calculate the actual nominal moment strength M_{oyn} for the equivalent uniaxial bending moment about the y axis when $M_{ox} = 0$.

5. Find M_{ny} by entering M_{nx}/M_{oxn} and the trial β value into the β factor contour plots of Fig. 9.36.

6. Make a second trial and adjustment, increasing the assumed β value if the M_{ny} value obtained from entering the chart is less than the required M_{ny}. Repeat this step until the two values of M_{ny} converge either through changing β or changing the section.

7. Design the lateral ties and detail the section.

9.16.4 Example 9.16: Design of a Biaxially Loaded Column

A corner column is subjected to a factored compressive axial load $P_u = 210,000$ lb (945 kN), a factored bending moment $M_{ux} = 1,680,000$ in.-lb (189.8 kN-m) about the x axis, and a factored bending moment $M_{uy} = 980,000$ in.-lb (110.7 kN-m) about the y axis, as shown in Fig. 9.37. Given:

$f'_c = 4000$ psi (27.6 MPa), normal-weight concrete
$f_y = 60,000$ psi (414 MPa)

Design a rectangular tied column section to resist the biaxial bending moments resulting from the given eccentric compressive load.

Solution   *Calculate the equivalent uniaxial bending moments assuming equal numbers of bars on all faces (step 1)*

Assume that $\phi = 0.70$ for tied columns.

$$\text{required nominal } P_n = \frac{210,000}{0.70} = 300,000 \text{ lb (1350 kN)}$$

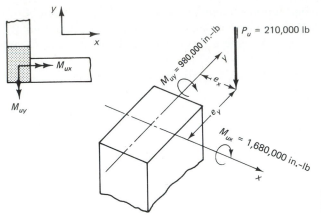

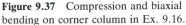

Figure 9.37   Compression and biaxial bending on corner column in Ex. 9.16.

$$\text{required nominal } M_{nx} = \frac{1,680,000}{0.70} = 2,400,000 \text{ in.-lb (271.2 kN-m)}$$

$$\text{required nominal } M_{ny} = \frac{980,000}{0.70} = 1,400,000 \text{ in.-lb (158.2 kN-m)}$$

Analyze for equivalent moment and equivalent eccentricity about the x axis since the larger of the two biaxial moments is $M_{nx} = 2,400,000$ in.-lb about the x axis.

$$\frac{M_{nx}}{M_{ny}} = \frac{2,400,000}{1,400,000} = 1.71$$

Since the column dimensions are proportional to the applied moments, assume that $h/b \simeq 1.71$ or $b = 12$ in. and $h = 20$ in. to give $h/b = 1.67$. Assume that the interaction contour factor $\beta = 0.61$.

$$\text{equivalent } M_{ox} = M_{nx} + M_{ny} \frac{h}{b} \frac{1 - \beta}{\beta}$$

or

$$M_{ox} = 2,400,000 + 1,400,000(1.67) \frac{1 - 0.61}{0.61} = 3,894,787 \text{ in.-lb (440.1 kN-m)}$$

($M_{ox} = 3,891,803$ in.-lb from exact solution.)

Verify capacity P_n of the assumed column section (step 2)

From step 1, $b = 12$ in. (304.8 mm) and $h = 20$ in. (508.0 mm). Assume that the steel ratio $\rho = \rho' \simeq 0.012$ and $d' = 2.5$ in. (63.5 mm). $d = 20.0 - 2.5 = 17.5$ in. (445.5 mm).

$$A_s = A_s' = 0.012 \times 12(20.0 - 2.5) = 2.52 \text{ in.}^2$$

Try $A_s = A_s' = 2.37$ in.^{2} (1528.7 mm^2) on each of the two 12-in. faces parallel to the x axis of bending in Fig. 9.38.

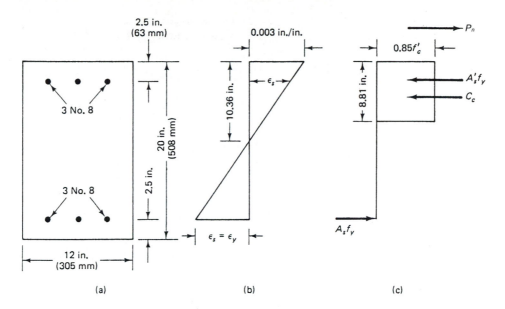

Figure 9.38 Equivalent column geometry: strain and stress diagrams (balanced failure): (a) cross section; (b) strains; (c) stresses.

Analyze the balanced condition

$$c_b = \frac{87,000}{87,000 + f_y} d = \frac{87,000}{87,000 + 60,000} 17.5 = 10.36 \text{ in.}$$

$$a_b = 0.85 \times 10.36 = 8.81 \text{ in. (223.7 mm)}$$

$$f'_s = \epsilon_c E_s \frac{c - d'}{c} = 0.003 \times 29 \times 10^6 \frac{10.36 - 2.5}{10.36}$$

$$= 66,006 \text{ psi} > f_y = 60,000 \text{ psi}$$

Hence $f'_s = 60,000$ psi.

$$P_{nb} = 0.85 f'_c b a_b + A'_s f_y - A_s f_y$$

$$= 0.85 \times 4000 \times 12.0 \times 8.81 + 2.37 \times 60,000 - 2.37 \times 60,000$$

$$= 359,448 \text{ lb} > P_n = 300,000 \text{ lb}$$

Therefore, tension controls with failure by initial yielding of the tension steel.

Evaluate the axial load capacity P_n

$$\rho = \frac{A_s}{bd} = \frac{2.37}{12 \times 17.5} = 0.0113$$

$$e = \frac{M_{ox}}{P_n} = \frac{3,894,787}{300,000} = 12.98 \text{ in. (329.8 mm)}$$

$$e' = e + \frac{d - d'}{2} = 12.98 + \frac{17.5 - 2.5}{2} = 20.48 \text{ in.}$$

$$1 - \frac{e'}{d} \quad \text{or} \quad \frac{h - 2e}{2d} = 1 - \frac{20.48}{17.5} = -0.17$$

$$1 - \frac{d'}{d} = 1 - \frac{2.5}{17.5} = 0.857$$

$$m = \frac{f_y}{0.85f'_c} = \frac{60,000}{0.85 \times 4000} = 17.65$$

The following equation applies for initial tension failure in columns if the compression steel A'_s is assumed to have yielded.

$$P_n = 0.85f'_cbd \left[\left(1 - \frac{e'}{d} \right) + \sqrt{\left(1 - \frac{e'}{d} \right)^2 + 2m\rho\left(1 - \frac{d'}{d} \right)} \right]$$

neglecting the area of concrete replaced by the steel; or

$$P_n = 0.85 \times 4000 \times 12 \times 17.5[-0.17$$

$$+ \sqrt{(-0.17)^2 + 2 \times 17.65 \times 0.0113 \times 0.857}]$$
$$= 714,000(0.4389) = 313,368 \text{ lb} > 300,000$$

Accept this for the first trial.

Check assumed value of $\phi = 0.70$

$$\phi = 0.90 - \frac{2.0P_u}{f'_cA_g} = 0.90 - \frac{2.0 \times 0.7 \times 313,368}{4000 \times 12 \times 20} = 0.44 < 0.70$$

Assume that $\phi = 0.70$. O.K.

Strain-compatibility check: $P_n = 0.85f'_cba$ for $A'_sf_y = A_sf_y$, or

$$a = \frac{P_n}{0.85f'_cb} = \frac{313,368}{0.85 \times 4000 \times 12} = 7.68 \text{ in.}$$

$$c = \frac{a}{\beta_1} = \frac{7.35}{0.85} = 9.04 \text{ in.}$$

$$f'_s = 87,000 \frac{9.04 - 2.5}{9.04} = 62,940 \text{ psi} > f_y = 60,000$$

Hence A'_s yielded a satisfying compatibility of strain requirement for the expression used to evaluate P_n.

Calculate the actual nominal resisting moment M_{oxn} *for equivalent uniaxial bending about the x axis when* $M_{oy} = 0$ *(step 3)*

Required $P_n = 300,000$ lb. Assuming that the compression steel has yielded (to be verified later), $f'_s = 60,000$ and $A_sf'_s - A_sf_y = 0$. Hence $P_n = 0.85f'_cba$,

or

$$a = \frac{P_n}{0.85f'_c b} = \frac{300{,}000}{0.85 \times 4000 \times 12} = 7.35 \text{ in.}$$

$$c = \frac{a}{\beta_1} = \frac{7.35}{0.85} = 8.65 \text{ in.}$$

$$f'_s = 87{,}000\left(\frac{8.65 - 2.5}{8.65}\right) = 61{,}855 \text{ psi} > 60{,}000 \qquad \text{O.K.}$$

A_s at the tension side has to yield since $a = 7.35$ in. $< a_b = 8.81$ in. with the neutral axis shallower than the balanced condition.

$$M_{oxn} = P_n e = 0.85f'_c ba\left(\bar{y} - \frac{a}{2}\right) + A'_s f_y(\bar{y} - d') + A_s f_y(d - \bar{y})$$

or

$$M_{oxn} = 0.85 \times 4000 \times 12 \times 7.35\left(\frac{20}{2} - \frac{7.35}{2}\right) + 2.37 \times 60{,}000\left(\frac{20}{2} - 2.5\right)$$

$$+ 2.37 \times 60{,}000\left(17.5 - \frac{20}{2}\right)$$

$$= 4{,}029{,}741 \text{ in.-lb } (455.4 \text{ kN-m}) > M_{ox}(3{,}894{,}787 \text{ in.-lb}) \qquad \text{O.K.}$$

If this calculation showed that $M_{oxn} < M_{ox}$ obtained in step 1, revise the assumed cross section by increasing the steel area or enlarging the section, or both.

Calculate the actual and nominal resisting moment M_{oyn} for the equivalent uniaxial bending moment about the y axis when $M_{ox} = 0$ (step 4)

In this condition, $b = 20$ in., $h = 12$ in., $d = 9.5$ in., and $A_s = A'_s = 2.37$ in.². By trial and adjustment, choose compressive block depth a such that the calculated P_n approximates the required P_n.

At the third trial, $a = 4.8$ in. and $c = 4.8/0.85 = 5.65$ in.

$$f'_s = 87{,}000\left(\frac{5.65 - 2.5}{5.65}\right) = 48{,}504 \text{ psi}$$

$$f_s = 87{,}000\left(\frac{d - c}{c}\right) = 87{,}000\left(\frac{9.5 - 5.65}{5.65}\right) = 59{,}283 \text{ psi}$$

$$P_n = 0.85 \times 4000 \times 20 \times 4.8 + 2.37 \times 48{,}504 - 2.37 \times 59{,}283$$

$$= 300{,}854 \simeq \text{required } P_n = 300{,}000 \qquad \text{O.K.}$$

Hence use $a = 4.8$ for calculating M_{oyn}.

$$M_{oyn} = 0.85 \times 4000 \times 20 \times 4.8\left(\frac{12}{2} - \frac{4.8}{2}\right) + 2.37 \times 48{,}504\left(\frac{12}{2} - 2.5\right)$$

$$+ 2.37 \times 59{,}283\left(9.5 - \frac{12}{2}\right) = 2{,}069{,}133 \text{ in.-lb}$$

Find M_{ny} by entering M_{nx}/M_{oxn} and trial β value into the β-factor contour plots in Fig. 9.36 (step 5)

First trial $\beta = 0.61$.   From step 3, $M_{oxn} = 4,029,741$ in.-lb

$$\frac{M_{nx}}{M_{oxn}} = \frac{2,400,000}{4,029,741} = 0.596$$

Enter into Fig. 9.36 the values of $\beta = 0.61$ and $M_{nx}/M_{oxn} = 0.596$ to get

$$\frac{M_{ny}}{M_{oyn}} = 0.62$$

But M_{oyn} from step 4 $= 2,069,133$ in.-lb, or

$$\frac{M_{ny}}{2,069,133} = 0.62$$

Hence

$$M_{ny} = 0.62 \times 2,069,133 = 1,282,862 \text{ in.-lb}$$
$$< \text{ required } M_{ny} = 1,400,000 \text{ in.-lb}$$

Revise the solution assuming a higher β value.   If adjusting β does not give the actual M_{ny} at least equal to the required M_{ny}, increase the reinforcement area or enlarge the section.

Second trial and adjustment (step 6)

Assume the same section but assume that $\beta = 0.64$.

$$M_{ox} = 2,400,000 + 1,400,000 \times 1.67 \times \frac{1 - 0.64}{0.64} = 3,715,125 \text{ in.-lb}$$

equivalent $e = \dfrac{3,715,125}{300,000} = 12.38$ in.

$$e' = 12.38 + 7.5 = 19.88$$

$$1 - \frac{e'}{d} = 1 - \frac{19.88}{17.5} = -0.136$$

$$1 - \frac{d'}{d} = 0.857 \text{ from step 2}$$

$$P_n = 714,000[-0.136 + \sqrt{(-0.136)^2 + 2 \times 17.65 \times 0.0113 \times 0.857}]$$
$$= 714,000(0.4642) = 331,472 \text{ lb} > 300,000 \qquad \text{O.K.}$$

Strain-compatibility check

$$a = \frac{P_n}{0.85 f'_c b} = \frac{331,472}{0.85 \times 4000 \times 12} = 8.12 \text{ in.} > \text{first trial} \quad (a = 7.68 \text{ in.}).$$

Hence the neutral axis would be lower, giving a larger value of strain ϵ'_s or $f'_s = f_y$ = yield strength assumed in calculating P_n.   From step 3, $M_{oxn} = 4,029,741$ in.-lb since this value does not change as long as the section and its reinforcement remain the same. $M_{nx}/M_{oxn} = 0.596$ from before. $\beta = 0.64$.   From contour plots in Fig. 9.36, $M_{ny}/M_{oyn} = 0.68$.   From step 4, $M_{oyn} = 2{,}069{,}133$.

$$M_{ny} = 0.68 \times 2{,}069{,}133 = 1{,}407{,}010 \text{ in.-lb} > \text{required } M_{ny} = 1{,}400{,}000$$

Adopt the design.

The use of hand-held computers, microcomputers, or charts in steps 2 to 6 reduces the calculation effort for biaxially loaded columns almost to that for the design of uniaxially loaded columns.

Select the longitudinal and lateral reinforcement (step 7)

Longitudinal bars: Provide three No. 8 bars (25.4-mm diameter) on each of the two 12-in.-wide faces.   Provide one No. 8 bar at the center of the 20-in.-wide face so that each face of this column would have an *equal* number of reinforcing bars.

Lateral ties: Try No. 3 bar lateral ties.   The spacing *s* should be the minimum of

$$16 \times \text{longitudinal bar diameter} = 16 \times \frac{8}{8} = 16 \text{ in.}$$

$$48 \times \text{lateral tie diameter} = 48 \times \frac{3}{8} = 18 \text{ in.}$$

The minimum lateral dimension = 12 in. Therefore, provide No. 3 (9.53-mm diameter) lateral ties at 12 in. (304.8 mm) center to center.   Reinforcing details are shown in Fig. 9.39.

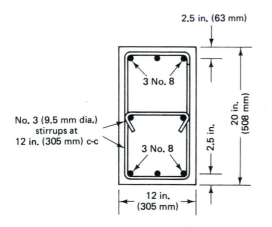

Figure 9.39   Biaxially loaded column section.

9.17 SI EXPRESSIONS AND EXAMPLE FOR THE DESIGN OF COMPRESSION MEMBERS

1. $c_b = \dfrac{600}{600 + f_y} d$

2. $f'_s = 0.003 E_s \dfrac{c_b - d'}{c_b} \leq f_y$

3. $E_c = w_c^{1.5}\, 0.043\sqrt{f_c'}$ MPa where $w_c = 1500$ to 2500 kg/m³ (90 to 1554 lb/ft³). For standard normal-weight concrete, $w_c = 2400$ kg/m³ to give $E_c = 29{,}700$ MPa.

4. Modulus of rupture $f_r = 0.7\sqrt{f_c'}$

5. $E_s = 200{,}000$ MPa

6. $\phi P_{n(\max)} = 0.80\phi[0.85\, f_c'(A_g - A_{st}) + f_y A_{st}]$, where $\phi = 0.80$ for tied columns and 0.85 for spirally reinforced columns

7. Ratio of spiral reinforcement should not be less than the value

$$\rho_s = 0.45 \left(\frac{A_g}{A_c} - 1\right)\frac{f_c'}{f_y}$$ where f_y is the specified yield strength of the spirals, but not to exceed 400 MPa.

8. $\dfrac{k\ell_n}{r} \le \left(34 - 12\dfrac{M_1}{M_2}\right)$

9. $EI = \dfrac{0.2\, E_c I_g + E_s I_{se}}{1 + \beta_d}$ or $EI = \dfrac{0.4\, E_c I_g}{1 + \beta_d}$

10. For compression members without transverse loads

$$c_m = 0.6 + 0.4\frac{M_1}{M_2} \ge 0.4$$

where M_1/M_2 is positive for single curvature. For members with transverse loads between supports, $c_m = 1.0$.

11. $M_{2,\min} = P_u(15 + 0.03h)$ about each axis separately, where 15 and h are in millimeters.

12. $M_1 = M_{1ns} + \delta_s M_{1s}$ and $M_2 = M_{2ns} + \delta_s M_{2s}$

13. $\delta_s M_s = \dfrac{M_s}{1 - \sum P_u\,(0.75\sum P_c)} \ge M_s$, where $\sum P_c = \sum \dfrac{\pi^2 E}{\ell r^2}$ as the Euler buckling load.

9.17.1 SI Example on Column Design

Solve Ex. 9.5 using SI units.

Data

$f_c' = 27.6$ MPa   (MPa = N/mm²)

$f_y = 414$ MPa

$e = 305$ mm

$b = 305$ mm

$h = 381$ mm

$A_s = A_s' =$ three No. 9 bars (28.7-mm diameter) $= 1936$ mm².

Use in this solution $A_s = A_s' =$ four No. 25 M $= 2000$ mm².².

Assume $d = 66$ mm to give $d = 381 - 66 = 315$ mm.

Solution

$$c_b = \left(\frac{600}{600 + f_y}\right) d = \left(\frac{600}{600 + 414}\right) 315$$

$$= 187 \text{ mm}$$

$$a_b = 0.85 \times 187 = 159 \text{ mm}$$

$$f_s' = 0.003 E_s \left(\frac{c_b - d'}{c_b}\right) < f_y$$

$$= 0.003 \times 200,000 \left(\frac{187 - 66}{187}\right) = 388 \text{ MPa}$$

$$< f_y = 414 \text{ MPa}$$

Hence compression reinforcement did not yield. And, trial and adjustment is to be applied to find the actual f_s'.

From Eq. 9.12,

$$P_{nb} = 0.85 f_c' b a_b + A_s' f_s' - A_s f_y$$

$$= 0.85 \times 27.6 \times 305 \times 159 + 2000 \times 388 - 2000 \times 414$$

$$= 1.09 \times 10^6 \text{ N} = 1090 \text{ kN}$$

From Eq. 9.13,

$$M_{nb} = P_{nb} e_b = 0.85 f_c' b a_b \left(\bar{y} - \frac{ab}{2}\right) + A_s' f_s' (\bar{y} - d') + A_s f_y (d - \bar{y})$$

where $\bar{y}$ for rectangular sections $= h/2 = 381/2 = 190.5$ mm.

$$M_{nb} = 0.85 \times 27.6 \times 305 \times 159 \left(190.5 - \frac{159}{2}\right)$$

$$+ 2000 \times 388(190.5 - 66) + 2000 \times 414(315 - 190.5)$$

$$= 3.26 \times 10^8 \text{ N-mm} = 326 \text{ kN-m}$$

$$e_b = \frac{M_{nb}}{P_{nb}} = \frac{326}{1090} \times 1000 \text{ mm} = 299 \text{ mm} < e = 305 \text{ mm}$$

Hence, initial failure will occur by initial yielding of the tension reinforcement.

Trial and adjustment procedure

Assume that $c = 180$ mm to give $a = 153$ mm. From similar triangles,

$$f_s' = 0.0003 E_s \left(\frac{c - d'}{c}\right) = 0.003 \times 200,000 \left(\frac{180 - 66}{180}\right)$$

$$= 380 \text{ MPa}$$

$$P_n = 0.85 \times 27.6 \times 305 \times 153 + 2000 \times 380 - 2000 \times 414$$

$$= 1.03 \times 10^6 \text{ N}$$

$$M_n = 0.85 \times 27.6 \times 305 \times 153 \left(190.5 - \frac{153}{2}\right) + 2000 \times 380(190.5 - 66)$$

$$+ 2000 \times 414(315 - 190.5) = 322 \times 10^6 \text{ N-mm}$$

$$e = \frac{322 \times 10^6}{1.03 \times 10^6} = 312 \text{ mm} > \text{actual } e = 305 \text{ mm}$$

Proceed to second cycle; assume that $c = 186$ mm to give $a = 158$ mm.

$$f'_s = 0.003 \times 200,000 \left(\frac{186 - 66}{186} \right) = 387 \text{ MPa}$$

$$P_n = 0.85 \times 27.6 \times 305 \times 158 + 2000 \times 387 - 2000 \times 414$$

$$= 1.07 \times 10^6 \text{ N}$$

$$M_n = 0.85 \times 27.6 \times 305 \times 158 \left(190.5 - \frac{158}{2} \right) + 2000 \times 387(190.5 - 66)$$

$$+ 2000 \times 414(315 - 190.5) = 326 \times 10^6 \text{ N-mm}$$

$$e = \frac{326 \times 10^6}{1.07 \times 10^6} \simeq 304 \text{ mm} \simeq \text{actual } e = 305 \text{ mm}$$

Hence, column capacity $P_n = 1070$ kN.

SELECTED REFERENCES

9.1. ACI Committee 318, *Building Code Requirements for Reinforced Concrete*, ACI Standard 318-95; and the *Commentary on Building Code Requirements for Reinforced Concrete*, American Concrete Institute, Detroit, 1995.

9.2. Hognestad, E., *A Study of Combined Bending and Axial Load in Reinforced Concrete Members*, Bulletin 399, University of Illinois, Urbana, Ill., November 1951.

9.3. ACI Committee 105, "Reinforced Concrete Column Investigation," *Journal of the American Concrete Institute*, Vol. 26, April 1930, pp. 601–612. Also see Vol. 27, pp. 675–676; Vol. 28, pp. 157–158; Vol. 29, pp. 53–56 and 274–284; Vol. 30, pp. 78–90 and 153–156.

9.4. Richart, F. E., Draffin, J. O., Olson, T. A., and Heitman, R. H., *The Effect of Eccentric Loading, Protective Shells, Slenderness Ratios, and Other Variables in Reinforced Concrete Columns*, Bulletin 368, Engineering Experiment Station, University of Illinois, Urbana, Ill., 1947, 130 pp.

9.5. Whitney, C. S., "Plastic Theory of Reinforced Concrete Design," *Transactions of the ASCE*, Vol. 107, 1942, pp. 251–326.

9.6. American Institute of Steel Construction, *Specifications for the Design, Fabrication and Erection of Structural Steel for Buildings* and the *Commentary on Their Specifications*, AISC, New York, 1969, and November 1978, 166 pp.

9.7. Johnston, B. G. (ed.), *Structural Stability Research Council Guide to Stability Design Criteria for Metal Structures*, Wiley, New York, 1966, 217 pp.

9.8. American Concrete Institute, *Design Handbook in Accordance with the Strength Design Method*, Vol. 2, *Columns*, Publication S-17A (91), ACI, Detroit, 1991.

9.9. Broms, B., and Viest, I. M., "Long Reinforced Concrete Columns," *Symposium Proceedings*, ASCE Transactions Paper 3155, January 1958, pp. 309–400.

9.10. Timoshenko, S. P., and Gere, J. M., *Theory of Elastic Stability*, 2nd ed., McGraw-Hill, New York, 1961, 541 pp.

9.11. Bresler, B., "Design Criteria for Reinforced Concrete Columns under Axial Load and Biaxial Bending," *Journal of the American Concrete Institute*, Proc. Vol. 57, November 1960, pp. 481–490.

9.12. Parme, A. L., Nieves, J. M., and Gouens, A., "Capacity of Reinforced Rectangular Columns Subjected to Biaxial Bending," *Journal of the American Concrete Institute*, Proc. Vol. 63, No. 9, September 1966, pp. 911–921.

PROBLEMS FOR SOLUTION

9.1. Calculate the axial load strength P_n for columns having the cross sections shown in Fig. 9.40. Assume zero eccentricity for all cases. Cases (a), (b), (c), and (d) are tied columns; case (e) is spirally reinforced.

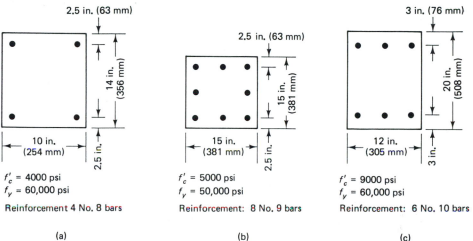

2.5 in. (63 mm)

14 in. (356 mm)

10 in. (254 mm)

2.5 in.

f'_c = 4000 psi
f_y = 60,000 psi

Reinforcement 4 No. 8 bars

(a)

2.5 in. (63 mm)

15 in. (381 mm)

15 in. (381 mm)

2.5 in.

f'_c = 5000 psi
f_y = 50,000 psi

Reinforcement: 8 No. 9 bars

(b)

3 in. (76 mm)

20 in. (508 mm)

12 in. (305 mm)

3 in.

f'_c = 9000 psi
f_y = 60,000 psi

Reinforcement: 6 No. 10 bars

(c)

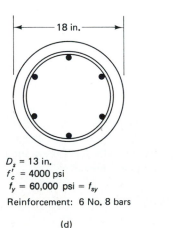

18 in.

D_s = 13 in.
f'_c = 4000 psi
f_y = 60,000 psi = f_{sy}

Reinforcement: 6 No. 8 bars

(d)

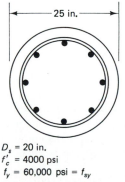

25 in.

D_s = 20 in.
f'_c = 4000 psi
f_y = 60,000 psi = f_{sy}

Reinforcement: 8 No. 10 bars

(e)

Figure 9.40 Column sections.

9.2. Calculate P_u and e in Fig. 9.40a and c of Problem 9.1.   Assume that the stresses in the tension steel are zero.

9.3. Determine P_{ub} and e_b for the rectangular cross sections of Fig. 9.40 in Problem 9.1.

9.4. For the cross section shown in Fig. 9.40a of Problem 9.1, determine the safe eccentricity e if $P_u = 20,000$ lb and the safe P_u if $e = 15$ in.

9.5. For the cross section shown in Fig. 9.40c of Problem 9.1, determine the safe eccentricity e if $P_u = 900,000$ lb.   Use the trial-and-adjustment method satisfying the compatibility of strains.

9.6. Repeat Problem 9.5 using Whitney's approximate procedure.   Compare the results.

9.7. Construct the load–moment interaction diagram for the cross sections shown in Fig. 9.40a and c of Problem 9.1.

9.8. Obtain the equivalent rectangular column for the cross sections of Fig. 9.40d and e of Problem 9.1.   Also calculate the balanced load P_{ub} and the balanced eccentricity e_b using the equivalent cross section.

9.9. For the cross section shown in Fig. 9.40d of Problem 9.1, calculate the design load P_u if $e = 6$ in. Repeat the calculation for $e = 20$ in.

9.10. Design the reinforcement for a nonslender 15 in. $\times$ 20 in. column to carry the following loading.   The factored ultimate axial force $P_u = 300,000$ lb.   The eccentricity e to plastic centroid $= 6$ in. Given:

$$f'_c = 4000 \text{ psi}$$
$$f_y = 60,000 \text{ psi}$$

9.11. Design a nonslender column to support the following service loads and moments. $P_L = 100$ kips, $P_D = 50$ kips, $M_L = 2500$ in.–kips, and $M_D = 1000$ in.–kips.   Given:

$$f'_c = 5000 \text{ psi}$$
$$f_y = 60,000 \text{ psi}$$

9.12. Design a nonslender circular column to support a factored ultimate load $P_u = 250,000$ lb and a factored moment $M_u = 5 \times 10^6$ in.-lb.   Given:

$$f'_c = 6000 \text{ psi, normal-weight concrete}$$
$$f_y = 60,000 \text{ psi}$$
$$d' = 2.50 \text{ in.}$$

9.13. Design the reinforcement for a 14 in. $\times$ 20 in. braced rectangular reinforced concrete column that can support a factored axial load $P_u = 500,000$ lb and a factored moment $M_u = 3,500,000$ in.-lb.   The unsupported length, l_u, of the column is 10 ft.   Assume that the end moments M_1 and M_2 are equal.   Given:

$$f'_c = 4000 \text{ psi, sand-lightweight concrete}$$
$$f_y = 60,000 \text{ psi}$$
$$d' = 2.5 \text{ in.}$$

9.14. A rectangular braced column of a multistory frame building has a floor height $l_u = 25$ ft. It is subjected to service dead-load moments $M_2 = 3,500,000$ on top and $M_1 = 2,500,000$ in.-lb at the bottom. The service live-load moments are 80% of the dead-load moments. The column carries a service axial dead load $P_D = 200,000$ lb and a service live load $P_L = 350,000$. Design the cross-section size and reinforcement for this column. Given:

$$f'_c = 7000 \text{ psi}$$
$$f_y = 60,000 \text{ psi}$$
$$\psi_A = 1.3, \quad \psi_B = 0.9$$
$$d' = 2.5 \text{ in.}$$

9.15. A rectangular unbraced exterior column of a multibay, multifloor frame system is subjected to $P_u = 500,000$ lb, factored end moments $M_1 = 2,500,000$ in.-lb, and $M_2 = 3,500,000$ in.-lb. The unbraced length, l_u, of the column $= 18$ ft. Design this column if
(a) it is subjected to gravity loads with sidesway considered as negligible.
(b) it is subjected to wind load resulting in a sidesway factored moment $M_u = 2,100,000$ in.-lb. Assume the total loading of all interior and exterior columns in a single floor is $\Sigma P_u = 20 \times 10^6$ lb and $\Sigma P_c = 44 \times 10^6$ lb. Use a section having a width $b = 14$. Given:

$$f'_c = 6500 \text{ psi, normal-weight concrete}$$
$$f_y = 60,000 \text{ psi}$$
$$\psi_A = 2, \quad \psi_B = 1.2$$
$$d' = 2.5 \text{ in.}$$

9.16. Design the column in Problem 9.15 using a circular cross section.

9.17. The columns of the first floor in a nine-story 7×3 bays office building have a clear height of 18 ft (5.49 m). They are not braced against sidesway, and the clear height above or below the first floor is 11 ft (3.35 m). Assume in your solution that the exterior columns have the same section as the interior columns. Design a typical interior column in that floor. Given:

$$\Sigma P_c = 38 \times 10^6 \text{ lb}, \quad \Sigma P_u = 16 \times 10^6 \text{ lb}$$

$\dfrac{EI}{l}$ for the connecting beams $= 450 \times 10^6$ in.-lb (50,850 kN-m)

Service loads for *interior columns* (lb) are
$$D = 360,000, \quad L = 130,000, \quad W = 0$$

Service loads for *exterior columns* (lb) are
$$D = 80,000, \quad L = 65,000, \quad W = 5000$$

Service moments for the *interior columns* (in.-lb) are

Top: $D = 200,000, \quad L = 160,000, \quad W = 600,000$
Bottom: $D = 500,000, \quad L = 400,000, \quad W = 600,000$

Service moments for *exterior columns* (in.-lb) are

Top: $D = 400,000$, $L = 240,000$, $W = 300,000$

Bottom: $D = 700,000$, $L = 360,000$, $W = 300,000$

$f'_c = 5000$ psi

$f_y = 60,000$ psi

$d' = 2.5$ in.

9.18. Design a typical exterior column for the structural framing system in Problem 9.15.

9.19. A nonslender square corner column is subjected to biaxial bending about its x and y axes. It supports a factored load $P_u = 200,000$ lb acting at eccentricities $e_x = e_y = 7$ in. Design the column size and reinforcement needed to resist the applied stresses. Given:

$f'_c = 5000$ psi

$f_y = 60,000$ psi

gross reinforcement percentage $\rho_g = 0.03$

$d' = 2.5$ in.

9.20. Design a nonslender rectangular end column to support a factored load $P_u = 200,000$ lb acting at eccentricities $e_x = 9.0$ in. and $e_y = 6.0$ in. Try a 12-in.-wide section with a total gross reinforcement percentage not to exceed $\rho_g = 0.025$. Given:

$f'_c = 5000$ psi, normal-weight concrete

$f_y = 60,000$ psi

$d' = 2.5$ in.

BOND DEVELOPMENT
OF REINFORCING BARS

10

10.1 INTRODUCTION

Reinforcement for concrete to develop the strength of a section in tension depends on the compatibility of the two materials to act together in resisting the external load. The reinforcing element, such as a reinforcing bar, has to undergo the same strain or deformation as the surrounding concrete in order to prevent the discontinuity or separation of the two materials under load. The modulus of elasticity, the ductility, and the yield or rupture strength of the reinforcement must also be considerably higher than those of the concrete in order to raise the capacity of the reinforced concrete section to a meaningful level. Consequently, materials such as brass, aluminum, rubber, or bamboo are not suitable for developing the bond or adhesion necessary between the reinforcement and the concrete. Steel and fiber glass do possess the principal factors necessary: yield strength, ductility, and bond value.

Bond strength results from a combination of several parameters, such as the mutual adhesion between the concrete and steel interfaces and the pressure of the

Photo 59 Ladd Canyon overpass, Oregon. (Courtesy of Portland Cement Association.)

Photo 60 Coronado High School, Scottsdale, Arizona. (Courtesy Portland Cement Association.)

hardened concrete against the steel bar or wire due to the drying shrinkage of the concrete.   Additionally, friction interlock between the bar surface deformations or projections and the concrete caused by the micro movements of the tensioned bar results in increased resistance to slippage.   The total effect of this is known as *bond*.   In summary, bond strength is controlled by the following major factors:

1. Adhesion between the concrete and the reinforcing elements
2. Gripping effect resulting from the drying shrinkage of the surrounding concrete and the shear interlock between the bar deformations and the surrounding concrete
3. Frictional resistance to sliding and interlock as the reinforcing element is subjected to tensile stress
4. Effect of concrete quality and strength in tension and compression
5. Mechanical anchorage effect of the ends of bars through development length, splicing, hooks, and crossbars
6. Diameter, shape, and spacing of reinforcement as they affect crack development

The individual contributions of these factors are difficult to separate or quantify.   Shear interlock, shrinking confining effect, and the quality of the concrete can be considered as major factors.

10.2 BOND STRESS DEVELOPMENT

Bond stress is primarily the result of the shear interlock between the reinforcing element and the enveloping concrete caused by the various factors previously enumerated.   It can be described as a local shearing stress per unit area of the bar surface.   This direct stress is transferred from the concrete to the bar interface so as to change the tensile stress in the reinforcing bar along its length.

Three types of tests can determine the bond quality of the reinforcing element: the pull-out test, the embedded-rod test, and the beam test.   Figure 10.1 shows the first two types of test.   The pull-out test can give a good comparison of the bond efficiency of the various types of bar surfaces and the corresponding embedment lengths.   It is, however, not truly representative of the bond stress development in a structural beam.   The concrete is subjected to compression and the reinforcing bar acts in tension in this test, whereas both the bar and the surrounding concrete in the beam are subjected to the same stress.

In the embedded-rod test (Fig. 10.1b), the number of cracks, their widths, and their spacing at the various loading levels are a measure of the bond stress development and bond strength.   The process resembles more closely the behavior in beams as the progressive increase in crack widths ultimately leads to bar slippage and beam failure.

The progressive slippage of the reinforcing bar in a beam and the redistribution of stresses is represented schematically in Fig. 10.2.   As the resistance to slippage over length l_1 becomes larger than the tensile strength of concrete, a new crack forms in that area and a new stress distribution develops around the newly formed crack.   The bond stress peak in Fig. 10.2a continues to progress to the right from position A to position B, passing the center line between the two potential cracks until a second crack forms at a distance a_c from crack 1.

Consequently, it is important to choose the appropriate length of the reinforcing bars that can minimize cracking and bond slippage.   As a result, the reinforcement can attain its full strength in tension, that is, its yield strength within the structural element without bond failure.

10.2.1 Anchorage Bond

Assume l_d in Fig. 10.3a to be the length of the bar embedded in the concrete subjected to a net pulling force dT.   If d_b is the diameter of the bar, μ is the average bond stress, and f_s is the stress in the reinforcing bar due to direct pull or bending stresses in a beam, the anchorage pulling force dT would be $\mu \pi d_b l_d$ and

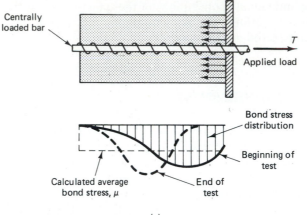

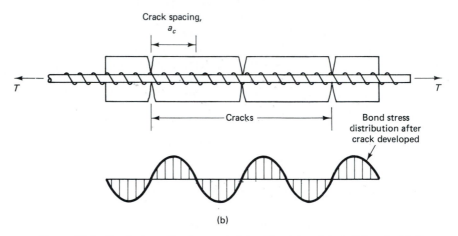

Figure 10.1   Bond stress development: (a) pull-out bond test; (b) embedded-rod test.

equal to the tensile force dT on the bar cross section; that is,

$$dT = \frac{\pi d_b^2}{4} f_s$$

Hence

$$\mu \pi d_b l_d = \pi \frac{d_b^2}{4} f_s$$

from which the average bond stress

$$\mu = \frac{f_s d_b}{4 l_d} \tag{10.1a}$$

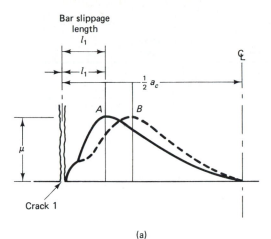

(a)

(b)

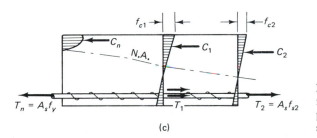

(c)

Figure 10.2 Stress redistribution with reinforcement slippage: (a) bond stress propagation; (b) reinforcement force or stress; (c) bending stress distribution.

and the development length

$$l_d = \frac{f_s}{4\mu} d_b$$ (10.1b)

10.2.2 Flexural Bond

The change in stress along the length of a bar in a beam due to the variation of moment along the span is shown schematically in Fig. 10.3b. If jd is the lever arm of the couple T due to moment M, then $T = M/jd$. In terms of the moment difference between cracked sections 1 and 2,

$$dT = \frac{dM}{jd}$$ (10.2a)

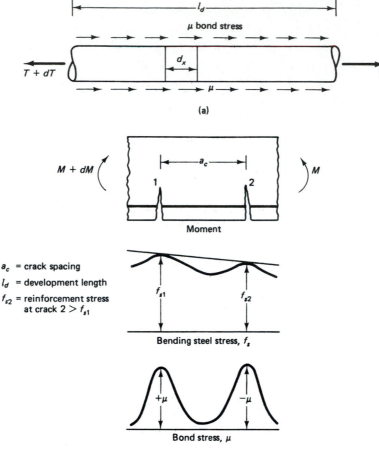

Figure 10.3 Bond stress across a reinforcing bar: (a) pull-out anchorage bond in a bar; (b) flexural bond.

Also,

$$dT = \mu d \times \Sigma\, o \qquad (10.2b)$$

where $\Sigma\, o$ is the total circumference of all the reinforcement subjected to the bond stress pull, to get $dM/dx = \mu\, \Sigma\, ojd$; since $dM/dx = $ shear V,

$$\mu = \frac{V}{\Sigma\, ojd} \qquad (10.2c)$$

Equation 10.2c is primarily of academic importance since it is indirectly accounted for in the development length approach given in Eq. 10.1b and the expressions to follow.

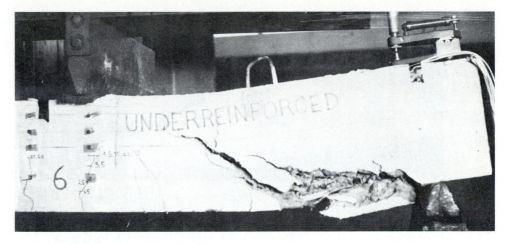

Photo 61   Bond failure at support of simply supported beam.   (Tests by Nawy et al.)

10.3 BASIC DEVELOPMENT LENGTH

From the discussion in the preceding section it can be concluded that the development length l_d as a function of the size and yield strength of the reinforcement determines the resistance of the bars to slippage and hence the magnitude of the beam's failure capacity.   It has been verified by tests that the bond strength μ is a function of the compressive strength of concrete such that

$$\mu = k\sqrt{f'_c} \tag{10.3a}$$

where k is a constant.   If the bond strength equals or exceeds the yield strength of a bar of cross-sectional area $A_b = \pi d_b^2/4$, then

$$\pi d_b l_d \mu \geq A_b f_y \tag{10.3b}$$

From Eqs. 10.1b, 10.3a, and 10.3b and considering l_{db} as the basic development length,

$$l_{db} = k_1 \frac{A_b f_y}{\sqrt{f'_c}} \tag{10.4}$$

or

$$\frac{l_{db}}{d_b} = k_2 \frac{d_b f_y}{\sqrt{f'_c}} \tag{10.5}$$

where k_2 is a function of the geometrical property of the reinforcing element and the relationship between bond strength and compressive strength of concrete.

   Equation 10.5 is consequently the basic model for defining the minimum development length of bars in structural elements, with the factor k_2 being the

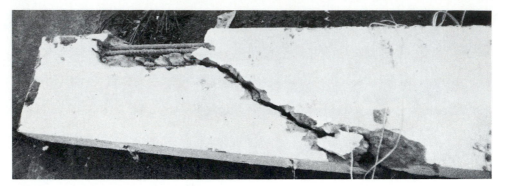

Photo 62 Bond failure and destruction of concrete cover at rupture load. (Tests by Nawy et al.)

experimental constant that covers the various factors affecting the development length. These factors include bar size, bar spacing, concrete cover, type of concrete, spacing and amount of transverse reinforcement, effect of use of excess main reinforcement, whether bars are coated, and the effect of bar splicing. These factors have been extensively investigated over the past 30 years, particularly by the group at the University of Texas at Austin.

10.3.1 Development of Deformed Bars in Tension

Equation 10.5 is transformed in the ACI 318-95 Code by replacing the coefficient $k_2 d_b$ by multipliers that reflect the effects of spacing of bars, cover, confinement

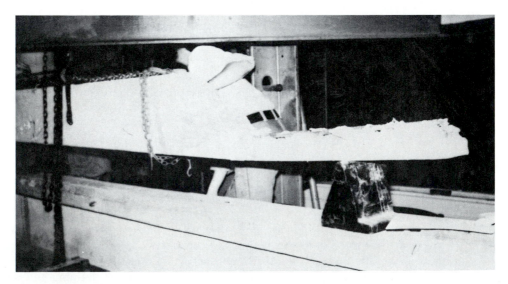

Photo 63 Bond failure in an overreinforced concrete beam. (Tests by Nawy et al.)

by transverse reinforcement, type of concrete, and whether the reinforcement is coated.

The full development length ℓ_d for deformed bars or wires obtained by applying these multipliers to the basic development length ℓ_{db} in Eq. 10.5 in terms of the bar diameter d_b is

$$\frac{\ell_d}{d_b} = \frac{3}{40} \frac{f_y}{\sqrt{f'_c}} \frac{\alpha\beta\gamma\lambda}{\left(\dfrac{c + K_{tr}}{d_b}\right)} \tag{10.6}$$

where the term $\dfrac{c + K_{tr}}{d_b}$ should not exceed a value of 2.5, but not less than 1.5 for usual structures and $\sqrt{f'_c}$ shall not exceed 100 psi ($\leq$6.9 MPa).

10.3.2 Modifying Multipliers of Development Length for Bars in Tension

$\alpha = $ *bar location factor*

> For horizontal reinforcement so placed that more than 12 in. of fresh concrete is below the development length or splice (top reinforcement)   1.3
>
> Other reinforcements   1.0

$\beta = $ *coating factor*

> Epoxy-coated bars or wires with cover less than $3d_b$ or clear spacing less than $6d_a$ 1.5
>
> All other epoxy-coated bars or wires   1.2
>
> Uncoated reinforcement   1.0

However, the product of $\alpha\beta$ should not exceed 1.7
$\gamma = $ *bar size factor*

> No 6 and smaller bars and deformed wires (No. 20 and smaller, SI)   0.8
>
> No. 7 and larger bars (No. 25 and larger, SI)   1.0

$c = $ *spacing or cover dimension, in.*
   Use the smaller of either the distance from the center of the bar to the nearest concrete surface or one-half center-to-center spacing of the bars being developed.

$K_{tr} = $ *transverse reinforcement index* $= \dfrac{A_{tr}f_{yt}}{1500sn}$, where constant 1500 carries
   units of psi
$A_{tr} = $ total cross-sectional area of all transverse reinforcement that is within the spacing s and that crosses the potential plane of splitting through to the reinforcement being developed, in. (mm^2)

f_{yt} = specified yield strength of transverse reinforcement, psi (MPa)

s = maximum spacing of transverse reinforcement within ℓ_d, center to center, in. (mm)

n = number of bars or wires being developed along the plane of splitting

The ACI Code permits using K_{tr} = 0 as a conservative design simplification even if transverse reinforcement is present.

λ = *lightweight aggregate concrete factor*

When lightweight aggregate concrete is used: 1.3

However, when f_{ct} is specified, use $\lambda = 6.7\sqrt{f'_c}/f_{ct}$ but not less than 1.0.

When normal-weight concrete is used: 1.0

The minimum development length in all cases is 12 in.

λ_s = *excess reinforcement factor*

The ACI Code permits the reduction of ℓ_d if the longitudinal flexural reinforcement is in excess of that required by analysis except where anchorage or development for f_y is specifically required or the reinforcement is designed for seismic effects.

Reduction multiplier $\lambda_s = A_s$ required/A_s provided and $\lambda_{s2} = f_y/60{,}000$ for cases where $f_y > 60{,}000$ psi. In lieu of using a refined computation of the development length of Eq. 10.6, Table 10.1 can be utilized for typical construction practices using a value of $\dfrac{c + K_{tr}}{d_b} = 1.5$ in such cases and $f'_c = 4000$ psi.

Table 10.2 is a general table for usual construction conditions giving the required development length for deformed bars of sizes Nos. 3 to 18.

Table 10.3 gives minimum beam width (inches) to satisfy two bar-diameter clear spacing, while Table 10.4 satisfies 1-bar-diameter or 1-in. clear spacing. In these two tables the following assumptions are made:

Side cover is 1.5 in on each side.

No. 3 stirrups for bars No. 11 or smaller.

No. 4 stirrups for bars No. 14 or No. 18.

Stirrups are bent around four bar diameters. Hence the distance from the centroid of the bar nearest the side face of the beam to the inside face of the No. 3 stirrup is taken as 0.75 in. for bars No. 11 or smaller and equal to the longitudinal bar radius for No. 14 and No. 18 bars.

10.3.3 Development of Deformed Bars in Compression and the Modifying Multipliers

Bars in compression require shorter development length than bars in tension. This is due to the absence of the weakening effect of the tensile cracks. Hence the

TABLE 10.1 *Simplified Development Length ℓ_d Equations*

	No. 6 and Smaller Bars and Deformed Wires (1)	No. 7 and Larger Bars (2)
Clear spacing of bars being developed or spliced not less than d_b, clear cover not less than d_b, and stirrups or ties throughout ℓ_d not less than the Code minimum	$\dfrac{\ell_d}{d_b} = \dfrac{f_y\alpha\beta\lambda}{25\sqrt{f'_c}}$	$\dfrac{\ell_d}{d_b} = \dfrac{f_y\alpha\beta\lambda}{20\sqrt{f'_c}}$
	*when $f'_c = 4000$ psi $\alpha, \beta, \lambda, \lambda_s = 1.0$ $\gamma = 0.8$	*when $f'_c = 4000$ psi $\alpha, \beta, \lambda, \lambda_s, \gamma = 1.0$
or		
Clear spacing of bars being developed or spliced not less than $2d_b$ and clear cover not less than d_b	$\ell_d = 38d_b$	$\ell_d = (38/0.8)d_b$ $= 48d_b$
Other cases (1.5 times the above values)	$\dfrac{\ell_d}{d_b} = \dfrac{3f_y\alpha\beta\lambda}{50\sqrt{f'_c}}$	$\dfrac{\ell_d}{d_b} = \dfrac{3f_y\alpha\beta\lambda}{40\sqrt{f'_c}}$
	*$\ell_d = 57d_b$	*$\ell_d = 72d_b$

TABLE 10.2 *Tension Reinforcement and Development Length (inches) for $f'_c = 4000$** psi Normal-weight Concrete, $f_y = 60,000$ psi Steel*

Bar Size (1)	Cross-sectional Area (in.²) (2)	Bar Diameter (in.) (3)	Development Length, $\ell_d^{c,d}$ (in.)	
			$s \geq 2d_b$ or d_b^b and Clear Cover $\geq d_b$ $\leq$ No. 6: $\ell_d = 38\,d_b$ $\geq$ No. 7: $\ell_d = 48\,d_b$ (4)	Other $\leq$ No. 6: $\ell_d = 57\,d_b$ $\geq$ No. 7: $\ell_d = 72\,d_b$ (5)
3	0.11	0.375	15	21
4	0.20	0.500	19	29
5	0.31	0.625	24	36
6	0.44	0.750	29	43
7	0.60	0.875	42	63
8	0.79	1.000	48	72
9	1.00	1.128	54	81
10	1.27	1.270	61	92
11	1.56	1.410	68	102
14	2.25	1.693	82	122
18	4.00	2.257	108	163

$\alpha, \beta, \lambda = 1.0$, $\gamma = 0.8$ for No. 6 bars or smaller and $= 1.0$ for No. 7 bars and larger.
[a]For f'_c values different from 4000 psi, multiply table values by $\sqrt{4000/f'_c}$. For $f_y = 40,000$ psi, multiply by $\frac{2}{3}$; $\sqrt{f'_c}$ should not exceed 100.
[b]Confined by stirrups.
[c]For compression development length, $\ell_d = $ multiplier $\times \ell_{db}$, where $\ell_{db} = 0.02d_bf_y/\sqrt{f'_c} \geq 0.0003d_bf_y$.
[d]Multiply table values by $\alpha = 1.3$ for top reinforcement; $\lambda = 1.3$ for lightweight aggregate; $\beta = 1.5$ for epoxy-coated bars with cover less than $3d_b$ or clear spacing less than $6d_b$ and $\beta = 1.2$ for other epoxy-coated bars. Minimum ℓ_d for all cases $= 12$ in.

TABLE 10.3 Minimum Beam Width (in.) to Satisfy Two-bar-diameter Clear Spacing

Bar Size	Number of Bars in Single Layer						
	2	3	4	5	6	7	8
4	6.8	8.3	9.8	11.3	12.8	14.3	15.8
5	7.1	9.0	10.9	12.8	14.6	16.5	18.4
6	7.5	9.8	12.0	14.3	16.5	18.8	21.0
7	7.9	10.5	13.1	15.8	18.4	21.0	23.6
8	8.3	11.3	14.3	17.3	20.3	23.3	26.3
9	8.6	12.0	15.4	18.8	22.2	25.6	28.9
10	9.1	12.9	16.7	20.5	24.3	28.1	31.9
11	9.5	13.7	17.9	22.2	26.4	30.6	34.9
14	12.2	15.9	20.9	26.0	31.1	36.2	41.2
18	15.0	19.8	26.6	33.3	40.1	46.9	53.7

expression for the basic development length is

$$l_{db} = 0.02 \frac{d_b f_y}{\sqrt{f_c'}} \tag{10.7a}$$

and
$$l_{db} \geq 0.0003 d_b f_y \tag{10.7b}$$

with the modifying multiplier for

1. Excess reinforcement: λ_s = required A_s/provided A_s.
2. Spirally enclosed reinforcement $\lambda_{s1} = 0.75$.

TABLE 10.4 Minimum Beam Width (in.) to Satisfy the Larger of One-bar-diameter or 1-Inch Clear Spacing

Bar Size	Number of Bars in Single Layer						
	2	3	4	5	6	7	8
4	6.8	8.3	9.8	11.3	12.8	14.3	15.8
5	6.9	8.5	10.1	11.8	13.4	15.0	16.6
6	7.0	8.8	10.5	12.3	14.0	15.8	17.5
7	7.1	9.0	10.9	12.8	14.6	16.5	18.4
8	7.3	9.3	11.3	13.3	15.3	17.3	19.3
9	7.5	9.8	12.0	14.3	16.5	18.8	21.0
10	7.8	10.3	12.9	15.4	18.0	20.5	23.0
11	8.1	10.9	13.7	16.5	19.3	22.2	25.0
14	9.1	12.5	15.9	19.2	22.6	26.0	29.4
18	10.8	15.3	19.8	24.3	28.8	33.3	37.9

10.3.4 Development of Bundled Bars in Tension and Compression

If bundled bars are used in tension or compression, l_d has to be increased by 20% for three-bar bundles and 33% for four-bar bundles. $\sqrt{f_c'}$ should not be taken greater than 100 psi. A unit of bundled bars is treated as a single bar of a diameter derived from the equivalent total area for the purpose of determining the modifying factors. However, although splice and development lengths of bundled bars are based on the diameter of individual bars increased by 20% or 33% as applicable, it is necessary to use an equivalent diameter of the entire bundle derived from the equivalent total area of bars when determining the factors that consider cover and clear spacing and represent the tendency of concrete to split.

10.3.5 Flow Chart for Reinforcement Development Length Computation

A flow chart for reinforcement development length computation is shown in Fig. 10.4.

10.3.6 SI Metric Conversion

$$K_{tr} = \frac{A_{tr}f_{yt}}{260 \; sn}$$

where f_{yt} is in MPa.

Equation 10.6:

$$\frac{\ell_d}{d_b} = \frac{15f_y\alpha\beta\gamma\lambda}{16 \sqrt{f_c'}\left(\dfrac{c + K_{tr}}{d_b}\right)}$$

TABLE 10.1a SI Development Length Simplified Expressions

	$\leq$ No. 20 (1)	$\geq$ No. 25 (2)
*main case	$\dfrac{\ell_d}{d_b} = \dfrac{f_y\alpha\beta\lambda}{2\sqrt{f_c'}}$	$\dfrac{\ell_d}{d_b} = \dfrac{5\,f_y\alpha\beta\lambda}{8\sqrt{f_c'}}$
*other case	$\dfrac{\ell_d}{d_b} = \dfrac{3\,f_y\alpha\beta\lambda}{4\sqrt{f_c'}}$	$\dfrac{\ell_d}{d_b} = \dfrac{15\,f_y\alpha\beta\lambda}{16\sqrt{f_c'}}$

*See table 10.1

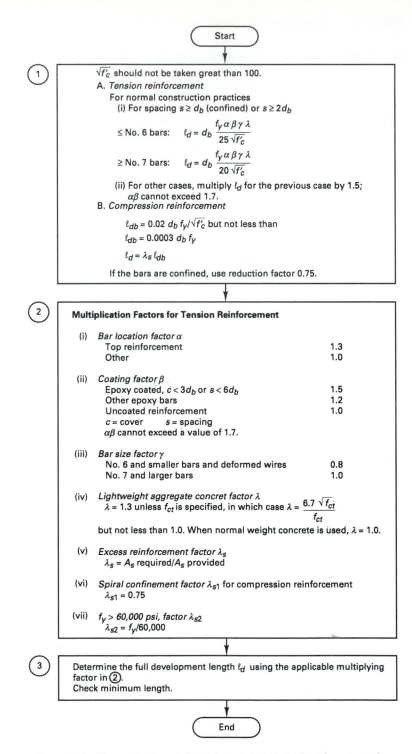

Start

1 $\sqrt{f'_c}$ should not be taken great than 100.

A. *Tension reinforcement*

For normal construction practices

(i) For spacing $s \geq d_b$ (confined) or $s \geq 2d_b$

$\leq$ No. 6 bars: $\quad \ell_d = d_b \dfrac{f_y \alpha \beta \gamma \lambda}{25 \sqrt{f'_c}}$

$\geq$ No. 7 bars: $\quad \ell_d = d_b \dfrac{f_y \alpha \beta \gamma \lambda}{20 \sqrt{f'_c}}$

(ii) For other cases, multiply ℓ_d for the previous case by 1.5;
$\alpha\beta$ cannot exceed 1.7.

B. *Compression reinforcement*

$\ell_{db} = 0.02\, d_b\, f_y / \sqrt{f'_c}$ but not less than

$\ell_{db} = 0.0003\, d_b\, f_y$

$\ell_d = \lambda_s\, \ell_{db}$

If the bars are confined, use reduction factor 0.75.

2 **Multiplication Factors for Tension Reinforcement**

(i) *Bar location factor α*

Top reinforcement	1.3
Other	1.0

(ii) *Coating factor β*

Epoxy coated, $c < 3d_b$ or $s < 6d_b$	1.5
Other epoxy bars	1.2
Uncoated reinforcement	1.0

$c = $ cover $\quad s = $ spacing
$\alpha\beta$ cannot exceed a value of 1.7.

(iii) *Bar size factor γ*

No. 6 and smaller bars and deformed wires	0.8
No. 7 and larger bars	1.0

(iv) *Lightweight aggregate concret factor λ*
$\lambda = 1.3$ unless f_{ct} is specified, in which case $\lambda = \dfrac{6.7 \sqrt{f_{ct}}}{f_{ct}}$

but not less than 1.0. When normal weight concrete is used, $\lambda = 1.0$.

(v) *Excess reinforcement factor λ_s*
$\lambda_s = A_s$ required$/A_s$ provided

(vi) *Spiral confinement factor λ_{s1} for compression reinforcement*
$\lambda_{s1} = 0.75$

(vii) $f_y > 60,000$ psi, factor λ_{s2}
$\lambda_{s2} = f_y/60,000$

3 Determine the full development length ℓ_d using the applicable multiplying factor in ②.
Check minimum length.

End

Figure 10.4 Flow chart for reinforcement development length computation.

10.3.7 Example 10.1: Development Length of Deformed Bars

Calculate the required embedment length of the deformed bars in the following four cases: (a) No. 7 bars (22.2-mm diameter), top reinforcement in single layer in a beam with No. 3 stirrups. Given:

$$f_y = 60,000 \text{ psi (414 MPa)}$$
$$f_c' = 4000 \text{ psi (27.6 MPa), normal weight concrete}$$
clear spacing between bars $= 2d_b$
clear side cover $= 1.5$ in. on each side
bars not spliced

(b) Same as part (a) except that clear spacing between bars $= d_b$ or 1 in. minimum. The bars are epoxy coated.

(c) Same as part (a) except that clear spacing between bars $= 3d_b$ and the bars are not top bars.

(d) Assume that the No. 7 bars in part (a) are in compression and the concrete is lightweight. Also assume that the provided A_s is 10% higher than the required A_s.

Solution

(a) Development length from Eq. 10.6 is

$$\ell_d = d_b \left[\frac{3}{40} \frac{f_y}{\sqrt{f_c'}} \frac{\alpha\beta\gamma\lambda}{\left(\dfrac{c + K_{tr}}{d_b} \right)} \right]$$

$\alpha = 1.3$ (top bar), $\beta = 1.0$, $\gamma = 1.0$ for No. 7, $\lambda = 1.0$, $d_b = 0.875$ in., and $c =$ smaller of distance from center of bar to the nearest concrete surface or one-half center to center spacing of bars.

$$c = 1.5 + \frac{0.875}{2} = 1.94 \quad \text{(cover)}$$

or

$$c = \text{bar spacing} = \frac{1.0 + 0.875}{2} = 0.938 \text{ in.} \quad \text{controls}$$

K_{tr} can be assumed zero as a design simplification even if transverse reinforcement is present, but the term $\dfrac{c + K_{tr}}{d_b}$ cannot be larger than 2.5 or less than 1.5. $=$ $0.938/0.875 = 1.072$. Assuming $K_{tr} = 0$, $\dfrac{c + K_{tr}}{d_b} = 1.072 < 1.5$; use 1.5

$$\sqrt{f_c'} = \sqrt{4000} = 64 < 100 \quad \text{O.K.}$$

$$\ell_d = d_b \left(\frac{3}{40} \times \frac{60,000}{\sqrt{4000}} \times \frac{1.3}{1.5} \right) = 61.7 d_b$$

$$\approx 54 \text{ in. (1370 mm)}$$

If $\ell_d = 48\alpha d_b$ from Table 10.1 is used with a value of $\dfrac{c + K_{tr}}{d_b} = 1.5$,

$\ell_d = 48\ \alpha d_b = 54$ in. (also from Table 10.2, $48 \times 0.875 \times 1.3 = 54$ in.).

(b) $\alpha = 1.3$ (top bar), $\beta = 1.2$, $\gamma = 1.0$, and $\lambda = 1.0$. Using Table 10.1, $\alpha\beta = 1.56 < 1.7$. Use $\alpha\beta = 1.56$.

$$\ell_d = 48\alpha\beta d_b = 48 \times 1.56 \times 0.875$$
$$= 66 \text{ in. (1680 mm)}$$

(c) $\alpha = 1.0$ (bottom bar), $\beta = 1.0$, $\gamma = 1.0$, and $\lambda = 1.0$. From Table 10.1

$$\ell_d = 48d_b = 42 \text{ in. (1070 mm)}$$

(d) $\lambda = 1.3$ for lightweight aggregate concrete. For compression steel, from Eq. 10.7a,

$$\ell_{db} = 0.02\ \frac{d_b f_y}{\sqrt{f_c'}} = \frac{0.02 \times 0.875 \times 60{,}000}{\sqrt{4000}}$$
$$= 16.6 \text{ in. (422 mm)}$$

From Eq. 10.7b,

$$\text{Min. } \ell_{db} = 0.0003 d_b f_y = 0.0003 \times 0.875 \times 60{,}000$$
$$= 15.8 \text{ in.}$$
$$\ell_{db} = 16.6 \quad \text{controls}$$
$$\lambda = 1.3$$

λ_{s2} for excess reinforcement $= 1/1.1$. Hence $\ell_d = 16.6 \times 1.3 \times 1/1.1 = 20$ in. (508 mm).

10.3.8 SI Example on Development Length Evaluation

Solve Ex. 10.1 using SI units.

Data

$$f_c' = 27.6 \text{ MPa}, \qquad d_b = \tfrac{7}{8} \text{ in. } = 22.2 \text{ mm} \quad \text{(soft conversion)}$$
$$f_y = 414 \text{ MPa}$$

Solution

Use closest in millimeters to No. 7 bars in column 2 of Table 10.1a.

(a)
$$\ell_d = d_b \left(\frac{5 f_y \alpha\beta\gamma\lambda}{8 \sqrt{f_c'}} \right)$$
$$f_y = 414 \text{ MPa}, \qquad f_c' = 27.6 \text{ MPa}$$
$$\alpha = 1.3 \quad \text{(top bar)}, \qquad \beta = \gamma = \lambda = 1.0$$
$$\ell_d = 22.2 \left(\frac{5 \times 414 \times 1.3}{8 \sqrt{27.6}} \right) = 1420 \text{ mm (55 in.)}$$

(b) $\alpha = 1.3$, $\beta = 1.2$, $\gamma = 1.0$, and $\lambda = 1.0$.

$$\ell_d = 1420 \times 1.2 = 1700 \text{ mm}$$

(c) $\alpha = \beta = \gamma = \lambda = 1.0$.

$$\ell_d = \frac{1420}{1.3} \cong 1090 \text{ mm}$$

(d) $\lambda = 1.3$ for lightweight aggregate concrete; $\lambda_{s2} = 1/1.1$. From before, $\ell_{db} = 16.6$ in. $= 422$ mm.

$$\ell_d = 422 \times 1.3 \times \frac{1}{1.1} = 498 \text{ mm}$$

10.3.9 Mechanical Anchorage and Hooks

Hooks are used when space limitation in a concrete section does not permit the necessary straight embedment length.   Hooks in structural members are placed relatively close to the free surface of a concrete element, where splitting forces proportional to the total bar force may determine the hook capacity.   The standard hook does *not* develop the tension yield strength of the bar.   If l_{hb} is the basic development length for the standard hook in tension, additional embedment length has to be incorporated to give a total length l_{dh} not less than $8d_b$ or 6 in., whichever is greater. l_{dh} length is shown in Fig. 10.5.   The length l_{hb} varies with the bar size, reinforcement yield strength, and compressive strength of concrete.   For $f_y = 60,000$ psi steel,

$$l_{hb} = 1200 \frac{d_b}{\sqrt{f'_c}} \tag{10.8}$$

where d_b is the diameter of the hook bar.

Modifying multipliers for hooks in tension

1. *Yield strength f_y effect:* for a yield-strength different than 60,000, $\lambda_{s2} = f_y/60,000$.
2. *Concrete cover effect:* for No. 11 bars and smaller, side cover normal to the plane of hook not less than $2\frac{1}{2}$ in. and for 90° hook with cover on bar extension beyond the hook not less than 2 in., $\lambda_d = 0.7$ (see Fig. 10.5c).
3. *Ties of stirrups:* for No. 11 bars and smaller, hook enclosed vertically or horizontally within ties or stirrup spaced not greater than $3d_b$, where d_b is diameter of hook bar, $\lambda_d = 0.8$.
4. *Excess reinforcement:* where anchorage or development for f_y is not specifically required but the reinforcement area A_s used is in excess of A_s required for analysis,

$$\lambda_d = \frac{\text{required } A_s}{\text{provided } A_s}$$

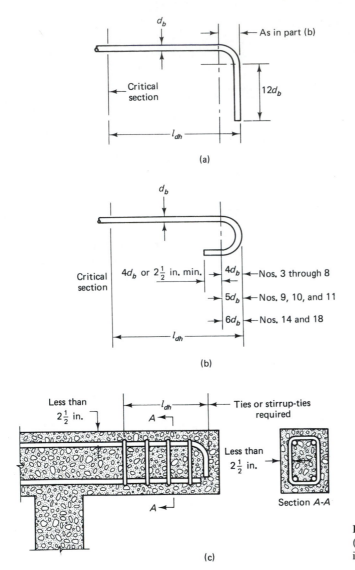

Figure 10.5   Standard bar hook details: (a) 90° hook; (b) 180° hook; (c) hook in small concrete cover.

5. *Bars developed by standard hooks at discontinuous ends:* if the concrete cover is less than $2\frac{1}{2}$ in., bars *should* be *enclosed* within ties or stirrups along the full development length l_{dh} spaced at no greater than $3d_b$; for this case $\lambda_{s2} = 0.8$ from item 3 above; modifying multiplier shall *not apply.*

6. *Lightweight concrete:* $\lambda = 1.3$. It should be noted that hooks cannot be considered effective in developing bars in compression. The total development or embedment length

$$l_{dh} = l_{hb} \times \lambda \tag{10.9}$$

Figure 10.5a and b shows details of standard 90° and 180° hooks used in axial tension or bending tension, and Fig. 10.5c gives details of bar hooks susceptible to concrete splitting when the cover is small, that is, less than $2\frac{1}{2}$ in. Confinement is enhanced through the use of closed ties or stirrups. No distinction is made between a top bar and a bottom bar if hooks are used.

10.3.10 Example 10.2: Embedment Length for a Standard 90° Hook

Compute the development length required for the top bars of a lightweight concrete beam extending into the column support shown in Fig. 10.6 assuming No. 9 reinforcing bars (28.6-mm diameter) hooked at the end. The concrete cover is 2 in. (50.8 mm). Given:

$$f'_c = 5000 \text{ psi (34.47 MPa)}$$
$$f_y = 60,000 \text{ psi (413.7 MPa)}$$

Solution   Top bars for hooks behave similarly to bottom bars; hence no modifier is needed. For No. 9 bars, $d_b = 1.128$ in. (28.65 mm).

$$\text{basic development length } l_{hb} = 1200 \frac{d_b}{\sqrt{f'_c}}$$

or

$$l_{hb} = \frac{1200 \times 1.128}{\sqrt{5000}} = 19.14 \text{ in.}$$

For lightweight concrete, $\lambda = 1.3$.

$$l_{dh} = 1.3 \times 19.14 = 24.88 \text{ in.} > 8d_b \quad \text{or} \quad 6 \text{ in.} \qquad \text{O.K.}$$

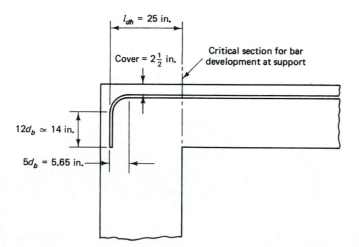

Figure 10.6   Hook embedment detail.

Use a 90° hook with embedment length $l_{dh} = 25$ in. (635 mm) beyond the critical section (face of support). Figure 10.6 shows the geometrical details of the hook.

10.3.11 Development of Web Reinforcement

1. For No. 5 bars and D31 wire and smaller, and for No. 6, 7, and 8 bars with $f_y = 40,000$ psi or less, a standard hook has to be used around the longitudinal reinforcement.
2. For Nos. 6, 7, and 8 stirrups with $f_y = $ greater than 40,000 psi, a standard stirrup hook around a longitudinal bar plus an embedment between midheight of the member and the outside end of the hook has to be used such that the length is equal to or greater than

$$0.014 d_b \frac{f_y}{\sqrt{f_c'}}$$

10.4 DEVELOPMENT OF FLEXURAL REINFORCEMENT IN CONTINUOUS BEAMS

As discussed earlier, reinforcing bars should be adequately embedded in order to prevent serious bar slippage resulting in bond pull-out failure. The critical locations for bar discontinuance are points along the structural member where there is a rapid drop in the bending moment or stress, such as the inflection points in a bending moment diagram of a continuous beam.

Tension reinforcement can be developed by bending the lower tension bars at a 45° inclination across the web of the beam and can be anchored or made continuous with the reinforcing bars on the top of the member. To ensure full development, reinforcement has to be extended beyond the point at which it is no longer required to resist flexure for a distance equal to the effective depth d or $12 d_b$, whichever is greater, except for supports of simple span beams or at the free end of a cantilever. Figure 10.7 shows details of flexural reinforcement development in typical continuous beams for both the positive and the negative steel reinforcement.

The following are general guidelines for full development of the reinforcement and for ensuring continuity in the case of continuous beams.

1. At least *one-third* of the positive moment reinforcement in simple beams and *one-fourth* of the positive moment reinforcement in continuous beams should be extended at least 6 in. into the support without being bent.
2. At simple supports as in Fig. 10.8a and at points of inflection as in Fig. 10.8b, the positive moment reinforcement should be limited to such a diameter that the development length

$$l_d \leq \frac{M_n}{V_u} + l_a \qquad (10.10)$$

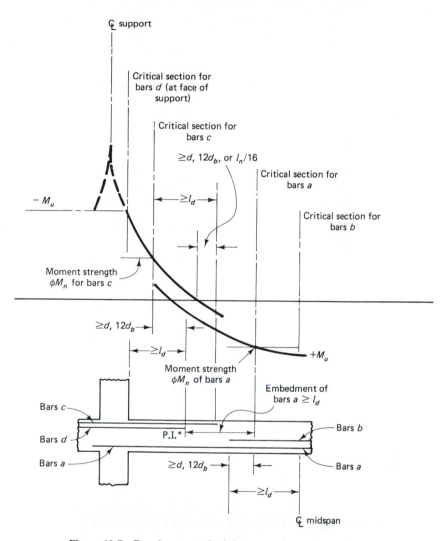

Figure 10.7   Development of reinforcement in continuous beams.

where M_n = nominal moment strength where all the reinforce-
                                   ment is stressed to f_y
                         V_u = factored shear force at the section under consid-
                                   eration
     inflection point l_a = effective depth d or $12d_b$, where d_b is the bar
                                   diameter, whichever is greater

Equation 10.10 imposes a design limitation on the flexural bond stress in areas
of large shear and small moment in order to prevent splitting. Such a condition exists in short-span, heavily loaded simple beams. Thus the bar diameter for positive moment is so chosen that, even if length AC to the critical

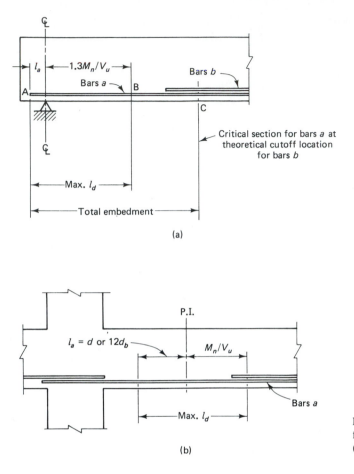

(a)

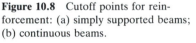

(b)

Figure 10.8 Cutoff points for reinforcement: (a) simply supported beams; (b) continuous beams.

section in Fig. 10.8a is larger than length AB, the bar size must be limited such that $l_d \leq 1.3 \, (M_n/V_u + l_a)$. For confining reactions such as at simple supports, the value M_n/V_u in Eq. 10.10 is increased by 30%.

3. At least *one-third* of the total tension reinforcement provided for negative bending moment at the support should be extended beyond the inflection point not less than the effective depth d of the member, $12d_b$, or $\frac{1}{16}$ of the clear span, whichever has the largest value.

4. Web stirrups have to be carried as close to the compression and tension surfaces of the member as the minimum concrete cover requirements allow. The ends of the stirrups without hooks should have an embedment of at least $d/2$ above or below the compression side of the member for full development length l_d, but not less than 12 in. or $24 \, d_b$. For stirrups with hook ends, the total embedment length should equal $0.5l_d$ plus the standard hook.

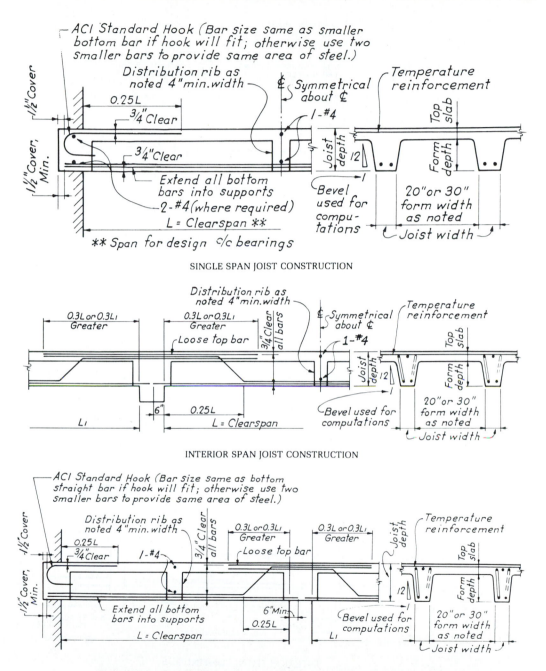

Figure 10.9 Cutoff point for one-way joist construction, Ref. 10.6. (*Note:* Continuing reinforcement shall have an embedment length not less than the required development length l_d beyond the point where bent or terminated tension reinforcement is no longer required to resist flexure.)

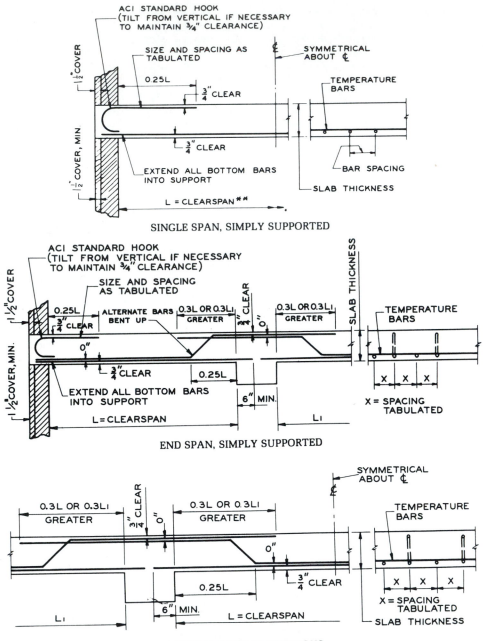

Figure 10.10 Cutoff points for one-way slabs, Ref. 10.6. (*Note:* Continuing reinforcement shall have an embedment length not less than the required development length l_d beyond the point where bent or terminated tension reinforcement is no longer required to resist flexure.)

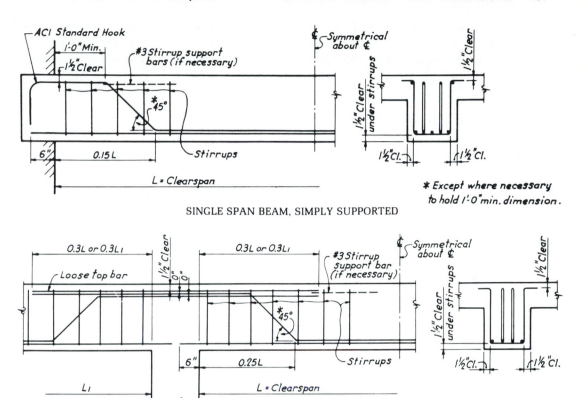

SINGLE SPAN BEAM, SIMPLY SUPPORTED

INTERIOR SPAN OF CONTINUOUS BEAM

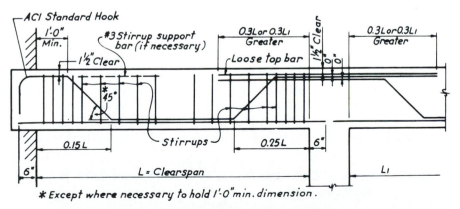

END SPAN OF SIMPLY SUPPORTED BEAM

Figure 10.11   Reinforcing details for continuous beams with diagonal tension steel, Ref. 10.6. (*Note:* Continuing reinforcement shall have an embedment length not less than the required development length l_d beyond the point where bent or terminated tension reinforcement is no longer required to resist flexure.)

A typical detail of cutoff points for continuous one-way beam and joist construction from Ref. 10.6 is given in Fig. 10.9. Typical cutoff points for one-way slabs are shown in Fig. 10.10 and for beams with diagonal tension stirrups are given in Fig. 10.11.

10.5 SPLICING OF REINFORCEMENT

Steel reinforcing bars are produced in standard lengths controlled by transportability and weight considerations. In general, 60-ft lengths are normally produced. But it is impractical in beams and slabs spanning over several supports to interweave bars of such lengths on site over several spans. Consequently, bars are cut to shorter lengths and lapped at the least critical bending moment locations for bar sizes No. 11 or smaller. A general rule of thumb for maximum bar length is about 40 ft for shipping purposes. The most effective means of continuity in reinforcement is to weld the cut pieces without reducing the mechanical or strength properties of the welded bar at the weld. However, cost considerations require alternatives. There are basically three types of splicing:

1. *Lap splicing:* depends on full bond development of the two lapping bars at the lap for bars of size not larger than No. 11.
2. *Welding by fusion of the two bars at the connection:* can be economically justifiable for bar sizes larger than No. 11 bars.
3. *Mechanical connecting:* can be achieved by mechanical sleeves threaded on the ends of the bars to be interconnected. Such connectors should have a yield strength at least 1.25 times the yield strength of the bars they interconnect. They are also more commonly used for large-diameter bars.

10.5.1 Lap Splicing

Figure 10.12a shows a bar lap splice and the force and stress distribution along the splice length l_s. Failure of the concrete at the splice region develops by a typical splitting mechanism as shown in Fig. 10.12b. At failure, one bar slips relative to the other. The *idealized* tensile stress distribution in the bars along the splice length l_s has a maximum value f_y at the splice end and $\frac{1}{2}f_y$ at $l_s/2$. At failure, the expected magnitude of slip is approximately $(0.5f_y/E_s) \times$ half splice length l_s in Fig. 10.12a.

10.5.2 Splices of Deformed Bars and Deformed Bars or
Wires in Tension

Two classes of lap splices are specified by the ACI Code. The minimum length l_1, but not less than 12 in., is

Class A: $1.0l_d$
Class B: $1.3l_d$

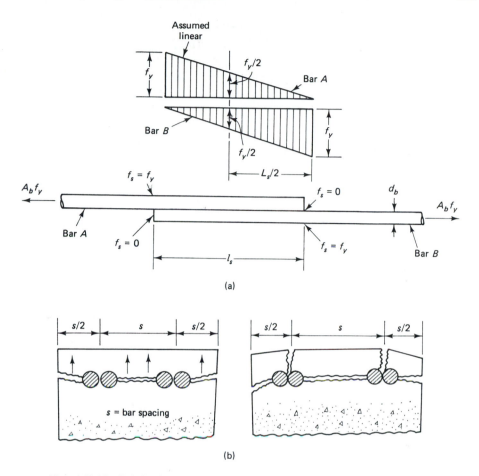

Figure 10.12   Reinforcing bars splicing: (a) lap splice idealized stress distribution; (b) splice splitting failure.

Table 10.5 gives the maximum percentage of tensile steel area A_s to be spliced. Splicing should be avoided at maximum tensile stress if at all possible; splicing may be by simple lapping of bars either in contact or separated by concrete. However, every effort should be made to stagger the splice, rather than having all the bars spliced within the required lap length.

10.5.3 Splices of Deformed Bars in Compression

The lap length l_s should be equal to at least the development length in compression as given in Section 10.3.3 and Eqs. 10.7a and 10.7b and the modifiers. l_s should also satisfy the following, but not be less than 12 in.:

$$f_y \leq 60,000 \text{ psi} \qquad l_s \geq 0.0005 f_y d_b \qquad (10.11a)$$

$$f_y > 60,000 \text{ psi} \qquad l_s \geq (0.0009 f_y - 24) d_b \qquad (10.11b)$$

TABLE 10.5   Tension Lap Slices

$\dfrac{A_s \text{ Provided}^a}{A_s \text{ Required}}$	Maximum Percent of A_s Spliced within Required Lap Length	
	50	100
Equal to or greater than 2	Class A	Class B
Less than 2	Class B	Class B

^aRatio of area of reinforcement provided to area of reinforcement required by analysis at splice location.

If the compressive strength f_c' of the concrete is less than 3000 psi, such as might occur in foundations, the splice length l_s has to be increased by one-third.

Modifying multipliers with values less than 1.0 are used in heavily reinforced tied compression members (0.83) and in spirally reinforced columns (0.75), but the lap length should not be less than 12 in.

10.5.4 Development of Welded Deformed Wire Fabric in Tension

The development length, ℓ_d, for deformed welded wire fabric should be taken as the ℓ_d value obtained from Eq. 10.6 or Table 10.1 multiplied by a fabric factor. The fabric factor, with at least one cross wire within the development length and not less than 2 in. from the point of the critical section, should be taken as the greater of the following two expressions:

$$\frac{f_y - 35,000}{f_y} \tag{10.12a}$$

or

$$\frac{5d_b}{s_w} \tag{10.12b}$$

but should not be taken greater than 1.0, where s_w = spacing of wire to be developed or spliced (in.).

10.5.5 Splices in Deformed Welded Wire Fabric

The minimum lap length l_s measured between the ends of the two lapped fabric sheets of welded deformed wire has to be $\geq 1.7l_d$ or 8 in. (204 mm), whichever is greater. Additionally, the overlap measured between the outermost cross wires of each fabric sheet should not be less than 2 in. (50.8 mm).

10.6 EXAMPLES OF EMBEDMENT LENGTH AND SPLICE DESIGN FOR BEAM REINFORCEMENT

10.6.1 Example 10.3: Embedment Length at Support of a Simply Supported Beam

Calculate the maximum development length that can be used for bars a at the support of the simply supported superstructure beam in Fig. 10.13 if the distance AC from the theoretical cutoff point of bars b is 42 in. (1067 mm). The beam is not integral with the support (nonconfining). Assume that the reinforcing bars used for moment strength are (a) No. 8 deformed (25.4 mm) and (b) No. 14 deformed (43.0 mm), if the beam was a raft foundation beam (a maximum No. 11 bar size is usually used in superstructure normal-size beams). Given:

$$S = \text{clear spacing between bars} = 3d_b$$
$$V_u = 100,000 \text{ lb (444.8 kN)}$$
$$M_n = 2,353,000 \text{ in.-lb (25.6 kNm)}$$
$$f'_c = 4000 \text{ psi (27.58 MPa), normal-weight concrete}$$
$$f_y = 60,000 \text{ psi (413.7 MPa)}$$
$$l_a = 12 \text{ in. (304.8 mm)}$$

Solution $\alpha = \beta = \lambda = 1$.
    (a) No. 8 bars: $d_b = 1.0$ in.

$$\sqrt{f'_c} = \sqrt{4000} = 63.2 < 100 \qquad \text{O.K.}$$

From column 2 of Table 10.1, $\ell_d = 48 \, d_b = 48$ in.

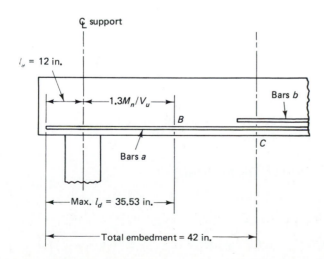

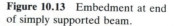

Figure 10.13   Embedment at end of simply supported beam.

From Eq. 10.10, $l_d \leq M_n/V_u + l_a$, where l_a = effective depth d or 12 d_b, whichever is greater, d_b being the bar diameter.

$$\frac{M_n}{V_u} = \frac{2,353,000}{100,000} = 23.53 \text{ in.}$$

l_a = embedment length beyond support center = 12 in.

maximum $l_d = 1.3 \times 23.53 + 12 = 42.59$ in.   say 45.0 in. < 48 in.

Use $\ell_d = 48$ in.

(b) No. 14 bars:

$$\alpha = \beta = \gamma = 1.0. \qquad d_b = 1.693 \text{ in.}$$

From column 2 of Table 10.1,

$$\ell_d = 48 \, d_b = 48 \times 1.693 = 82 \text{ in.}$$

$$\frac{M_n}{V_u} + l_d = 45 \text{ in.}$$

Use $l_d = 82$ in. (2110 mm).

10.6.2 Example 10.4: Embedment Length at Support of a Continuous Beam

A continuous reinforced concrete beam has clear spans l_{nr} = 36 ft (10.97 m) and l_{nl} = 22 ft (6.7 m) and the bending moment diagram segment at an interior support as shown in Fig. 10.14. Calculate the cutoff lengths of the negative moment top reinforcement bars to satisfy the development length requirements at the cutoff points. The beam is singly reinforced and has the dimensions h = 27 in. (685.8 mm), d = 23.5 in. (596.9 mm), and b = 15 in. (381.0 mm). It is subjected to a factored negative bending at the center of the intermediate support.

$$-M_u = 6,127,000 \text{ in.-lb (692.4 kNm)}$$

Given:

s = clear spacing between bars = $3d_b$

f'_c = 4000 psi (27.58 MPa), normal-weight concrete

f_y = 60,000 psi (413.7 MPa)

Required A_s = 5.62 in.²

Provided A_s = 6.00 in.² (six No. 9 bars)

Solution

$$\sqrt{f'_c} = \sqrt{4000} = 63.2 < 100 \qquad \text{O.K.} \qquad d_b = 1.128 \text{ in.}$$

$$\lambda_d = 1.0 \qquad \alpha \text{ for top bars} = 1.3; \qquad \beta = \gamma = \lambda = 1.0$$

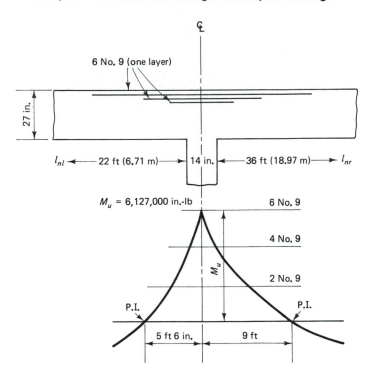

Figure 10.14 Continuous beam support moment.

From Table 10.1, column 2,

$$l_d = \alpha 48 d_b = 1.3 \times 48 \times 1.128 = 70 \text{ in.}$$

$$l_d = \frac{\text{required } A_s}{\text{provided } A_s} \times 50 = \frac{5.62}{6.00} \times 70 = 66 \text{ in.}$$

Use $l_d = 66$ in. (1630 mm) for the six No. 9 bars.

Cutoff points

At least one-third of the bars have to extend beyond the point of inflection by the largest of $\frac{1}{16}$ (span l_n), d, or $12d_b$.

$$\tfrac{1}{3}A_s = \text{two No. 9 bars} \qquad 12d_b = 12 \times 1\tfrac{1}{8} = 13.5 \text{ in.}$$

1. Right span $l_{nr} = 36$ ft:

$$\tfrac{1}{16} l_{nr} = \tfrac{36}{16} \times 12 = 27.0 \text{ in.} \quad \text{controls}$$

2. Left span $l_{nl} = 22$ ft:

$$\tfrac{1}{16}l_{nl} = \tfrac{22}{16} \times 12 = 16.5 \text{ in.} \qquad d = 23.5 \text{ in.} \quad \text{controls} \qquad \text{say 24 in.}$$

As given in Fig. 10.7, details of the development length dimensions at all cutoff points for this continuous beam are shown in Fig. 10.15.

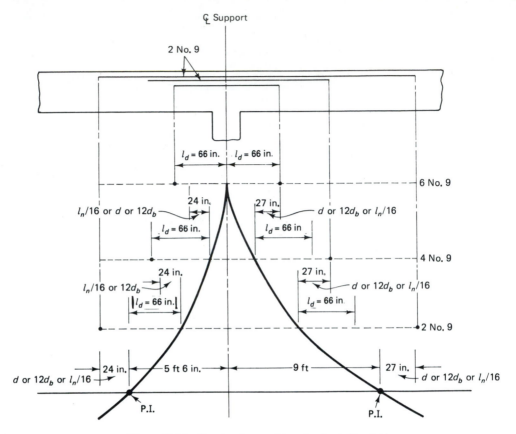

Figure 10.15   Bar development length of cutoff points.

10.6.3 Example 10.5: Splice Design for Tension Reinforcement

Calculate the lap splice length for No. 7 tension bottom bars (22.2-mm diameter) spaced at $2d_b$, minimum spacing.   The ratio of the provided A_s to the required A_s is (a) >2.0, (b)<2.0, and the maximum percentage of A_s spliced within the section is 75%.   Given:

$$f'_c = 5000 \text{ psi (34.47 MPa)}$$

$$f_y = 60{,}000 \text{ psi (413.7 MPa)}$$

$$s = \text{clear spacing between bars} = 2\tfrac{1}{2}d_b.$$

Solution

$$d_b = 0.875 \text{ in. for No. 7 bar}$$

$$\alpha = \beta = \gamma = \lambda = 1.0$$

$$\text{Check } \sqrt{5000} = 70.7 < 100 \qquad \text{O.K.}$$

From Table 10.1, column 2, $l_d = 48\,d_b = 48 \times 0.875 = 42$ in.

(a) For provided A_s/required $A_s > 2.0$, class A splice: $l_s = 1 \times l_d = 21$ in.
(b) For provided A_s/required $A_s < 2.0$, class B splice:

$$\text{lap splice length } l_s = 1.3 l_d = 1.3 \times 42 = 55 \ (1400 \text{ mm})$$

10.6.4 Example 10.6: Splice Design for Deformed Compression Reinforcement

Calculate the lap splice length for a No. 9 compression deformed bars (28.7-mm diameter) in a normal-weight concrete beam at clear spacing $3d_b$. Given:

$$f_c' = 7000 \text{ psi (48.26 MPa)}$$
$$f_y = 80,000 \text{ psi (551.6 MPa)}$$

Solution

$$d_b = 1.128 \text{ in. for No. 9 bar}$$

$$\sqrt{7000} = 83.7 < 100 \qquad \text{O.K.}$$
$$\alpha = \beta = \gamma = \lambda = 1.0$$

Splice length l_s: From Eq. 10.11b, for $f_y > 60,000$ psi,

$$l_s = (0.0009 f_y - 24) d_b = (0.0009 \times 80,000 - 24)1.128 = 54.14 \text{ in.}$$

Use lap splice length $l_s = 55$ in. (1400 mm).

10.7 TYPICAL DETAILING OF REINFORCEMENT AND BAR SCHEDULING

The design examples for bond development length, lap splicing, and spacing reinforcement are applied in Figs. 10.16 to 10.20. Additional examples from the author's parking-garage working drawing details are given in Figs. 10.21 to 10.25. These representative examples can serve as a good guideline for producing correct engineering working drawings. It should be recognized that successful execution of a designed system is directly dependent on the availability of clear and correct detailing and the avoidance of any congestion of the reinforcement. Such congestion can only lead to honeycombing in the concrete, resulting in possible cracking, capacity reduction, and even failure. Consequently, equal attention has to be given to detailing as to design if a constructed system is to perform the structural functions for which it is intended.

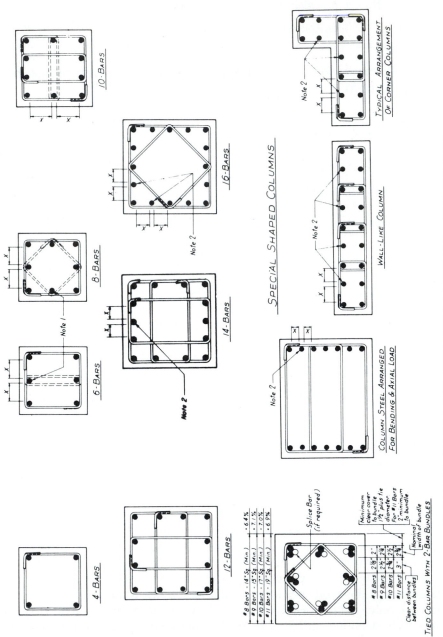

Figure 10.16 Column ties for preassembled lap-spliced cages. (From Ref. 10.6.)

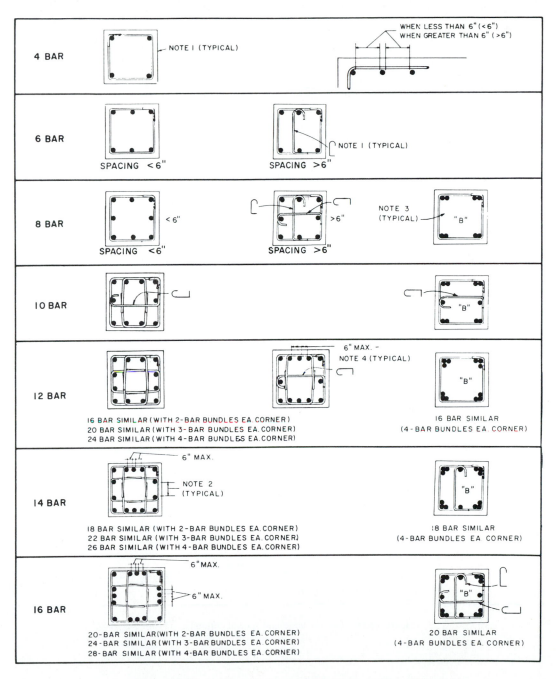

Figure 10.17 Column ties for standard columns. (From Ref. 10.6.)

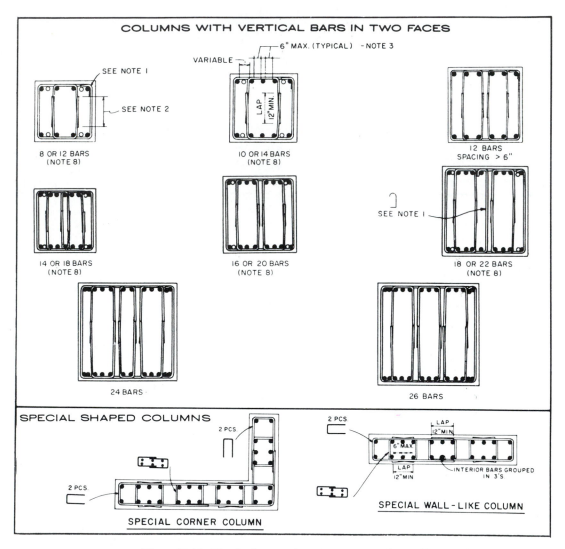

Figure 10.18   Ties for large and special columns.   (From Ref. 10.6.)

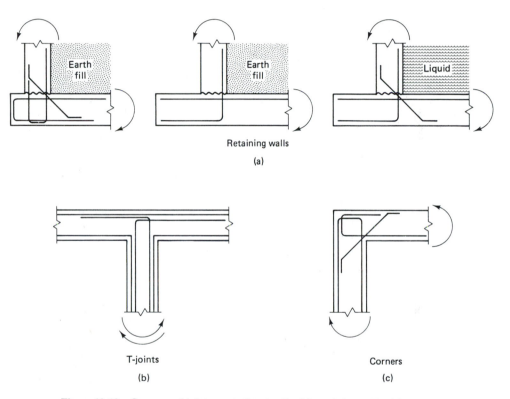

Figure 10.19   Corner and joint connection details: (a) retaining walls; (b) T joints; (c) corners.   (From Ref. 10.6.)

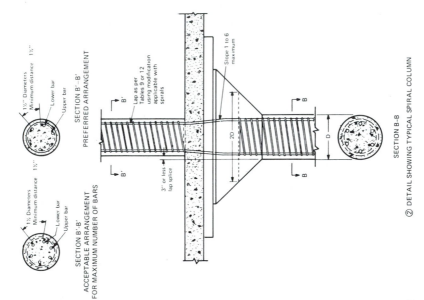

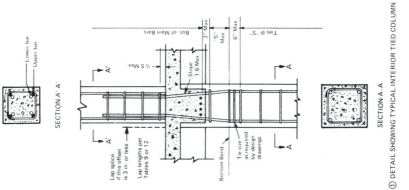

Figure 10.20 Column splice details. (From Ref. 10.6.)

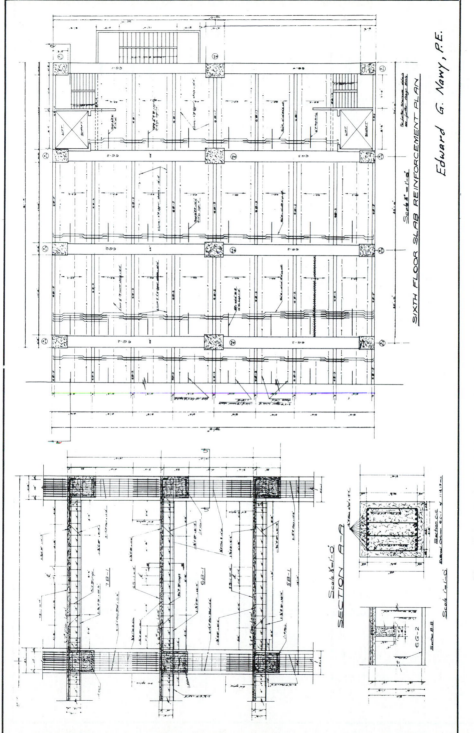

Figure 10.21 Typical beam and slab reinforcing working drawing. (Design by E. G. Nawy.)

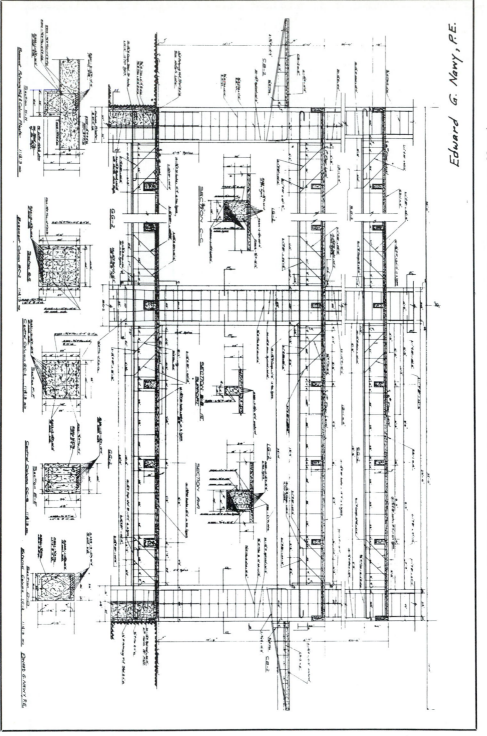

Figure 10.22 Typical working drawing of column reinforcement details. (Design by E. G. Nawy.)

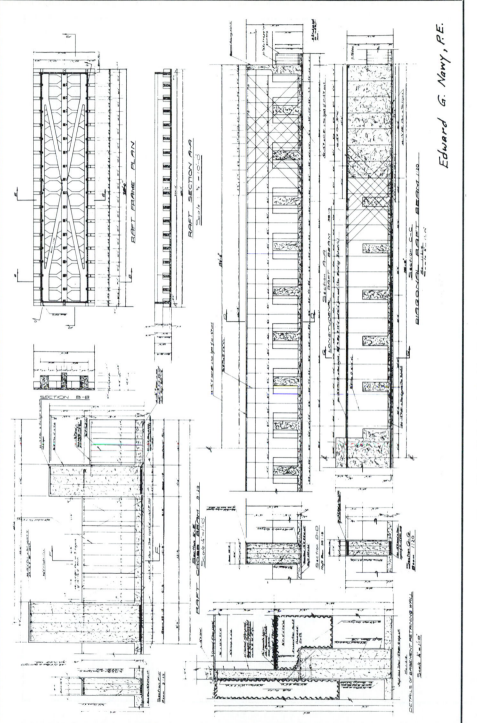

Figure 10.23　Raft foundation details.　(Design by E. G. Nawy.)

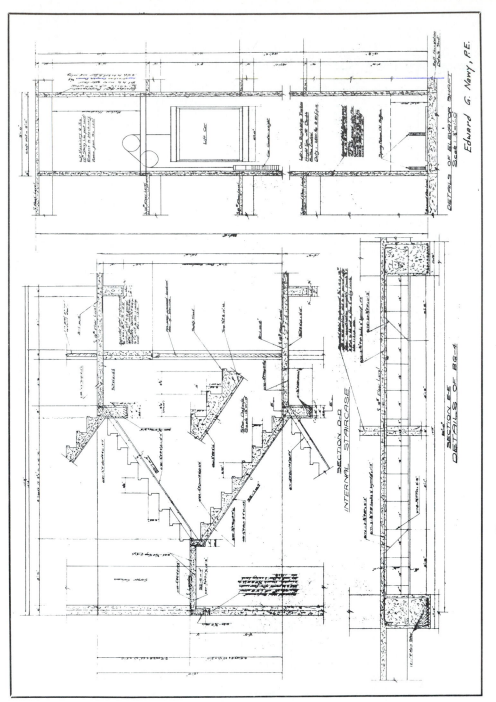

Figure 10.24 Elevator and stairwell details. (Design by E. G. Nawy.)

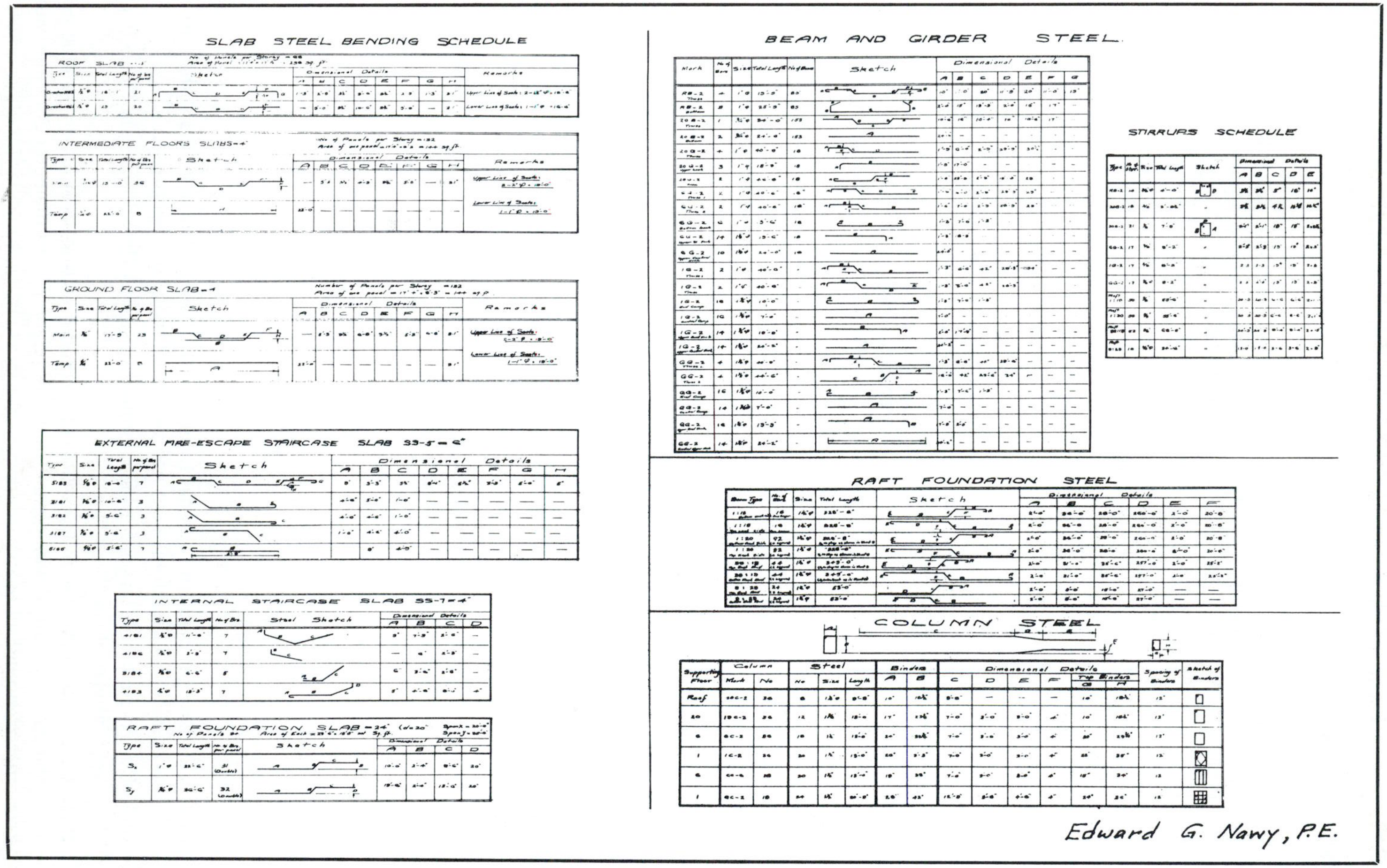

Figure 10.25 Typical reinforcement bar bending schedule. (Design by E. G. Nawy.)

SELECTED REFERENCES

10.1. Mathey, R. G., and Watstein, D., "Investigation of Bond in Beam Pull-out Specimens with High Yield Strength Deformed Bars," *Journal of the American Concrete Institute*, Proc. Vol. 57, No. 9, March 1961, pp. 1071–1090.

10.2. Jirsa, J. O., Lutz, L. A., and Gergely, P., "Rationale for Suggested Development, Splice and Standard Hook Provisions for Deformed Bars in Tension," *Concrete International: Design and Construction*, American Concrete Institute, Vol. 1, No. 7, July 1979, pp. 47–61.

10.3. Ferguson, P. M., Breen, J. E., and Thompson, J. N., "Pullout Tests on High Strength Reinforcing Bars," Part 1, *Journal of the American Concrete Institute*, Proc. Vol. 62, No. 8, August 1965, pp. 933–950.

10.4. Ferguson, P. M., and Breen, J. E., "Lapped Splices for High Strength Reinforcing Bars," *Journal of the American Concrete Institute*, Proc. Vol. 62, No. 9, September 1965, pp. 1063–1078.

10.5. ACI Committee 408, *Suggested Development, Splice, and Hook Provisions for Deformed Bars in Tension*, ACI 408-1 R-90, American Concrete Institute, Detroit, 1990, 3 pp.

10.6. ACI Committee 315, *ACI Detailing Manual—1980*, Special Publication SP-66, American Concrete Institute, Detroit, 1980, 206 pp.

10.7. Wire Reinforcement Institute, *Reinforcement Anchorages and Splices*, 3rd ed., WRI Publication, McLean, Va., 1979, 32 pp.

PROBLEMS FOR SOLUTION

10.1. Calculate the basic development lengths in tension for the following deformed bars embedded in normal-weight concrete.
(a) No. 5, No. 8. Given:

$$f'_c = 5000 \text{ psi (34.47 MPa)}$$
$$f_y = 60,000 \text{ psi (413.7 MPa)}$$

(b) No. 14, No. 18. Given:

$$f'_c = 4000 \text{ psi}$$
$$f_y = 60,000 \text{ psi}$$
$$f_y = 80,000 \text{ psi}$$

10.2. Calculate the total embedment length for the bars in Problem 10.1 if they are used as compression reinforcement and the concrete is sand-lightweight.

10.3. Design the cutoff length for the continuous beam in Ex. 10.4 if eight No. 8 bars are used instead of six No. 9 bars.

10.4. Design the compression lap splice for a column section 16 in. × 16 in. (406 mm × 406 mm) reinforced with eight No. 9 bars (eight bars of diameter 28.7 mm) equally spaced around all faces.

(a) f'_c = 5000 psi (34.47 MPa)
   f_y = 60,000 psi (413.7 MPa)
(b) f'_c = 7000 psi (48.26 MPa)
   f_y = 80,000 psi (551.6 MPa)

10.5. An 18-ft (5.49-m) normal-weight concrete cantilever beam is subjected to a factored M_u = 3,500,000 in.-lb (396 kN-m) and a factored shear V_u = 32,400 lb (144 kN) at the face of the support. Design the top reinforcement and the appropriate embedment of 90° hook into the concrete wall to sustain the external shear and moment. Given:

$$f'_c = 4500 \text{ psi}$$
$$f_y = 60,000 \text{ psi}$$

10.6. Design the beam reinforcement in Problem 10.5 if it was simply supported having a span l_n = 36 ft (10.97 m) and subjected to the same factored M_u value at midspan and the shear V_u at the face of the support. Evaluate the required embedment length at the support to ensure that no bond failure due to slippage can develop. Assume (a) confining beam reaction and (b) beam not monolithic with its support.

DESIGN OF TWO-WAY SLABS AND PLATES

11.1 INTRODUCTION: REVIEW OF METHODS

Supported floor systems are usually constructed of reinforced concrete cast in place. Two-way slabs and plates are those panels in which the dimensional ratio of length to width is less than 2. The analysis and design of framed floor slab systems represented in Fig. 11.1 encompasses more than one aspect. The present state of knowledge permits reasonable evaluation of (1) the moment capacity, (2) the slab–column shear capacity, and (3) serviceability behavior as determined by deflection control and crack control. Flat plates are slabs supported directly on columns without beams, as shown in Fig. 11.1a, compared to Fig. 11.1b for slabs on beams, or Fig. 11.1c for waffle slab floors. Lift slabs are another form of construction but mostly in prestressed concrete.

The evolution of the state of knowledge in slab design in the last 50 years will be briefly reviewed. The analysis of slab behavior in flexure up to the 1940s and early 1950s followed the classical theory of elasticity, particularly in the United States. The small deflections theory of plates, assuming the material to be ho-

Photo 64 Sydney Opera House, Sydney, Australia. (Courtesy of Australian Information Service.)

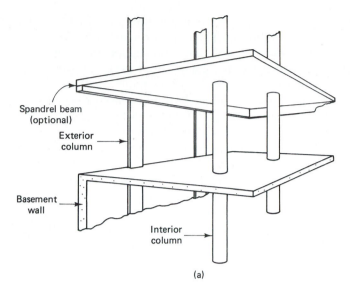

(a)

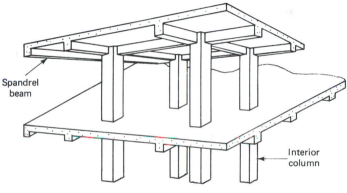

(b)

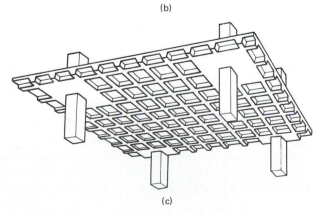

(c)

Figure 11.1   Two-way-action floor systems: (a) two-way flat-plate floor; (b) two-way slab floor on beams; (c) waffle slab floor.

mogeneous and isotropic, formed the basis of ACI Code recommendations with moment coefficient tables. The work, principally by Westergaard, that empirically allowed limited moment redistribution guided the thinking of the code writers. Hence the elastic solutions, complicated even for simple shapes and boundary conditions when no computers were available, made it mandatory to idealize and sometimes empiricize conditions beyond economic bounds.

In 1943, Johansen presented his yield-line theory for evaluating the collapse capacity of slabs. Since that time, extensive research into the ultimate behavior of reinforced concrete slabs has been undertaken. Studies by many investigators, such as those of Ockleston, Mansfield, Rzhanitsyn, Powell, Wood, Sawczuk, Gamble-Sozen-Siess, Park and the author contributed immensely to further understanding of the limit-state behavior of slabs and plates at failure as well as serviceable load levels.

The various methods that are used for the analysis (design) of two-way action slabs and plates are summarized in the following.

11.1.1 Semielastic ACI Code Approach

The ACI approach gives two alternatives for the analysis and design of a framed two-way action slab or plate system: the direct design method and the equivalent frame method. Both methods are discussed in more detail in Sections 11.3 and 11.6.

11.1.2 Yield-line Theory

Whereas the semielastic code approach applies to standard cases and shapes and has an inherent, excessively large safety factor with respect to capacity, the yield-line theory is a plastic theory that is easy to apply to irregular shapes and boundary conditions. Provided that serviceability constraints are applied, Johansen's yield-line theory is the simplest approach that the designer can use, representing the true behavior of reinforced concrete slabs and plates. It permits evaluation of the bending moments from an assumed collapse mechanism that is a function of the type of external load and the shape of the floor panel. This topic will be discussed in more detail in Section 11.9.

11.1.3 Limit Theory of Plates

The interest in developing a limit solution became necessary due to the possibility of finding a variation in the collapse field that can give a lower failure load. Hence an upper-bound solution requiring a valid mechanism when supplying the work equation was sought, as well as a lower-bound solution requiring that the stress field satisfies everywhere the differential equation of equilibrium; that is,

$$\frac{\partial^2 M_x}{\partial x^2} - 2\,\frac{\partial^2 M_{xy}}{\partial x\,\partial y} + \frac{\partial^2 M_y}{\partial y^2} = -w \tag{11.1}$$

where M_x, M_y, and M_{xy} are the bending moments and w is the unit intensity of load. Variable reinforcement permits the lower-bound solution still to be valid. Wood, Park, and other researchers have given more accurate semiexact predictions of the collapse load.

For limit-state solutions, the slab is assumed to be completely rigid until collapse. Further work at Rutgers by the author incorporated the deflection effect at high load levels as well as the compressive membrane force effects in predicting the collapse load.

11.1.4 Strip Method

This method was proposed by Hillerborg, attempting to fit the reinforcement to the strip fields. Since practical considerations require the reinforcement to be placed in orthogonal directions, Hillerborg set twisting moments equal to zero and transformed the slab into intersecting beam strips; hence the name strip method.

Except for Johansen's yield-line theory, most of the other solutions are lower bound. Johansen's upper-bound solution can give the highest collapse load as long as a valid failure mechanism is used in predicting the collapse load.

11.1.5 Summary

Both the direct design method (DDM) and the equivalent frame method (EFM) will be discussed with appropriate examples. Both methods are based on the concept of an equivalent frame, except that the DDM has several limitations, is less refined, and is suitable for gravity loads only, whereas the EFM is more general, can be utilized for horizontal loading, and is adaptable for computer programming.

11.2 FLEXURAL BEHAVIOR OF TWO-WAY SLABS AND PLATES

11.2.1 Two-way Action

A single rectangular panel supported on all four sides by unyielding supports such as shear walls or stiff beams is first considered. The purpose is to visualize the physical behavior of the panel under gravity load. The panel will deflect in a dishlike form under the external load, and its corners will lift if it is not monolithically cast with the supports. The contours shown in Fig. 11.2a indicate that the curvatures and consequently the moments at the central area C are more severe in the shorter direction y with its steep contours than in the longer direction x.

Evaluation of the division of moments in the x and y directions is extremely complex because the behavior is highly statically indeterminate. The discussion of the simple case of the panel in Fig. 11.2a is expanded further by taking strips AB and DE at midspan as in Fig. 11.2b such that the deflection of both strips at central point C is the same.

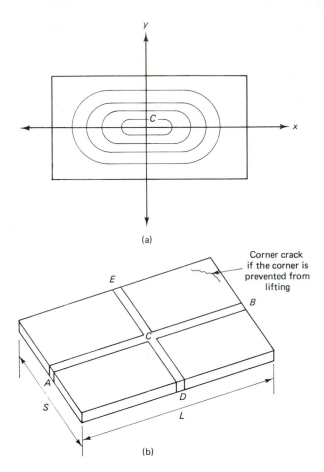

(a)

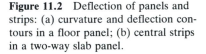

(b)

Figure 11.2   Deflection of panels and strips: (a) curvature and deflection contours in a floor panel; (b) central strips in a two-way slab panel.

The deflection of a simply supported uniformly loaded beam is $5wl^4/384EI$; that is, $\Delta = kwl^4$, where k is a constant. If the thickness of the two strips is the same, the deflection of strip AB would be $kw_{AB}L^4$, and the deflection of strip DE would be $kw_{DE}S^4$, where w_{AB} and w_{DE} are the portions of the total load intensity w transferred to strips AB and DE, respectively; that is, $w = w_{AB} + w_{DE}$. Equating the deflections of the two strips at the central point C, we get

$$w_{AB} = \frac{wS^4}{L^4 + S^4} \tag{11.2a}$$

and

$$w_{DE} = \frac{wL^4}{L^4 + S^4} \tag{11.2b}$$

It is seen from the two relationships w_{AB} and w_{DE} in Eqs. 11.2a and 11.2b that the shorter span S of strip DE carries the heavier portion of the load. Hence the shorter span of such a slab panel on unyielding supports is subjected to the larger moment, supporting the foregoing discussion of the steepness of the curvature contours in Fig. 11.2a.

11.2.2 Relative Stiffness Effects

Alternatively, we must consider a slab panel supported by flexible supports such as beams and columns or flat plates supported by a grid of columns. The distribution of moments in the short and long directions is considerably more complex. The complexity arises from the fact that the degree of stiffness of the yielding supports determines the intensity of steepness of the curvature contours in Fig. 11.2a in both the x and y directions and the redistribution of moments.

The ratio of the stiffness of the beam supports to the slab stiffness can result in curvatures and moments in the long direction larger than those in the short direction, because the total floor behaves as an orthotropic *plate* supported on a grid of columns without beams. The moment values in the long and short directions in Exs. 11.1 and 11.2 arithmetically illustrate this discussion. If the long span L in such floor systems of slab panels without beams is considerably larger than the short span S, the maximum moment at the center of a plate panel would approximate the moment at the middle of a uniformly loaded strip of span L and clamped at both ends.

In summary, because slabs are flexible and highly underreinforced, redistribution of moments in both the long and short directions depends on the relative stiffnesses of the supports and the supported panels. Overstress in one region is reduced by such redistribution of moments to the lesser stressed regions.

11.3 THE DIRECT DESIGN METHOD

The following discussion of the direct design method (DDM) of analysis for two-way systems summarizes the ACI Code approach for evaluation and distribution of the total moments in a two-way slab panel. The various moment coefficients are taken directly from the ACI Code provisions. An assumption is made that vertical planes cut through an entire rectangle in plan multistory building along lines AB and CD in Fig. 11.3a midway between columns. A rigid frame results in the x direction. Similarly, vertical planes EF and HG result in a rigid frame in the y direction. A solution of such an idealized frame consisting of horizontal beams or equivalent slabs and supporting columns enables the design of the slab as the beam part of the frame. Approximate determinations of the moments and shears using simplified coefficients are presented throughout the direct design method. The equivalent frame method treats the idealized frame in a manner similar to an actual frame, and hence is more exact and has fewer limitations than the direct design method. It basically involves a full moment distribution of many cycles, compared to the direct design method, which involves a one-cycle-moment distribution approximation.

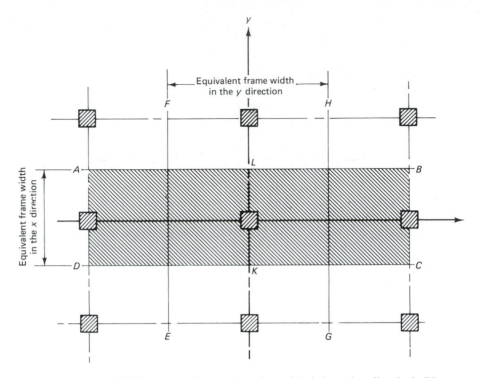

Figure 11.3 (a) Floor plan with equivalent frame (shaded area in x direction): (b) column and middle strips of the equivalent frame (y direction).

11.3.1 Limitations of the Direct Design Method

The following are the limitations of this method:

1. A minimum of three continuous spans in each direction.
2. The ratio of the longer to the shorter span within a panel should not exceed 2.0.
3. Successive span lengths in each direction should not differ by more than one-third of the longer span.
4. Columns may be offset a maximum of 10% of the span in the direction of the offset from either axis between center lines of successive columns.
5. All loads shall be due to gravity only and uniformly distributed over the entire panel. The live load shall not exceed three times the dead load.
6. If the panel is supported by beams on all sides, the relative stiffness of the beams in two perpendicular directions shall not be less than 0.2 nor greater than 5.0.

It should be noted that the majority of normal floor systems satisfy these conditions.

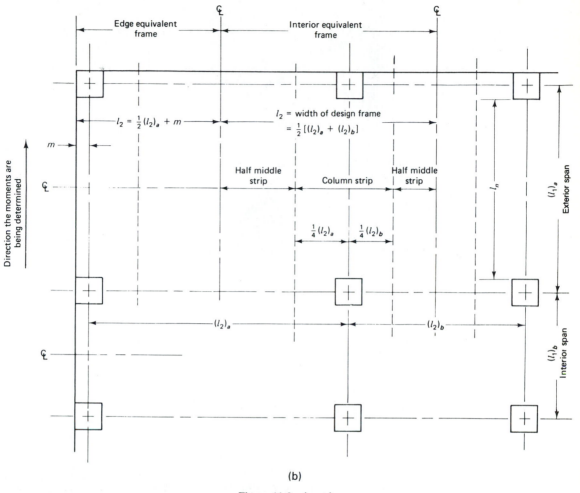

(b)

Figure 11.3 (*cont.*)

11.3.2 Determination of the Factored Total Statical Moment M_0

There are basically four major steps in the design of the floor panels.

1. Determination of the total factored statical moment in each of the two perpendicular directions.
2. Distribution of the total factored design moment to the design of sections for negative and positive moment.
3. Distribution of the negative and positive design moments to the column and middle strips and to the panel beams, if any. A column strip is a width = 25% of the equivalent frame width on each side of the column center line, and the middle strip is the balance of the equivalent frame width.

Photo 65 Sydney Opera House during construction.

4. Proportioning of the size and distribution of the reinforcement in the two perpendicular directions.

Hence correct determination of the values of the distributed moments becomes a principal objective. Consider typical interior panels having center-line dimensions l_1 in the direction of the moments being considered and dimensions l_2 in the direction perpendicular to l_1, as shown in Fig. 11.3b. The clear span l_n extends from face to face of columns, capitals, or walls. Its value should not be less than $0.65l_1$, and circular supports should be treated as square supports having the same cross-sectional area. The total statical moment of a uniformly loaded simply supported beam as a one-dimensional member is $M_0 = wl^2/8$. In a two-way slab panel as a two-dimensional member, the idealization of the structure through conversion to an equivalent frame makes it possible to calculate M_0 once in the x direction and again in the orthogonal y direction. If we take as a free-body diagram the typical interior panel shown in Fig. 11.4a, symmetry reduces the shears and twisting moments to zero along the edges of the cut segment. If no restraint existed at ends A and B, the panel would be considered simply supported in span l_n direction. If we cut at midspan as in Fig. 11.4b and consider half the panel as a free-body diagram, the moment M_0 at midspan would be

$$M_0 = \frac{w_u l_2 l_{n1}}{2}\frac{l_{n1}}{2} - \frac{w_u l_2 l_{n1}}{2}\frac{l_{n1}}{4} \qquad (11.3)$$

or

$$M_0 = \frac{w_u l_2 (l_{n1})^2}{8}$$

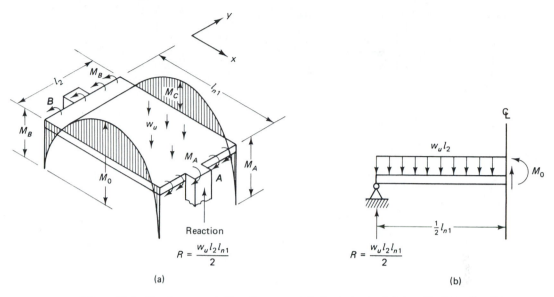

Figure 11.4   Simple moment M_0 acting on an interior two-way slab panel in the x direction: (a) moment on panel; (b) free-body diagram.

Due to the actual existence of restraint at the supports, M_0 in the x direction would be distributed to the supports and midspan such that

$$M_0 = M_C + \tfrac{1}{2}(M_A + M_B) \tag{11.4}$$

The distribution would depend on the degree of stiffness of the support. In a similar manner, M_0 in the y direction would be the sum of the moments at midspan and the average of the moments at the supports in that direction.

The distribution of the statical factored moment M_0 to the column strip of the equivalent frame leads to the proportioning of the reinforcement in those strips.

11.4 DISTRIBUTED FACTORED MOMENTS AND SLAB REINFORCEMENT BY THE DIRECT DESIGN METHOD

11.4.1 Negative and Positive Factored Design Moments

From Fig. 11.5a, the negative factored moment factor in interior spans is 0.65 and the positive factor is 0.35 of the total statical moment M_0. For end spans of flat-plate floor panels, the M_0 factors are given in Table 11.1.

11.4.2 Factored Moments in Column Strips

A column strip is a design strip with a width on each side of the column equal to $0.25l_2$ or $0.25l_1$, whichever is *less*, as shown in Figs. 11.3b and 11.5. The strip includes beams, if any. The middle strip is a design strip bound by the two column strips of the panel being analyzed.

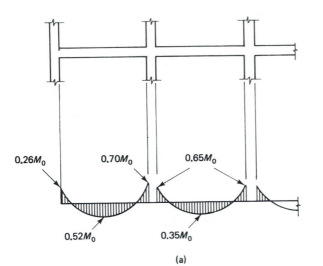

(a)

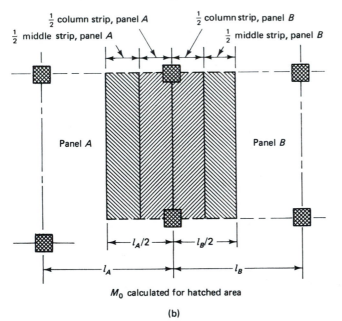

M_0 calculated for hatched area

(b)

Figure 11.5 Distribution of the statical factored moments M_0 for slab without beams into negative and positive moments: (a) moment coefficients for multispans; (b) slab areas for which M_0 is calculated.

Interior panels. For interior negative moments, column strips have to be proportioned to resist the following portions in percent of the *interior* negative factored moments, with linear interpolation made for intermediate values.

l_2/l_1	0.5	1.0	2.0
$\alpha_1(l_2/l_1) = 0$	75	75	75
$\alpha_1(l_2/l_1) \geq 1.0$	90	75	45

TABLE 11.1 Moment Factors for M_0 Distribution in Exterior Spans

	Exterior Edge Unrestrained	Slab with Beams between All Supports	Slab without Beams between Interior Supports		Exterior Edge Fully Restrained
			Without Edge Beam	With Edge Beam	
Interior negative factored moment	0.75	0.70	0.70	0.70	0.65
Positive factored moment	0.63	0.57	0.52	0.50	0.35
Exterior negative factored moment	0	0.16	0.26	0.30	0.65

α_1 in these tables is α in the direction of span l_1 for cases of two-way slabs on beams and is equal to the ratio of flexural stiffness of the beam section to the flexural stiffness of a width of slab bound laterally by center lines of adjacent panels, if any, on each side of the beam $\alpha_1 = E_{cb}I_b/E_{cs}I_s$, where E_{cb} and E_{cs} are the modulus values of concrete, and I_b and I_s are the moments of inertia of the beam and the slab, respectively.   The factored moments in beams between supports have to be proportioned to resist 85% of the column strip moment when $\alpha_1(l_2/l_1)$ ≥ 1.0.   Linear interpolation between 85% and 0% needs to be made for cases where $\alpha_1(l_2/l_1)$ varies between 1.0 and 0.

Exterior panels.   For exterior negative moments, the column strips should be proportioned to resist the following portions in percent of the *exterior* negative factored moments with linear interpolation made for the intermediate values, where β_t is the torsional stiffness ratio. β_t = ratio of torsional stiffness of the edge beam section to the flexural stiffness of a width of a slab equal to the span length of beam center to center of supports.

l_2/l_1		0.5	1.0	2.0
$\alpha_1(l_2/l_1) = 0$	$\beta_t = 0$	100	100	100
	$\beta_t \geq 2.5$	75	75	75
$\alpha_1(l_2/l_1) \geq 1.0$	$\beta_t = 0$	100	100	100
	$\beta_t \geq 2.5$	90	75	45

Positive moments. For positive moments, the column strips have to be proportioned to resist the following portions in percent of the positive factored moments with linear interpolation being made for intermediate values.

l_2/l_1	0.5	1.0	2.0
$\alpha_1(l_2/l_1) = 0$	60	60	60
$\alpha_1(l_2/l_1) \geq 1.0$	90	75	45

11.4.3 Factored Moments in Middle Strips

That portion of the negative and positive factored moments not resisted by the column strips would have to be proportionately assigned to the corresponding half of the middle strips. Adjacent spans do not necessarily have to be equal so that the two halves of the column strip flanking a row of columns need not be equal in width. Hence each middle strip has to be proportioned to resist the sum of the moments assigned to its two half middle strips. A middle strip adjacent to and parallel with an edge supported by a wall must be proportioned to resist twice the moment assigned to the half middle strip corresponding to the first row of interior columns.

11.4.4 Pattern Loading Consideration

In the analysis of continuous members, pattern loading has to be considered. As a result, the maximum moments are obtained. Pattern loading can cause reversal of stress, as seen in Fig. 11.6, as a function of the relative stiffness of the beams and columns intersecting at the joint. Pattern loading analysis is cumbersome. By limiting the applicability of the Direct Design Method of slabs with "Live load not exceeding two times dead load", there is no longer a need to check for pattern load effects. Slab and column dimensions for virtually all practical cases will meet the values for α_{min} specified in Table 11.2.

11.4.5 Shear–Moment Transfer to Columns Supporting Flat Plates

11.4.5.1. Shear strength. The shear behavior of two-way slabs and plates is a three-dimensional stress problem. The critical shear failure plane follows the perimeter of the loaded area and is located at a distance that gives a minimum shear perimeter b_0. Based on extensive analytical and experimental verification, the shear plane should not be closer than a distance $d/2$ from the concentrated load or reaction area.

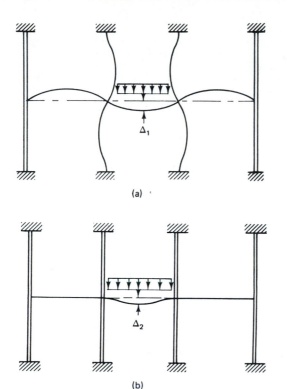

(a)

(b)

Figure 11.6 Pattern loading effect on deflection and cracking: (a) large deflection Δ_1 with more flexible columns; (b) small deflection Δ_2 with stiffer columns.

If no special shear reinforcement is provided, the maximum allowable nominal shear strength V_c of the section as required by the ACI is the smallest of the values from Eqs. 11.5:

$$V_c = \left(2 + \frac{4}{\beta_c} \right) \sqrt{f_c'}\, b_0 d \qquad (11.5a)$$

where β_c is the ratio of the long side to the short side of the column, concentrated load, or reaction areas, and b_0 is the perimeter of the critical section:

$$V_c = \left(\frac{\alpha_s d}{b_0} + 2 \right) \sqrt{f_c'}\, b_0 d \qquad (11.5b)$$

where α_s is 40 for interior columns, 30 for edge columns, and 20 for corner columns and

$$V_c = 4 \sqrt{f_c'}\, b_0 d \qquad (11.5c)$$

Equations 11.5(a) and (b) are the results of tests that indicate that as the ratio b_0/d increases the available nominal shear strength V_c decreases so that in such situations Eq. 11.5(c) would not control because it becomes unsafe. It is clear from Eq.

TABLE 11.2 Values of α_{min}[a]

β_a	Aspect Ratio, l_2/l_1	Relative Beam Stiffness, α				
		0	0.5	1.0	2.0	4.0
2.0	0.5–2.0	0	0	0	0	0
1.0	0.5	0.6	0	0	0	0
	0.8	0.7	0	0	0	0
	1.0	0.7	0.1	0	0	0
	1.25	0.8	0.4	0	0	0
	2.0	1.2	0.5	0.2	0	0
0.5	0.5	1.3	0.3	0	0	0
	0.8	1.5	0.5	0.2	0	0
	1.0	1.6	0.6	0.2	0	0
	1.25	1.9	1.0	0.5	0	0
	2.0	4.9	1.6	0.8	0.3	0
0.33	0.5	4.8	0.5	0.1	0	0
	0.8	2.0	0.9	0.3	0	0
	1.0	2.3	0.9	0.4	0	0
	1.25	2.8	1.5	0.8	0.2	0
	2.0	13.0	2.6	1.2	0.5	0.3

$$^a\beta_a = \frac{\text{unfactored dead load per unit area}}{\text{unfactored live load per unit area}}$$

$$\alpha_c = \frac{\Sigma K_c}{\Sigma(K_b + K_s)}$$

$$= \frac{\text{sum of stiffness of columns above and below slab}}{\text{sum of stiffness of beams and slabs framing into}}$$
the joint in the direction of the span for which the
moments are being determined

11.5c that the shear strength provided by the plain concrete cannot exceed $4\sqrt{f_c'}$, which is almost double the shear strength allowed in one-way members such as beams and one-way slabs.

If special shear reinforcement is provided, the maximum nominal shear strength V_n cannot exceed $6\sqrt{f_c'}\,b_0 d$, provided that the value used for V_c in the term $V_s = V_n - V_c$ does not exceed $2\sqrt{f_c'}\,b_0 d$.

11.4.5.2. Shear–moment transfer.   The unbalanced moment at the column face support of a slab without beams is one of the more critical design considerations in proportioning a flat plate or a flat slab.   To ensure adequate shear strength requires moment transfer to the column by flexure across the perimeter of the column and by eccentric shearing stress such that approximately 60% is transferred by flexure and 40% by shear.

The fraction γ_v of the moment transferred by eccentricity of the shear stress decreases as the width of the face of the critical section resisting the moment increases such that

$$\gamma_v = 1 - \frac{1}{1 + \frac{2}{3}\sqrt{b_1/b_2}} \tag{11.6a}$$

where $b_2 = c_2 + d$ is the width of the face of the critical section resisting the moment and $b_1 = c_1 + d$ for interior columns and $b_1 = c_1 + d/2$ for edge columns and $b_1 = c_1 + d/2$ and $b_2 = c_2 + d/2$ for corner columns. Side b_1 is the width of the face at right angles to side b_2.

The remaining portion γ_f of the unbalanced moment transferred by flexure is given by

$$\gamma_f = \frac{1}{1 + \frac{2}{3}\sqrt{b_1/b_2}} \quad \text{or} \quad \gamma_f = 1 - \gamma_v \tag{11.6b}$$

acting on an effective slab between lines that are $1\frac{1}{2}$ times the total slab thickness h on both sides of the column support.

For the exterior support, provided that V_u at the support is less than $0.75\phi V_c$, the value of γ_f can be increased to 1.0. *At the interior support*, γ_f can be increased by 25% provided that V_u is equal to or less than $0.4\phi V_c$ and ρ is equal to or less than $0.375\,\rho_b$.

The distribution of shear stresses around the column edges is as shown in Fig. 11.7. It is considered to vary linearly about the centroid of the critical section. The factored shear force V_u and the unbalanced factored moment M_u, both assumed acting at the column face, have to be transferred to the centroidal axis c–c of the critical section. Thus the axis position has to be located, thereby obtaining the shear force arm g (distance from the column face to the centroidal axis plane) of the critical section c–c for the shear moment transfer.

For calculating the maximum shear stress sustained by the plate in the edge column region, the ACI Code requires using the full nominal moment strength M_n provided by the column strip in Eqs. 11.7 as the unbalanced moment, multiplied by the transfer fraction factor γ_v. This unbalanced moment $M_n \geq M_{ue}/\phi$ is composed of two parts: the negative end panel moment $M_{ne} = M_e/\phi$ at the face of the column and the moment $(V_u/\phi)g$, due to the eccentric factored perimetric shear force V_u. The limiting value of the shear stress intensity is expressed as

$$\frac{v_{u(AB)}}{\phi} = \frac{V_u}{\phi A_c} + \frac{\gamma_v M_{ue} c_{AB}}{\phi J_c} \tag{11.7a}$$

$$\frac{v_{u(CD)}}{\phi} = \frac{V_u}{\phi A_c} - \frac{\gamma_v M_{ue} c_{CD}}{\phi J_c} \tag{11.7b}$$

where the nominal shear strength intensity is

$$v_n = \frac{v_u}{\phi} \tag{11.7c}$$

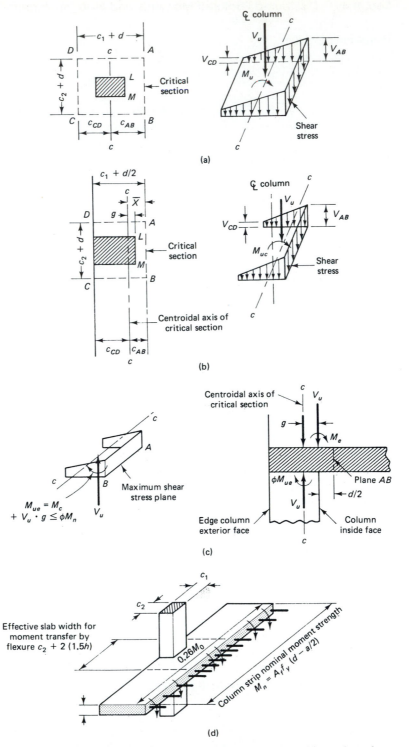

Figure 11.7 Shear stress distribution around column edges: (a) interior column; (b) end column; (c) critical surface; (d) transfer nominal moment strength M_n.

and where A_c = area of concrete of assumed critical section
$\qquad\qquad$ = $2d(c_1 + c_2 + 2d)$ for an interior column
$\qquad J_c$ = property of assumed critical section analogous
$\qquad\qquad$ to polar moment of inertia

The value of J_c for an interior column is

$$J_c = \frac{d(c_1 + d)^3}{6} + \frac{d^3(c_1 + d)}{6} + \frac{d(c_2 + d)(c_1 + d)^2}{2} \qquad (11.7d)$$

The value of J_c for an edge column with bending parallel to the edge is

$$J_c = \frac{(c_1 + d/2)(d)^3}{6} + \frac{2(d)}{3}(c^3{}_{AB} + c^3{}_{CD}) + (c_2 + d)(d)(c_{AB})^2 \qquad (11.7e)$$

It can be recognized from basic principles of mechanics of materials that the shearing stress

$$v_u = \frac{V_u}{A_c} + \gamma_v \frac{Mc}{J}$$

where the second part of the term is the shearing stress resulting from the torsional moment at the column face.

If the nominal moment strength M_n of the shear moment transfer zone after the design of the reinforcement results in a larger value than M_{ue}/ϕ, the M_n value should be used in Eqs. 11.7a and b in lieu of M_{ue}/ϕ. In such a case, where the moment strength value $M_n = M_{ne} + (V_u/\phi)g$ is increased because of the use of flexural reinforcement in excess of what is needed to resist M_{ue}/ϕ, the slab stiffness is raised, thereby increasing the transferred shear stress v_u calculated from Eqs. 11.7a and b for development of full moment transfer. Consequently, it is advisable to maintain a design moment M_{ue} with a value close to the factored moment value M_{ue} if an increase in the shear stress due to additional moment transfer needs to be avoided and a possible resulting need for additional increase in the plate design thickness prevented.

Numerical Ex. 11.1 illustrates the procedure for calculating the limit perimeter shear stress in the plate at the edge column region.

A higher perimetric shearing stress v_u can occur than evaluated by Eq. 11.7a or b when adjoining spans are unequal or unequally loaded in the case of an interior column. The ACI Code stipulates in the slab section pertaining to factored moments in columns and walls that the supporting element, such as a column or a wall, has to resist an unbalanced moment M', such that

$$M' = 0.07[(w_{nd} + 0.5w_{nl})l_2 l_{n2} - w'_{nd} l'_2 (l'_n)^2]$$

where w'_{nd}, l'_2, and l'_n refer to the shorter span.

Hence, an additional term is added to Eq. 11.7a or b in such cases so that

$$v_u = \frac{V_u}{A_c} + \frac{\gamma_v M_u c_{AB}}{J_c} + \frac{\gamma_v M'_C}{J'_c} \tag{11.8}$$

where J'_c is the polar moment of inertia with moment areas taken in a direction perpendicular to that used for J_c.

11.4.6 Deflection Requirements for Minimum Thickness: An Indirect Approach

The serviceability of a floor system can be maintained through deflection control and crack control. Since deflection is a function of the stiffness of the slab as a measure of its thickness, a minimum thickness has to be provided irrespective of the flexural thickness requirement. Table 11.3 gives the minimum thickness of slabs without interior beams. This occurs when $\alpha_m \cong 0$. Table 11.4 gives the maximum permissible computed deflections to safeguard against plaster cracking and to maintain esthetic appearance. Deflection computations for two-way-action slabs can be made using the analytical procedures described in Section 11.8 in order to determine whether the analysis gives long-term deflections within the limitations of Table 11.4.

Approximate empirical limitation on deflection through determination of the minimum thickness of the slab on beams or drop panels or bands can be obtained from Table 11.3 if the stiffness ratio $\alpha_m < 0.2$.
For $\alpha_m > 0.2$ but not greater than 2.0,

$$h = \frac{l_n (0.8 + f_y/200{,}000)}{36 + 5\beta(\alpha_m - 0.2)} \tag{11.9}$$

and need not be less than 5 in.
For $\alpha_m > 2.0$,

$$h = \frac{l_n (0.8 + f_y/200{,}000)}{36} \tag{11.10}$$

where l_n is the length of clear span in the long direction for deflection determination. For moment computation, l_n is the length of the clear span in the direction that moments are being computed.

For slabs without beams, but with drop panels having a width in each direction from center line of support a distance not less than one-sixth the span length in that direction measured center to center of supports and a projection below the slab at least one-fourth the slab thickness beyond the drop, the thickness required by Eq. 11.9 or 11.10 may be reduced by 10%. At discontinuous edges, an edge beam shall be provided with a stiffness ratio α not less than 0.80; or the minimum thickness required by Eq. 11.9 or 11.10 shall be increased by at least 10% in the panel with a discontinuous edge.

TABLE 11.3 Minimum Thickness of Slab Without Interior Beams

Yield Stress, $f_{y,0}$ (psi)[b]	Without Drop Panels[a]			With Drop Panels[a]		
	Exterior Panels			Exterior Panels		
	Without Edge Beams	With Edge Beams[c]	Interior Panels	Without Edge Beams	With Edge Beams[c]	Interior Panels
40,000	$\dfrac{l_n}{33}$	$\dfrac{l_n}{36}$	$\dfrac{l_n}{36}$	$\dfrac{l_n}{36}$	$\dfrac{l_n}{40}$	$\dfrac{l_n}{40}$
60,000	$\dfrac{l_n}{30}$	$\dfrac{l_n}{33}$	$\dfrac{l_n}{33}$	$\dfrac{l_n}{33}$	$\dfrac{l_n}{36}$	$\dfrac{l_n}{36}$

[a]Drop panel as defined by the ACI Code.

[b]For values of reinforcement yield stress between 40,000 and 60,000 psi, minimum thickness shall be obtained by linear interpolation.

[c]Slabs with beams between columns along exterior edges. The value of α for the edge beam shall not be less than 0.8.

TABLE 11.4 Minimum Permissible Ratios of Span (l) to Deflection (a) (l = Longer Span)

Type of Member	Deflection, a, to Be Considered	$(l/a)_{min}$
Flat roofs not supporting and not attached to nonstructural elements likely to be damaged by large deflections	Immediate deflection due to live load L	180[a]
Floors not supporting and not attached to nonstructural elements likely to be damaged by large deflections	Immediate deflection due to live load L	360
Roof or floor construction supporting or attached to nonstructural elements likely to be damaged by large deflections	That part of total deflection occurring after attachment of nonstructural elements: sum of long-term deflection due to all sustained loads (dead load plus any sustained portion of live load) and immediate deflection due to any additional live load[b]	480[c]
Roof or floor construction supporting or attached to nonstructural elements not likely to be damaged by large deflections		240[c]

[a]Limit not intended to safeguard against ponding. Ponding should be checked by suitable calculations of deflection, including added deflections due to ponded water and considering long-term effects of all sustained loads, camber, construction tolerances, and reliability of provisions for drainage.

[b]Long-term deflection has to be determined, but may be reduced by the amount of deflection calculated to occur before attachment of nonstructural elements. This reduction is made on the basis of accepted engineering data relating to time–deflection characteristics of members similar to those being considered.

[c]Ratio limit may be lower if adequate measures are taken to prevent damage to supported or attached elements, but should not be lower than tolerance of nonstructural elements.

Figure 11.8 gives a plot of the thickness ratio h/l_n to aspect ratio β for the two equations for various stiffness ratios α. Note from the plots that Eq. 11.9 is an upper-limit expression applicable to limited conditions when the stiffnesses of the panel beam supports are so low that the stiffness ratio α has a value close to 0.2, gradually approaching the condition of a flat plate. It is not applicable when $\alpha = 0$. If it is to be used in the latter condition, however, we can assume that part of the slab in the column region acts as a beam. Thickness h cannot be less than the following values:

Slabs without beams or drop panels	5 in.
Slabs without beams, but with drop panels	4 in.
Slabs with beams on all four edges with a value of α_m at least equal to 2.0	3.5 in.

h also has to be increased by at least 10% for flat-plate floors if the end panels have no edge beams and by 45% for corner panels.

In addition, in the equations above,

α = ratio of flexural stiffness of beam section to flexural stiffness of a width of slab bounded laterally by the center line of the adjacent panel (if any) on each side of the beam

α_m = average value of α for all beams on edges of a panel

β = ratio of clear spans in long to short direction of two-way slabs

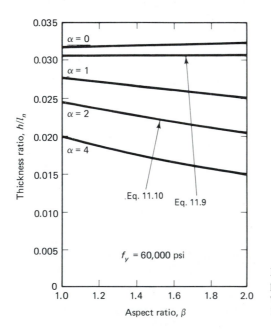

Figure 11.8 Thickness ratio versus aspect ratio for deflection control of two-way slabs on beams, Eqs. 11.9 and 11.10 limits.

It has to be emphasized that a deflection check is critical for the construction loading condition. Shoring and reshoring patterns can result in dead-load deflection in excess of the normal service-load state at a time when the concrete has only a 7-day strength or less and not the normal design 28-day strength. The stiffness *EI* in such a state is less than the design value. Flexural cracking lowers further the stiffness values of the two-way slab or plate, with a possible increase in long-term deflection several times the anticipated design deflection. Consequently, reinforced concrete two-way slabs and plates have to be constructed with a camber of $\frac{1}{8}$ in. in 10-ft span or more and crack control exercised as in Section 11.9 in order to counter the effects of excessive deflection at the construction loading stage.

11.5 DESIGN AND ANALYSIS PROCEDURE: DIRECT DESIGN METHOD

11.5.1 Operational Steps

Figure 11.9 gives a logic flow chart for the following operational steps:

1. Determine whether the slab geometry and loading allow the use of the direct design method as listed in Section 11.3.1.
2. Select slab thickness to satisfy deflection and shear requirements. Such calculations require a knowledge of the supporting beam or column dimensions. A reasonable value of such a dimension of columns or beams would be 8% to 15% of the average of the long and short span dimensions, that is, $\frac{1}{2}(l_1 + l_2)$. For shear check, the critical section is at a distance $d/2$ from the face of the support. If the thickness shown for deflection is not adequate to carry the shear, use one or more of the following:
   (a) Increase the column dimension.
   (b) Increase concrete strength.
   (c) Increase slab thickness.
   (d) Use special shear reinforcement.
   (e) Use drop panels or column capitals to improve shear strength.
3. Divide the structure into equivalent design frames bound by center lines of panels on each side of a line of columns.
4. Compute the total statical factored moment $M_0 = (w_u l_2 l_{n1}^2)/8$.
5. Select the distribution factors of the negative and positive moments to the exterior and interior columns and spans as in Fig. 11.3b and Table 11.1 and calculate the respective factored moments.
6. Distribute the factored equivalent frame moments from step 4 to the column and middle strips.
7. Determine whether the trial slab thickness chosen is adequate for moment-shear transfer in the case of flat plates at the interior column junction com-

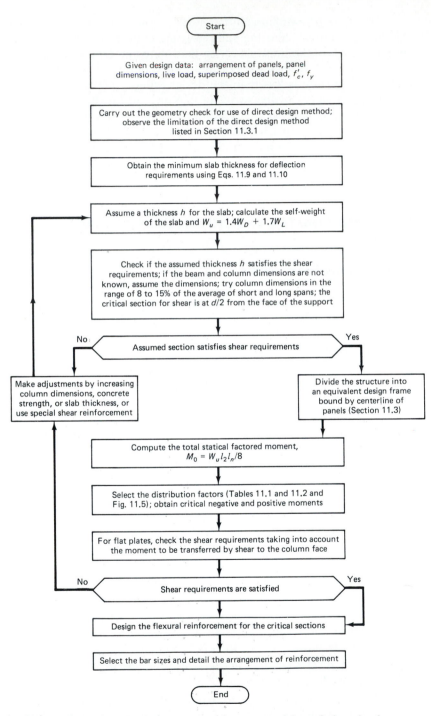

Figure 11.9   Flow chart for design sequence in two-way slabs and plates by the direct design method.

puting that portion of the moment transferred by shear and the properties of the critical shear section at distance $d/2$ from column face.

8. Design the flexural reinforcement to resist the factored moments in step 6.
9. Select the size and spacing of the reinforcement to fulfill the requirements for crack control, bar development lengths, and shrinkage and temperature stresses.

11.5.2 Example 11.1: Design of Flat Plate without Beams by the Direct Design Method

A three-story building is four panels by four panels in plan.   The clear height between the floors is 12 ft, and the floor system is a reinforced concrete flat-plate construction with no edge beams.   The dimensions of the end panels as well as the size of the supporting columns are shown in Fig. 11.10.   Given:

> live load $= 50$ psf (2.39 kPa)
>
> $f_c' = 4000$ psi (27.6 MPa), normal-weight concrete
>
> $f_y = 60,000$ psi (414 MPa)

The building is not subject to earthquake; consider gravity loads only.   Design the end panel and the size and spacing of the reinforcement needed.   Consider flooring weight to be 10 psf in addition to the floor self-weight.

Solution   *Geometry check for use of direct design method (step 1)*

(a) Ratio $\dfrac{\text{longer span}}{\text{shorter span}} = \dfrac{24}{18} = 1.33 < 2.0$, hence two-way action

(b) More than three spans in each direction and successive spans in each direction the same and columns are not offset.

(c) Assume a thickness of 9 in. and flooring of 10 psf.

$$w_d = 10 + \frac{9}{12} \times 150 = 122.5 \text{ psf} \qquad 2w_d = 245 \text{ psf}$$

$$w_l = 50 \text{ psf} < 2w_d \qquad \text{O.K.}$$

Hence the direct design method is applicable.

Minimum slab thickness for deflection requirement (step 2)

$$\text{E–W direction } l_{n1} = 24 \times 12 - \frac{18}{2} - \frac{20}{2} = 269 \text{ in (6.83 m)}$$

$$\text{N–S direction } l_{n2} = 18 \times 12 - \frac{20}{2} - \frac{20}{2} = 196 \text{ in. (4.98 m)}$$

$$\text{ratio of longer to shorter clear span } \beta = \frac{269}{196} = 1.37$$

Minimum preliminary thickness h from Table 11.3 for a flat plate without edge beams or drop panels using $f_y = 60,000$-psi steel is $h = l_n/30$, to be increased by at least 10% when no edge beam is used.

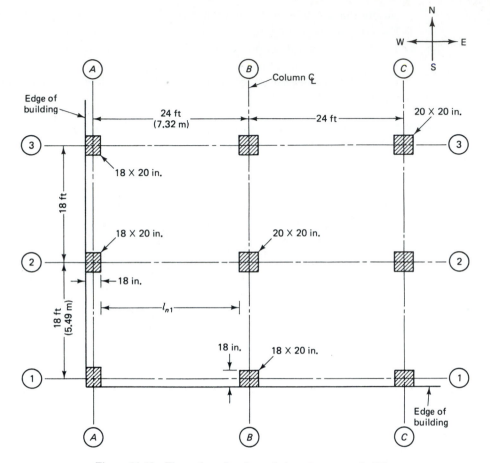

Figure 11.10 Floor plan of end panels in a three-story building.

$$\text{E--W:} \quad l_n = 269 \text{ in.}$$

$$h = \frac{269}{30} \times 1.1 = 9.86 \text{ in.}$$

Try a slab thickness $h = 10$ in. This thickness is larger than the absolute minimum thickness of 5 in. required in the code for flat plates; hence O.K. Assume that $d \simeq h - 1$ in. $= 9$ in.

$$\text{new } w_d = 10 + \frac{10}{12} \times 150 = 135.0 \text{ psf}$$

Therefore,

$$2w_d = 270 \text{ psf}$$

$$w_l = 50 \text{ psf} < 3w_d \qquad \text{O.K.}$$

Shear thickness requirement (step 2)

$$w_u = 1.7L + 1.4D = 1.7 \times 50 + 1.4 \times 135.0$$
$$= 274 \text{ psf } (13.12 \text{ kPa})$$

Interior column: The controlling critical plane of maximum perimetric shear stress is at a distance $d/2$ from the column faces; hence, the net factored perimetric shear force is

$$V_u = [(l_1 \times l_2 - (c_1 + d)(c_2 + d)]w_u$$

$$= \left(18 \times 24 - \frac{20 + 9}{12} \times \frac{20 + 9}{12}\right) 274 = 116{,}768$$

$$V_n = \frac{V_u}{\phi} = \frac{116{,}768}{0.85} = 137{,}374 \text{ lb}$$

From Fig. 11.11, the perimeter of the critical shear failure surface is

$$b_0 = 2(c_1 + d + c_2 + d) = 2(c_1 + c_2 + 2d)$$

perimetric shear surface $A_c = b_0 d = 2d(c_1 + c_2 + 2d) = 2 \times 9.0(20 + 20 + 18)$

$$= 116 \times 9 = 1044 \text{ in.}^2 \ (673{,}400 \text{ mm}^2)$$

Since moments are not known at this stage, only a preliminary check for shear can be made.

$$\beta_c = \text{ratio of longer to shorter side of columns} = \frac{20}{20} = 1.0$$

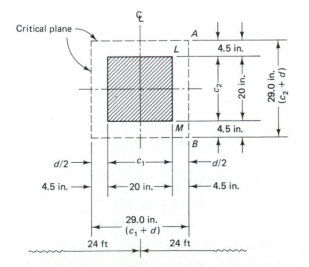

Figure 11.11   Critical plane for shear moment transfer in Ex. 11.1 interior column (line B–B, Fig. 11.10).

Available nominal shear V_c is the least of

$$V_c = \left(2 + \frac{4}{\beta_c}\right)\sqrt{f_c'}\,b_0 d = \left(2 + \frac{4}{1}\right)\sqrt{4000} \times 1044 = 369{,}170\,\text{lb}\,(1.64 \times 10^3\,\text{kN})$$

or

$$V_c = \left(\frac{\alpha_s d}{b_0} + 2\right)\sqrt{f_c'}\,b_0 d = \left(\frac{40 \times 9}{116} + 2\right)\sqrt{4000} \times 1044 = 336{,}972\,\text{lb}\,(1.5 \times 10^3\,\text{kN})$$

or

$$V_c = 4\sqrt{f_c'}\,b_0 d = 4\sqrt{4000} \times 1044 = 264{,}113\,\text{lb}\,(1.16 \times 10^3\,\text{kN})$$

$$\text{controlling } V_c = 264{,}113\,\text{lb} > \text{required } V_n = 137{,}374\,\text{lb}$$

Hence the floor thickness is adequate.

Exterior column: Include weight of exterior wall, assuming its service weight to be 270 plf. Net factored perimetric shear force is

$$V_u = \left[18 \times \left(\frac{24}{2} + \frac{18}{2 \times 12}\right) - \frac{(18 + 4.50)(20 + 9.0)}{144}\right] 274$$
$$+ \left(18 - \frac{20}{12}\right) \times 270 \times 1.4 = 67{,}815\,\text{lb} \qquad V_n = \frac{67{,}815}{0.85} = 79{,}782\,\text{lb}$$

Consider the line of action of V_u to be at the column face LM in Fig. 11.12 for shear moment transfer to the centroidal plane c–c. This approximation is adequate since V_u acts *perimetrically* around the column faces and not along line AB only. From Fig. 11.12,

$$A_c = d(2c_1 + c_2 + 2d) = 9.0(2 \times 18 + 20 + 18) = 9 \times 74$$
$$= 666\,\text{in}^2\,(429{,}700\,\text{mm}^2)$$

Available nominal shear V_c is the least of

$$V_c = \left(2 + \frac{4}{\beta_c}\right)\sqrt{f_c'}\,b_0 d = \left(2 + \frac{4}{20/18}\right)\sqrt{4000} \times 666 = 235{,}881\,\text{lb}\,(1.05 \times 10^3\,\text{kN})$$

or

$$V_c = \left(\frac{\alpha_s d}{b_0} + 2\right)\sqrt{f_c'}\,b_0 d = \left(\frac{30 \times 9}{74} + 2\right)\sqrt{4000} \times 666 = 237{,}930\,\text{lb}\,(1.06 \times 10^3\,\text{kN})$$

where $\alpha_s = 30$ for edge column

or

$$V_c = 4\sqrt{f_c'}\,b_0 d = 4\sqrt{4000} \times 666 = 168{,}486\,\text{lb}\,(0.75 \times 10^3\,\text{kN}) \qquad \text{controls}$$
$$> \text{required } V_n = 79{,}782\,\text{lb} \qquad \text{O.K.}$$

Statical moment computation (steps 3 to 5)

$$\text{E–W: } l_{n1} = 269\,\text{in.} = 22.42\,\text{ft}$$
$$\text{N–S: } l_{n2} = 196\,\text{in.} = 16.33\,\text{ft}$$

$$0.65l_1 = 0.65 \times 24 = 15.6\,\text{ft} \qquad \text{Use } l_{n1} = 22.4\,\text{ft}$$
$$0.65l_2 = 0.65 \times 18 = 11.7\,\text{ft} \qquad \text{Use } l_{n2} = 16.33\,\text{ft}$$

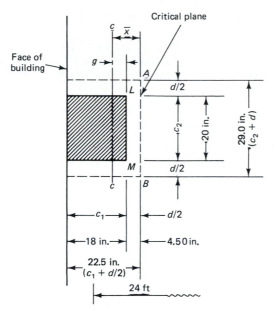

Figure 11.12 Centroidal axis for shear moment transfer in Ex. 11.1 end column (line A–A or 1–1, Fig. 11.10).

(a) *E–W direction*

$$M_0 = \frac{w_n l_2 l_{n1}^2}{8} = \frac{274 \times 18(22.42)^2}{8} = 309,888 \text{ ft-lb (420 kN-m)}$$

For end panel of a flat plate without end beams, the moment distribution factors as in Table 11.1 are

$$-M_u \text{ at first interior support} = 0.70 M_0$$
$$+M_u \text{ at midspan of panel} = 0.52 M_0$$
$$-M_u \text{ at exterior face} = 0.26 M_0$$
$$\text{negative design moment} -M_u = 0.70 \times 309,888$$
$$= 216,922 \text{ ft-lb (295 kN-m)}$$
$$\text{positive design moment} +M_u = 0.52 \times 309,888$$
$$= 161,142 \text{ ft-lb (218 kN-m)}$$
$$\text{negative moment at exterior} -M_u = 0.26 \times 309,888$$
$$= 80,571 \text{ ft-lb (109 kN-m)}$$

(b) *N–S direction*

$$M_0 = \frac{w_u l_1 l_{n2}^2}{8} = \frac{274 \times 24(16.33)^2}{8} = 219,202 \text{ ft-lb (298 kN-m)}$$

$$\text{negative design moment} -M_u = 0.70 \times 219,202$$
$$= 153,411 \text{ ft-lb (208 kN-m)}$$
$$\text{positive design moment} +M_u = 0.52 \times 219,202$$
$$= 113,985 \text{ ft-lb (77 kN-m)}$$

negative design moment at exterior face $-M_{u1} = 0.26 \times 219{,}202$

$$= 56{,}993 \text{ ft-lb (75 kN-m)}$$

Note that the smaller moment factor 0.35 could be used for the positive factored moment in the N–S direction in this example if the exterior edge is fully restrained. For the N–S direction, panel BC12 was considered.

Moment distribution in the column and middle strips (steps 6 and 7)

At the exterior column there is no torsional edge beam; hence the torsional stiffness ratio β_t of an edge beam to the columns is zero. Hence $\alpha_1 = 0$. From the *exterior* factored moments tables for the column strip in Section 11.4.2, the distribution factor for the negative moment at the exterior support is 100%, the positive midspan moment is 60%, and the interior negative moment is 75%. Table 11.5 gives the moment values resulting from the moment distributions to the column and middle strips.

Check the shear moment transfer capacity at the exterior column supports

$$-M_c \text{ at interior column } 2\text{-}B = 216{,}922 \text{ ft-lb}$$

$$-M_e \text{ at exterior column } 2\text{-}A = 80{,}571 \text{ ft-lb}$$

$$V_u = 67{,}815 \text{ lb acting at the face of the column}$$

The ACI Code stipulates that the nominal moment strength be used in evaluating the unbalanced transfer moment at the edge column; that is, use M_n based on $-M_e = 80{,}571$ ft-lb. Factored shear force at the edge column adjusted for the interior moment is

$$V_u = 67{,}815 - \frac{216{,}922 - 80{,}571}{24 - \dfrac{9 + 10}{12}} = 61{,}732 \text{ lb}$$

$V_n = 61{,}732/0.85 = 72{,}626$ lb, assuming that the design M_u has the same value as the factored M_u.

$$A_c \text{ from before } = 666 \text{ in.}^2$$

From Figs. 11.7c and 11.12, taking the moment of area of the critical plane about axis AB,

$$d(2c_1 + c_2 + 2d)\bar{x} = d\left(c_1 + \frac{d}{2}\right)^2$$

where $\bar{x}$ is the distance to the centroid of the critical section or

$$(2 \times 18 + 20 + 18)\bar{x} = \left(18 + \frac{9.0}{2}\right)^2$$

$$\bar{x} = \frac{506.25}{74} = 6.84 \text{ in. (173.8 mm)}$$

and $g = 6.84 - \dfrac{9.0}{2} = 2.34$ in, where g is the distance from the column face to the centroidal axis of the section.

TABLE 11.5 Moment Distribution Operations Table

	E–W Direction l_2/l_1: 18/24 = 0.75 $\alpha_1(l_2/l_1)$: 0			N–S Direction 24/18 = 1.33 0		
Column strip	Interior negative moment	Positive midspan moment	Exterior negative moment	Interior negative moment	Positive midspan moment	Exterior negative moment
M_u (ft-lb)	216,922	161,142	80,571	153,441	113,985	56,993
Distribution factor (%)	75	60	100	75	60	100
Column strip design moments (ft-lb)	0.75 × 216,922 / 162,692	0.60 × 161,142 / 96,685	1.0 × 80,571 / 80,571	0.75 × 153,441 / 115,081	0.60 × 113,985 / 68,391	1.0 × 56,993 / 56,993
Middle strip design moments (ft-lb)	216,922 − 162,692 / 54,230	161,142 − 96,685 / 64,457	80,571 − 80,571 / 0	153,441 − 115,081 / 38,360	113,985 − 68,391 / 45,594	56,993 − 56,993 / 0

To transfer the shear V_u from the face of column to the centroid of the critical section adds an additional moment to the value of $M_e = 80{,}571$ ft-lb. Therefore, the total external factored moment $M_{ue} = 80{,}571 + 61{,}732(2.34/12) = 92{,}609$ ft-lb. Total required minimum unbalanced moment strength is

$$M_n = \frac{M_{ue}}{\phi} = \frac{92{,}609}{0.90} = 102{,}899 \text{ ft-lb}$$

The fraction of nominal moment strength M_n to be transferred by shear is

$$\gamma_v = 1 - \frac{1}{1 + \frac{2}{3}\sqrt{b_1/b_2}} = 1 - \frac{1}{1 + 0.59} = 0.37$$

where $b_1 = c_1 + d/2 = 18 + 4.5 = 22.5$ in. and $b_2 = c_2 + d = 20 + 9 = 29$ in. It should be noted that the dimension $c_1 + d$ for the end column in the above expression becomes $c_1 + d/2$. Hence $M_{nv} = 0.37 M_n$. Moment of inertia of sides parallel to the moment direction about N–S axis is

$$I_1 = \left(\frac{bh^3}{12} + Ar^2 + \frac{hb^3}{12} \right)2 \qquad \text{for both faces}$$

$$I_1 = \left[\frac{9.0(22.5)^3}{12} + (9.0 \times 22.5)\left(\frac{22.50}{2} - 6.84 \right)^2 + \frac{22.5(9.0)^3}{12} \right]2$$

$$= (8543 + 3938 + 1367)2 = 27{,}696 \text{ in.}^4$$

Moment of inertia of sides perpendicular to the moment direction about N–S axis is

$$I_2 = A(\bar{x})^2 = [(20 + 9.0)9.0](6.84)^2 = 12{,}211 \text{ in.}^4$$

Therefore,

$$\text{torsional moment of inertia } J_c = 27{,}696 + 12{,}211$$

$$= 39{,}907 \text{ in.}^4$$

If Eq. 11.7e is used instead from first principle calculations, as shown above, the same value $J_c = 39{,}907$ in.^{4} is obtained.

Shearing stress due to perimeter shear, effect of M_n, and weight of wall is

$$v_n = \frac{V_u}{\phi A_c} + \frac{\gamma_v c_{AB} M_n}{J_c}, \qquad \text{where } M_{nv} = \gamma_v \times M_n$$

$$= \frac{61{,}732}{0.85 \times 666} + \frac{0.37 \times 6.84 \times 102{,}899 \times 12}{39{,}907}$$

$$= 109.05 + 78.31 = 187.36 \text{ psi}$$

From before, maximum allowable $v_c = 4\sqrt{f_c'} = 4\sqrt{4000} = 253.0$ psi and

$$v_n < v_c$$

Therefore, accept plate thickness. For the corner panel column, special shear-head provision or an enlarged column or capital might be needed to resist the high shear stresses at that location.

Design of reinforcement in the slab area at column face for the unbalanced moment transferred to the column by flexure

From Eq. 11.6b,

$$\gamma_f = 1 - \gamma_v = 1 - 0.37 = 0.63$$

$$M_{nf} = \gamma_f M_n = 0.63 \times 102,899 \times 12 = 777,916 \text{ in.-lb}$$

This moment has to be transferred within $1.5h$ on each side of the column as in Fig. 11.7d.

$$\text{transfer width} = (1.5 \times 10.0)2 + 20 = 50.0 \text{ in.}$$

$$M_{nf} = A_s f_y \left(d - \frac{a}{2} \right) \qquad \text{assume that } \left(d - \frac{a}{2} \right) \approx 0.9d$$

or $777,916 = A_s \times 60,000(9.0 \times 0.9)$ gives

$$A_s = 1.60 \text{ in.}^2 \text{ over a strip width} = 50.0 \text{ in.}$$

Verifying A_s,

$$a = \frac{1.60 \times 60,000}{0.85 \times 4,000 \times 50.0} = 0.56 \text{ in.}$$

Therefore,

$$777,916 = A_s \times 60,000 \left(9.0 - \frac{0.56}{2} \right)$$

$$A_s = 1.49 \text{ in.}^2 \approx 5 \text{ No. 5 bars at 4 in. c-c}$$

are to be used in the 20-in.-column-width strip and anchored into the column as required for bond length development.

This additional steel will have to be used to effect the moment transfer. Reinforcement to carry the total edge column strip moment $M_e = 80,571$ ft-lb ($M_{ne} = 89,523$ ft-lb) is proportioned in the next section.

Checks have to be made in a similar manner for the shear-moment transfer at the face of the interior column C. As also described in Section 11.4.5.2, checks are sometimes necessary for pattern loading conditions and for cases where adjoining spans are not equal or not equally loaded.

Proportioning of the plate reinforcement (steps 8 and 9)

(a) *E–W direction (long span)*

1. *Summary of moments in column strip (ft-lb)*

$$\text{interior column negative } M_n = \frac{162,692}{\phi = 0.9} = 180,769$$

$$\text{midspan positive } M_n = \frac{96,685}{0.9} = 107,428$$

$$\text{exterior column negative } M_{ne} = \frac{80,571}{0.9} = 89,523$$

2. *Summary of moments in middle strip (ft-lb)*

$$\text{interior column negative } M_n = \frac{54,248}{0.9} = 60,276$$

$$\text{midspan positive } M_n = \frac{64,457}{0.9} = 71,619$$

$$\text{exterior column negative } M_n = 0$$

3. *Design of reinforcement for column strip*

$$-M_n = 180,769 \text{ ft-lb acts on a strip width of } 2(0.25 \times 18) = 9.0 \text{ ft}$$

$$\text{unit } -M_n \text{ per 12-in.-wide strip} = \frac{180,769 \times 12}{9.0} = 241,025 \text{ in.-lb}$$

$$\text{unit } + M_n = \frac{107,428 \times 12}{9.0} = 143,427 \text{ in.-lb/12-in.-wide strip}$$

minimum A_s for two-way plates using $f_y = 60,000$-psi steel $= 0.0018bh$

$$= 0.0018 \times 10 \times 12 = 0.216 \text{ in.}^2 \text{ per 12-in. strip}$$

Negative steel:

$$M_n = A_s f_y \left(d - \frac{a}{2} \right) \quad \text{or} \quad 241,025 = A_s \times 60,000 \left(9.0 - \frac{a}{2} \right)$$

Assume that moment arm $d - a/2 \simeq 0.9d$ for first trial and $d = h - \frac{3}{4}$ in. $- \frac{1}{2}$ diameter of bar $\simeq 9.0$ in. for all practical purposes. Therefore,

$$A_s = \frac{241,025}{60,000 \times 0.9 \times 9.0} = 0.50 \text{ in.}^2$$

$$a = \frac{A_s f_y}{0.85 f'_c b} = \frac{0.50 \times 60,000}{0.85 \times 4000 \times 12} = 0.74 \text{ in.}$$

For the second trial-and-adjustment cycle,

$$241,025 \doteq A_s \times 60,000 \left(9.0 - \frac{0.74}{2} \right)$$

Therefore, required A_s per 12-in.-wide strip $= 0.47$ in.^{2}. Try No. 5 bars (area per bar $= 0.305$ in.^{2}).

$$\text{spacing } s = \frac{\text{area of one bar}}{\text{required } A_s \text{ per 12-in. strip}}$$

Therefore,

$$s \text{ for negative moment} = \frac{0.305}{0.47/12} = 7.79 \text{ in. c-c (194 mm)}$$
$$\text{(No. 5 bars)}$$

$$s \text{ for positive moment} = 7.79 \times \frac{241,025}{143,237} = 13.11 \text{ in. c-c (326 mm)}$$

The maximum allowable spacing $= 2h = 2 \times 10 = 20$ in. (508 mm).   Try No. 4 bars for positive moment ($A_s = 0.20$ in.²).

$$A_s = \frac{143{,}237}{241{,}025} \times 0.47 = 0.28 \text{ in.}^2 \text{ per 12-in. strip}$$

minimum temperature reinforcement $= 0.0018 \, bh = 0.0018 \times 12 \times 10$

$$= 0.216 \text{ in.}^2/\text{f} < 0.28 \text{ in.}^2 \qquad \text{O.K.}$$

$$s = \frac{0.20}{0.28/12} = 8.57 \text{ in. c-c (218 mm)}$$

For an external negative moment, use No. 4 bars.

$$s = 8.57 \times \frac{143{,}237}{119{,}364} = 10.28 \text{ in. c-c}$$

Use 14 No. 5 bars at $7\frac{1}{2}$ in. center to center for negative moment at interior column side, 12 No. 4 bars at 8 in. center to center for positive moment; and 10 No. 4 bars at 10 in. center to center for the exterior negative moment M_e with 8 of these bars to be placed outside the shear moment transfer band width 50 in., as seen in Fig. 11.13b.

   4. *Design of reinforcement for middle strip*

$$\text{unit} - M_n = \frac{54{,}230}{0.9} = 60{,}256 \text{ acting on a strip width of } 18.0 - 9.0 = 9.0 \text{ ft}$$

$$\text{unit} - M \text{ per 12-in.-width strip} = \frac{60{,}256 \times 12}{9} = 80{,}341 \text{ lb-in.}$$

$$80{,}341 = A_s \times 60{,}000(9.0 \times 0.9)$$

$$A_s = 0.17 \text{ in.}^2 \qquad a = \frac{0.17 \times 60{,}000}{0.85 \times 4000 \times 12} = 0.25 \text{ in.}$$

Second cycle:

$$80{,}341 = A_s \times 60{,}000\left(9.0 - \frac{0.25}{2}\right)$$

$$A_s = 0.15 \text{ in.}^2/\text{12-in. strip} \qquad \text{minimum } A_s = 0.194$$

$$\text{use } A_s = 0.194 \text{ in.}^2/\text{12-in. strip}$$

Try No. 3 bars ($A_s = 0.11$ in.² per bar).

$$\text{unit} + M = \frac{64{,}457 \times 12}{9 \times 0.9} = 95{,}492 \text{ in.-lb per 12-in. strip}$$

$$A_s = \frac{95{,}492}{60{,}000 \times 8.875} = 0.18 \text{ in.}^2 \qquad \text{use minimum } A_s = 0.216 \text{ in.}^2/12 \text{ in.}$$

Hence use negative and positive steel spacing $s = 0.11/(0.216/12) = 6.11$ in. (No. 3 at 6 in. center to center).   Use No. 3 bars at 6 in. center to center for both the negative moment and positive moments.

   (b) *N–S direction (short span):* The same procedure has to be followed as for the E–W direction.   The width of the column strip on one side of the column =

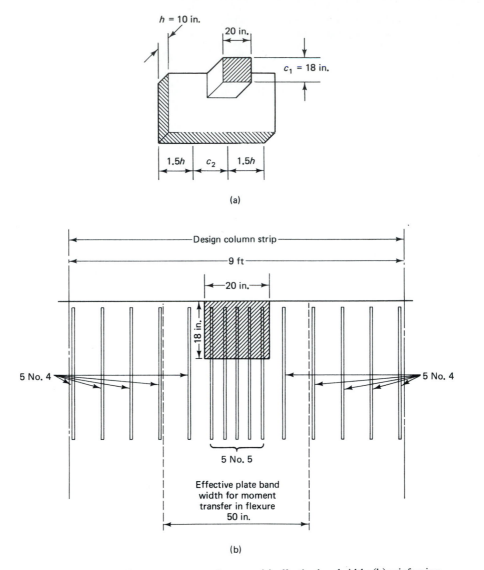

Figure 11.13 Shear moment transfer zone: (a) effective bandwidth; (b) reinforcing details.

$0.25l_1 = 0.25 \times 24 = 6$ ft, which is greater than $0.25l_2 = 4.5$ ft; hence a width of 4.5 ft controls. The total width of the column strip in the N–S direction $= 2 \times 4.5 = 9.0$ ft. The width of the middle strip $= 24.0 - 9.0 = 15.0$ ft. Also, the effective depth d_2 would be smaller; $d_2 = (h - 3/4\text{-in. cover} - 0.5\text{ in.} - 0.5/2) = 8.5$ in. The moment values and the bar size and distribution for the panel in the N–S direction as well as the E–W directions are listed in Table 11.6. It is recommended for crack-control purposes that a minimum of No. 3 bars at 12 in. center to center be used and

TABLE 11.6 Moments, Bar Sizes, and Distribution

Strip	Moment Type	E-W			N-S		
		Moment (lb-in./12 in.)	A_s Req'd	Bar Size and Spacing	Moment (lb-in./12 in.)	A_s Req'd	Bar Size and Spacing
Column	Interior negative	241,025	0.47	No. 5 at $7\frac{1}{2}$	170,490	0.37	No. 4 at 6
	Exterior negative	119,364	0.23	No. 4 at 10	84,434[b]	0.18	No. 3 at 6
	Midspan positive	143,237	0.28	No. 4 at $8\frac{1}{4}$	101,320	0.22	No. 3 at 6
Middle	Interior negative	80,341	0.15	No. 3 at 8	34,098	0.07	No. 3 at 6
	Exterior negative	0	0	No. 3 at 6[a]	0	0	No. 3 at 6
	Midspan positive	95,492	0.18	No. 3 at 6	40,528	0.09	No. 3 at 6

[a]Minimum temperature steel = $0.0018bh$ where it controls.
[b]For panel BC12 (see comment on page 000).

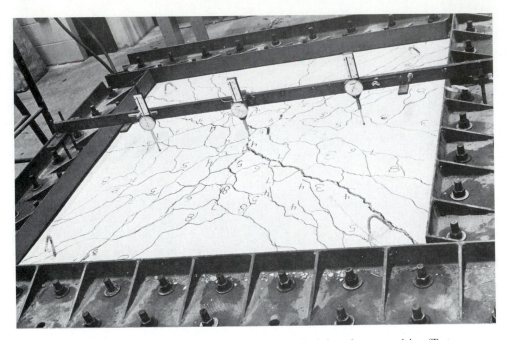

Photo 66 Flexural cracking in restrained one-panel reinforced concrete slab. (Tests by Nawy et al.)

that bar spacing not exceed 12 in. center to center. In this case the minimum reinforcement required by the ACI Code for slabs reinforced with $f_y = 60,000$-psi steel $= 0.0018 \, bh =$ No. 3 at $6\frac{1}{2}$ in. on centers. Space at 6 in. on centers.

The choice of size and spacing of the reinforcement is a matter of engineering judgment. As an example, the designer could have chosen for the positive moment in the middle strip No. 4 bars at 12 in. center to center, instead of No. 3 bars at 6 in. center to center, as long as the maximum permissible spacing is not exceeded and practicable bar sizes are used for the middle strip.

The placing of the reinforcement is schematically shown in Fig. 11.14. The minimum cutoff of reinforcement for bond requirements in flat-plate floors is given in Fig. 11.15. The exterior panel negative steel at outer edges, if no edge beams are used, has to be bent into full hooks in order to ensure sufficient anchorage of the reinforcement. The floor reinforcement plan gives the E–W steel for panel AB23 and N–S steel for panel BC12 of Fig. 11.10.

11.5.3 Example 11.2: Design of Two-way Slab on Beams by Direct Design Method (DDM)

A two-story factory building is three panels by three panels in plan, monolithically supported on beams. Each panel is 18 ft (5.49 m) center to center in the N–S direction and 24 ft (7.32 m) center to center in the E–W direction, as shown in Fig. 11.16. The clear height between the floors is 16 ft. The dimensions of the supporting beams

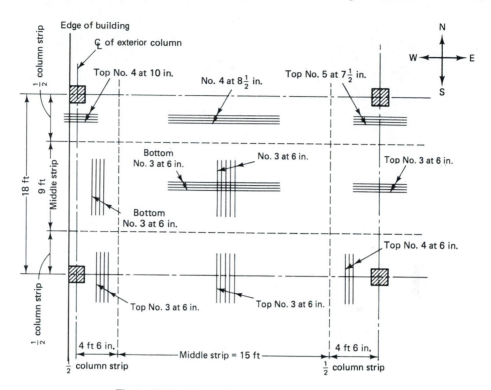

Figure 11.14 Schematic reinforcement distribution.

and columns are also shown in Fig. 11.16, and the building is subject to gravity loads only. Given:

> live load = 115 psf (5.45 kPa)
>
> f'_c = 4000 psi (27.6 MPa), normal-weight concrete
>
> f_y = 60,000 psi (414 MPa)

Assume that $\beta_c > 2.5$.

Design the interior panel and the size and spacing of reinforcement needed. Consider flooring weight to be 14 psf in addition to the slab self-weight.

Solution   *Geometry check for use of direct design method (step 1)*

(a) Ratio $\dfrac{\text{longer span}}{\text{shorter span}} = \dfrac{24}{18} = 1.33 < 2.0$; hence two-way action.

(b) More than three panels in each direction.

(c) Assume a thickness of 7 in.

$$w_d = 14 + \frac{7}{12} \times 150 = 101.5 \text{ psf}$$

$$2w_d = 203 \text{ psf}$$

$$w_l = 115 \text{ psf} < 2w_d$$

Hence the direct design method is applicable.

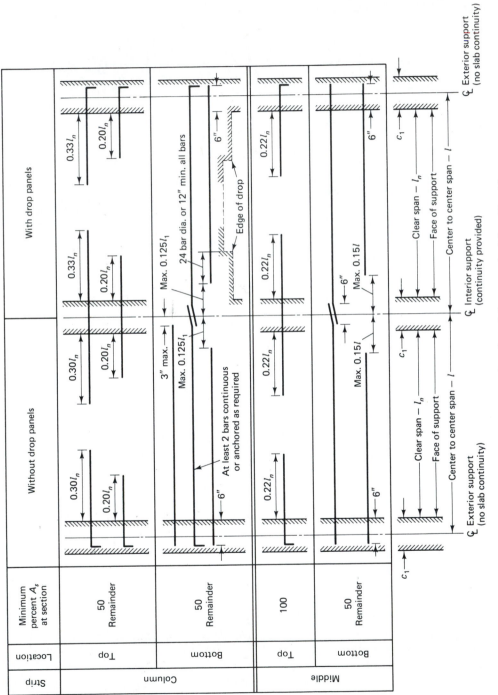

Figure 11.15 Minimum extensions for reinforcement in slabs without beams.

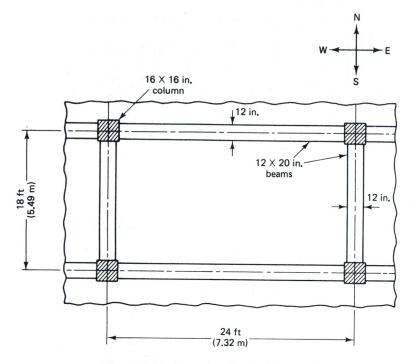

Figure 11.16 Floor plan of an interior panel.

Minimum slab thickness for deflection requirement (step 2)

$$l_n \, (\text{E--W}) = 24 \times 12 - 2 \times 8 = 272 \text{ in.}$$

$$l_n \, (\text{N--S}) = 18 \times 12 - 2 \times 8 = 200 \text{ in.}$$

$$\beta = \frac{272}{200} = 1.36$$

For a preliminary estimate of the required total thickness h, using Eq. 11.9,

$$h = \frac{l_n\left(0.8 + \dfrac{f_y}{200,000}\right)}{36 + 9\beta}$$

$$= \frac{272(0.8 + 60,000/200,000)}{36 + 9 \times 1.36}$$

$$= 6.21 \text{ in. } (15.8 \text{ cm}) > 5 \text{ in.} \qquad \text{O.K.}$$

To check h from Eq. 11.9, the stiffness ratio α_m is needed. Since this is an interior panel, the end and corner adjacent panel would necessitate larger thickness. Try $h = 7$ in.

To locate the beam centroid for the section in Fig. 11.17,

$$(38 \times 7)(\bar{y} + 3.5) + \frac{12(\bar{y}^2)}{2} = \frac{12(13 - \bar{y})^2}{2}$$

$$\bar{y} = 0.20 \text{ in.}$$

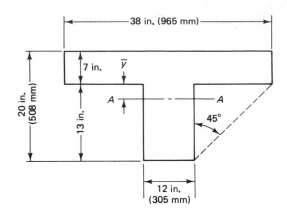

Figure 11.17   Effective flanged beam section.

$I_b = \frac{1}{3} \times 12(0.20)^3 + \frac{1}{12} \times 38(7)^3 + 38 \times 7(0.20 + 3.5)^2$

$$+ \frac{1}{3} \times 12(13 - 0.20)^3 = 13{,}116.4 \text{ in.}^4$$

$I_s = h^3/12 \times$ width of slab bound laterally by the center line of the adjacent panel on each side of the beam section shown in Fig. 11.16

$$I_{s1} \text{ (N–S)} = \frac{(7)^3}{12} \times 24 \times 12 = 8232 \text{ in.}^4$$

$$I_{s2} \text{ (E–W)} = \frac{(7)^3}{12} \times 18 \times 12 = 6174 \text{ in.}^4$$

Therefore,

$$\alpha_1 = \frac{13{,}116.4}{8232} = 1.59 \qquad \alpha_2 = \frac{13{,}116.4}{6174} = 2.12$$

$$\alpha_m = \frac{1.59 \times 2 + 2.12 \times 2}{4} = 1.86$$

From Eq. 11.9,

$$h = \frac{l_n(0.8 + f_y/200{,}000)}{36 + 5\beta[\alpha_m - 0.2]}$$

$$= \frac{272(0.8 + 60{,}000/200{,}000)}{36 + 5 \times 1.35[1.86 - 0.2]} = 6.29 \text{ in. (16 cm)}$$

Minimum h in this case from Eq. 11.9 = 6.29 in.   Therefore, for deflection, use $h = 7$ in. (17.8 cm) assumed at the beginning.

Statical moment computation (steps 3 to 5)

Given the flooring weighs 14 psf.

$$w_u = 1.4D + 1.7L$$

$$= 1.4\left(\frac{7}{12} \times 150 + 14\right) + 1.7 \times 115 = 338 \text{ psf}$$

$$\text{E--W } l_{n1} = 272 \text{ in.} = 22.7 \text{ ft}$$

$$\text{N--S } l_{n2} = 200 \text{ in.} = 16.7 \text{ ft}$$

$$0.65l_1 = 15.6 \text{ ft} \quad \text{use } l_{n1} = 22.7 \text{ ft}$$

$$0.65l_2 = 11.7 \text{ ft} \quad \text{use } l_{n2} = 16.7 \text{ ft}$$

(a) *E--W direction*

$$M_0 = \frac{w_u l_2 l_{n1}^2}{8} = \frac{338 \times 18.0(22.7)^2}{8} = 391{,}878 \text{ lb-ft } (532 \text{ kN-m})$$

Moment distribution factors for interior panels from Fig. 11.5 are

$$-M_u = 0.65M_0 = 0.65 \times 391{,}878 = 254{,}720 \text{ lb-ft}$$

$$+M_u = 0.35M_0 = 0.35 \times 391{,}878 = 137{,}157 \text{ lb-ft}$$

(b) *N--S direction*

$$M_0 = \frac{w_u l_1 l_{n2}^2}{8} = \frac{338 \times 24.0(16.7)^2}{8} = 282{,}794 \text{ lb-ft } (384 \text{ kN-m})$$

Moment distribution factors for interior panel from Fig. 11.5 or Table 11.1 are

$$-M_u = 0.65M_0 = 0.65 \times 282{,}794 = 183{,}816 \text{ lb-ft } (249 \text{ kN-m})$$

$$+M_u = 0.35M_0 = 0.35 \times 282{,}794 = 98{,}978 \text{ lb-ft } (134 \text{ kN-m})$$

Moment distribution in the column and middle strips (steps 5 to 7)

(a) *E--W stiffness ratio (long span)*

$$\alpha = \frac{E_{cb}I_{b2}}{E_{cs}I_{s2}} = \frac{13{,}116.4}{6174} = 2.12$$

$$\frac{l_2}{l_1} = \frac{18}{24} = 0.75 \qquad \alpha\frac{l_2}{l_1} = 1.59 > 1.0$$

Moment factors for the column strip for this panel from the factored moment coefficient for the column strip of an interior panel (Section 11.4.2, interior panels and positive moments) are linearly interpolated to give the following:

$$-M: \quad 0.75 + \frac{0.90 - 0.75}{2} = 0.83$$

$$+M: \quad 0.75 + \frac{0.90 - 0.75}{2} = 0.83$$

(b) *N–S stiffness ratio* α

$$\alpha = \frac{E_{cb}I_{b1}}{E_{cs}I_{s1}} = \frac{13,116.4}{8232} = 1.59$$

$$\frac{l_2}{l_1} = \frac{24}{18} = 1.33 \qquad \alpha\frac{l_2}{l_1} = 2.12 > 1.0$$

Hence moment factors are in this case using the same tables by linear interpolation.

$$-M: \quad 0.75 - (0.75 - 0.45)\tfrac{1}{3} = 0.65$$
$$+M: \quad 0.75 - (0.75 - 0.45)\tfrac{1}{3} = 0.65$$

The distributed moments are then evaluated using the interpolated factors above to produce a moment distribution operations table (Table 11.7). Note from this table that the stiffness ratio of the slab to the supporting beams for the span ratio in this example has resulted in middle strip moments in the N–S direction larger than the moments in the E–W direction.

Check slab thickness for shear capacity

$$\alpha_1 \frac{l_2}{l_1} = 1.59 > 1.0$$

Hence shear will be transferred to the beams surrounding the slab according to a tributary area bound by 45° lines drawn from the corners of the panel and the center line of the panel parallel to the long side.

TABLE 11.7 Moment Distribution Operations Table

Column Strip	E–W Direction l_2/l_1: 18/24 = 0.75 $\alpha_1(l_2/l_1)$: 2.12 × 0.75 = 1.59		N–S Direction 24/18 = 1.33 1.59 × 1.33 = 2.12	
	Negative Moment	Positive Moment	Negative Moment	Positive Moment
M_u (ft-lb)	254,720	137,157	183,816	92,978
Distribution factor (%)	83	83	65	65
Total column strip design moment (ft-lb)	211,418	113,840	119,480	60,436
Beam moment, 85%	179,705	96,764	101,558	51,371
Column strip slab moment (ft-lb)	31,713	17,076	17,922	9,065
Total middle strip design moment (ft-lb)	254,720 × 0.17 = 43,403	137,157 × 0.17 = 23,317	183,816 × 0.35 = 64,336	92,978 × 0.35 = 32,542

The largest part of the load has to be carried in the short direction with the largest value at the face of the first interior support. The factored shear on a 12-in.-wide strip spanning in the short direction can be approximated as

$$V_u = 1.15 \frac{w_u l_{n2}}{2} = \frac{1.15 \times 338.0(16.7 \times 12)}{2 \times 12} = 3246 \text{ lb/ft width}$$

where the value 1.15 is the continuity factor.

$$\text{slab effective } d = 7 - 0.75 - 0.25 = 6.0 \text{ in. (152.4 mm)}$$

$$\phi V_c = \phi(2\sqrt{f_c'}\, bd)$$

$$= 0.85 \times 2\sqrt{4000} \times 12 \times 6 = 7741 \text{ lb}$$

$$V_u < \phi V_c \qquad \text{hence safe}$$

Proportioning of the slab reinforcement (steps 7 and 8)

Minimum A_s using $f_y = 60{,}000$-psi steel is $0.0018bh$ or minimum $A_s = 0.0018 \times 12 \times 7 = 0.15$ in.^{2}/12-in. strip. For No. 3 steel, $S = 0.11/(0.15/12) = 8.8$ in. on centers; use No. 3 at $8\frac{1}{2}$ in. center to center. As was done in Ex. 11.1, the moments per 12-in.-wide strip have to be evaluated.

(a) *E–W direction*

Column strip

$$-M_n = \frac{31{,}713}{\phi = 0.9} = 35{,}237$$

$$0.25 l_2 = 0.25 \times 18 \text{ ft} = 4.5 \text{ ft} < 0.25 \times 24 \text{ ft}$$

Hence the half column strip $= 4.5$ ft controls. The net width of the slab in the column strip on which moments act $= 2 \times 4.5 - 38$ in./$12 = 5.83$ ft.

$$\text{required unit} - M \text{ per 12-in. strip} = \frac{35{,}237 \times 12}{5.83} = 72{,}529 \text{ in.-lb}$$

$$\text{required unit} + M \text{ per 12-in. strip} = \frac{17{,}076 \times 12}{0.9 \times 5.83} = 39{,}053 \text{ in.-lb}$$

Middle strip

$$\text{width of strip} = 18 - 9.0 = 9.0 \text{ ft}$$

$$\text{required unit} - M \text{ per 12-in. strip} = \frac{43{,}403 \times 12}{0.9 \times 9.0} = 64{,}301 \text{ in.-lb}$$

$$\text{required unit} + M \text{ per 12-in. strip} = \frac{23{,}317 \times 12}{0.9 \times 9.0} = 34{,}544 \text{ in.-lb}$$

(b) *N–S direction (short span):* From before, the maximum allowable width of the half column strip $= 4.5$ ft.

Column strip

net width of slab in column strip on which moments act $= 2 \times 4.5 - \dfrac{38}{12} = 5.83$ ft.

$$\text{required unit} - M \text{ per 12-in. strip} = \frac{17{,}922 \times 12}{0.9 \times 5.83} = 40{,}988 \text{ in.-lb}$$

$$\text{required unit} + M \text{ per 12-in. strip} = \frac{9065 \times 12}{0.9 \times 5.83} = 20{,}732 \text{ in.-lb}$$

Middle strip

$$\text{width of strip} = 24 - 9.0 = 15.0 \text{ ft}$$

$$\text{required unit} - M \text{ per 12-in. strip} = \frac{64{,}336 \times 12}{0.9 \times 15.0} = 57{,}188 \text{ in.-lb}$$

$$\text{required unit} + M \text{ per 12-in. strip} = \frac{32{,}542 \times 12}{0.9 \times 15.0} = 28{,}926 \text{ in.-lb}$$

Selection of size and spacing of reinforcement (step 9)

  The maximum unit moment in the negative moment region of the column strip in the E–W direction $= 72{,}529$ lb-in. per 12-in.-wide strip.

$$M_n = A_s f_y \left(d - \frac{a}{2} \right)$$

Hence

$$72{,}529 = A_s \times 60{,}000 \; (\approx 0.9d)$$

$$A_s = \frac{72{,}529}{60{,}000 \times 0.9 \times 6.0} = 0.22 \text{ in.}^2$$

Adjustment trial

$$a = \frac{A_s f_y}{0.85 f_c' b} = \frac{0.22 \times 60{,}000}{0.85 \times 4000 \times 12} = 0.32 \text{ in.}$$

Hence

$$72{,}529 = A_s \times 60{,}000 \left(6.0 - \frac{0.32}{2} \right)$$

Therefore, required $A_s = 0.21$ in.² per 12-in. strip. Try No. 4 bars (0.20 in.²) (12.7-mm diameter)

$$s = \frac{\text{area of one bar}}{\text{required area per 12-in. strip}} = \frac{0.20}{0.21/12} = 11.43 \text{ in. c-c}$$

Hence, use No. 4 bars at 11 in. center to center (12.7-mm diameter at 280 mm center to center).

  In the same manner, calculate the area of steel needed in each direction for both the column and middle strips (Table 11.8). Note that the effective depth d in

TABLE 11.8 Column and Middle Strip Calculations

| | Column Strip | | | | Middle Strip | | | |
| | Support | | Midspan | | Support | | Midspan | |
Direction	$\dfrac{A_s}{12\text{ in.}}$	Bar Size and Spacing (in. c-c)	$\dfrac{A_s}{12\text{ in.}}$	Bar Size and Spacing (in. c-c)[a]	$\dfrac{A_s}{12\text{ in.}}$	Bar Size and Spacing (in. c-c)	$\dfrac{A_s}{12\text{ in.}}$	Bar Size and Spacing (in. c-c)[a]
E–W ($d = 6.0$)	0.21	No. 4 at 11	0.11	No. 3 at $8\frac{1}{2}$	0.19	No. 4 at 12	0.10	No. 3 at $8\frac{1}{2}$
N–S ($d = 5.5$)	0.14	No. 3 at 9	0.06	No. 3 at $8\frac{1}{2}$	0.20	No. 4 at 12	0.11	No. 3 at $8\frac{1}{2}$

[a]Maximum spacing of bars should not exceed 12 in. center to center for crack control.

the N–S direction would be $= 7.0 - (0.75 + 0.5 + 0.25) = 5.5$ in. since it is assumed in this design that the E–W grid of reinforcement is closest to the concrete surface.

Compare the reinforcement areas obtained in this example with those of Ex. 11.1 in conjunction with the discussion in Section 11.2.1 on two-way action and moment redistribution as a function of stiffness ratios. It should be noted that, when the slab or plate panel is either supported on flexible supports or on columns only, the moments are not necessarily more severe in the shorter direction.

Carry the reinforcement at the same spacing for each respective strip up to the webs of the supporting beams. Also, as the next step, design (analyze) the supporting beams in the usual manner as discussed in Chapter 5.

More refinements could have been obtained using the equivalent frame method for moment calculations. Also, for cases where limitations exist, such as horizontal loads and others discussed in Section 11.3.1, the equivalent frame method would have to be used for moment calculations.

11.6 EQUIVALENT FRAME METHOD FOR FLOOR SLAB DESIGN

11.6.1 Applicability

The equivalent frame method for the design of two-way slab and plate systems is a more rigorous form of the direct design method presented in Sections 11.3 and 11.4. It differs only in the means of computing the longitudinal variation of bending moments along the design frame such that it would be applicable to a wide range of applications. Its main features can be summarized in the following:

1. Moments are distributed to critical sections by an elastic analysis such as moment distribution, rather than by general factors. Pattern loadings have to be considered for the most critical loading conditions.
2. There are no limitations on dimensions or loadings.
3. Contrary to the simplifications in the direct design method, variations in the moment of inertia along the axes of numbers have to be considered, such as the effects of column capitals.
4. Effects of lateral loading can be accounted for in the analysis.
5. In contrast to the direct design method, use of a computer facilitates the analysis in this method through evaluation of the various stiffnesses.
6. Because of the refinement possible in its use, the total statical moment need not exceed the statical moment M_0 required by the direct design method.

11.6.2 Stiffness Coefficients

The structure, divided into continuous frames as shown in Fig. 11.3 for frames in both orthogonal directions, would have the row of columns and a wide continuous beam (slab) *ABCD*. Each floor is analyzed separately whereby the columns are assumed fixed at the floors above and below. To satisfy statics and equilibrium, each equivalent frame must carry the total applied load. Alternate span loading has to be used for the worst live-load condition.

It is necessary to account for the rotational resistance of the column at the joint when running a moment relaxation or distribution except when the columns are very slender, with very low rigidity compared to the rigidity of the slab at the joint. In such cases, such as in lift slab construction, only a continuous beam is necessary. A schematic illustration of the constituent elements of the equivalent frame is given in Fig. 11.18. The slab strips are assumed to be supported by transverse slabs. The column provides a resisting torque M_T equivalent to the applied torsional moment intensity m_t. The exterior ends of the slab strip rotate *more* than the central section because of the torsional deformation. To account for this rotation and deformation, the actual column and the transverse slab strip are replaced by an *equivalent* column, such that the flexibility of the equivalent column is *equal* to the *sum* of the flexibilities of the actual column and the slab strip. This assumption is represented as follows:

$$\frac{1}{K_{ec}} = \frac{1}{\Sigma \, K_c} + \frac{1}{K_t} \qquad (11.11)$$

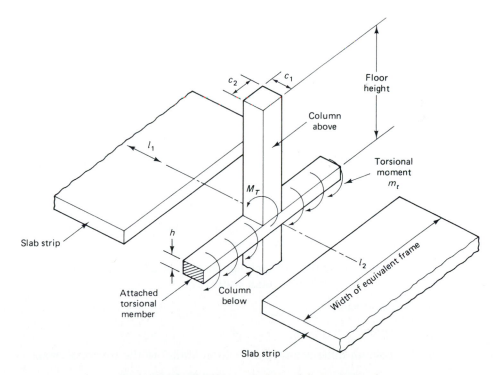

Figure 11.18   Constituent elements of the equivalent frame.

where K_{ec} = flexural stiffness of the equivalent column, moment per unit rotation

  $\Sigma\,K_c$ = sum of flexural stiffnesses of the upper and lower columns at the joint, moment per unit rotation

  K_t = torsional stiffness of the torsional beam, moment per unit rotation

Alternatively, the flexibility equation 11.11 can be written as a stiffness equation

$$K_{ec} = \frac{\Sigma\,K_c}{1 + \Sigma\,K_c/K_t} \qquad (11.12)$$

The column stiffness for an equivalent frame (Ref. 11.7) can be defined as

$$K_c = \frac{EI}{l'}\left[1 + 3\left(\frac{l}{l'}\right)^2\right] \qquad (11.13)$$

where I = column moment of inertia

  l = center-line span

  l' = clear span of the equivalent beam

The carry-over factors are approximated by $-\frac{1}{2}\left(1 + \dfrac{1}{3h}\right)$. A simpler expression for K_c (Ref. 11.8) gives results within 5% of the more refined values from Eq. 11.13 as follows:

$$K_c = \frac{4EI}{L - 2h} \qquad (11.14)$$

where h is the slab thickness.

The torsional stiffness of the slab in the column line is

$$K_t = \Sigma\,\frac{9E_{cs}C}{L_2[1 - (c_2/L_2)]^3} \qquad (11.15)$$

where the torsional constant is

$$C = \frac{\Sigma(1 - 0.63x/y)x^3y}{3}$$

and x = shorter dimension of rectangular part of cross section at column junction (such as slab depth)

  y = longer dimension of rectangular part of cross section at column junction (such as column width)

  L_2 = band width

  L_n = span

  c_2 = column dimensions in direction parallel to the torsional beam

$$K_s = \frac{4E_{cs}I_s}{L_n - c_1/2} \qquad (11.16)$$

As the effective stiffness K_{ec} of the column and the slab stiffness K_s are established, the analysis of the equivalent frame can be performed by any applicable methods, such as relaxation or moment distribution. The distribution factor for fixed-end moment (FEM) is

$$DF = \frac{K_s}{\Sigma K} \tag{11.17}$$

where $\Sigma K = K_{ec} + K_{s(\text{left})} + K_{s(\text{right})}$.

For carry-over factors, COF $\simeq \frac{1}{2}$ can be used without loss of accuracy since the nonprismatic section causes only very small effects on fixed-end moments and carry-over factors. The fixed-end moments for a uniformly distributed load are $w_u l_2 (l_n)^2 / 12$ at the supports, such that after moment redistribution the sum of the negative distributed moment at the support and the midspan would always be equal to the static moment $M_0 = w_u l_2 (l_n)^2 / 8$.

11.6.3 Pattern Loading of Spans

Loading all spans simultaneously does not necessarily produce the maximum positive and negative flexural stresses. Consequently, it is advisable to analyze the multispan frame also using alternate span loading patterns for the *live* load. For a three-span frame, the suggested patterns for the live load are as shown in Fig. 11.19. The ACI Code, however, permits the full factored live load to be used on the entire slab system if the live load is less than 75% of the dead load.

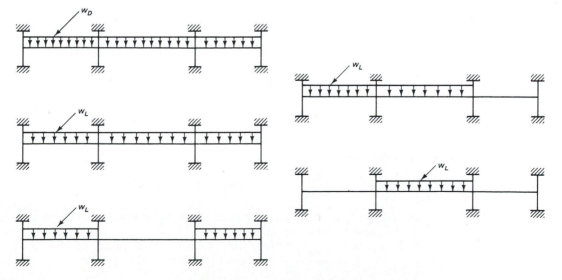

Figure 11.19 Loading patterns for live load.

11.6.4 Operational Flow Chart for the Design (Analysis) of Two-way Floor Slabs by the Equivalent Frame Method

The flow chart for this procedure is shown in Fig. 11.20.

11.6.5 Example 11.3: Design of Flat Plate without End Beams by the Equivalent Frame Method (EFM)

A reinforced concrete flat-plate apartment floor system without end beams and without drop panels is shown in Fig. 11.21. The end panel center-line dimensions are 17'-6" × 20'-0" (5.33 m × 6.10 m) and the interior panel dimensions are 24'-0" × 20'-0" (7.32 m × 6.10 m). The floor heights l_u of intermediate floors are typically 8'-9" (2.67 m). Evaluate the required nominal moment strengths M_n for a typical floor panel in the north–south direction to withstand a working live load $w_L = 40$ psf (1.92 kPa) and a superimposed dead load $w_D = 20$ psf due to partitions and flooring. Assume that all panels are simultaneously loaded by the live load in your solution. Given:

$$f'_c = 4000 \text{ psi (27.58 MPa), normal-weight concrete}$$

$$f_y = 60,000 \text{ psi (413.7 MPa)}$$

Solution   *Equivalent frame characteristics*

Deflection thickness check: From Table 11.3, the minimum thickness h for interior panels using steel having $f_y = 60,000$ psi and without drop panels is

$$h = \frac{l_n}{33}$$

$$l_n = \text{clear span} = 24 \text{ ft less } 20 \text{ in.} = 22.33 \text{ ft}$$

$$\text{required } h = \frac{22.33}{33} \times 12 = 8.12 \text{ in.} \quad \text{say 8.25 in. (21.0 cm)}$$

Thickness of exterior panel without edge beams $= l_n/30$ to be increased at discontinuous edges by at least 10%. Exterior panel $l_n = 20'-0'' - 14$ in. $= 18.82$ ft. $h = 18.82/30 \times 12 \times 1.10 = 8.07$ in. Use $h = 8.25$ in. (21 cm) for all panels of the floor system.

Take the equivalent frame in the N–S direction whose plan is shown in the shaded portion in Fig. 11.21.

$$w_u = 1.4\left(\frac{8.25}{12} \times 150 + 20\right) + 1.7 \times 40 = 240 \text{ psf}$$

Approximate flexural stiffness of column above and below the floor joint (moment per unit rotation), from Eq. 11.14, is

$$K_c = \frac{4E_c I_c}{L_n - 2h} \quad \text{where } L_n = 8'-9'' = 105 \text{ in.}$$

Start

1 Input: Slab plan dimensions L_{EW}, L_{NS}, l_u, W_L, W_D, f'_c, f_y, E_c, E_s

2 Select preliminary thickness h for E–W, N–S using Eq. 11.8, 11.9, and 11.10 for slabs with beams or slabs with drop panels, and Table 11.3 for flat plates. Chosen thicknesses h should always satisfy the minimum code requirements for thickness h. Increase h from Table 11.3 by at least 10% if no edge beams are used.

3 Compute E_{cc}, E_{cs}, E_{cb}

where c = column, s = slab, b = beam

$$K_c \simeq \frac{4EI}{L - 2h}$$

$$K_t = \Sigma \frac{9E_{cs}C}{L_2(1 - c_2/L_2)^3}$$

where $C = \Sigma(1 - 0.63x/y)\dfrac{x^3 y}{3}$

Find $K_{ec} = \left(\dfrac{1}{K_c} + \dfrac{1}{K_t}\right)^{-1}$

4 Compute $K_s \simeq \dfrac{4E_{cs}I}{(L_1 - c_1/2)}$

where L_1 = centerline span

c_1 = column depth

Find $DF = \dfrac{K_s}{\Sigma K}$

where $\Sigma K = K_{ec} + K_{s(left)} + K_{s(right)}$

5 Compute FEM for load intensity w_u and run a moment distribution where

$\text{FEM} = \pm \dfrac{w_u (L_n)^2}{12}$. Adjust distributed M_u to support face values such that

the adjusted M_u = distributed $M_u - \dfrac{V_c}{3}$

6 Convert the final M_o values to required nominal moment strengths $M_n = \dfrac{M_u}{\phi}$

7 Consider inelastic moment redistribution if necessary, using factor R, not exceeding 10%,

$R = 20 \left(1 - \dfrac{\rho - \rho'}{\rho_b}\right)$

such that $(\rho - \rho') \leq \frac{1}{2}\rho_b$.

Figure 11.20 Flow chart for the design (analysis) of reinforced concrete two-way slabs and plates by the equivalent frame method.

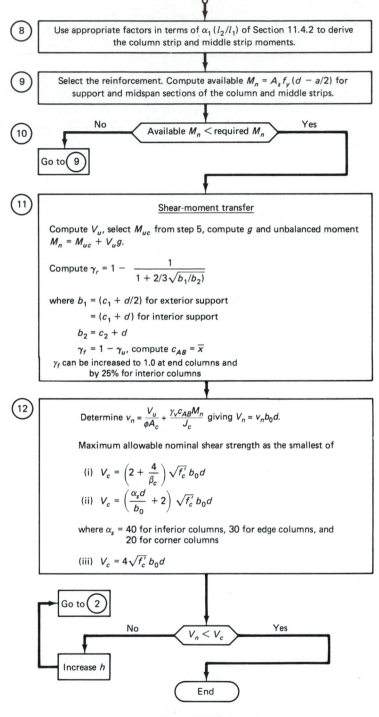

8 Use appropriate factors in terms of $\alpha_1 (l_2/l_1)$ of Section 11.4.2 to derive the column strip and middle strip moments.

9 Select the reinforcement. Compute available $M_n = A_s f_y (d - a/2)$ for support and midspan sections of the column and middle strips.

10 No — Available $M_n <$ required M_n — Yes

Go to **9**

11 Shear-moment transfer

Compute V_u, select M_{uc} from step 5, compute g and unbalanced moment $M_n = M_{uc} + V_u g$.

Compute $\gamma_r = 1 - \dfrac{1}{1 + 2/3\sqrt{b_1/b_2)}}$

where $b_1 = (c_1 + d/2)$ for exterior support

$\quad = (c_1 + d)$ for interior support

$b_2 = c_2 + d$

$\gamma_f = 1 - \gamma_u$, compute $c_{AB} = \overline{x}$

γ_f can be increased to 1.0 at end columns and by 25% for interior columns

12 Determine $v_n = \dfrac{V_u}{\phi A_c} + \dfrac{\gamma_v c_{AB} M_n}{J_c}$ giving $V_n = v_n b_0 d$.

Maximum allowable nominal shear strength as the smallest of

(i) $V_c = \left(2 + \dfrac{4}{\beta_c} \right) \sqrt{f_c'} \, b_0 d$

(ii) $V_c = \left(\dfrac{\alpha_s d}{b_0} + 2 \right) \sqrt{f_c'} \, b_0 d$

where $\alpha_s = 40$ for inferior columns, 30 for edge columns, and 20 for corner columns

(iii) $V_c = 4\sqrt{f_c'} \, b_0 d$

Go to **2**

No — $V_n < V_c$ — Yes

Increase h

End

Figure 11.20 *(cont.)*

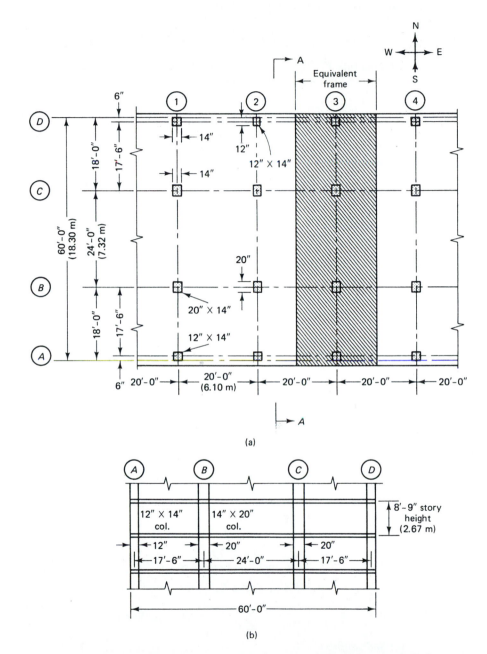

(a)

(b)

Figure 11.21 Flat-plate apartment structure (a) plan; (b) section $A-A$, N–S.

(a) *Exterior column (14 in. × 12 in.) stiffness*

$$b = 14 \text{ in.} \qquad I_c = \frac{14(12)^3}{12} = 2016 \text{ in.}^4$$

Assume that $E_{col}/E_{slab} = E_{cc}/E_{cs} = 1.0$. Use $E_{cc} = E_{cs} = 1.0$ in the calculations as E_{cs} drops out in the equation for K_{ec}.

$$\text{total } K_c = \frac{4 \times 1 \times 2016}{105 - (2 \times 8.25)} \times 2 \text{ (for top and bottom columns)}$$

$$= 182.2 \text{ in.-lb/rad}/E_{cc}$$

Torsional constant C from Eq. 11.15 is

$$C = \sum \left(1 - 0.63 \frac{x}{y}\right) \frac{x^3 y}{3}$$

$$= \left(1 - 0.63 \times \frac{8.25}{12}\right)(8.25)^3 \times \frac{12}{3} = 1273$$

Torsional stiffness of the slab at the column line is

$$K_t = \sum \frac{9 E_{cs} C}{L_2[1 - (c_2/L_2)]^3}$$

$$= \frac{9 \times 1 \times 1273}{20 \times 12 \, [1 - 14/(12 \times 20)]^3}$$

$$= 57.1 \text{ in.-lb/rad}/E_{cs}$$

From Eq. 11.12, the equivalent column stiffness is

$$K_{ec} = \left(\frac{1}{K_c} + \frac{1}{K_t}\right)^{-1} = \left(\frac{1}{182.2} + \frac{1}{57.2}\right)^{-1} = 43.5 \text{ in.-lb/rad}/E_{cc}$$

(b) *Interior column (14 in. × 20 in.) stiffness*

$$b = 14 \text{.in.} \qquad I = \frac{14(20)^3}{12} = 9333 \text{ in.}^4$$

$$\text{total } K_c = \frac{4 \times 1 \times 9333}{105 - 2 \times 8.25} \times 2 = 843.7 \text{ in.-lb/rad}/E_{cc}$$

$$C = (1 - 0.63 \times 8.25/20) \times (8.25)^3 \times 20/3 = 2770$$

$$K_t = \frac{9 \times 2770}{20 \times 12[1 - 14/(12 \times 20)]^3} + \frac{9 \times 2770}{20 \times 12[1 - 14/(12 \times 20)]^3}$$

$$= 248.8 \text{ in.-lb/rad}/E_{cs}$$

$$K_{ec} = \left(\frac{1}{843.7} + \frac{1}{248.8}\right)^{-1} = 192.1 \text{ in.-lb/rad}/E_{cc}$$

(c) *Slab stiffness*

$$h = 8.25 \text{ in.}$$

From Eq. 11.16,

$$K_s = \frac{4 E_{cs} I_s}{L_n - c_1/2}$$

where L_n = center-line span

c_1 = column depth

Slab band width in E–W direction = 20/2 + 20/2 = 20 ft.

$$I_s = 20 \times \frac{12(8.25)^3}{12} = 11,230 \text{ in.}^4$$

Slab at right of exterior column A:

$$K_s = \frac{4 \times 1 \times 20(8.25)^3}{12 \times 17.5 - 12/2} = 220.2 \text{ in.-lb/rad}/E_{cs}$$

Slab at left of interior column B:

$$K_s = \frac{4 \times 1 \times 20(8.25)^3}{12 \times 17.5 - 20/2} = 224.6 \text{ in.-lb/rad}/E_{cs}$$

Slab at right of interior column B:

$$K_s = \frac{4 \times 1 \times 20(8.25)^3}{12 \times 14 - 20/2} = 161.6 \text{ in.-lb/rad}/E_{cs}$$

From Eq. 9.12, slab distribution factor at joints: DF = $K_s/\Sigma K$, where

$$\Sigma K = K_{ec} + K_{s(\text{left})} + K_{s(\text{right})}$$

Outer joint A slab:

$$\text{DF} = \frac{220.2}{43.5 + 220.2} = 0.835$$

Left joint B slab:

$$\text{DF} = \frac{224.6}{192.1 + 224.6 + 161.6} = 0.388$$

Right joint B slab:

$$\text{DF} = \frac{161.6}{192.1 + 224.6 + 161.6} = 0.279$$

Moment distribution of factored moments M_u

From before, w_u = 240 psf.

Joint A, span AB:

$$\text{factored fixed-end moment } M_u = \frac{w_u(L_{AB})^2}{12}$$

$$= \frac{240(17.5)^2}{12} \times 12$$

$$= \begin{array}{l} 73,500 \text{ in.-lb/ft} \\ (27.2 \text{ kNm/m}) \end{array}$$

Joint B, span BC:

$$\text{factored fixed-end moment } M_u = \frac{w_u (L_{BC})^2}{12}$$

$$= \frac{240(24.0)^2}{12} \times 12 = 138{,}240 \text{ in.-lb/ft}$$

$$(51.2 \text{ kNm/m})$$

Run a moment distribution for the factored moments as shown in Table 11.9.

Factored and required nominal moment strengths M_n

     Joint A (span AB) moment $-M_u$

moment reduction to face of column $= Vc/3$

$$V_{AB} = \frac{w_u L}{2} - \frac{M_{\bar{u}@B} - M_{\bar{u}@A}}{L_n} = \frac{240 \times 17.5}{2} - \frac{120{,}910 - 10{,}060}{17.5 \times 12}$$

$$= 2100.0 - 527.9 = 1572.1 \text{ lb/ft}$$

$$c = 12 \text{ in.}$$

$$\text{required column face } M_u = 10{,}060 - \frac{1572.1 \times 12}{3}$$

$$= 10{,}060 - 6288 = 3772 \text{ in.-lb/ft } (1.40 \text{ kNm/m})$$

$$\text{required } -M_n = \frac{M_u}{\phi} = \frac{3772}{0.9} = 4191 \text{ in.-lb/ft } (1.55 \text{ kNm/m})$$

TABLE 11.9   Moment Distribution of Factored Loads

	Ⓐ		Ⓑ		Ⓒ	
DF	0.835	0.388	0.279		0.279	0.388
COF	0.5	0.5	0.5		0.5	0.5
FEM$_u \times 10^3$ in.-lb per foot	−73.50	+73.50	−138.24		138.24	73.50
distribution	+61.37	+25.12	+18.06		18.06	25.12
CO distribution	+12.56 −10.49	+30.69 −8.40	−9.03 −6.04			
Final $M_u \times 10^3$ per foot	−10.06	+120.91	−135.25		Line of symmetry	

Joint B (span BA) moment $-M_u$

$$\text{center line } M_u = 120{,}910 \text{ in.-lb/ft}$$

$$V_{BA} = 2100.0 + 527.9 = 2627.9 \text{ lb/ft}$$

$$c = 20 \text{ in.}$$

$$\text{required column face } M_u = 120{,}910 - \frac{2627.9 \times 20}{3}$$

$$= 120{,}910 - 17{,}519 = 103{,}391 \text{ in.-lb/ft (38.3 kNm/m)}$$

$$\text{required } -M_n = \frac{M_u}{\phi} = \frac{103{,}391}{0.9} = 114{,}879 \text{ in.-lb/ft (42.6 kNm/m)}$$

Joint B (span BC) moment $-M_u$

$$\text{center line } M_u = 135{,}250 \text{ in.-lb/ft}$$

$$V_{BC} = \frac{240 \times 24}{2} = 2880 \text{ lb/ft}$$

$$\text{required column face } -M_u = 135{,}250 - \frac{2880 \times 20}{3}$$

$$= 135{,}250 - 19{,}200 = 116{,}050 \text{ in.-lb/ft (43.0 kNm/m)}$$

$$\text{required } -M_n = \frac{M_u}{\phi} = \frac{116{,}050}{0.9} = 128{,}944 \text{ in.-lb/ft (47.8 kNm/m)}$$

Span AB maximum positive moment $+M_u$: Assume that point of zero shear and maximum moment is x ft from face A.

$$x = \frac{V_{AB}}{w_u} = \frac{1570.4}{240} = 6.54 \text{ ft}$$

End M_u at A (Table 11.9) = 9700 in.-lb/ft

$$\text{maximum } +M_u = V_{AB} \times x - \frac{w_u x^2}{2} - M_u$$

$$= 1570.4 \times 6.54 \times 12 - \frac{240(6.54)^2}{2} \times 12 - 9700$$

$$= 123{,}245 - 61{,}591 - 10{,}600 = 51{,}594 \text{ at } 6.60 \text{ ft from } A$$

$$\text{required } +M_n = \frac{M_u}{\phi} = \frac{51{,}594}{0.9} = 57{,}327 \text{ in.-lb/ft (21.3 kNm/m)}$$

Span BC maximum positive moment $+M_u$

$$V_{BC} = 2880 \text{ lb/ft} \qquad x = \frac{L_n}{2} = \frac{24}{2} = 12 \text{ ft}$$

$$\text{simple span midspan moment } M = V_{BC} \times \frac{L_n}{2} - \left(w_u \times \frac{L}{2} \right)\left(\frac{L}{4} \right)$$

$$= 2880 \times \frac{24}{2} - \frac{240(24)^2}{8}$$

$$= 17{,}280 \text{ ft-lb/ft} = 207{,}360 \text{ in.-lb/ft}$$

Alternatively,

$$\text{simple span moment } M_0 = \frac{w_u L^2}{8} = \frac{240(24)^2}{8} \times 12$$

$$= 207,360 \text{ in.-lb/ft}$$

$$+ M_u = M_0 - (-M_u)$$

$$M_{\tilde{u}} \text{ from Table 11.9} = -135,250 \text{ in.-lb/ft}$$

$$\text{required maximum } + M_u = 207,360 - 135,250$$

$$= 72,110 \text{ in.-lb/ft (15.93 kNm/m) at midspan}$$

$$\text{required } + M_n = \frac{M_u}{\phi} = \frac{72,110}{0.9} = 80,122 \text{ in.-lb/ft (18.90 kNm/m)}$$

Figure 11.22 gives a plot of the required moment strengths M_n across the continuous spans at the column face.

For a complete design, a similar analysis has to be performed in the E–W direction. From there on, the nominal moment strength values are split into column strip moments and middle strip moments in both the N–S and E–W directions in a manner identical to the procedure in Ex. 11.1. An operations table is then developed similar to that of Table 11.5 of the direct design method. Shear–moment transfer check at the column faces also has to be made and shear–moment reinforcement designed following the same steps as in Ex. 11.1.

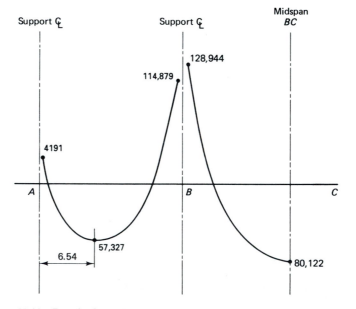

Figure 11.22 Required nominal moment strengths M_n (in.-lb/ft) by equivalent frame method in N–S direction.

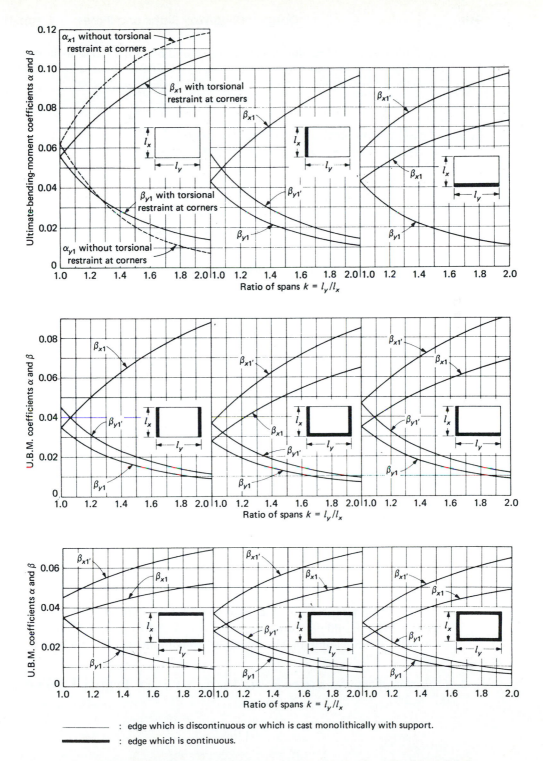

Figure 11.23 Ultimate load moment coefficients in two-way-action slabs and plates. From Ref. 11.9.)

No congestion of reinforcement results from the flexural moments in this example (No. 5 at 12 in. center to center maximum). Hence no inelastic redistribution of flexural moments from support to midspan (up to 10%) was applied.

11.6.6 Coefficients Method of Ultimate Moment Evaluation for Slabs on Continuous Supports

For slabs continuous over some supports along some edges and discontinuous at others, Fig. 11.23 can be used for a rapid evaluation check of the support moment coefficients at ultimate load.   A provision is made in the chart at the discontinuous edge for possible moment restraint of the slab as cast monolithically with the edge beams.

$$\text{short span:}\quad \text{support } M_x = -\beta'_{x1} w_u l_x^2$$

$$\text{span}\quad M_x = +\beta_{x1} w_u l_x^2$$

$$\text{long span:}\quad \text{support } M_y = -\beta'_{y1} w_u l_y^2$$

$$\text{span}\quad M_y = +\beta_{y1} w_u l_y^2$$

where w_u is the unit intensity of factored load per unit area of the slab.

11.7 SI TWO-WAY SLAB DESIGN EXPRESSIONS AND EXAMPLE

1. *Material properties*

$$E_c = w_c^{1.5} 0.0043 \sqrt{f'_c} \text{ MPa}$$

$$E_s = 2000,000 \text{ MPa}$$

$$c_b = \frac{600}{600 + f_y} d$$

$$\beta_1 = 0.85 - 0.008 (f'_c - 30)$$

Minimum temperature steel = $0.0018bh$. The value of β_1 for strengths above 30 MPa should be reduced at the rate of 0.008 for each 1 MPa in excess of 30 MPa, but β_1 cannot be less than 0.65. Maximum spacing of reinforcment = $2\,h$.

2. *Deflection of two-way slab on beams*: For $\alpha_m \leq 0.2$, use Table 11.3.   For $\alpha_m > 0.2 < 2.0$,

$$h = \frac{l_n\left(0.8 + \dfrac{f_y}{1500}\right)}{36 + 5\beta[\alpha_m - 0.2]} \tag{11.9}$$

For $\alpha_m > 2.0$,

$$h = \frac{l\left(0.8 + \dfrac{f_y}{1500}\right)}{36 + 9\beta}$$

3. *Shear*

(a) Slabs on beams

$$V_c = \frac{\sqrt{f_c'}}{6} b_o d$$

(b) Flat plates: The smaller of

$$V_c = \left(2 + \frac{4}{\beta_c}\right) \frac{\sqrt{f_c'}}{12} b_o d$$

$$V_c = \left(\frac{\alpha_s d}{b_o} + 2\right) \frac{\sqrt{f_c'}}{12} b_o d$$

$$V_c = \frac{4\sqrt{f'}_c}{12} b_o d$$

where b_o = perimetric length of critical section in slabs and footings (at $d/2$ from column face in two-way action)

   β = ratio of longer side to shorter side of the panel

   α_s = 40 for interior columns

   = 30 for edge columns

   = 20 for corner columns

4. *Unbalanced moments*

$$\gamma_f = \frac{1}{1 + \frac{2}{3}\sqrt{b_1/b_2}}$$

$$\gamma_v = 1 - \gamma_f$$

(a) Exterior supports: The value of γ_f can be increased up to 1.0 if V_u at support $< 0.75 \phi\sqrt{f_c'}$.

(b) Interior supports: The value of γ_f can be increased by 25% if $V_u \leq 0.4 \phi V_c$ and $\rho < 0.375\rho_b$.

11.7.1 SI Design of Two-way Slabs and Plates

Solve Ex. 11.1 using SI units.

Data

$$f_c' = 27.6 \text{ MPa} \qquad\qquad \text{MPa} = \text{N/mm}^2$$
$$f_y = 414 \text{ MPa} \qquad\qquad \text{Pa} = \text{N/m}^2$$
$$\text{live load } w_w = 2.40 \text{ kPa} \qquad\qquad 1 \text{ kg} = 9.81 \text{ N}$$
$$\text{unit weight of concrete} = 240 \text{ kg/m}^3$$

Geometry from Fig. 11.10

$$l_{E-W} = 7.32 \text{ m}$$

$$l_{N-S} = 5.49 \text{ m}$$

additional flooring $w_D = 0.5$ kPa

Solution

Geometry check for use of the direct design method (step 1)

(a) Ratio $\dfrac{\text{longer span}}{\text{shorter span}} = \dfrac{7.32}{5.49} = 1.33 < 2.0$; hence, two-way action.

(b) More than three spans in each direction and successive spans in each direction are the same, with columns not offset.

(c) Assume a thickness of 230 mm, flooring of 500 N/m².

$$w_d = 500 + \frac{230}{1000}(\text{m}) \times 2400 \, \text{kg/m}^3 \times \frac{9.81 \, \text{N}}{\text{kg}}$$

$$\approx 6000 \, \text{N/m}^2 = 6.0 \, \text{kN/m}^2$$

$$2w_d = 12.0 \, \text{kN/m}^2 > w_l = 2.4 \, \text{kN/m}^2 \qquad \text{O.K.}$$

From items a, b, and c, the DDM is applicable.

Minimum slab thickness for deflection requirement (step 2)

$$\text{E-W direction:} \quad l_{n1} = 7.32 - 0.225 - 0.254 = 6.84 \text{ (m)}$$

$$\text{N-S direction:} \quad l_{n2} = 5.49 - 0.254 - 0.254 = 4.98 \text{ (m)}$$

Ratio of longer to shorter clear span $\beta = 6.84/4.98 = 1.37$. Minimum preliminary thickness h from Table 11.3 for a flat plate without edge beams or drop panels using $f_y = 414$-MPa steel is $h = l_n/30$, to be increased by at least 10% when no edge beam is used.

$$\text{E-W:} \quad \ell_n = 6.84 \text{ m}$$

$$h = \frac{6.48}{30} \times 1.1 = 0.251 \text{ m}$$

Try a slab thickness $h = 260$ mm. This thickness is larger than the absolute minimum thickness of 5 in. $= 120$ mm required in the code for flat plates; hence O.K. Assume that $d = h - 30$ mm $= 230$ mm. New $w_D = 6.7$ kN/m² (6.7 kPa). Therefore,

$$2w_D = 13.4 \text{ kPa}$$

$$w_\ell = 2.4 \text{ kPa} < 2 \, w_D \qquad \text{O.K.}$$

Shear thickness requirement (step 3)

$$w_U = 1.7L + 1.4D = 1.7 \times 2.4 + 1.4 \times 6.7 = 13.5 \text{ kPa}$$

Interior column: The controlling critical plane of maximum perimetric shear stress is at a distance $d/2$ from the column face; hence, the net factored perimetric shear force is

$$V_u = [l_1 \times l_2 - (c_1 + d)(c_2 + d)]w_u$$

$$= (7.32 \times 5.49 - 0.74 \times 0.74)13.5 \times 10^3 = 535 \times 10^3 \text{ N} = 535 \text{ kN}$$

$$V_n = \frac{V_u}{\phi} = \frac{535}{0.85} = 629 \text{ kN}$$

From Fig. 11.11, perimeter of the critical shear failure surface is

$$b_o = 2(c_1 + d + c_2 + d) = 2(c_1 + c_2 + 2d)$$

$$= 2(0.51 + 0.51 + 0.45) = 2.94 \text{ m}^3 = 2.94 \times 10^6 \text{ mm}^3$$

Perimetric shear surface is

$$A_c = b_o d = 2.94 \times 0.23 = 0.66 \text{ m}^2 = 0.66 \times 10^6 \text{ mm}^2$$

Since moments are not known at this stage, only a preliminary check for shear can be made

$$\beta_c = \text{ratio of longer to shorter side of columns} = 510 \text{ mm}/510 \text{ mm} = 1.0$$

Available nominal shear V_c is the least of

$$V_c = \left(2 + \frac{4}{\beta_c}\right) \frac{\sqrt{f'_c}}{12} b_o d = (2 + 4) \frac{\sqrt{27.6}}{12} \times 0.66 \times 10^6 = 1.70 \times 10^6 \text{N}$$

$$V_c = \left(\frac{\alpha_s d}{b_0} + 2\right) \frac{\sqrt{f'_c}}{12} b_o d = \left(\frac{40 \times 0.23}{2.94} + 2\right) \frac{\sqrt{27.6}}{12} \times 0.66 \times 10^6 \text{ N}$$

$$= 1.50 \times 10^6 \text{ N}$$

$$V_c = \frac{4 \sqrt{f'_c}}{12} b_o d = 1.16 \times 10^6 \text{ N}$$

Controlling $V_c = 1.16 \times 10^6$ N > required $V_n = 0.629 \times 10^6$ N; hence the floor thickness is adequate at this design stage

Exterior column: Include weight of exterior wall, assuming its service weight to be 13 kPa. Net factored perimetric shear force is

$$V_u = \left[5.49 \times \left(\frac{7.32}{2} + \frac{0.457}{2}\right) - \left(0.457 + \frac{0.23}{2}\right)(0.508 + 0.23)\right] 13.49 \times 10^3$$

$$+ (5.49 - 0.508) \times 13 \times 0.43 \times 10^3 = 310 \times 10^3 \text{ N} = 310 \text{ kN}$$

$$V_n = \frac{V_u}{\phi} = \frac{310}{0.85} = 364 \text{ kN}$$

Consider the line of action of V_u to be at the column face *LM* in Fig. 11.12 for shear moment transfer to the centroidal plane c-c. This approximation is adequate since

V_u acts perimetrically around the column faces and not along line AB only. From Fig. 11.12,

$$A_c = d(2c_1 + c_2 + 2d) = 0.23(2 \times 0.457 + 0.508 + 2 \times 0.23)$$

$$= 0.430 \text{ m}^2 = 0.430 \times 10^6 \text{ mm}^2$$

$$\beta_c = \frac{508}{457} = 1.11$$

Available nominal shear V_c is the least of

$$V_c = \left(2 + \frac{4}{\beta_c}\right) \frac{\sqrt{f'_c}}{12} b_0 d = 1.05 \times 10^6 \text{ N}$$

$$V_c = \left(\frac{\alpha_s d}{b_0} + 2\right) \frac{\sqrt{f'_c}}{12} b_0 d = 1.06 \times 10^6 \text{ N}$$

$$V_c = \frac{4\sqrt{f'_c}}{12} b_0 d = 0.75 \times 10^6 \text{ N} > \text{required} \quad V_n = 0.364 \times 10^6 \text{ N} \qquad \text{O.K.}$$

Statical moment computation (steps 3 to 5)

$$\text{E–W:} \quad \ell_{n1} = 6.83 \text{ m}$$
$$\text{N–S:} \quad \ell_{n2} = 4.98 \text{ m}$$
$$0.65\ell_1 = 0.65 \times 7.32 = 4.76 \text{ m} \qquad \text{Use } \ell_{n1} = 6.83 \text{ m}$$
$$0.65\ell_2 = 0.65 \times 5.49 = 3.57 \text{ m} \qquad \text{Use } \ell_{n2} = 4.98 \text{ m}$$

(a) *E–W direction*

$$M_0 = \frac{w_u \ell_2 \ell_{n1}^2}{8} = \frac{13.49 \times 10^3 \times 5.49(6.83)^2}{8} = 432 \text{ kN-m}$$

For end panel of a flat plate without end beams, the moment distribution factors as in Table 11.1 are

$$-M_u \text{ at first interior support} = 0.7M_0$$
$$+M_u \text{ at midspan of panel} = 0.52M_0$$
$$-M_u \text{ at exterior face} = 0.26M_0$$
$$\text{negative design moment } -M_u = 0.70 \times 432 = 302 \text{ kN-m}$$
$$\text{positive design moment } +M_u = 0.52 \times 306 = 159 \text{ kN-m}$$
$$\text{negative moment at exterior } -M_u = 0.26 \times 432 = 112 \text{ kN-m}$$

(b) *N–S direction*

$$M_0 = \frac{w_u \ell_1 \ell_{n2}^2}{8} = \frac{13.49 \times 10^3 \times 7.32(4.98)^2}{8} = 306 \text{ kN-m}$$

$$\text{negative design moment } -M_n = 0.70 \times 306 = 214 \text{ kN-m}$$
$$\text{positive design moment } +M_n = 0.52 \times 306 = 159 \text{ kN-m}$$
$$\text{negative at exterior } -M_n = 0.26 \times 306 = 80 \text{ kN-m}$$

Moment distribution in the column and middle strips (steps 6 and 7)

Check the shear–moment transfer capacity at the exterior column supports:

$$-M_c \text{ at interior column 2-B } = 302 \text{ kN-m}$$

$$-M_e \text{ at exterior column 2-A } = 112 \text{ kN-m}$$

$$V_u = 310 \text{ kN acting at the face of the column}$$

Factored shear force at the edge column adjusted for the interior moment is

$$V_u = 310 - \frac{302 - 112}{7.32 - 0.483} = 282 \text{ kN}$$

$$V_n = \frac{V_u}{\phi} = \frac{V_u}{0.85} = 332 \text{ kN}$$

Assuming that the design M_u has the same value as the factored M_u,

$$A_c \text{ from before } = 0.433 \text{ m2}$$

From Figs. 11.7c and 11.12, taking the moment of area of the critical plane about axis AB,

$$d(2c_1 + c_2 + 2d)\,\bar{x} = d\left(c_1 + \frac{d}{2}\right)^2$$

where $\bar{x}$ is the distance to the centroid of the critical section, or

$$(2 \times 0.457 + 0.508 + 0.457)\,\bar{x} = \left(0.457 + \frac{0.23}{2}\right)^2$$

$$\bar{x} = 0.174 \text{ m}$$

$g = 0.174 - 0.23/2 = 0.059 \text{ m} = 59 \text{ mm}$, where g is the distance from the column face to the centroidal axis of the section.

The total external factored moment is

$$M_{ue} = 112 + 282 \times 0.059 = 129 \text{ kN-m}$$

Total required minimum unbalanced moment strength is

$$M_n = \frac{M_{ue}}{\phi} = \frac{129}{0.9} = 143 \text{ kN-m}$$

The fraction of nominal moment strength M_n to be transferred by shear is

$$\gamma_v = \frac{1}{1 + \frac{2}{3}\sqrt{b_1/b_2}} = 0.37$$

$$\text{where } b_1 = c_1 + d/2 = 0.457 + 0.23/2 = 0.572 \text{ m}$$

$$b_2 = c_2 + d = 0.508 + 0.23 = 0.738 \text{ m}$$

Moment of inertia of sides parallel to the moment direction about n–s axis is

$$I_1 = \left(\frac{bh^3}{12} + Ad^2 + \frac{hb^3}{12}\right)2 \quad \text{for all faces}$$

$$= \left[\frac{0.23 \times (0.572)^3}{12} + 0.23 \times 0.572 \times \left(\frac{0.572}{2} - 0.174\right)^2 + \frac{0.572(0.23)^3}{12}\right]2$$

$$= [358{,}702 + 165{,}029 + 57{,}996]2 = 1{,}163{,}450 \text{ cm}^4$$

Moment of inertia of sides perpendicular to the moment direction about $n{-}s$ axis is

$$I_2 = A(\bar{x})^2$$
$$= [(50.8 + 23)23](17.4)^2 = 513,900 \text{ cm}^2$$

Therefore, the torsional moment of inertia is

$$J_c = 1,163,450 + 513,900 = 1,677,350 \text{ cm}^2$$

Shearing stress due to perimeter shear effect on M_n is

$$V_n = \frac{V_u}{\phi A_c} + \frac{\gamma_v C_{AB} M_n}{J_c}$$
$$= \frac{282}{0.85 \times 0.433} + \frac{0.37 \times 0.174 \times 143}{1,677,350 \times 10^{-8}} = 766 \text{ kPa} + 550 \text{ kPa}$$
$$= 1.32 \text{ MPa}$$

From before, maximum allowable $J_c = 1.34$ MPa:

$$J_n < J_c$$

Therefore, accept plate thickness.

Design of reinforcement in the slab area at column face for the unbalanced moment transferred to the column by flexure
From Eq. 11.6b,

$$\gamma_f = 1 - \gamma_v = 1 - 0.37 = 0.63$$
$$M_{nf} = \gamma_f M_n = 0.63 \times 14.3 = 90.1 \text{ kN-m}$$

This moment has to be transferred within $1.5\,h$ on each side of the column as in Fig. 11.7d.

$$\text{transfer width} = (1.5 \times 0.254) \times 2 + 0.508 = 1.27 \text{ m}$$
$$M_{nf} = A_s f_y \left(d - \frac{a}{2} \right), \quad \text{assume that } \left(d - \frac{a}{2} \right) \simeq 0.9d$$

or

$$90.1 \times 10^6 = A_s \times 414 \times 0.9 \times 230$$
$$A_s = 1050 \text{ mm}^2 \text{ over strip width} = 1300 \text{ mm}$$

Verify A_s.

$$a = \frac{1050 \times 414}{0.85 \times 27.6 \times 1300} = 14.3 \text{ mm}$$

Therefore,

$$90.1 \times 10^6 = A_s \times 414 \left(230 - \frac{14.3}{2} \right)$$
$$A_s \simeq 1000 \text{ mm}^2$$

Use five No. 15 M bars (1000 mm²) at 100 mm c-c to be used in the 510-mm column width at the top and anchor into the column as required for bond length development.

Proportioning of the plate reinforcement (steps 8 and 9)

(a) *E–W direction (long span)*

1. *Summary of moments in column strip*

$$\text{interior column negative } M_n = \frac{0.75 \times 302}{\phi = 0.9} = 252 \text{ kN-m}$$

$$\text{midspan } M_n = \frac{0.6 \times 225}{0.9} = 150 \text{ kN-m}$$

$$\text{exterior column negative } M_{ne} = \frac{1 \times 112}{0.9} = 124 \text{ kN-m}$$

2. *Summary of moment in middle strip*

$$\text{interior column negative } M_n = \frac{302 - 0.75 \times 302}{0.9} = 84 \text{ kN-m}$$

$$\text{midspan positive } M_n = 225 - 0.6 \times 225 = 90 \text{ kN-m}$$

$$\text{exterior column negative } M_n = 0$$

3. *Design of reinforcement for column strip*

$$-M_n = 252 \text{ kN-m acts on a strip width of } 2(0.25 \times 5.49) = 2.745 \text{ m}$$

$$\text{unit } -M_n \text{ per meter wide strip} = \frac{252}{2.745} = 91.8 \text{ kN-m}$$

$$\text{unit } +M_n \text{ per meter wide strip} = \frac{150}{2.745} = 54.6 \text{ kN-m}$$

Minimum temperature steel reinforcement A_s for two-way plates

Using f_y = 414 MPa steel, A_s = $0.0018bh$ = $0.0018 \times 1000 \times 260$ = 468 mm² per 1-m strip width. Try No. 10 M bars (A_s = 100 mm²).

$$s = \frac{100 \text{ mm}^2}{468 \text{ mm}^2/1000 \text{ mm}} \approx 215 \text{ mm c-c}$$

Negative steel

$$91.8 \times 10^6 = A_s \times 414 \times \left(230 - \frac{a}{2}\right)$$

Assuming $(d - a/2)$ = $0.9d$ for the first trial,

$$A_s = \frac{91.8 \times 10^6 \text{ N-mm}}{414 \times 0.9 \times 230} = 1070 \text{ mm}^2$$

$$a = \frac{A_s f_y}{0.85 f'_c b} = \frac{1070 \times 414}{0.85 \times 27.6 \times 1000} = 18.8 \text{ mm}$$

For the second trial and adjustment cycle,

$$91.8 \times 10^6 = A_s \times 414 \left(230 - \frac{18.8}{2}\right)$$

Negative A_s = 1010 mm²/m width strip.

Try No. 15 M bars from Table B.2b in the appendix (area per bar = 200 mm²).

$$\text{spacing } s = \frac{200}{1010/1000} = 200 \text{ mm c-c for negative moment}$$

Positive steel

Spacing s as required by moment = $200 \times 91.8/54.6 = 330$ mm c-c. The maximum allowable spacing = $2h = 2 \times 260 = 520$ mm. Try No. 10 M bars for positive moment ($A_s = 100$ mm²) from Table B.2b in the appendix.

$$A_s = \frac{54.6}{91.8} \times 1010 = 600 \text{ mm}^2/\text{1-m strip}$$

$$s = \frac{100}{600/1000} \simeq 170 \text{ mm c-c}$$

4. *Design of reinforcement for middle strip*

Unit $-M_n = 840/0.9 = 93.3$ kN-m acting on a strip width of 2.745 m.

unit $-M_n$ per meter width strip = $93.3/2.745 = 34.0$ kN-m = 34×10^6 N-mm

$$34 \times 10^6 = A_s \times 414 \times 0.9 \times 230$$

$$A_s = 400 \text{ mm}^2/\text{1-m strip}$$

$$a = \frac{400 \times 414}{0.85 \times 27.6 \times 1000} = 5.6 \text{ mm}$$

Second cycle using a = 5.6 mm

$$34 \times 10^6 = A_s \times 414 \left(230 - \frac{5.6}{2}\right)$$

$$A_s = 363 \text{ mm}^2/\text{1-m strip}$$

From Table B.2b, try No. 10 M bars ($A_s = 100$ mm²).

$$\text{unit } +M_n = \frac{90 \text{ kN-m}}{0.9 \times 2.745} = 36.4 \text{ kN-m} = 36.4 \times 10^6 \text{ N-mm/meter strip}$$

$$A_s = \frac{36.4 \times 10^6}{414 \times 0.9 \times 230} = 425 \text{ mm}^2$$

Minimum $A_s = 0.0018bh = 0.0018 \times 100 \times 260 = 468$ mm², controls.
For positive reinforcement, the spacing using No. 10 M bars is

$$s = \frac{\text{area of bar}}{\text{required unit area}} = \frac{100}{463/1000} = 213 \text{ mm c-c}$$

Use No. 15 M bars at 200 mm c-c for the negative reinforcement and No. 10 M bars at 200 mm c-c for positive reinforcement.

(b) *N–S direction (short span)*: The same procedure is followed as for the E–W direction for both the negative and positive moments. The effective width $d_2 = 260 - 30 - 10 = 250$ mm is to be used for the N–S direction. A table for moments

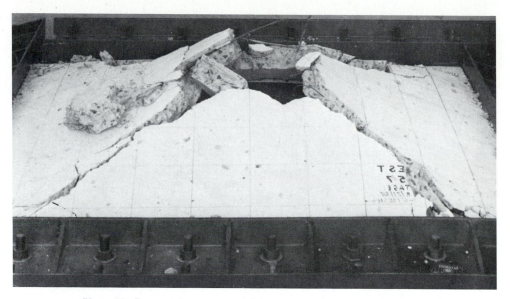

Photo 67   Rectangular concrete slab at rupture.   (Tests by Nawy et al.)

and bar sizes is then constructed as in Table 11.6 and a plan for schematic distribution of reinforcement in SI units is provided similar to Fig. 11.14.

11.8 DIRECT METHOD OF DEFLECTION EVALUATION

11.8.1 The Equivalent Frame Approach

As in the equivalent frame method discussed in detail in the preceding sections, the structure is divided into continuous frames centered on the column lines in each of the two perpendicular directions. Each frame would be composed of a row of columns and a *broad* band of slab together with column line beams, if any, between panel center lines.

By the requirement of statics, the applied load must be accounted for in each of the two perpendicular (orthogonal) directions. In order to account for the torsional deformations of the support beams, an *equivalent* column is used whose flexibility is the *sum* of the flexibilities of the actual column and the torsional flexibility of the transverse beam or slab strips (stiffness is the inverse of flexibility). In other words,

$$\frac{1}{K_{ec}} = \frac{1}{\Sigma\,K_c} + \frac{1}{K_t} \tag{11.18}$$

where K_{ec} = flexural stiffness of the equivalent column; bending moment per unit rotation

$\Sigma\, K_c$ = sum of flexural stiffnesses of upper and lower columns; bending moment per unit rotation

K_t = torsional stiffness of the transverse beam or slab strip; torsional moment per unit rotation

The value of K_{ec} would thus have to be known in order to calculate the deflection by this procedure.

The slab–beam strips are considered to be supported *not* on the columns but on *transverse* slab–beam strips on the column center lines. Figure 11.24a illustrates this point. Deformation of a typical panel is considered in *one direction at a time*. Thereafter, the contribution in each of the two directions, *x* and *y*, is added to obtain the total deflection at any point in the slab or plate.

First, the deflection due to bending in the *x* direction is computed (Fig. 11.24b). Then the deflection due to bending in the *y* direction is found. The midpanel deflection can now be obtained as the sum of the center-span deflections of the column strip in one direction and that of the middle strip in the orthogonal direction (Fig. 11.24c).

The deflection of each panel can be considered as the sum of three components:

1. Basic midspan deflection of the panel, assumed fixed at both ends, given by

$$\delta' = \frac{wl^4}{384 E_c I_{\text{frame}}}$$

This has to be proportioned to separate deflection δ_c of the column strip and δ_s of the middle strip, such that

$$\delta_c = \delta'\,\frac{M_{\text{col strip}}}{M_{\text{frame}}}\,\frac{E_c I_{cs}}{E_c I_c}$$

$$\delta_s = \delta'\,\frac{M_{\text{slab strip}}}{M_{\text{frame}}}\,\frac{E_c I_{cs}}{E_c I_s}$$

where I_{cs} is the moment of inertia of the total frame, I_c the moment of inertia of the column strip, and I_s the moment of inertia of the middle slab strip.

2. Center deflection, $\delta''_{\theta L} = \frac{1}{8}\theta L$, due to rotation at the left end while the right end is considered fixed, where $\theta_L = $ left M_{net}/K_{ec} and K_{ec} is the flexural stiffness of equivalent column (moment per unit rotation).

3. Center deflection, $\delta''_{\theta R} = \frac{1}{8}\theta L$ due to rotation at the right end while the left end is considered fixed, where $\theta_R = $ right M_{net}/K_{ec}. Hence

$$\delta_{cx} \quad\text{or}\quad \delta_{cy} = \delta_c + \delta''_{\theta L} + \delta''_{\theta R} \tag{11.19a}$$

$$\delta_{sx} \quad\text{or}\quad \delta_{sy} = \delta_s + \delta''_{\theta L} + \delta''_{\theta R} \tag{11.19b}$$

(Use in Eqs. 11.19a and 11.19b the values of δ_c, $\delta''_{\theta L}$, and $\delta''_{\theta R}$ that correspond to the applicable span directions. From Fig. 11.24b and c, the total deflection

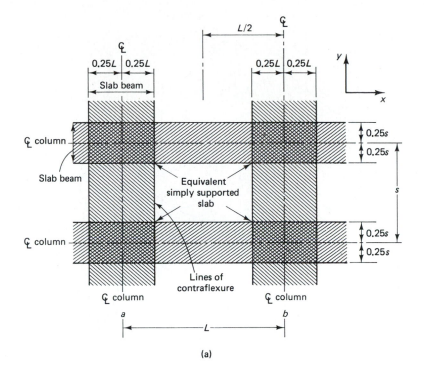

(a)

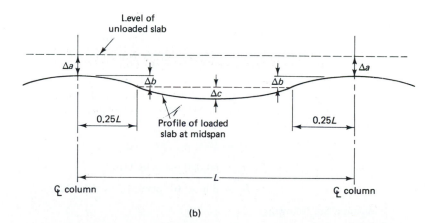

(b)

Figure 11.24 Equivalent frame method for deflection analysis: (a) plate panel transferred into equivalent frames; (b) profile of deflected shape at center line; (c) deflected shape of panel.

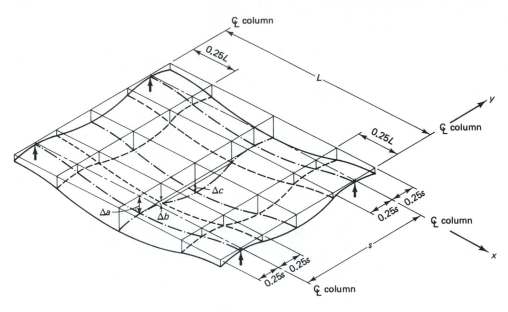

(c)

Figure 11.24   (*cont.*)

is

$$\Delta = \delta_{sx} + \delta_{cy} = \delta_{sy} + \delta_{cx} \tag{11.20}$$

11.8.2 Example 11.4: Central Deflection Calculations of a Slab Panel on Beams

A 7-in. (177.8-mm) slab of a five-panel by five-panel floor system spanning 25 ft in the E–W direction (7.62 m) and 20 ft in the N–S direction (6.10 m) is shown in Fig. 11.25a.   The panel is monolithically supported by beams 15 in. × 27 in. in the E–W direction (381 mm × 686 mm) and 15 in. × 24 in. in the N–S direction (381 mm × 610 mm).   The floor is subjected to a time-dependent deflection due to an equivalent uniform working load intensity $w = 450$ psf (21.5 kPa).   Material properties of the floor are

$$f'_c = 4000 \text{ psi (27.6 MPa)}$$

$$f_y = 60{,}000 \text{ psi (414 MPa)}$$

$$E_c = 3.6 \times 10^6 \text{ psi (24.8} \times 10 \text{ kPa)}$$

Assume the following:

1. Net moment M_w from adjacent spans (ft-lb):

E–W	N–S
Support 1: 20×10^3	Support 1: 40×10^3
Support 2: 5×10^3	Support 4: 20×10^3

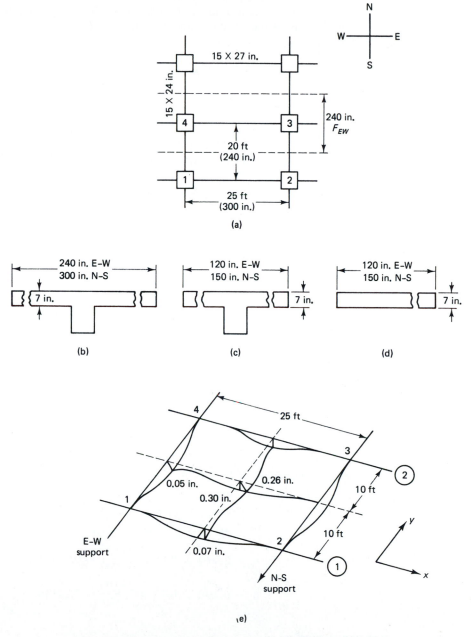

Figure 11.25 Example on equivalent frame deflection evaluation.

2. Equivalent column stiffness $K_{ec} \simeq 400E_c$ lb-in. per radian in both directions. Find the maximum central deflection of the panel due to the long-term loading and determine if its magnitude is acceptable if the floor supports sensitive equipment that can be damaged by large deflections.

3. Cracked moment of inertia:

$$\text{E--W:} \quad I_{cr} = 45{,}500 \text{ in.}^4$$

$$\text{N--S:} \quad I_{cr} = 32{,}500 \text{ in.}^4$$

Solution Calculate the gross moments of inertia (in.^{4}) of the sections in Fig. 11.25: the total equivalent frame I_{cs} in part (b), the column strip beam I_c in part (c), and the middle strip slab I_s in part (d). These values are

	I_{cs}	I_c	I_s
E--W	63,600	53,700	3430
N--S	47,000	40,000	4288

Next, calculate factors $\alpha_1 l_2/l_1$ and $\alpha_2 l_1/l_2$ as in Ex. 11.2. In both cases they are greater than 1.0. Hence the factored moments coefficients (percent) obtained from the tables in Section 11.4.2 are as follows:

	Column Strip (+ and −)	Middle Strip (+ and −)
E--W	81.0	19.0
N--S	67.5	32.5

E--W direction deflections (span = 25 ft)

long-term $w_w = 450$ psf

$$\delta'_{25} = \frac{450 \times 20(25)^4 \times 1728}{384 \times 3.6 \times 10^6 \times 63{,}600} = 0.0691 \text{ in.}$$

$$\delta_c = 0.0691 \times 0.81 \times \frac{63{,}600}{53{,}700} = 0.0663 \text{ in.}$$

$$\delta_s = 0.0691 \times 0.19 \times \frac{63{,}600}{3{,}430} = 0.243 \text{ in.}$$

Rotation at end 1 is

$$\theta_1 = \frac{M_1}{K_{ec}} = \frac{20 \times 10^3 \times 12}{400 \times 3.6 \times 10^6} = 1.67 \times 10^{-4} \text{ rad}$$

and rotation at end 2 is

$$\theta_2 = \frac{M_2}{K_{ec}} = \frac{5 \times 10^3 \times 12}{400 \times 3.6 \times 10^6} = 0.42 \times 10^{-4} \text{ rad}$$

where θ is the rotation at one end if the other end is fixed.

$$\delta'' = \text{deflection adjustment due to rotation at supports}$$
$$1 \text{ and } 2 = \frac{\theta l}{8}$$
$$\delta'' = \frac{(1.67 + 0.42) \times 10^{-4} \times 300}{8} = 0.0078 \text{ in.}$$

Therefore,

$$\text{net } \delta_{cx} = 0.0663 + 0.0078 = 0.0741 \qquad \text{say 0.07 in.}$$
$$\text{net } \delta_{sx} = 0.243 + 0.0078 = 0.2508 \qquad \text{say 0.25 in.}$$

N–S direction deflections (span = 20 ft)

$$\delta'_{20} = \frac{450 \times 25(20)^4 \times 1728}{384 \times 3.6 \times 10^6 \times 47,000} = 0.0479 \text{ in.}$$

$$\delta_c = 0.0479 \times 0.675 \times \frac{47,000}{40,000} = 0.038 \text{ in.}$$

$$\delta_s = 0.0479 \times 0.325 \times \frac{47,000}{4288} = 0.171 \text{ in.}$$

$$\text{rotation } \theta_1 = \frac{M_1}{K_{ec}} = \frac{40 \times 10^3 \times 12}{400 \times 3.6 \times 10^6} = 3.3 \times 10^{-4} \text{ rad}$$

$$\text{rotation } \theta_4 = \frac{M_4}{K_{ec}} = \frac{20 \times 10^3 \times 12}{400 \times 3.6 \times 10^6} = 1.67 \times 10^{-4} \text{ rad}$$

$$\delta'' = \frac{\theta l_2}{8} = \frac{(3.3 + 1.67)10^{-4} \times 240}{8} = 0.0149 \text{ in.}$$

Therefore,

$$\text{net } \delta_{cy} = 0.038 + 0.0149 = 0.0529 \qquad \text{say 0.05 in.}$$
$$\delta_{sy} = 0.171 + 0.0149 = 0.1859 \qquad \text{say 0.19 in.}$$
$$\text{total central deflection } \Delta = \delta_{sx} + \delta_{cy} = \delta_{sy} + \delta_{cx}$$
$$\Delta_{E-W} = \delta_{sx} + \delta_{cy} = 0.25 + 0.05 = 0.30 \text{ in.}$$
$$\Delta_{N-S} = \delta_{sy} + \delta_{cx} = 0.19 + 0.07 = 0.26 \text{ in.}$$

Hence the average deflection at the center of the interior panel is

$$\tfrac{1}{2}(\Delta_{E-W} + \Delta_{N-S}) = 0.28 \text{ in. (7.1 mm)}$$

Adjustment for cracked section: Use Branson's effective moment of inertia equation,

$$I_e = \left(\frac{M_{cr}}{M_a}\right)^3 I_g + \left[1 - \left(\frac{M_{cr}}{M_a}\right)^3\right] I_{cr}$$

as discussed in Chapter 8. Calculation of ratio M_{cr}/M_a:

$$M_{cr} = \frac{f_r I_g}{y_t}$$

where f_r = modulus of rupture of concrete

y_t = distance of center of gravity of section from outer tension fibers

E–W (240-in. flange width): y_t = 21.54 in.

N–S (300-in. flange width): y_t = 19.20 in.

$$f_r = 7.5\sqrt{f'_c} = 7.5\sqrt{4000} = 474 \text{ psi}$$

Hence

$$M_{cr} \quad (\text{E–W}) = \frac{474 \times 63,600}{21.54} \times \frac{1}{12} = 1.17 \times 10^5 \text{ ft-lb}$$

$$M_{cr} \quad (\text{N–S}) = \frac{474 \times 47,000}{19.20} \times \frac{1}{12} = 0.97 \times 10^5 \text{ ft-lb}$$

$$\text{interior panel } M_a = \frac{w_w l^2}{16} = \frac{20 \times 450(25)^2}{16} \quad \text{for E–W}$$

$$= 3.52 \times 10^5 \text{ ft-lb}$$

$$= \frac{25 \times 450(20)^2}{16} \quad \text{for N–S}$$

$$= 2.81 \times 10^5 \text{ ft-lb}$$

Note that the moment factor $\frac{1}{16}$ is used to be on the safe side, although the actual moment coefficients for two-way action would have been smaller.

E–W effective moment of inertia I_e

$$\frac{M_{cr}}{M_a} = \frac{1.17 \times 10^5}{3.52 \times 10^5} = 0.332$$

$$\left(\frac{M_{cr}}{M_a}\right)^3 = 0.037$$

$$I_e = 0.037 \times 63,600 + (1 - 0.037)45,500 = 46,170 \text{ in.}^4$$

N–S effective moment of inertia I_e

$$\frac{M_{cr}}{M_a} = \frac{0.97 \times 10^5}{2.81 \times 10^5} = 0.345$$

$$\left(\frac{M_{cr}}{M_a}\right)^3 = 0.041$$

$$I_e = 0.041 \times 47,000 + (1 - 0.041)32,500 = 33,095 \text{ in.}^4$$

$$\text{average } \frac{I_g}{I_e} = \frac{1}{2}\left(\frac{63,600}{46,170} + \frac{47,000}{33,095}\right) = 1.40$$

adjusted central deflection for cracked section effect

$$= 1.40 \times 0.28 = 0.39 \text{ in. (9.9 mm)}$$

$$\frac{l}{\Delta} = \frac{25 \text{ ft} \times 12}{0.39} = 769 > 480 \qquad \text{allowed in Table 11.3}$$

Hence the long-term central deflection is acceptable.

11.9 CRACKING BEHAVIOR AND CRACK CONTROL IN TWO-WAY-ACTION SLABS AND PLATES

11.9.1 Flexural Cracking Mechanism and Fracture Hypothesis

Flexural cracking behavior in concrete structural floors under two-way action is significantly different from that in one-way members. Crack-control equations for beams underestimate the crack widths developed in two-way slabs and plates and do not tell the designer how to space the reinforcement. Cracking in two-way slabs and plates is controlled primarily by the steel stress level and the spacing of the reinforcement in the two perpendicular directions. In addition, the clear concrete cover in two-way slabs and plates is nearly constant [$\frac{3}{4}$ in. (19 mm) for interior exposure], whereas it is a major variable in the crack-control equations for beams. The results from extensive tests on slabs and plates by Nawy et al. demonstrate this difference in behavior in a fracture hypothesis on crack development and propagation in two-way plate action. As seen in Fig. 11.26, stress concentration develops initially at the points of intersection of the reinforcement in the reinforcing bars and at the welded joints of the wire mesh, that is, at grid nodal points, thereby dynamically generating fracture lines along the paths of least resistance: A_1B_1, A_1A_2, A_2B_2, and B_2B_1. The resulting fracture pattern is a total repetitive cracking grid, provided that the spacing of the nodal points A_1, B_1, A_2, and B_2 is close enough to generate this preferred initial fracture mechanism of orthogonal cracks narrow in width as a fracture mechanism.

If the spacing of the reinforcing grid intersections is too large, the magnitude of the stress concentration and the energy absorbed per unit grid are too low to generate cracks along the reinforcing wires or bars. As a result, the principal cracks follow diagonal yield-line cracking in the plain concrete field away from the reinforcing bars early in the loading history. These cracks are wide and few.

This hypothesis also leads to the conclusion that surface deformations of the individual reinforcing elements have little effect in arresting the generation of the cracks or controlling their type or width in a two-way-action slab or plate. In a

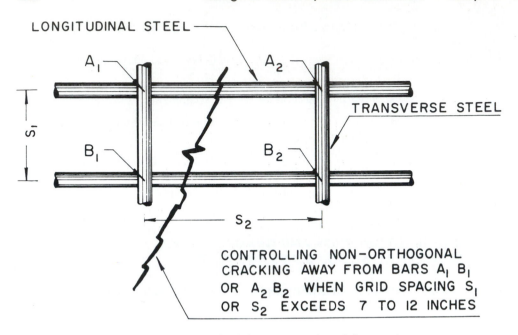

Figure 11.26 Grid unit in two-way-action reinforcement.

similar manner, we may conclude that the scale effect on two-way-action cracking behavior is insignificant, since the cracking grid would be a reflection of the reinforcement grid if the preferred orthogonal narrow cracking widths develop. Therefore, to control cracking in two-way-action floors, the major parameter to be considered is the reinforcement spacing in two perpendicular directions. Concrete cover has only a minor effect, since it is usually a small, constant value of 0.75 in. (20 mm).

For a constant area of steel determined for bending in one direction, that is, for energy absorption per unit slab area, the smaller the spacing of the transverse bars or wires, the smaller should be the diameter of the longitudinal bars. The reason is that less energy has to be absorbed by the individual longitudinal bars. If we consider that the magnitude of fracture is determined by the energy imposed per specific volume of reinforcement acting on a finite element of the slab, a proper choice of the reinforcement grid size and bar size can control cracking into preferred orthogonal grids.

It must be emphasized that this hypothesis is important for serviceability and reasonable overload conditions. In relating orthogonal cracks to yield-line cracks, the failure of a slab ultimately follows the generally accepted rigid-plastic yield-line criteria.

11.9.2 Crack Control Equation

The basic equation (Section 8.11) for relating crack width to strain in the reinforcement is

$$w = \alpha a_c^\beta \, \epsilon_s^\gamma \tag{11.21}$$

The effect of the tensile strain in the concrete between the cracks is neglected as insignificant. a_c is the crack spacing, ϵ_s the unit strain in the reinforcement, and α, β, and γ are constants. As a result of this fracture hypothesis, the mathematical model in Eq. 11.21, and the statistical analysis of the data of 90 slabs tested to failure, the following crack-control equation emerged:

$$w = K\beta f_s \sqrt{\frac{d_{b_1} s_2}{Q_{i_1}}} \tag{11.22}$$

where the quantity under the radical, $G_1 = d_{b_1} s_2 / Q_{i_1}$, is termed the grid index and can be transformed into

$$G_1 = \frac{s_1 s_2 d_c}{d_{b1}} \frac{8}{\pi} \tag{11.23}$$

where K = fracture coefficient, having a value of $K = 2.8 \times 10^{-5}$ for uniformly loaded, restrained, two-way-action square slabs and plates. For concentrated loads or reactions or when the ratio of short to long span is less than 0.75, but larger than 0.5, a value of $K = 2.1 \times 10^{-5}$ is applicable. For a span aspect ratio of 0.5, $K = 1.6 \times 10^{-5}$. Units of coefficient K are in.^{2}/lb.

β = ratio of the distance from the neutral axis to the tensile face of the slab to the distance from the neutral axis to the centroid of the reinforcement grid (to simplify the calculations use $\beta = 1.25$, although it varies between 1.20 and 1.35)

f_s = actual average service load stress level, or 40% of the design yield strength, f_s (ksi)

d_{b1} = diameter of the reinforcement in direction 1 closest to the concrete outer fibers (in.)

d_c = concrete cover to centroid of reinforcement (in.)

s_1 = spacing of the reinforcement in direction 1

s_2 = spacing of the reinforcement in perpendicular direction 2

1 = direction of the reinforcement closest to the outer concrete fibers; this is the direction for which crack control check is to be made

Q_{i_1} = active steel ratio

$= \dfrac{\text{area of steel } A_s \text{ per foot width}}{12(d_{b1} + 2c_1)}$

where c_1 is clear concrete cover measured from the tensile face of the concrete to the nearest edge of the reinforcing bar in direction 1

w = crack width at face of concrete caused by flexural load (in.)

Subscripts 1 and 2 pertain to the directions of reinforcement. Detailed values of the fracture coefficients for various boundary conditions are given in Table 11.10. A graphical solution of Eq. 11.22 is given in Fig. 11.27 for

$$f_y = 60,000 \text{ psi } (414 \text{ MPa})$$

$$f_s = 40\% \, f_y = 24,000 \text{ psi } (165.5 \text{ MPa})$$

for rapid determination of the reinforcement size and spacing needed for crack control. Equation 11.22 in SI units is

$$w_{\max} \text{ (mm)} = 0.145 \, K\beta f_s \sqrt{G_I} \qquad (11.22a)$$

where f_s = MPa, $G_I = \dfrac{s_1 s_2 d_c}{d_{b1}} \times \dfrac{8}{\pi}$, and s_1, s_2, d_c, and d_{b1} are in millimeters.

The grid index, G_1, specifies the size and spacing of the bars in the two perpendicular directions of any concrete floor system, and $w_{\max}$ is the maximum allowable crack width.

The crack control equation and guidelines presented are important not only for the control of corrosion in the reinforcement but also for deflection control. The reduction of the stiffness EI of the two-way slab or plate due to orthogonal cracking when the limits of permissible crack widths in Table 8.5 are exceeded can

TABLE 11.10   Fracture Coefficients for Slabs and Plates

Loading Type[a]	Slab Shape	Boundary Condition[b]	Span Ratio,[c] S/L	Fracture Coefficient, 10^{-5} K
A	Square	4 edges r	1.0	2.1
A	Square	4 edges s	1.0	2.1
B	Rectangular	4 edges r	0.5	1.6
B	Rectangular	4 edges r	0.7	2.2
B	Rectangular	3 edges r, 1 edge h	0.7	2.3
B	Rectangular	2 edges r, 2 edges h	0.7	2.7
B	Square	4 edges r	1.0	2.8
B	Square	3 edges r, 1 edge h	1.0	2.9
B	Square	2 edges r, 2 edges h	1.0	4.2

[a]Loading type: A, concentrated; B, uniformly distributed.
[b]Boundary condition: r, restrained; s, simply supported; h, hinged.
[c]Span ratio: S, clear short span; L, clear long span.

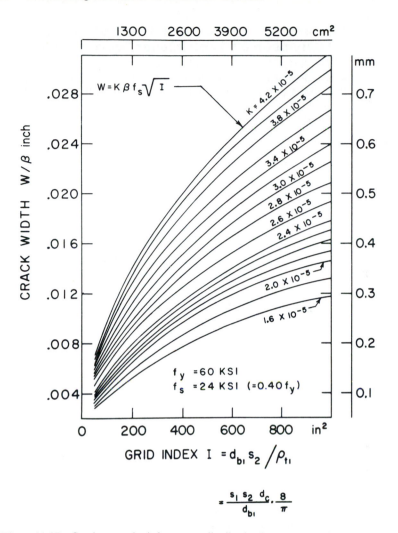

Figure 11.27 Crack-control reinforcement distribution in two-way-action slabs and plates for all exposure conditions: $f_y = 60,000$ psi, $f_s = 24,000$ psi ($= 0.40 f_y$).

lead to excessive deflection in both the short and long term.   Deflection values several times those anticipated in the design, including deflection due to construction loading, can be reasonably controlled through camber and control of the flexural crack width in the slab or plate.   Proper selection of the *reinforcement spacing* s_1 and s_2 in both perpendicular directions, as discussed in this section, and not exceeding 12 in. center to center can maintain good serviceability performance of a slab system under normal and reasonable overload conditions.   The 1985 Australian Code and other codes generally follow the principles presented on the selection of reinforcement size and spacing in slabs and plates, and good engineering practice mandates such caution.

11.9.3 Example 11.5: Crack-Control Evaluation for Serviceability in an Interior Two-way Panel

Check the bar size and spacing used in an interior panel to determine if it satisfies serviceability through crack control. As shown in Fig. 11.28, the panel has a width/length ratio $l_s/l_l = 1.0$, and its thickness is 5 in. (125 mm). The floor is subjected to normal weather conditions. Reinforcement used for flexure is No. 4 bars at 9 in. center to center in each direction. Given:

$$\beta = 1.25$$
$$w_{max} = 0.016 \text{ in.}$$
$$f_y = 60 \text{ ksi } (60,000 \text{ psi})$$
$$K = 2.8 \times 10^{-5} \text{ in.}^2/\text{lb}$$

Solution

$$f_s = 0.40f_y = 0.40(60) = 24 \text{ ksi}$$

The fracture coefficient K for this panel aspect ratio is $K = 2.8 \times 10^{-5}$. The maximum permissible crack width for normal interior conditions is $w_{max} = 0.016$ in. (0.4 mm) (Table 8.4).

$$w_{max} = K\beta f_s \sqrt{G_I}$$

$$\text{grid index } G_I = \frac{s_1 s_2 d_c}{d_{b1}} \times \frac{8}{\pi}$$

$$0.016 = 2.8 \times 10^{-5} \times 1.25 \times 24\sqrt{G_I}$$

to give

$$G_I = 363 \text{ in.}^2 = \frac{s_2 s_1 d_c}{d_{b1}} \times \frac{8}{\pi}$$

If $s_1 = s_2$ for this square panel, cover $d_c = 0.75 + 0.25 = 1.0$ in. to the center of the first reinforcement layer and $d_{b1} = 0.5$ in. = diameter of the No. 4 bar.

$$363 = \frac{s^2 \times 1.0}{0.5} \times \frac{8}{\pi} \qquad \text{giving } s = 8.4 \text{ in. or } 8.5 \text{ in. maximum}$$

Hence the 9-in. center-to-center spacing specified for flexure is not satisfactory. Reduce the spacing of reinforcement to No. 4 bars at $8\frac{1}{2}$ in. (216 mm) center to center for crack control (12.7-mm diameter at 216 mm center to center).

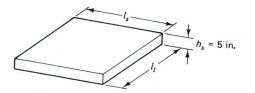

Figure 11.28   Square panel.

11.9.4 Example 11.6: Crack-control Evaluation for Serviceability in a Rectangular Panel Subjected to Severe Exposure Conditions

Select the bar size and spacing necessary for crack control at the column reaction region of the 7-in.-thick slab shown in Fig. 11.29 that is unformly loaded. Select the bar size for two conditions:

> *Condition A:* Floor is subjected to severe exposure of humidity and moist air.
> *Condition B:* Floor sustains an aggressive chemical environment where the design working stress level in the reinforcement is limited to 15 ksi (15,000 psi).

Given:

$$\beta = 1.20$$
$$l_s/l_l = 0.8$$
$$f_y = 60 \text{ ksi (414 MPa)}$$

Solution   *Condition A: Humidity and moist air*

Permissible $w_{max} = 0.012$ in. (0.3 mm) (Table 8.4). Try No. 4 bars $d_b = 0.5$, $d_c = 0.75 + 0.25 = 1.0$ in. Assume that $s_1 = s_2 = s$ for the given panel. The aspect ratio $l_s/l_l = 0.8$. $K = 2.1 \times 10^{-5}$ for concentrated reaction at the column support (Table 11.10).

$$0.012 = 2.1 \times 10^{-5} \times 1.20 \times 0.4 \times 60\sqrt{G_I}$$

to give $G_I = 394$ in.^{2}. Therefore,

$$394 = \frac{s^2 d_c}{d_{b1}} \times \frac{8}{\pi} = \frac{s^2 \times 1.0}{0.5} \times \frac{8}{\pi}$$

$$s = 8.8 \text{ in.}$$

Hence use No. 4 bars at $8\frac{1}{2}$ in. center to center each way for crack control.

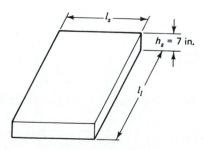

Figure 11.29   Rectangular panel.

Condition B: Aggressive chemical environment

Permissible $w_{max} = 0.007$ in. (0.18 mm) (Table 8.2). $f_s = 15$ ksi to be used as a low stress level for sanitary or water-retaining structures instead of $0.4 f_y$. Try No. 5 bars ($d_{b1} = 0.625$ in.).

$$0.007 = 2.1 \times 10^{-5} \times 1.20 \times 15.0 \sqrt{G_I}$$

to give a grid index $G_I = 343$ in.^{2}.

$$d_{c1} = 0.75 + 0.312 = 1.06 \text{ in.}$$

$$G_I = 343 = \frac{s^2 \times 1.06}{0.625} \times \frac{8}{\pi} \quad \text{to get } s = 8.9 \text{ in.}$$

Use No. 5 bars at 9-in. (229-mm) center-to-center spacing each way for crack control.

Reinforcement summary

Condition A: No. 4 bars at $8\frac{1}{2}$ in. c-c (12.7-mm diameter at 216 mm c-c)
Condition B: No. 5 bars at 9 in. c-c (15.9-mm diameter at 229 mm c-c)

11.9.5 Example 11.7: SI Example on Crack Control in Two-way Slabs and Plates

Solve Ex. 11.5 using SI units.

Data

$$f_s = 166 \text{ MPa}$$
$$\text{max allowable } w_{max} = 0.40 \text{ mm}$$
$$K = 2.8 \times 10^{-5}, \qquad \beta = 1.25$$
$$d_c = 25 \text{ mm}, \qquad d_{b1} = 12.7 \text{ mm}$$

Solution

$$w_{max} \text{ (mm)} = 0.145 \, K\beta f_s \, \sqrt{G_I}$$

where

$$G_I = \frac{s_1 s_2 d_c}{d_{b1}} \times \frac{8}{\pi}$$

$$0.40 = 0.145 \times 2.8 \times 10^{-5} \times 1.25 \times 166 \sqrt{G_i}$$

$$\sqrt{G_I} = \frac{0.4 \times 10^5}{0.145 \times 2.8 \times 1.25 \times 166} = 475$$

If $s_1 = s_2$ for this square panel,

$$(474.8)^2 = \frac{s^2 \times 25}{12.7} \times \frac{8}{\pi}$$

$$s = \left[\frac{(475)^2 \times 12.7 \times \pi}{25 \times 8}\right]^{1/2} = 212 \text{ mm}$$

Hence space the bars at 20 cm c-c.

11.10 YIELD-LINE THEORY FOR TWO-WAY ACTION PLATES

A study of the hinge-field mechanism in a slab or plate at loads close to failure aids the engineering student in developing a feel for the two-way-action behavior of plates. Hinge fields are successions of hinge bands that are idealized by lines; hence the name *yield-line theory* by K. W. Johansen.

To do justice to this subject, an extensive discussion over several chapters or a whole textbook is necessary. The intention of this chapter is only to introduce the reader to the fundamentals of the yield-line theory and its application.

The yield-line theory is an upper-bound solution to the plate problem. This means that the predicted moment capacity of the slab has the highest expected value in comparison with test results. Additionally, the theory assumes a totally rigid-plastic behavior; that is, the plate stays planar at collapse, producing rigid planar failure systems. Consequently, deflection is not accounted for, nor are the compressive membrane forces that will act in the plane of the slab or plate considered. The plates are assumed to be considerably underreinforced such that the maximum reinforcement percentage ρ does not exceed 1/2% of the section *bd*.

Since the solutions are upper bound, the slab thickness obtained by this process is in many instances thinner than what is obtained by the other lower-bound solutions, such as the direct design method. Consequently, it is important to apply rigorously the serviceability requirements for deflection control and for crack control in conjunction with the use of the yield-line theory as given in Sections 11.8 and 11.9.

One distinct advantage in this theory is that solutions are possible for any shape of a plate, whereas most other approaches are applicable only to the rectangular shapes with rigorous computations for boundary effects. The engineer can, with ease, find the moment capacity for a triangular, trapezoidal, rectangular, circular, and any other conceivable shape provided that the failure mechanism is known or predictable. Since most failure patterns are presently identifiable, solutions can be readily obtained, as seen in Section 11.10.2.

11.10.1 Fundamental Concepts of Hinge-Field Failure Mechanisms in Flexure

Under the action of a two-dimensional system of bending moments, yielding of a rigid-plastic plate occurs when the principal moments satisfy Johansen's square yield criterion, as shown in Fig. 11.30. In this criterion, yielding is considered to

Photo 68   Testing setup of four-panel prestressed concrete floor.   (Tests by Nawy et al.)

have occurred when the numerically greater of the principal moments reaches the value of $\pm M$ at the yield-line cracks.   The directions of the principal curvature rates are considered to coincide with the curvatures of the principal moments.   The idealized moment–curvature relationship is shown as the solid line in Fig. 11.31.

Line OA is considered almost vertical at point O and strain hardening is neglected.

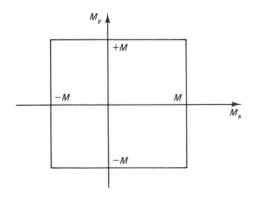

Figure 11.30 Johansen's square yield criterion.

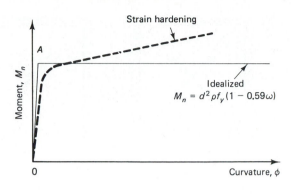

Figure 11.31 Moment–curvature relationship.

If we consider the simplest case of a square slab with supports, with degree of fixity i varying from $i = 0$ for simply supported to $i = 1.0$ for fully restrained on all four sides, the failure mechanism would be as shown in Fig. 11.32, when a uniformly distributed load is applied.

Take the simply supported case (a). The yield-line moments along the yield lines are the principal moments.   Hence the twisting moments are zero in the yield lines and in most cases the shearing forces are also zero.   Consequently, only moment m per unit length of the yield line acts about the lines AD and BE in Fig. 11.33.   The total moments can be represented by a vector in the direction of the yield line whose value is $M \times$ length of the yield line, that is, $M\,(a/2 \cos \theta)$ in Fig. 11.33c.   The virtual work of the yield moments of the shaded triangular segment ABO is the scalar product of the two moment vectors $Ma/2 \cos \theta$ on fracture lines AO and BO and a rotation θ.   In other words, the internal work

$$E_I = \sum \overline{M\theta}$$

If the displacement of the shaded segment at its center of gravity c is δ, the external work

$$E_E = \text{force} \times \text{displacement}$$

$$= \sum \int \int w_u \, dx \, dy \, \delta$$

where w_u is the intensity of external load per unit area.   But $E_I = E_E$; hence

$$\sum \overline{M\theta} = \sum \int \int w_u \, dx \, dy \, \delta \tag{11.24}$$

Applying Eq. 11.24 to the particular case under discussion gives us

$$\overline{M\theta} = Ma \, \frac{\Delta}{a/2}$$

since angle θ in Fig. 11.33b is small, where $\theta = \Delta/(a/2)$.

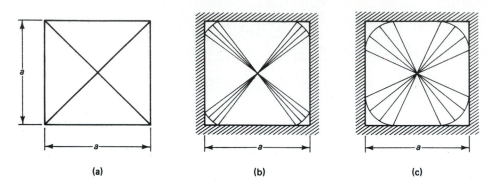

Figure 11.32   Failure mechanism of a square slab: (a) $i = 0$; (b) $i = 0.5$; (c) $i = 1.0$.

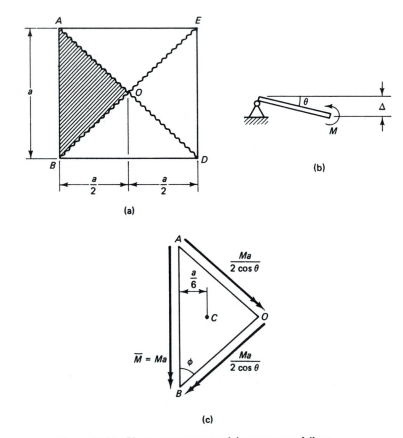

Figure 11.33   Vector moments on slab segment at failure.

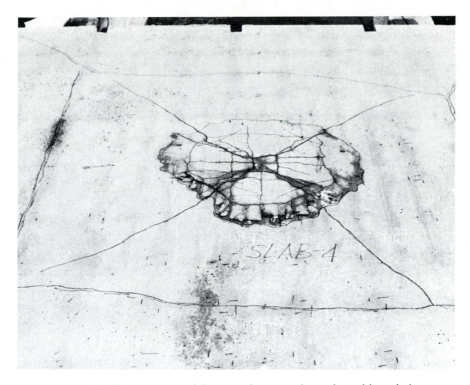

Photo 69 Yield-line pattern at failure at column reaction and panel boundaries of a two-way multipanel floor. (Tests by Nawy, Chakrabarti, et al.)

Work per one triangular segment:

$$E_I = \overline{M\theta} = 2M\Delta$$

$$E_E = \frac{w_u a^2}{4} \times \frac{\Delta}{3}$$

where deflection at center of gravity of the triangle = $\Delta/3$. Therefore,

$$4(2M\Delta) = 4\left(\frac{w_u a^2}{12}\Delta\right)$$

or
$$\text{unit } M = \frac{w_u a^2}{24} \tag{11.25}$$

If the square slab was fully fixed on all four sides, $E_I = 4(4M\Delta)$ since fracture lines develop around not only the diagonals but also the four edges, as shown in Fig. 11.32c. Hence

$$\text{unit } M = \frac{w_u a^2}{48} \tag{11.26}$$

It is to be noted that a lower-bound solution as proposed by Mansfield's failure pattern in Fig. 11.32c gives a value $M = w_u a^2/42.88$. Hence, for a uniformly loaded square slab with load intensity w_u per unit area and degree of support fixity i on all sides,

$$w_u a^2 = M[24(1 + i)] \tag{11.27}$$

The general equation for the yield-line moment capacity of a rectangular isotropic slab on beams and having dimensions $a \times b$ as shown in Fig. 11.34, with side a being the shorter dimension, is

$$\text{unit } M \frac{\text{ft-lb}}{\text{ft}} = \frac{w_u a_r^2}{24} \left[\sqrt{3 + \left(\frac{a_r}{b_r}\right)^2} - \frac{a_r}{b_r} \right]^2 \tag{11.28}$$

where $a_r = \dfrac{2a}{\sqrt{1 + i_2} + \sqrt{1 + i_4}}$

$b_r = \dfrac{2b}{\sqrt{1 + i_1} + \sqrt{1 + i_3}}$

i = degree of restraint depending on stiffness ratios as discussed in Section 11.2

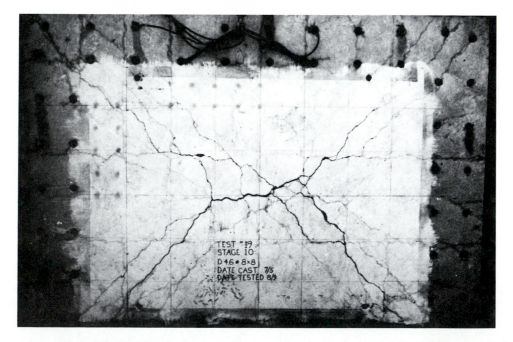

Photo 70 Yield-line patterns at failure at tension face of rectangular restrained panel. (Tests by Nawy et al.)

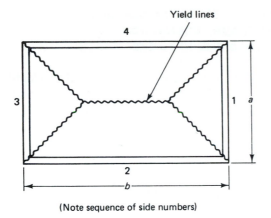

(Note sequence of side numbers)

Figure 11.34   Rectangular slab.   (Note the sequence of side numbers.)

Note that Eq. 11.28 reduces to the simplified form of Eq. 11.26 or 11.27 for the case of a square slab restrained on all four sides ($i = 1.0$).

Affine slabs.   Slabs that are reinforced differently in the two perpendicular directions are called *orthotropic slabs* or *plates*.   The moment in the x direction equals M and in the y direction equals μM, where μ is a measure of the degree of orthotropy or the ratio

$$\frac{M_y}{M_x} = \frac{(A_s)_y}{(A_s)_x}$$

To simplify the analysis, the slab should be converted to an affine (isotropic) slab where the strength and reinforcement area in both the x and y directions are the same.   Such conversion can be made as follows:

1. *Divide* the linear dimension by $\sqrt{\mu}$ in the M direction of the positive moment for a slab to be reinforced for a moment M in both directions using the same unit load intensity w_u per unit area.

2. In the case of concentrated loads or total loads, also divide such loads by $\sqrt{\mu}$.

3. In the case of line loads, the line load has to be divided by $\sqrt{\mu \cos^2 \theta + \sin^2 \theta}$, where θ is the angle between the line load and the M direction.

If the slab is to be analyzed as an affine slab with the moment μM in both directions, the dimension in the μM direction would have to be *multiplied* by $\sqrt{\mu}$.   In either case, the result would of course have to be the same (see Ex. 11.7).

11.10.2 Failure Mechanisms and Moment Capacities of Slabs of Various Shapes Subjected to Distributed or Concentrated Loads

The preceding concise introduction to the virtual-work method of yield-line moments evaluation should facilitate good understanding of the mathematical procedures of most standard rectangular shapes subjected to uniform loading. More complicated slab shapes and other types of symmetrical and nonsymmetrical loading require additional and more advanced knowledge of the subject as discussed in the introduction. Also, the assumed failure shape and minimization energy principles can give values for particular cases that can differ slightly from one author to another depending on the mathematical assumptions made with respect to the failure shape.

The following summary of failure patterns and the respective moment capacities in terms of load, many of them due to Mansfield (Ref. 11.13), should give

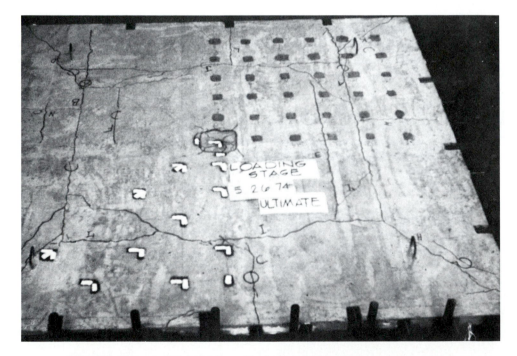

Photo 71 Four-panel slab at failure showing the yield-line patterns at the negative compression face of the supports. (Tests by Nawy and Chakrabarti.)

the reader adequate coverage in a capsule of solutions to most cases expected in today's and tomorrow's structures.

1. Point load to corner of rectangular cantilever plates:

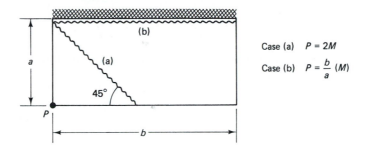

Case (a) $P = 2M$

Case (b) $P = \dfrac{b}{a} (M)$

2. Square plate centrally loaded having boundaries simply supported against both downward and upward movements:

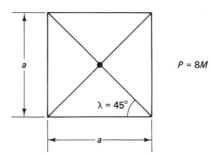

$P = 8M$

3. Regular n-sided plate with simply supported edges and centrally loaded ($n > 4$):

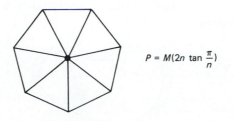

$P = M(2n \tan \dfrac{\pi}{n})$

4. Square plates centrally loaded, having boundaries simply supported against downward movement but free for upward movement:

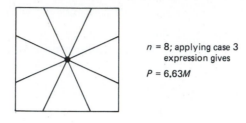

$n = 8$; applying case 3 expression gives

$P = 6.63M$

5. Circular centrally loaded plate simply supported along the edges:

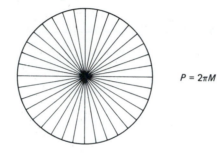

$P = 2\pi M$

6. Circular plate with fully restrained edges and centrally loaded by point load P:

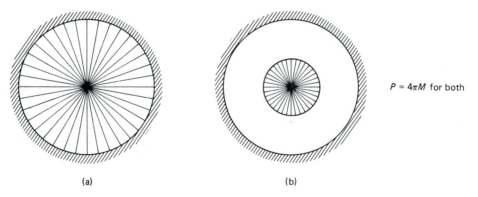

$P = 4\pi M$ for both

(a)   (b)

7. Point load P applied anywhere in arbitrarily shaped plate fully restrained on all boundaries:

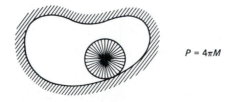

$P = 4\pi M$

8. Equilateral triangular plate with simply supported edges and centrally loaded by point load *P:*

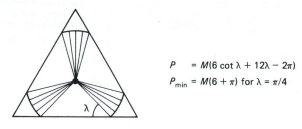

$$P = M(6 \cot \lambda + 12\lambda - 2\pi)$$
$$P_{min} = M(6 + \pi) \text{ for } \lambda = \pi/4$$

9. Acute-angled triangular plate on simply supported edges loaded with point load *P* at the center of the inscribed circle:

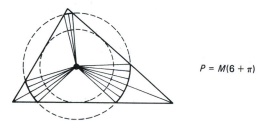

$$P = M(6 + \pi)$$

10. Obtuse-angled triangular plate with simply supported edges and load *P* at the center of the inscribed circle:

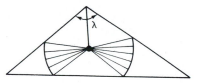

$$P = M(4 + 2\lambda + 2 \cot 1/2\lambda),$$
where λ is in radians

As λ approaches π, the plate degenerates into case 11

11. A long strip simply supported along edges and load with point *P* midway between edges;

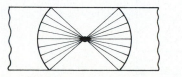

$$P = M(4 + 2\pi)$$

12. Simply supported strip with equal loads P between edges:

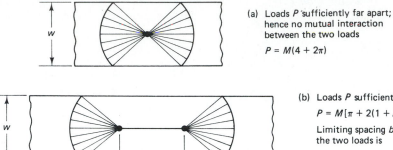

(a) Loads P sufficiently far apart; hence no mutual interaction between the two loads

$$P = M(4 + 2\pi)$$

(b) Loads P sufficiently close

$$P = M[\pi + 2(1 + b/w)]$$

Limiting spacing b between the two loads is

$$b_{lim} = (1 + 1/2\pi)w$$

13. Strip simply supported with unequal loads P and kP midway between the edges, where $k < 1.0$ and the loads are sufficiently apart:

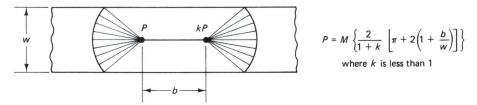

$$P = M\left\{\frac{2}{1+k}\left[\pi + 2\left(1 + \frac{b}{w}\right)\right]\right\}$$

where k is less than 1

14. Uniformly loaded square slab with degree of fixity i varying between zero and 1.0:

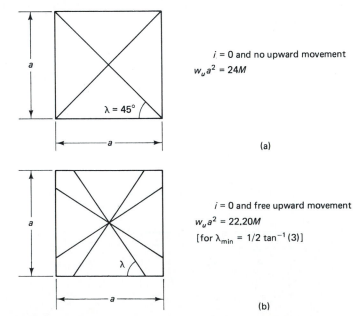

$i = 0$ and no upward movement

$$w_u a^2 = 24M$$

(a)

$i = 0$ and free upward movement

$$w_u a^2 = 22.20M$$

$$[\text{for } \lambda_{min} = 1/2 \tan^{-1}(3)]$$

(b)

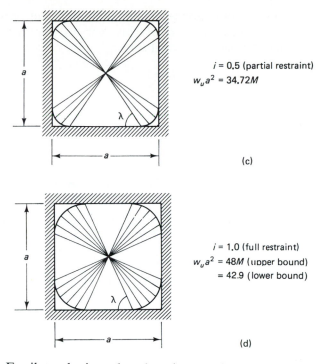

$i = 0.5$ (partial restraint)
$w_u a^2 = 34.72M$

(c)

$i = 1.0$ (full restraint)
$w_u a^2 = 48M$ (upper bound)
       $= 42.9$ (lower bound)

(d)

15. Equilateral triangular plate ($\lambda = 60°$) uniformly loaded:

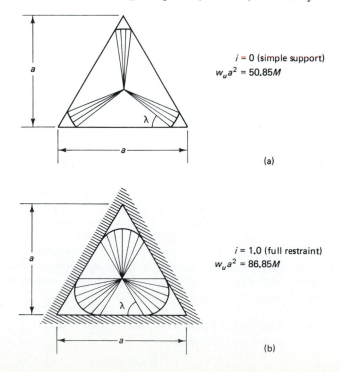

$i = 0$ (simple support)
$w_u a^2 = 50.85M$

(a)

$i = 1.0$ (full restraint)
$w_u a^2 = 86.85M$

(b)

16. Rectangular slab uniformly loaded with unit load of intensity w_u supported on all four sides with degree of fixity i varying from zero to 1.0 (note sequence of numbers assigned to panel sides):

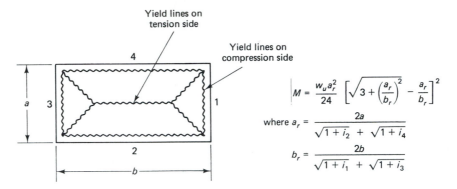

$$\left|M\right| = \frac{w_u a_r^2}{24}\left[\sqrt{3 + \left(\frac{a_r}{b_r}\right)^2} - \frac{a_r}{b_r}\right]^2$$

where $a_r = \dfrac{2a}{\sqrt{1 + i_2} + \sqrt{1 + i_4}}$

$b_r = \dfrac{2b}{\sqrt{1 + i_1} + \sqrt{1 + i_3}}$

General note: Load P is assumed in the foregoing expression to act at a point. To adjust for the fact that P acts on a finite area, assume that it acts over a circular area of radius ρ. For a slab fully restrained on all boundaries, the hinge field would be bound by a circle touching the slab boundary (circle radius = r). In such a case,

$$M + M' = \frac{P}{2\pi}\left(1 - \frac{2\rho}{3r}\right) \tag{11.29}$$

where M = positive unit moment
       M' = negative unit moment

Reaction of columns supporting flat plates can be similarly considered for analyzing the flexural local capacity of the plate in the column area. For rectangular supports, an approximation to equivalent circular support can be made in the use of Eq. 11.29.

11.10.3 Example 11.8: Rectangular Slab Yield-line Design

The reinforced concrete slab shown in Fig. 11.35 is 14 ft 6 in. × 24 ft in plan (4.42 m × 7.32 m). It carries an external factored ultimate uniform load w_u = 220 psf (10.5 kPa), including its self-weight. It is simply supported on one long edge and the adjacent short edge and built in on the opposite edges. Let the reinforcement spanning the short direction be twice the reinforcement spanning the long direction. Also assume the reinforcement on the built-in edges to be equal to the strong reinforcement. Design the slab structure for flexure, including the reinforcement needed and its spacing, using the yield-line theory. Given:

     f_c' = 4000 psi (27.6 MPa), normal-weight concrete
     f_y = 60,000 psi (414 MPa)

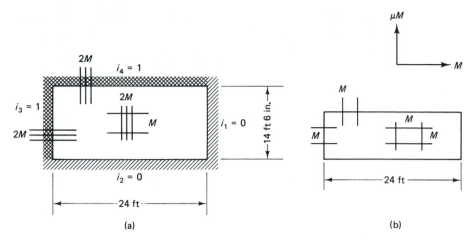

Figure 11.35 Slab geometry: (a) orthotropic; (b) affine.

Solution μ = ratio of reinforcement in the strong direction to the weak direction = 2. From Eq. 11.28, the expression for the unit moment in an affine rectangular slab is

$$M_u = \frac{w_u a_r^2}{24}\left[\sqrt{3 + \left(\frac{a_r}{b_r}\right)^2} - \frac{a_r}{b_r}\right]^2$$

Change to affine slab converting the span dimension:

$$a = 14.5 \times \frac{1}{\sqrt{\mu}} = 14.5 \times \frac{1}{\sqrt{2}} = 10.25 \text{ ft}$$

$$a_r = \frac{2a}{\sqrt{i_2 + 1} + \sqrt{i_4 + 1}} = \frac{2 \times 10.25}{\sqrt{0 + 1} + \sqrt{1 + 1}}$$

$$= \frac{20.50}{2.414} = 8.492$$

$$b_r = \frac{2b}{\sqrt{1 + i_1} + \sqrt{1 + i_3}} = \frac{2 \times 24.0}{\sqrt{1 + 0} + \sqrt{1 + 1}}$$

$$= \frac{48.0}{2.414} = 19.884$$

$$\frac{a_r}{b_r} = \frac{8.492}{19.884} = 0.427$$

$$w_n = \frac{w_u}{\phi = 0.9} = 244 \text{ psf}$$

$$M_n = \frac{w_n(8.492)^2}{24}[\sqrt{3 + (0.427)^2} - 0.427]^2$$

$$= \frac{w_n \times 72.11}{24}(1.841) = 5.532w_n = 1350 \text{ lb}$$

$$\mu M_n = 2 \times 1350 = 2700 \text{ lb or ft-lb/ft}$$

Assume that $d = 4$ in. $(h = 5$ in.). $M_n = d^2\rho f_y(1 - 0.59\omega)$, where $\omega = \rho(f_y/f_c')$ (see Chapter 5); or

$$2700 = (4)^2\rho \times 60,000\left(1 - 0.59\rho\frac{60,000}{4000}\right)$$

to get $\rho = 0.00289 = 0.289\%$ in the short direction.

reinforcement A_s on 12-in. (305-mm) strip $= 0.00289 \times 4 \times 12 = 0.139$ in.^{2}/12 in.

This steel area is less than that of the balanced ratio $0.75\bar{\rho}_b$; hence O.K.   No. 3 bars at $9\frac{1}{2}$ in. center to center $\simeq 0.139$ in^2.   Hence use No. 3 bars at $9\frac{1}{2}$ in. center to center in the short direction and on the tension top face of the fixed supports (9.53-mm bars at 241 mm center to center).

maximum allowable spacing $s = 2h = 2 \times 5 = 10$ in. (254 mm)

Percentage of reinforcement in the long direction gives $\rho = 0.00171 = 0.171\%$. Use No. 3 bars at 10 in. center to center for the long direction.

Alternate affine slab in the perpendicular direction

Multiply the M direction by $\sqrt{\mu}$ to get μM in both directions.   Affine $b = 24.0\sqrt{2} = 33.9$ ft.

$$a_r = \frac{2 \times 14.5}{2.414} = 12.0 \text{ ft} \qquad b_r = \frac{2 \times 33.9}{2.414} = 28.1 \text{ ft}$$

$$\frac{a_r}{b_r} = \frac{12.0}{28.1} = 0.427 \qquad \text{(same in the preceding solution)}$$

Hence

$$M_n = \frac{w_n(12.0)^2}{24}[\sqrt{3 + (0.427)^2} - 0.427]^2$$

$$= 6w_n(1.839) = 2700 \text{ lb}   \text{(as before)}$$

A check of the thickness for minimum deflection and crack-control requirements would have to be made before the design is complete.

11.10.4 Example 11.9: Moment Capacity and Yield-line Design of a Triangular Balcony Slab

The balcony floor in Fig. 11.36 is triangular in shape and is supported at the two perpendicular sides and carries a factored uniform line load of intensity $p = 400$ lb per linear foot (5.84 kN/m) acting on the triangle hypotenuse.   The reinforcement in the short direction is three times the reinforcement in the long direction ($\mu = 3.0$). Analyze this floor moment capacity by the yield-line theory and design the thickness and reinforcement spacing in both directions if the longer of the two perpendicular supports is 16 ft (4.88 m) and it subtends an angle of 30° with the hypotenuse.   The shorter side is 9.24 ft (2.82 m).   Given:

$f_c' = 4000$ psi (27.58 kPa), normal-weight concrete
$f_y = 60,000$ psi (413.7 kPa)

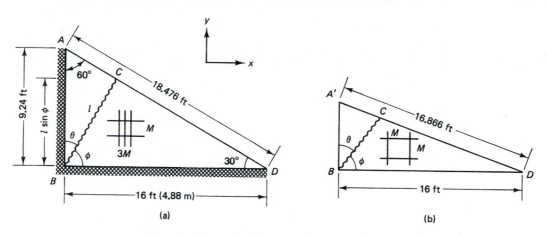

Figure 11.36 Balcony floor plate: (a) floor geometry; (b) affine slab.

Assume that the self-weight of the plate can be neglected in the solution on the assumption that it is small compared to the line load.

Solution

$$AB = 16 \tan 30° = 9.24 \text{ ft}$$

$$AD = \sqrt{16^2 + 9.24^2} = 18.48 \text{ ft}$$

The yield line at failure is expected to be line BC subtending an angle ϕ with side BD. Its components in the x and y directions are $l \cos \phi$ and $l \sin \phi$, respectively.

Assume that point C deforms by a magnitude Δ, causing the center of gravity of the load on segment DC at its center line to deflect $\frac{1}{2}\Delta$. Summing vectorially, the work in the x and y directions in Fig. 11.36 gives

$$E_I = 3Ml \cos \phi \frac{\Delta}{l \sin \phi} + Ml \sin \phi \frac{\Delta}{l \cos \phi}$$

$$E_E = 18.48p \times \frac{\Delta}{2} = 9.24p$$

Since $E_I = E_E$,

$$\frac{p}{M} = 0.324 \cot \phi + 0.108 \tan \phi$$

$$\frac{d(p/M)}{d\phi} = -0.324 \csc^2 \phi + 0.108 \sec^2 \phi = 0$$

$$\frac{\sec^2 \phi}{\csc^2 \phi} = 3.0 \quad \text{or} \quad \tan \phi = 1.732 \quad \phi_{\min} = 60°$$

Hence

$$p = M\left(0.324 \times \frac{1}{1.732} + 0.108 \times 1.732\right) = 0.375M$$

Affine slab solution

$A'B$ in Fig. 11.36b becomes $9.24/\sqrt{3} = 5.335$ ft.   Free edge $A'D$ becomes 16.866 ft.   If

$$p' = \text{affine linear load} = \frac{p}{\sqrt{\mu \cos^2 \theta + \sin^2 \theta}}$$

where $\theta = $ angle between the line load and M direction, or

$$p' = \frac{p}{\sqrt{3} \cos^2 30 + \sin^2 30} = 0.632p$$

$$E_I = \Delta M \cot \phi + \Delta M \tan \phi$$

$$E_E = p' \times 16.866 \times \frac{\Delta}{2} = 8.433p' \, \Delta$$

But $E_I = E_E$; therefore,

$$\frac{p'}{M} = 0.1186 \cot \phi + 0.1186 \tan \phi$$

$$\phi_{\min} = 45° \qquad \tan \phi = 1.0$$

Therefore,

$$p' = 2 \times 0.1186M = 0.237M$$

or
$$p = \frac{0.237}{0.632} M = 0.375M \quad \text{(as before)}$$

Design of reinforcement

$400 = 0.375M$ to give $M = 1067$ lb or $\mu M = 3 \times 1067 = 3200$ lb or ft-lb/ft. Assume that $h = 5$ in. ($d = 4$ in. $= 100$ mm).

$$3200 = d^2 \rho f_y (1 - 0.59\omega)$$

or
$$3200 = 4^2 \rho \times 60{,}000 \left(1 - 0.59\rho \times \frac{60{,}000}{4000} \right)$$

to give $\rho = 0.0034$.   The required $A_s = 0.0034 \times 4 \times 12 = 0.163$ in.^{2}.   Use No. 3 bars at 8 in. center to center in the short direction $\simeq 0.165$ in.^{2} (9.52-mm diameter at 165 mm center to center) and No. 3 bars at $2h = 9$ in. center to center in the long direction to satisfy moment requirements.

SELECTED REFERENCES

11.1. ACI Committee 318, *Building Code Requirements for Reinforced Concrete*, ACI Standard 318–95; and the *Commentary on Building Code Requirements for Reinforced Concrete*, American Concrete Institute, Detroit, 1995.

11.2. CEB-FIP, *Concrete Design—U.S. and European Practices*, Joint ACI-CEB Symposium, Bulletin d'Information 113, Comité Euro-International du Béton, Paris, February 1990.

11.3. Gamble, W. L., Sozen, M. A., and Siess, C. P., *An Experimental Study of Reinforced Concrete Two-Way Floor Slab*, Civil Engineering Studies, Structural Research Series 211, University of Illinois, Urbana, Ill., 1961, 304 pp.

11.4. Corley, W. G., and Jirsa, J. D., "Equivalent Frame Analysis for Slab Design," *Journal of the American Concrete Institute*, Proc. Vol. 67, No. 11, November 1970, pp. 875–884.

11.5. Wang, C. K., and Salmon, C. G., *Reinforced Concrete Design*, 3rd ed., Harper & Row, New York, 918 pp.

11.6. Branson, D. E., *Deformation of Concrete Structures*, McGraw-Hill, New York, 1977, 546 pp.

11.7. Cross, H., and Morgan, N., *Continuous Frames of Reinforced Concrete*, Wiley, New York, 1954.

11.8. Rice, P. F., Hoffman, E. S., Gustafson, D. P., and Gouwens, A. J., *Structural Design Guide to the ACI Building Code*, 3rd ed., Van Nostrand Reinhold, New York, 1985, 477 pp.

11.9. Reynolds, C. E., and Steedman, J. C., *Reinforced Concrete Designer's Handbook*, 9th ed., Viewpoint Publications, London, 1981, 505 pp.

11.10. Nawy, E. G., *Prestressed Concrete: A Fundamental Approach*, 2nd ed., Prentice Hall, Englewood Cliffs, N.J., 1996, 800 pp.

11.11. Nilson, A. H., and Walters, D. B., "Deflection of Two-Way Floor Systems by the Equivalent Frame Method," *Journal of the American Concrete Institute*, Proc. Vol. 72, No. 5, May 1975, pp. 210–218.

11.12. Nawy, E. G., and Chakrabarti, P., "Deflection of Prestressed Concrete Flat Plates," *Journal of the Prestressed Concrete Institute*, Vol. 21, No. 2, March–April 1976, pp. 86–102.

11.13. Mansfield, E. H., "Studies in Collapse Analysis of Rigid-Plastic Plates with a Square Yield Diagram," *Proceedings of the Royal Society*, Vol. 241, August 1957, pp. 311–338.

11.14. Wood, R. H., *Plastic and Elastic Design of Slabs and Plates*, Thames and Hudson, London, 1961, pp. 225–261.

11.15. Johansen, K. W., *Yield-Line Theory*, Cement and Concrete Association, London, 1962, 181 pp.

11.16. Hognestad, E., "Yield-Line Theory for the Ultimate Flexural Strength of Reinforced Concrete Slabs," *Journal of the American Concrete Institute*, Proc. Vol. 49, No. 7, March 1953, pp. 637–655.

11.17. Hung, T. Y., and Nawy, E. G., "Limit Strength and Serviceability Factors in Uniformly Loaded, Isotropically Reinforced Two-Way Slabs," *Symposium on Cracking, Deflection, and Ultimate Load of Concrete Slab Systems*, Special Publication SP-30, American Concrete Institute, Detroit, 1972, pp. 301–324.

11.18. Nawy, E. G., and Blair, K. W., "Further Studies on Flexural Crack Control in Structural Slab Systems," *Symposium on Cracking, Deflection, and Ultimate Load of Concrete Slab Systems*, Special Publication SP-30, American Concrete Institute, Detroit, 1972, pp. 1–41.

11.19. Nawy, E. G., "Crack Control through Reinforcement Distribution in Two-Way Acting Slabs and Plates," *Journal of the American Concrete Institute*, Proc. Vol. 69, No. 4, April 1972, pp. 217–219.

11.20. ACI Committee 340, *Design Hand Book: Beams, Slabs, Etc.*, Special Publication SP-17 (81), Vol. 1, American Concrete Institute, Detroit, 1991, 508 pp.

11.21. Hawkins, N., and Corley, W. G., "Transfer of Unbalanced Moment and Shear from Plates to Columns," *Symposium on Cracking, Deflection, and Ultimate Load of Concrete Slab Systems*, Special Publication SP-30, American Concrete Institute, Detroit, 1972, pp. 147–176.

11.22. Park, R., and Gamble, W. L., *Reinforced Concrete Slabs*, Wiley, 1980, 618 pp.

11.23. Warner, R. F., Rangan, B. V., and Hall, H. S., *Reinforced Concrete*, Pitman Australia, London, 1982, 471 pp.

11.24. Standards Association of Australia, *SAA Concrete Structures Code, 1985*, SAA, Sydney, 1985, pp. 1–158.

11.25. Nawy, E. G., "Strength, Serviceability and Ductility," Chapter 12 in *Handbook of Structural Concrete*, Pitman Books, London/McGraw-Hill, 1983, 1968 pp.

11.26. Simmonds, S. H., and Janko, M., "Design Factors for Equivalent Frame Method," *Journal of the American Concrete Institute*, Proc. Vol. 68, No. 11, November 1971, pp. 825–831.

11.27. Nawy, E. G., *High Strength, High Performance Concrete*, Longman, London, Wiley, New York 1996, pp. 400.

PROBLEMS FOR SOLUTION

11.1. An end panel of a floor system supported by beams on all sides carries a uniform service live load $w_L = 75$ psf and an external dead load $w_D = 20$ psf in addition to its self-weight. The center-line dimensions of the panel are 18 ft × 20 ft (the dimension of the discontinuous side is 18 ft). Design the panel and the size and spacing of the reinforcement using the ACI Code direct design method. Given:

$f'_c = 4000$ psi, normal-weight concrete

$f_y = 60,000$ psi

column sizes at each corner 12 in. × 12 in.

interior column sizes are 12 in. × 14 in.

width of the supporting beam webs = 12 in.

assume reinforcement ratio $\rho \simeq 0.4\, \rho_b$ for the supporting beams.

11.2. Solve Problem 11.1 by the direct design method and by the equivalent frame method of the ACI Code assuming that the floor is a flat plate system with no edge beams.

11.3. Determine the size and spacing of reinforcement in the N–S direction for the floor slab in Ex. 11.3 for both the column and the middle strips.

11.4. Solve for the factored moments for the floor slab on beams in Ex. 11.2 using Fig. 11.23.

11.5. An interior flat-plate panel is supported on columns spaced 18 ft × 20 ft.   The panel dimensions, loading, and material properties are the same as those in Problem 11.1. Design the panel and size and spacing of the reinforcement by the ACI Code direct design method.

11.6. Calculate the time-dependent deflection at the center of the panel in (a) Problem 11.1 and (b) Problem 11.5.   Check also if the panels satisfy the serviceability requirements for deflection control and crack control for aggressive environment. Assume that $K_{ec}/E_c = 350$ in.^{3} per radian for part (a) and $= 225$ in.^{3} per radian for part (b).

11.7. Use the yield-line theory to evaluate the slab thickness needed in the column zone of the flat plate in Problem 11.5 for flexure, assuming that the hinge field would have a radius of 24 in.

11.8. An isotropically reinforced long strip is simply supported on the edges.   A concentrated load P acts on the minor axis of the slab midway between the long edges. Prove that the magnitude of the collapse load is $P = M(4 + 2\pi)$.

11.9. A slab 21 ft × 13 ft 6 in. carries an external factored ultimate load of 200 lb per square foot, including its self-weight.   It is simply supported on one long edge and the adjacent short edge and built in on the opposite edges.   Let the reinforcement spacing the short way be three times the reinforcement spanning the long way.   Also assume the reinforcement on the built-in edges to be equal to the strong reinforcement.   Design the slab structures and the reinforcement needed and its spacing using the yield-line theory.   Given:

$$f_c' = 4000 \text{ psi, normal-weight concrete}$$
$$f_y = 60,000 \text{ psi}$$

11.10. Calculate the maximum crack width in a two-way interior panel of a reinforced concrete floor system.   The slab thickness is 8 in. (203.2 mm) and the panel size is 20 ft × 28 ft (6.10 m × 8.53 m).   Also design the size and spacing of the reinforcement necessary for crack control assuming that (a) the floor is exposed to normal environment; (b) the floor is part of a parking garage.   Given: $f_y = 60.0$ ksi (414 MPa).

FOOTINGS

12

12.1 INTRODUCTION

Cumulative floor loads of a superstructure are supported by foundation substructures in direct contact with the soil. The function of the foundation is to transmit safely the high concentrated column and/or wall reactions or lateral loads from earth-retaining walls to the ground without causing unsafe differential settlement of the supported structural system or soil failure.

If the supporting foundations are not adequately proportioned, one part of a structure can settle more than an adjacent part. Various members of such a system become overstressed at the column–beam joints due to *uneven* settlement of the supports leading to large deformations. The additional bending and torsional moments in excess of the resisting capacity of the members can lead to excessive cracking due to yielding of the reinforcement and ultimately to failure.

If the total structure undergoes *even* settlement, little or no overstress occurs. Such behavior is observed when the foundation is excessively rigid and the supporting soil highly yielding such that a structure behaves similar to a floating body

Photo 72 One Shell Plaza, New Orleans. (Courtesy of Portland Cement Association.)

564

that can sink or tilt without breakage. Numerous examples of such structures can be found in such locations as Mexico City with buildings on mat foundations or rigid supports that sank several feet over the years due to the high consolidation of the supporting soil. Examples of other famous cases of very slow and relatively uneven consolidation process can be cited. Gradual loss of stability of a structure undergoing tilting with time, like the leaning Tower of Pisa, is an example of foundation problems resulting from uneven bearing support.

Layouts of structural supports vary widely and soil conditions differ from site to site and within a site. As a result, the type of foundation to be selected has to be governed by these factors and by optimal cost considerations. In summary, the structural engineer has to acquire the maximum economically feasible soil data on the site before embarking on a study of the various possible alternatives for site layout.

Basic knowledge of soil mechanics and foundation engineering is assumed in presenting the topic of design of footings in this chapter. Background knowledge

TABLE 12.1 Presumptive Bearing Capacity (tons/ft²)

Type of Soil	Bearing Capacity
Massive crystalline bedrock, such as granite, diorite, gneiss, and trap rock	100
Foliated rocks, such as schist or slate	40
Sedimentary rocks, such as hard shales, sandstones, limestones, and siltstones	15
Gravel and gravel–sand mixtures (GW and GP soils)	
Densely compacted	5
Medium compacted	4
Loose, not compacted	3
Sands and gravely sands, well graded (SW soil)	
Densely compacted	$3\frac{3}{4}$
Medium compacted	3
Loose, not compacted	$2\frac{1}{4}$
Sands and gravely sands, poorly graded (SP soil)	
Densely compacted	3
Medium compacted	$2\frac{1}{4}$
Loose, not compacted	$1\frac{3}{4}$
Silty gravels and gravel–sand–silt mixtures (GM soil)	
Densely compacted	$2\frac{1}{4}$
Medium compacted	2
Loose, not compacted	$1\frac{1}{2}$
Silty sand and silt-sand mixtures (SM soil)	2
Clayey gravels, gravel–sand–clay mixtures, clayey sands, sand–clay mixtures (GC and SC soils)	2
Inorganic silts, and fine sands; silty or clayey fine sands and clayey silts, with slight plasticity; inorganic clays of low to medium plasticity; gravely clays; sandy clays; silty clays; lean clays (ML and CL soils)	1
Inorganic clays of high plasticity, fat clays; micaceous or diatomaceous fine sand or silty soils, elastic silts (CH and MH soils)	1

of the methodology of determining the resistance of cohesive and noncohesive soils is necessary to select the appropriate bearing capacity value for the particular site and the particular foundation system under consideration.

The bearing capacity of soils is usually determined by borings, test pits, or other soil investigations. If these are not available for the preliminary design, representative values at the footing level can normally be used from Table 12.1.

12.2 TYPES OF FOUNDATIONS

There are basically six types of foundation substructures, as shown in Fig. 12.1. The foundation area must be adequate to carry the column loads, the footing weight, and any overburden weight within the permissible soil pressure.

1. *Wall footings.* Such footings comprise a continuous slab strip along the length of the wall having a width larger than the wall thickness. The projection of the slab footing is treated as a cantilever loaded up by the distributed soil pressure. The length of the projection is determined by the soil bearing pressure, with the critical section for bending being at the face of the wall. The main reinforcement is placed perpendicular to the wall direction.

2. *Independent isolated column footings.* These consist of rectangular or square slabs of either constant thickness or sloping toward the cantilever tip. They are reinforced in both directions and are economical for relatively small loads or for footings on rock.

3. *Combined footings.* Such footings support two or more column loads. They are necessary when a wall column has to be placed on a building line and the footing slab cannot project outside the building line. In such a case, an independent footing would be eccentrically loaded, causing apparent tension on the foundation soil.

    In order to achieve a relatively uniform stress distribution, the footing for the exterior wall column can be combined with the footing of the adjoining interior column. Additionally, combined footings are also used when the distance between adjoining columns is relatively small, such as in the case of corridor columns, when it becomes more economical to build a combined footing for the closely spaced columns.

4. *Cantilever or strap footings.* These are similar to the combined footings, except that the footings for the exterior and interior columns are built independently. They are joined by a strap beam to transmit the effect of the bending moment produced by the eccentric wall column load to the interior column footing area.

5. *Pile foundations.* This type of foundation is essential when the supporting ground consists of structurally unsound layers of material to large depths. The piles may be driven either to solid bearing on rocks or hardpan or deep enough into the soil to develop the allowable capacity of the pile through

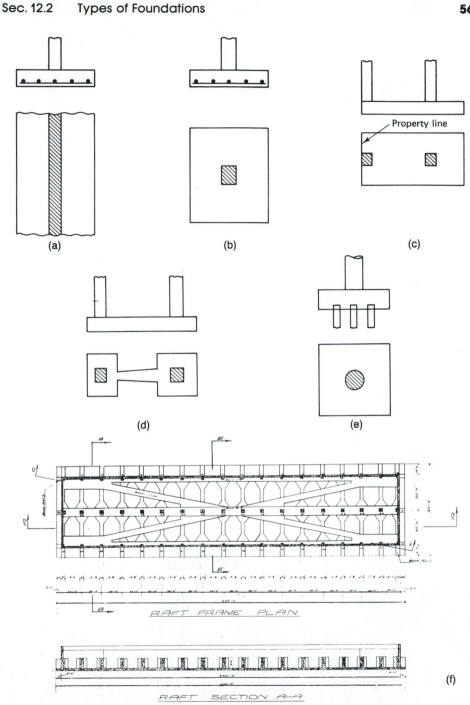

Figure 12.1 Types of foundations: (a) wall footing; (b) isolated footing; (c) combined footing; (d) strap footing; (e) pile foundation; (f) raft foundation.

skin frictional resistance or a combination of both. The piles could be either precast, and hence driven into the soil, or cast in place by drilling a caisson and subsequently filling it with concrete. The precast piles could be reinforced or prestressed concrete. Other types of piles are made of steel or treated wood. In all types, the piles have to be provided with appropriately designed concrete caps reinforced in both directions.

6. *Raft, mat, or floating foundations.* Such foundation systems are necessary when the allowable bearing capacity of soil is very low to great depths, making pile foundations uneconomical. In this case it becomes necessary to have a deep enough excavation with sufficient depth of soil removed that the net bearing pressure of the soil on the foundation is almost equivalent to the structure load. It becomes necessary to spread the foundation substructure over the entire area of the building such that the superstructure is considered to be theoretically floating on a raft. Continuously consolidating soils require such a substructure, which is basically an inverted floor system. Otherwise, friction piles or piles driven to rock become mandatory.

12.3 SHEAR AND FLEXURAL BEHAVIOR OF FOOTINGS

To simplify foundation design, footings are assumed to be rigid and the supporting soil layers elastic. Consequently, uniform or uniformly varying soil distribution can be assumed. The net soil pressure is used in the calculation of bending moments and shears by subtracting the footing weight intensity and the surcharge from the total soil pressure. If a column footing is considered as an *inverted* floor segment where the intensity of net soil pressure is considered to be acting as a column-supported cantilever slab, the slab would be subjected to both bending and shear in a similar manner to a floor slab subjected to gravity loads.

Photo 73 Reinforced concrete footing after excavation. (Tests by F. E. Richart.)

When heavy concentrated loads are involved, it has been found that shear rather than flexure controls most foundation designs. The mechanism of shear failure in footing slabs is similar to that in supported floor slabs. However, the shear capacity is considerably higher than that of beams, as will be discussed in the next section. Since the footing in most cases bends in double curvature, shear and bending about both principal axes of the footing plan have to be considered.

The state of stress at any element in the footing is due primarily to the combined effects of shear, flexure, and axial compression. Consequently, a basic understanding of the fundamental behavior of the footing slab and the cracking mechanism involved is essential. It enables developing a background feeling for the underlying hypothesis used in the analysis and design requirements of footings both in shear and in flexure.

12.3.1 Failure Mechanism

The inclined shear cracks develop in essentially the same manner as in beams, stabilizing at approximately 65% of the ultimate load and extending rapidly toward the neutral axis. Thereafter, the cracks propagate slowly toward the compression zone such that a very shallow depth in compression remains at failure.

The inclined cracks always form close to the concentrated load or column reaction in two-way slabs or footings, as seen in Fig. 12.2a. This is due partly to the heavy concentration of bending moments in the region close to the column face, forming a truncated pyramid at the foot of the column region. The column can perimetrically punch through the slab in this failure form if the slab is not adequately designed to resist shear failure (also called *diagonal tension* or *punching shear*). The action of the confining surrounding punched slab on the column base interface punching through the slab in Fig. 12.2b can be represented by the resulting shear forces V_1 and V_2, the compressive forces C_1 and C_2, and the tensile forces T_1 and T_2, in addition to the internal dowel and membrane action of the slab.

Figure 12.2c shows an infinitesimal element taken from the compression zone above the inclined crack. The element is subjected to the following four stress components: (1) vertical shear stress v_0, (2) direct compressive stress f_c, (3) vertical compressive stress f_3, and (4) lateral compressive stress f_2.

The vertical shearing stress v_0 is the result of the total shear that has to be entirely transmitted by the compression zone above the inclined crack. The direct compressive stress f_c, which varies along the length of the critical section, results from the bending moments. The vertical compressive force f_3 is due to the heavy concentrated column load. It has a major influence on increasing the shear capacity of the slab, as demonstrated in Ref. 12.2 for pressure in an infinite semielastic solid loaded at the surface. The lateral compressive stress f_2 is the result of the bending moment about an axis perpendicular to the critical section. It contributes further to the increase in the compressive strength of the concrete as a result of the triaxial state of stress. Consequently, the existence of the multiaxial forces

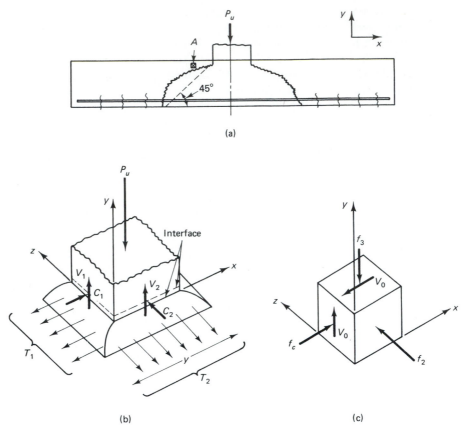

Figure 12.2 Two-way-action failure mechanism in slabs and footings: (a) footing elevation; (b) failure pyramid; (c) element *A* in the compression zone.

and stresses in Fig. 12.2c explains why the shear capacity of a slab subjected to concentrated loads in considerably higher than that of a beam.

In addition, the inclined crack generating *close* to the critical section in two-way slabs and footings due to the high moment concentration justifies considering the critical section to be at a distance of $d/2$ from the face of the column in slabs and footings, while in beams and one-way slabs and footings, the ACI Code specifies the critical section at a distance d from the face of the column support. The nominal shearing stress at failure varies between $6\sqrt{f_c'}$ and $9\sqrt{f_c'}$ for the footing slabs, whereas it does not exceed $2\sqrt{f_c'}$ to $4\sqrt{f_c'}$ in beams. The Code, however, allows a maximum nominal resisting shear strength of plain concrete not to exceed $v_c = 4\sqrt{f_c'}$ for the supported two-way slab or the footing and $v_c = 2\sqrt{f_c'}$ for beams and one-way-action footings. For plain concrete footings cast against soil, the effective thickness used in computing stresses is taken as the overall thickness minus 3 in. The overall thickness should not be less than 8 in.

12.3.2 Loads and Reactions

Based on the foregoing discussion, it is essential to make the correct assumptions for evaluating all the combined forces acting on the foundation. The footing slab has to be proportioned to sustain all the applied factored loads and induced reactions, which include axial loads, shears, and moments to be resisted at the base of the footing.

After the permissible soil pressure is determined from the available site data and the principles of soil mechanics and the local codes, the footing area size is computed on the basis of the *unfactored* (service) loads, such as dead, live, wind, or earthquake loads in whatever combination governs the design.

The minimum eccentricity requirement for column slenderness considerations is neglected in the design of footings or pile caps and only the computed end moments that exist at the base of a column are considered to have been transferred to the footing. In cases where eccentric loads or moments exist due to any loading combinations, the extreme soil pressure resulting from such loading conditions has to be within such permissible bearing values such as those in Table 12.1 or as determined by actual soil tests.

Once the size of a footing or pile cap for a single pile or a group of piles is determined, the design of the footing geometry becomes possible using the principles and methodologies presented in the preceding chapters for shear and flexure design. The external *service* loads and moments used to determine the size of the foundation area are converted to their ultimate *factored* values using the appropriate load factors and strength reduction factors ϕ for determining the nominal resisting values to be used in the analysis and proportioning the size and reinforcement distribution in the footing.

12.4 SOIL BEARING PRESSURE AT BASE OF FOOTINGS

The distribution of soil bearing pressure on the footing depends on the manner in which the column or wall loads are transmitted to the footing slab and the degree of rigidity of the footing. The soil under the footing is assumed to be a homogeneous elastic material, and the footing is assumed to be rigid as a most common type of foundation. Consequently, the soil bearing pressure can be considered uniformly distributed if the reaction load acts through the axis of the footing slab area. If the load is not axial or symmetrically applied, the soil pressure distribution becomes trapezoidal due to the combined effects of axial load and bending.

12.4.1 Eccentric Load Effect on Footings

As indicated in Section 12.2, exterior column footings and combined footings can be subjected to eccentric loading. When the eccentric moment is very large, tensile stress on one side of the footing can result, since the bending stress distribution depends on the magnitude of load eccentricity. It is always advisable to proportion the area of these footings such that the load falls within the middle kern, as shown in Figs. 12.3 and 12.4. In such a case, the location of the load is in the middle third of the footing dimension in each direction, thereby avoiding tension in the soil that can theoretically occur prior to stress redistribution.

1. *Eccentricity case $e < L/6$* (Fig. 12.3a). In this case, the direct stress P/A_f is larger than the bending stress M_c/I. The stress

$$p_{max} = \frac{P}{A_f} + \frac{Pe_1 c}{I} \tag{12.1a}$$

$$p_{min} = \frac{P}{A_f} - \frac{Pe_1 c}{I} \tag{12.1b}$$

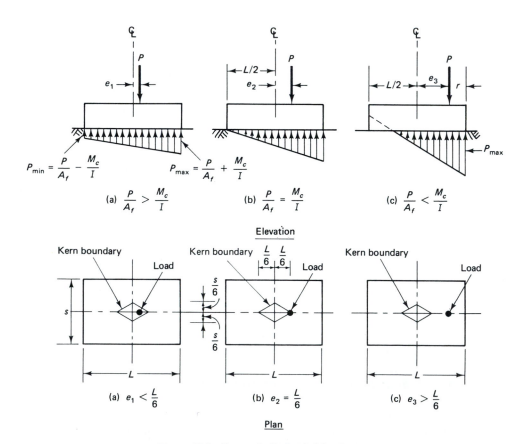

Figure 12.3   Eccentrically loaded footings.

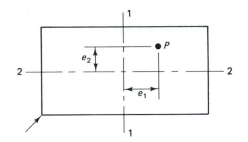

Figure 12.4 Biaxial loading of footing.

2. *Eccentricity case $e_2 = L/6$ (Fig. 12.3b):*

$$\text{direct stress} \quad = \frac{P}{A_f} = \frac{P}{sL} \tag{12.2a}$$

$$\text{bending stress} = \frac{M_c}{I} = Pe_2 \times \frac{c}{I} \tag{12.2b}$$

$$\frac{c}{I} \qquad = \frac{L/2}{s(L^3/12)} = \frac{1}{s(L^2/6)} = \frac{6}{sL^2} \tag{12.2c}$$

where s and L are the width and the length of the footing, respectively. In order to find the limiting case where *no* tension exists on the footing, the direct stress P/A_f has to be equivalent to the bending stress so that

$$\frac{P}{A_f} - Pe_2\frac{c}{I} = 0 \tag{12.2d}$$

Substituting for P/A_f and C/I from Eqs. 12.2a and c into Eq. 12.2d,

$$\frac{P}{sL} - Pe_2 \times \frac{6}{sL^2} = 0 \quad \text{or} \quad e_2 = \frac{L}{6}$$

Consequently, the eccentric load has to act within the middle third of the footing dimension to avoid tension on the soil.

3. *Eccentricity case $e_3 > L/6$ (Fig. 12.3c).* As the load acts outside the middle third, tensile stress results at the left side of the footing, as shown in Fig. 12.3c. If the maximum bearing pressure p_{max} due to load P does not exceed the allowable bearing capacity of the soil, no uplift is expected at the left end of the footing, and the center of gravity of the triangular bearing stress distribution *coincides* with the point of action of load p in Fig. 12.3c.

The distance from the load P to the tip of footing is $r = (L/2) - e_3 =$ distance of the centroid of the stress triangle from the base of the triangle. Therefore, the width of the triangle is $3r = 3[(L/2) - e_3]$. Hence the maximum compressive bearing stress is

$$p_{max} = \frac{P}{\dfrac{3r \times s}{2}} = \frac{2P}{3s\left(\dfrac{L}{2} - e_3\right)} \tag{12.3a}$$

4. *Eccentricity about two axes, biaxial loading* (Fig. 12.4).   In the case where a concentrated load has an eccentricity in two directions (both within their respective kern points), the stresses are

$$p_{max} = \frac{P}{A_f} \pm \frac{Pe_1c_1}{I_1} \pm \frac{Pe_2c_2}{I_2} \qquad (12.3b)$$

12.4.2 Example 12.1: Concentrically Loaded Footings

A column support transmits axially a total service load of 400,000 lb (1779 kN) to a square footing at the frost line (3 ft below grade), as shown in Fig. 12.5.   The frost line is the subgrade soil level below which the groundwater does not freeze throughout the year.   Test borings indicate a densely compacted gravel–sand soil.   Determine the required area of the footing and the net soil pressure intensity p_n to which it is subjected.   Given:

> unit weight of soil γ = 135 lb/ft^3 (21.1 kN/m^3)
> footing slab thickness = 2 ft (0.61 m)

Solution   Since the footing is concentrically loaded, the soil bearing pressure is considered uniformly distributed assuming the footing is rigid.   From the soil test borings and Table 12.1, the presumptive bearing capacity of the soil is 5 tons/ft^2 at the level of the footing, that is, 10,000 lb/ft^2 (478.8 kPa).   Assume that the average weight of the soil and concrete above the footing is 135 pcf.   Since the top of the footing has to be below the frost line (minimum 3 ft below grade), the net allowable pressure is

$$p_n = 10,000 - (5 \times 135 + 100 \text{ psf for surcharge paving}) = 9225 \text{ psf}$$

$$\text{minimum area of footing } A_f = \frac{400,000}{9,225} = 43.36 \text{ ft}^2$$

Use square footing 6 ft 8 in. × 6 ft 8 in. (2.03 m × 2.03 m):

$$A_f = 44.44 \text{ ft}^2 (4.13 \text{ m}^2) > 43.36 \text{ ft}^2$$

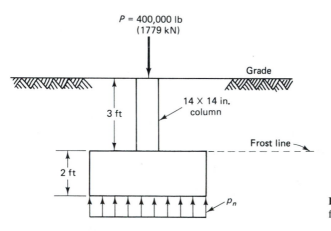

Figure 12.5   Concentrically loaded footing.

12.4.3 Example 12.2: Eccentrically Loaded Footings

A reinforced concrete footing supports a 14 in. × 14 in. column reaction $P = 400,000$ lb (1779 kN) at the frost line (3 ft below grade). The load acts at an eccentricity $e_1 = 0.4$ ft, $e_2 = 1.3$ ft, and $e_3 = 2.2$ ft. Select the necessary area of footing assuming that it is rigid and has a thickness $h = 2\frac{1}{2}$ ft. Soil test borings have indicated that the bearing area is composed of layers of shale and clay to a considerable depth below the foundation. Use a unit weight $\gamma = 140$ lb/ft³.

Solution From Table 12.1, assume an allowable bearing capacity $p_g = 6.5$ tons/ft² (13,000 lb/ft²) at the footing base level.

Eccentricity $e_1 = 0.4$ ft

By trial and adjustment, assume a footing 5 ft × 9 ft (1.52 m × 2.74 m), $A_f = 45$ ft². Assume that the footing base is 6 ft below grade and that a slab on grade surcharge weighs 120 psf. Assume that the average weight of the soil and footing is ≈ 140 pcf.

$$\text{net allowable bearing pressure } p_n = 13,000 - (6 \times 140 + 120)$$
$$= 12,040 \text{ lb/ft}^2 \text{ (576.5 KPa)}$$

Stress due to the service eccentric column load is

$$p = \frac{P}{A_f} \pm \frac{P \times e}{I/c} = \frac{400,000}{45} \pm \frac{400,000 \times 0.4 \times 6}{5(9)^2}, \qquad \text{where } \frac{I}{c} = \frac{bh^2}{6}$$
$$= 8889 \pm 2370 = 11,259 \text{ lb/ft}^2 \text{ (C) and } 6519 \text{ lb/ft}^2 \text{ (C)} < 12,040 \text{ lb/ft}^2$$

The distribution of the bearing pressure is as shown in Fig. 12.6a; therefore, O.K.

Eccentricity $e_2 = 1.3$ ft

By trial and adjustment, assume a footing 6 ft × 10 ft (1.83 m × 3.05 m), $A_f = 60$ ft² (5.57 m²). The actual service load-bearing pressure is

$$p = \frac{400,000}{60.0} \pm \frac{400,000 \times 1.3 \times 6}{6(10)^2}$$
$$= 6667 \pm 5200 = 11,867 \text{ lb/ft}^2 \text{ (C) and } 1467 \text{ lb/ft}^2 \text{ (C)}$$
$$< 12,040 \text{ lb/ft}^2 \qquad \text{therefore, O.K.}$$

Notice in comparing the two cases that as the moment increases leading to larger eccentricities, the minimum bearing pressure decreases, as seen from Fig. 12.6a and b.

Eccentricity $e_3 = 2.2$ ft

By trial and adjustment, try a footing 7 ft × 11 ft (2.13 m × 3.35 m), $A_f = 77$ ft² (7.15 m²).

$$p = \frac{400,000}{77.0} \pm \frac{400,000 \times 2.2 \times 6}{7(11)^2}$$
$$= 5195 \pm 6234 = 11,429 \text{ lb/ft}^2 \text{ (C) and } -1039 \text{ lb/ft}^2 \text{ (T)}$$

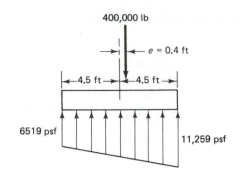

(a)

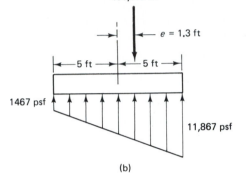

(b)

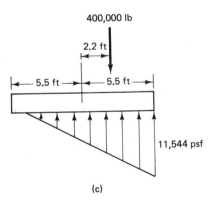

(c)

Figure 12.6   Bearing area and bearing
stress distribution in Ex. 12.2.

Check by Eq. 12.3 for $e > L/6 > 11.0/6 = 1.83$ ft:

$$p = \frac{2P}{3S[(L/2) - e_3]} = \frac{2 \times 400{,}000}{3 \times 7[(11/2) - 2.2]}$$

$$= 11{,}544 \text{ lb/ft } 2 < 12{,}040 \text{ lb/ft}^2 \qquad \text{O.K.}$$

Figure 12.6c shows that as the load acts outside the middle third of the base only part
of the footing is subjected to compressive bearing stress.

12.5 DESIGN CONSIDERATIONS IN FLEXURE

The maximum external moment on any section of a footing is determined on the basis of computing the factored moment of the forces acting on the entire area of footing on *one side* of a vertical plane assumed to pass through the footing.   This plane is taken at the following locations:

1. At the face of column, pedestal, or wall for an isolated footing, as in Fig. 12.7a
2. Halfway between the middle and edge of wall for footing supporting a masonry wall, as in Fig. 12.7b
3. Halfway between face of column and edge of steel base for footings supporting a column with steel base plates

12.5.1 Reinforcement Distribution

In one-way footings and in two-way square footings, the flexural reinforcement should be uniformly distributed across the entire width of the footing.   This recommendation is conservative, particularly if the soil bearing pressure is not uniform.

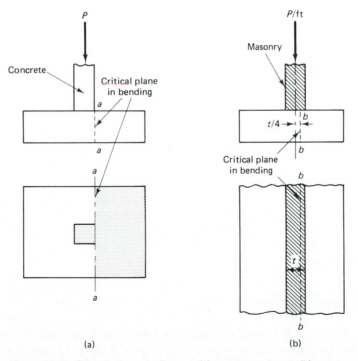

Figure 12.7   Critical planes in flexure: (a) concrete column; (b) masonry wall.

However, no meaningful saving can be accomplished if refinement is made in the bending moment assumptions.

In two-way rectangular footings supporting one column, the bending moment in the short direction is taken as equivalent to the bending moment in the long direction. The distribution of reinforcement differs in the long and short directions. The effective depth is assumed without meaningful loss of accuracy to be equal in both the short and long directions, although it differs slightly because of the two-layer reinforcing mats. The following is the recommended reinforcement distribution:

1. Reinforcement in the long direction is to be uniformly distributed across the entire width of the footing.
2. For reinforcement in the short direction, a central band of width equal to the width of footing in the short direction shall contain a major portion of the reinforcement total areas as in Eq. 12.4 uniformly distributed along the band width:

$$\frac{\text{reinforcement in band width}}{\text{total reinforcement in short direction, } A_s} = \frac{2}{\beta + 1} \qquad (12.4)$$

where β is the ratio of long to short side of footing. The remainder of the reinforcement required in the short direction is uniformly distributed outside the center band of the footing.

In all cases, the depth of the footing above the reinforcement has to be at least 6 in. (152 mm) for footings on soil and at least 12 in. (305 mm) for footings on piles (footings on piles must always be reinforced). A practical depth for column footings should not be less than 9 in. (229 mm).

12.6 DESIGN CONSIDERATIONS IN SHEAR

As discussed in Section 12.3.1, the behavior of footings in shear is not different from that of beams and supported slabs. Consequently, the same principles and expressions as those used in Chapter 6 on shear and diagonal tension are applicable to the shear design of foundations. The shear strength of slabs and footings in the vicinity of column reactions is governed by the more severe of the following two conditions.

12.6.1 Beam Action

The critical section for shear in slabs and footings is assumed to extend in a plane across the entire width and located at a distance d from the face of the concentrated load or reaction area. In this case, if only shear and flexure act, the nominal shear strength of the section is

$$V_c = 2\sqrt{f'_c}\, b_w d \qquad (12.5)$$

where b_w is the footing width. V_c must always be larger than the nominal shear force $V_n = V_u/\phi$ unless shear reinforcement is provided.

12.6.2 Two-way Action

The plane of the critical section perpendicular to the plane of the slab is assumed to be so located that it has a minimum perimeter b_0. This critical section need not be closer than $d/2$ to the perimeter of the concentrated load or reaction area. The fundamental shear failure mechanism in two-way action as presented in Section 12.3.1 demonstrates that the critical section occurs at a distance $d/2$ from the face of the support and not at d as in beam action. Maximum allowable nominal shear strength is the smallest of

$$\text{(i)}\quad V_c = \left(2 + \frac{4}{\beta_c}\right)\sqrt{f'_c}\, b_0 d \qquad (12.6a)$$

$$\text{(ii)}\quad V_c = \left(\frac{\alpha_s d}{b_0} + 2\right)\sqrt{f'_c}\, b_0 d \qquad (12.6b)$$

where α_s is 40 for interior columns, 30 for edge columns, and 20 for corner columns.

$$\text{(iii)}\quad V_c = 4\sqrt{f'_c}\, b_0 d \qquad (12.6c)$$

where β_c = long side c_l/short side c_s of the concentrated load or reaction area

b_0 = perimeter of the critical section, that is, the length of the idealized failure plane

Figure 12.8 gives the relationship of the column side ratio β_c to the shear strength V_c of the footing. V_c must always be larger than the nominal shear force $V_n = V_n/\phi$ unless shear reinforcement is provided.

In cases of both one- and two-way action, if shear reinforcement consisting of bars or wires is used,

$$V_n = V_c + V_s \le 6\sqrt{f'_c}\, b_0 d \qquad (12.7)$$

where $V_c = 2\sqrt{f'_c}\, b_0 d$ and V_s is based on the shear reinforcement size and spacing as described in Chapter 6, unless shear heads made from steel I or channel shapes are used.

It is worthwhile to keep in mind that in most footing slabs, as in most supported superstructure slabs or plates, the use of shear reinforcement is not popular, due to practical considerations and the difficulty of holding the shear reinforcement in position.

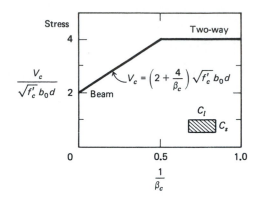

Figure 12.8   Shear strength in footings.

12.6.3 Force and Moment Transfer at Column Base

The forces and moments at the base of a column or wall are transferred to the footing by bearing on the concrete and by reinforcement, dowels, and mechanical connectors. Such reinforcement can transmit the compressive forces that exceed the concrete bearing strength of the footing or the supported column as well as any tensile force across the interface.

The permissible bearing stress on the actual loaded area of the column base or footing top area of contact is

$$f_b = \phi(0.85f_c'), \qquad \text{where } \phi = 0.70 \tag{12.8a}$$

or

$$f_b = 0.60f_c' \tag{12.8b}$$

Hence the permissible bearing stress on the column can normally be considered $0.60f_c'$ for the column concrete. The compressive force that *exceeds* that developed by the permissible bearing stress at the base of the column or at the top of the footing has to be carried by dowels or extended longitudinal bars.

If the footing supporting surface is wider on all sides than the loaded area, the code allows the design bearing strength on the loaded area to be multiplied by $\sqrt{A_2/A_1}$, but the value of $\sqrt{A_2/A_1}$ cannot exceed 2.0. A_1 is the loaded area and A_2 is the maximum area of the supporting surface that is geometrically similar and concentric with the loaded area.

A minimum area of reinforcement of $0.005A_g$ (but not less than four bars) has to be provided across the interface of the column and the footing even when the concrete bearing strength is not exceeded, A_g (in.^{2}) being the gross area of the column cross section.

Lateral forces due to horizontal normal loads, wind, or earthquake can be resisted by shear-friction reinforcement, as described in Section 6.10.

12.7 OPERATIONAL PROCEDURE FOR THE DESIGN OF FOOTINGS

The following sequence of steps can be used for the selection and geometrical proportioning of the size and reinforcement spacing in footings.

1. Determine the allowable bearing capacity of the soil based on site boring test data and soil investigations.

2. Determine the service loads and bending moments acting at the base of the columns supporting the superstructure. Select the controlling service load and moment combinations.

3. Calculate the required area of the footing by dividing the total controlling service load by the selected allowable bearing capacity of the soil if the load is concentric or by also taking into account the controlling bending stress if combined load and bending moments exist.

4. Calculate the factored loads and moments for the controlling loading condition and find the required nominal resisting values by dividing the factored loads and moments by the applicable strength reduction factors ϕ.

5. By trial and adjustment, determine the required effective depth d of the section that has adequate punching shear capacity at a distance d from the support face for one-way action and at a distance $d/2$ for two-way action such that $V_c = 2\sqrt{f_c'}\, b_w d$ for one-way action and $V_c =$ smallest of values from Eqs. 12.6 for two-way action, where b_w is the footing width for one-way action and b_0 is the perimeter of the failure planes in two-way action.   Use an average value of d, since there are two reinforcing mats in the footing.   If the footing is rectangular, check the beam shear capacity in each direction on planes at a distance d from the face of the column support.

6. Calculate the factored moment of resistance M_u on a plane at the face of the column support due to the controlling factored loads from that plane to the extremity of the footing. Find $M_n = M_u/(\phi = 0.9)$.   Select a total reinforcement area A_s based on M_n and the applicable effective depth.

7. Determine the size and spacing of the flexural reinforcement in the long and short directions:

    (a) Distribute the steel uniformly across the width of the footing in the long direction.

    (b) Determine the portion A_{s1} of the total steel area A_s determined in step 6 for the short direction to be uniformly distributed over the central band:

$$A_{s1} = \frac{2}{\beta + 1} A_s$$

Distribute uniformly the remainder of the reinforcement $(A_s - A_{s1})$ outside the center band of the footing. Verify that the area of steel in each principal direction of the footing plan exceeds the minimum value required for temperature and shrinkage: $A_s = 0.0018b_w d$ for sections reinforced with grade 60 steel and $0.0020b_w d$ with grade 40 steel.

8. Check the development length and anchorage available to verify that bond requirements are satisfied (see Chapter 10).

9. Check the bearing stresses on the column and the footing at their area of contact such that the bearing strength P_{nb} for both is larger than the nominal value of column reaction $P_n = P_u/(\phi = 0.70)$. For footing bearing $P_{nb} = \sqrt{A_2/A_1}\,(0.85f_c'A)$, $\sqrt{A_2/A_1}$ not to exceed 2.0.

10. Determine the number and size of the dowel bars that transfer the column load to the footing slab.

Figure 12.9 presents a flow chart for the sequence of calculation operations.

12.7.1 SI Footing Design Expressions

$$E_c = w_c^{1.5}\, 0.043\, \sqrt{f_c'}\ \text{MPa}$$

$$f_r = 0.7\, \sqrt{f_c'}$$

$$K_{tr} = \frac{A_{tr}f_{yt}}{260sn} \qquad \text{where } f_{yt} \text{ is in MPa}$$

$$\ell_d = d_b \left[\frac{15 f_y \alpha \beta \gamma \lambda}{16\, \sqrt{f_c'}\, \left(\dfrac{c + K_{tr}}{d_b} \right)} \right]$$

If $\alpha = \beta = \gamma = \lambda = 1.0$ and $f_c' = 27.6$ MPa,

$$\text{bars} \leq \text{No. 20 M,} \quad s \geq 2d_b: \quad \ell_d = 38d_b$$
$$s < 2d_b: \quad \ell_d = 57d_b$$
$$\text{bars} \geq \text{No. 25 M,} \quad s > 2d_b: \quad \ell_d = 48d_b$$
$$s \leq 2d_b: \quad \ell_d = 72d_b$$

Shear in beam action

$$V_c = 2\, \sqrt{f_c'}\, b_0 d$$

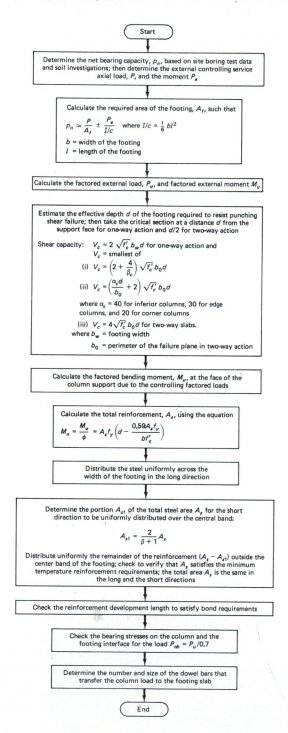

Figure 12.9 Flow chart for footing design.

Shear in two-way action

The smallest of

$$V_c = \left(2 + \frac{4}{\beta_c}\right) \sqrt{f'_c}\, b_o d/12$$

$$V_c = \left(\frac{\alpha_s d}{b_o} + 2\right) \sqrt{f'_c}\, b_o d/12$$

$$V_c = 4\sqrt{f'_c}\, b_o d/12$$

$\alpha_s = 40$ for interior columns, $\alpha_s = 30$ for edge columns, and $\alpha_s = 20$ for corner columns.

12.8 EXAMPLES OF FOOTING DESIGN

12.8.1 Example 12.3: Design of Two-way Isolated Footing

Design the footing thickness and reinforcement distribution for the isolated square footing in Ex. 12.1 if the total service load $P = 400{,}000$ comprises 230,000 lb (1023 kN) dead load and 170,000 lb (756 kN) live load. Given:

$f'_c = 3000$ psi (20.68 MPa), normal-weight concrete (footing)
$f'_c = 5500$ psi (37.91 MPa) in column
$f_y = 60{,}000$ psi (413.7 MPa)

Solution   *Factored load intensity (step 4)*

Data from Ex. 12.1:

column size $= 14$ in. $\times$ 14 in. (355.6 mm $\times$ 355.6 mm)

footing area $= 6$ ft 8 in. $\times$ 6 ft 8 in. (2.03 m $\times$ 2.03 m), $A_f = 44.49$ ft²

assumed footing slab thickness $h = 2$ ft

factored load $U = 1.4 \times 230{,}000 + 1.7 \times 170{,}000 = 611{,}000$ lb

factored load intensity $= q_s = \dfrac{U}{A_f} = \dfrac{611{,}000}{44.49} = 13{,}733$ lb/ft² (657.6 kPa)

Shear capacity (step 5)

Assume that the thickness of the footing slab $\simeq 2$ ft. The average depth $d = h - 3$ in. minimum clear cover minus steel diameter $\simeq 20$ in.

Beam action (at d from support face): The area to be considered for factored shear V_u is shown as *ABCD* in Fig. 12.10.

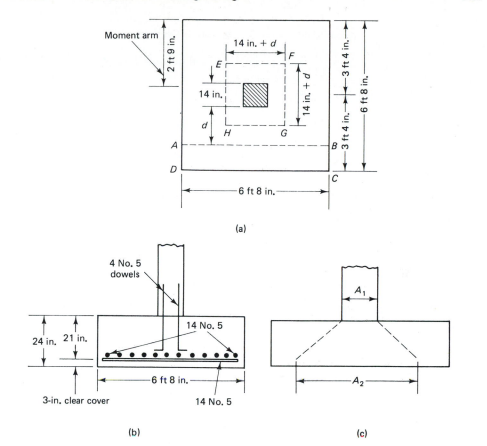

(a)

(b)

(c)

Figure 12.10   Details of footing in Ex. 12.3.

$$\text{factored } V_u = 13,733 \left(\frac{6 \text{ ft } 8 \text{ in.}}{2} - \frac{14}{2 \times 12} - \frac{20}{12} \right)(6 \text{ ft } 8 \text{ in.}) = 99,340 \text{ lb}$$

$$\text{required } V_n = \frac{V_u}{\phi} = \frac{99,340}{0.85} = 116,871 \text{ lb}$$

$$b_w = 6 \text{ ft } 8 \text{ in.} = 80 \text{ in.} \quad (2.03 \text{ m})$$

$$\text{available } V_c = 2\sqrt{f_c'} \, b_w d = 2\sqrt{3000} \times 80 \times 20 = 175,271 \text{ lb}$$

Two-way action (at d/2 from support face): The area to be considered for factored shear V_u is equal to the total area of footing less area *EFGH* of the failure zone.

$$\text{factored } V_u = 13,733 \left[44.49 - \left(\frac{14 + 20}{12} \right)^2 \right] = 500,736 \text{ lb}$$

$$\text{required } V_n = \frac{V_u}{\phi} = 589,100 \text{ lb (2620 kN)}$$

$$b_0 = \text{perimeter of failure zone } EFGH = (14 + 20)4 = 136 \text{ in.}$$

$$\beta_c = \frac{14}{14} = 1.0$$

The available nominal shear strength from Eqs. 12.6 is the smallest of

$$V_c = \left(2 + \frac{4}{\beta_c}\right)\sqrt{f'_c}\, b_0 d \leq 4\sqrt{f'_c}\, b_0 d$$

or

$$V_c = \left(2 + \frac{4}{1}\right)\sqrt{3000} \times 136 \times 20 = 893,882 \text{ lb}$$

$$V_c = \left(\frac{\alpha_s d}{b_0} + 2\right)\sqrt{f'_c}\, b_0 d = \left(\frac{40 \times 20}{136} + 2\right)\sqrt{3000} \times 136 \times 20$$

$$= 1,174,317 \text{ lb}$$

where $\alpha = 40$ used for interior column value

$$V_c = 4\sqrt{3000} \times 136 \times 20 = 595,922 \text{ lb} \qquad \text{controls}$$

Since available $V_c = 595,922 \text{ lb}, >$ required $V_c = 589,100 \text{ lb}$. Therefore, $d = 20 \text{ in.}$ is adequate for shear.

Bending moment capacity (steps 6 and 7)

The critical section is at the face of the column.

$$\text{moment arm} = \frac{6 \text{ ft } 8 \text{ in.}}{2} - \frac{14}{2 \times 12} = 2 \text{ ft } 9 \text{ in.}$$

$$\text{factored moment } M_u = 13,733 \times 6.67 \left[\frac{(2 \text{ ft } 9 \text{ in.})^2}{2}\right]$$

$$= 346,359.1 \text{ ft-lb} = 4,156,310 \text{ in.-lb}$$

$$M_n = \frac{M_u}{\phi} = \frac{4,156,310}{0.90} = 4,618,122 \text{ in.-lb}$$

$$(521.8 \text{ kN-m})$$

$$M_n = A_s f_y \left(d - \frac{a}{2}\right)$$

Assume that $(d - a/2) \approx 0.9d$. Use average $d = 20 \text{ in.}$

$$4,618,122 = A_s \times 60,000 \times 0.9 \times 20$$

or

$$A_s = \frac{4,618,122}{60,000 \times 0.9 \times 20} = 4.28 \text{ in.}^2/80\text{-in. band}$$

$$a = \frac{A_s f_y}{0.85 f'_c b} = \frac{4.28 \times 60,000}{0.85 \times 3000 \times 80} = 1.26 \text{ in.}$$

$$4{,}618{,}122 = A_s \times 60{,}000\left(20.0 - \frac{1.26}{2}\right)$$

$$A_s = 3.98 \text{ in.}^2 \qquad \rho = \frac{A_s}{bd} = \frac{3.98}{80 \times 20} = 0.0025$$

Minimum allowable shrinkage steel

$$\rho_{min} = 0.0018 < \rho \qquad \text{O.K.}$$

Use 14 No. 5 bars ($A_s = 4.27$ in.^{2}) each way spaced at $\simeq 5\frac{1}{2}$ in. (139.7 mm) center to center.

Development of reinforcement (step 8)

The critical section for development-length determination is the same as the critical section in flexure, that is, at the face of the column. From Table 10.2, $\ell_d = 24$ in. for No. 5 bars (bottom bars).

$$\text{check } \sqrt{f_c'} = \sqrt{3000} = 54.8 < 100 \quad \text{O.K.;} \qquad s = 5\frac{1}{2} \text{ in.} > 2d_b$$

Use $l_d = 24$ in. The projection length of each bar beyond the column face is

$$\frac{1}{2}(6 \text{ ft } 8 \text{ in.} - 14 \text{ in.}) - 3 \text{ in. cover} = 30 \text{ in.} > 24 \text{ in.} \qquad \text{O.K.}$$

Force transfer at interface of column and footing (step 9)

Column $f_c' = 5500$ psi. Factored $P_u = 611{,}000$ lb.
(a) Bearing strength on column using Eq. 12.8b:

$$\phi P_{nb} = 0.70 \times 0.85 f_c' A_1 = 0.60 f_c' A_1$$

or $\qquad \phi P_{nb} = 0.60 f_c' A_1 = 0.60 \times 5500 \times 14 \times 14$

$$= 646{,}800 \text{ lb} > 611{,}000 \qquad \text{O.K.}$$

From step 9 of the design operational procedure on bearing strength on footing concrete,

$$\sqrt{\frac{A_2}{A_1}} = \sqrt{\frac{(6 \text{ ft } 8 \text{ in.}) \times (6 \text{ ft } 8 \text{ in.})}{(14 \times 14)/144}} = 5.714 > 2.0 \qquad \text{use } 2.0$$

$$\phi P_{nb} = 2.0(0.60 f_c' A_1) = 2.0 \times 0.60 \times 3000 \times 14 \times 14 = 705{,}600 \text{ lb}$$

$$> 611{,}000 \qquad \text{O.K.}$$

Dowel bars between column and footing (step 10)

Even though the bearing strength at the interface between the column and the footing slab is adequate to transfer the factored P_u, a minimum area of reinforcement is necessary across the interface. The minimum $A_s = 0.005 (14 \times 14) = 0.98$ in.^{2}, but not less than four bars. Use four No. 5 bars as dowels ($A_s = 1.22$ in.^{2}).

Development of dowel reinforcement in compression: From Eqs. 10.7 a and b for No. 5 bars and Section 10.3.5,

$$l_{db} = \frac{0.02 d_b f_y}{\sqrt{f_c'}}$$

and $l_{db} \geq 0.0003 d_b f_y$, where d_b is the dowel bar diameter.   Within column,

$$l_d = \frac{0.02 \times 0.625 \times 60{,}000}{\sqrt{5500}} = 10.11 \text{ in.}$$

$$0.0003 \times 0.625 \times 60{,}000 = 11.25 \text{ in.} \qquad \text{controls}$$

Within footing,

$$l_d = \frac{0.02 \times 0.625 \times 60{,}000}{\sqrt{3000}} = 13.69 \text{ in.}$$

Available length for development above the footing reinforcement assuming column bars size to be the same as the dowel bars size:

$$l = 24 - 3 \text{ (cover)} - 2 \times 0.625 \text{ (footing bars)} - 0.625 \text{ (dowels)}$$
$$= 19.13 > 13.69 \text{ in.} \qquad \text{O.K.}$$

12.8.2 Example 12.4: Design of Two-way Rectangular Isolated Footing

Determine the size and distribution of the bending reinforcement of an isolated rectangular footing subjected to a concentrated concentric factored column load $P_u = 770{,}000$ lb (3425 kN) and having an area 10 ft × 15 ft (3.05 m × 4.57 m).   Given:

$f_c' = 3000$ psi (20.68 MPa), footing
$f_y = 60{,}000$ psi (413.7 MPa)
column size = 14 in. × 18 in.

Solution   factored load intensity $q_s = \dfrac{770{,}000}{10 \times 15} = 5134$ lb/ft²

Shear capacity (step 5)

Through trial and adjustment, assume that the footing slab is 2 ft 4 in. thick.
Beam action (at distance d from column face): Average effective depth ≈ 2 ft 4 in. − 3 in. (cover) − ¾ in. (diameter of bars in first layer) ≈ 24 in.
From Fig. 12.11, length *CD* subjected to bearing intensity q_s in one-way beam action:

$$\frac{15 \text{ ft}}{2} - \frac{18 \text{ in.}}{2 \times 12} - \frac{24 \text{ in.}}{12} = 4 \text{ ft 9 in.} = 57 \text{ in.}$$

$$\text{factored } V_u = 5134 \times 10 \text{ ft} \times 4 \text{ ft 9 in.} = 243{,}865 \text{ lb}$$

$$\text{required } V_n = \frac{V_u}{\phi} = \frac{243{,}865}{0.85} = 286{,}900 \text{ lb}$$

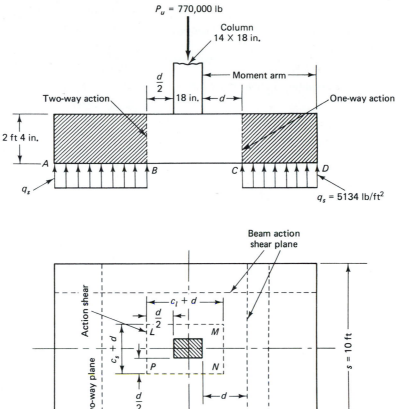

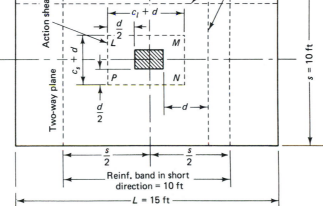

Figure 12.11   Beam action and two-way action planes in Ex. 12.4.

$$\text{available } V_c = 2\sqrt{f_c'}\, b_w d = 2\sqrt{3000} \times 120 \times 24$$

$$= 315{,}488 \text{ lb} > 286{,}900 \qquad \text{O.K.}$$

Notice that the shorter side length was used for b_w to give the lower available V_n value.

Two-way action (at distance $d/2$ from column face):

loaded area outside the failure zone $LMNP$ in Fig. 12.11

$$= 15 \times 10 - (c_l + d)(c_s + d)$$

$$= 150 - \frac{(18 + 24)(14 + 24)}{144}$$

$$= 138.92 \text{ ft}^2$$

$$\text{factored } V_u = 5134 \times 138.92 = 713,215 \text{ lb}$$

$$\text{required } V_n = \frac{713,215}{0.85} = 839,077 \text{ lb (3732 kN)}$$

$$\text{perimeter of shear failure plane } b_0 = 2[(c_l + d) + (c_s + d)]$$

$$= 2[(18 + 24) + (14 + 24)] = 160 \text{ in.}$$

From Eqs. 12.6,

$$V_c = \left(2 + \frac{4}{\beta_c}\right) \sqrt{f_c'}\, b_0 d \le 4\sqrt{f_c'}\, b_0 d$$

$$\beta_c = \frac{18}{14} = 1.286$$

or

$$V_c = \left(2 + \frac{4}{1.286}\right) \sqrt{3000} \times 160 \times 24 = 1,074,851 \text{ lb.}$$

$$V_c = \left(\frac{\alpha_s d}{b_0} + 2\right) \sqrt{f_c'}\, b_0 d = \left(\frac{40 \times 24}{160} + 2\right) \sqrt{3000} \times 160 \times 24$$

$$= 1,682,604 \text{ lb}$$

and

$$V_c = 4\sqrt{f_c'}\, b_0 d = 4\sqrt{3000} \times 160 \times 24$$

$$= 841,302 \text{ lb} \quad \text{controls}$$

Design of two-way reinforcement

The critical section for bending is at the face of the column. The controlling moment arm is in the long direction:

$$\frac{15 \text{ ft}}{2} - \frac{18 \text{ in.}}{2 \times 12} = 6.75 \text{ ft (2.06 m)}$$

$$\text{factored moment } M_u = 5134 \times \frac{10(6.75)^2}{2}$$

$$= 1,169,589 \text{ ft-lb} = 14,035,073 \text{ in.-lb (1585.96 kN-m)}$$

$$M_n = \frac{14,035,073}{0.9} = 15,594,526 \text{ in.-lb (1762.18 kN-m)}$$

Assume that $(d - a/2) \simeq 0.9d$.

$$M_n = A_s f_y \left(d - \frac{a}{2}\right) \quad \text{or} \quad 15,594,526 = A_s \times 60,000 \times 0.9 \times 24$$

$$A_s = \frac{15,594,526}{60,000 \times 0.9 \times 24} = 12.03 \text{ in.}^2/\text{10-ft-wide strip}$$

Check:

$$a = \frac{A_s f_y}{0.85 f'_c b} = \frac{12.03 \times 60,000}{0.85 \times 3000 \times 120} = 2.36 \text{ in.}$$

$$15,594,526 = A_s \times 60,000\left(24 - \frac{2.36}{2}\right)$$

$$A_s = 11.39 \text{ in.}^2 = \frac{11.39}{10 \text{ ft}} = 1.14 \text{ in.}^2 \text{ ft/width}$$

Try No. 8 bars, $A_s = 0.79$ in.^{2} per bar.

$$\text{number of bars in the short direction} = \frac{11.39}{0.79} = 14.42$$

Use 15 bars.

Reinforcement in the short direction

The band width $= s = 10$ ft (Fig. 12.11). From Eq. 12.4,

$$\beta = \frac{15}{10} = 1.5$$

$$\frac{A_{s1}}{A_s} = \frac{2}{\beta + 1} \quad \text{or} \quad \frac{A_{s1}}{11.39} = \frac{2}{1 + 1.5}$$

Therefore,

$$A_{s1} = \frac{2 \times 11.39}{2.5} = 9.11 \text{ in.}^2$$

to be placed in the central 10-ft-wide band and the balance ($11.39 - 9.11 = 2.28$ in.^{2}) to be placed in the remainder of the footing. Use 12 No. 8 bars in the central band $= 9.48$ in.^{2} and two No. 8 bars at each side of the band, as in Fig. 12.12. To complete the design, a check of the development length, bearing stress at the column–footing interface, and dowel action has to be made, as in Ex. 12.3.

12.8.3 Example 12.5: Proportioning of a Combined Footing

A combined footing has the layout shown in Fig. 12.13. Column L at the property line is subjected to a total service axial load $P_L = 200,000$ lb (889.6 kN), and the internal column R is subjected to a total service load $P_R = 350,000$ lb (1556.8 kN). The live load is 35% of the total load. The bearing capacity of the soil at the level of the footing base is 4000 lb/ft^2 (191.5 kPa), and the average value of the soil and footing unit weight $\gamma = 120$ pcf (1922 kg/m^3). A surcharge of 100 lb/ft^2 results from the slab on grade. Proportion the footing size and select the necessary size and distribution of the footing slab reinforcement and verify the development length required. Given:

$f'_c = 3000$ psi (20.68 MPa)
$f_y = 60,000$ psi (413.7 MPa)
base of footing at 7 ft below grade

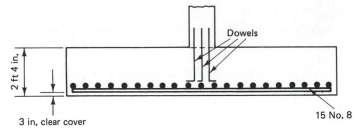

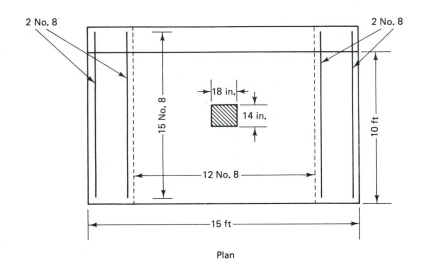

Figure 12.12   Footing reinforcement details of Ex. 12.4.

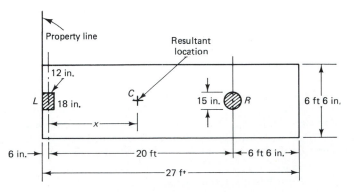

Figure 12.13   Combined footing plan geometry in Ex. 12.5.

Solution   total columns load $= 200,000 + 350,000 = 550,000$ lb (2446.4 kN)

net allowable soil capacity $p_n = p_g - 120(7$ ft height to base of footing$) - 100$

or

$$p_n = 4000 - 120 \times 7 - 100 = 3060 \text{ lb/ft}^2$$

$$\text{minimum footing area } A_f = \frac{P}{p_n} = \frac{550,000}{3060} = 179.8 \text{ ft}^2$$

Center of gravity of column loads from the property line:

$$\bar{x} = \frac{200,000 \times 0.5 + 350,000 \times 20.5}{550,000} = 13.23 \text{ ft}$$

length of footing $L = 2 \times 13.23 = 26.46$ ft

Use $L = 27$ ft.

$$\text{width of footing } S = \frac{179.8}{27.0} = 6.66 \text{ ft}$$

Use $S = 6$ ft 6 in. as shown in Fig. 12.13.

Factored shears and moments

$$\text{column } L: \quad P_D = 0.65 \times 200,000 = 130,000 \text{ lb}$$
$$P_L = 200,000 - 130,000 = 70,000 \text{ lb}$$
$$P_U = 1.4 \times 130,000 + 1.7 \times 70,000 = 301,000 \text{ lb}$$
$$\text{column } R: \quad P_D = 227,500 \text{ lb}$$
$$P_L = 122,500 \text{ lb}$$
$$P_U = 1.4 \times 227,500 + 1.7 \times 122,500 = 526,750 \text{ lb}$$

The net factored soil bearing pressure for footing structural design is

$$q_s = \frac{P_u}{A_f} = \frac{301,000 + 526,750}{6.5 \times 27.0} = 4716.5 \text{ lb/ft}^2$$

Assume that the column loads are acting through their axes.

factored bearing pressure per foot width $= q_s \times S = 4,716.5 \times 6.5 = 30,658$ lb/ft

V_u at center line of column $L = 301,000 - 30,658 \times \dfrac{6}{12} = 285,671$ lb

V_u at center line of column $R = 526,750 - 30,658 \times 6.5 = 327,473$ lb

The maximum moment is at the point C of zero shear in Fig. 12.14 x (ft) from the center of the left column L.

$$x = \frac{285,671 \text{ lb}}{30,658 \text{ plf}} = 9.32 \text{ ft}$$

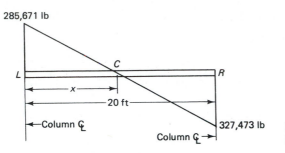

285,671 lb

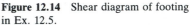

Figure 12.14   Shear diagram of footing in Ex. 12.5.

327,473 lb

Taking a free-body diagram to the left of a section through C, the factored moment at point C is

$$M_{uc} = \frac{w_u l^2}{2} - P_{ul}x$$

$$M_u \text{ from left side} = 30{,}658\frac{(9.32 + 0.50)^2}{2} - 301{,}000 \times 9.32$$

$$= -1{,}327{,}108\,\text{ft-lb} = -15{,}925{,}293\,\text{in.-lb} \qquad \text{(Fig. 12.15)}$$

$$M_u \text{ from right side} = 30{,}658\frac{(27.0 - 9.82)^2}{2} - 526{,}750(20.0 - 9.32)$$

$$= 1{,}101{,}299\,\text{ft-lb} = -13{,}215{,}588\,\text{in.-lb less than } -15{,}925{,}293\,\text{in.-lb}$$

Hence M_u from the left side controls.   Note that M_u from the right side differs from M_u from the left side because the footing length of 27 ft is used instead of the computed length of 26.46 ft and because x is rounded off.   Therefore, the load is not exactly uniform due to the small eccentricity.

Design of the footing in the longitudinal direction

(a) *Shear:* The combined footing is considered as a beam in the shear computations.   Hence the critical section is at a distance d from the face of the support. Controlling V_n at the column center line

$$\frac{V_u}{\phi} = \frac{327{,}473}{0.85} = 385{,}262\,\text{lb}$$

Assume that the total footing thickness = 3 ft (0.92 m).   The effective footing depth $d = 32$ in. for minimum steel cover of ≈ 4 in. For the controlling interior column R, the equivalent rectangular column size $\sqrt{\pi(15)^2/4} = 13.29$ in.

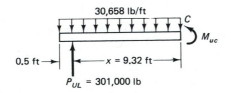

30,658 lb/ft

0.5 ft → | ← $x = 9.32$ ft →

$P_{UL} = 301{,}000$ lb

Figure 12.15   Free-body diagram.

$$\text{required } V_n \text{ at } d \text{ section} = 385,262 - \frac{(13.29/2 + d)}{12} \times \frac{30,658}{\phi}$$

$$= 385,262 - \frac{38.65 \times 30,658}{12 \times 0.85} = 269,107 \text{ lb (1196.9 kN)}$$

$$V_c = 2\sqrt{f_c'}\, b_w d = 2\sqrt{3000} \times 6.5 \times 12 \times 32$$

$$= 273,423 \text{ lb (1216.2 kN)} > 269,107 \qquad \text{O.K.}$$

(b) *Moment and reinforcement in the longitudinal direction (step 4):* The distribution of shear and moment in the longitudinal direction is shown in Fig. 12.16. The critical section for moment is taken at the face of the columns.

$$\text{controlling moment } M_n = \frac{M_u}{\phi} = \frac{15,925,293}{0.9} = 17,694,770 \text{ in.-lb (1999.5 kNm)}$$

$$M_n = A_s f_y \left(d - \frac{a}{2} \right)$$

Assume that $(d - a/2) \simeq 0.9d$.

$$17,694,770 = A_s \times 60,000(0.9 \times 32)$$

or

$$A_s = \frac{17,694,770}{60,000 \times 0.9 \times 32} = 10.24 \text{ in.}^2$$

$$a = \frac{A_s f_y}{0.85 f_c' b} = \frac{10.24 \times 60,000}{0.85 \times 3,000 \times 6.5 \times 12} = 3.09 \text{ in.}$$

$$17,694,770 = A_s \times 60,000 \left(32 - \frac{3.09}{2} \right)$$

$$A_s = 9.68 \text{ in.}^2 \text{ (6245 mm}^2\text{)}$$

Use 22 No. 6 bars at the top for the middle span.

$$A_s = 9.68 \text{ in.}^2 \text{ (22 bars, 19.1-mm diameter)}$$

Design of footing in the transverse direction

Both columns are treated as isolated columns. The width of the band should not be larger than the width of the column plus half the effective depth d on *each* side of the column. This assumption is on the safe side since the actual bending stress distribution is highly indeterminate. It is, however, possible to assume that the flexural reinforcement in the transverse direction can raise the shear punching capacity within the $d/2$ zone from the face of the rectangular left column L and the *equivalent* rectangular right column R. Figure 12.17 shows the transverse band widths for both columns L and R determined on the basis of this discussion.

$$\text{band width } b_L = 12 + \frac{32}{2} = 28 \text{ in.} = 2.33 \text{ ft}$$

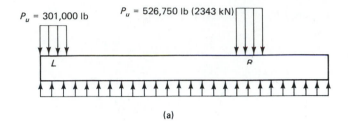

(a)

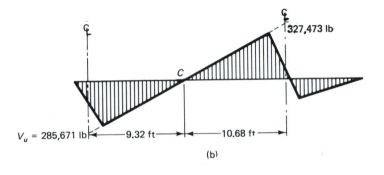

(b)

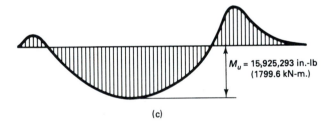

(c)

Figure 12.16 Longitudinal shear and moment distribution: (a) elevation; (b) shear; (c) moment.

The rectangular column size equivalent to the circular interior 15-in.-diameter column = 13.29 in.

$$\text{band width } b_R = 13.29 + 2\left(\frac{32}{2}\right) = 45.3 \text{ in.} = 3.77 \text{ ft}$$

Column L transverse band reinforcement

$$\text{moment arm} = \frac{6 \text{ ft } 6 \text{ in.}}{2} - \frac{18}{2 \times 12} = 2.50 \text{ ft} = 30.0 \text{ in.}$$

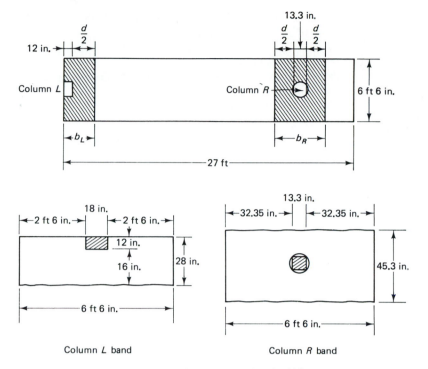

Figure 12.17 Footing transverse band widths.

The net factored bearing pressure in the transverse direction is

$$q_s = \frac{301,000}{6.5} = 46,308 \text{ lb/ft}$$

$$M_u = q_s \frac{l^2}{2} = 46,308 \frac{(2.50)^2}{2} = 144,713 \text{ ft-lb} = 1,736,550 \text{ in.-lb}$$

$$M_n = \frac{M_u}{\phi} = \frac{1,736,550}{0.90} = 1,929,500 \text{ in.-lb (218.0 kN-m)}$$

$$M_n = A_s f_y \left(d - \frac{a}{2} \right)$$

or

$$1,929,500 = A_s \times 60,000 \times 0.9 \times 32$$

$$A_s = 1.12 \text{ in.}^2$$

$$a = \frac{A_s f_y}{0.85 f'_c b} = \frac{1.12 \times 60,000}{0.85 \times 3000 \times 28} = 0.94 \text{ in.}$$

$$1,929,500 = A_s \times 60,000 \left(32 - \frac{0.94}{2} \right)$$

$$A_s = 1.02 \text{ in.}^2 \text{ (658 mm}^2\text{)}$$

$$\text{min. } A_s = 0.0018b_w d = 0.0018 \times 28 \times 32 = 1.62 \text{ in.}^2$$

$$\rho = \frac{1.02}{28 \times 32} = 0.00114$$

Use six No. 5 bars, $A_s = 1.86$ in.^{2} (six bars, 15.9-mm diameter) equally spaced in the band, which is to be centered under the column.

Column R transverse band reinforcement Equivalent square column size = 13.3 in. × 13.3 in.

$$\text{moment arm} = \frac{6 \text{ ft } 6 \text{ in.}}{2} - \frac{13.3 \text{ in.}}{2 \times 12} = 2.69 \text{ ft} = 32.35 \text{ in.}$$

Net factored bearing pressure in the transverse direction is

$$q_s = \frac{526,750}{6.50} = 81,038 \text{ lb/ft}$$

$$M_u = q_s \frac{l^2}{2} = 81,038 \frac{(2.69)^2}{2} = 293,200 \text{ ft-lb} = 3,518,400 \text{ in.-lb}$$

$$M_n = \frac{M_u}{\phi} = \frac{3,518,400}{0.90} = 3,909,333 \text{ in.-lb (441.8 kN-m)}$$

$$M_n = A_s f_y \left(d - \frac{a}{2} \right) \qquad \text{assume that } d - \frac{a}{2} \approx 0.90d$$

or

$$3,909,333 = A_s \times 60,000 \times 0.9 \times 32$$

$$A_s = 2.26 \text{ in.}^2 \qquad a = \frac{A_s f_y}{0.85 f'_c b} = \frac{2.26 \times 60,000}{0.85 \times 3000 \times 45.3} = 1.17 \text{ in.}$$

$$3,909,333 = A_s \times 60,000 \left(32 - \frac{1.17}{2} \right)$$

$$A_s = 2.07 \text{ in.}^2 \text{ (3347 mm}^2\text{)}$$

$$\rho = \frac{2.07}{45.3 \times 32} = 0.0014 < \rho_{\min}$$

where $\rho_{\min} = 0.0018$ (shrinkage temperature reinforcement).

$$\text{minimum } A_s = 0.0018 \times 45.3 \times 32 = 2.61 \text{ in.}^2$$

Use nine No. 5 bars, $A_s = 2.79$ in.^{2} (nine bars, 15.9-mm diameter) equally spaced.

Development length check for top bars in tension

$$f'_c = 3000 \text{ psi (27.6 MPa)}$$

Top reinforcement more than 12-in. concrete below bars, $\alpha = 1.3$. From Eq. 10.6,

$$\frac{\ell_d}{d_b} = \frac{3}{40} \frac{f_y}{\sqrt{f'_c}} \frac{\alpha\beta\gamma\lambda}{\left(\dfrac{c + K_{tr}}{d} \right)}$$

(a) *Longitudinal top bars*: Assume transverse reinforcement index $K_{tr} \approx 0$.   Spacing $c = 4.91$ in. $> 2d_b$. $c/d_b = 4.91/0.75 = 6.5 > 2.5$; use 2.5.   No. 6 bar $d_b = 0.75$ in. (19.1 mm).

$$\alpha = 1.3, \quad \beta = 1.0, \quad \gamma = 0.8, \quad \lambda = 1.0, \quad K_{tr} = 0$$

$$\ell_d = 0.75 \left(\frac{3}{40} \frac{60{,}000}{\sqrt{3000}} \times \frac{1.3 \times 1.0 \times 0.8 \times 1.0}{2.5} \right) = 25.7 \text{ in.}$$

$$> \min \ell_d = 12 \text{ in.}$$

Use $\ell_d = 26$ in. $= 2.2$ ft. (660 mm).

The distance from c at the maximum moment in Fig. 12.14 to the center of the left column $= 9.32 + 0.50 = 9.82$ ft. > 2.3,   O.K.

(b) *Transverse bottom steel*: No. 5 bars, $d_b = 0.625$

$$\alpha = 1.0 \text{ (bottom steel)}, \quad \beta = 1.0, \quad \gamma = 0.8, \quad \lambda = 1.0, \quad K_{tr} \approx 0$$

$$c = 3.8 \text{ in.}, \quad c/d_b = 3.8/0.625 = 6.1 > 2.5, \qquad \text{use } 2.5$$

$$\ell_d = 0.625 \left(\frac{3}{40} \times \frac{60{,}000}{\sqrt{3000}} \times \frac{1.0 \times 1.0 \times 0.8 \times 1.0}{2.5} \right) = 16.4 \text{ in.}$$

$$\lambda_s = \frac{A_s \text{ required}}{A_s \text{ provided}}$$

$$\text{column } L \text{ steel modifier: } \quad \lambda_s = \frac{1.62}{1.86} = 0.87$$

$$\text{column } R \text{ steel modifier: } \quad \lambda_s = \frac{2.61}{2.79} = 0.94$$

modified $\ell_d = 16.4 \times 0.94 = 15.4$ in.

available development length $= (32.35 - 3.0)$ in. > 15.4 in.   O.K.

Therefore, adopt reinforcement as in Fig. 12.18.   Check for dowel steel from the columns to the footing slab.

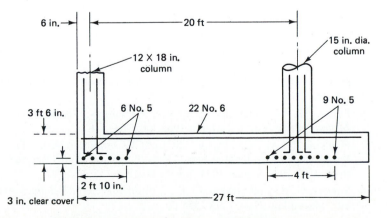

Figure 12.18   Combined footing reinforcement.

12.9 STRUCTURAL DESIGN OF OTHER TYPES OF FOUNDATIONS

From the discussion and the examples given in the foregoing sections, it is clear that the design of foundation substructures follows all the hypotheses and procedures used in proportioning the superstructures once the intensity and distribution of the soil bearing pressure is determined. If a cluster of piles supports a very heavy reaction through a pile cap, the analysis reduces to determining the punching load for each pile and determining the corresponding thickness of the cap. A determination of the center of gravity of the resultant of all pile forces has to be made if the system is subjected to bending in addition to axial load in order to choose the appropriate pile cap layout.

When raft foundations are necessary in poor soil conditions and deep excavations, the design of such a substructure is not too different from the design of any heavily loaded floor system. Once the soil pressure distribution is determined, the design becomes that of an inverted floor supported by deep beams longitudinally and transversely.

Variations are to be expected in the described foundation types, particularly in cases of specialized or unique structures. Through an understanding of the basic principles presented, the student and the designer should have no difficulty in utilizing the soil data developed by the geotechnical engineer in selecting and proportioning the appropriate foundation substructure.

SELECTED REFERENCES

12.1. Richart, F. E., "Reinforced Concrete Walls and Column Footings," *Journal of the American Concrete Institute*, Proc. Vol. 45, October and November 1948, pp. 97–127 and 237–245.

12.2. Timoshenko, S., and Woinowsky-Kreiger, *Theory of Plates and Shells*, 2nd ed., McGraw Hill, New York, 1968, 580 pp.

12.3. Balmer, G. G., Jones, V., and McHenry, D., *Shearing Strength of Concrete under High Triaxial Stress*, Structural Research Laboratory Report SP–23, U.S. Department of Interior, Bureau of Reclamation, Washington, D.C., 1949, 26 pp.

12.4. American Insurance Association, *The National Building Code*, 1976 ed., AIA, New York, December 1977, 767 pp.

12.5. Moe, J., *Shearing Strength of Reinforced Concrete Slabs and Footings under Concentrated Load*, Bulletin D47, Portland Cement Association, Skokie, Ill., April 1961, 134 pp.

12.6. Furlong, R. W., "Design Aids for Square Footings," *Journal of the American Concrete Institute*, Proc. Vol. 62, March 1965, pp. 363–371.

12.7. Hawkins, N. M., Chairman, ASCE-ACI Committee 426, "The Shear Strength of Reinforced Concrete Members—Slabs," *Journal of the Structural Division, ASCE*, Proc. Vol. 100, August 1974, pp. 1543–1591.

12.8. Sowers, G. B., and Sowers, G. F., *Introductory Soil Mechanics and Foundations*, 3rd ed., Macmillan, New York, 556 pp.

12.9. Bowles, J. E., *Foundations Analysis and Design*, McGraw–Hill, New York, 1982, 816 pp.

12.10. Winterkorn, H. F., and Fang, H. Y., *Foundation Engineering Handbook*, Van Nostrand Reinhold, New York, 1975, 751 pp.

12.11. Baker, A. L. L., *Raft Foundations—The Soil Line Method of Design*, Concrete Publications, London, 1948, 141 pp.

PROBLEMS FOR SOLUTION

12.1. Design a reinforced concrete, square, isolated footing to support an axial column service live load P_L = 300,000 lb (1334 kN) and service dead load P_D = 625,000 lb (2780 kN). The size of the column is 30 in. × 24 in. (0.76 m × 0.61 m). The soil test borings indicate that it is composed of medium compacted sands and gravely sands, poorly graded. The frost line is assumed to be 3 ft below grade. Given:

> average weight of soil and concrete above the footing, γ = 130 pcf (20.41 kN/m³)
>
> footing f_c' = 3000 psi (20.68 MPa)
>
> column f_c' = 4000 psi (27.58 MPa)
>
> f_y = 60,000 psi (413.7 MPa)
>
> surcharge = 120 psf (5.7 kPa)

12.2. Design a reinforced concrete wall footing for (a) a 10-in. (0.25-m) reinforced concrete wall and (b) a 12-in. (0.30-m) masonry wall. The intensity of service linear dead load is W_D = 20,000 lb/ft (292.0 kN/m) and a service linear live load W_L = 15,000 lb/ft (219.0 kN/m) of wall length. Assume an evenly distributed soil bearing pressure and that the average soil bearing pressure at the base of the footing is 3 tons/ft² (87.6 kN/m). The frost line is assumed to be 2 ft below grade. Given:

> average weight of soil and footing above base = 125 pcf (19.6 kN/m³)
>
> footing f_c' = 3000 psi (20.68 MPa)
>
> column f_c' = 5000 psi (34.47 MPa)
>
> f_y = 60,000 psi (413.7 MPa)

12.3. A combined footing is subjected to an exterior 16 in. × 16 in. (0.4 m × 0.4 m) column abutting the property line carrying a total service load P_W = 300,000 lb (1334.4 kN) and an interior column 20 in. × 20 in. (0.5 m × 0.5 m) carrying a total factored load P_W = 400,000 lb (1779.2 kN). The live load is 30% of the total load. The center-line distance between the two columns is 22′–0″ (6.71 m). Design the appropriate reinforced concrete footing on a soil weighing 135 pcf (21.2 kN/m³). The bearing capacity of the soil at the level of the footing base is 6000 lb/ft². The frost line is assumed to be at 3′–6″ (1.07 m) below grade. Assume a surcharge of 125 psf (19.62 kN/m³) at grade level. Given:

> footing f_c' = 3500 psi (24.13 MPa)
>
> column f_c' = 5500 psi (37.42 MPa)
>
> f_y = 60,000 psi (413.7 MPa)

12.4. Redesign the isolated reinforced concrete footing in Problem 12.1 if the load is applied at an eccentricity (a) $e = 0.5$ ft (0.15 m) and (b) $e = 1.8$ ft (0.55 m).

12.5. Redesign the combined reinforced concrete footing in Problem 12.3 if the center-line distance between the two columns is 15′–0″ (4.6 m).

CONTINUOUS REINFORCED CONCRETE STRUCTURES

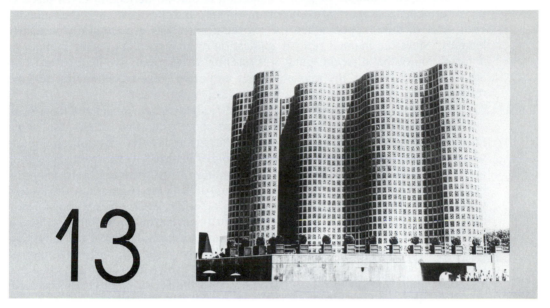

13.1 INTRODUCTION

Preceding chapters have covered the design and analysis of individual elements and isolated reinforced concrete sections. However, except for precast construction, concrete structural systems are constructed in continuous monolithic pours with the reinforcement well developed through adjoining spans and overlapping columns. Consequently, support sections and joints transmit flexural moments, thereby rendering a structure statically indeterminate in view of the continuity of the components. The three equations of equilibrium of forces and moments are no longer sufficient to solve for the unknown moments and reactions, and equations of geometry have to be developed to eliminate the indeterminacy.

Equations of geometry consider the *deformation* of the structure under load or stress, because the deflected shape of individual elements depends not only on load but also on the rotations and slopes of the element at its ends. The magnitudes of the slope and rotation depend on the rigidity (or flexibility) of the joint, that

Photo 74   New York Hall of Science, Queens, New York. (Courtesy of Ammann & Whitney).

is, the relative rigidities of the adjoining members, which normally comprise horizontal beam elements and vertical column elements.

The equations of geometry ensure the compatibility of the deformations with the geometry of the structure, hence the name *geometry conditions* or *compatibility conditions*. An example of such a condition is that no deflection takes place at the intermediate support of a continuous beam, and the rotation is the same on both sides of that support when the two adjacent spans are equal and similarly loaded. Since the manner of load application on one span affects the manner of deformation of the adjacent spans, it is essential to alternate live-load patterns on adjacent spans to give the maximum and minimum (reverse) moments and the resulting stresses in the various parts of the structural system.

Several methods of analysis of statically indeterminate structures have been developed over the years. Some of them are more exact than others, and some include approximations and simplifications that facilitate relatively quick solutions when computer utilization is not readily possible or justified. The classical methods are based primarily on a physical understanding of the structural deformational behavior and are particularly important in interpreting the response of the system to the type and sense of applied load. The availability of computers transformed this need for understanding physical behavior to formatting a problem so that it can be understood by the computer through matrix formulation of the computations. In this manner it becomes possible to keep track of a larger number of calculations and hence to be able to analyze more complex structural systems.

It is important for historical purposes to list the major methods of structural analysis. The common basic concept in these methods is either the *force method* with consistent deformation, for which the redundant forces are the primary unknowns, or the *displacement method*, for which displacements are the primary unknowns. The former is also termed the *flexibility method* and the latter is termed the *stiffness method*.

Emile Clapeyron's *three moments theorem* (1897) is a "force method" in which bending moments at supports are considered as the redundants to be evaluated by solving simultaneous equations whose number is equivalent to the number of indeterminacies. Carlo Castigliano's *energy method*, embodying his second theorem of differential coefficients of internal work (1858), postulates that the partial differential of the internal work of an elastic structure as a function of one of the external forces gives the relative displacement at the point of application of that force. Both force methods, while very powerful at the time, have the limitation of applicability to few spans with essentially unyielding supports and the necessity for exceedingly tedious computational effort. Another application of the force method is the elastic center and the analogous column. In this method the redundants are chosen at a point called the elastic center, involving computations similar to those for stresses in a column subjected to combined bending and direct stress.

The slope deflection concept is an example of the displacement method, where deflections and slopes are taken as the primary unknowns. It was developed independently by Axel Bendixen in Germany (1914) and George Manney in the

United States (1915) and is the precursor of the moment distribution method of Hardy Cross (1929). It is worthwhile noting that it was originally developed because of the need to consider the secondary effects, that is, the internal bending moments and shears resulting from rotational restraints that develop in the bolted or welded truss joints.

The availability of computers for speedy solution of complex problems has decreased interest in use of the classical methods discussed thus far. The approaches here are more in the direction of formatting problems in such a manner that the computer can keep track of large quantities of numbers. Such formatting or bookkeeping can be achieved by the use of matrix methods. Matrix formulations can be used for the "force method–method of consistent deformation" solutions through the use of the *flexibility matrix*. The displacement method can be applied through the use of the *stiffness matrix*. The unknown joint displacements are obtained by solving an equal number of simultaneous equations in matrix form. It should be noted at the outset that the matrix displacement method using stiffness matrices is the most powerful of the various methods of analysis hitherto discussed when computers are used, and in its generalization as the *finite-element method* it is capable of analyzing any elastic structural system and most plastic systems.

13.2 LONGHAND DISPLACEMENT METHODS

13.2.1 Slope Deflection Method

This displacement method involves writing two equations for each span of a continuous structure, one at each end, expressing the end moments as the sum of the following *four* contributions:

1. The restraining moment resulting from an assumed fixed-end condition of the loaded span
2. The moment due to the rotation of the tangent to the elastic curve at one end of the span
3. The moment due to the rotation of the tangent to the elastic curve at the other end of the span
4. The moment resulting from the translation of one end of the span with respect to the other

The equations have to be set to conform to the requirements of equilibrium and compatability at each joint of a continuous beam or a frame.

Consequently, a *large* set of simultaneous linear algebraic equations results for a total structural system, with displacements as the unknowns. For a structure with several spans or a high-rise building, the computational effort needed to solve the equations could be staggering and the probability of errors great. Use of this method is limited today, because other faster methods are available. Example 13.2 with two redundancies is given as an illustration.

13.2.2 Moment Distribution Method

This method is a *numerical* application of the displacement method in which the desired moments, shears, or stresses are obtained by a method of successive approximations suitable for longhand computations. The method lends itself to simple physical interpretation. Hence it can be used for quick approximate as well as exact solutions, depending on the number of successive cycles chosen. It is essentially an iterative solution of the slope deflection equations and has been used extensively since its development by Hardy Cross in the United States in 1929. The student and the designer are assumed to be well acquainted with this hand computational method, and the reader is referred to the various texts on the subject of structural analysis given in the list of references to supplement the examples presented in this chapter using moment distribution.

13.3 FORCE METHOD OF ANALYSIS

13.3.1 The Force Method and the Flexibility Matrix

In this method, earlier forces or moments can be used as redundants. Moments will be used in this book as redundants. They are more convenient than forces in the analysis, particularly at the limit state of failure, as shown in A. L. L. Baker's method of "imposed rotations" presented in Section 13.7.2. Hence a somewhat more detailed discussion with an analysis example is presented here to fit the sequence of analysis methods. It is also assumed that the reader has a background from other courses in structural analysis in both the force method and the matrix force method such that this discussion will serve only as a refresher on the topic.

In this method, cuts or hinges are inserted at suitable points in an indeterminate frame or continuous beam. As many supporting reactions or moments as necessary are removed until the structure becomes statically determinate, thereby facilitating the analysis. If the total strain energy U with respect to the elastic redundant moment X_i at any hinge i is made equal to the *elastic* rotation at the hinge, that is,

$$\frac{\partial U}{\partial X_i} = -\theta_i \qquad (13.1)$$

and if δ_{ik} is assumed to represent the relative rotation of the ith hinge due to a moment at the kth hinge, then $\delta_{ik} = \delta_{ki}$ from Maxwell's reciprocal theorem. The coefficients δ_{ik} are called *influence coefficients* because they represent the displacement or rotation at a particular section due to a unit moment at *another* section, that is, $\delta_{ik} = -\theta_i$.

From the principle of virtual work,

$$\delta_{ik} = \Sigma \int_0^l \frac{M_i M_k}{EI}\, ds \qquad (13.2)$$

The right side of Eq. 13.2 represents the integration of the products of the areas of the M_i diagrams and the ordinates of the M_k diagrams at the *centroids* of the M_i diagrams along the horizontal distances along the span.   In other words,

$$\delta_{ik} = \frac{A_i}{EI} \eta \tag{13.3}$$

where A_i is the area under the primary M_i bending moment diagram and η is the ordinate of the M_k moment diagram under the centroid of the M_i diagram (Ref. 13.2).   As an example, in Fig. 13.1 the influence coefficient δ_{ki} is obtained by superposing the moment diagram M_0 of the primary structure on the moment diagram X_1 of the redundant structure created by the assumed developed hinge 1.

$$A_i = \frac{2}{3} la$$

η under the centroid of M_0 diagram $= c/2$.

$$\delta_{01} = -\frac{1}{EI} \left(\frac{2}{3} la \right) \left(\frac{c}{2} \right) = -\frac{1}{3EI} lac$$

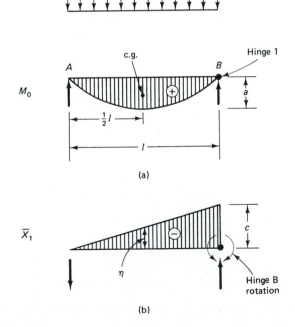

(a)

(b)

Figure 13.1 Influence coefficient determination from superposing M_0 and X_1: (a) primary structure moment; (b) redundant structure moment.

δ_{11} is obtained by superposing the redundant structure X_1 on itself.

$$A_i = -\frac{1}{2} la$$

$$\delta_{11} = -\frac{1}{EI}\left(\frac{1}{2} la \times \frac{2}{3}c\right) = -\frac{1}{3EI} lac$$

Table 13.1 gives the product integral values $\int M_i M_k ds$ for evaluating the influence coefficients δ_{ik} for various combinations of primary and redundant moment diagrams. It can aid the designer in easily getting sets of equations 13.6 to follow.

Equation 13.2 can be rewritten as

$$\Sigma \int_0^l \frac{M_i M_k}{E_c I} ds = -\theta_i \qquad (13.4)$$

TABLE 13.1 Product Integral Values $\int_0^l M_i M_k\, ds$ for Various Moment Combinations $(EI\delta_{ik})$

M_k \ M_i	rectangle (l, a)	right triangle (a, left)	right triangle (a, right)	Parabolic (a)	triangle (a)	trapezoid (d, b)
rectangle (l, c)	lac	$\frac{1}{2} lac$	$\frac{1}{2} lac$	$\frac{2}{3} lac$	$\frac{1}{2} lac$	$\frac{1}{2} l\,(a+b)c$
right triangle (c, left)	$\frac{1}{2} lac$	$\frac{1}{3} lac$	$\frac{1}{6} lac$	$\frac{1}{3} lac$	$\frac{1}{4} lac$	$\frac{1}{6} l\,(2a+b)c$
right triangle (c, right)	$\frac{1}{2} lac$	$\frac{1}{6} lac$	$\frac{1}{3} lac$	$\frac{1}{3} lac$	$\frac{1}{4} lac$	$\frac{1}{6} l\,(a+2b)c$
Parabolic (c)	$\frac{2}{3} lac$	$\frac{1}{3} lac$	$\frac{1}{3} lac$	$\frac{8}{15} lac$	$\frac{5}{12} lac$	$\frac{1}{3} l\,(a+b)c$
triangle (c)	$\frac{1}{2} lac$	$\frac{1}{4} lac$	$\frac{1}{4} lac$	$\frac{5}{12} lac$	$\frac{1}{3} lac$	$\frac{1}{4} l\,(a+b)c$
trapezoid (c, d)	$\frac{1}{2} la\,(c+d)$	$\frac{1}{6} la\,(2c+d)$	$\frac{1}{6} la\,(c+2d)$	$\frac{1}{3} la\,(c+d)$	$\frac{1}{4} la\,(c+d)$	$\frac{1}{6} l\,[a(2c+d)+b(2d+c)]$

Substituting δ_{i0} and δ_{ik} for M_k in Eq. 13.4, the following expression is obtained:

$$\delta_{i0} + \sum_{k=1}^{k=n} \delta_{ik}X_k = -\theta_i \qquad (13.5)$$

where $\theta_i = 0$ is the net *elastic* rotation.

Hence, to solve for a structure having n degrees of indeterminacy, it should satisfy the following condition

$$\delta_{10} + \delta_{11}X_1 + \delta_{12}X_2 + \cdots + \delta_{1n}X_n = -\theta_1 = 0$$
$$\delta_{20} + \delta_{21}X_1 + \delta_{22}X_2 + \cdots + \delta_{2n}X_n = -\theta_2 = 0 \qquad (13.6)$$
$$\delta_{n0} + \delta_{n1}X_1 + \delta_{n2}X_2 + \cdots + \delta_{nn}X_n = -\theta_n = 0$$

The number of equations in a set is equal to the number of redundancies. In matrix form, the structure flexibility matrix $[\theta]$ can be defined for n loading conditions as

$$-[\theta] = \begin{bmatrix} \delta_{11} & \delta_{12} & \cdots & \delta_{1n} \\ \delta_{21} & \delta_{22} & \cdots & \delta_{2n} \\ \vdots & \vdots & & \vdots \\ \delta_{n1} & \delta_{n2} & \cdots & \delta_{nn} \end{bmatrix} \begin{bmatrix} X_{11} & X_{12} & \cdots & X_{1n} \\ X_{21} & X_{22} & \cdots & X_{2n} \\ \vdots & \vdots & & \vdots \\ X_{n1} & X_{n2} & \cdots & X_{nn} \end{bmatrix}$$

$$+ \begin{bmatrix} \delta_{11}^{\circ} & \delta_{12}^{\circ} & \cdots & \delta_{1n}^{\circ} \\ \delta_{21}^{\circ} & \delta_{22}^{\circ} & \cdots & \delta_{2n}^{\circ} \\ \vdots & \vdots & & \vdots \\ \delta_{n1}^{\circ} & \delta_{n2}^{\circ} & \cdots & \delta_{nn}^{\circ} \end{bmatrix} \qquad (13.7a)$$

or in shorter form,

$$[\delta][X] + [\delta^{\circ}] = -[\theta] \qquad (13.7b)$$

Solving for the unknown redundants by inversion of the $[\delta]$ matrix,

$$[X] = [\delta]^{-1}[\theta] - [\delta^{\circ}] \qquad (13.8)$$

The parameters $\delta_{11} \ldots \delta_{nn}$, $X_{11} \ldots X_{nn}$, $\delta_{11}^{\circ} \ldots \delta_{nn}^{\circ}$ can best be described in Fig. 13.2 for a typical two-span continuous beam ABC with a hinge introduced at interior support B due to moment X_1 causing a rotation θ_1 at this support. The beam is subjected to a single external loading condition of uniformly distributed load.

13.3.2 Example 13.1: Analysis of Two-span Continuous Beam by the Force and Matrix Methods

Consider the two-span continuous prismatic beam ABC in Fig. 13.3 having a fixed moment at the right end C. Solve for the moments at B and C due to a uniform load of intensity w per unit length of span using (a) the force method (the method of consistent deformations) and (b) the matrix force method.

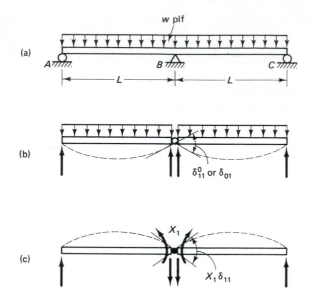

Figure 13.2 Reduction of indeterminate beam through introduction of hinge as redundant: (a) continuous beam elevation; (b) primary structure; (c) redundant structure.

Solution   **(a) Method of consistant deformations.** Since the beam is prismatic, the EI values are not included, for simplification,

$$\delta_{01} + X_1\delta_{11} + X_2\delta_{12} = 0 \tag{a}$$

$$\delta_{02} + X_1\delta_{12} + X_2\delta_{22} = 0 \tag{b}$$

From Eq. 13.3, $\delta_{ik} = (A_i/EI)\,\eta$; hence

$$\delta_{01} = +2\left[\left(\frac{wL^2}{8} \times L \times \frac{2}{3}\right)\left(-\frac{1}{2}\right)\right] = -\frac{wL^3}{12}$$

$$\delta_{02} = \left[\left(\frac{wL^2}{8} \times L \times \frac{2}{3}\right)\left(-\frac{1}{2}\right)\right] = -\frac{wL^3}{24}$$

$$\delta_{11} = -2\left[\left(\frac{1 \times L}{2}\right)\left(-\frac{2}{3}\right)\right] = +\frac{2L}{3}$$

$$\delta_{12} = \left(-\frac{1 \times L}{2}\right)\left(-\frac{1}{3}\right) = +\frac{L}{6} = \delta_{21}$$

$$\delta_{22} = \left(-\frac{1 \times L}{2}\right)\left(-\frac{2}{3}\right) = +\frac{L}{3}$$

Substituting for the values of δ_{01} through δ_{22} in Eqs. (a) and (b),

$$-\frac{wL^3}{12} + \frac{2L}{3}X_1 + \frac{L}{6}X_2 = 0 \tag{c}$$

$$-\frac{wL^3}{24} + \frac{L}{6}X_1 + \frac{L}{3}X_2 = 0 \tag{d}$$

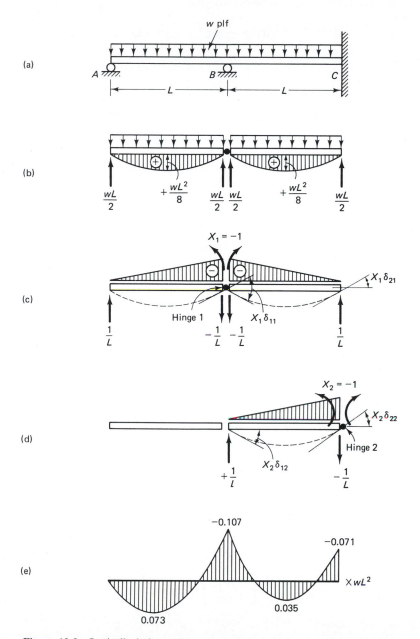

Figure 13.3 Statically indeterminate beam in Ex. 13.1: (a) structure elevation; (b) primary structure (M_0); (c) redundant structure $X_1 = -1$; (d) redundant structure $X_2 = -1$; (e) final bending moments.

Solving Eqs. (c) and (d) simultaneously gives

$$X_1 = \frac{3}{28} wL^2 = 0.107wL^2$$

$$X_2 = \frac{2}{28} wL^2 = 0.071wL^2$$

Hence

$$M_B = -0.107wL^2 \quad \text{and} \quad M_C = -0.071wL^2$$

The balance of moments at midspan and the support reactions can simply be found from statics.

(b) Matrix force method. The structure flexibility matrix $[F]$ from part (a) of the solution is

$$[F] = \frac{L}{6EI} \begin{bmatrix} 4 & 1 \\ 1 & 2 \end{bmatrix}$$

The matrices $[\Delta]$ and $[\Delta°]$ are, respectively,

$$[\Delta] = \begin{bmatrix} 0 \\ 0 \end{bmatrix}$$

$$[\Delta°] = \begin{bmatrix} -\dfrac{wL^3}{12EI} \\ -\dfrac{wL^3}{24EI} \end{bmatrix} = -\frac{wL^3}{24EI} \begin{bmatrix} 2 \\ 1 \end{bmatrix}$$

The flexibility matrix is inverted to yield

$$[F]^{-1} = \frac{6EI}{7L} \begin{bmatrix} 2 & -1 \\ -1 & 4 \end{bmatrix}$$

$$[X] = [F]^{-1} \left[[\Delta] - [\Delta°] \right]$$

or

$$\begin{bmatrix} X_{11} & X_{12} \\ X_{21} & X_{22} \end{bmatrix} = \frac{6EI}{7L} \begin{bmatrix} 2 & -1 \\ -1 & 4 \end{bmatrix} \left[\begin{bmatrix} 0 \\ 0 \end{bmatrix} - \frac{wL^3}{24EI} \begin{bmatrix} 2 \\ 1 \end{bmatrix} \right]$$

$$= \begin{bmatrix} -\dfrac{3wL^2}{28} \\ -\dfrac{2wL^2}{28} \end{bmatrix}$$

Hence

$$M_B = -\frac{3wL^2}{28} = -0.107wL^2$$

$$M_C = -\frac{2wL^2}{28} = -0.071wL^2$$

13.4 DISPLACEMENT METHOD OF ANALYSIS

13.4.1 Displacement Method and the Stiffness Matrix

The displacement method is analogous to the force (deformation) method except that the nodal displacements are considered as the unknowns instead of the redundant forces or moments. It is essentially the slope deflection method and can be considered as the direct link to computer methods of structural analysis. Since the joint displacements represent the freedom to move or rotate, the term "degrees of freedom" represents the joint displacements as a measure of the *kinematic* degrees of indeterminacy.

A set of equilibrium equations equal to the number of unknown displacements has to be solved in order to determine these unknown displacements. The computational operation involves (1) computation of the force–displacement or moment–rotation relationships, that is, the stiffness; (2) setting up the geometrical relationships; (3) setting up the equilibrium equations in order to determine the unknown *displacements* or *rotations*; and (4) calculation of the forces or moments by substituting the displacements or rotations computed in (3) in the force–displacement or moment–rotation relationship established in (1).

In matrix form, we must establish the kinematic degrees of freedom n to be used in the solution. Next, the static matrix $[A]$ and the deformation matrix $[B]$ have to be established using basic concepts, to be followed by a visual check to ensure that the matrix $[B] = [A^T]$; that is, $[B]$ is the transpose of $[A]$. The member stiffness matrix $[S]$ is then computed. The fixed-end moments $\{M_0\}$ are also computed, and the external force (moment) matrix $\{P\}$ is established in which the elements of the $\{P\}$ matrix are the reversals of the forces (moments) acting on the member ends in the fixed condition.

Combining the equilibrium conditions as in Ref. 13.3,

$$\{P\}_{np \times 1} = [A]_{nP \times nM} \cdot \{M\}_{nM \times 1} \tag{13.9}$$

the moment–rotation relationships,

$$\{M\}_{nM \times 1} = [S]_{nM \times nP} \cdot \{\theta\}_{nP \times 1} \tag{13.10}$$

and the compatibility conditions,

$$\{\theta\}_{nM \times 1} = [B]_{nM \times nP} \cdot \{X\}_{nP \times 1} \tag{13.11}$$

and using the inverse of the global stiffness

$$[K] = [ASA^T]_{nP \times nP} \tag{13.12}$$

where the inverse of the matrix is

$$[K]^{-1} = [ASA^T]^{-1}$$

the following two equations of the joint displacement (rotation) matrix $[X]$ and internal force (moment) matrix $[M]$, respectively, are obtained:

$$\{X\}_{nP \times 1} = [ASA^T]^{-1}_{nP \times nP} \cdot \{P\}_{nP \times 1} \qquad (13.13)$$

$$\{M\}_{nM \times 1} = [SA^T]_{nM \times nP} \cdot \{X\}_{nP \times 1} \qquad (13.14)$$

The final end moments $\{M_F\}$ are the *sum* of the moments for the joint fixed condition $\{M_0\}$ and joint rotation condition $\{M\}$ of Eq. 13.14, so

$$\{M_F\}_{nM \times 1} = \{M_0\}_{nM \times 1} + \{M\}_{nM \times 1} \qquad (13.15)$$

13.4.2 Example 13.2: Analysis of Two-span Continuous Beam by the Displacement and Matrix Displacement Methods

Solve Ex. 13.1 for the moments M_B and M_C at the supports of the continuous beam using (a) the displacement method and (b) the matrix displacement method.

Solution (a) Displacement method. This method is the longhand slope deflection method and hence is expected to be cumbersome if the number of redundancies is large. As seen from Fig. 13.3, the structure is statically indeterminate to the second degree; it has two redundancies. The longhand solution would have been considerably quicker if the second displacement method of moment distribution could be used. Figure 13.4 shows the displacement (rotation) and the free-body diagrams of segments AB and BC of the structure. The basic slope deflection equations are

$$M_A = M_{FAB} + \frac{4EI}{L}\theta_A + \frac{2EI}{L}\theta B - \frac{6EI}{L}\Delta \qquad (13.16a)$$

$$M_B = M_{FBA} + \frac{2EI}{L}\theta_A + \frac{4EI}{L}\theta_B - \frac{6EI}{L}\Delta \qquad (13.16b)$$

Since the joints are not rotating, $\Delta = 0$. Hence Eqs. 13.16a and b become

$$M_A = M_{FAB} + \frac{4EI}{L}\theta_A + \frac{2EI}{L}\theta_B \qquad (13.17a)$$

$$M_B = M_{FBA} + \frac{2EI}{L}\theta_A + \frac{4EI}{L}\theta_B \qquad (13.17b)$$

Writing the two joint equilibrium equations corresponding to two degrees of freedom, we have

$$M_1 = 0 \qquad M_2 + M_3 = 0 \qquad M_4 = 0 \qquad (13.18)$$

As a sign convention in Fig. 13.4, M_1, M_2, M_3, and M_4 as unknowns are considered to act clockwise on member ends and counterclockwise on the joints.

(1) *Fixed-end moments*

$$M_{F1} = -\frac{wL^2}{12} \qquad M_{F2} = +\frac{wL^2}{12}$$

$$M_{F3} = -\frac{wL^2}{12} \qquad M_{F4} = +\frac{wL^2}{12}$$

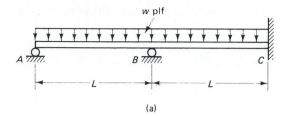

(a)

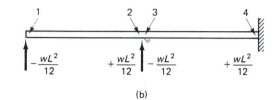

(b)

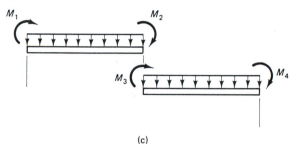

(c)

(d)

Figure 13.4 Slope deflection solution of Ex. 13.2: (a) continuous beam elevation; (b) fixed-end moments; (c) free-body diagrams with unknown moments; (d) free-body diagrams representing degrees of freedom θ_A and θ_B rotations.

(2) *Slope deflection moments:* From Eqs. 13.17a and b,

$$M_1 = -\frac{wL^2}{12} + \frac{4EI}{L}\theta_A + \frac{2EI}{L}\theta_B \tag{13.19a}$$

$$M_2 = +\frac{wL^2}{12} + \frac{2EI}{L}\theta_A + \frac{4EI}{L}\theta_B \tag{13.19b}$$

$$M_3 = -\frac{wL^2}{12} + \frac{4EI}{L}\theta_B + \frac{2EI}{L} \quad (\theta_C = 0) \tag{13.19c}$$

$$M_4 = +\frac{wL^2}{12} + \frac{2EI}{L}\theta_B + \frac{4EI}{L} \quad (\theta_C = 0) \tag{13.19d}$$

where $\theta_C = 0$ because support C is fixed.

(3) *Joint equilibrium moments:* Substituting the slope deflection moments of Eqs. 13.19 into Eq. 13.18 yields

$$M_1 = -\frac{wL^2}{12} + \frac{4EI}{L}\theta_A + \frac{2EI}{L}\theta_B = 0 \tag{13.20a}$$

$$M_2 + M_3 = \frac{2EI}{L}\theta_A + \frac{8EI}{L}\theta_B = 0 \tag{13.20b}$$

$$M_4 = +\frac{wL^2}{12} + \frac{2EI}{L}\theta_B = 0 \tag{13.20c}$$

Solving for Eqs. 13.20a and b, we get

$$EI\theta_A = \frac{1}{42}wL^3$$

$$EI\theta_B = \frac{1}{4 \times 42}wL^3$$

(4) *Final end moments M_F:* Substituting for the values of θ_A and θ_B into Eqs. 13.19, the following final moment values are obtained:

$$M_1 = 0$$

$$M_2 = +\frac{3}{28}wL^2$$

$$M_3 = -\frac{3}{28}wL^2$$

$$M_4 = +\frac{2}{28}wL^2$$

Hence $M_A = 0$, $M_B = 0.107wL^2$, and $M_C = 0.073wL^2$.

(b) Stiffness matrix solution. Figure 13.5 gives the moments and forces on the joints of the continuous beam *ABC* identifying the matrix notations used in the solution.

(1) *Static matrix [A]:* From Fig. 13.5e, the equilibrium of joints where $P_1 = M_1$ and $P_2 = M_2 + M_3$ gives static equilibrium in matrix form as follows:

$$[A]_{2 \times 4} =$$

P \ M	1	2	3	4
1	+1.0			
2		+1.0	+1.0	

(2) *Deformation matrix [B]:* This matrix relates the rotations (internal deformations) to joint displacements. From the geometry of Fig. 13.5d, the deformation

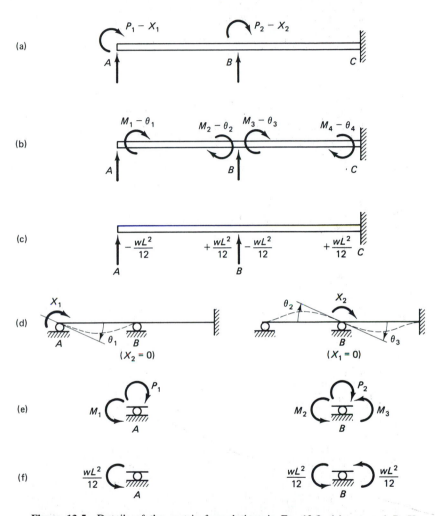

Figure 13.5 Details of the matrix formulations in Ex. 13.2: (a) external $P-X$ diagram; (b) internal $M-\theta$ diagram; (c) fixed-end moments M_0; (d) compatibility of internal deformations (rotations) with joint displacements; (e) equilibrium moments at joints; (f) fixed-end moments at joints. Sign convention: Moment causing compression at bottom face is positive.

matrix is

$$[B]_{4 \times 2} = \begin{array}{c|c|c} & 1 & 2 \\ \hline 1 & +1.0 & \\ \hline 2 & & +1.0 \\ \hline 3 & & +1.0 \\ \hline 4 & & \end{array}$$

where $\theta_2 = X_1$ and $\theta_3 = X_1$ when $X_1 \neq 0$ and $X_2 = 0$
$\theta_4 = X_2$ when $X_2 \neq 0$ and $X_1 = 0$

It should be emphasized that after establishing matrix $[B]$ independently from matrix $[A]$ a check is made that one is the transpose of the other.

(3) *Member stiffness matrix* $[S]$: This matrix is obtained from the pair of Eqs. 13.17a and b and the slope deflection solutions given in Eqs. 13.19 of the previous solution, where

$$M_1 = \frac{4EI}{L} \theta_A + \frac{2EI}{L} \theta_B - \frac{wL^2}{12}$$

$$M_2 = \frac{2EI}{L} \theta_A + \frac{4EI}{L} \theta_B + \frac{wL^2}{12}$$

$$M_3 = \frac{4EI}{L} \theta_B - \frac{wL^2}{12}$$

$$M_4 = \frac{2EI}{L} \theta_B + \frac{wL^2}{12}$$

Hence

$$[S]_{4 \times 4} = \begin{array}{c|c|c|c|c} & 1 & 2 & 3 & 4 \\ \hline 1 & \dfrac{4EI}{L} & \dfrac{2EI}{L} & & \\ \hline 2 & \dfrac{2EI}{L} & \dfrac{4EI}{L} & & \\ \hline 3 & & & \dfrac{4EI}{L} & \\ \hline 4 & & & & \dfrac{2EI}{L} \end{array}$$

(4) *External force (moment) matrix* $\{P\}$: From the fixed-end moments in Fig. 13.5c and the rotational equilibrium of joints A and B in Fig. 13.5f, the net *reverse* moments on the joints are

$$P_1 = -\left(-\frac{wL^2}{12}\right) = +\frac{wL^2}{12}$$

$$P_2 = -\left(\frac{wL^2}{12} - \frac{wL^2}{12}\right) = 0$$

$$\{P\}_{2\times 1} = \begin{array}{c|c}
\diagdown \, M_0 \\
P \diagdown & 1 \\
\hline
1 & +\dfrac{wL^2}{12} \\
\hline
2 & 0
\end{array}$$

(5) $[SA^T]$ *matrix* $= [S] \times [A^T]$: Transpose $[A^T] = [B]$; hence

$$[SA^T]_{4\times 2} = \begin{array}{c|c|c}
\diagdown \, X \\
M \diagdown & 1 & 2 \\
\hline
1 & \dfrac{4EI}{L} & \dfrac{2EI}{L} \\
\hline
2 & \dfrac{2EI}{L} & \dfrac{4EI}{L} \\
\hline
3 & 0 & \dfrac{4EI}{L} \\
\hline
4 & 0 & \dfrac{2EI}{L}
\end{array}$$

(6) *Global stiffness matrix* $[K]$: The complete stiffness matrix from Eq. 13.12 $[K] = [ASA^T]$; hence

$$[K]_{2\times 2} = \begin{array}{c|c|c}
\diagdown \, X \\
P \diagdown & 1 & 2 \\
\hline
1 & \dfrac{4EI}{L} & \dfrac{2EI}{L} \\
\hline
2 & \dfrac{2EI}{L} & \dfrac{8EI}{L}
\end{array}$$

Note that the global stiffness matrix $[K]$ is the same as the factors of the pair of equilibrium Eqs. 13.20a and b in the longhand slope deflection solution of part (a) in this problem.

(7) *Inverse of the stiffness matrix* $[K]^{-1}$:

$$[K]^{-1}_{2\times2} = \begin{array}{c|c|c} \diagdown^{P}_{X} & 1 & 2 \\ \hline 1 & \dfrac{2L}{7EI} & -\dfrac{L}{14EI} \\ \hline 2 & -\dfrac{L}{14EI} & +\dfrac{L}{7EI} \end{array}$$

(8) *Joint displacement matrix* $\{X\}$: The joint displacement matrix $\{X\}$ from Eq. 13.13 is the product of the premultiplier inverse stiffness matrix $[K]^{-1}$ and the postmultiplier external moment matrix $\{P\}$: $\{X\} = [K]^{-1} \times \{P\}$.

The matrix operation gives

$$X_1 = \frac{2L}{7EI} \times \frac{wL^2}{12} - 0 = +\frac{wL^3}{42EI}$$

$$X_2 = -\frac{L}{14EI} \times \frac{wL^2}{12} = -\frac{wL^3}{4 \times 42EI}$$

and in matrix format, the joint displacement matrix would thus be

$$\{X\}_{2\times1} = \begin{array}{c|c} 1 & +\dfrac{wL^3}{42EI} \\ \hline 2 & -\dfrac{wL^3}{4 \times 42EI} \end{array}$$

(9) *Internal moment matrix* $\{M\}$: From Eq. 13.14, the internal moment matrix is the product of the $[SAT]$ premultiplier matrix and the joint displacement $\{X\}$ postmultiplier matrix, giving

$$M_1 = +\frac{1}{12} wL^2$$

$$M_2 = +\frac{1}{42} wL^2$$

$$M_3 = -\frac{1}{42} wL^2$$

$$M_4 = -\frac{1}{84} wL^2$$

and in matrix form

$$\{M\}_{4 \times 1} = \begin{array}{|c|c|}\hline 1 & +\dfrac{1}{12}\, wL^2 \\\hline 2 & +\dfrac{1}{42}\, wL^2 \\\hline 3 & -\dfrac{1}{42}\, wL^2 \\\hline 4 & -\dfrac{1}{84}\, wL^2 \\\hline\end{array}$$

The final end moments $[M_F]$ from Eq. 13.15 in matrix form are the *sum* of the fixed-end moments $[M_0]$ and the internal moments $\{M\}$. In matrix form, the final moment values are as follows:

$$\{M_F\}_{4 \times 1} = [M_0]_{4 \times 1} + \{M\}_{4 \times 1}$$

or

$$\{M_F\}_{4 \times 1} = \begin{array}{|c|}\hline -\dfrac{1}{12}\, wL^2 \\\hline +\dfrac{1}{12}\, wL^2 \\\hline -\dfrac{1}{12}\, wL^2 \\\hline +\dfrac{1}{12}\, wL^2 \\\hline\end{array} \;+\; \begin{array}{|c|}\hline +\dfrac{1}{12}\, wL^2 \\\hline +\dfrac{1}{42}\, wL^2 \\\hline -\dfrac{1}{42}\, wL^2 \\\hline -\dfrac{1}{84}\, wL^2 \\\hline\end{array} \;=\; \begin{array}{|c|}\hline 0 \\\hline +\dfrac{3}{28}\, wL^2 \\\hline -\dfrac{3}{28}\, wL^2 \\\hline +\dfrac{2}{28}\, wL^2 \\\hline\end{array}$$

Hence

$$M_A = 0$$

$$M_B = \frac{3}{28}\, wL^2 = 0.107 wL^2$$

$$M_C = \frac{2}{28}\, wL^2 = 0.073 wL^2$$

13.5 FINITE-ELEMENT METHODS AND COMPUTER USAGE

The finite-element method is an extension of the matrix displacement method in which the body or structure to be analyzed is modeled as an assembly of finite elements interconnected at specified nodal points. The difference between the two methods is the choice of the stiffness matrix, permitting the inclusion of *different* types of elements into the analysis. As a result, the solution is greatly facilitated through convergence using the computer. The accuracy of the stiffness matrices can be distinctly improved by the introduction of additional nodes along the length of the member or in the plane of a planar element, depending on the degree of accuracy needed.

Numerous canned computer programs for structural analysis are available using the matrix displacement method or the finite-element method. FORTRAN standard programming language is widely used in such problem-oriented programs as STRESS, discussed in Ref. 13.4; PSCST finite-element programs, presented in Ref. 13.6; ANSYS finite-element program for three-dimensional analysis (Ref. 13.7); the PCA ADOSS program in BASIC language for the design of slabs and plate systems as part of continuous frames (Ref. 13.8); and others. The reader would do well to become familiar with the available programs and acquire the background knowledge for use of computer methods in the solution of highly indeterminate continuous structures and high-rise building frames.

It should be noted, however, that the designer must always execute computational checks on the computer output. Such checks can be accomplished through the use of the classical structural analysis methods on small subassemblages of elements or individual elements. Hence the preceding presentations in Sections 13.1 to 13.4 are a necessary refresher for conducting such checks for which hand-held computers can often be adequate.

13.6 APPROXIMATE ANALYSIS OF CONTINUOUS BEAMS AND FRAMES

13.6.1 Idealization Principle

The use of computers has facilitated the rapid analysis of continuous structures with high degrees of indeterminacy, giving relatively exact solutions. This advantage has come through the digital computational process, applying matrix methods and finite-element techniques and utilizing the revolutionary advances in the hardware and software capabilities and speed of today's desktop computers. However, preceding a detailed computer analysis, the elastic properties of the members have to be assumed as an input requirement. These include modulus of elasticity, cross-sectional area, cross-sectional moment of inertia, and the length of members. All these parameters are needed for establishing preliminary stiffness values for the beams and columns. Also, the pattern of load distribution that can give the worst loading conditions has to be set by the design engineer. Hence *approximate* structural analysis has to be initially performed with the appropriate idealizations *prior* to embarking on an "exact" solution using the computer. In many instances for moderate-sized structures, the approximate solution is often sufficient since the input of stiffness values into a computer involves idealizations and assumptions

Photo 75   Chicago Mercantile Exchange; a high-strength concrete unique cantilever supports an office tower. (Courtesy of Robert B. Johnson, Alfred Benesch and Co., Chicago.)

based on engineering judgment and isolation of controlling segments of an indeterminate structure to arrive at a preliminary stiffness.

Taking the simple case of a fixed-end beam as in Fig. 13.6a, the elastic curve of the beam changes slope at a distance of $0.211L$ from the fixed supports, thereby creating *inflection points*: points of zero moments at points C and D in the span. Consequently, AC and BD can be treated as cantilever beams, and segment CD can be considered a simply supported beam of span $L_1 = 0.578L$ and solved by simple statics.

Frames can be treated in a similar manner through location of the inflection points. Figure 13.7 shows two portal frames, one with a hinged base and the other with a fixed base subjected to gravity loading. Note that the bending moment diagrams are consistently drawn on the *tension* side of the members. The bending moment at midspan of the horizontal top member would be the difference between the moment at B or C and the total static moment $wL_B^2/8$. If the same frame is subjected to horizontal wind forces, the inflection points and the resulting bending moments are as shown in Fig. 13.8.

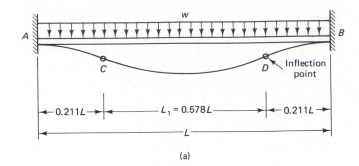

(a)

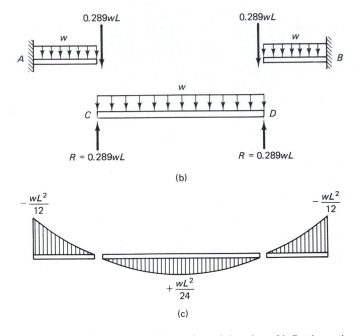

(c)

Figure 13.6 Idealization of fixed-end beam through location of inflection points: (a) fixed-end beam elevation; (b) idealization through location of inflection points; (c) bending moment diagram.

On the basis of the foregoing discussion for gravity and wind loads, a multistory frame can thus be idealized as shown in Fig. 13.9a for gravity loading and in Fig. 13.9b for wind loading, with the inflection points located at the sections where a change in curvature direction takes place. Figure 13.9b is uniquely suitable for approximate analysis due to horizontal wind loads assumed concentrated at joints.

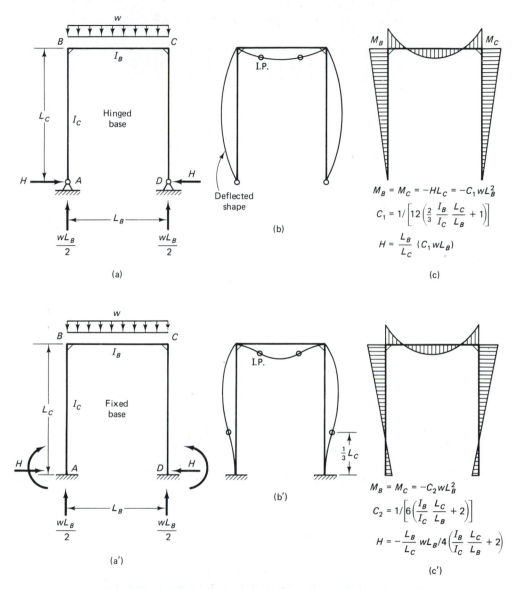

Figure 13.7 Idealization of portal frame through location of inflection points, gravity loading: (a, a′) frame elevation; (b, b′) deflected shape; (c, c′) bending moment diagrams.

13.6.1.1. Portal method of wind loading frame analysis. It is seen from Fig. 13.9(b) that the structure is divided into portals because moments are taken as zero at midspans of the horizontal members and at midheight of the vertical members, rendering the entire structure statically determinate. This method

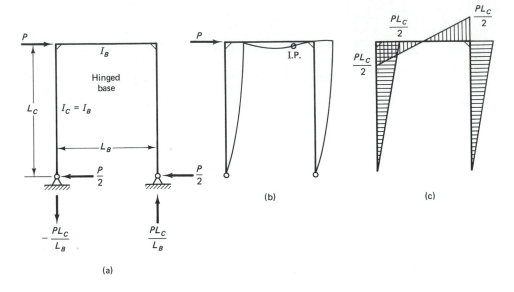

(a)

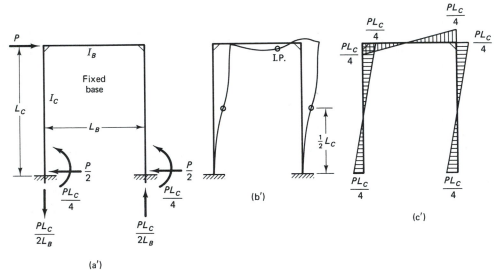

(a')

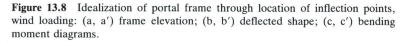

Figure 13.8 Idealization of portal frame through location of inflection points, wind loading: (a, a') frame elevation; (b, b') deflected shape; (c, c') bending moment diagrams.

of analysis for wind on frames is termed the *portal method*. It is based on the following assumptions:

1. All wind loads are transferred to the joint.
2. Shear resisted by each exterior column is assumed to be one-half that resisted by each interior column.

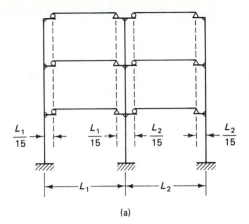

(a)

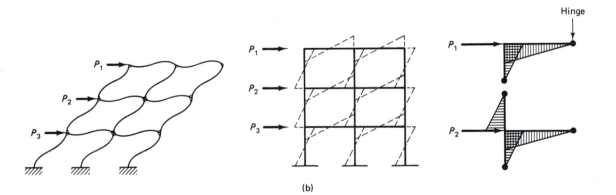

(b)

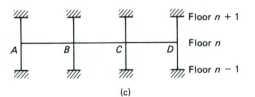

(c)

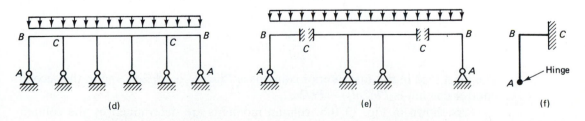

(d)   (e)   (f)

Figure 13.9 Idealization of continuous structures for appoximate analysis: (a) gravity loading; hinges assumed at $\frac{1}{15}$ of span from column support for preliminary analysis; (b) wind loading; (c) alternative idealization of multistory frame for gravity loading; (d) single-floor multispan symmetrical portals; (e) portal idealization of structure in (d); (f) portal unit *ABC*.

3. The total horizontal shear in the columns of a given story is equal to the total lateral force above that story.

4. Inflection points, equivalent to hinges, occur at midspan of beams and at column midheight except in basement levels. In that case, it can be assumed to occur at about one-fourth to one-third of the column height above the foundation.

These assumptions are based on the fact that the total shearing force due to wind at any floor level can be divided among the columns at that level in proportion to their stiffnesses, and the vertical reactions on the columns due to wind can be considered to be proportional to their distance from the *center* of the building. Such assumptions are true only if the beams are infinitely stiff relative to the columns. Yet they are nearly correct in most cases and do provide an adequate safety factor, considering that I_c values for columns are constant between any two floors and the I_b values for beams are in most cases constant throughout the frame. It must also be remembered that the ratio of stiffnesses of the beams to the columns does not significantly affect the value of the ultimate load because excessive stiffness in the columns is eliminated by plastic yield before failure.

Figure 13.9c shows the idealized portion of a high-rise building for approximate analysis where fixity of columns can be assumed at the $n + 1$ and $n - 1$ floors. The end moment values of the beam $AB-BC-CD$ are not drastically affected by this approximation, and the moment coefficients for continuous beams on knife-edge supports can be used in the approximate analysis. Figure 13.9d and e shows an approximation procedure for a symmetrical one-story frame under gravity loading. Note that the replacement of the portal intermediate columns by fixed ends at column locations C transformed the structure into simple and essentially single bay portals.

Figure 13.10 shows a portal subjected to wind load P and having equal spans. Inflection points are at midheights and at midspans. It is seen that axial loads due to horizontal wind force P occur at the exterior columns only, since the combined tension and compression due to the portal effect results in zero axial load in the interior columns. If the spans are unequal, wind loading P would cause axial loads not only on the exterior columns but also on the columns between the unequal spans.

The general expression for axial load in the exterior columns of an n-bays frame with unequal spans is

$$\frac{Ph}{2nl_1} \quad \text{and} \quad \frac{Ph}{2nl_n}$$

The axial load in the first interior columns is $Ph/2nl_1 - Ph/2nl_2$ and in the second interior column is $Ph/2nl_2 - Ph/2nl_3$.

As shown in Fig. 13.10b, column moments are determined by the column shear times one-half the column height. Consequently, for joint B of the portal frame, the column moment becomes $(P/3) \cdot (h/2)$ to give a moment value of

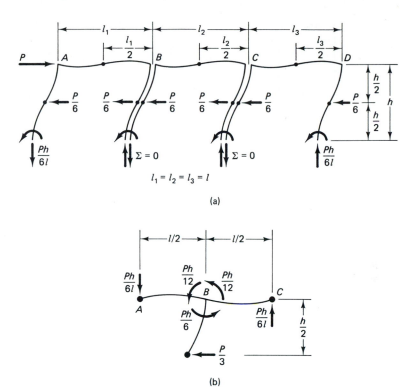

Figure 13.10   Three-bay, equal-span portal frame wind analysis: (a) horizontal shears and axial loads; (b) moments at joint B (inflection points at midheights and midspans).

$Ph/6$.   This moment must be balanced by equal moments in BA and BC of a magnitude $Ph/12$ without considering their relative stiffnesses.   The shear in beams AB and BC is then determined by dividing the beam end moments by one-half the beam length.   In this case, the end shear becomes $(Ph/12)/(l/2)$, giving a value of $(Ph/6l)$ at each floor level.   Summation is then made of the beam shear and column load values as we proceed from the top floor to the foundation floor level.

Example 13.5 illustrates the use of the portal frame method for the analysis of forces in an indeterminate frame due to wind loading.

13.6.2 Indeterminate Frames and Portals

13.6.2.1. General properties.   Concrete frames are indeterminate structures consisting of horizontal and vertical or inclined members joined in such a manner that the connection can withstand the stresses and bending moments that act on it.   The degree of indeterminacy depends on the number of spans, number of vertical members, and type of end reactions.   Typical frame configurations are shown in Fig. 13.11.   If n is the number of joints, b the number of members, r

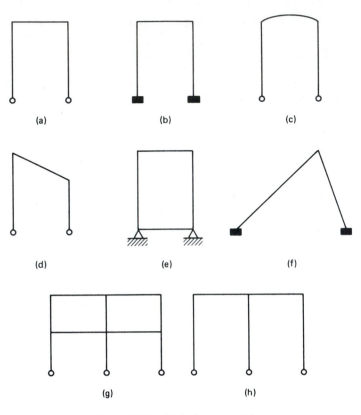

Figure 13.11 Typical structural frames.

the number of reactions, and s the number of indeterminacies, the degree of indeterminacy is determined from the following inequalities:

$$3n + s > 3b + r \qquad \text{unstable} \qquad\qquad (13.21a)$$

$$3n + s = 3b + r \qquad \text{statically determinate} \qquad\qquad (13.21b)$$

$$3n + s < 3b + r \qquad \text{statically indeterminate} \qquad\qquad (13.21c)$$

The degree of indeterminacy is

$$s = 3b + r - 3n \qquad\qquad (13.22)$$

where $3n$ equations of static equilibrium are always available and the total number of unknowns is $3b + r$.

As an example, the degree of indeterminacy of the frame in Fig. 13.11a is

$$s = 3 \times 3 + 2 \times 2 - 3 \times 4 = 1$$

For the frame in Fig. 13.11g,

$$s = 3 \times 10 + 2 \times 3 - 3 \times 9 = 9$$

For a frame to perform satisfactorily, the following conditions must be satisfied:

1. The design should be based on the most unfavorable moment and shear combinations. If moment reversal is possible due to reversal of live-load direction, the highest values of positive and negative bending moments have to be considered in the design.

2. Proper foundation support for horizontal thrust has to be provided. If the frame is designed as hinged, which is an expensive construction procedure, an actual hinge system has to be provided.

13.6.2.2. Forces and moments in portal frames.

The behavior of concrete frames before cracking can reasonably be considered elastic, as was done in the case of continuous beam at service-load and slight overload conditions. Consequently, well before the development of plastic hinging, the bending moment diagrams shown in Figs. 13.12 and 13.13 can easily be used in the design of in-

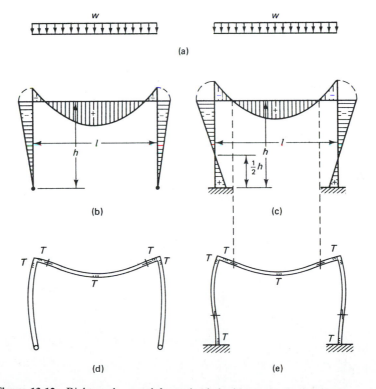

Figure 13.12 Right-angle portal frame loaded with gravity load intensity w (T indicates tension fibers): (a) load intensity; (b) bending moment (hinged base frame); (c) bending moment (fixed base frame); (d) deformation of frame (b); (e) deformation of frame (c).

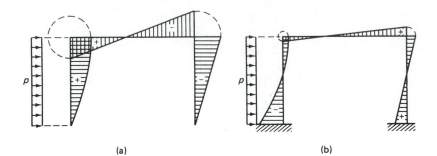

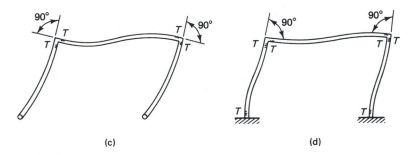

Figure 13.13 Right-angle portal frame loaded with wind load intensity p (T indicates tension fibers): (a) bending moment (hinged base frame); (b) bending moment (fixed base frame); (c) deformation of frame (a); (d) deformation of frame (b).

determinate reinforced concrete frames. It has to be assumed that the student or the design engineer is well versed in these procedures as a basic background, and only the minimum guidelines and simplifications are presented in this book.

13.6.2.3. Uniform gravity loading on single-bay portal. Assuming that the moments of inertia I_c of the vertical columns and I_b of the horizontal beam of the portal in Fig. 13.14a are not equal, the following values of the moments and thrusts can be deduced:

End shear in beam

$$V_B = V_C = -\frac{1}{2} wl \qquad (13.23a)$$

horizontal thrust $H = \frac{l}{h} C_1 wl$ \qquad (13.23b)

where

$$C_1 = \frac{1}{12\left(\frac{2}{3}\frac{I_b}{I_c}\frac{h}{l} + 1\right)} \qquad (13.23c)$$

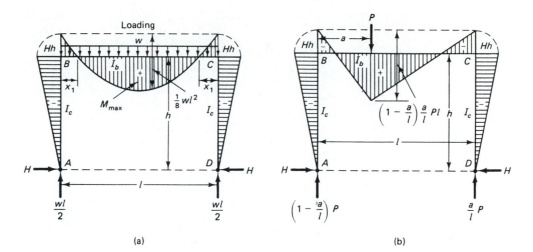

(a)   (b)

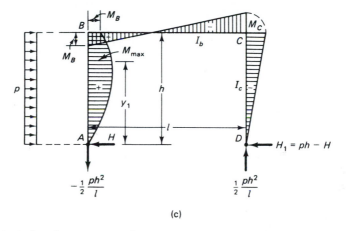

(c)

Figure 13.14 Bending moment ordinates in single-bay frame: (a) uniform gravity loading; (b) concentrated gravity loading; (c) uniform horizontal pressure.

Maximum negative moment at beam-column junction

$$M_B = M_C = -Hh = -C_1 wl^2 \tag{13.23d}$$

Maximum positive moment at midspan

$$M_{max} = \frac{1}{8} wl^2 - Hh = \left(\frac{1}{8} - C_1\right) wl^2 \tag{13.23e}$$

Bending moments at any point X

$$M_X = \frac{1}{2} l(l - x)w - C_1 wl^2 \tag{13.23f}$$

Photo 76 River City Apartment Complex, Chicago, Illinois. The marina area has room for 70 moors, (Courtesy Bertrand Goldberg Associates, Architects and Engineers, Chicago, Illinois.)

where points of contraflexure from either corner of the portal are

$$x_1 = \frac{1}{2}[1 - \sqrt{(1 - 8C_1)}]\, l = C_2 l \qquad (13.23g)$$

and

$$C_2 = \frac{1}{2}(1 - \sqrt{1 - 8C_1}) \qquad (13.23h)$$

13.6.2.4. Concentrated gravity loading on single-bay portal.

Since the concentrated load P does not have to act at midspan, nonsymmetry of shears results.   The end shears from Fig. 13.14b are

$$V_B = \left(1 - \frac{a}{l}\right)P \quad \text{and} \quad V_C = \frac{a}{l}P \qquad (13.24a)$$

Horizontal thrust

$$H = C_3 \frac{a}{l}\left(1 - \frac{a}{l}\right)P\frac{l}{h} \qquad (13.24b)$$

where

$$C_3 = \frac{1}{2\left(\dfrac{2}{3}\dfrac{I_b}{I_c}\dfrac{h}{l} + 1\right)} \qquad (13.24c)$$

Bending moments at corners

$$M_B = M_C = -Hh = -C_3 \frac{a}{l}\left(1 - \frac{a}{l}\right)Pl \qquad (13.24d)$$

Bending moments at any point along BC: For $x < a$,

$$M_x = \left(1 - \frac{a}{l}\right)\left(\frac{x}{l} - \frac{a}{l}C_3\right)Pl \qquad (13.24e)$$

For $x > a$,

$$M_x = \frac{a}{l}\left[1 - \frac{x}{l} - \left(1 - \frac{a}{l}\right)C_3\right]Pl \qquad (13.24f)$$

Maximum postive moment at $x = a$

$$M_{max} = \frac{a}{l}\left(1 - \frac{a}{l}\right)Pl - Hh = (1 - C_3)\frac{a}{l}\left(1 - \frac{a}{l}\right)Pl \qquad (13.24g)$$

Horizontal thrust for several constructed gravity loads

$$H = \frac{l}{h}C_3\left[P_1\frac{a_1}{l}\left(1 - \frac{a_1}{l}\right) + P_2\frac{a_2}{l}\left(1 - \frac{a_2}{l}\right) + \cdots\right] \qquad (13.24h)$$

or

$$H = \frac{l}{h}C_3 \Sigma\, P\frac{a}{l}\left(1 - \frac{a}{l}\right) \qquad (13.24i)$$

13.6.2.5. Uniform horizontal pressure on single-bay portal.
From Fig. 13.14c,

Vertical reactions at supports

$$R_A = -\frac{1}{2}ph\frac{h}{l} \quad \text{and} \quad R_D = +\frac{1}{2}ph\frac{h}{l} \qquad (13.25a)$$

Horizontal reactions: For windward hinge A,

$$H_A = \frac{1}{8}\frac{11\frac{I_b}{I_c}\frac{h}{l} + 18}{2\frac{I_b}{I_c}\frac{h}{l} + 3}ph = C_4 ph \qquad (13.25b)$$

where

$$C_4 = \frac{1}{8}\frac{11\frac{I_b}{I_c}\frac{h}{l} + 18}{2\frac{I_b}{I_c}\frac{h}{l} + 3} \qquad (13.25c)$$

For leeward hinge D,

$$H_D = ph - H_A = (1 - C_4)ph \tag{13.25d}$$

Bending moments at any point y along the column height due to horizontal pressure, with y being measured from the *bottom*

$$M_y = H_A y - \frac{1}{2}py^2 \tag{13.25e}$$

Maximum moment at windward column

$$M_{\max} = \frac{1}{2}\left(\frac{1}{8}\frac{11\frac{I_b}{I_c}\frac{h}{l} + 18}{2\frac{I_b}{I_c}\frac{h}{l} + 3}\right)ph^2 = \frac{1}{2}(C_4)\,ph^2 \tag{13.25f}$$

Point of maximum bending moment above support A

$$y_1 = \frac{1}{8}\left(\frac{11\frac{I_b}{I_c}\frac{h}{l} + 18}{2\frac{I_b}{I_c}\frac{h}{l} + 3}\right)h = C_4 h \tag{13.25g}$$

Bending moments in corners of portal

$$M_B = H_A h - \frac{1}{2}ph^2 = \frac{3}{8}\frac{\frac{I_b}{I_c}\frac{h}{l} + 2}{2\frac{I_b}{I_c}\frac{h}{l} + 3}ph^2$$

$$= (C_4 - 0.5)ph^2 \tag{13.25h}$$

$$M_C = -H_D h = -(1 - C_4)ph^2 \tag{13.25i}$$

The constants C_1, C_2, C_3, and C_4 in Eqs. 13.23, 13.24, and 13.25 can be graphically represented as in Fig. 13.15. Canned computer programs for the analysis of indeterminate beams and frames render the use of charts such as Fig. 13.15 unnecessary except for a quick check of numerical values.

13.6.3 Example 13.3: Forces and Moments in a Warehouse Portal Frame

A warehouse structure is constructed of a single-bay portal frame hinged at the base. The frame has a clear span of 80 ft (24.4 m) and is built in 8-ft segments. It is subjected to a uniform gravity load intensity $w_L = 240$ plf (3.5 kN/m) and a horizontal uniform wind pressure of intensity $p_w = 65$ plf (0.95 kN/m) at the windward side and a suction of intensity $p_L = 40$ plf (0.58 kN/m) at the leeward side, as shown in Fig. 13.16. Compute the shears, moments, and reactions that would be needed for the

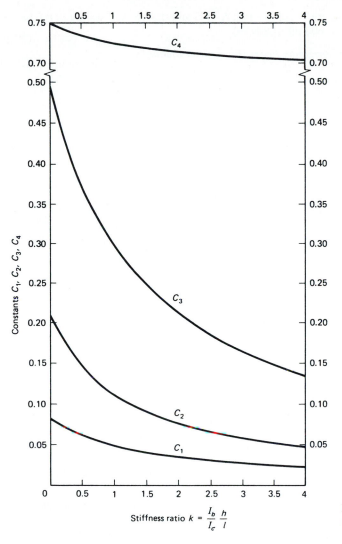

Figure 13.15   Constants C_1 to C_4 in Eqs. 13.23, 13.24, and 13.25.

Stiffness ratio $k = \dfrac{I_b}{I_c} \dfrac{h}{l}$

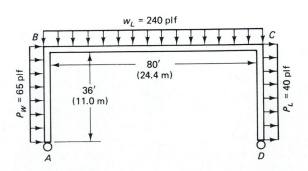

Figure 13.16   Warehouse frame elevation.

design of the structure.   Assume that $I_b = I_c$ and that the self-weight of the horizontal top member $w_D = 600$ plf (8.8 N/m).

Solution   *Frame horizontal beam BC, gravity loading*

From Eq. 13.23c,

$$\text{stiffness coefficient } C_1 = \frac{1}{12 \left(\frac{2}{3}k + 1 \right)}$$

where

$$k = \frac{I_b}{I_c} \frac{h}{l} = \frac{36}{80} \times 1.0 = 0.45$$

From Fig. 13.15, $C_1 = 0.064$.   Assume a 2-in. concrete topping and 5-psf insulation and water proofing.

$$w_{s3} = \left(\frac{2}{12} \times 150 + 5 \right)8 = 240 \text{ plf over 8-ft-wide strip}$$

$$w_u = 1.4(600 + 240) + 1.7 \times 240 = 1584 \text{ plf (23.1 kN/m)}$$

From Eq. 13.23d,

$$M_{uB} = M_{uC} = -Hh = -C_1wl^2 = -0.064 \times 1584(80)^2 \times 12$$
$$= -7.79 \times 10^6 \text{ in.-lb (380 kNm)}$$

From Eq. 13.23e,

$$\text{maximum } M_u \text{ at midspan} = \left(\frac{1}{8} - C_1 \right)wl^2$$

$$= \left(\frac{1}{8} - 0.064 \right)1584(80)^2 \times 12$$

$$= +7.42 \times 10^6 \text{ in.-lb (839 kN/m)}$$

$$\text{column vertical reactions } R_C = R_D = \frac{wL}{2} = \frac{1584 \times 80}{2} = 63{,}360 \text{ lb (282 kN)}$$

Column top moments and reactions: wind loading

$$\text{windward } p_u = 65 \times 1.7 = 110.5 \text{ plf}$$
$$\text{leeward } p_u = 40 \times 1.7 = 68.0 \text{ plf}$$

From Eqs. 13.25h and 13.25i,

$$M_B = (C_4 - 0.5)ph^2$$
$$M_C = -(1 - C_4)ph^2$$

From before,

$$k = \frac{I_b}{I_c} \frac{h}{l} = 0.45 \quad \text{for } I_b = I_c$$

From Fig. 13.15, $C_4 = 0.73$.

Windward side moment M_B

$$M_{B1} = (0.73 - 0.50)110.5(36)^2 \times 12 = 395,254 \text{ in.-lb}$$

$$M_{B2} = (1 - 0.73)68.0(36)^2 \times 12 = 285,535 \text{ in.-lb}$$

$$\text{total } M_B = 395,254 + 285,535 = 680,789 \text{ in.-lb (76.9 kNm)}$$

Leeward side moment M_C

$$M_{C1} = -(1 - 0.73)110.5(36)^2 \times 12 = -463,994 \text{ in.-lb}$$

$$M_{C2} = -(0.73 - 0.50)68.0(36)^2 \times 12 = -243,233 \text{ in.-lb}$$

$$\text{total } M_C = -463,994 - 243,233 = -707,227 \text{ in.-lb (79.9 kNm)}$$

Controlling wind moment $M_{uw} = 707,277$ in.-lb, since wind can blow from either left or right.   By superposition of moments due to gravity on the moment due to wind,

$$\text{maximum } M_B = \text{maximum } M_C = 7.79 \times 10^6 - 707,227$$

$$= 7.08 \times 10^6 \text{ in.-lb}$$

$$\text{maximum midspan moment} = 7.42 \times 10^6 \text{ in.-lb}$$

Vertical support reactions: From Eq. 13.25a,

$$R_{cA} = -\frac{1}{2}ph\,\frac{h}{l} = -\frac{(110.5 + 68.0)(36)^2}{2 \times 80} = -1446 \text{ lb}$$

$$R_{cD} = -R_{cA} = +1446 \text{ lb (6.4 kN)}$$

Total R_u due to gravity and wind load $= 63,360 + 1446 = 64,806$ lb (288 kN).   Hence design the vertical supports for a combined $P_u = 64,806$ lb and $M_u = 7.08 \times 10^6$ in.-lb.   Note that the axial load value is so small compared to the moment magnitude that the design of the vertical supports would be governed by flexure.   The design of the beam *BC* and verticals *AB* and *CD* for flexure and shear would be accomplished in accordance with the discussions in Chapters 5 and 6.   Check also for moment redistribution from support to midspan if necessary, as discussed in the ACI Code section on moment coefficients.

13.6.4 Loading

The first step in the analysis of frames is the determination of the *service* loads and wind stresses, as required in the general building code under which the project is to be designed and constructed, such as ANSI standard A581 (Ref. 13.25). Dead load includes member self-weight, weight of fixed service equipment, such as electrical and plumbing, and the weight of built-in partitions, which is normally taken as 20 psf at service level. Live loads include loads due to movable objects and movable partitions. The uniformly distributed live loads range between 40 psf at service for residential use and 450 psf for havy manufacturing and warehouse storage. Portions of buildings such as library stacks and film rooms require substantially heavier live loads. Live loads on roofs include maintenance equipment, snow loads, ponding of water, and landscaping where applicable. If concentrated live loads have to be included, they would more likely affect individual supporting members and do not generally need to be included in the frame analysis.

Live-load reduction for the design of beams, slabs, and columns is generally allowed in most building codes to account for the probability that the total floor area *influencing* the load on an individual element may not be fully loaded. For example, the influencing area for an interior column is the total area of the four surrounding bays, while for an edge column, the influencing area is the two adjacent bays, and for a corner column, it is one bay only.

The reduced live load L_r per square foot of floor area supported on columns, beams, and two-way slabs having an influence area of more than 400 ft^2 is

$$L_r = L\left(0.25 + \frac{15}{\sqrt{A_I}}\right) \tag{13.26}$$

where L = unreduced service live load
$\quad\quad A_I$ = influencing area

The reduced live load cannot be taken less than 50% for members supporting one floor or less than 40% of the unit live load L.

13.6.5 ACI Moment Coefficients

13.6.5.1. Center-line moments.   The ACI building code allows moment and shear coefficients for rapid analysis of standard buildings of usual types of spans, construction, and story heights. Table 13.2 represents these coefficients. The limitations are as follows:

1. Maximum allowable ratio of live to dead load is 2:1.
2. Maximum allowable span difference should be such that the larger of the two adjacent spans does not exceed the shorter by more than 20%.

It should be noted that the values in Table 13.2 are somewhat conservative, and economy can be achieved by more precise analysis of the multistory structural system. Figure 13.17 gives a graphical representation of the moment and shear coefficients of Table 13.2 for a typical three-span structure under various support conditions. Moment coefficients for continuous beams and alternative loadings where ACI moment coefficients cannot be used are given in Table 13.3.

13.6.5.2. Redistribution of negative moments in continuous non-prestressed flexural members.   Except where approximate values for moments are used, negative moments calcualted by the elastic theory at supports of continuous flexural members on each be increased or decreased by not more than the value obtained from the following expression:

$$p_d = 20\left(1 - \frac{\rho - \rho'}{\rho_b}\right) \quad \text{percent} \tag{13.27}$$

where $\rho = A_s/bd$
$\quad\quad \rho' = A_s'/bd$

TABLE 13.2   ACI Moment and Shear Coefficients[a]

Positive moment	
End spans	
If discontinuous end is unrestrained	$\frac{1}{11}w_u l_n^2$
If discontinuous end is integral with the support	$\frac{1}{14}w_u l_n^2$
Interior spans	$\frac{1}{16}w_u l_n^2$
Negative moment at exterior face of first interior support	
Two spans	$\frac{1}{9}w_u l_n^2$
More than two spans	$\frac{1}{10}w_u l_n^2$
Negative moment at other faces of interior supports	$\frac{1}{11}w_u l_n^2$
Negative moment at face of all supports for (1) slabs with spans not exceeding 10 ft and (2) beams and girders where ratio of sum of column stiffness to beam stiffness exceeds 8 at each end of the span	$\frac{1}{12}w_u l_n^2$
Negative moment at interior faces of exterior supports for members built integrally with their supports	
Where the support is a spandrel beam or girder	$\frac{1}{24}w_u l_n^2$
Where the support is a column	$\frac{1}{16}w_u l_n^2$
Shear in end members at first interior support	$1.15\dfrac{w_u l_n}{2}$
Shear at all other supports	$\dfrac{w_u l_n}{2}$

Source: Ref. 13.13.
[a] w_u = total factored load per unit length of beam or per unit area of slab
  l_n = clear span for positive moment and shear and the average of the two adjacent clear spans for negative moment

The redistribution of negative moments can be made only when the section at which the moment is reduced is so designed that

$$\rho \quad \text{or} \quad (\rho - \rho') \le 0.50\rho_b \tag{13.28}$$

where

$$\rho_b = 0.85 \,\beta_1 \frac{f_c'}{f_y} \times \frac{87,000}{87,000 + f_y}$$

is defined as the balanced steel percentage for a singly reinforced section (Chapter 5).

From the foregoing discussion it can be expected that a *decrease* in the negative moment at the supports requires a comparable *increase* in the postive moment in the span, such that the support reinforcement percentage $\rho - \rho' \le 0.50\rho_b$. Similarly, an increase in the negative moment at the support requires a comparable decrease in the positive mometn such that the midspan $(\rho - \rho') \le 0.50\rho_b$.

Analysis of Eq. 13.27 indicates that, for singly reinforced beams ($\rho' = 0$) and using a reinforcement percentage $\rho = 0.50\rho_b$ to prevent steel congestion, the redistribution factor $p_D = 10\%$. Hence a flat redistribution reduction factor of 10% for the elastic negative moment can easily be applied. In actual practice the

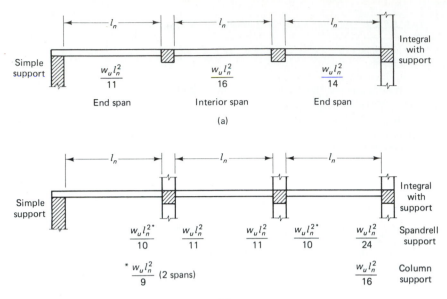

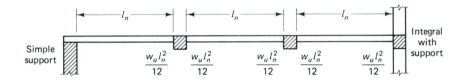

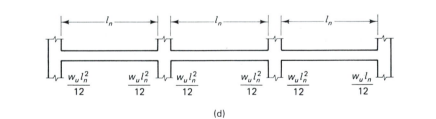

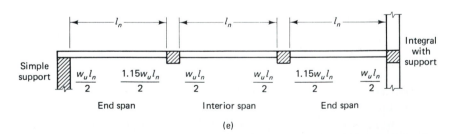

Figure 13.17 ACI moment and shear coefficients: (a) positive moments, all cases; (b) negative moments, beams and slabs; (c) negative moments, slabs with span ≤10 ft; (d) negative moments, beams with stiff columns ($\Sigma\, K_C/\Sigma\, K_B > 8$); (e) support shears, all cases.

TABLE 13.3 Bending Moments and Shear Diagrams for Continuous Beams

1. CONTINUOUS BEAM—THREE EQUAL SPANS—ONE END SPAN UNLOADED

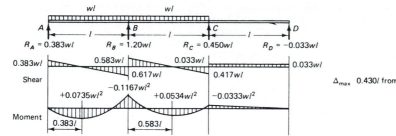

Δ_{max} 0.430l from A) = 0.0059wl^4/EI

2. CONTINUOUS BEAM—THREE EQUAL SPANS—END SPANS LOADED

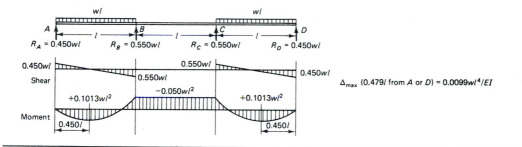

Δ_{max} (0.479l from A or D) = 0.0099wl^4/EI

3. CONTINUOUS BEAM—THREE EQUAL SPANS—ALL SPANS LOADED

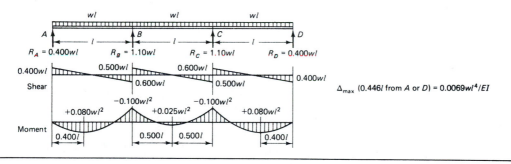

Δ_{max} (0.446l from A or D) = 0.0069wl^4/EI

4. CONTINUOUS BEAM—FOUR EQUAL SPANS—THIRD SPAN UNLOADED

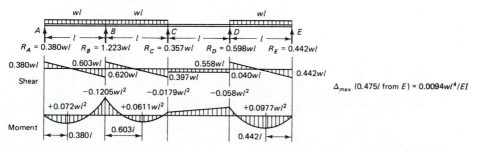

Δ_{max} (0.475l from E) = 0.0094wl^4/EI

TABLE 13.3 (continued)

5. CONTINUOUS BEAM—FOUR EQUAL SPANS—LOAD FIRST AND THIRD SPANS

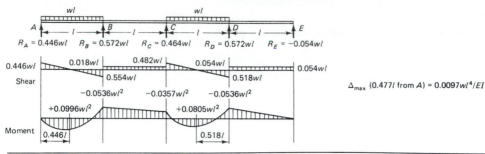

$R_A = 0.446wl$ $R_B = 0.572wl$ $R_C = 0.464wl$ $R_D = 0.572wl$ $R_E = -0.054wl$

Δ_{max} (0.477l from A) = 0.0097wl^4/EI

6. CONTINUOUS BEAM—FOUR EQUAL SPANS—ALL SPANS LOADED

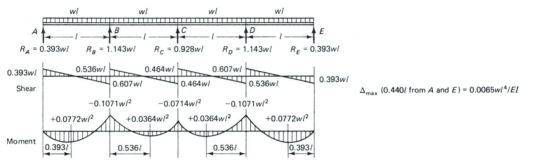

$R_A = 0.393wl$ $R_B = 1.143wl$ $R_C = 0.928wl$ $R_D = 1.143wl$ $R_E = 0.393wl$

Δ_{max} (0.440l from A and E) = 0.0065wl^4/EI

7. CONTINUOUS BEAM—TWO EQUAL SPANS—CONCENTRATED LOAD AT CENTER OF ONE SPAN

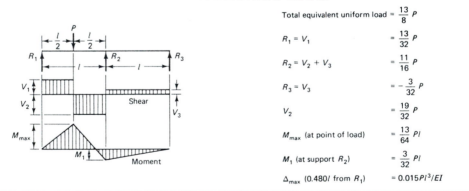

Total equivalent uniform load = $\dfrac{13}{8} P$

$R_1 = V_1$ $= \dfrac{13}{32} P$

$R_2 = V_2 + V_3$ $= \dfrac{11}{16} P$

$R_3 = V_3$ $= -\dfrac{3}{32} P$

V_2 $= \dfrac{19}{32} P$

M_{max} (at point of load) $= \dfrac{13}{64} Pl$

M_1 (at support R_2) $= \dfrac{3}{32} Pl$

Δ_{max} (0.480l from R_1) $= 0.015 Pl^3/EI$

8. CONTINUOUS BEAM—TWO EQUAL SPANS—CONCENTRATED LOAD AT ANY POINT

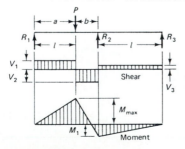

$R_1 = V_1$ $= \dfrac{Pb}{4l^3} [4l^2 - a(l + a)]$

$R_2 = V_2 + V_3$ $= \dfrac{Pa}{2l^3} [2l^2 + b(l + a)]$

$R_3 = V_3$ $= -\dfrac{Pab}{4l^3} (l + a)$

V_2 $= \dfrac{Pa}{4l^3} [4l^2 + b(l + a)]$

M_{max} (at point of load) $= \dfrac{Pab}{4l^3} [4l^2 - a(l + a)]$

M_1 (at support R_2) $= \dfrac{F}{4l^2}$

application of *computed* redistribution percentage from the high negative moment regions to the lower positive regions is limited. The computational effort can be warranted, however, if reasonable saving is achieved in construction costs through reinforcement modification instead of using a flat 10% value for p_D.

13.6.5.3. Effective span moments. The usual assumption in frame analysis postulates that the members are prismatic, having constant moment of inertia between center lines. In reality, a beam stops to have constant section at the face of the column support, and its moment of inertia is greatly increased as it approaches the column center line. To account for this discrepancy in the analysis, the moments and shears have to be adjusted by the change in moment values between the support center line and the support face, as shown in Fig. 13.18. The adjusted moment value can be determined by mapping the area of the shear diagram between the column face and its center line. This area can be taken as $\frac{1}{2}Val$ for the knife-edge support assumption of $\frac{1}{3}Val$ for the finite support area as usually exists.

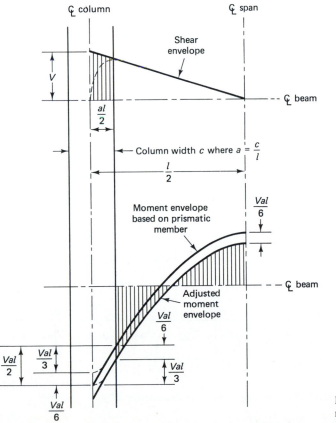

Figure 13.18 Modification of frame moments at support and midspan.

13.6.6 Example 13.4: Analysis for Continuity by Moment Distribution

For continuous beams and one-way slabs not meeting the ACI moment coefficient conditions of Table 13.2 and for two-way-action slabs and plates as part of a structural framing system, analysis by the other methods discussed in Sections 13.1 to 13.5 becomes necessary. The moment distribution method discussed in Section 13.2.2 will be used in the following illustrative example. Except in unique cases, building frame and continuous beam moments computed by adjusting the fixed-end moments by *two cycles* of moment distribution are sufficiently accurate for design purposes.

Problem statement. A flat-plate floor of a multifloor framing system is shown in Fig. 13.19. The end panels center-line dimensions are 17'-6" × 20'-0" (5.33 m × 6.10 m), and the interior panel dimensions are 24'-0" × 20' 0" (7.52 m × 6.10 m). The floor heights l_n of intermediate floors are typically 8'-9" (2.67 m). Compute the factored controlling moments needed for the design of a typical floor panel to withstand service live load $w_L = 40$ psf (1.92 kPa) and a superimposed dead load $w_{SD} = 14$ psf due to partitions and flooring. Assume that the slab thicknesses $h = 6\frac{1}{2}$ in. (16.5 cm) and that all panels are simultaneously loaded by the live load in the solution. Use

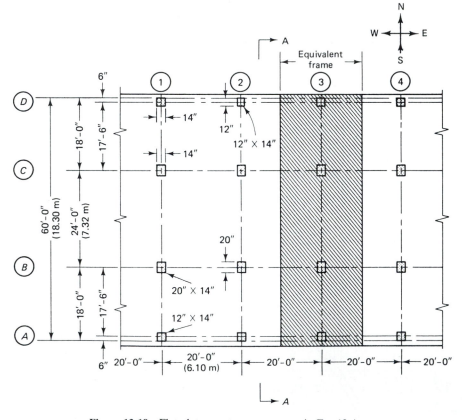

Figure 13.19 Flat-plate apartment structure in Ex. 13.4.

the equivalent frame method for determining the stiffness of the members intersecting at the panel supports.

Solution   *Equivalent frame characteristics*

Take the equivalent frame in the N–S direction whose plane is shown in the shaded portion in Fig. 13.19.   The approximate stiffness of the column above and below the floor joint (moment per unit rotation) can be approximated by

$$K_c = \frac{4E_c I_c}{L_n - 2h} \tag{13.29}$$

where $l_n = 8'\text{-}9'' = 105$ in.

Exterior column (14 in. × 12 in.) stiffness

$$b = 14 \text{ in.} \qquad I_c = 14(12)^3/12 = 2016 \text{ in.}^4$$

Assume that

$$\frac{E_{col}}{E_{slab}} = \frac{E_{cc}}{E_{cs}} = 1.0$$

Use $E_{cc} = E_{cs} = 1.0$ in the calculations because E_{cs} drops out in the equation for K_{ec}.

$$\text{total } K_c = \frac{4 \times 1 \times 2016}{105 - (2 \times 6.5)} \times 2 \quad \text{(for top and bottom columns)}$$

$$= 175.3 \text{ in.-lb/rad}/E_{cc}$$

$$\text{torsional constant } C = \Sigma \left(1 - 0.63 \frac{x}{y}\right) x^3 \frac{y}{3}$$

$$= \left(1 - 0.63 \times \frac{6.5}{12}\right) 6.5^3 \times \frac{12}{3} = 724$$

Torsional stiffness of the slab at the column line is

$$K_t = \Sigma \frac{9 E_{cs} C}{L_2 (1 - c_2/L_2)^3}$$

$$= \frac{9 \times 1 \times 724}{20 \times 12[1 - 14/(12 \times 20)]^3} + \frac{9 \times 1 \times 724}{20 \times 12[1 - 14/(12 \times 20)]^3}$$

$$= 65 \text{ in.-lb/rad}/E_{cs}$$

The equivalent column stiffness is

$$K_{ec} = \left(\frac{1}{K_c} + \frac{1}{K_t}\right)^{-1} = \left(\frac{1}{175.3} + \frac{1}{65}\right)^{-1} = 47 \text{ in.-lb/rad}/E_{cc}$$

Interior column (14 in. × 20 in.) stiffness

$$b = 14 \text{ in.} \qquad I = \frac{14(20)^3}{12} = 9333 \text{ in.}^4$$

$$\text{total } K_c = \frac{4 \times 1 \times 9333}{105 - 2 \times 6.5} \times 2 = 812 \text{ in.lb/rad/}E_{cc}$$

$$C = (1 - 0.63 \times 6.5/20) \times (6.5)^3 \times 20/3 = 1456$$

$$K_t = \frac{9 \times 1456}{20 \times 12[1 - 14/(12 \times 20)]^3} + \frac{9 \times 1456}{20 \times 12[1 - 14/(12 \times 20)]^3}$$

$$= 131 \text{ in.-lb/rad/}E_{cs}$$

$$K_{ec} = \left(\frac{1}{812} + \frac{1}{131}\right)^{-1} = 113 \text{ in.-lb/rad/}E_{cc}$$

Slab stiffness: From Eq. 13.29,

$$K_s = \frac{4E_{cs}I_s}{L_n - c_1/2}$$

where L_n = center-line span
c_n = column depth

Slab band width in E–W direction = 20/2 + 20/2 = 20 ft.

$$I_s = 20 \times \frac{12(6.5)^3}{12} = 5493 \text{ in.}^4$$

Slab at right of exterior column A:

$$K_s = \frac{4 \times 1 \times 20(6.5)^3}{12 \times 17.5 - 12/2} = 108 \text{ in.-lb/rad/}E_{cs}$$

Slab at left of interior column B:

$$K_s = \frac{4 \times 1 \times 20(6.5)^3}{12 \times 17.5 - 20/2} = 110 \text{ in.-lb/rad/}E_{cs}$$

Slab at right of interior column B:

$$K_s = \frac{4 \times 1 \times 20(6.5)^3}{12 \times 24 - 20/2} = 79 \text{ in.-lb/rad/}E_{cs}$$

Slab distribution factor at joints: DF = $K_s/\Sigma K$, where $\Sigma K = K_{ec} + K_{s(\text{left})} + K_{s(\text{right})}$.

$$\text{outer joint } A \text{ slab DF} = \frac{108}{47 + 108} = 0.697$$

$$\text{left joint } B \text{ slab DF} = \frac{110}{113 + 110 + 79} = 0.364$$

$$\text{right joint } B \text{ slab DF} = \frac{79}{113 + 110 + 79} = 0.262$$

Two-cycle moment distribution

$$\text{slab self-weight } w_D = \frac{6.5}{12} \times 150 = 81 \text{ psf}$$

$$w_u = 1.4(81 + 14 + 20) + 1.7 \times 40 = 201 \text{ psf}$$

Exterior spans AB and CD

$$\text{fixed-end moment, FEM} = \frac{wL_n^2}{12}$$

$$= \frac{201(17.5)^2}{12} \times 12 = 61,556 \simeq 61.56 \times 10^3 \text{ in.-lb/ft}$$

Interior span BC

$$\text{fixed-end moment, FEM} = \frac{201(24)^2}{12} \times 12 = 115,776 \simeq 115.78 \times 10^3 \text{ in.-lb/ft}$$

Running a moment distribution analysis as shown in Table 13.4, a carryover factor COF $= \frac{1}{2}$ can be used for all spans. This assumption is justified because the effect of nonprismatic sections would have negligible bearing on the fixed-end moments and carry-over factors. It can also be assumed in multistory, multispan frames that the frame at a joint two spans away from the left joint (joint C) can be considered fixed in the distribution of moments. Hence the slabs at interior supports B and C have to be designed for factored moments $M_u = 97,540$ in.-lb/ft for spans BA and CD and $M_u = 112,440$ in.-lb/ft for span BC. These moments can be adjusted for values at column face.

Proportioning of the reinforcement after similar evaluation for the E–W direction can be made similar to the discussion presented in Chapter 11, including proportioning for shear and moment transfer at column junctions.

TABLE 13.4 Two-cycle Moment of Distribution of Factored Moments M_v

	A		B	℄	C	
DF	0.697		0.364	0.262	0.262	0.364
COF	0.5		0.5	0.5	0.5	0.5
FEM ($\times 10^3$ in.-lb/ft)	−61.56		+61.56	−115.78	+115.78	−61.56
Distribution	+42.91		+19.74	+14.21	−14.21	
CO	+9.87		+21.46	−7.11		
Distribution	−6.88		−5.22	−3.76		
Final M_u ($\times 10^3$ in.-lb/ft)	−15.66		+97.54	−112.44		

13.6.7 Example 13.5: Analysis of Wind Moments and Shears in a Multistory Frame by the Portal Method

The structure in Ex. 13.4 is four stories in height plus a basement, as shown in Fig. 13.20. It is subjected to a wind intensity of 35 psf (1676 Pa). Assume all floor heights as well as the basement height to be the same: 8'-9" (2.67 m). Compute the moments

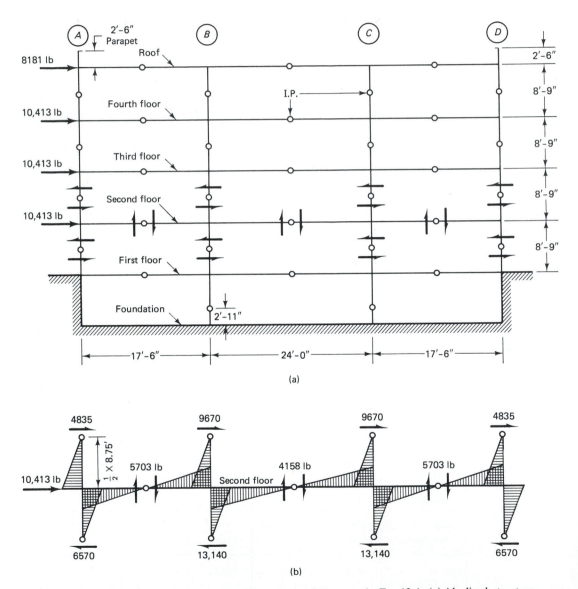

(a)

(b)

Figure 13.20 Portal frame wind analysis of structure in Ex. 13.4: (a) idealized structure frame with inflection points; (b) shear forces on second-story portals.

and shears in the first- and second-story columns and the second-floor interior beams on lines 1, 2, 3, or 4.   Assume that the building roof has a 2'-6" (76 cm) parapet wall.

Solution   *External forces*

Figure 13.20 shows the idealized structure and the inflection points for a portal frame analysis due to wind.   The inflection points (virtual hinges) are at midheights of columns and at midspans of beams except for the basement, where the column hinges are one-third basement height above the foundation.

The wind forces are based on the tributary areas and are concentrated at the joints.   Hence portal frame analysis is applicable.

$$\text{half-floor height} = \frac{8'\text{-}9''}{2} = 4.375 \text{ ft}$$

load factor = 1.7

wind force at roof joint = 1.7(4.375 + 2.5)20 ft × 35 psf = 8181 lb

wind force at each floor joint = 1.7(8.75 × 20 × 35) = 10,413 lb

total shear in the second story is equivalent to the sum of the horizontal

shear forces *above* that level = 8181 + 2 × 10,413 = 29,007 lb (129 kN)

total shear in the first story = 8181 + 3 × 10,413 = 39,420 lb (175 kN)

Second-story columns: shears and moment

$$\text{shear in exterior columns, } V_{2e} = \frac{29,007}{6} = 4835 \text{ lb}$$

$$\text{shear in interior columns, } V_{2i} = 2 \times 4835 \text{ or } \frac{29,007}{3} = 9670 \text{ lb}$$

$$\text{moment in exterior columns, } M_{2e} = \frac{4835 \times 8.75}{2} = 21,153 \text{ ft.-lb}$$

$$\text{moment in interior columns, } M_{2i} = \frac{9670 \times 8.75}{2} = 42,306 \text{ ft.-lb}$$

First-story columns: shears and moments

$$\text{shear in exterior columns, } V_{1e} = \frac{39,420}{6} = 6570 \text{ lb}$$

shear in interior columns, $V_{1i} = 2 \times 6570 = 13,140$ lb

$$\text{moment in exterior columns, } M_{1e} = \frac{6570 \times 8.75}{2} = 28,744 \text{ ft.-lb}$$

$$\text{moment in interior columns, } M_{1i} = \frac{13,140 \times 8.75}{2} = 57,488 \text{ ft.-lb}$$

Moments and shears in interior beams on line 1, 2, 3, or 4

Beams in bay AB or CD

$$M_{ue} = M_{2e} + M_{1e} = 21,153 + 28,744 = 49,897 \text{ ft.-lb/bay}$$

$$V_{ue} = \frac{49,897}{17.5 \times 0.5} = 5703 \text{ lb/bay}$$

Beams in bay BC

$$M_{ui} = (M_{2i} + M_{1i}) - M_{ue} = (42{,}306 + 57{,}488) - 49{,}897 = 44{,}061 \text{ ft.-lb/bay}$$

$$V_{ui} = \frac{44.061}{24 \times 0.5} = 36.72 \text{ lb/bay}$$

These shears and moments would have to be combined with the gravity load moments and shears computed by the frame analysis of Ex. 13.4. Axial loads on columns are computed similarly as discussed in Section 13.6.1.1. Since bay spans are not equal, net *small* axial loads on the interior columns due to wind loading have to be computed and added to those due to gravity loads and adjustments made in shears and moments. It should be noted that the numerical computations can become exceedingly cumbersome for tall buildings with a large number of bays. In such cases, the use of computers in evaluating the axial forces, shears, and moments becomes necessary.

The horizontal and vertical shear forces computed in this example for the *second*-floor portals are shown in Fig. 13.20b. For wind blowing from left to right, the total axial tensile force due to wind in column *A* and the axial compressive force due to wind in column *D* are equal to the sum of all the vertical shears in the beams in all the floors and the roof in bays *AB* and *CD*, respectively. Note that the bending moment diagram in Fig. 13.20b due to wind is drawn on the *tension* face of the members.

NOTE: The portal method is an *approximate* and rapid method of analyzing the moments and shears due to wind in multistory frames. Exact analysis taking into account the $P-\Delta$ effects on the columns can be achieved by the utilization of canned computer programs, as explained in Section 9.13.

13.7 LIMIT DESIGN (ANALYSIS) OF INDETERMINATE BEAMS AND FRAMES

The discussions presented in this chapter thus far entail the *elastic* analysis of indeterminate beam and frame systems with examples for "exact" solutions as well as by approximate methods based on the geometrical behavior of the structure. Also, redistribution factors p_d for continuity are introduced where permissible provided that *adequate* longitudinal reinforcement is provided at the critical continuity zones to control the cracking levels of such zones.

These procedures do not necessarily give the most efficient solution to a statically indeterminate continuous beam or frame, since full redistribution at ultimate load is not considered. As the applied load is gradually increased until the structure as a whole reaches its limit capacity, the critical sections, such as the supports or corners of frames, develop severe cracking, and rotations become so large that for all practical purposes rotating *plastic hinges* develop. If the number of plastic hinges that develops equals the number of the indeterminacies, the structure becomes determinate, as *full redistribution* of moments would have taken place throughout the structure. With the development of an additional hinge, the structure becomes a mechanism resulting in a collapse.

Analysis of the structure at *full* moment redistribution is termed *plastic* or *limit* analysis. Since concrete cracks severely at high overloads, it is possible for the designer to *impose the desirable* locations of the plastic hinges by making the concrete member fail or making it adequately strong at any section by decreasing or increasing the reinforcement percentage without appreciably altering the stiffness of the member. This flexibility in proportioning is not available in the plastic design of steel structures where the resulting locations of the plastic hinges are obtained from mechanisms determined by upper- and lower-bound solutions. Details of the *theory of imposed rotations* by A. L. L. Baker are well presented in Refs. 13.9, 13.10, and 13.11.

13.7.1 Method of Imposed Rotations

The *imposed* locations of the plastic hinges coincide with the locations of the maximum elastic moments for combined gravity loads and horizontal wind loads. These locations occur at intermediate supports of continuous beams and beam–column corners of frames as seen in the portal frame of Fig. 13.21. By super-imposing Fig. 13.21a and b, the maximum elastic moment occurs at corner C. As plastic moments are a magnification of the elastic moments, the natural location for the development of a plastic hinge is corner C. Since the structure is inde-terminate to the first degree, only one hinge develops, resulting in a *basic* frame ABC, which is the fundamental frame for the imposed hinges seen in Fig. 13.21e, numbered in the order in which they are expected to form. This structure has nine indeterminacies; hence nine plastic hinges are formed. A tenth hinge reduces the structure to a mechanism resulting in a collapse. Note that no plastic hinges are permitted to form at midspan of the horizontal members. The plastic moments corresponding to assumed hinges 1, 2, 3, . . . , n are denoted by $\overline{X}_1, \overline{X}_2, \overline{X}_3, \ldots,$ $\overline{X}_n$ and are assumed to remain constant throughout the progressive deformation of the structure.

Hence the derivative of the total strain energy U with respect to the assumed plastic moments $\overline{X}_i$ at any hinge i is made equal to the plastic rotation at the hinge:

$$\frac{\partial U}{\partial X_i} = -\theta_i \tag{13.30}$$

Equation 13.30 is similar to Eq. 13.1 except that plastic moment $\overline{X}_i$ is used instead of the elastic moment X_i. If δ_{ik} is assumed to represent the relative rotation of the ith hinge due to a unit moment at the kth hinge, $\delta_{ik} = \delta_{ki}$ from Maxwell's reciprocal theorem. The coefficients δ_{ik} are called *influence coefficients* because they represent the displacement or rotation at a particular section due to a unit moment at *another* section; that is, $\delta_{ik} = -\theta_i$.

From the principle of virtual work,

$$\delta_{ik} = \Sigma \int_0^l \frac{M_i M_k}{E_c I}\, ds = -\theta_i \tag{13.31}$$

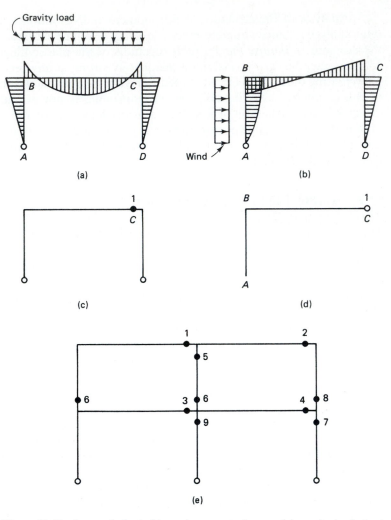

Figure 13.21 Imposed plastic hinges in concrete frames: (a) gravity load elastic moment; (b) wind load intensity moments; (c) hinge 1 at C reducing frame to statically determinate; (d) basic plastic frame; (e) succession of plastic hinges in two spans, two-level frame.

where θ_i has a finite value *not* equal to zero, as was the case in the elastic analysis represented by Eq. 13.4.

By substituting δ_{io} and δ_{ik} for M_k in Eq. 13.31, the following expression is obtained:

$$\delta_{i0} + \sum_{k=1}^{k=n} \delta_{ik}\overline{X}_k = -\theta_i \qquad (13.32)$$

Photo 77   Rotation measurement setup using inclinometers in flanged beam (Nawy et al.).

Hence a structure should develop n plastic hinges to reduce it to statically determinate

$$\delta_{10} + \delta_{11}\overline{X}_1 + \delta_{12}\overline{X}_2 + \cdots + \delta_{in}\overline{X}_n = -\theta_1$$
$$\delta_{20} + \delta_{21}\overline{X}_1 + \delta_{22}\overline{X}_2 + \cdots + \delta_{2n}\overline{X}_n = -\theta_2 \qquad (13.33)$$
$$\delta_{n0} + \delta_{n1}\overline{X}_1 + \delta_{n2}\overline{X}_2 + \cdots + \delta_{nn}\overline{X}_n = -\theta_n$$

The number of equations in a set is equal to the number of redundancies or indeterminacies. By trial and adjustment of the redundant plastic moments $\overline{X}_1$, . . . , $\overline{X}_n$ in the solution of the set of Eqs. 13.33 for controlled maximum allowable rotation of the largest rotating hinge θ_1, the plastic moments at beam supports and column ends are obtained for the plastic design of the concrete structure. It has to be emphasized that the arbitrary plastic moment values $\overline{X}_1, \overline{X}_2, \ldots, \overline{X}_n$ are chosen in Eqs. 13.33 to result in plastic rotations $\theta_1, \theta_2, \ldots, \theta_n$ that give *full redistribution* of moments throughout the structure.

As shown in Section 13.3.1, the influence coefficient δ_{ik} in Eqs. 13.33 is

$$\delta_{ik} = \frac{A_i}{EI} \eta \qquad (13.34)$$

where A_i is the area under the primary M_i bending moment diagram and η is the ordinate of the M_k moment diagram under the centroid of the M_i diagram (Ref. 13.9). Table 13.1 and the example accompanying it give solutions for products of integral values $\int_0^l M_i M_k$ for various moment combinations $EI\delta_{ik}$.

13.7.2 Example 13.6: Determination of Plastic Hinge Rotations in Continuous Beams

Determine the required plastic hinge rotation in the four-span beam of Fig. 13.22. The beam is subjected to a simple span plastic moment M_0 so that midspan moment = support moment = $\frac{1}{2}M_0$ before full rotation of the hinges and full moment redistribution takes place.

Solution   The structure is statically indeterminate to the third degree, so three hinges will develop at the plastic limit.   Assume the maximum ordinate c of the redundant moment at hinge location to be unity.   From Table 13.1 and Fig. 13.23

$$EI\delta_{10} = -\frac{2}{3}\,M_0 l \qquad EI\delta_{11} = \frac{2}{3}\,l$$

$$EI\delta_{12} = \frac{1}{6}\,l \qquad EI\delta_{13} = 0$$

From Eq. 13.33,

$$-\theta_1 = \delta_{10} + \delta_{11}\overline{X}_1 + \delta_{12}\overline{X}_2 + \delta_{13}\overline{X}_3$$

$$-EI\theta_1 = -\frac{2}{3}\,M_0 l + 0.5M_0\left(\frac{2l}{3}\right) + 0.5M_0\left(\frac{l}{6}\right) + 0 = -\frac{M_0 l}{4}$$

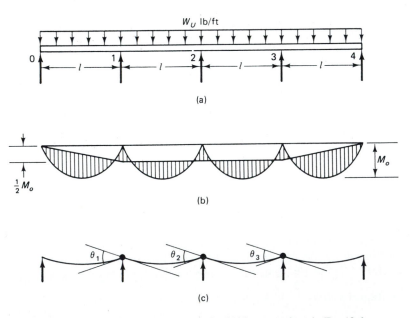

Figure 13.22   Primary moments and plastic hinge rotations in Ex. 13.6.

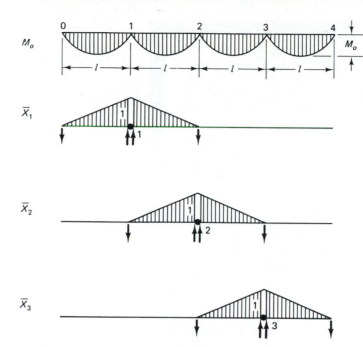

Figure 13.23   Primary and redundant moments in Ex. 13.6.

Similarly, from Table 13.1 and Fig. 13.23

$$EI\delta_{20} = \frac{2}{3} M_0 l\left(-\frac{1}{2}\right) + \frac{2}{3} M_0 l\left(-\frac{1}{2}\right) = -\frac{2}{3} M_0 l$$

$$EI\delta_{21} = \left(-\frac{l}{2}\right)\left(-\frac{1}{3}\right) = +\frac{l}{6}$$

$$EI\delta_{22} = 2\left(-\frac{l}{2}\right)\left(-\frac{2}{3}\right) = +\frac{2l}{3}$$

$$EI\delta_{23} = \left(-\frac{l}{2}\right)\left(-\frac{1}{3}\right) = +\frac{l}{6}$$

From Eq. 13.33,

$$-\theta_2 = \delta_{20} + \delta_{21}\overline{X}_1 + \delta_{22}\overline{X}_2 + \delta_{23}\overline{X}_3$$

$$-EI\theta_2 = -\frac{2}{3}M_0 l + 0.5M_0\left(+\frac{l}{6}\right) + 0.5M_0\left(+\frac{2}{3}l\right)$$

$$+0.5M_0\left(+\frac{l}{6}\right) = -\frac{M_0 l}{6}$$

From symmetry, $\theta_3 = \theta_1$. Therefore, the required plastic hinge rotations at the support are

$$\theta_1 = \frac{M_0 l}{4EI} = \theta_3 \quad \text{and} \quad \theta_2 = \frac{M_0 l}{6EI}$$

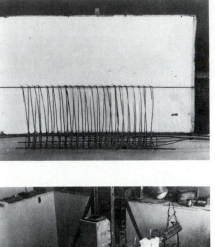

Photo 78   I-beams loaded to failure
with spiral confining reinforcement: (a)
rectangular spirals; (b) test setup after
posttensioning; (c) postfailure condition.
(Nawy and Salek.)

Since θ_2 is less than θ_1, the first hinge to develop and the controlling one in the design are $\theta_1 = M_0 l/4EI$.   Note that the same procedure used in Ex. 13.1 can be used in the limit design of any continuous beam or multistory frame.   It is also important to maintain the correct sign convention by drawing all moments at the *tension* side of the member, as noted earlier in this chapter.

The preceding discussion gives the basic *imposed rotations* approach embodied in A. L. L. Baker's theory.   Other modified approaches have been proposed by Cohn (Ref. 13.21), Sawyer (Ref. 13.22), and Furlong (Ref. 13.23).   Cohn's method

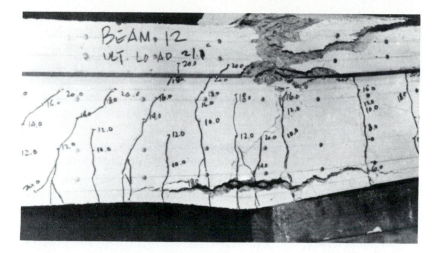

Photo 79 Tensile hinge at failure in a T beam with rectangular confining reinforcement. (Nawy and Potyondy.)

is based on the requirement of limit equilibrium and serviceability with subsequent check of rotation compatibility.   Sawyer's method is based on simultaneous requirement of limit equilibrium and rotational compatibility with subsequent check of serviceability requirement.

Furlong's method is based on assigning ultimate moments for various loading patterns on the continuous spans that would satisfy serviceability and limit equilibrium for the worst case.   The sections are reinforced in such a manner that the ultimate moment strengths for each span are equal to or greater than the *product* of the maximum ultimate moment M_0 in the span when the ends are free to rotate and using a moment coefficient k_1 for various boundary conditions as listed in Table 13.5.

13.7.3 Rotational Capacity of Plastic Hinges

Rotation is the *total* change in slope along the short plasticity length concentrated at the hinge zone.   It can also be described as the angle of discontinuity between the plastic parts of the member on either side of the plastic hinge.   There are two

TABLE 13.5 Beam Moment Coefficients for Assigned Moments

Boundary Condition	Moment Type	Beam Loaded by One Concentrated Load at Midspan	All Other Beams
Span with ends restrained	Negative	0.37	0.50
	Positive	0.42	0.33
Span with one end restrained	Negative	0.56	0.75
	Positive	0.50	0.46

types of hinges, as shown in Fig. 13.24, tensile hinges and compressive hinges. So that the first hinge that develops in the structure, usually the critical hinge, can rotate *without* rupture until the nth hinge develops, the concrete section at this hinge has to be made ductile enough through section core confinement to be able to sustain the necessary rotation. This is equally applicable to both tension and compression hinges, where confinement of the concrete core is obtained through concentration of closed stirrups at the supports and column ends. Figure 13.24 shows typical tensile and compressive hinges.

The plasticity length l_p determines the extent of the severe cracking and the rotation magnitude of the hinge. Therefore, it is important to limit the magnitude of l_p through use of *closely* spaced ties or closed stirrups. In this manner the strain capacity of the concrete at the confined section can be significantly raised, as demonstrated experimentally by several investigators, including the author's work in Refs. 13.17, 13.18, and 13.19. Several empirical expressions have been developed: Baker (Ref. 13.9), Corley (Ref. 13.16), Nawy and Potyondy (Ref. 13.19), Sawyer (Ref. 13.22), Mattock (13.24), and others.

The following simplified expressions for plasticity length l_p and concrete strain ϵ_c (Ref. 13.24) are presented:

$$l_p = 0.5d + 0.05Z \tag{13.35}$$

$$\epsilon_c = 0.003 + 0.02\frac{b}{Z} + 0.2\rho_s \tag{13.36}$$

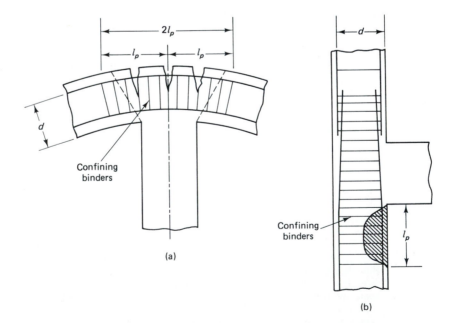

Figure 13.24 Plasticity zones l_p in plastic hinges: (a) tensile hinge; (b) compressive hinge.

where d = effective depth of beam, in.

b = beam width, in.

Z = distance from the critical section to the point of contraflexure

ρ_s = ratio of volume of confining binder steel (including the compression steel) to the volume of the concrete core

l_p = *half* the plasticity length on each side of the center line of the plastic hinge

Equation 13.35 can be more conservative for high values of ρ_s.

Additionally, such a ρ_s should be chosen that the necessary confinement is achieved as detailed in Ref. 13.20. The maximum spacing of the confining hoops should not exceed the *smallest* of the following: d/4, 8 times the diameter of the smallest longitudinal bars, 24 times the diameter of the hoop bars, or 12 in. in beams and 4 in. in columns.

Once the concrete strain ϵ_c is determined, the angle of rotation of the plastic hinge is readily determined from the expression

$$\theta_p = \left(\frac{\epsilon_c}{c} - \frac{\epsilon_{ce}}{kd} \right) l_p \tag{13.37}$$

where c = neutral axis depth at the limit state at failure

ϵ_{ce} = strain in the concrete at the extreme compression fibers when the yield curvature is reached

kd = natural axis depth corresponding to ϵ_{ce}

ϵ_c = concrete compressive strain at the end of the inelastic range or at the limit state at failure

The strain ϵ_{ce} can usually be taken at the load level where the strain in the tension reinforcement reaches the yield strain $\epsilon_y = f_y/E_s$. The strain ϵ_{ce} can be taken = 0.001 in./in. or higher, depending on whether the tension steel yields before the concrete crushes at the extreme compression fibers in cases of over-reinforced beams as is the case in some prestressed beams. If concrete crushes first, the value of ϵ_{ce} would have to be higher than 0.001 in./in. A limit of allowable $\epsilon_c = 1.0\%$ is recommended in determining the maximum allowable plastic rotation θ_p, although strains of confined concrete as high as 13% could be obtained, as shown in the author's work (Ref. 13.19). A typical comparison of plastic rotations obtained by several authors for various degrees of confinement is shown in Fig. 13.25.

It should be stated that the discussions presented are equally applicable to reinforced and prestressed concrete indeterminate structures at the plastic loading range, where full redistribution of moment has taken place. As the load reaches the limit state at failure, the flexural behavior of the prestressed concrete elements is expected to resemble closely that of reinforced concrete elements.

Discussion and design examples on the confinement of members by the ACI and UBC Codes for resisting seismic loading are presented in detail in Chapter 15.

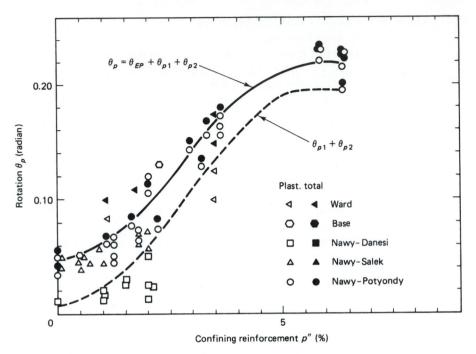

Figure 13.25   Comparison of plastic rotation with results of other authors.

13.7.4 Example 13.7: Calculation of Available Rotational Capacity

Determine the required and the available rotational capacities of the critical plastic hinges in the continuous beam of Ex. 13.6 for both confined and unconfined concrete. Given:

$$M_u = \tfrac{1}{2}M_0 = 400bd^2$$

$$c = 0.28d$$

$$kd = 0.375d$$

$$\epsilon_{ce} = \begin{cases} 0.001 \text{ in./in. at end of the elastic range (unconfined)} \\ 0.004 \text{ in./in. at end of the inelastic range for unconfined sections} \end{cases}$$
maximum allowable $\epsilon_c = 0.01$ in./in. (confined)

$$E_c I_c = 150{,}000bd^3 \text{ in.}^2\text{-lb}$$

$$\frac{Z}{d} = 5.5$$

$$f'_c = 5000 \text{ psi}$$

$$f_y = 60{,}000 \text{ psi for the mild steel}$$

Also calculate the maximum allowable span-to-depth ratio l/d for the beam if full redistribution of moments is to occur at the limit state at failure.

Solution $M_0 = 2 \times 400bd^2 = 800bd^2$

From Ex. 13.6,

$$\text{required } \theta_1 = \theta_3 = \frac{M_0 l}{4E_c I_c} = \frac{800bd^2 l}{4 \times 150,000bd^3} = \frac{1}{750}\frac{l}{d} \text{ rad}$$

$$\text{required } \theta_2 = \frac{M_0 l}{6EI} = \frac{800bd^2 l}{6 \times 150,000bd^3} = \frac{1}{1125}\frac{l}{d} \text{ rad}$$

From Eq. 13.35,

$$l_p = 0.5d + 0.05Z = 0.5d + 0.05 \times 5.5d = 0.775d$$

Total plasticity length on both sides of the hinge center line $= 2 \times 0.775d = 1.55d$.

(a) *Unconfined sections:* From Eq. 13.37,

$$\text{available } \theta_p = \left(\frac{\epsilon_c}{c} - \frac{\epsilon_{ce}}{kd}\right)l_p = \left(\frac{0.004}{0.28d} - \frac{0.001}{0.375d}\right)1.55d = 0.018 \text{ rad}$$

For full moment redistribution,

$$\frac{1}{750}\frac{l}{d} \leq 0.018 \quad \text{and} \quad \frac{1}{1125}\frac{l}{d} \leq 0.018$$

or

$$\frac{l}{d} \leq 13.5 \quad \text{and} \quad \frac{l}{d} \leq 20.3$$

(b) *Confined sections*

maximum allowance $\epsilon_c = 0.01$ in./in.

$$\text{available } \theta_p = \left(\frac{0.01}{0.28d} - \frac{0.001}{0.375d}\right)1.55d = 0.051 \text{ rad}$$

For full moment redistribution,

$$\frac{1}{750}\frac{l}{d} \leq 0.051 \quad \text{and} \quad \frac{1}{1125}\frac{l}{d} \leq 0.051$$

or

$$\frac{l}{d} \leq 38.3 \quad \text{and} \quad \frac{l}{d} \leq = 57.4$$

It can be seen from comparing the results of the unconfined sections in case (a) to the confined sections in case (b) that confinement of the concrete at the plastic hinging zone permits more slender sections for full plasticity and hence a more economical indeterminate structural system.

13.7.5 Example 13.8: Check for Plastic Rotation Serviceability

If closed stirrup binders are used in Ex. 13.7 with binders ratio $\rho_s = 0.025$ and $l/d = 35$ with c at failure $= 0.25d$, verify whether the continuous beam satisfies rotation serviceability criteria. Given: $b = \frac{1}{2}d$.

Solution $\dfrac{Z}{d} = 5.5$

Hence

$$\frac{b}{z} = \frac{1}{11}$$

$$\text{available } \epsilon_c = 0.003 + 0.2\,\frac{b}{Z} + 0.2\rho_s$$

$$= 0.003 + 0.02 \times \frac{1}{11} + 0.2 \times 0.025 = 0.01 \text{ in./in.}$$

Maximum allowable strain to be utilized is $\epsilon_c = 0.01$ in./in. For $\epsilon_c = 0.01$, the corresponding available plastic rotation is

$$\theta_p = \left(\frac{0.01}{0.25d} - \frac{0.001}{0.375d}\right) 1.55d = 0.058 \text{ rad}$$

$$\text{required } \theta_1 = \frac{1}{750}\frac{l}{d} = \frac{35}{750} = 0.046$$

$$\text{required } \theta_2 = \frac{1}{1125}\frac{l}{d} = \frac{35}{1125} = 0.031 \text{ rad}$$

Available $\theta_p = 0.058$ rad $>$ requires $\theta = 0.046$ rad. The beam satisfies serviceability criteria for plastic rotation.

The foregoing discussion for the limit design of reinforced and prestressed concrete indeterminate beams and frames permits the design engineer to provide ductile connections at beam–column supports and generate full moment redistribution throughout the structure, resulting in full utilization of the strength of the structural system. Also, continuity to withstand seismic loading can be effectively utilized through the appropriate confinement of the beam–column zones utilizing the procedures presented in this section, as discussed in Chapter 15 on design for seismic loading.

SELECTED REFERENCES

13.1. Gartner, R., *Statically Indeterminate Structures*, 2nd ed., Concrete Publications, London, 1974, 114 pp.

13.2. Gerstle, K. H., *Basic Structural Analysis*, Civil Engineering and Engineering Mechanics Series, Prentice Hall, Englewood Cliffs, N.J., 1974, 498 pp.

13.3. Wang, C. K., and Salmon, C. G., *Introductory Structural Analysis*, Civil Engineering and Engineering Mechanics Series, Prentice Hall, Englewood Cliffs, N.J., 1984, 591 pp.

13.4. Fenves, S. J., *Computer Methods in Civil Engineering*, Prentice Hall, Englewood Cliffs, N.J., 1967.

13.5. Vanderbilt, M. D., *Matrix Structural Analysis*, Quantum Publications, New York, 1974, 396 pp.

13.6. Weaver, W., Jr., and Johnston, P. R., *Finite Elements for Structural Analysis*, Civil Engineering and Engineering Mechanics Series, Prentice Hall, Englewood Cliffs, N.J., 1984, 403 pp.

13.7. DeSalvo, G. J., and Swanson, J. A., *Engineering Analysis System User's Manual: ANSYS Revision 4.1*, Vol. 1, Swanson Analysis Systems, Houston, Pa., 1983, pp. 1.1–6.32.

13.8. Portland Cement Association, *ADOSS Basic Language Program for the Analysis and Design of Slab Systems*, PCA, Skokie, Ill., 1987, pp. 1.1–6.8.

13.9. Baker, A. L. L., *The Ultimate Load Theory Applied to the Design of Reinforced and Prestressed Concrete Frames*, Concrete Publications, London, 1956, 91 pp.

13.10. Baker, A. L. L., *Limit State Design of Reinforced Concrete*, Cement and Concrete Association, London, 1970, 345 pp.

13.11. Ramakrishnan, V., and Arthur, P. D., *Ultimate Strength Design of Structural Concrete*, Wheeler Publications, London, 1977, 264 pp.

13.12. Taylor, F. W., Thompson, S. E., and Smulski, E., *Concrete Plain and Reinforced*, Vol. II, Wiley, New York, 1947, 688 pp.

13.13. ACI Committee 318, *Building Code Requirements for Reinforced Concrete*, ACI Standard 318–95, American Concrete Institute, Detroit, 1989.

13.14. ACI Committee 318, *Commentary on Building Code Requirements for Reinforced Concrete*, ACI 318 R-89, American Concrete Institute, Detroit, 1995.

13.15. Cronin, R. C., "Simplified Frame Analysis," Chapter 2 in *Simplified Design*, Gerald B. Neville, editor, Portland Cement Association, Skokie, Ill., 1984, pp. 2–1 to 2–31.

13.16. Corley, W. G., "Rotational Capacity of Reinforced Concrete Beams," *Journal of the Structural Division*, *ASCE*, Vol. 92, No. ST 5, October 1966, pp. 121–146.

13.17. Nawy, E. G., Danesi, R., and Grosco, J., "Rectangular Spiral Binders Effect on the Rotation Capacity of Plastic Hinges in Reinforced Concrete Beams," *Journal of the American Concrete Institute*, December 1968, pp. 1001–1010.

13.18. Nawy, E. G., and Salek, F., "Moment–Rotation Relationships of Non-bonded Prestressed Flanged Sections Confined with Rectangular Spirals," *Journal of the Prestressed Concrete Institute*, August 1968, pp. 40–55.

13.19. Nawy, E. G., and Potyondy, J. G., *Moment Rotation, Cracking and Deflection of Spirally Bound Pretensioned Prestressed Concrete Beams*, Engineering Research Bulletin 51, Bureau of Engineering Research, Rutgers University, New Brunswick, N.J., 1970, pp. 1–97.

13.20. Park, R., and Paulay, J., *Reinforced Concrete Structures*, Wiley, New York, 1975, 769 pp.

13.21. Cohn, M. Z., "Rotational Compatibility in the Limit Design of Reinforced Concrete Beams," *Proceedings of the International Symposium on the Flexural Mechanics of Reinforced Concrete*, ASCE-ACI, Miami, Fla., November 1964, pp. 359–382.

13.22. Sawyer, H. A., "Design of Concrete Frames for Two Failure Stages," *Proceedings of the International Symposium on the Flexural Mechanisms of Reinforced Concrete*, ASCE-ACI, Miami, Fla., November 1964, pp. 405–431.

13.23. Furlong, R. W., "Design of Concrete Frames by Assigned Limit Moments," *Journal of the American Concrete Institute*, Vol. 67, No. 4, April 1970, pp. 341–353.

13.24. Mattock, A. H., "Discussion of Rotational Capacity of Reinforced Concrete Beams by W. G. Corley," *Journal of the Structural Division, ASCE,* Vol. 93, No. ST 2, April 1967, pp. 519–522.

13.25. American National Standards Institute, *ANSI A58.1—1982*, ANSI, New York, 1982, 100 pp.

13.26. Nawy, E. G., *Prestressed Concrete: A Fundamental Approach*, 2nd ed., International Series in Civil Engineering and Engineering Mechanics, Prentice Hall, Englewood Cliffs, N.J., 1996, 800 pp.

PROBLEMS FOR SOLUTION

13.1. A continuous reinforced concrete beam is shown in Fig. 13.26. Find the moments and reactions at supports B, C, and D using the following methods of analysis: (a) flexibility matrix; (b) stiffness matrix; (c) slope deflection; (d) moment distribution. Assume that $I_{AB} = I_{BC} = 2I_{CD}$.

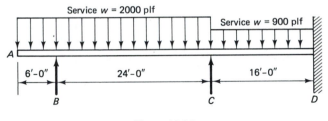

Figure 13.26

13.2. Solve Ex. 13.4 for service live-load intensity of 60 psf (2.87 MPa). Use a height between floors $l_u = 12'\text{-}6''$ (3.81 m). Use in your solution the appropriate slab thickness as required by moment, shear, and deflection. Given:

$$f'_c = 5000 \text{ psi, normal weight (34.47 MPa)}$$
$$f_y = 60,000 \text{ psi (413.7 MPa)}$$

13.3. Find the moments and shears caused by a wind intensity of 25 psf acting on the structural system in Problem 13.2 using the portal method of wind analysis. Assume that the wind load factor is 1.7.

13.4. Assume that the continuous structure in Problem 13.1 is subjected to a constant intensity of dead load $w_D = 800$ plf (11.7 kN/m) and live load $w_L = 2000$ plf (29.2 kN/m) across all spans and the cantilever segment AB. Analyze the structure by the limit theory of imposed rotations, and design the section for a maximum rotation capability for limit strain in confined concrete of value $\epsilon_c = 0.01$ in./in. Design the confining reinforcement to achieve such ductility.

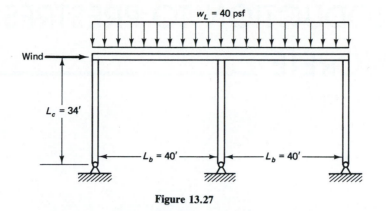

Figure 13.27

13.5. A two-bay warehouse portal frame has the dimensions shown in Fig. 13.27.   Analyze the structure for limit moments by the method of imposed rotations assuming that the portal is hinged at the foundation.   Design the portal beam and column elements for service gravity live load of $w_L = 40$ psf (1915 Pa) and wind and suction load concentrated at the upper joints having a service-level intensity $w_W = 125$ plf of height.   Use a degree of confinement that permits a limit strain $\epsilon_c = 0.01$ in./in. Select the confining steel.   Use Ref. 13.10 expressions for two-bay portal frame redundants

$$\overline{X}_1 = 0.1m + 0.5M$$
$$\overline{X}_2 = 0.175m + 0.4M$$
$$\overline{X}_3 \approx 0.1m - 0.01M$$

where $m =$ wind moment and $M =$ gravity load moment.   Assume:

$$f'_c = 5000 \text{ psi, normal weight (34.47 MPa)}$$
$$f_y = 60{,}000 \text{ psi (413.7 MPa)}$$

The portals are spaced 20 ft on centers.

INTRODUCTION TO PRESTRESSED CONCRETE

14.1 BASIC CONCEPTS OF PRESTRESSING

Reinforced concrete is weak in tension but strong in compression. In order to maximize utilization of its material properties, an internally or externally compressive force P is induced on the structural element through the use of stressed high-strength prestressing wires or tendons prior to loading. As a result, the concrete section is generally stressed only in compression under service and sometimes overload conditions. Such a system of construction is termed as prestressed concrete.

The prestressing force P that satisfies the particular conditions of geometry and loading of a given element (Fig. 14.1) is determined from the principles of mechanics and of stress–strain relationships. Sometimes simplification is necessary, as when a prestressed beam is assumed to be homogeneous and elastic. A

Photo 80 Sunshine Skyway Bridge, Tampa Bay, Florida. Designed by Figg and Muller Engineers, Inc., the bridge has a 1200-ft cable-stayed main span with a single pylon, 175-ft vertical clearance, and total length of 21,878 ft. It has twin 40-ft roadways and has 135-ft spans in precast segmental bridge with trestle and high approaches to elevation +130 ft. (Courtesy of Figg and Muller Engineers, Inc.)

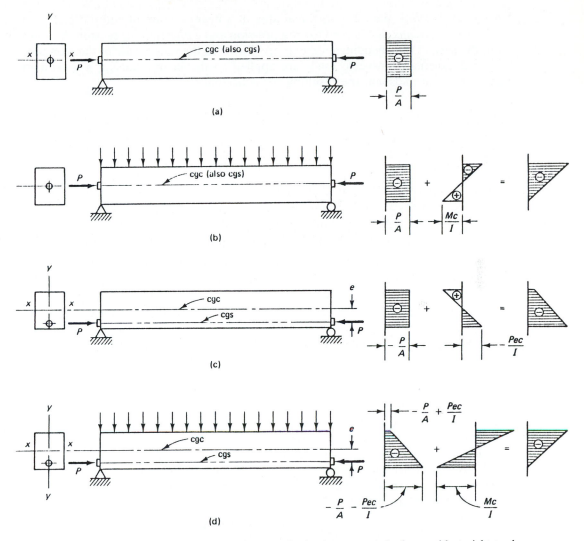

Figure 14.1 Concrete fiber stress distribution in a rectangular beam with straight tendon: (a) concentric tendon, prestress only; (b) concentric tendon, self-weight added; (c) eccentric tendon, prestress only; (d) eccentric tendon, self-weight added.

comprehensive treatment of the subject of prestressed concrete may be found in Ref. 14.1.

Consider, then, a simply supported rectangular beam subjected to a *concentric* prestressing force P, as shown in Figure 14.1(a). The compressive stress on the beam cross section is uniform and has an intensity

$$f = -\frac{P}{A_c} \tag{14.1}$$

where $A_c = bh$ is the cross-sectional area of a beam section of width b and total depth h. A *minus* sign is used for compression and a *plus* sign for tension throughout the text. Also, bending moments are drawn on the tensile side of the member.

If external transverse loads are applied to the beam, causing a maximum moment M at midspan, the resulting stress becomes

$$f^t = -\frac{P}{A} - \frac{Mc}{I_g} \tag{14.2a}$$

and

$$f_b = -\frac{P}{A} + \frac{Mc}{I_g} \tag{14.2b}$$

where f^t = stress at the top fibers
  f_b = stress at the bottom fibers
  $c = \frac{1}{2} h$ for the rectangular section
  I_g = gross moment of inertia of the section ($bh^3/12$ in this case)

Equation 14.2b indicates that the presence of prestressing-compressive stress $-P/A$ is reducing the tensile flexural stress Mc/I to the extent intended in the design, either eliminating tension totally (even inducing compression) or permitting a level of tensile stress within allowable code limits. The section is then considered uncracked and behaves elastically: the concrete's inability to withstand tensile stresses is effectively compensated for by the compressive force of the prestressing tendon.

The compressive stresses in Eq. 14.2a at the top fibers of the beam due to prestressing are compounded by the application of the loading stress $-Mc/I$, as seen in Fig. 14.1b. Hence the compressive stress capacity of the beam to take a substantial external load is reduced by the *concentric* prestressing force. In order to avoid this limitation, the prestressing tendon is placed *eccentrically* below the neutral axis at midspan, to induce tensile stresses at the top fibers due to prestressing (Fig. 14.1c and d). If the tendon is placed at eccentricity e from the center of gravity of the concrete, termed the *cgc line*, it creates a moment Pe, and the ensuing stresses at midspan become

$$f^t = -\frac{P}{A_c} + \frac{Pec}{I_g} - \frac{Mc}{I_g} \tag{14.3a}$$

$$f_b = -\frac{P}{A_c} - \frac{Pec}{I_g} + \frac{Mc}{I_g} \tag{14.3b}$$

Since the support section of a simply supported beam carries no moment from the external transverse load, high tensile fiber stresses at the top fibers are caused by the eccentric prestressing force. To limit such stresses, the eccentricity of the prestressing tendon profile, the *cgs line*, is made less at the support section than at the midspan section, or eliminated altogether, or else a negative eccentricity above the cgc line is used. The cgs line is the profile of the center of gravity of the prestressing tendon and the cgc line is the profile of the center of gravity of the concrete.

In designing prestressed concrete elements, the concrete fiber stresses are *directly* computed from the external forces applied to the concrete by longitudinal prestressing and the external transverse load. Equations 14.3a and 14.3b can be modified and simplified for use in calculating stresses at the initial prestressing stage and at service load levels. If P_i is the initial prestressing force before stress losses and P_e is the effective prestressing force after losses, then

$$\gamma = \frac{P_e}{P_i}$$

can be defined as the residual prestress factor. Substituting r^2 for I_g/A_c in Eqs. 14.3, where r is the radius of gyration of the gross section, the expressions for stress can be rewritten as follows:

1. *Prestressing force only*

$$f^t = -\frac{P_i}{A_c}\left(1 - \frac{ec_t}{r^2}\right) \tag{14.4a}$$

$$f_b = -\frac{P_i}{A_c}\left(1 + \frac{ec_b}{r^2}\right) \tag{14.4b}$$

where c_t and c_b are the distances from the center of gravity of the section (the cgc line) to the extreme top and bottom fibers, respectively.

2. *Prestressing plus self-weight.* If the beam self-weight causes a moment M_D at the section under consideration, Eqs. 14.4a and 14.4b, respectively, become

$$f^t = -\frac{P_i}{A_c}\left(1 - \frac{ec_t}{r^2}\right) - \frac{M_D}{S^t} \tag{14.5a}$$

and

$$f_b = -\frac{P_i}{A_c}\left(1 + \frac{ec_b}{r^2}\right) + \frac{M_D}{S_b} \tag{14.5b}$$

where S^t and S_b are the moduli of the sections for the top and bottom fibers, respectively.

The change in eccentricity from the midspan to the support section is obtained by raising the prestressing tendon either abruptly from the midspan to the support, a process called *harping*, or gradually in a parabolic form, a process called *draping*. Figure 14.2a shows a harped profile usually used for pretensioned beams and for concentrated transverse loads. Figure 14.2b shows a draped tendon usually used in posttensioning.

Subsequent to erection and installation of the floor or deck, live loads act on the structure, causing a superimposed moment M_s. The full intensity of such loads normally occurs after the building is completed and some time-dependent losses in prestress have already taken place. Hence the prestressing force used in the stress equations would have to be the effective prestressing force P_e. If the total

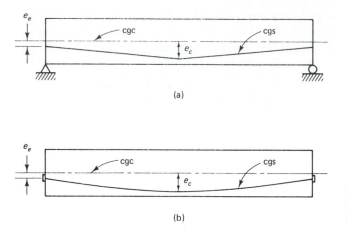

(a)

(b)

Figure 14.2 Prestressing tendon profile: (a) harped tendon; (b) draped tendon.

moment due to gravity loads is M_T, then

$$M_T = M_D + M_{SD} + M_L \tag{14.6}$$

where M_D = moment due to self-weight
M_{SD} = moment due to superimposed dead load, such as flooring
M_L = moment due to live load, including impact and seismic loads if any

Equations 14.5 then become

$$f^t = -\frac{P_e}{A_c}\left(1 - \frac{ec_t}{r^2}\right) - \frac{M_T}{S^t} \tag{14.7a}$$

$$f_b = -\frac{P_e}{A_c}\left(1 + \frac{ec_b}{r^2}\right) + \frac{M_T}{S_b} \tag{14.7b}$$

The tensile stress in the concrete permitted at the extreme fibers of the section cannot exceed the maximum permissible in the code (e.g., $f_t = 6\sqrt{f_c'}$ in the ACI Code). If it is exceeded, bonded nonprestressed reinforcement proportioned to resist the total tensile force has to be provided to control cracking at service loads so that $f_t = 12\sqrt{f_c'}$ can be used.

14.1.1 Example 14.1: Computation of Fiber Stresses in a Prestressed Beam by Basic Principles

A pretensioned simply supported T beam has a span of 64 ft (19.51 m) and the geometry shown in Fig. 14.3. It is subjected to a uniform intensity of superimposed gravity dead-load intensity W_{SD} and live-load intensity W_L summing to 420 plf (6.13 kN/m). The initial prestressing stress before losses is $f_{pi} \approx 0.70 f_{pu} = 189,000$ psi (1303 MPa), and the effective prestress after losses is $f_{pe} = 150,000$ psi (1034 MPa). Compute the extreme fiber stresses at the midspan section due to (a) the initial full prestress and no external gravity load and (b) the final service-load conditions when prestress losses have taken place. Allowable stress data are as follows:

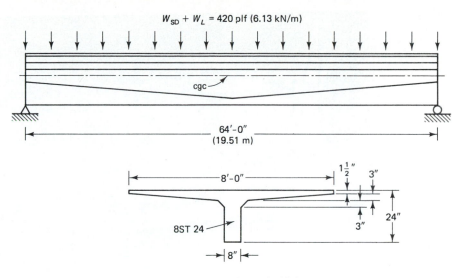

Figure 14.3   Example 14.1

f'_c = 6000 psi, normal weight (41.3 MPa)

f_{pu} = 270,000 psi, stress relieved (1862 MPa) = specified tensile strength of the tendons

f_{py} = 220,000 psi (1517 MPa) = specified yield strength of the tendons

f_{pe} = 150,000 psi (1034 MPa)

f_t = 12 $\sqrt{f'_c}$ = 930 psi (6.4 MPa) = maximum allowable tensile stress in concrete

f'_{ci} = 4800 psi (33.1 MPa) = concrete compressive strength at time of initial prestress

f_{ci} = 0.6f'_{ci} = 2880 psi (19.9 MPa) = maximum allowable stress in concrete at initial prestress

f_c = 0.45f'_c = maximum allowable compressive stress in concrete at service-load level

Assume that twelve $\frac{1}{2}$in.-diameter seven-wire strand (twelve 12.7-mm-diameter strand) tendons are used to prestress the beam, and take

$$A_c = 474 \text{ in.}^2 \ (3058 \text{ cm}^2)$$

$$I_c = 21{,}550 \text{ in.}^4 \ (896{,}978 \text{ cm}^4)$$

$$r^2 = \frac{I_c}{A_c} = 45.44 \text{ in.}^2$$

$$c_b = 18.06 \text{ in. } (459 \text{ mm})$$

$$c_t = 5.94 \text{ in. } (151 \text{ mm})$$

$$e_c = 14.6 \text{ in. } (371 \text{ mm})$$

$$e_e = 0$$

Photo 81 Prestressed lightweight concrete midbody for Arctic offshore drilling platform, Global Marine Development.   (Courtesy of Ben C. Gerwick.)

$$S_b = 1193 \text{ in.}^3 \ (19{,}550 \text{ cm}^3)$$
$$S^t = 3626 \text{ in.}^3 \ (59{,}419 \text{ cm}^3)$$
$$W_D = 494 \text{ plf } (6.13 \text{ kN/m})$$

Solution   *Initial conditions at prestressing*

$$A_{ps} = 12 \times 0.153 = 1.836 \text{ in.}^2$$
$$P_i = A_{ps}f_{pi} = 1.836 \times 189{,}000 = 347{,}000 \text{ lb } (1544 \text{ kN})$$
$$P_e = 1.836 \times 150{,}000 = 275{,}400 \text{ lb } (1225 \text{ kN})$$

The midspan self-weight dead-load moment is

$$M_D = \frac{W_D l^2}{8} = \frac{494(64)^2}{8} \times 12 = 3{,}035{,}136 \text{ in.-lb } (343 \text{ kNm})$$

From Eqs. 14.5a and 14.5b,

$$f^t = -\frac{P_i}{A_c}\left(1 - \frac{ec_t}{r^2}\right) - \frac{M_D}{S^t}$$
$$= -\frac{347{,}004}{474}\left(1 - \frac{14.6 \times 5.94}{45.44}\right) - \frac{3{,}035{,}136}{3626}$$
$$= +665.8 - 837.0 = -171.2 \text{ psi } (C)$$
$$f_b = -\frac{P_i}{A_c}\left(1 + \frac{ec_b}{r^2}\right) + \frac{M_D}{S_b}$$

$$= -\frac{347{,}000}{474}\left(1 + \frac{14.6 \times 18.06}{45.44}\right) + \frac{3{,}035{,}136}{1193}$$

$$= -4980.1 + 2544.1 = -2436.0 \text{ psi } (C)$$

$$<f_{ci} = -2880 \text{ psi allowed} \qquad \text{O.K.}$$

Final condition at service load

Midspan moment due to superimposed dead and live load is

$$M_{SD} + M_L = \frac{420(64)^2}{8} \times 12 = 2{,}580{,}480 \text{ in.-lb}$$

total moment $M_T = 3{,}035{,}136 + 2{,}580{,}480$

$$= 5{,}615{,}616 \text{ in.-lb } (635 \text{ kNm})$$

$$f^t = \frac{P_e}{A_c}\left(1 - \frac{ec_t}{r^2}\right) - \frac{M_T}{S^t}$$

$$= -\frac{275{,}400}{474}\left(1 - \frac{14.6 \times 5.94}{45.44}\right) - \frac{5{,}615{,}616}{3626}$$

$$= +527.9 - 1548.7 = -1020.8 \text{ psi } (C) \text{ (7 MPa)}$$

$$<f_c = 0.45 \times 6000 = 2700 \text{ psi} \qquad \text{O.K.}$$

$$f_b = -\frac{P_e}{A_c}\left(1 + \frac{ec_b}{r^2}\right) + \frac{M_T}{S_b}$$

$$= -\frac{275{,}400}{474}\left(1 + \frac{14.6 \times 18.06}{45.44}\right) + \frac{5{,}615{,}616}{1193}$$

$$= -3952.5 + 4707.1 = +755 \text{ psi } (T) \text{ (5.2 MPa)}$$

$$<f_t = 12\sqrt{f_c'} = 930 \text{ psi} \qquad \text{O.K.}$$

14.2 PARTIAL LOSS OF PRESTRESS

It is a well-established fact that the initial prestressing force applied to the concrete element undergoes a progressive process of reduction over a period of approximately 5 years. Consequently, it is important to determine the level of the prestressing force at each loading stage, from the stage of transfer of the prestressing force to the concrete, to the various stages of prestressing available at service load, up to the ultimate. Essentially, the reduction in the prestressing force can be grouped into two categories:

1. Immediate elastic loss during the fabrication or construction process, including elastic shortening of the concrete, anchorage losses, and frictional losses
2. Time-dependent losses such as creep, shrinkage, and those due to temperature effects and steel relaxation, all of which are determinable at the service-load limit state of stress in the prestressed concrete element

TABLE 14.1 AASHTO LUmp-Sum Losses

Type of Ptrestressing Steel	Total Loss [psi (N/mm²)]	
	f'_c = 4000 psi (27.6 N/mm²)	f'_c = 5000 psi (34.5 N/mm²)
Pretensioning strand		45,000 psi (310)
Posttensioning[a] wire or strand	32,000 psi (221)	33,000 psi (228)
Bars	22,000 psi (152)	23,000 psi (159)

[a]Losses due to friction are excluded. Such losses should be computed according to Section 6.5 of the AASHTO specifications.

An exact determination of the magnitude of these losses, particularly the time-dependent ones, is not feasible, since they depend on a multiplicity of inter-related factors. Empirical methods of estimating losses differ with the different codes of practice or recommendations, such as those of the Prestressed Concrete Institute, the ACI–ASCE joint committee approach, the AASHTO lump-sum approach, the Comité Eurointernationale du Béton (CEB), and the FIP (Federation Internationale de la Précontrainte). The degree of rigor of these methods depends on the approach chosen and the accepted practice of record.

A very high degree of refinement of loss estimation is neither desirable nor warranted, because of the multiplicity of factors affecting the estimate. Consequently, lump-sum estimates of losses are more realistic, particularly in routine designs and under average conditions. Such lump-sum losses are summarized in Table 14.1 of AASHTO and Table 14.2 of PTI. They include elastic shortening, relaxation in the prestressing steel, creep, and shrinkage and are applicable only to routine, standard conditions of loading; normal concrete, quality control, construction procedures, and environmental conditions; and the importance and magnitude of the system. Detailed analysis has to be performed if these standard conditions are not fulfilled.

A summary of the sources of the separate prestressing losses and the stages of their occurrence is given in Table 14.3, in which the subscript i denotes "initial" and the subscript j denotes the loading stage after jacking. From this table, the total loss in prestress can be calculated for pretensioned and posttensioned members as follows:

1. *Pretensioned members*

$$\Delta f_{pT} = \Delta f_{pES} + \Delta f_{pR} + \Delta f_{pCR} + \Delta f_{pSH} \qquad (14.8a)$$

TABLE 14.2 Approximate Prestress Loss Values for Posttensioning[a]

Posttensioning Tendon Material	Prestress Loss [psi (N/mm²)]	
	Slabs	Beams and Joists
Stress-relieved 270K strand and stress-relieved 240K wire	30,000 (207)	35,000 (241)
Bar	20,000 (138)	25,000 (172)
Low-relaxation 270K strand	15,000 (103)	20,000 (138)

Source: Post-Tensioning Institute.

[a]This table of approximate prestress losses was developed to provide a common posttensioning industry basis for determining tendon requirements on projects in which the magnitude of prestress losses is not specified by the designer. These loss values are based on use of normal-weight concrete and on average values of concrete strength, prestress level, and exposure conditions. Actual values of losses may vary significantly above or below the table values where the concrete is stressed at low strengths, where the concrete is highly prestressed, or in very dry or very wet exposure conditions. The table values do not include losses due to friction.

TABLE 14.3 Types of Prestress Loss

Type of Prestress Loss	Stage of Occurrence		Tendon Stress Loss	
	Pretensioned Members	Post-tensioned Members	During Time Interval (t_i, t_j)	Total or During Life
Elastic shortening of concrete (ES)	At transfer	At sequential jacking	—	Δf_{pES}
Relaxation of tendons (R)	Before and after transfer	After transfer	$\Delta f_{pR}(t_i, t_j)$	Δf_{pR}
Creep of concrete (CR)	After transfer	After transfer	$\Delta f_{pC}(t_i, t_j)$	Δf_{pCR}
Shrinkage of concrete (SH)	After transfer	After transfer	$\Delta f_{pS}(t_i, t_j)$	Δf_{pSH}
Friction (F)	—	At jacking	—	Δf_{pF}
Anchorage seating loss (A)	—	At transfer	—	Δf_{pA}
Total	Life	Life	$\Delta f_{pT}(t_i, t_j)$	Δf_{pT}

where $\Delta f_{pR} = \Delta f_{pR}(t_0, t_{tr}) + \Delta f_{pR}(t_{tr}, t_s)$
t_0 = time at jacking
t_{tr} = time at transfer
t_s = time at stabilized loss

Hence, computations for steel relaxation loss have to be performed for the time interval t_1 through t_2 of the respective loading stages.

As an example, the transfer stage, say, at 18 hours would result in t_{tr} = t_2 = 18 hours and t_0 = t_1 = 0. If the next loading stage is between transfer and 5 years (17,520 hours), when losses are considered stabilized, then t_2 = t_s = 17,520 hours and t_1 = 18 hours. Then, if f_{pi} is the initial prestressing stress that the concrete element is subjected to and f_{pJ} is the jacking stress in the tendon,

$$f_{pi} = f_{pJ} - \Delta f_{pR}(t_0, t_{tr}) - \Delta f_{pES} \qquad (14.8b)$$

2. *Post-tensioned members*

$$\Delta f_{pT} = \Delta f_{pA} + \Delta f_{pF} + \Delta f_{pES} + \Delta f_{pR} + \Delta f_{pCR} + \Delta f_{pSH} \qquad (14.8c)$$

where Δf_{pES} is applicable only when tendons are jacked sequentially, and not simultaneously.

In the posttensioned case, computation of relaxation loss starts between the transfer time $t_1 = t_{tr}$ and the end of the time interval t_2 under consideration. Hence

$$f_{pi} = f_{pJ} - \Delta f_{pES} - \Delta f_{pF} \qquad (14.8d)$$

14.2.1 Example 14.2: Step-by-Step Computation of all Time-dependent Losses in a Post-tensioned Beam

A simply supported post-tensioned, 70-ft-span-lightweight, steam-cured, double T-beam as shown in Fig. 14.4 is prestressed by twelve $\frac{1}{2}$-in. (12.7-mm) diameter 270-K grade stress-relieved strands. The tendons are harped, and the eccentricity at midspan is 18.73 in. (476 mm) and at the end 12.98 in. (330 mm). Compute the prestress loss at the critical section in the beam at 0.40 span at (a) stage I at transfer, (b) stage II after concrete topping is placed, and (c) 2 years after concrete topping is placed. Suppose the topping is 2-in. (51-mm) normal-weight concrete cast at 30 days. Suppose also that prestress transfer occurred 18 hours after casting the section and tensioning the strands. Assume that stress increase due to topping = 5048 psi (34.8 MPa). Given:

f'_c = 5000 psi (34.5 MPa), lightweight　　f_{pi} = 189,000 psi (1303 MPa)
f'_{ci} = 3500 psi (24.1 MPa)　　　　　　　　f_{py} = 230,000 psi

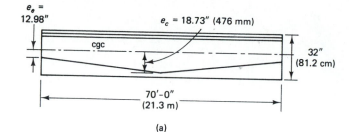

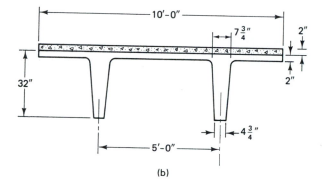

Figure 14.4 Double-T pretensioned beam: (a) elevation; (b) pretensioned section.

and the following noncomposite section properties:

$A_c = 615$ in.^{2} (3968 cm^2) $E_s = 28 \times 10^6$ psi

$I_c = 59{,}720$ in.^{4} (2.49 $\times$ 10^6 cm^4) $S_b = 2717$ in.^{3}

$c_b = 21.98$ in. (55.8 cm) $S^t = 5960$ in.^{3}

$c_t = 10.02$ in. (25.5 cm)

shrinkage loss at transfer $= 6190$ psi (42.7 MPa)

Solution   (1) *Anchorage seating loss*

$$\Delta_A = \frac{1''}{4} = 0.25'' \qquad L = 70 \text{ ft}$$

From ACI Code, the anchorage slip stress loss is

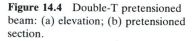

$$\Delta f_{pA} = \frac{\Delta_A}{L} E_{Ps} = \frac{0.25}{70 \times 12} \times 28 \times 10^6 \approx 8333 \text{ psi (40.2 MPa)}$$

(2) *Elastic shortening:* Since all jacks are simultaneously tensioned, the elastic shortening will precipitate during jacking. As a result, no elastic shortening stresses takes place in the strands. Hence $\Delta f_{pES} = 0$.

Photo 82 Linn Cove Viaduct, Grand-father Mountain, North Carolina. A 90° cantilever and a 10% superelevation in one direction to a full 10% in the opposite direction within 180 ft. Designed by Figg and Muller Engineers, Inc., Tallahassee, Florida (Courtesy of Figg and Muller Engineers, Inc.)

(3) *Frictional loss:* Assume that the parabolic tendon approximates the shape of an arc of a circle. Then, from the equation of the parabola (see Ref. 14.1),

$$\alpha = \frac{8y}{x} = \frac{8(18.73 - 12.98)}{70 \times 12} = 0.0548 \text{ rad}$$

From ACI Code, use $K = 0.001$ and $\mu = 0.25$. Then

$$f_{pi} = 189,000 \text{ psi (1303 MPa)}$$

From ACI Code, the stress loss in prestress due to friction is

$$\begin{aligned}
\Delta f_{pF} &= f_{pi}(\mu\alpha - KL) \\
&= 189,000(0.25 \times 0.0548 + 0.001 \times 70) \\
&= 15,819 \text{ psi (104.3 MPa)}
\end{aligned}$$

The stress remaining in the prestressing steel afater all initial instantaneous losses is

$$f_{pi} = 189,000 - 8333 - 0 - 15,819 = 164,848 \text{ psi (1136 MPa)}$$

Hence the net prestressing force is

$$P_i = 164,848 \times 12 \times 0.153 = 296,727 \text{ lb}$$

Stage I: Stress at transfer

    (1) *Anchorage seating loss*

$$\text{anchorage loss} = 8333 \text{ psi}$$
$$\text{net stress} = 164{,}838 \text{ psi}$$

    (2) *Relaxation loss*

$$\Delta f_{pR} = 164{,}848 \frac{\log 18}{10}\left(\frac{164{,}838}{230{,}000} - 0.55\right)$$
$$\approx 3450 \text{ psi } (238 \text{ MPa})$$

    (3) *Creep loss*

$$\Delta f_{pCR} = 0$$

    (4) *Shrinkage loss*

$$\Delta f_{pSH} = 0$$

So the tendon stress f_{pi} at the end of stage I is

$$164{,}848 - 3450 = 161{,}398 \text{ psi } (1113 \text{ MPa})$$

Stage II: Transfer to placement of topping after 30 days

    (1) *Creep loss*

$$P_i = 161{,}398 \times 12 \times 0.153 = 296{,}327 \text{ lb}$$

$$\bar{f}_{cs} = -\frac{P_i}{A_c}\left(1 + \frac{e^2}{r^2}\right) + \frac{M_D e}{I_c}$$

$$= -\frac{296{,}327}{615}\left[1 + \frac{(17.58)^2}{97.11}\right] + \frac{3{,}464{,}496 \times 17.58}{59{,}720}$$

$$= -2016.2 + 1020.0 = 996.2 \text{ psi } (6.94 \text{ MPa})$$

Hence the creep loss is

$$\Delta f_{pCR} = nK_{CR}(\bar{f}_{cs} - \bar{f}_{csd})$$
$$= 9.72 \times 1.6(996.2 - 519.3) \approx 7417 \text{ psi } (51.9 \text{ MPa})$$

    (2) *Shrinkage loss:* Given

$\Delta f_{pSH} = 3{,}590 \text{ psi } (24.8 \text{ MPa})$ at 30 days using a shrinkage reduction coefficient
$\qquad = 0.58$

    (3) *Steel relaxation loss at 30 days*

$$f_{ps} = 161{,}398 \text{ psi}$$

The relaxation loss in stress becomes

$$\Delta f_{pR} = 161{,}398 \frac{\log 720 - \log 18}{10}\left(\frac{161{,}398}{230{,}000} - 0.55\right)$$
$$\approx 3923 \text{ psi } (27.0 \text{ MPa})$$

Stage II: Total losses

$$\Delta f_{pT} = \Delta f_{pCR} + \Delta f_{pSH} + \Delta f_{pR}$$
$$= 7417 + 3590 + 3923 = 14,930 \ (103.0 \ \text{MPa}) \ \text{psi} \ (122.2 \ \text{MPa})$$

The increase in stress in the strands due to the addition of topping is $f_{SD} = 5048$ psi (34.8 MPa); hence the strand stress at the end of stage II is

$$f_{pe} = f_{pi} - \Delta f_{pT} + \Delta f_{SD} = 161,398 - 14,930 + 5048 = 151,516 \ \text{psi} \ (1045 \ \text{MPa})$$

Stage III: At end of 2 years

$$f_{pe} = 151,516 \ \text{psi}$$
$$t_1 = 720 \ \text{hours}$$
$$t_2 = 17,520 \ \text{hours}$$

The steel relaxation stress loss is

$$\Delta f_{pR} = 151,516 \left(\frac{\log 17,520 - \log 720}{10} \right) \left(\frac{151,516}{230,000} - 0.55 \right)$$
$$\simeq 2248 \ \text{psi} \ (15.8 \ \text{MPa})$$

Assuming that the creep shrinkage losses were maintained till stage III, the strand stress f_{pe} at the end of stage III is approximately

$$151,516 - 2248 = 149,232 \ \text{psi} \ (1029 \ \text{MPa})$$

Summary of stresses

Stress Level at Various Stages	Steel Stress (psi)	Percent
After tensioning ($0.70 f_{pu}$)	189,000	100.0
Elastic shortening loss	0	0.0
Anchorage loss[a]	−8,333	−4.4
Frictional loss[a]	−15,819	−8.4
Creep loss	−7,417	−3.9
Shrinkage loss	−3,590	−1.9
Relaxation loss (3528 + 4008 + 2056)	−9,657	−5.1
Increase due to topping	+5,048	+2.7
Final net stress f_{pe}	149,232	79.0

Percentage of total losses = $100 - 77.9 = 22.1\%$, say, 22% for this posttensioned beam

[a]Frictional and anchorage seating losses are included in this table since the total jacking stress is given as 189,000 psi; otherwise, the tendons would have to be jacked an additional stress of such magnitude as to neutralize the frictional and anchorage seating losses.

14.2.2 Example 14.3: Lump-sum Computation of Time-Dependent Losses in Prestressing

Solve Example 14.2 by the approximate lump-sum method and compare the results.

Solution   From Table 14.2, the total loss $\Delta_{pT} = 35,000$ psi (241.3 MPa).   So the net final strand stress by the lump-sum method is

$$f_{pe} = 189,000 - 45,000 = 144,000 \text{ psi (993 MPa)}$$

$$\text{step-by-step } f_{pe} \text{ value} = 149,232 \text{ psi}$$

$$\text{percent difference} = \frac{149,232 - 144,000}{189,000} = 2.8\%$$

The difference between the step-by-step "exact" method and the approximate lump-sum method is quite small, indicating that in normal, standard cases both methods are equally reliable.

14.3 FLEXURAL DESIGN OF PRESTRESSED CONCRETE ELEMENTS

Flexural stresses are the result of external, or imposed, bending moments.   In most cases, they control the selection of the geometrical dimensions of the prestressed concrete section regardless of whether it is pretensioned or posttensioned. The design process starts with the choice of a preliminary geometry, and by trial and adjustment it converges to a final section with geometrical details of the concrete cross section and the sizes and alignments of the prestressing strands.   The section satisfies the flexural (bending) requirements of concrete stress and steel stress limitations.   Thereafter, other factors, such as shear and torsion capacity, deflection, and cracking, are analyzed and satisfied.

While the input data for the analysis of sections differ from the data needed for design, every design is essentially an analysis.   We assume the geometrical properties of the section to be prestressed and then proceed to determine whether the section can safely carry the prestressing forces and the required external loads. Hence a good understanding of the fundamental principles of analysis and the alternatives presented thereby significantly simplifies the task of designing the section.   The basic mechanics of materials, principles of equilibrium of internal couples, and elastic principles of superposition have to be adhered to in all stages of loading.

It suffices in the flexural design of reinforced concrete members to apply only the limit states of stress at failure for the choice of the section, provided that other requirements such as serviceability, shear capacity, and bond are met.   In the design of prestressed members, however, additional checks are needed at the load transfer and limit state at service load, as well as the limit state at failure, with the failure load indicating the reserve strength for overload conditions.   All these checks are necessary to ensure that at service load cracking is negligible and the long-term effects on deflection or camber are well controlled.

In view of the preceding, this section covers the major aspects of both the service-load flexural design and the ultimate-load flexural design check. Note that a logical sequence in the design process entails *first* the service-load design of the section required in flexure and then the analysis of the available moment strength M_n of the section for the limit state at failure. A negative sign $(-)$ is used to denote compressive stress, and a positive sign $(+)$ is used to denote tensile stress in the concrete section. A convex or hogging shape indicates negative bending moment and a concave or sagging shape positive, bending moment, as shown in Fig. 14.5.

Unlike the case of reinforced concrete members, the external dead load and partial live load are applied to the prestressed concrete member at varying concrete strengths at various loading stages. These loading stages can be summarized as follows:

1. Initial prestress force P_i is applied; then, at transfer, the force is transmitted from the prestressing strands to the concrete.
2. The full self-weight W_D acts on the member together with the initial prestressing force, provided that the member is simply supported, that is, there is no intermediate support.
3. The full superimposed dead load W_{SD}, including topping for composite action, is applied to the member.
4. Most short-term losses in the prestressing force occur initially, leading to a reduced prestressing force P_{eo}.
5. The member is subjected to the full service load, with long-term losses due to creep, shrinkage, and steel strand relaxation taking place and leading to a net prestressing force P_e.
6. Overloading of the member occurs under certain conditions up to the limit state at failure.

A typical loading history and corresponding stress distribution across the depth of the critical section are shown in Fig. 14.6, while a schematic plot of load versus deformation (camber or deflection) is shown in Fig. 14.7 for the various loading stages from the self-weight effect up to rupture.

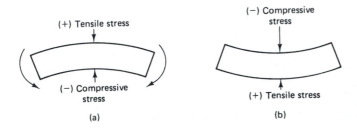

(+) Tensile stress

(−) Compressive stress

(a)

(−) Compressive stress

(+) Tensile stress

(b)

Figure 14.5 Sign convention for flexure stress and bending moment: (a) negative bending moment; (b) positive bending moment.

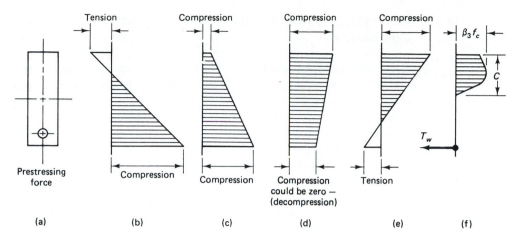

Figure 14.6 Flexural stress distribution throughout loading history: (a) beam section; (b) initial prestressing stage; (c) self-weight and effective prestress; (d) full dead load plus effective prestress; (e) full service load plus effective prestress; (f) limit state of stress at ultimate load for underreinforced beam.

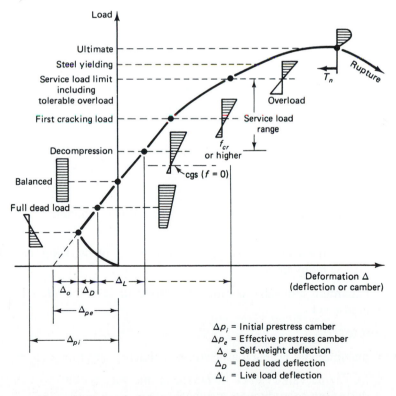

Figure 14.7 Load–deformation curve of typical prestressed beam.

14.3.1. Selection of Geometrical Properties of Section Components

14.3.1.1. General guidelines. Under service-load conditions, the beam is assumed to be homogeneous and elastic. Since it is also assumed (because expected) that the prestress compressive force transmitted to the concrete closes the crack that might develop at the tensile fibers of the beam, beam sections are considered uncracked. Stress analysis of prestressed beams under these conditions is no different from stress analysis of a steel beam or, more accurately, a beam–column. The axial force due to prestressing is always present regardless of whether bending moments do or do not exist due to other external or self-loads.

It is advantageous to have the alignment of the prestressing tendons eccentric at the critical sections, such as the midspan section in a simple beam and the support section in a continuous beam. As compared to a rectangular solid section, a nonsymmetrical flanged section has the advantage of efficiently using the concrete material and of concentrating the concrete in the compressive zone of the section where it is most needed.

Equations 14.9 to 14.11, to be subsequently presented, are stress equations that are convenient in the analysis of stresses in the section once the section is chosen. For design, it is necessary to transpose the three equations into geometrical equations so that the student and the designer can readily choose the concrete section. A logical transposition is to define the minimum section modulus that can withstand all the loads after losses.

14.3.1.2. Minimum section modulus. To design or choose the section, a determination of the required minimum section modulus S_b' has to be made first. If

f_{ci} = maximum allowable compressive stress in concrete immediately after transfer and prior to losses

  = $0.60 f_{ci}'$

f_{ti} = maximum allowable tensile stress in concrete immediately after transfer and prior to losses

  = $3\sqrt{f_{ci}'}$ (the value can be increased to $6\sqrt{f_{ci}'}$ at the supports for simply supported members)

f_c = maximum allowable compressive stress in concrete after losses at service-load level

  = $0.45 f_c'$

f_t = maximum allowable tensile stress in concrete after losses at service-load level

  = $6\sqrt{f_c'}$ (the value can be increased in one-way systems to $12\sqrt{f_c'}$ if long-term deflection requirements are met)

then the *actual* extreme fiber stresses in the concrete cannot exceed the values listed.

Using the uncracked unsymmetrical section, a summary of the equations of stress for the various loading stages is as follows:

1. *Stress at transfer*

$$f^t = -\frac{P_i}{A_c}\left(1 - \frac{ec_t}{r^2}\right) - \frac{M_D}{S^t} \leq f_{ti} \tag{14.9a}$$

$$f_b = -\frac{P_i}{A_c}\left(1 + \frac{ec_b}{r^2}\right) + \frac{M_D}{S_b} \leq f_{ci} \tag{14.9b}$$

where P_i is the initial prestressing force. While a more accurate value to use would be the horizontal component of P_i, it is reasonable for all practical purposes to disregard such refinement.

2. *Effective stresses after losses*

$$f^t = -\frac{P_e}{A_c}\left(1 - \frac{ec_t}{r^2}\right) - \frac{M_D}{S^t} \leq f_t \tag{14.10a}$$

$$f_b = -\frac{P_e}{A_c}\left(1 + \frac{ec_b}{r^2}\right) + \frac{M_D}{S_b} \leq f_c \tag{14.10b}$$

3. *Service-load final stresses*

$$f^t = -\frac{P_e}{A_c}\left(1 - \frac{ec_t}{r^2}\right) - \frac{M_T}{S^t} \leq f_c \tag{14.11a}$$

$$f_b = -\frac{P_e}{A_c}\left(1 + \frac{ec_b}{r^2}\right) + \frac{M_T}{S_b} \leq f_t \tag{14.11b}$$

where $M_T = M_D + M_{SD} + M_L$
P_i = initial prestress
P_e = effective prestress after losses;
    t denotes the top, and b denotes the bottom fibers
e = eccentricity of tendons from the concrete section center of gravity, cgc
r^2 = square of radius of gyration
S^t/S_b = top/bottom section section modulus value of concrete section

The *decompression stage* denotes the increase in steel strain due to the increase in load from the stage when the effective prestress P_e acts *alone* to the stage when the additional load causes the compressive stress in the concrete at the cgs level to reduce to zero (see Fig. 14.7). At this stage, the *change* in concrete stress due to decompression is

$$f_{\text{decomp}} = \frac{P_e}{A_c}\left(1 + \frac{e^2}{r^2}\right) \tag{14.11c}$$

This relationship is based on the assumption that the strain between the concrete and the prestressing steel bonded to the surrounding concrete is such that the gain in the steel stress is the same as the decrease in the concrete stress.

14.3.1.3. Beams with variable tendon eccentricity.

Beams are prestressed with either draped or harped tendons. The maximum eccentricity is usually at the midspan controlling section for the simply supported case. Assuming that the effective prestressing force is

$$P_e = \gamma P_i$$

where γ is the residual prestress ratio, the loss of prestress is

$$P_i - P_e = (1 - \gamma)P_i \tag{a}$$

If the actual concrete extreme fiber stress is equivalent to the maximum allowable stress, the change in this stress after losses, from Eqs. 14.9a and 14.9b, is given by

$$\Delta f^t = (1 - \gamma)\left(f_{ti} + \frac{M_D}{S^t}\right) \tag{b}$$

$$\Delta f_b = (1 - \gamma)\left(-f_{ci} + \frac{M_D}{S_b}\right) \tag{c}$$

From Fig. 14.8, as the superimposed dead-load moment M_{SD} and live-load moment M_L act on the beam, the net stress at the top fibers is

$$f_n^t = f_{ti} - \Delta f^t - f_c \tag{d}$$

or $$f_n^t = \gamma f_{ti} - (1 - \gamma)\frac{M_D}{S^t} - f_c$$

The net stress at the bottom fibers is

$$f_{bn} = f_t - f_{ci} - \Delta f_b$$

or $$f_{bn} = f_t - \gamma f_{ci} - (1 - \gamma)\frac{M_D}{S_b} \tag{e}$$

From Eqs. (d) and (e), the chosen section should have section moduli values

$$S^t \geq \frac{(1 - \gamma)M_D + M_{SD} + M_L}{\gamma f_{ti} - f_c} \tag{14.12a}$$

and $$S_b \geq \frac{(1 - \gamma)M_D + M_{SD} + M_L}{f_t - \gamma f_{ci}} \tag{14.12b}$$

The required eccentricity of the prestressing tendon at the critical section, such as the midspan section, is

$$e_c = (f_{ti} - \bar{f}_{ci})\frac{S^t}{P_i} + \frac{M_D}{P_i} \tag{14.12c}$$

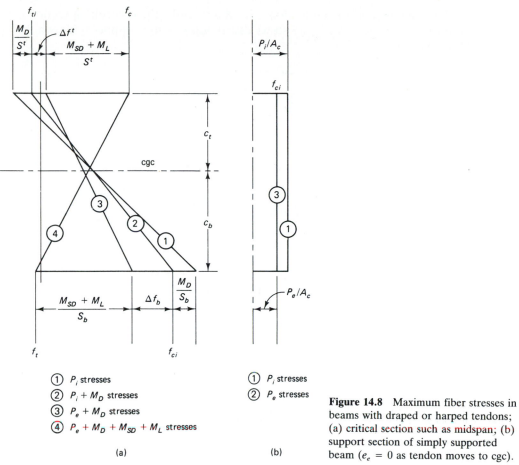

① P_i stresses
② $P_i + M_D$ stresses
③ $P_e + M_D$ stresses
④ $P_e + M_D + M_{SD} + M_L$ stresses

(a)

① P_i stresses
② P_e stresses

(b)

Figure 14.8   Maximum fiber stresses in beams with draped or harped tendons; (a) critical section such as midspan; (b) support section of simply supported beam ($e_e = 0$ as tendon moves to cgc).

where $\bar{f}_{ci}$ is the concrete stress at transfer at the level of the centroid cgc of the concrete section and

$$P_i = \bar{f}_{ci} A_c$$

Thus

$$\bar{f}_{ci} = f_{ti} - \frac{c_t}{h}\,(f_{ti} - f_{ci}) \tag{14.12d}$$

14.3.1.4. Beams with constant tendon eccentricity.   Beams with constant tendon eccentricity are beams with straight tendons, as is normally the case in precast, moderate-span, simply supported beams. Because the tendon has a large eccentricity at the support, creating large tensile stresses at the top fibers without any reduction due to superimposed $M_D + M_{SD} + M_L$, in such beams smaller eccentricity of the tendon at midspan has to be used as compared to a similar beam with a draped tendon. In other words, the controlling section is the support section, for which the stress distribution at the support is shown in Fig. 14.9. Hence:

$$\Delta f^t = (1 - \gamma)(f_{ti}) \tag{a'}$$

and

$$\Delta f_b = (1 - \gamma)(-f_{ci}) \tag{b'}$$

The net stress at the service-load condition after losses at the top fibers is

$$f_n^t = f_{ti} - \Delta f^t - f_c$$

or

$$f_n^t = \gamma f_{ti} - f_{cs} \tag{c'}$$

where f_{cs} is the actual service-load stress in concrete. The net stress at service load after losses at the bottom fibers is

$$f_{bn} = f_t - f_{ci} - \Delta f_b$$

or

$$f_{bn} = f_t - \gamma f_{ci} \tag{d'}$$

From Eqs. (c) and (d), the chosen section should have section moduli values

$$S^t \geq \frac{M_D + M_{SD} + M_L}{\gamma f_{ti} - f_c} \tag{14.13a}$$

and

$$S_b \geq \frac{M_D + M_{SD} + M_L}{f_t - \gamma f_{ci}} \tag{14.13b}$$

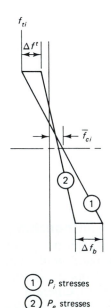

Figure 14.9 Maximum fiber stresses at support section of beams with straight tendons (stress distribution at midspan section similar to that of Fig. 14.8).

① P_i stresses
② P_e stresses

TABLE 14.4   Section Properties and Moduli of Standard PCI Rectangular Sections

Designation	12RB16	12RB20	12RB24	12RB28	12RB32	12RB36	16RB32	16RB36	16RB40
Section modulus, S (in.³)	512	800	1152	1568	2048	2592	2731	3456	4267
Width, b (in.)	12	12	12	12	12	12	16	16	16
Depth, h (in.)	16	20	24	28	32	36	32	36	40

The required eccentricity value at the critical section, such as the support for an ideal beam section having properties close to those required by Eqs. 14.13a and 14.13b is

$$e_e = (f_{ti} - \bar{f}_{ci})\frac{S^t}{P_i} \qquad (14.13c)$$

Table 14.4 gives the section moduli of standard PCI rectangular sections. Tables 14.5 and 14.6 give the geometrical outer dimensions of standard PCI T sections and AASHTO I sections, respectively, as well as the top-section moduli of those sections needed in the preliminary choice of the section in the service-load analysis. Table 14.7 gives dimensional details of the *actual* as-built geometry of the standard PCI and AASHTO sections.

14.3.2 Example 14.4: Service-load Design Example

Design a simply supported pretensioned beam with harped tendon and with a span of 65 ft (19.8 m) using the ACI 318 Building Code allowable stresses. The beam has to carry a superimposed service load of 1100 plf (16.1 kN/m) and superimposed dead load of 100 plf (1.5 kN/m) and has no concrete topping. Assume that beam is made of normal-weight concrete with $f_c' = 5000$ psi (34.5 MPa) and the concrete strength f_{ci}' at transfer is 75% of the cylinder strength. Assume also that the time-dependent losses of the initial prestress are 18% of the initial prestress and $f_{pu} = 270,000$ psi (1862 MPa) for stress-relieved tendons.

TABLE 14.5   Geometrical Outer Dimensions and Section Moduli of Standard PCI T Sections

Designation	Top-/Bottom-Section Modulus (in.³)	Flange Width b_f (in.)	Flange Width t_f (in.)	Total Depth h (in.)	Web Width b_w (in.)
8DT12	1001/315	96	2	12	9.5
8DT14	1307/429	96	2	14	9.5
8DT16	1630/556	96	2	16	9.5
8DT20	2320/860	96	2	20	9.5
8DT24	3063/1224	96	2	24	9.5
8DT32	5140/2615	96	2	32	9.5
10DT32	5960/2717	120	2	32	12.5
8ST36	6899/3650	96	3	36	8
10ST48	13,194/4803	120	3	48	8

TABLE 14.6 Geometrical Outer Dimensions and Section Moduli of Standard AASHTO Bridge Sections

Designation	AASHTO Sections					
	Type 1	Type 2	Type 3	Type 4	Type 5	Type 6
Area A_c, in.^{2}	276	369	560	789	1,013	1,085
Moment of inertia I_g, in.^{4}	22,750	50,979	125,390	260,741	521,180	733,320
Top-/bottom-section modulus, in.^{3}	1,476 1,807	2,527 3,320	5,070 6,186	8,908 10,544	16,790 16,307	20,587 20,157
Top flange width, b_f (in.)	12	12	16	20	42	42
Top flange average thickness, t_f (in.)	6	8	9	11	7	7
Bottom flange width, b_2 (in.)	16	18	22	26	28	28
Bottom flange average thickness, t_2 (in.)	7	9	11	12	13	13
Total depth, h (in.)	28	36	45	54	63	72
Web width, b_w (in.)	6	6	7	8	8	8
c_t/c_b (in.)	15.41 12.59	20.17 15.83	24.73 20.27	29.27 24.73	31.04 31.96	35.62 36.38
r^2, in.^{2}	82	132	224	330	514	676

Solution

$$\gamma = 100 - 18 = 82\%$$

$$f'_{ci} = 0.75 \times 5000 = -3750 \text{ psi } (25.9 \text{ MPa})$$

$$f_{ci} = 0.60 \times 3750 = -2250 \text{ psi } (15.5 \text{ MPa})$$

$$f_{ti} = 3\sqrt{3750} = 184 \text{ psi } \quad (\text{midspan})$$

$$= 6\sqrt{3750} = 368 \text{ psi } \quad (\text{support})$$

$$f_c = 0.45 \times 5000 = -2250 \text{ psi } (15.5 \text{ MPa})$$

Use $f_t = 6\sqrt{5000} = 425$ psi (2.95 MPa) as the maximum stress in tension, and assume a self-weight of approximately 850 plf (12.4 kN/m). Then the self-weight moment is given by

$$M_D = \frac{wl^2}{8} = \frac{850(65)^2}{8} \times 12 = 5,386,875 \text{ in.-lb } (608.7 \text{ kN-m})$$

and the superimposed load moment is

$$M_{SD} + M_L = \frac{(1100 + 100)(65)^2}{8} \times 12 = 7,605,000 \text{ in.-lb } (859.4 \text{ kN-m})$$

TABLE 14.7   Geometrical Details of As-bUILT PCI and AASHTO Sections

Designation	b_f (in.)	h_f (in.)	b_{w1} (in.)	b_{w2} (in.)	h (in.)	b (in.)
8DT12	96	2	5.75	3.75	12	48
8DT14	96	2	5.75	3.75	14	48
8DT16	96	2	5.75	3.75	16	48
8DT18	96	2	5.75	3.75	18	48
8DT20	96	2	5.75	3.75	20	48
8DT24	96	2	5.75	3.75	24	48
8DT32	96	2	7.75	4.75	32	48
10DT32	120	2	7.75	4.75	32	60

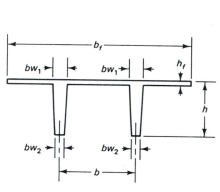

Actual double-T sections

Designation	b_f (in.)	x_1 (in.)	x_2 (in.)	b_w (in.)	h (in.)
8ST36	96	1.5	3	8	36
10ST48	120	1.5	4	8	48

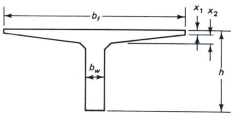

Actual T sections

Designation	b_f (in.)	x_1 (in.)	x_2 (in.)	b_2 (in.)	x_3 (in.)	x_4 (in.)	b_w (in.)	h (in.)
AASHTO 1	12	4	3	16	5	5	6	28
AASHTO 2	12	6	3	18	6	6	6	36
AASHTO 3	16	7	4.5	22	7.5	7	7	45
AASHTO 4	20	8	6	26	9	8	8	54
AASHTO 5	42	5	7	28	10	8	8	63
AASHTO 6	42	5	7	28	10	8	8	72

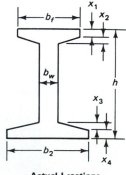

Actual I sections

Since the tendon is harped, the critical section is close to the midspan, where dead-load and superimposed dead-load moments reach their maximum. The critical section is in many cases taken at $0.40L$ from the support, where L is the beam span. From Eqs. 14.12a and 14.12b,

$$S^t \geq \frac{(1 - \gamma)M_D + M_{SD} + M_L}{\gamma f_{ti} - f_c}$$

$$\geq \frac{(1 - 0.82)5,386,875 + 7,605,000}{0.82 \times 184 + 2250} = 3572 \text{ in.}^3 \ (58,535 \text{ cm}^3)$$

$$S_b \geq \frac{(1 - \gamma)M_D + M_{SD} + M_L}{f_t - \gamma f_{ci}}$$

$$\geq \frac{(1 - 0.82)5,386,875 + 7,605,000}{425 + (0.82 \times 2250)} = 3777 \text{ in.}^3 \ (61.892 \text{ cm}^3)$$

required $S^t = 3572$ in.^{3} (58,535 cm^3)

required $S_b = 3777$ in.^{3} (61,872 cm^3)

Since the section moduli at the top and bottom fibers are almost equal, a symmetrical section is adequate. Next, analyze the section in Fig. 14.10 chosen by trial and adjustment.

Photo 83 Crack development is prestressed T beam. (Test by Nawy et al.)

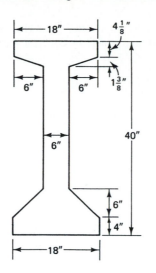

Figure 14.10   I-beam section in Ex. 14.1.

Analysis of stresses at transfer

From Eq. 14.12d,

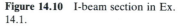

$$\bar{f}_{ci} = f_{ti} - \frac{c_t}{h}(f_{ti} - f_{ci})$$

$$= +184 - \frac{21.16}{40}(+184 + 2250) \simeq -1104 \text{ psi } (C) \text{ (7.6 MPa)}$$

$$P_i = A_c\bar{f}_{ci} = 377 \times 1104 = 416,208 \text{ lb (1851 kN)}$$

$$M_D = \frac{393(65)^2}{8} \times 12 = 2,490,638 \text{ in.-lb (281 kN-m)}$$

From Eq. 14.12c, the eccentricity required at the section of maximum moment at midspan is

$$e_c = (f_{ti} - \bar{f}_{ci})\frac{S^t}{P_i} + \frac{M_D}{P_i}$$

$$= (184 + 1104)\frac{3572}{416,208} + \frac{2,490,638}{416,208}$$

$$= 11.05 + 5.98 = 17.04 \text{ in. (433 mm)}$$

Since $c_b = 18.84$ in. and assuming a cover of 3.75 in., try $e_c = 18.84 - 3.75 \cong 15.0$ in. (381 mm).

$$\text{required area of tendons } A_p = \frac{P_i}{f_{pi}} = \frac{416,208}{189,000} = 2.2 \text{ in.}^2 \text{ (14.2 cm}^2\text{)}$$

$$\text{number of strands} = \frac{2.2}{0.153} = 14.38$$

Try thirteen $\frac{1}{2}$-in. strands, $A_p = 1.99$ in.^{2} (12.8 cm), and an actual $P_i = 189,000 \times 1.99 = 376,110$ lb (1673 kN), and check the concrete extreme fiber stresses. From Eq. 14.9a,

$$f^t = -\frac{P_i}{A_c}\left(1 - \frac{ec_t}{r^2}\right) - \frac{M_D}{S^t}$$

$$= -\frac{376,110}{377}\left(1 - \frac{15.0 \times 21.16}{187.5}\right) - \frac{2,490,638}{3340}$$

$$= +691.2 - 745.7 = -54.5 \text{ psi } (C), \text{ no tension at transfer} \qquad \text{O.K.}$$

From Eq. 14.9b,

$$f_b = -\frac{P_i}{A_c}\left(1 + \frac{ec_b}{r^2}\right) + \frac{M_D}{S_b}$$

$$= -\frac{376,110}{377}\left(1 + \frac{15 \times 18.84}{187.5}\right) + \frac{2,490,638}{3750}$$

$$= -2501.3 + 664.2 = -1837.1 \text{ psi } (C) < f_{ci} = 2250 \text{ psi} \qquad \text{O.K.}$$

Analysis of stresses at service load

From Eq. 14.11a,

$$f^t = -\frac{P_e}{A_c}\left(1 - \frac{ec_t}{r^2}\right) - \frac{M_T}{S^t}$$

$$P_e = 13 \times 0.153 \times 154,980 = 308,255 \text{ lb (1372 kN)}$$

total moment $M_T = M_D + M_{SD} + M_L = 2,490,638 + 7,605,000$
$$= 10,095,638 \text{ in.-lb (1141 kN-m)}$$

$$f^t = -\frac{308,225}{377}\left(1 - \frac{15.0 \times 21.16}{187.5}\right) - \frac{10,095,638}{3340}$$

$$= +566.5 - 3022.6 = -2456.1 \text{ psi } (C) > f_c = -2250 \text{ psi}$$

Hence either enlarge the depth of the section or use higher-strength concrete. Using $f_c' = 6000$ psi,

$$f_c = 0.45 \times 6000 = -2700 \text{ psi} \qquad \text{O.K.}$$

$$f_b = -\frac{P_e}{A_c}\left(1 + \frac{ec_b}{r^2}\right) + \frac{M_T}{S_b} = \frac{308,255}{377}\left(1 + \frac{15.0 \times 18.84}{187.5}\right) + \frac{10,095,638}{3750}$$

$$= 2050 + 2692.2 = 642 \text{ psi } (T) \qquad \text{O.K.}$$

Check support section stresses

$$f_{ci}' = 0.75 \times 6000 = 4500 \text{ psi}$$

$$f_{ci} = 0.60 \times 4500 = 2700 \text{ psi}$$

$$f_{ti} = 3\sqrt{f_{ci}'} = 201 \text{ psi for span}$$

$$f_{ti} = 6\sqrt{f_{ci}'} = 402 \text{ psi for support}$$

$$f_c = 0.45 f'_c = 2700 \text{ psi}$$

$$f_{t1} = 6\sqrt{f'_c} = 465 \text{ psi}$$

$$f_{t2} = 12\sqrt{f'_c} = 930 \text{ psi}$$

(a) *At transfer:* Support section compressive fiber stress,

$$f_b = -\frac{P_i}{A_c}\left(1 + \frac{ec_b}{r^2}\right) + 0$$

$$P_i = 376,110 \text{ lb}$$

or

$$-2700 = -\frac{376,110}{377}\left(1 + \frac{e \times 18.84}{187.5}\right)$$

so that

$$e_e = 16.98 \text{ in.}$$

Accordingly, try $e_e = 12.49$ in.:

$$f' = -\frac{376,110}{377}\left(1 - \frac{12.49 \times 21.16}{187.5}\right) - 0$$

$$= 409 \text{ psi } (T) > f_{ti} = 402 \text{ psi}$$

Thus, use mild steel at the top fibers at the suport section to take all tensile stresses in the concrete, or use a higher-strength concrete for the section, or reduce the eccentricity.

(b) *At service load*

$$f' = -\frac{308,225}{377}\left(1 - \frac{12.49 \times 21.16}{187.5}\right) - 0 = 334.9 \text{ psi } (T) < 465 \text{ psi} \qquad \text{O.K.}$$

$$f_b = -\frac{308,255}{377}\left(1 + \frac{12.49 \times 18.84}{187.5}\right) + 0 = -1843 \text{ psi } (C) < -2700 \text{ psi} \qquad \text{O.K.}$$

Hence adopt the 40-in. (102-cm)-deep I-section prestressed beam of f'_c equal to 6000 psi (41.4 MPa) normal-weight concrete with thirteen $\frac{1}{2}$-in. strands having midspan eccentricity $e_c = 15.0$ in. (381 mm) and end section eccentricity $e_e = 12.5$ in. (318 m). An alternative to this solution is to continue using $f'_c = 5000$ psi, but change the number of strands and eccentricities.

14.3.3 Flow Chart for Service-load Flexural Design of Prestressed Beams

Figure 14.11 shows a flow chart for the service-load flexural design of prestressed beams.

14.4 ULTIMATE-STRENGTH FLEXURAL DESIGN OF PRESTRESSED BEAMS

14.4.1 Rectangular Sections

The actual distribution of the compressive stress in a section at failure has the form of a rising parabola, as shown in Fig. 14.12c. It is time consuming to evaluate the volume of the compressive stress block if it has a parabolic shape. An equivalent rectangular stress block due to Whitney can be used with ease and without loss of accuracy to calculate the compressive force and hence the flexural moment strength of the section. This equivalent stress block has a depth a and an average compressive strength $0.85f'_c$. As seen from Fig. 14.12d, the value of $a = \beta_1 c$ is determined by using a coefficient β_1 such that the area of the equivalent rectangular block is approximately the same as that of the parabolic compressive block, resulting in a compressive force C of essentially the same value in both cases.

The value $0.85f'_c$ for the average stress of the equivalent compressive block is based on the core test results of concrete in the structure at a minimum age of 28 days. Based on exhaustive experimental tests, a maximum allowable strain of 0.003 in./in. was adopted by the ACI as a safe limiting value. Even though several

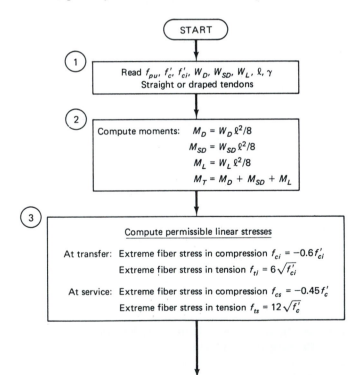

Figure 14.11 Flow chart for service-load flexural design of prestressed beams.

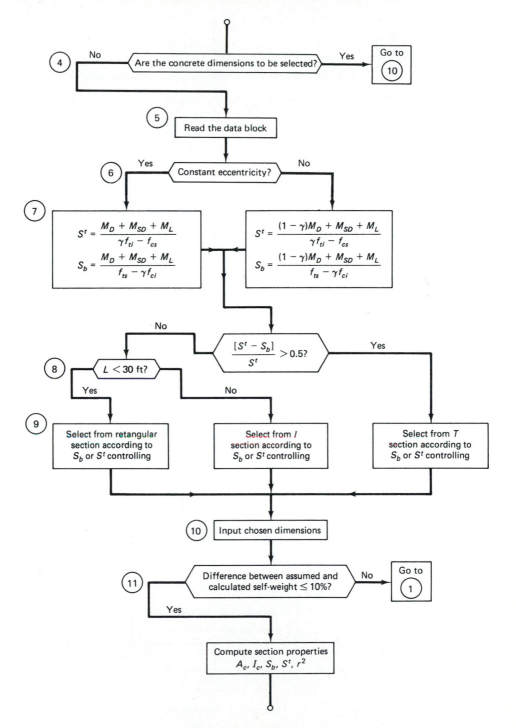

Figure 14.11 (*continued*)

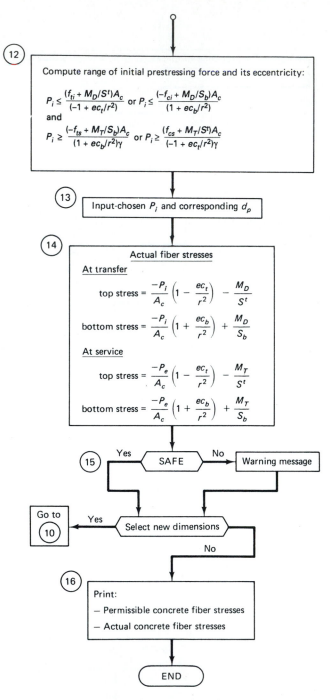

Figure 14.11 *(continued)*

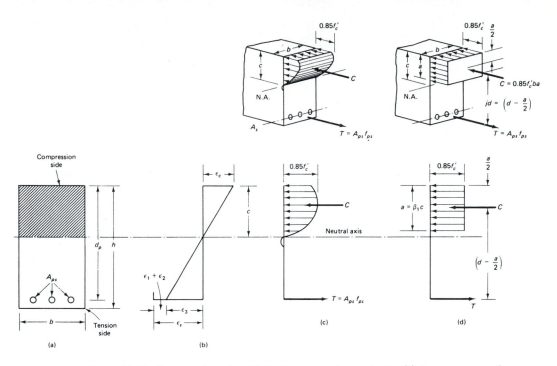

Figure 14.12 Stress and strain distribution across beam depth: (a) beam cross section; (b) strains; (c) actual stress block; (d) assumed equivalent stress block.

forms of stress blocks, including the trapezoidal, have been proposed to date, the simplified equivalent rectangular block is accepted as the standard in the analysis and design of reinforced concrete. The behavior of the steel is assumed to be elastoplastic.

Using all the preceding assumptions, the stress distribution diagram shown in Fig. 14.12c can be redrawn as shown in Fig. 14.12d. We can easily deduce that the compression force C can be written $0.85f'_cba$, that is, the *volume* of the compressive block at or near the ultimate when the tension steel has yielded ($\epsilon_s > \epsilon_y$). The tensile force T can be written $A_{ps}f_{ps}$; thus the equilibrium equation can be written as

$$A_{ps}f_{ps} = 0.85f'_cba \tag{14.14}$$

A little algebra yields

$$a = \beta_1 c = \frac{A_{ps}f_{ps}}{0.85f'_cb}$$

The nominal moment strength is obtained by multiplying C or T by the moment arm $(d_p - a/2)$, yielding

$$M_n = A_{ps}f_{ps}\left(d_p - \frac{a}{2}\right) \tag{14.15a}$$

where d_p is the distance from the compression fibers to the center of the prestressed reinforcement. The steel percentage $p_p = A_{ps}/bd_p$ gives nominal strength of the prestressing steel only as follows

$$M_n = p_p f_{ps} bd_p^2 \left(1 - 0.59 p_p \frac{f_{ps}}{f_c'} \right) \tag{14.15b}$$

If ω_p is the reinforcement index $= p_p(f_{ps}/f_c')$, Eq. 14.15b becomes

$$M_n = p_p f_{ps} bd_p^2 (1 - 0.59 \omega_p) \tag{14.15c}$$

The contribution of the mild steel tension reinforcement should be similarly treated, so that the depth a of the compressive block is

$$a = \frac{A_{ps} f_{ps} + A_s f_y}{0.85 f_c' b} \tag{14.16a}$$

If $c = a/\beta_1$, the strain at the level of the mild steel is (Fig. 14.12)

$$\epsilon_s = \epsilon_c \frac{d - c}{c} \tag{14.16b}$$

Equation 14.15b, for rectangular sections but with mild tension steel and no compression steel accounted for, becomes

$$M_n = p_p f_{ps} bd_p^2 \left(1 - 0.59 p_p \frac{f_{ps}}{f_c'} \right) + p f_y bd^2 \left(1 - 0.59 \frac{f_y}{f_c'} \right) \tag{14.17a}$$

or can be rewritten as either

$$M_n = A_{ps} f_{ps} \left\{ 1 - 0.59 \left(\omega_p + \frac{d}{d_p} \omega \right) \right\} + A_s f_y \left\{ 1 - 0.59 \left(\frac{d_p}{d} \omega_p + \omega \right) \right\} \tag{14.17b}$$

where $\omega = p(f_y/f_c')$, or

$$M_n = A_{ps} f_{ps} \left(d_p - \frac{a}{2} \right) + A_s f_y \left(d - \frac{a}{2} \right) \tag{14.17c}$$

The contribution from compression reinforcement can be taken into account provided it has been found to have yielded

$$a = \frac{A_{ps} f_{ps} + A_s f_y - A_s' f_y}{0.85 f_c' b} \tag{14.18}$$

where b is the section width at the compression face of the beam.

Taking moments about the center of gravity of the compressive block in Fig. 14.13, the nominal moment strength in Eq. 14.17c becomes

$$M_n = A_{ps} f_{ps} \left(d_p - \frac{a}{2} \right) + A_s f_y \left(d - \frac{a}{2} \right) + A_s' f_y \left(\frac{a}{2} - d' \right) \tag{14.19}$$

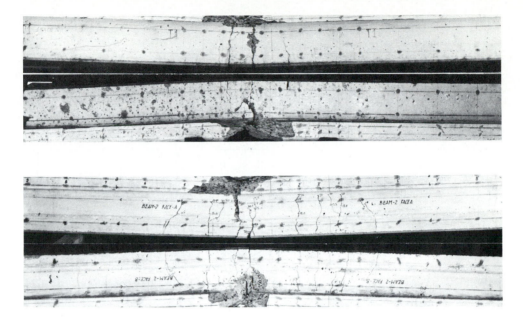

Photo 84   Flexural cracks at failure of prestressed T beams.   (Tests by Nawy and Potyondy.)

14.4.2 Nominal Moment Strength of Flanged Sections

When the compression flange thickness h_f is less than the neutral axis depth c and equivalent rectangular block depth a, the section can be treated as a flanged section as in Fig. 14.14.   From the figure,

$$T_p + T_s = T_{pw} + T_{pf} \tag{14.20}$$

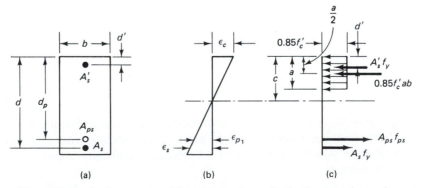

Figure 14.13   Strain, stress, and forces across beam depth of rectangular section: (a) beam section; (b) strain; (c) stresses and forces.

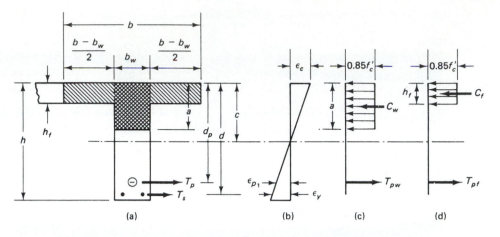

Figure 14.14 Strain, stress, and forces in flanged sections: (a) beam section; (b) strain; (c) web stress and forces; (d) flange stress and force.

where T_p = total prestressing force = $A_{ps}f_{ps}$
 T_s = ultimate force in the nonprestressed steel = $A_s f_y$
T_{pw} = part of the total force in the tension reinforcement required to de-
$\phantom{T_{pw} =}$ velop the web = $A_{pw}f_{ps}$
A_{pw} = total reinforcement area corresponding to the force T_{pw}
T_{pf} = part of the total force in the tension reinforcement required to de-
$\phantom{T_{pf} =}$ velop the flange = C_f = $0.85f'_c(b - b_w)h_f$
C_w = $0.85f'_c b_w a$

Substituting in Eq. 14.20, we obtain

$$T_{pw} = A_{ps}f_{ps} + A_s f_y - 0.85f'_c(b - b_w)h_f \tag{14.21}$$

Summing up all forces in Fig. 14.14c and d, we have

$$T_{pw} + T_{pf} = C_w + C_f$$

giving

$$a = \frac{A_{pw}f_{ps}}{0.85f'_c b_w} \tag{14.22a}$$

or $\qquad\qquad a = \dfrac{A_{ps}f_{ps} + A_s f_y - 0.85f'_c(b - b_w)h_f}{0.85f'_c b_w} \tag{14.22b}$

Equation 14.19 for a beam with compression reinforcement can be rewritten to give the nominal moment strength for a flanged section where the neutral axis falls outside the flange and $a > h_f$ as follows, taking moments about the center of the prestressing steel:

$$M_n = A_{pw}f_{ps}\left(d_p - \frac{a}{2}\right) + A_s f_y(d - d_p) + 0.85f'_c(b - b_w)h_f\left(d_p - \frac{h_f}{2}\right) \tag{14.23}$$

The design moment in all cases would be

$$M_u = \phi M_n \tag{14.24}$$

where $\phi = 0.90$ for flexure.

In order to determine whether the neutral axis falls outside the flange, requiring a flanged section analysis, we must determine, as discussed in Chapter 5, whether the total compressive force C_n is larger or smaller than the total tensile force T_n. If $T_p + T_s$ in Fig. 14.14 is larger than C_f, the neutral axis falls outside the flange and the section has to be treated as a flanged section. Otherwise, it should be treated as a rectangular section of the width b of the compression flange.

Another method of determining whether the section can be considered flanged is to calculate the value of the equivalent rectangular block depth a from Eq. 14.22b, thereby determining the neutral-axis depth $c = a/\beta_1$.

14.4.3 Determination of Prestressing Steel Nominal Failure Stress f_{ps}

The value of the stress f_{ps} of the prestressing steel at failure is not readily available. However, it can be determined by *strain compatibility* through the various loading stages up to the limit state at failure, as defined in Eqs. 14.15. Such a procedure is required if

$$f_{pe} = \frac{P_e}{A_{ps}} < 0.50 f_{pu} \tag{14.25a}$$

Approximate determination is allowed by the ACI 318 building code provided that

$$f_{pe} = \frac{P_e}{A_{ps}} \geq 0.50 f_{pu} \tag{14.25b}$$

with separate equations for f_{ps} given for bonded and nonbonded members.

Bonded tendons.   The empirical expression for bonded members is

$$f_{ps} = f_{pu}\left(1 - \frac{\gamma_p}{\beta_1}\left[\rho_p \frac{f_{pu}}{f'_c} + \frac{d}{d_p}(\omega - \omega')\right]\right) \tag{14.26}$$

where the reinforcement index for the compression nonprestressed reinforcement is $\omega' = \rho'(f_y/f'_c)$. If the compression reinforcement is taken into account when calculating f_{ps} by Eq. 14.26, the term $[\rho_p(f_{pu}/f'_c) + (d/d_p)(\omega - \omega')]$ should not be less than 0.17 and d' should not be greater than $0.15d_p$. Also,

$$\gamma_p = 0.55 \text{ for } f_{py}/f_{pu} \text{ not less than } 0.80$$
$$= 0.40 \text{ for } f_{py}/f_{pu} \text{ not less than } 0.85$$
$$= 0.28 \text{ for } f_{py}/f_{pu} \text{ not less than } 0.90$$

The value of the factor γ_p is based on the criterion that $f_{py} = 0.80f_{pu}$ for high-strength prestressing bars, 0.85 for stress-relieved strands, and 0.90 for low-relaxation strands.

Unbonded tendons. For a span-to-depth ratio of 35 or less,

$$f_{ps} = f_{pe} + 10,000 + \frac{f'_c}{100\rho_p} \qquad (14.27a)$$

where f_{ps} shall not be greater than f_{py} or $f_{pe} + 60,000$.

For a span-to-depth ratio greater than 35,

$$f_{ps} = f_{pe} + 10,000 + \frac{f'_c}{300\rho_D} \qquad (14.27b)$$

where f_{ps} shall not be greater than f_{py} or $f_{pe} + 30,000$.   Code requirements for maximum and minimum reinforcement index ω have to be observed.

14.5 FLOW CHART FOR STRENGTH FLEXURAL ANALYSIS OF PRESTRESSED BEAMS

A flow chart for strength flexural analysis is shown in Fig. 14.15.

14.6 EXAMPLE 14.5: ULTIMATE-STRENGTH DESIGN OF PRESTRESSED SIMPLY SUPPORTED BEAM BY STRAIN COMPATIBILITY

Design the bonded beam in Ex. 14.4 by the ultimate-load theory using nonprestressed reinforcement to *partially* carry part of the factored loads.   Use strain compatibility to evaluate f_{ps}, given

$$f_{pu} = 270,000 \text{ psi (1862 MPa)}$$

$$f_{py} = 0.85 f_{pu} \text{ for stress-relieved strands}$$

$$f_y = 60,000 \text{ psi (414 MPa)}$$

$$f'_c = 5,000 \text{ psi (34.5 MPa), normal-weight concrete}$$

Use seven-wire $\frac{1}{2}$-in.-diameter strands.   The nonprestressed partial mild steel is to be placed with a $1\frac{1}{2}$-in. clear cover, and no compression steel is to be accounted for.   No wind or earthquake is taken into consideration.   The stress–strain diagram of the prestressing tendons is given in Fig. 14.16.

Solution   From Ex. 14.4,

$$\text{service } W_L = 1100 \text{ plf (16.1 kN/m)}$$

$$\text{service } W_{SD} = 100 \text{ plf (1.46 kN/m)}$$

$$\text{assumed } W_D = 393 \text{ plf (5.74 kN/m)}$$

$$\text{beam span} = 65 \text{ ft (19.8 m)}$$

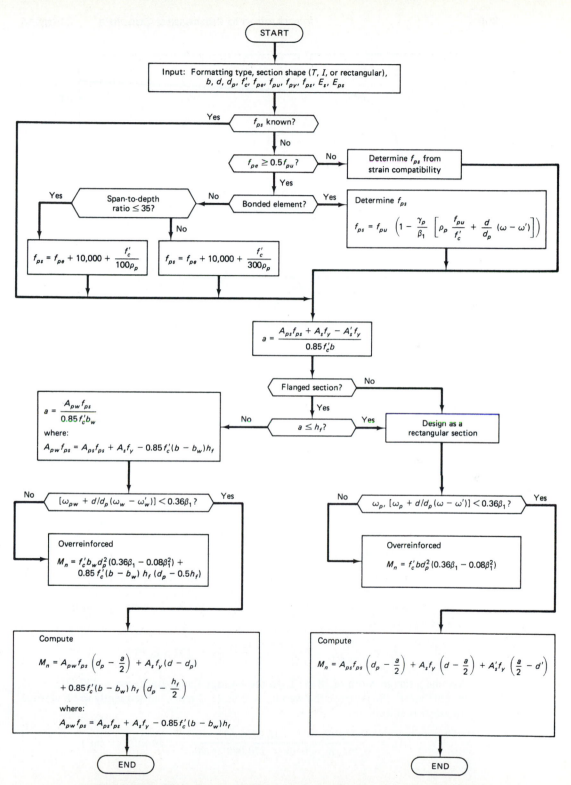

Figure 14.15 Flow chart for ultimate load flexural analysis of rectangular and flanged prestressed sections based on cgs profile depth.

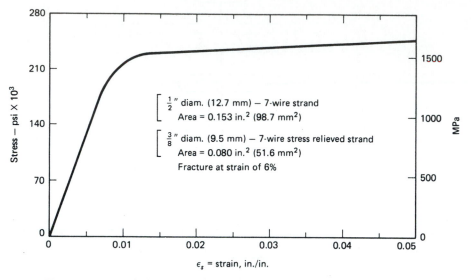

Figure 14.16   Typical stress–strain relationship of seven-wire 270-K prestressing strand.

Factored moment

$$W_u = 1.4(W_D + W_{SD}) + 1.7W_L$$

$$= 1.4(100 + 393) + 1.7(1100) = 2560 \text{ plf } (37.4 \text{ kN/m})$$

The factored moment is given by

$$M_u = \frac{w_u l^2}{8} = \frac{2560(65)^2 12}{8} = 16,224,000 \text{ in.-lb } (1833 \text{ kN-m})$$

and the required nominal moment strength is

$$M_n = \frac{M_u}{\phi} = \frac{16,224,000}{0.9} = 18,026,667 \text{ in.-lb } (2037 \text{ kN-m})$$

Choice of preliminary section

Assuming a depth of 0.6 in. per foot of span, we can have a trial section depth $h = 0.6 \times 65 \simeq 40$ in. (102 cm). Then assume a mild partial steel 4 No. 6 = 4 × 0.44 = 1.76 in.² (11.4 cm²). Empirical expressions for A'_c and A_{ps} can be used as follows (see Ref. 14.1):

$$A'_c = \frac{M_n}{0.68 f'_c h} = \frac{18,026,667}{0.68 \times 5000 \times 40} = 132.5 \text{ in.}^2 \ (855 \text{ cm}^2)$$

Assume a flange width of 18 in. Then the average flange thickness = 132.5/18 ≈ 7.5 in. (191 mm). So suppose the web $b_w = 6$ in. (152 mm), subsequently to be verified for shear requirements.

$$A_{ps} = \frac{M_n}{0.76 f_{pu} h} = \frac{18,026,667}{0.76 \times 270,000 \times 40} = 2.19 \text{ in.}^2 \ (14 \text{ cm}^2)$$

and the number of $\frac{1}{2}$-in. stress-relieved wire strands $= 2.19/0.153 = 14.31$. So try thirteen $\frac{1}{2}$-in. strands:

$$A_{ps} = 13 \times 0.153 = 1.99 \text{ in.}^2 \ (12.8 \text{ cm}^2)$$

Calculate the stress f_{ps} in the prestressing tendon at nominal strength using the strain-compatibility approach

The geometrical properties of the trial section are very close to the assumed dimensions for the depth h and the top flange width b. Hence use the following data for the purpose of the example:

$$A_c = 377 \text{ in.}^2$$

$$c_t = 21.16 \text{ in.}$$

$$d_p = 15 + c_t = 15 + 21.16 = 36.16 \text{ in.}$$

$$r^2 = 187.5 \text{ in.}^2$$

$$e = 15 \text{ in. at midspan}$$

$$e^2 = 225 \text{ in.}^2$$

$$\frac{e^2}{r^2} = \frac{225}{187.5} = 1.20$$

$$E_c = 57{,}000\sqrt{5000} = 4.03 \times 10^6 \text{ psi} \ (27.8 \times 10^3 \text{ MPa})$$

$$E_p = 28 \times 10^6 \text{ psi} \ (193 \times 10^3 \text{ MPa})$$

The maximum allowable compressive strain ϵ_c at failure $= 0.003$ in./in. Assume that the effective prestress at service load is $f_{pe} \approx 155{,}000$ psi (1069 MPa).

$$\epsilon_1 = \epsilon_{pe} = \frac{f_{pe}}{E_p} = \frac{155{,}000}{28 \times 10^6} = 0.0055 \text{ in./in.}$$

$$P_e = 13 \times 0.153 \times 155{,}000 = 308{,}295 \text{ lb}$$

The increase in prestressing steel strain as the concrete is decompressed by the increased external load (see Fig. 14.7 and Eq. 14.11c) is given as

$$\epsilon_2 = \epsilon_{decomp} = \frac{P_e}{A_c E_c}\left(1 + \frac{e^2}{r^2}\right)$$

$$= \frac{308{,}295}{377 \times 4.03 \times 10^6}(1 + 1.20) = 0.0004 \text{ in./in.}$$

Assume that the stress $f_{ps} \approx 205{,}000$ psi as a first trial. Suppose the neutral axis inside the flange is verified. Then, from Eq. 14.16a,

$$a = \frac{A_{ps}f_{ps} + A_s f_y}{0.85 f'_c b} = \frac{1.99 \times 205{,}000 + 1.76 \times 60{,}000}{0.85 \times 5000 \times 18}$$

$$= 6.71 \text{ in. (17 cm)} < h_f = 7.5 \text{ in.}$$

Hence the equivalent compressive block is inside the flange and the section is to be treated as rectangular.

Accordingly, for 5000-psi concrete,

$$\beta_1 = 0.85 - 0.05 = 0.8$$

$$c = \frac{a}{\beta_1} = \frac{6.71}{0.80} = 8.39 \text{ in. (22.7 cm)}$$

$$d = 40 - (1.5 + \tfrac{1}{2} \text{ in. for stirrups} + \tfrac{6}{16} \text{ in. for bar}) \simeq 37.6 \text{ in.}$$

The increment of strain due to overload to the ultimate, from Eq. 14.16b, is

$$\epsilon_3 = \epsilon_c \frac{d - c}{c} = 0.003 \frac{37.6 - 8.39}{8.39} = 0.0104 \text{ in./in.}$$

and the total strain is

$$\epsilon_{ps} = \epsilon_1 + \epsilon_2 + \epsilon_3$$
$$= 0.0055 + 0.0004 + 0.0104 = 0.0163 \text{ in./in.}$$

From the stress–strain diagram in Fig. 14.16, the f_{ps} corresponding to $\epsilon_{ps} = 0.0163$ is 230,000 psi.

Second trial for f_{ps} value: Assume that

$$f_{ps} = 229{,}000 \text{ psi}$$

$$a = \frac{1.99 \times 229{,}000 + 1.76 \times 60{,}000}{0.85 \times 5000 \times 18} = 7.34 \text{ in.}$$

$$c = \frac{7.34}{0.80} = 9.17 \text{ in.}$$

$$\epsilon_3 = 0.003 \frac{37.6 - 9.17}{9.17} = 0.0093$$

Then the total strain is $\epsilon_{ps} = 0.0055 + 0.0004 + 0.0093 = 0.0152$ in./in. From Fig. 14.16, $f_{ps} = 229{,}000$ psi (1,579 MPa); use

$$A_s = 4 \text{ No. } 6 = 1.76 \text{ in.}^2$$

Available moment strength

From Eq. 14.17c, for the neutral axis falling within the flange,

$$M_n = 1.99 \times 229{,}000 \left(36.16 - \frac{7.34}{2} \right) + 1.76 \times 60{,}000 \left(37.6 - \frac{7.34}{2} \right)$$

$$= 14{,}806{,}017 + 3{,}583{,}008 = 18{,}389{,}025 \text{ in.-lb (2078 kN-m)}$$
$$> \text{ required } M_n = 18{,}026{,}667 \text{ in.-lb} \qquad \text{O.K.}$$

The percentage of moment resisted by the nonprestressed steel is

$$\frac{3{,}583{,}008}{18{,}026{,}667} \cong 20\%$$

Check for minimum and maximum reinforcement

Minimum $A_s = 0.004A$, where A is the area of the part of the section between the tension face and the cgc. From the cross section of Fig. 14.10,

$$A = 377 - 18\left(4.125 + \frac{1.375}{2}\right) - 6(21.16 - 5.5) \approx 201 \text{ in.}^2$$

minimum $A_s = 0.004 \times 201 = 0.80 \text{ in.}^2 < 1.76$ used   O.K.

The maximum steel index, from Ref. 14.1, is

$$\omega_p + \frac{d}{d_p}(\omega - \omega') \le 0.36\beta_1 \le 0.29 \text{ for } \beta_1 = 0.80$$

and the actual total reinforcement index is

$$\omega_T = \frac{1.99 \times 229,000}{18 \times 36.16 \times 5000} + \frac{37.6}{36.16}\frac{1.76 \times 60,000}{18 \times 37.6 \times 5000}$$

$$= 0.14 + 0.03 = 0.17 < 0.29 \quad \text{O.K.}$$

Choice of section for ultimate load

The section in Ex. 14.4 with the modifications shown in Fig. 14.17 has the normal moment strength M_n that can carry the factored load, provided that four No. 6 nonprestressed bars are used at the tension side as a partially prestressed section. So we can adopt the section for flexure, as it also satisfies the service-load flexural stress requirements both at midspan and at the support. Note that the section could only develop the required nominal strength $M_n = 18,026,667$ in.-lb by the addition

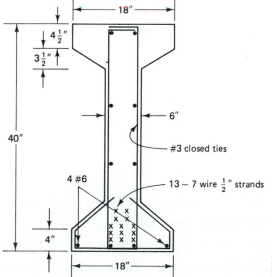

Figure 14.17 Midspan section of the beam in Ex. 14.5.

of the nonprestressed bars at the tension face to resist 20% of the total *required* moment strength. Note also that this section is adequate with a concrete $f'_c = 5000$ psi, while the section in Ex. 14.4 has to have $f'_c = 6000$ psi strength in order not to exceed the allowable service-load concrete stresses. Hence ultimate-load computations are necessary in prestressed concrete design to ensure that the constructed elements can carry all the factored load and are thus an integral part of the total design.

14.7 WEB REINFORCEMENT DESIGN PROCEDURE FOR SHEAR

The following is a summary of a recommended sequence of design steps:

1. Determine the required nominal shear strength value $V_n = V_u/\phi$ at a distance $h/2$ from the face of the support.
2. Calculate the nominal shear strength V_c that the web has by one of the following two methods.
   (a) *ACI short method if $f_{pe} > 0.40f_{pu}$*

$$V_c = \left(0.60\lambda\sqrt{f'_c} + \frac{700V_u d_p}{M_u} \right) b_w d_p$$

where $2\lambda\sqrt{f'_c}\, b_w d_p \le V_c \le 5\lambda\sqrt{f'_c}\, b_w d_p$ and where $V_u d_p/M_u \le 1.0$ and V_u is calculated at the same section for which M_u is calculated.

   If the average tensile splitting strength f_{ct} is specified for lightweight concrete, then $\lambda = f_{ct}/6.7\sqrt{f'_c}$ with $\sqrt{f'_c}$ not to exceed a value of 100.
   (b) *Detailed analysis where V_c is the lesser of V_{ci} and V_{cw}*

$$V_{ci} = 0.60\lambda\sqrt{f'_c}\, b_w d_p + V_d + \frac{V_i}{M_{max}}(M_{cr}) \ge 1.7\lambda\sqrt{f'_c}\, b_w d_p$$

$$V_{cw} = (3.5\lambda\sqrt{f'_c} + 0.3\bar{f}_c)b_w d_p + V_p$$

using d_p or $0.8h$, whichever is larger, and where

$$M_{cr} = (I_c/y_t)(6\lambda\sqrt{f'_c} + f_{ce} - f_d) \text{ or}$$

$$M_{cr} = S_b(6\lambda\sqrt{f'_c} + f_{ce} - f_d)$$

$\qquad V_i =$ factored shear force at section due to externally applied loads occurring simultaneously with M_{max}

$\qquad f_{ce} =$ compressive stress in concrete after occurrence of all losses at extreme fibers of section where external load causes tension

3. If $V_u/\phi \leq \frac{1}{2}V_c$, no web steel is needed. If $V_u/\phi > \frac{1}{2}V_c < V_c$, provide minimum reinforcement. If $V_u/\phi > V_c$ and $V_s = V_u/\phi - V_c \leq 8\lambda\sqrt{f_c'}\, b_w d_p$, design the web steel. If $V_s = V_u/\phi - V_c > 8\lambda\sqrt{f_c'}\, b_w d_p$, or if $V_u > \phi(V_c + 8\lambda\sqrt{f_c'}\, b_w d_p)$, enlarge the section.

4. Calculate the required minimum web reinforcement. The spacing is $s \leq 0.75h$ or 24 in., whichever is smaller.

$$\text{minimum } A_v = \frac{50 b_w s}{f_y} \qquad \text{(conservative)}$$

If $f_{pe} \geq 0.40 f_{pu}$, a less conservative minimum A_v is the smaller of

$$\frac{A_{ps} f_{pu} s}{80 f_y d_p} \sqrt{\frac{d_p}{b_w}}$$

where $d_p \geq 0.80h$, and

$$A_v = \frac{50 b_w s}{f_y}$$

5. Calculate the required web reinforcement size and spacing. If $V_s = (V_u/\phi - V_c) \leq 4\lambda\sqrt{f_c'}\, b_w d_p$, then the stirrup spacing s is as required by the design expressions in step 6, to follow. If $V_s = (V_u/\phi - V_c) > 4\lambda\sqrt{f_c'}\, b_w d_p$, the stirrup spacing s is half the spacing required by the design expressions in step 6.

6. $s = \dfrac{A_v f_y d_p}{(V_u/\phi) - V_c} = \dfrac{A_v \phi f_y d_p}{V_u - \phi V_c} \leq 0.75h \leq 24$ in. $\geq$ maximum s from step 4

7. Draw the shear envelope over the beam span, and mark the band requiring web steel.

8. Sketch the size and distribution of web stirrups along the span using No. 3 or No. 4 size stirrups as preferable, but no larger size than No. 6 stirrups.

9. Design the vertical dowel reinforcement in cases of composite sections.
  (a) $V_{nh} \leq 80 b_v d_{pc}$ for both roughened contact and no vertical ties or dowels, and nonroughened but with minimum vertical ties, use

$$A_v = \frac{50 b_w s}{f_y} = \frac{50 b_v l_{vh}}{f_y}$$

  (b) $V_{nh} \leq 350 b_v d_{pc}$ for a roughened contact surface with full amplitude $\frac{1}{4}$ in.
  (c) For cases where $V_{nh} > 350 b_v d_{pc}$, design vertical ties for $V_{nh} = A_{vf} f_y \mu$,

where A_{vf} = area of frictional steel dowels
         μ = coefficient of friction = 1.0λ for intentionally roughened surface, where $\lambda = 1.0$ for normal-weight concrete. In all cases, $V_n \leq V_{nh} \leq 0.2 f_c' A_c \leq 800 A_{cc}$, where $A_{cc} = b_v l_{vh}$.

An alternative method of determining the dowel reinforcement area A_{vf} is by computing the horizontal force F_h at the concrete contact surface such that

$$F_h \le \mu_e A_{vf} f_y \le V_{nh}$$

where

$$\mu_e = \frac{1000\lambda^2 b_v l_{vh}}{F_h} \le 2.9$$

14.7.1 Example 14.6: Shear Strength and Web–Shear Steel Design in a Prestressed Beam

Design the bonded beam of Ex. 14.4 to be safe against shear failure, and proportion the required web reinforcement by the ACI short method.

Solution   *Data and nominal shear strength determination*

f_{pu} = 270,000 psi (1862 MPa)	d_p = 36.16 in. (91.8 cm)
f_y = 60,000 psi (414 MPa)	d = 37.6 in. (95.5 cm)
f_{pe} = 155,000 psi (1069 MPa)	b_w = 6 in. (15 cm)
f_c' = 5000 psi, normal-weight concrete	e_c = 15 in. (31 cm)
A_{ps} = 13 seven-wire $\frac{1}{2}$-in. strands = 1.99 in.^{2}	e_e = 12.5 in. (32 cm)
(12.8 cm^2)	I_c = 70,700 in.^{4} (2.94 × 10^6 cm^4)
A_s = 4 No. 6 bars = 1.76 in.^{2} (11.4 cm^2)	A_c = 377 in.^{2} (2432 cm^2)
span = 65 ft (19.8 m)	r^2 = 187.5 in.^{2} (1210 cm^2)
service W_L = 1100 plf (16.1 kN/m)	c_b = 18.84 in. (48 cm)
service W_{SD} = 100 plf (1.46 kN/m)	c_t = 21.16 in. (54 cm)
service W_D = 393 plf (5.7 kN/m)	P_e = 308,255 lb (1371 kN)

$h = 40$ in. (101.6 cm)

factored load $W_U = 1.4D + 1.7L$

$$= 1.4(100 + 393) + 1.7 \times 1100 = 2560 \text{ plf}$$

factored shear force at face of support $= V_u = \dfrac{W_U L}{2}$

$$= \frac{2560 \times 65}{2} = 83,200 \text{ lb}$$

required $V_n = \dfrac{V_u}{\phi} = \dfrac{83,200}{0.85} = 97,882$ lb at support

Plane at $\frac{1}{2}d_p$ from face of support

Nominal shear strength V_c of web (steps 2, 3)

$$\frac{1}{2}d_p = \frac{36.16}{2 \times 12} \simeq 1.5 \text{ ft}$$

$$V_n = 97,882 \times \frac{(65/2) - 1.5}{65/2} = 93,364 \text{ lb}$$

$$V_n \text{ at } \tfrac{1}{2}d_p = 0.85 \times 93,364 = 79,359 \text{ lb}$$

$$f_{pe} = 155,000 \text{ psi}$$

$$0.40f_{pu} = 0.40 \times 270,000 = 108,000 \text{ psi (745 MPa)} < f_{pe}$$

$$= 155,000 \text{ psi (1069 MPa)}$$

Use ACI short method. Since $d_p > 0.8h$, use $d_p = 36.16$ in., assuming that part of the prestressing strands continue straight to the support.

$$V_c = \left(0.60\lambda\sqrt{f_c'} + 700\frac{V_u d_p}{M_u}\right)b_w d_p \ge 2\lambda\sqrt{f_c'}\,b_w d_p \le 5\lambda\sqrt{f_c'}\,b_w d_p$$

$$\lambda = 1.0 \text{ for normal-weight concrete}$$

$$M_u \text{ at } \frac{d}{2} \text{ from face} = \text{reaction} \times 1.5 - \frac{W_u(1.5)^2}{2} = 83,200 \times 1.5 - \frac{2560(1.5)^2}{2}$$

$$= 121,920 \text{ ft-lb} = 1,463,400 \text{ in.-lb}$$

$$\frac{V_u d_p}{M_u} = \frac{79,359 \times 36.16}{1,463,040} = 1.96 > 1.0$$

So use $V_u d_p / M_u = 1.0$. Then

$$\text{Minimum } V_c = 2\lambda\sqrt{f_c'}\,b_w d_p = 2 \times 1.0\sqrt{5000} \times 6 \times 36.16 = 30,683 \text{ lb}$$

$$\text{Maximum } V_c = 5\lambda\sqrt{f_c'}\,b_w d_p = 76,707 \text{ lb}$$

$$V_c = (0.60 \times 1.0\sqrt{5000} + 700 \times 1.0)6 \times 36.16 = 161,077 \text{ lb}$$
$$> \max V_c = 76,707$$

Then $V_c = 76,707$ lb controls (341 kN). Also, $V_u/\phi > \frac{1}{2}V_c$; hence web steel is needed. Accordingly,

$$V_s = \frac{V_u}{\phi} - V_c = 97,882 - 76,707 = 21,175 \text{ lb}$$

$$8\lambda\sqrt{f_c'}\,b_w d_p = 8 \times 1.0\sqrt{5000} \times 6 \times 36.16 = 122,713 \text{ lb } (546 \text{ kN}) > V_s$$
$$= 21,175 \text{ lb}$$

So the section depth is adequate.

Minimum web steel (step 4): From Equation 5.22b,

$$\text{min. } \frac{A_v}{s} = \frac{A_{ps}f_{pu}}{80f_y d_p}\sqrt{\frac{d_p}{b_w}} = \frac{1.99 \times 270,000}{90 \times 60,000 \times 36.16}\sqrt{\frac{36.16}{6}} = 0.0076 \text{ in.}^2/\text{in.}$$

Required web steel (steps 5, 6):

$$s = \frac{A_v f_y d_p}{(V_u/\phi) - V_c} \le 0.75h \le 24 \text{ in.}$$

or

$$\frac{A_v}{s} = \frac{V_s}{f_v d_p} = \frac{21,175}{60,000 \times 36.16} = 0.0098 \text{ in.}^2/\text{in.}$$

Use a minimum required web shear steel $A_v/s = 0.0098$ in.^{2}/in., although 0.0076 in.^{2}/in. could be used since $f_{pe} > 40\% \, f_{pu}$. So, trying No. 3 U stirrups, $A_c = 2 \times 0.11$

$= 0.22$ in.², and we get $0.0098 = 0.22/s$, so that the maximum spacing is

$$s = \frac{0.22}{0.0098} = 22.4 \text{ in. (5.7 cm)}$$

and

$$4\lambda\sqrt{f'_c}\, b_w d_p = 4 \times 1.0\sqrt{5000} \times 6 \times 36.16 = 61{,}366 \text{ lb} > V_s$$

Hence we do not need to use $\frac{1}{2}s$.   Now

$$0.75h = 0.75 \times 40.0 = 30.0 \text{ in.}$$

Thus use No. 3 U web–shear reinforcement at 22 in. center to center (9.5-mm diameter at 56 cm center to center).

Plane at which no web steel is needed.

Assume that such a plane is at distance x from support.   By similar triangles,

$$V_c = 76{,}707 = 97{,}882 \times \frac{65/2 - x}{65/2}$$

or

$$\frac{65}{2} - x = \frac{76{,}707}{97{,}882} \times \frac{65}{2}$$

giving

$$x = 7.03 \text{ ft (2.14 m)} \approx 84 \text{ in.}$$

Therefore, adopt the design in question, using No. 3 U at 22 in. center to center over a stretch length of approximately 84 in., with the first stirrup to start at 18 in. from the face of support.   Extend the stirrups to the midspan if composite action doweling is needed.

SELECTED REFERENCES

14.1. Nawy, E. G., *Prestressed Concrete: A Fundamental Approach*, 2nd ed., Prentice Hall, Englewood Cliffs, N.J., 1996, 800 pp.

14.2. Prestressed Concrete Institute, *PCI Design Handbook*, PCI, Chicago, 1993.

14.3. Post-Tensioning Institute, *Post-Tensioning Manual*, 4th ed., PTI, Phoenix, Ariz., 1985.

14.4. Lin, T. Y., "Cable Friction in Post-Tensioning," *Journal of the Structural Division, ASCE*, November 1956, pp. 1107–1 to 1107–13.

14.5. Lin, T. Y., and Burns, N. H., *Design of Prestressed Concrete Structures*, 3d ed., Wiley, New York, 1981.

14.6. ACI Committee 318, *Building Code Requirements for Reinforced Concrete, ACI Standard 318–95*, ACI, Detroit, 1995.

14.7. Nilson, A. H., *Design of Prestressed Concrete*, Wiley, New York, 1987.

14.8. Nawy, E. G., *Simplified Reinforced Concrete*, Prentice Hall, Englewood Cliffs, N.J., 1986, 318 pp.

PROBLEMS FOR SOLUTION

14.1. An AASHTO prestressed simply supported I beam has a span of 34 ft (10.4 m) and is 36 in. (91.4 cm) deep. Its cross section is shown in Fig. 14.18. It is subjected to a live-load intensity $W_L = 3600$ plf (52.6 kN/m). Determine the required $\frac{1}{2}$-in.-diameter, stress-relieved, seven-wire strands to resist the applied gravity load and the self-weight of the beam, assuming that the tendon eccentricity at midspan is $e_c = 13.12$ in. (333 mm). Maximum permissible stresses are as follows:

$$f'_c = 6000 \text{ psi (41.4 MPa)}$$

$$f_c = 0.45f'_c$$

$$= 2700 \text{ psi (26.7 MPa)}$$

$$f_t = 12\sqrt{f'_c} = 930 \text{ psi (6.4 MPa)}$$

$$f_{pu} = 270{,}000 \text{ psi (1862 MPa)}$$

$$f_{pi} = 189{,}000 \text{ psi (1303 MPa)}$$

$$f_{pe} = 145{,}000 \text{ psi (1000 MPa)}$$

The section properties, given these stresses, are

$$A_c = 369 \text{ in.}^2$$

$$I_g = 50{,}979 \text{ in.}^4$$

$$r^2 = \frac{I_c}{A_c} = 138 \text{ in.}^2$$

$$c_b = 15.83 \text{ in.}$$

$$S_b = 3220 \text{ in.}^3$$

$$S^t = 2527 \text{ in.}^3$$

$$W_D = 384 \text{ plf}$$

$$W_L = 3600 \text{ plf}$$

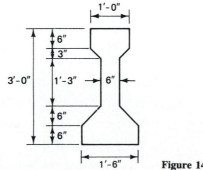

Figure 14.18

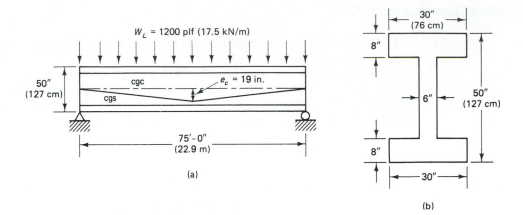

Figure 14.19 (a) elevation; (b) section.

14.2. A simply supported pretensioned beam has a span of 75 ft (22.9 m) and the cross section shown in Fig. 14.19. It is subjected to a uniform gravitational live-load intensity $W_L = 1200$ plf (17.5 kN/m) in addition to its self-weight and is prestressed with 20 stress-relieved $\frac{1}{2}$-in. (12.7-mm) diameter seven-wire strands. Compute the total prestress losses by the step-by-step method, and compare them with the values obtained by the lump-sum method. Take the following values as given:

$$f'_c = 6000 \text{ psi (41.4 MPa), normal-weight concrete}$$

$$f'_{ci} = 4500 \text{ psi (31 MPa)}$$

$$f_{pu} = 270,000 \text{ psi (1862 MPa)}$$

$$f_{pi} = 0.70 f_{pu}$$

$$\text{relaxation time } t = 15 \text{ years}$$

$$e_c = 19 \text{ in. (483 mm)}$$

$$\text{relative humidity, RH} = 75\%$$

$$\frac{V}{S} = 3.0 \text{ in. (7.62 cm)}$$

14.3. For service-load and ultimate-load conditions, design a pretensioned symmetrical I-section beam to carry a superimposed dead load of 750 plf (10.95 kN/m) and a service live load of 1500 plf (21.90 kN/m) on a 50-ft (15.2-m) simply supported span. Assume that the sectional properties are $b = 0.5h$, $h_f = 0.2h$, and $b_w = 0.40b$, and suppose the following data are given:

$$f_{pu} = 270,000 \text{ psi (1862 MPa)}$$

$$E_{ps} = 28.5 \times 10^6 \text{ psi (196} \times 10^3 \text{ MPa)}$$

$$f'_c = 5000 \text{ psi (34.5 MPa) normal-weight concrete}$$

$$f'_{ci} = 3500 \text{ psi (24.1 MPa)}$$

$$f_t = 12\sqrt{f'_c} \text{ assuming that deflection is not critical}$$

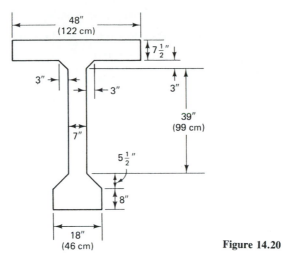

Figure 14.20

Sketch the design details, including the anchorage zone reinforcement, and arrangement of strands for (a) the straight-tendon case and (b) a harped tendon at the third span points with end eccentricity zero.   Assume total prestress losses of 22%.

14.4. A post-tensioned bonded prestressed beam has the cross section shown in Fig. 14.20. It has a span of 75 ft (22.9 m) and is subjected to a service superimposed dead load W_{SD} = 450 plf (6.6 kN/m) and a superimposed service live load W_L = 2300 plf (33.6 kN/m).   Design the web reinforcement necessary to prevent shear cracking (a) by the detailed design method and (b) by the alternative method at a section 15 ft (4.6 m) from the face of the support.   The profile of the prestressing tendon is parabolic. Use No. 3 stirrups in your design, and detail the section.   The following data are given:

$$A_c = 876 \text{ in.}^2 \ (5652 \text{ cm}^2)$$

$$I_c = 433,350 \text{ in.}^4 \ (18.03 \times 10^6 \text{ cm}^4)$$

$$r^2 = 495 \text{ in.}^2 \ (3194 \text{ cm}^2)$$

$$c_t = 25 \text{ in. } (63.5 \text{ cm})$$

$$S^t = 17,300 \text{ in.}^3 \ (2.83 \times 10^5 \text{ cm}^3)$$

$$c_b = 38 \text{ in. } (96.5 \text{ cm})$$

$$S_b = 11,400 \text{ in.}^3 \ (1.86 \times 10^5 \text{ cm}^3)$$

$$W_d = 910 \text{ plf } (13.3 \text{ kN/m})$$

$$e_c = 32 \text{ in. } (81.3 \text{ cm})$$

$$e_e = 2 \text{ in. } (5 \text{ cm})$$

$$f'_c = 5000 \text{ psi } (44.5 \text{ MPa}), \text{ normal-weight concrete}$$

$$f'_{ci} = 3500 \text{ psi } (24.1 \text{ MPa})$$

f_y for stirrups $= 60,000$ psi (41.8 MPa)

$f_{pu} = 270,000$ psi (1862 MPa) low-relaxation strands

$f_{ps} = 243,000$ psi (1675 MPa)

$f_{pe} = 157,500$ psi (1086 MPa)

$A_{ps} =$ twenty-four $\frac{1}{2}$-in. (12.7-mm) diameter seven-wire strands

SEISMIC DESIGN OF CONCRETE STRUCTURES

15

Photo 85 Northridge, California, 1994 earthquake structural failure. (Courtesy Dr. Murat Saatcioglu.)

15.1 INTRODUCTION: MECHANISM OF EARTHQUAKES

Earth's crust is composed of several layers of hard tectonic plates, called *lithospheres*, that float on the softer, underpinning, fluid medium called *mantle*. These plates or rock masses, when fractured, *form fault lines*. The adjoining plates or rock masses are prevented by the interacting frictional forces from moving past one another most of the time. However, when this frictional ultimate resistance is reached because of the continuous motion of the underlying fluid, any two plates can impact on one another, generating seismic waves that can cause large horizontal and vertical ground motions. These ground motions translate into inertia forces in structures.

The length and width of a fault are interrelated to the magnitude of the earthquake. The fault is the cause rather than the result of the earthquake. A fault can cause an earthquake due to the following reasons (Ref. 15.4):

1. Cumulative strain in the fault over a long period of time reaches the rupture strain.

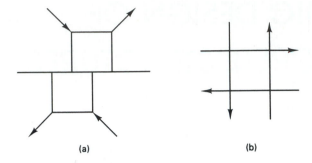

(a)   (b)

Figure 15.1 Mechanism of earthquakes: (a) slip of tectonic plates; (b) reverse moment couples.

2. Slip of the tectonic plates at the fault zones causes a rebound, as in Fig. 15.1a.

3. Sudden push and pull forces at the fault lead to reverse moment couples, as in Fig. 15.1b. The moment caused by these couples as a measure of earthquake size can be termed the *seismic moments*. Their magnitude is equal to rock rigidity × fault area × amount of slip. The range of slip velocity in such faults as the San Andreas fault in California is 30 to 100 mm per year. On this basis, a slippage or horizontal motion of 3 m at such faults in one single earthquake is expected to occur at intervals of 30 to 100 years.

15.2 ELASTIC AND INELASTIC RESPONSE

Earthquakes may be placed in three categories: low, moderate, and high intensity. Seismic risk level is designated by zones or areas of nearly equal risk or probability of occurrence, leading to seismic regionalization. Five seismic zones are identified for the United States by the Uniform Building Code (Ref. 15.2):

Zone 0 = little or no seismic risk
Zone 1 = minor seismic risk
Zone 2 = moderate seismic risk
Zone 3 = major seismic risk
Zone 4 = areas within zone 3 that are close to a major fault

A structure is expected to respond essentially elastically to low-intensity earthquakes, where the stresses are expected to remain within the elastic range, with a slight possibility of developing limited inelasticity that is not expected to cause structural or nonstructural damage. Structural response is expected to be inelastic under high-intensity earthquakes having an intensity of 5 or higher on the Richter scale and in regions close to the epicenter.

No special detailing of a reinforced concrete structure is needed in regions of low seismic risk; standard design in accordance with the ACI 318-95 Code

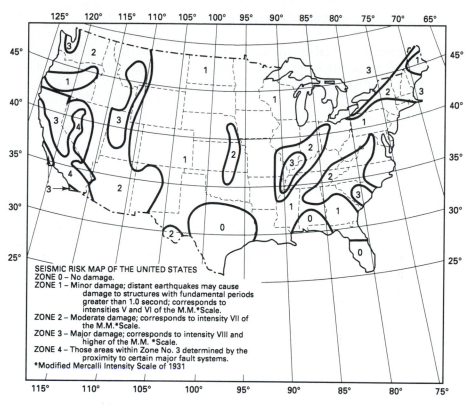

Figure 15.2 Seismic zones map of the United States.

provides the necessary strength and ductility in the structure. In moderate seismic risk zones, special reinforcing details may be needed in the moment-resistant frames in order to accommodate the degree of inelastic deformation expected in such zones. Structures located in high seismic risk zones have to be designed for inelastic response with special detailing of the plastic hinging regions. This can be achieved through section confinement in order to ensure high ductility capacity and load resistances that prevent collapse. Figure 15.2 provides a seismic zone map of the United States as a guide for the extent of structural detailing needed in a specific seismic risk zone.

15.3 SEISMIC DESIGN CONCEPTS AND PROCEDURES

Building codes generally recommend the static analysis for the design of structural frames and shear walls to resist earthquakes. This can be justified in view of the uncertainties associated with earthquake loading and all the assumptions that have to be made in modeling the structure and its degrees of freedom for rigorous dynamic analysis. The static approach is based on structural analysis under equivalent lateral loads, such as those specified in the Uniform Building Code.

15.3.1 Design Seismic Base Shear

A realistic evaluation of the base horizontal shear force due to seismic effects should assume that the structure undergoes several cycles of inelastic deformations during a major seismic ground motion. This implies that the design seismic base shear can be lower than that expected for elastic response. Hence the load level is a function of the type of structural system and its ability to deform inelastically while dissipating seismically induced energy without a collapse (Ref. 15.6). The design base shear associated with the expected inelastic deformation can be obtained from the following equation (UBC, Sec. 1628, Ref. 15.2):

$$V = \frac{ZIC}{R_w} W \qquad (15.1)$$

where I = occupancy importance coefficient, equal to 1.25 for essential and hazardous facilities, and 1.0 for all other cases, including assembly buildings.

Z = seismic zone factor indicative of the effective peak ground acceleration:

zone 1: $Z = 0.075$
zone 2A: $Z = 0.15$
zone 2B: $Z = 0.20$
zone 3: $Z = 0.30$
zone 4: $Z = 0.40$

C = seismic response coefficient, a function of the fundamental period of vibration T and the site–structure resonance coefficient S. The value of C need not exceed 2.75, which is the value that may be used for any structure without regard to soil type or structure period. Except when forces are scaled up by $3(R_w/8)$, the minimum value of the ratio C/R_w is 0.075:

$$C = \frac{1.25S}{T^{2/3}} \le 2.75 \qquad (15.2)$$

S = site response coefficient, ranging between 1.0 (rocklike soil where shear wave velocity is greater than 2500 ft/s) and 2.0 (soil profile containing more than 40 ft of soft clay soil with shear wave velocity less than 500 ft/s).

W = total seismic dead weight of the building, which includes the weights of all permanent structural and nonstructural components, such as walls, floors, roofs, and fixed service equipment. In storage and warehouse occupancies, a minimum of 25% of the floor live load must be included.

R_w = system quality factor; it idealizes the inelastic behavior of the structures under the spectral representation of a major seismic ground

Photo 86 311 S. Wacker Street, Chicago, 12,000 concrete (Courtesy Portland Cement Association.)

motion (CZ). Its value is a function of the constituent material and components and type of structural system. The value of R_w used in the design of any story should be less than or equal to the R_w of the story above. The following are the values recommended by the UBC provisions for concrete systems (Ref. 15.3, Table 16N):

(a) Concrete and masonry shear walls in bearing wall systems ($h \leq$ 160 ft.): $R_w = 6$

(b) Concrete and masonry shear walls in building frame systems ($h \leq 200$ ft): $R_w = 8$

(c) Special ductile moment-resistant concrete frames (h = not limited): $R_w = 12$

(d) Dual systems (h = not limited): $R_w = 12$

Photo 87   Bridge girder collapse in the San Francisco 1989 earthquake. (Courtesy Portland Cement Association.)

15.3.2 Fundamental Period of Vibration

The basic natural period T of a simple one-degree-of-freedom system is the time required to complete one whole cycle during dynamic loading. In other words, it is the time required for a phase angle ωt to travel from 0 to 2π, where ω is the angular frequency of the system. Hence $\omega t = 2\pi$, leading to the expression

$$T = \frac{2\pi}{\omega} = 2\pi \left(\frac{m}{k}\right)^{1/2} \qquad (15.3)$$

where m = mass of system
k = spring constant and damping is not considered

Most reinforced concrete structures are multidegrees-of-freedom systems, as in Fig. 15.3. In this case the structural mass can be assumed to be concentrated in the vertical spring element at the floor level, resulting in multiple modes with frequencies (periods) for each mode. The compound natural period T is then evaluated with due consideration given to the distribution of mass and stiffness. The UBC requires that T be established using the structural properties and de-

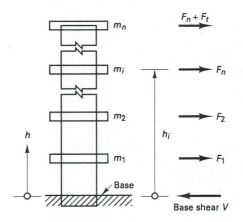

Figure 15.3   Modeling multistory structures.

formation characteristics of the resisting elements in a properly substantiated analysis using an expression such as Eq. 15.6 to follow.

Two methods are allowed by the UBC provisions for calculation of the fundamental period of an idealized structural system.

Method A

$$T = C_t(h_n)^{3/4} \tag{15.4}$$

where C_t can be taken $= 0.030$ for reinforced concrete moment-resisting frames, and $C_t = 0.020$ for all other buildings except steel moment resisting, for which $C_t = 0.035$. For buildings stabilized with concrete or masonry shear walls,

$$C_t = \frac{0.10}{\sqrt{A_c}} \tag{15.5a}$$

$$A_c = \sum A_e \left[0.2 + \left(\frac{D_e}{h_n} \right)^2 \right] \tag{15.5b}$$

where A_c = combined effective area (ft²) of all the shear walls in the first story of the structure

A_e = effective horizontal cross-sectional area (ft²) of a shear wall in the first story of the structure

D_e = wall length (ft) of a shear wall element in the first story in the direction parallel to the applied force

h_n = height of wall (ft) above the base level

The value of D_e/h_n should not exceed 0.9.

Photo 88   Northridge, California, 1994 earthquake structural failure. (Courtesy Dr. Murat Saatcioglu.)

Method B.   The fundamental period T may be calculated by detailed analysis using the structural and deformational properties of the resisting elements. This requirement is satisfied using the following expression:

$$T = 2\pi \sqrt{\left(\sum_{i=1}^{n} w_i \delta_i^2 \right) \div \left(g \sum_{i=1}^{n} f_i \delta_i \right)} \qquad (15.6)$$

where W_i = portion of W that is located at or assigned to level i
$\quad\quad\ f_i$ = any lateral force distributed in accordance with Eqs. 15.7 and 15.8
$\quad\quad\ \delta_i$ = lateral elastic deflection calculated using the applied lateral forces f_i
$\quad\quad\ g$ = acceleration due to gravity

The value of the fundamental period T from this method should not exceed the T value obtained by method A by more than 30% the T in seismic zone 4 and 40% in seismic zones 1, 2, and 3.

15.3.3 Vertical Distribution of Design Base Shear *V*

The design base shear V should be distributed over the height of the structure as follows:

$$V = F_t + \sum_{i=1}^{n} F_i \tag{15.7a}$$

$$F_i = \frac{W_i h_i}{\sum_{i=1}^{n} W_i h_i} V \tag{15.8a}$$

$$F_t = 0.07TV \leq 0.25V \tag{15.8b}$$

where F_t = that portion of V considered concentrated at the top of the structure in addition to F_n. The value of T used for the purpose of calculating F_t can be considered as the period corresponding to the base shear V in Eq. 15.1. Use $F_t = 0$ if $T = 0.7$ s or less.

The remaining portion of the base shear has to be distributed on the height of the structure, including level n, on the basis of the following expression:

$$F_x = \frac{(V - F_t)W_x h_x}{\sum_{i=1}^{n} W_i h_i} \tag{15.8c}$$

At each level x, the force F_x should be applied over the area of the building in accordance with the mass distribution at that level.

Additional information on the dynamic behavior of structures and dynamic modeling and detailing of high-rise structures in seismic zones can be found in the selected list of references at the end of the chapter.

15.3.4 Horizontal Distribution of Story Shear V_x

The story shear V_x in any story = sum of the forces F_t plus the F_x above that story. It should be distributed to the various structural elements according to their rigidities, taking into account the rigidity of the diaphragm. If the diaphragms are not flexible, the mass at each level must be displaced from the calculated center of mass in each direction a distance equal to 5% of the building dimension at that level *perpendicular* to that direction of the force.

15.3.5 Lateral Shear *V* in Rigid Structures

Rigid nonframe building structures are those whose fundamental period T is less than 0.06 seconds. Such structures should be designed for a lateral force

$$V = 0.5ZIW \tag{15.9}$$

The force V should be distributed similarly to the distribution of the mass and should be assumed to act in any horizontal direction.

Photo 89 Skybridge, Vancouver, Canada, a 2020-ft longlength cable-stayed bridge and the world's longest transit bridge. (Courtesy Portland Cement Association.)

15.4 FLEXURAL DESIGN OF BEAMS AND COLUMNS

Moment-resisting ductile frames of reinforced concrete structures are designed for strength and ductility. During a strong earthquake, it is anticipated that the critical regions of frame members will develop plastic hinges to dissipate seismically induced energy. In a well-designed frame structure, the energy dissipation occurs in the plastic hinges that form at the ends of the beams, while the columns remain elastic and provide overall strength and stability to the stories above. This can be achieved if the sequence of plastification in the structure can be controlled.

In an effort to control the seismic response of structures, the ACI 318-95 building code calls for "strong columns and weak beams," although it may be difficult to prevent hinging of the lower-story columns. All the potential hinging regions are required to be detailed by special confining reinforcement for improved ductility and energy absorption capacity.

15.4.1 Seismic Shear Forces in Beams and Columns

Shear failure in reinforced concrete members is regarded as brittle failure. Therefore, in designing earthquake-resistant structures, it is important to provide excess shear capacity over and above that corresponding to flexural failure. The ACI 318-

95 requirements are based on the strong column–weak beam concept subsequently discussed. Hence plastification of the critical regions at the ends of the beams will have to be considered as a possible loading condition.

The shear force is then computed based on the moment resistances in the developed plastic hinges, labeled as probable moment resistance M_{pr}, developed when the longitudinal flexural steel enters into the hardening stage. Consequently, the computation of the probable moment resistance, $1.25f_y$, is used as the stress in the longitudinal reinforcement. In order to absorb the energy that can cause plastic hinging, the earthquake–resistant frame has to be ductile in part through confinement of the longitudinal reinforcement of the columns and the beam–column joints and in part through the provision of the excess shear capacity previously discussed.

Figure 15.4 shows the deformed geometry of and the moment and shear forces for a beam subjected to gravity loading and reversible sidesway. If the intensity of gravity load is w_u, then ACI 318-95 stipulates

$$w_u = 0.75 (1.4D + 1.7L + 1.87E) \qquad (15.10a)$$

The UBC (Sec. 1921.2.7) stipulates

$$w_u = 1.4(D + L + E)$$
$$\text{and} \qquad (15.10b)$$
$$w_u = 0.9D \pm 1.4E$$

The seismic shear forces are

$$V_L = \frac{M_{prL}^- + M_{prR}^+}{\ell} + 0.75 \frac{1.4D + 1.7L}{2} \qquad (15.11a)$$

$$V_R = \frac{M_{prL}^+ + M_{prR}^-}{\ell} - 0.75 \frac{1.4D + 1.7L}{2} \qquad (15.11b)$$

where ℓ = span, L and R subscripts = left and right ends, and M_{pr} = probable moment strength at the end of the beam based on steel reinforcement tensile

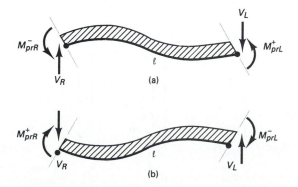

Figure 15.4   Seismic moments and shears at beam ends: (a) sidesway to the left; (b) sidesway to the right.

strength of $1.25f_y$ and strength reduction factor $\phi = 1.0$. These instantaneous moments M_{pr} should be computed on the basis of equilibrium of moments at the joint where the beam moments are equal to the probable moments of resistance.

The shear forces in the columns are computed in a similar manner, so the horizontal V_e at top and bottom of the column is

$$V_e = \frac{M_{pr1} + M_{pr2}}{h} \tag{15.12}$$

except that end moments for columns (M_{pr1} and M_{pr2}) need not be greater than the moments generated by the M_{pr} of beams framing into the beam–column joint. h = column height, and the subscripts 1 and 2 indicate the top and bottom column end moments, respectively, as seen in Fig. 15.5.

15.4.2 Strong Column–Weak Beam Concept

As previously stated, U.S. seismic codes require that earthquake-induced energy be dissipated by plastic hinging of the beams, rather than the columns. This hypothesis is due to the fact that compression members such as columns have lower ductility than flexure-dominant beams. Furthermore, the consequence of a column failure is far more severe than a local beam failure. Therefore, the ACI 318-95 Code as well as the UBC stipulate "strong columns and weak beams." This is ensured by the following inequality:

$$\sum M_{\text{col}} \geq \left(\frac{6}{5}\right) \sum M_{\text{bm}} \tag{15.13}$$

where $\sum M_{\text{col}}$ = sum of moments at the center of the joint corresponding to the design flexural strength of the columns framing into that joint
$\sum M_{\text{bm}}$ = sum of moments at the center of the joint corresponding to the design flexural strengths of the beams framing into that joint

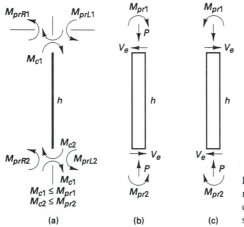

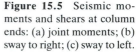

Figure 15.5 Seismic moments and shears at column ends: (a) joint moments; (b) sway to right; (c) sway to left.

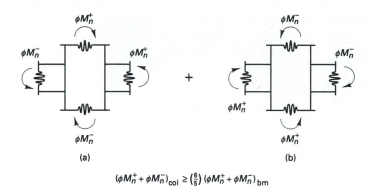

$$(\phi M_n^+ + \phi M_n^-)_{\text{col}} \geq \left(\tfrac{6}{5}\right)(\phi M_n^+ + \phi M_n^-)_{\text{bm}}$$

Figure 15.6 Seismic moment summation at beam–column joint: (a) sidesway to left; (b) sidesway to right.

For a joint subjected to reversible base shear forces, as shown in Fig. 15.6, Eq. 15.13 becomes

$$(\phi M_n^+ \; + \; \phi M_n^-)_{\text{col}} \geq \frac{6}{5}(\phi M_n^+ \; + \; \phi M_n^-)_{\text{bm}} \tag{15.14}$$

where $\phi = 0.90$ for beams, 0.70 for tied columns, and 0.75 for spiral columns. For beam–columns, $\phi = 0.90$ to 0.7.

15.5 SEISMIC DETAILING REQUIREMENTS FOR BEAMS AND COLUMNS

15.5.1 Longitudinal Reinforcement

1. In seismic design, when the factored axial load P_u is negligible or significantly less than $A_g f_c'/10$, the member is considered a flexural member (beam). If $P_u > A_g f_c'/10$, the member is considered a beam–column, because it is subjected to both axial and flexural loads as columns and shear walls are.

2. The shortest cross-sectional dimension ≥ 12 in. (300 mm).

3. The limitation on the longitudinal reinforcement ratio in the beam–column element is $0.01 \leq \rho_g = \dfrac{A_s}{A_g} \leq 0.06$. For practical considerations, an upper limitation of 6% is too excessive, because it results in impractical congestion of longitudinal reinforcement. A practical maximum total percentage ρ_g of 3.5% to 4.0% should be a reasonable limit.

4. A minimum percentage of longitudinal reinforcement in flexural members (beams) is
   (a) For positive reinforcement,

$$\rho \geq \frac{3\sqrt{f_c'}}{f_y} \geq \frac{200}{f_y} \tag{15.15a}$$

Photo 90 Column localized damage in a high-rise frame building, Los Angeles 1994 earthquake. (Courtesy Portland Cement Association.)

(b) For negative reinforcement,

$$\rho \geq \frac{6\sqrt{f_c'}}{f_y} \geq \frac{200}{f_y} \qquad (15.15b)$$

But under no condition should ρ exceed 0.025. The stresses f_c' and f_y in these expressions are in psi units. All reinforcement has to be continued through the joint.

5. Main reinforcement should be chosen on the basis of the strong column–weak beam concept of the ACI Code: $\sum M_{col} \geq \frac{6}{5} \sum M_{bm}$.

6. The nominal moment strength requirements are
   (a) M_n^+ at joint face $\geq \frac{1}{2}M_n^-$ at that face.
   (b) Neither the negative nor the positive moment strength at *any* section along the span can be less than *one-quarter* the maximum moment strength provided at the face of either joint. Hence at joint face

$$M_n^+ \geq \frac{1}{2}M_a \qquad (15.16a)$$

At any section,

$$M_a^+ \geq \frac{1}{4}(M_a^-)_{\max} \tag{15.16b}$$

$$M_a^- \geq \frac{1}{4}(M_a^-)_{\max} \tag{15.16c}$$

15.5.2 Transverse Confining Reinforcement

Transverse reinforcement in the form of closely spaced hoops (ties) or spirals has to be adequately provided. The aim is to produce adequate rotational capacity within the elastic hinges that may develop as a result of the seismic forces.

1. For column spirals, the minimum volumetric ratio of the spiral hoops needed for the concrete core confinement is

$$\rho_s \geq \frac{0.12f_c'}{f_{yh}} \tag{15.17a}$$

or

$$\rho_s \geq 0.45 \left(\frac{A_g}{A_{ch}} - 1\right) \frac{f_c'}{f_{yh}} \tag{15.17b}$$

whichever is greater, where

ρ_s = ratio of volume of spiral reinforcement to the core volume measured out-to-out

A_g = gross area of the column section

A_{ch} = core area of section measured to the outside of the transverse reinforcement (in.²)

f_{yh} = specified yield strength of transverse reinforcement, psi

2. For column rectangular hoops, the total cross-sectional area within spacing s is

$$A_{sh} \geq 0.09 sh_c \frac{f_c'}{f_{yh}} \tag{15.18a}$$

or $\tag{15.18b}$

$$A_{sh} \geq 0.3 sh_c \left(\frac{A_g}{A_{ch}} - 1\right) \frac{f_c'}{f_{yh}}$$

whichever is greater, where

A_{sh} = total cross-sectional area of transverse reinforcement (including cross ties) within spacing s and perpendicular to dimension h_c

h_c = cross-sectional dimension of column core measured c–c of confining reinforcement, in.

A_{ch} = cross-sectional area of structural member, measured out-to-out of transverse reinforcement

s = spacing of transverse reinforcement measured along the longitudinal axis of the member, in.

s_{max} = one-quarter of the smallest cross-sectional dimension of the member or 4 in., whichever is smaller (UBC requires 4 in.)

3. The confining transverse reinforcement in *columns* should be placed on *both* sides of a potential hinge over a distance ℓ_0. The largest of the following three conditions governs:
   (a) depth of member at joint face
   (b) one-sixth of the clear span
   (c) 18 in.

4. For beam confinement, the confining transverse reinforcement at *beam* ends should be placed over a length equal to *twice* the member depth h from the face of the joint on either side or of any other location where plastic hinges can develop. The maximum hoop spacing should be the smallest of the following four conditions:
   (a) One-fourth effective depth d
   (b) 8 × diameter of longitudinal bars
   (c) 24 × diameter of the hoop
   (d) 12 in. (300 mm)
   Figure 15.7 (Ref. 15.8) summarizes typical detailing requirements for a confined column.

5. Reduction in confinement at joints: a 50% reduction in confinement and an increase in the minimum tie spacing to 6 in. are allowed by the ACI Code if a joint is confined on all *four* faces by adjoining beams with each beam wide enough to cover three-quarters of the adjoining face.

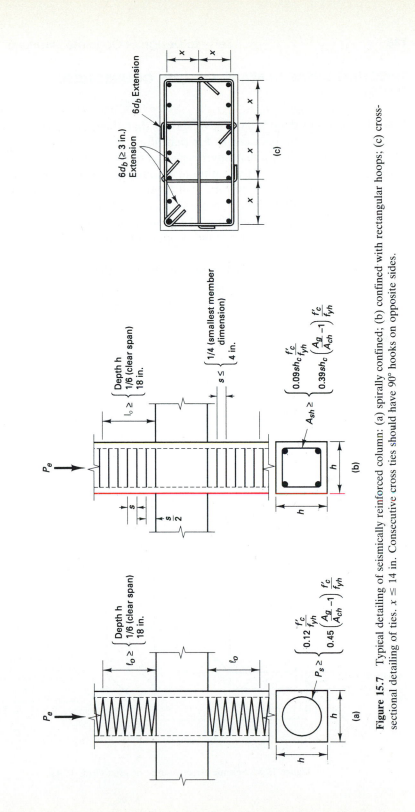

Figure 15.7 Typical detailing of seismically reinforced column: (a) spirally confined; (b) confined with rectangular hoops; (c) cross-sectional detailing of ties. $x \leq 14$ in. Consecutive cross ties should have 90° hooks on opposite sides.

15.6 HORIZONTAL SHEAR IN BEAM—COLUMN CONNECTIONS (JOINTS)

Test of joints and deep beams have shown that shear strength is not as sensitive to joint (shear) reinforcement as for that along the span. On this basis, the ACI Code has assumed the joint strength as a function of only the compressive strength of the concrete and requires a minimum amount of transverse reinforcement in the joint. The effective area A_j in Fig. 15.8 from the ACI 318 Commentary should in no case be greater than the column cross-sectional area.

The minimal shear strength of the joint should not be taken greater than the forces V_n specified below for normal-weight concrete.

1. Confined on all faces by beams framing into the joint:

$$V_n \leq 20\sqrt{f_c'}\, A_j \qquad\qquad (15.19a)$$

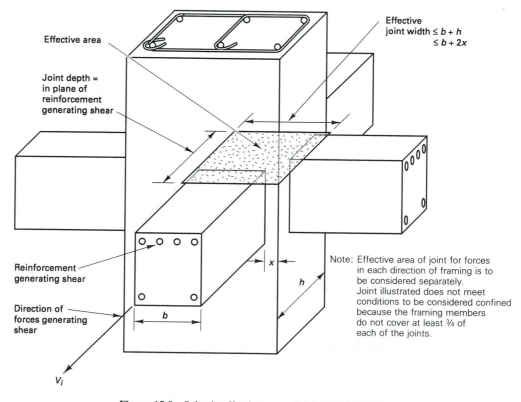

Figure 15.8   Seismic effective area of joint (Ref. 15.1).

2. Confined on three faces or on two opposite faces:

$$V_n \leq 15\sqrt{f_c'} \, A_j \tag{15.19b}$$

3. All other cases:

$$V_n \leq 12\sqrt{f_c'} \, A_j \tag{15.19c}$$

A framing beam is considered to provide confinement to the joint only if at least three-quarters of the joint is covered by the beam.

The value of allowable V_n should be reduced by 25% if lightweight concrete is used. Also, test data indicate that the value in Eq. 15.19c is unconservative when applied to corner joints. A_j = effective cross-sectional area within a joint, as in Fig. 15.8, in a plane parallel to the plane of reinforcement generating shear at the joint. The ACI Code assumes that the horizontal shear in the joint is determined on the basis that the stress in the flexural tensile steel = $1.25f_y$. Figure 15.9 shows the forces acting on a beam–column connection at the joint. These forces are the result of the equilibrium forces shown in the deformed wall of Fig. 15.10.

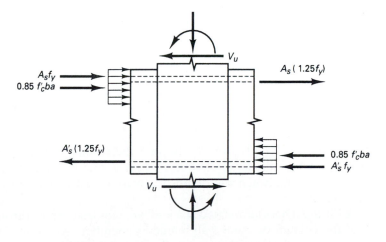

Figure 15.9   Reversible forces at beam–column joint connection. (V_u = horizontal shear at joint).

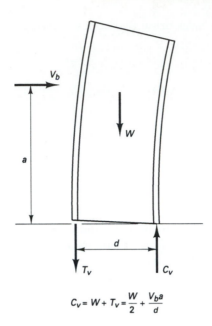

$$C_v = W + T_v = \frac{W}{2} + \frac{V_b a}{d}$$

Figure 15.10
Forces and loads on
deformed shear wall.

15.6.1 Development of Reinforcement at the Joint

For bars of sizes Nos. 3 through 11 terminating at an exterior joint with standard 90° hooks in normal concrete, the development length ℓ_{dh} beyond the column face, as required by the ACI 318 Code, should not be less than the following:

$$\ell_{dh} \geq f_y d_b / 65\sqrt{f'_c} \qquad (15.20a)$$

$$\ell_{dh} \geq 8 d_b \qquad (15.20b)$$

where d_b = bar diameter.

$$\ell_{dh} \geq 6 \text{ in.} \qquad (15.20c)$$

The development length provided beyond the column face must be no less than $\ell_d = 2.5\ell_{dh}$ when the depth of concrete cast in one lift beneath the bar ≤ 12 in. or $\ell_d = 3.5 \times \ell_{dh}$ when the depth of concrete cast in one lift beneath the bar exceeds 12 in.

All straight bars terminated at a joint are required to pass through the confined core of the column or shear wall boundary member. Any portion of the straight embedment length not within the confined core should be increased by a factor of 1.6.

15.7 DESIGN OF SHEAR WALLS

Structural walls in a frame building should be so proportioned that they possess the necessary stiffness needed to reduce the relative interstory distortions caused by seismic-induced motions. Such walls are termed shear walls. Their additional function is to reduce the possibility of damage to the nonstructural elements that most buildings contain.

Buildings stiffened by shear walls are considerably more effective than rigid-frame buildings with regard to damage control, overall safety, and integrity of the structure. This performance is due to the fact that shear walls are considerably stiffer than regular frame elements and thus can respond to or absorb greater lateral forces induced by the earthquake motions, while controlling interstory drift.

15.7.1 Forces and Reinforcement in Shear Walls and Diaphragms

Shear walls, that is, structural walls, with height-to-depth ratio in excess of 2.0 essentially act as vertical cantilever beams. As a result, their strength is determined by flexure rather than shear. If they are subjected to factored in-plane seismic shear forces $V_{uh} > A_{cv} \sqrt{f'_c}$, they have to be reinforced with a minimum percentage $\rho_v \geq 0.0025$.

At least two curtains of reinforcement are needed in the wall if the in-plane factored shear forces exceed a value of $2A_{cv}\sqrt{f'_c}$.

$\rho_v = A_{sv}/A_{cv}$

A_{cv} = net area of concrete cross section = thickness × length of section in direction of shear considered

A_{sv} = projection on A_{cv} of area of distributed shear reinforcement crossing the plane of A_{cv}

If the extreme fiber compressive stresses exceed $0.2f'_c$, the shear walls have to be provided with boundary elements along their vertical boundaries and around the edges of openings.

The nominal shear strength V_n of structural walls and diaphragms of high-rise buildings with aspect ratio greater than 2 should not exceed the shear force calculated from

$$V_n = A_{cv}(2\sqrt{f'_c} + \rho_n f_y) \qquad (15.21a)$$

where ρ_n = ratio of distributed shear reinforcement on a plane perpendicular to the plane of A_{cv}.

For low-rise walls with aspect ratio h_w/ℓ_w less than 2, the ACI Code requires that the coefficient 2 in Eq. 15.21a be increased linearly up to a value of 3 when

Photo 91   Northridge, California, 1994 earthquake structural failure. (Courtesy Dr. Murat Saatcioglu.)

the h_w/ℓ_w ratio reaches 1.5 in order to account for the higher shear capacity of low-rise walls. In other words,

$$V_N = A_{cv}(\alpha_c \sqrt{f_c'} + \rho_n f_y) \tag{15.21b}$$

where $\alpha_c = 2$ when $h_w/\ell_w \geq 2$ and $\alpha_c = 3$ when $h_w/\ell_w = 1.5$; $V_u = \phi V_n$

         $\phi = 0.6$ for designing the joint, if nominal shear is less than the shear corresponding to the development of the nominal flexural strength of the member.

The nominal flexural strength is determined considering the most critical factored axial loads including earthquake effects. The maximum allowable nominal unit shear strength in structural walls is $8A_{cv}\sqrt{f_c'}$ where A_{cv} is the total cross-sectional area (in.²) previously defined and f_c' is in psi. However, the nominal shear strength of any one of the individual wall piers can be permitted to have a maximum value of $10A_{cp}\sqrt{f_c'}$, where A_{cp} is the cross-sectional area of the individual pier.

      Figure 15.10 shows the forces acting on a shear wall (structural wall). Figure 15.11 shows typical failure modes of low- and high-rise structural walls.

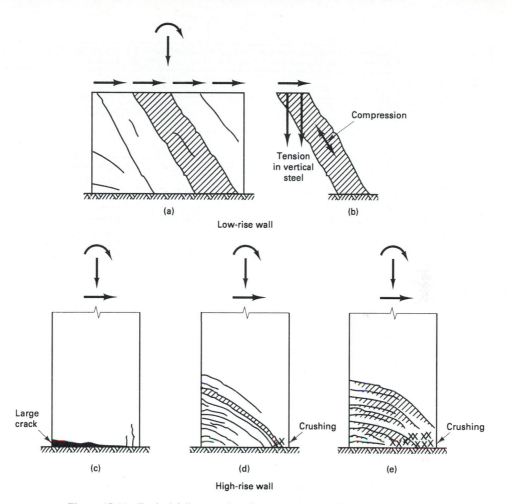

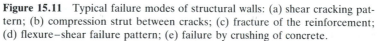

Figure 15.11 Typical failure modes of structural walls: (a) shear cracking pattern; (b) compression strut between cracks; (c) fracture of the reinforcement; (d) flexure–shear failure pattern; (e) failure by crushing of concrete.

15.8 DESIGN PROCEDURE FOR EARTHQUAKE-RESISTANT STRUCTURES

Figure 15.12 gives a logic flow chart for the following operational steps.

1. Determine the earthquake zone. If the Uniform Building Code is used, select the UBC seismic coefficients based on the idealized building properties for determining the base shear:

$$V = \frac{ZIC}{R_w}$$

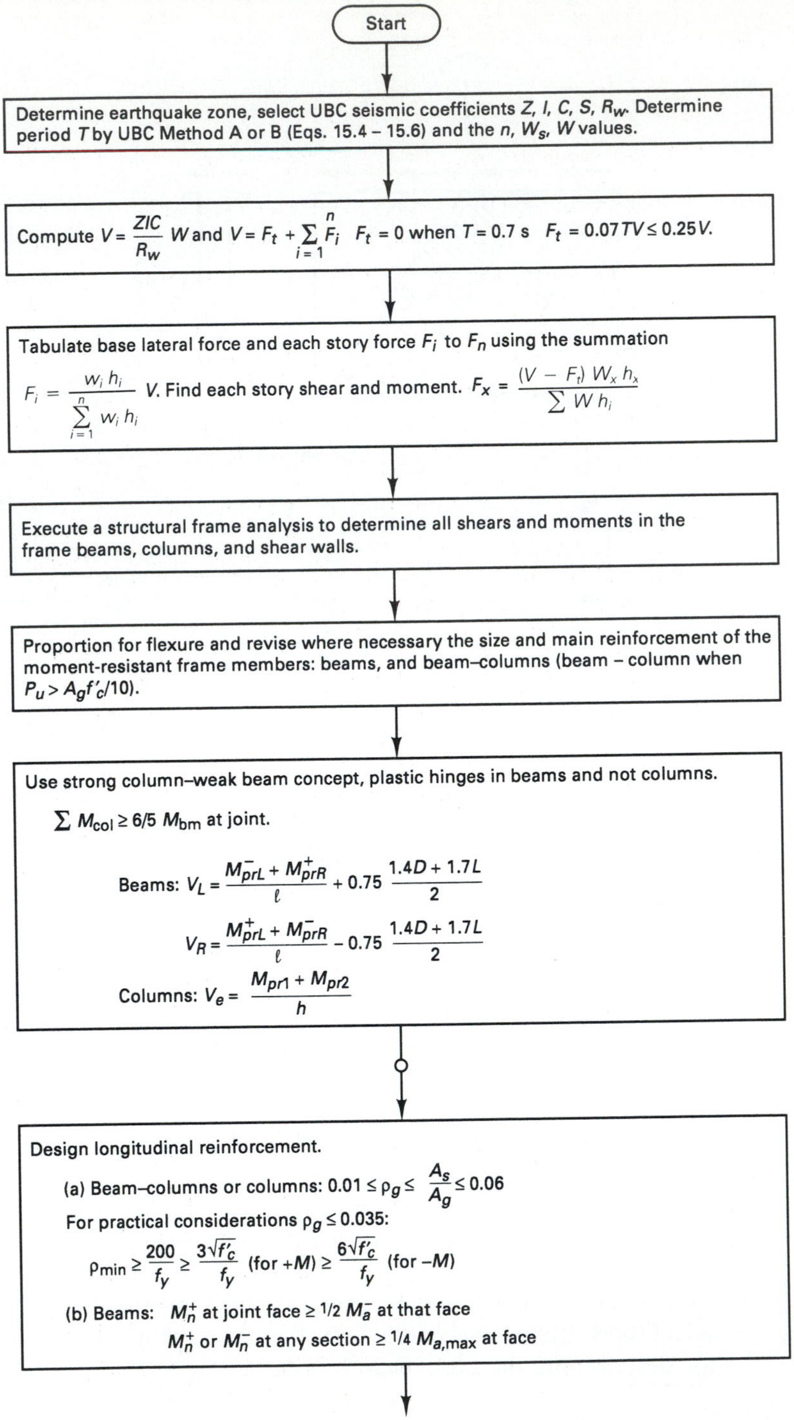

Start

Determine earthquake zone, select UBC seismic coefficients Z, I, C, S, R_w. Determine period T by UBC Method A or B (Eqs. 15.4 – 15.6) and the n, W_s, W values.

Compute $V = \dfrac{ZIC}{R_w} W$ and $V = F_t + \sum\limits_{i=1}^{n} F_i$ $F_t = 0$ when $T = 0.7$ s $F_t = 0.07 TV \leq 0.25 V$.

Tabulate base lateral force and each story force F_i to F_n using the summation

$F_i = \dfrac{w_i h_i}{\sum\limits_{i=1}^{n} w_i h_i} V$. Find each story shear and moment. $F_x = \dfrac{(V - F_t) W_x h_x}{\sum W h_i}$

Execute a structural frame analysis to determine all shears and moments in the frame beams, columns, and shear walls.

Proportion for flexure and revise where necessary the size and main reinforcement of the moment-resistant frame members: beams, and beam–columns (beam – column when $P_u > A_g f'_c/10$).

Use strong column–weak beam concept, plastic hinges in beams and not columns.

$\sum M_{col} \geq 6/5\, M_{bm}$ at joint.

Beams: $V_L = \dfrac{M^-_{prL} + M^+_{prR}}{\ell} + 0.75 \dfrac{1.4D + 1.7L}{2}$

$V_R = \dfrac{M^+_{prL} + M^-_{prR}}{\ell} - 0.75 \dfrac{1.4D + 1.7L}{2}$

Columns: $V_e = \dfrac{M_{pr1} + M_{pr2}}{h}$

Design longitudinal reinforcement.

(a) Beam–columns or columns: $0.01 \leq \rho_g \leq \dfrac{A_s}{A_g} \leq 0.06$

For practical considerations $\rho_g \leq 0.035$:

$\rho_{min} \geq \dfrac{200}{f_y} \geq \dfrac{3\sqrt{f'_c}}{f_y}$ (for +M) $\geq \dfrac{6\sqrt{f'_c}}{f_y}$ (for –M)

(b) Beams: M^+_n at joint face $\geq 1/2\, M^-_a$ at that face

M^+_n or M^-_n at any section $\geq 1/4\, M_{a,max}$ at face

Figure 15.12   Flow chart for seismic design of ductile structures.

Transverse confining reinforcement.

(a) Spirals for columns: $\rho_s \geq \dfrac{0.12f'_c}{f_{yh}}$ or $\geq 0.45\left(\dfrac{A_g}{A_{ch}}-1\right)\dfrac{f'_c}{f_{yh}}$

Whichever is greater.

(b) hoops for columns: $A_{sh} \geq 0.09\, s\, h_c\, \dfrac{f'_c}{f_{yh}}$

$\geq 0.3\, s\, h_c\left(\dfrac{A_g}{A_{ch}}-1\right)\dfrac{f'_c}{f_{yh}}$

s = 1/4 of smallest cross-sectional dimension or 4 in., whichever is smaller. Use standard tie spacing for the balance of the length.

(c) Beams: Place hoops over a length = $2h$ from face of columns. Maximum spacing: smaller of s = 1/4d, 8d_b main bar, 24d_b hoop, or 12 in.
If joint confined on all four sides, 50% reduction in confining steel and increase in minimum spacing of ties to 6 in. in columns is allowed.
Use the standard size and spacing of stirrups for the balance of the span as needed for shear.

Beam–column connection (joint)

Available nominal shear strength ≥ applied V_u

Confined on all faces: $V_n = 20\sqrt{f'_c}\ A_j$

Confined on three faces
or two opposite faces: $V_n \leq 15\sqrt{f'_c}\ A_j$

All other cases: $V_n \leq 12\ \sqrt{f'_c}\ A_j$

Check development length, normal-weight concrete,

$\ell_{dh} \geq f_y\, d_b/65\sqrt{f'_c} \geq 8d_b \geq 6$in.
$\ell_d = 2.5\ \ell_{dh}$ for 12 in. or less concrete below straight bar
$\ell_d = 3.5\ \ell_{dh}$ for > 12 in. in one pour

If bars have 90° hooks, $\ell_d = \ell_{dh}$. For lightweight concrete, adjust as in the ACI Code.

Design shear wall.

$V_{uh} > 2A_{cv}\sqrt{f'_c}$; use two reinforcement curtains in wall.
If wall $f_c > 0.2\ f'_c$, provide boundary elements.

Available $V_n = A_{cv}\left(\alpha_s\sqrt{f'_c} + \rho_n\, f_y\right)$
For $h_w/\ell_w \geq 2.0$, $\alpha_s = 2.0$
For $h_w/\ell_w = 1.5$, $\alpha_s = 3.0$
Interpolate intermediate values of h_w/ℓ_w.

Maximum allowance: $v_n = 8A_{cv}\sqrt{f'_c}$ for total wall
$v_n = 10A_{cp}\sqrt{f'_c}$ for individual pier

End

where Z (zone) $= 0.3$ for zone 3 and 0.4 for zone 4

I (occupation factor) $= 1.25$ to 1.0

C (seismic response) $= (1.25/T)^{2/3}$, $S \leq 2.75$

S (site-structure response): ranges between 1.0 for rocklike soil and shear wave velocity $v > 2500$ ft/s and 2.0 for soft clay substrata and $v < 500$ ft/s.

R_w (system quality factor) $= 4$ for shear walls, 6 for building frame systems, 12 for fully ductile moment-resistant concrete frame systems and dual systems

W_s = weight of one story, n = number of stories

W = total vertical weight of building $= nW_s$ if all stories have equal idealized weight

$$V = F_t + \sum_{i=1}^{n} F_i \text{ and } F_t = 0.07TV. \text{ But } F_t = 0 \text{ if } T = 0.7 \text{ or less.}$$

$T_n = C_t(h_w)^{3/4}$, where h_n = building height and $C_t = 0.10/\sqrt{A_c}$ (see Eq. 15.5)

2. Tabulated base lateral force and each story level force F_1 to F_n (Ex. 15.1) using UBC:

$$F_i = \frac{W_i h_i}{\sum\limits_{i=1}^{n} W_i h_i} V$$

Find the story shear and story moment. Execute a structural frame analysis to determine all shears and moments in the frame beams, columns, and shear walls.

3. Proportion members of the ductile moment-resistant frame, that is, all beams, columns, and beam–columns. If the frame is not a ductile moment-resisting frame, the designer has the uneconomical and inefficient alternative of choosing a brittle system using a low R_w factor.

4. Using the strong column–weak beam concept, plastic hinges are assumed to form in the beams.

Seismic beam shear forces

$$V_L = \frac{M_{prL}^- + M_{prR}^+}{\ell} + 0.75 \frac{1.4D + 1.7L}{2}$$

$$V_R = \frac{M_{prL}^+ + M_{prR}^-}{\ell} - 0.75 \frac{1.4D + 1.7L}{2}$$

ℓ = beam span, M_{pr} = probable moment of resistance, and L, R = left and right.

Seismic column shear force

$$V_e = \frac{M_{pr1} + M_{pr2}}{h}$$

where h = column height.

$$\sum M_{\text{col}} \geq \frac{6}{5} \sum M_{\text{bm}}$$

at joint to ensure hinges form in the beams; hence

$$(\phi M_n^+ + \phi M_n^-)_{\text{col}} \geq \frac{6}{5}(\phi M_n^+ + \phi M_n^-)_{\text{bm}}$$

The nominal moment strengths M_n have to be evaluated and the member proportioned prior to evaluating the seismic beam shear forces.

Beam: flexural design, P_u insignificant
Column: combined bending and axial load P_u
Beam–column: $P_u > A_g f_c'/10$
Shortest cross-sectional dimension ≥ 12 in.

5. *Longitudinal reinforcement*
   Beam–column or columns

$$0.01 \leq \rho_g = \frac{A_s}{A_g} \leq 0.06$$

For practical considerations, $\rho_g \leq 0.035$.

Beam (positive reinforcement):

$$\rho_{\text{min}} \geq \frac{200}{f_y} \geq \frac{3\sqrt{f_c'}}{f_y}$$

Beam (negative reinforcement):

$$\rho_{\text{min}} \geq \frac{200}{f_y} \geq \frac{6\sqrt{f_c'}}{f_y}$$

where f_y is in psi units. ρ should never exceed 0.025.
  For proportioning reinforcement in beams, the nominal moment strength requirements are
  (a) M_n^+ at face of joint $\geq \frac{1}{2}M_n^-$ at that face.
  (b) M_n^+ or M_n^- at any section $\geq \frac{1}{4}M_{a,\text{max}}$ at the face.

6. *Transverse confining reinforcement*
   (a) *Spirals*

$$\rho_s \geq \frac{0.12 f_c'}{f_{yh}} \quad \text{or} \quad \rho_s \geq 0.45\left(\frac{A_g}{A_{ch}} - 1\right)\frac{f_c'}{f_{yh}}$$

whichever is greater.

$$A_g = \text{gross area}$$

$$A_{ch} = \text{core area to outside of spirals}$$

$$f_{yh} = \text{specified yield strength}$$

(b) *Rectangular hoops in columns*: Total cross sectional within spacing s:

$$A_{sh} \geq 0.09 s h_c \frac{f'_c}{f_{yh}}$$

$$\geq 0.3 s h_c \left(\frac{A_g}{A_{ch}} - 1 \right) \frac{f'_c}{f_{yh}}$$

whichever is greater.

A_{sh} = total cross-sectional area of transverse reinforcement (including

cross ties) within spacing s and perpendicular to dimension h_c

h_c = cross-sectional dimension of column core, in.

s = spacing of transverse hoops

$s_{\max}$ = one-quarter of the smallest cross-sectional dimension or 4 in.,

whichever is smaller

Placement of confining reinforcement: Place confining reinforcement on either side of potential hinge over a distance the largest of

  (i) Depth of member at joint face
 (ii) One-sixth clear span
(iii) 18 in.

The spacing of the ties in the balance of column height follows normal column tie requirements.

(c) *Confining reinforcement in beam ends*: Should be placed on a length = $2h$ on both sides of the joint if it is internal; otherwise, maximum hoop spacing, smallest of

  (i) One-quarter effective depth d
 (ii) 8 × diameter of longitudinal bar
(iii) 24 × diameter of hoop
(iv) 12 in.

The ties in the balance of the beam span follow the standard shear web reinforcement requirements. If the joint is confined on all four sides, 50% reduction in confinement and increase in minimum tie spacing to 6 in. in the columns are allowed. No smooth bar reinforcement is allowed in seismic structures.

7. *Beam–column connections (joints)*: Normal concrete nominal shear strength V_n at a joint:

   (a) Confined on all faces: $V_n \leq 20\sqrt{f'_c}\, A_j$
   (b) Confined on three faces or two opposite faces: $V_n \leq 15\sqrt{f'_c}\, A_j$
   (c) All other cases: $V_n \leq 12\sqrt{f'_c}\, A_j$

   where A_j is effective area at joint (Fig. 15.8). The value of allowable V_n should be reduced by 25% for lightweight concrete. Note from Fig. 15.9 that the horizontal shear in the joint is determined by assuming a stress = $1.25f_y$ in the tensile reinforcement.

8. *Development length of reinforcing bars*: For bar sizes Nos. 3 to 11 without hooks, the largest of

$$\ell_d = 2.5\ell_{dh} \text{ when concrete below bars} \leq 12 \text{ in.}$$

$$\ell_d = 3.5\ell_{dh} \text{ when concrete below bars} \geq 12 \text{ in.}$$

where for normal-weight concrete

$$\ell_{dh} \geq f_y d_b / 65\sqrt{f'_c}$$

$$\geq 8d_b$$

$$\geq 6 \text{ in.}$$

When standard 90° hooks are used, $\ell_d = \ell_{dh}$. Any portion of straight embedment length not within the confined core should be increased by a factor of 1.6.

9. *Shear walls*: *height/depth* > 2.0

    (i) Minimum $\rho_v = 0.0025$ if $V_{uh} > A_{cv}\sqrt{f'_c}$. At least two curtains of reinforcement needed if in-plane factored shear force $V_{uh} > 2A_{cv}\sqrt{f'_c}$, where A_{cv} = net area of concrete cross section = thickness × length of section in direction of the considered shear.

    (ii) If extreme fiber compressive stresses exceed $0.2f'_c$, shear walls have to be provided with boundary elements along their vertical boundaries and around the edges of openings.

    (iii) Available $V_n = A_{cv} (2\sqrt{f'_c} + \rho_n f_y)$ for $h_w/\ell_w \geq 2.0$. For $h_w/\ell_w < 2$, the factor of 2 inside the parentheses varies linearly from 3.0 for $h_w/\ell_w = 1.5$ to 2.0 for $h_w/\ell_w = 2.0$; $V_u = \phi V_n$, where $\phi = 0.60$.

    (iv) Maximum allowable nominal unit shear $V_n = 8A_{cv}\sqrt{f'_c}$ for total wall, but can be increased to $V_n = 10A_{cp}\sqrt{f'_c}$ for an individual pier, where A_{cp} is the cross-sectional area of the individual pier.

15.8.1 SI Seismic Design Expressions

compressive strength $f'_c \geq 20$ MPa

$$E_c = w_c^{1.5}\, 0.043\, \sqrt{f'_c} \text{ MPa}$$

$$E_s = 200,000 \text{ MPa}$$

Equation 15.10: $w_u = 0.75\,(1.4D + 1.7L)$

Equation 15.11a: $V_L = \dfrac{M^-_{prL} + M^+_{prR}}{\ell} + 0.75\left(\dfrac{1.4D + 1.7L}{2}\right)$

Equation 15.11b: $V_R = \dfrac{M^+_{prL} + M^-_{prR}}{\ell} - 0.75\left(\dfrac{1.4D + 1.7L}{2}\right)$

Equation 15.12: $V_e = \dfrac{M_{pr1} + M_{pr2}}{h}$

Equation 15.14: $(\phi M^+_n + \phi M^-_n)_{col} \geq \dfrac{6}{5}(\phi M^+_n + \phi m^-_n)_{bm}$

$\phi = 0.9$ for beams and 0.7 or 0.75 for columns.

Photo 93   Masonry collapse in Los Angeles earthquake, 1994. (Courtesy Portland Cement Association.)

Equation 15.15

For positive moment: $\rho \geq \dfrac{\sqrt{f_c'}}{4f_y} \geq \dfrac{1.4}{f_y}$

where f_c, f_y are in MPa

For negative moment: $\rho \geq \dfrac{\sqrt{f_c'}}{2f_y} \geq \dfrac{1.4}{f_y}$

Equation 15.16a

At joint face: $M_n^+ \geq \dfrac{1}{2}M_a^-$

At any section:

$$M_a^+ \geq \frac{1}{4}(M_a^-)_{\max}$$

$$M_a^- \geq \frac{1}{4}(M_a^-)_{\max}$$

Equation 15.17a: $\rho_s \geq 0.12 \dfrac{f_c'}{f_{yh}}$

or Eq. 15.17b: $\rho_s \geq 0.45 \left(\dfrac{A_g}{A_{ch}} - 1\right)\dfrac{f_c'}{f_{yh}}$,   whichever is greater

Equation 15.18a: $A_{sh} \geq 0.09 s h_c \dfrac{f_c'}{f_{yh}}$

or Equation 15.18b: $A_{sh} \geq 0.3\ s h_c \left(\dfrac{A_g}{A_{ch}} - 1\right)\dfrac{f_c'}{f_{yh}}$,   whichever is greater
(same stipulations for max s as in Section 15.5.2).

Equation 15.19a: $V_n \leq 1.7\sqrt{f_c'}A_j$

Equation 15.19b: $V_n \leq 1.25\sqrt{f_c'}A_j$

Equation 15.19c: $V_n \leq 1.0\sqrt{f_c'}A_j$

Stress in the flexural reinforcement at the joint has to be taken as $1.25f_y$ (see Fig. 15.9).

Equation 15.20a: $\ell_{dh} = \dfrac{f_y d_b}{5.4\ \sqrt{f_c'}}$

for bar sizes No. 10 M through No. 35 M.

$\ell_d = 2.5\ell_{dh}$, when concrete depth below bar cast is one lift $\leq$ 300 mm

$\ell_d = 3.5\ell_{dh}$, when $\geq$ 300 mm

Equation 15.21: $V_n = A_{cv}\dfrac{\sqrt{f_c'}}{6} + \rho_n f_y$

For walls or diaphragms with $h_w/\ell_w < 2.0$,

$$V_n = A_{cv}(\alpha_c\sqrt{f_c'} + \rho_n f_y)$$

where α_c varies linearly from 3.0 for $h_w/\ell_w = 1.5$ to 2.0 for $h_w/\ell_w = 2.0$.

Modulus of rupture: $f_r = 0.7\sqrt{f_c'}$.

V_c for beams $= \lambda\left(1 + \dfrac{N_u}{14A_g}\right)\dfrac{\sqrt{f_c'}}{6}b_w d$. When seismic force is not applicable,
λ = type of concrete factors.

15.9 EXAMPLE 15.1: SEISMIC BASE SHEAR AND LATERAL FORCES AND MOMENTS BY THE UBC APPROACH

A building system with shear walls has a ground floor and five upper floors as in Fig. 15.3. Each floor has a weight W_s. Calculate the seismic base shear V and the overturning moment at each story level in terms of the single floor weight W_s assuming that all story heights are equal to h and their idealized weights are the same. The fundamental period $T = 0.15$ seconds.

Given:

$$\text{system quality factor } R_w = 6$$

$$\text{zone coefficient } Z = 0.4$$

$$\text{occupancy importance coefficient } I = 1.0$$

$$\text{site–structure resonance coefficient } S = 1.5$$

Solution From Eq. 15.1,

$$V = \left(\frac{ZIC}{R_w}\right)W$$

From Eq. 15.2,

$$C = \left(\frac{1.25}{T^{2/3}}\right) S \leq 2.75$$

$$T = 0.15 \quad \text{s}$$

Hence

$$C = \frac{1.25}{(0.15)^{2/3}} \times 1.5 = 6.7 > 2.75$$

Use $C = 2.75$.

Since there are five floors, each having a total vertical weight W_s, $W = 5W_s$; therefore,

$$V = \frac{0.4 \times 1 \times 2.75}{6} \times 5W_s$$

$$= 0.92W_s$$

Because $T < 0.7$ seconds, the horizontal force F_t in Eq. 15.7a can be assumed equal to zero. From Eq. 15.8a, the lateral earthquake force at the ith level is

$$F_i = \left(\frac{W_i h_i}{\sum\limits_{i=1}^{n} W_i h_i}\right) V \quad \text{or} \quad F_i = C_i V = 0.92 C_i W_s$$

where $i = 5$ at the *top* floor and height h is the same for all floors. The total cumulative load is

$$\sum_{i=1}^{n} = 1W_s + 2W_s + 3W_s + 4W_s + 5W_s = 15W_s$$

Photo 94 Overpass collapse in 1971 Los Angeles earthquake. (Courtesy Portland Cement Association.)

Floor	C_i	Lateral force $F_i = 0.92W_sC_i$	Story Shear	Story Moment
(1)	(2)	(3)	(4)	(5)
5	$C_5 = \dfrac{5W_s}{15W_s} = 0.333$	$0.3064W_s$	0	0
4	$C_4 = \dfrac{4}{15} = 0.267$	$0.2456W_s$	$0.3064W_s$	$0.3064W_sh$
3	$C_3 = \dfrac{3}{15} = 0.200$	$0.1840W_s$	$0.5520W_s$	$0.8584W_sh$
2	$C_2 = \dfrac{2}{15} = 0.133$	$0.1224W_s$	$0.7360W_s$	$1.5944W_sh$
1	$C_1 = \dfrac{1}{15} = 0.006$	$0.0055W_s$	$0.8584W_s$	$2.4528W_sh$
Wall base	$C_0 = 0$	0	$0.8639W_s$	$3.3167W_sh$

hence seismic base shear $V = 0.8639W_s$. The moments at each story level are tabulated in column (5).

15.10 EXAMPLE 15.2: DESIGN OF CONFINING REINFORCEMENT FOR BEAM–COLUMN CONNECTIONS

Design the transverse confining reinforcement of joint A in a ductile moment-resistant frame of a building as shown in Fig. 15.13.   The structure is located in seismic zone 3.   Given:

$$\text{working:} \quad W_D = 60{,}000 \text{ lb (269 kN)}$$

$$W_L = 75{,}000 \text{ lb (337 kN)}$$

$$f_c' = 4000 \text{ psi (27.6 MPa)}$$

$$f_y = 60{,}000 \text{ psi (414 MPa)}$$

All beams are 12 in. × 27 in. (305 mm × 686 mm) with four No. 9 bars (28.6-mm diameter) top and bottom and columns 15 in. × 24 in. (381 mm × 610 mm).

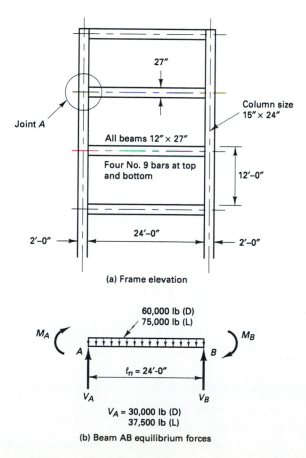

(a) Frame elevation

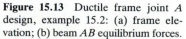

(b) Beam AB equilibrium forces

Figure 15.13 Ductile frame joint A design, example 15.2: (a) frame elevation; (b) beam AB equilibrium forces.

Solution

1. *Web shear reinforcement along beam span*

$$d = 24.0 - 2.5 = 21.5 \text{ in (622 mm)}$$

$$A_s = 4 \times 1.0 = 4.0 \text{ in.}^2 \text{ (2600 mm}^2)$$

$$\rho = \frac{4.0}{12 \times 21.5} = 0.016 < 0.025 \text{ allowed by ACI318   O.K.}$$

$$M_n = A_s f_y \left(d - \frac{a}{2} \right)$$

$$a = \frac{A_s f_y}{0.85 f_c' b} = \frac{4.0 \times 60,000}{0.85 \times 4000 \times 12} = 5.9 \text{ in.}$$

$$M_n = 4.0 \times 60,000 \left(21.5 - \frac{5.9}{2} \right) = 4,452,000 \text{ in.-lb. (5030 kN-m)}$$

Use 4,452,000 in this example as the required M_n value that would have been obtained from the structural analysis.

From Eq. 15.11,

$$V_L = \frac{M_A + M_B}{\ell_n} + 0.75 \left(\frac{1.4D + 1.7L}{2} \right)$$

$$= \frac{4,452,000 + 4,452,000}{24 \times 12} + 0.75 \left(\frac{1.4 \times 60,000 + 1.7 \times 75,000}{2} \right)$$

$$= 30,917 + 79,313 = 110,230 \text{ lb (490 kN)}$$

$$V_c = 2\sqrt{f_c'} b_w d = 2\sqrt{4000} \times 12 \times 24.5 = 37,188 \text{ lb (165 kN)}$$

Note that V_c can be assumed zero under certain ACI 318-95 Code conditions.
V_n at d from face of support is

$$V_n = 110,230 \left(\frac{12 - 27/12}{12} \right) = 89,562 \text{ lb (398 kN)}$$

$$V_s = V_n - V_c = 89,562 - 37,188 = 52,374 \text{ lb (233 kN)}$$

Try No. 4 hoop, $A_v = 2 \times 0.20 = 0.40 \text{ in.}^2$.

$$s = \frac{A_v f_y d}{V_s} = \frac{0.40 \times 60,000 \times 21.5}{52,374} = 9.85 \text{ in. (250 m)}$$

Use No. 4 closed hoops at 9¾ in c-c at critical section. Increase s to $d/2 = 10.75$ in., say 10 in. c-c, as the midspan is approached. Stop stirrups at $V_c/2$. See Chapter 6 for computational steps.

2. *Confining reinforcement in the beam at beam–column joint*: From Fig. 5.14,

$$V_{\text{col}} = \frac{M_n}{h_1/2 + h_2/2} = \frac{4,452,000}{(6.0 + 6.0)12} = 30,917 \text{ lb} (138 \text{ kN})$$

$$V_n \text{ at each joint} = A_s f_y - V_{\text{col}} = 4.0 \times 60,000 - 30,917 = 209,080 \text{ lb} (929 \text{ kN})$$

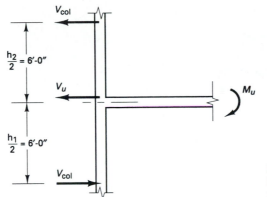

Figure 15.14 Shear forces at beam–column joint of example 15.2.

From Eq. 15.19b,

$$\text{allowable } V_n \leq 15\sqrt{f_c'}A_j, \qquad A_j = 15 \times 24 = 360 \text{ in.}^2 \ (232{,}000 \text{ mm}^2)$$

$$\text{allowable } V_u = 0.85 \ (15\sqrt{4000} \times 360 = 290{,}297 \text{ lb (1291 kN)}$$

> actual $V_u = 177{,}720$ lb, O.K.   Column $d = 24 - 2.5 = 21.5$ in.
V_c at the A_j plane is

$$V_c = 2\sqrt{4000} \times 15 \times 21.5 = 40{,}790 \text{ lb}$$

$$V_s = \frac{209{,}080}{0.85} - 40{,}790 = 168{,}290 \text{ lb (749 kN)}$$

$$s = \frac{A_v f_y d}{V_s} = \frac{0.40 \times 60{,}000 \times 21.5}{168{,}290} = 3.07 \text{ in (78 m)}$$

Maximum allowable $s = d/4 = 21.5/4 = 5.5$, 4 in., $8d_b = 8 \times \frac{9}{8} = 9$ in., 24 d_b hoop = $24 \times \frac{1}{2} = 12$ in., or 12 in. minimum.   Hence, use No. 4 hoops plus two No. 4 crossties at 3 in. c-c.   Place the confining hoops in the beam over a distance $\ell_0 = 2h = 2 \times 27 = 54$ in. (1372 mm).
    Use $\ell_0 = 54$ in., spacing the No. 4 hoops and crossties at 3 in. c-c over this distance.

3. *Confining reinforcement in the column at beam–column joint*: From Eqs. 15.18a and b,

$$A_{sh} \geq 0.09 sh_c \frac{f_c'}{f_{yh}}$$

or

$$A_{sh} \geq 0.3 sh_c \left(\frac{A_g}{A_{ch}} - 1\right) \frac{f_c'}{F_{yh}}$$

whichever is greater.

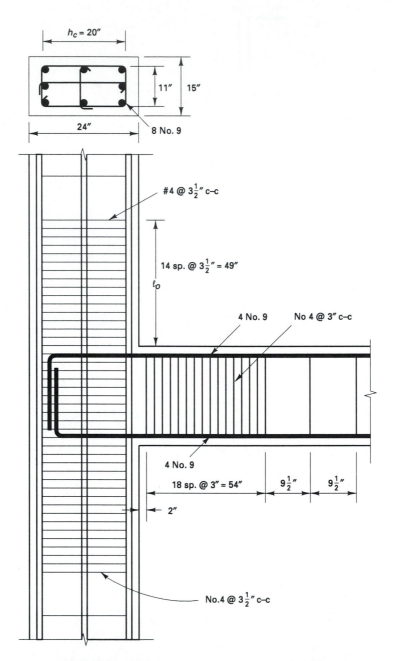

Figure 15.15 Confining hoops at joint in Example 15.2.

$$h_c = \text{column core dimension} = 24 - 2(1.5 + 0.5) = 20 \text{ in.}$$

Trying $s = 3\frac{1}{2}$ in.

$$A_{sh} = 0.09 \times 3.5 \times 20 \left(\frac{4000}{60,000}\right) = 0.42 \text{ in.}^2$$

$$A_{sh} = 0.3 \times 3.5 \times 200 \left(\frac{15 \times 24}{11 \times 20} - 1\right)\left(\frac{4000}{60,000}\right) = 0.89 \text{ in.}^2 \qquad \text{controls}$$

max. allowable $s = \frac{1}{4}b = \frac{1}{4}$ smallest column dimension or 4 in.

$$= 0.25 \times 15 = 3.75 \text{ in.}$$

Use No. 4 hoops plus two No. 4 crossties at $3\frac{1}{2}$ in. c-c. Place the confining hoops in the column on both sides of potential hinge over a distance ℓ_0 being the largest of

(a) Depth of member = 24 in. (610 mm)
(b) $\frac{1}{6} \times$ clear span = $(24 \times 12)/6 = 48$ in. (1220 mm)
(c) 18 in. (450 mm)

Use $\ell_0 = 48$ in. (1220 mm), spacing the No. 4 hoops and crossties at 3.5 in. c-c over this distance (12.7-mm-diameter bars at 89 mm c-c). Figure 15.15 shows detailing of the confining hoops at the beam–column joint at each floor level.

15.11 EXAMPLE 15.3: SEISMIC SHEAR WALL DESIGN AND DETAILING

Design by the ACI 318 Code the reinforcement for a shear wall in a multibay, ductile frame, 12-story structure (adapted from Ref. 15.8) having a total height $h_w = 148$ ft (45 m) and having equal spans of 22 ft (6.7 m). Except for the ground story, which is 16 ft (4.88 m) high, all other stories have 12 ft (3.67 m) heights. The total gravity factored load on the shear wall is $W_u = 4,800,000$ lb (21.4 MN). The factored moment at the base of the wall due to seismic loads from the lateral load analysis of the transverse frames is $M_u = 554 \times 10^6$ in.-lb (62.6 MN-m). The maximum axial force on the boundary element is $P_u = 4,500,000$ lb (20 MN). The horizontal shear force at the base is 885,000 lb (3940 kN).
Given:

wall length (horizontally) = $26' - 2'' = 26.17$ ft = 314 in. (7980 mm)

thickness t = 20 in. = 1.67 ft (508 mm)

boundary element width   = 32 in. (813 mm)

depth   = 50 in. (1270 mm)

A_s = 30 No. 11 bars (30 bars of 35-mm diameter) in each boundary element

$f_c' = 4000$ psi (27.6 MPa), normal weight
$f_y = 60,000$ psi (414 MPa)

Use $\phi = 0.60$ as the strength reduction factor for shear in this example.

Photo 95 Northridge, California, 1994 earthquake structural failure. (Courtesy Dr. Murat Saatcioglu.)

Solution

1. *Wall geometry and forces:* ℓ_n = 22 ft (6.7 m), ℓ_w (horizontal dimension) = 26.17 ft, b_{web} = 20 in. = 1.67 ft, and b_{bound} = 32 in. = 2.67 ft.

$$\text{factored } W_u = 4{,}800{,}000 \text{ lb (21.4 MN)}$$

$$M_u = 554 \times 10^6 \text{ in. lb (62.6 MN-m)}$$

$$P_u = 4{,}500{,}000 \text{ lb (20 MN)}$$

2. *Boundary element check:* ℓ_w = 26.17 ft, b = 1.67 ft, P_u = 4,500,000 lb, and M_u = 550 × 10⁶ in.-lb. Assume that the wall will not be provided with confinement over its entire section.

$$\text{gross } I_g = \frac{bh^3}{12} = \frac{1.67(26.17)^3}{12} = 2495 \text{ ft}^4$$

$$A_g = 1.67 \times 26.17 = 43.7 \text{ ft}^2$$

$$f_c = \frac{P}{A} \pm \frac{M_c}{I}, \qquad c = \frac{26.17}{2} \times 12 = 157 \text{ in. (3990 mm)}$$

Concrete compressive stress in the wall is

$$f_c = \frac{4,500,000}{43.7\,(12)^2} - \frac{554 \times 10^6 \times 157}{2494(12)^4}$$

$$= -715 - 1682 = -2400 \text{ psi (C) (16.5 MPa)}$$

Maximum allowable $f_c = 0.2f'_c = 0.2 \times 4000 = 800$ psi (5.52 MPa) in compression if a boundary element is not required. Hence boundary elements are needed subject to the confinement and loading requirements of Section 15.7.

3. *Longitudinal and transverse reinforcement:* Check if two curtains of reinforcement are needed, that is, if in-plane factored shear $> 2A_{cv}\sqrt{f'_c}$ (Section 15.7.1).

$V_u = 885,000$ lb

A_{cv} = area bound by web thickness and length of section in direction of shear force

$\quad = 20 \times 314 = 6280$ in.^{2}

$$2A_{cv}\sqrt{f'_c} = 2 \times 6280 \sqrt{4000} = 799,400 \text{ lb (353 kN)} < V_u = 885,000 \text{ lb}$$

Hence two curtains of reinforcement are required.

$$\min \rho_v = \frac{A_{sv}}{A_{cv}} = \rho_n = 0.0025 \quad \text{and} \quad \max s = 18 \text{ in.}$$

A_{cv} per ft of wall $= 20 \times 12 = 240$ in.^{2}

required A_s in each direction $= 0.0025 \times 240 = 0.60$ in.^{2}/ft

Trying No. 5 bars (15.8-mm diameter), $A_s = 2(0.31) = 0.62$ in.^{2} in two curtains.

$$s = \frac{\text{one bar area}}{\text{required } A_s/12 \text{ in.}} = \frac{0.62}{0.60/12}$$

$$= 12.4 \text{ in. (315 mm)} < 18 \text{ in. limit} \qquad \text{O.K.}$$

Use $s = 12$ in.

Check for shear reinforcement capacity

A check is needed in order to determine that the No. 5 bars in two curtains at 12 in. c-c both ways are adequate for the wall section to sustain the applied shear force at the base. The shear wall aspect ratio is

$$\frac{h_w}{\ell_w} = \frac{148}{26.17} = 5.66 > 2$$

Hence from Eq. 15.21a

$$\phi V_n = \phi A_{cv}(2\sqrt{f'_c} + \rho_n f_y)$$

where $\phi = 0.60$ in this example; otherwise, refer to the ACI 318-95 Code for other conditions.

$$A_{cv} = 20(26.17 \times 12) = 6280 \text{ in.}^2$$

$$\rho_n = \frac{2(0.31)}{20 \times 12} = 0.0026$$

available $\phi V_n = 0.60 \times 6280 \ (2\sqrt{4000} + 0.0026 \times 60,000)$

$$= 1,065,000 \ \text{lb} > V_u = 885,000 \ \text{lb}$$

$$(4.7 \ \text{MN} > \text{required} \ 3.9 \ \text{MN})$$

Hence the wall section is adequate. Therefore, use two curtains of No. 5 bars spaced at 12 in. c-c in both horizontal and vertical directions.

4. *Boundary element check if acting as a short column under factored vertical forces due to gravity and lateral loads*: P_u acting on wall = 4,500,000 lb. From before, $b = 32$ in., $h = 50$ in., $A_s = 30$ No. 11 bars = $30 \times 1.56 = 46.8$ in.² (30,190 mm²) in each boundary element.

$$\rho_{st} = \frac{A_s}{A_g} = \frac{46.8}{32 \times 50 = 1600} = 0.0293$$

$$\rho_{min} = 0.01 < \rho_{st} < \rho_{max} = 0.06 \qquad \text{O.K.}$$

The axial load capacity of the boundary element acting as a short column is

$$\phi P_{n(max)} = 0.80 \ \phi[0.85 f'_c \ (A_g - A_{st}) + A_s f_y]$$

$$= 0.80 \times 0.70[0.85 \times 4000(1600 - 46.8) + 46.8 \times 60,000]$$

$$= 4,530,000 \ \text{lb} > P_u = 4,500,000 \ \text{lb} \qquad \text{O.K.}$$

5. *Boundary element transverse confining reinforcement*: $b_w = 20$ in., $b_b = 32$ in., h or $\ell_w = 314$ in., and $A_g = 1600$ in.². From Eqs. 15.17a and b,

$$\rho_s \geq \frac{0.12 f'_c}{f_{yh}}$$

and

$$A_{sh} \geq 0.3 s h_c \left(\frac{A_g}{A_{ch}} - 1\right) \frac{f'_c}{f_{yh}}$$

Assume No. 5 hoops and crossties spaced at 4 in. c-c.

(a) *Short direction*

$$h_c = 50 - 2\left(1.5 + \frac{5}{16}\right) = 46.37 \ \text{in.}$$

$$b_c = 32 - 2\left(1.5 + \frac{5}{16}\right) = 28.37 \ \text{in.}$$

$$A_{ch} = 46.33 \times 28.37 = 1314 \ \text{in.}^2 \quad \text{(core area)}$$

$$A_{sh} = \frac{0.12 \ f'_c s h_c}{f_{yh}} = \frac{0.12 \times 4000 \times 4 \times 46.37}{60,000} = 1.48 \ \text{in.}^2$$

$$A_{sh} = 0.3 \times 4 \times 46.37 \left(\frac{1600}{1314} - 1\right) \frac{4000}{60,000} = 0.80 \ \text{in.}^2$$

$A_{sh} = 1.48$ in.² (955 mm²) governs.

Use three No. 5 crossties, for a total of five legs being provided including the hoop every 4 in. along the boundary length (wall length ℓ_w). A_{sh} provided = $5 \times 0.31 = 1.55$ in.², O.K.

Photo 96 Interfirst Plaza, Dallas, Texas, 10,000-psi concrete. (Courtesy Portland Cement Association.)

(b) *Longitudinal direction*

$$h_c = 28.37 \text{ in.,} \qquad A_{ch} = 1314 \text{ in.}^2$$

$$A_{sh} = \frac{0.12 \, f'_c h_c s}{f_{yh}} - \frac{0.12 \times 4000 \times 4 \times 28.37}{60,000} = 0.91 \text{ in.}^2$$

or

$$A_{sh} = 0.3 \times 4 \times 28.37 \left(\frac{1600}{1314} - 1 \right) \frac{4000}{60,000} = 0.49 \text{ in.}^2$$

$A_{sh} = 0.91$ in.^{2} (587 mm^2) controls.   With one No. 5 crosstie, a total of three legs is provided every 4 in. c-c. A_{sh} provided $= 3 \times 0.31 = 0.93$ in.^{2} (600 mm^2).

6. *Development of reinforcement*: Development length of No. 5 horizontal bars assuming no hooks are used within the boundary element: From Eqs. 15.20a, b, and c,

$$\ell_{dh} \geq \frac{f_y d_b}{65\sqrt{f'_c}} = \frac{60,000 \times 0.625}{65\sqrt{4000}} = 9 \text{ in.}$$

$$\geq 8d_b = 8 \times 0.625 = 5 \text{ in.}$$

$$\geq 6 \text{ in.}$$

$$\ell_{dh} = 9 \text{ in. (229 mm)} \qquad \text{governs}$$

$$\ell_d = 3.5\ell_{dh} = \times 9 \cong 32 \text{ in. (815 mm)}$$

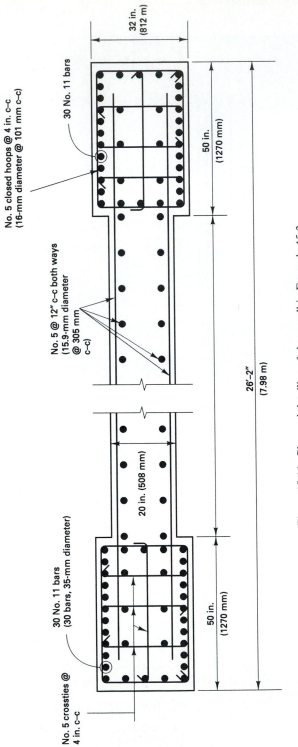

No. 5 closed hoops @ 4 in. c–c
(16-mm diameter @ 101 mm c–c)

30 No. 11 bars

32 in.
(812 mm)

50 in.
(1270 mm)

No. 5 @ 12" c–c both ways
(15.9-mm diameter
@ 305 mm
c–c)

26'–2"
(7.98 m)

20 in. (508 mm)

30 No. 11 bars
(30 bars, 35-mm diameter)

50 in.
(1270 mm)

No. 5 crossties @
4 in. c–c

Figure 15.16 Plan and detailing of shear wall in Example 15.3.

If bars are straight as in this example, ensure that development length is provided. If 90° hooks are used, $\ell_d = \ell_{dh} = 9$ in. Note that no lap splices should be allowed for the No. 5 horizontal bars.

7. *Verify adequacy of shear wall section at its base under combined axial load and bending in its plane*: From before,

$P_u = 4,800,000$ lb (total gravity factored load)

$M_u = 554 \times 10^6$ in.-lb, $e = \dfrac{M_u}{P_u} = 115$ in., $b_{web} = 20$ in., $b_{bound} = 32$ in.

$\ell_w = 26.17$ ft $= 314$ in.,   wall height $h_w = 148$ ft $= 1776$ in. (45 m)

column action $\phi = 0.70$,   beam action $\phi = 0.90$

no. of longitudinal bars in wall plane $= 110$ composed of two (30 No. 11 bars)

for *both* boundary elements and two curtains of No. 5 bars at 12 in.

c-c over $\ell_w = 314$ in.

total A_{st} in the lateral cross section $= 2 \times 46.8 + 2 (25 \times 0.31) = 109.1$ in.^{2}

$$A_g = 2(32 \times 50) + 20(314 - 2 \times 50) = 7480 \text{ in.}^2 \ (4,830,000 \text{ mm}^2)$$

$$\rho = \frac{109.1}{7480} = 0.015 > 0.01 \quad \text{and} \quad < 0.06 \quad \text{O.K.}$$

$$\frac{e}{\ell_w} = \frac{115}{314} = 0.366$$

$$\frac{\phi M_n}{A_g \ell_w} = \frac{554 \times 10^6}{7480 \times 314} = 236 \text{ psi}$$

From Fig. 9.20 with $h_e/h \equiv 1.0$, enter the plot with $\phi M_n/A_g\ell_w = 236$ and $e/\ell_w = 0.366$ coordinates. This gives a value of $\phi P_n/A_g = 0.85$ ksi or available $P_u = 850\,A_g = 850 \times 7480 = 6,360,000$ lb $>$ actual $P_u = 4,800,000$ lb (available 28.2 MN $>$ required 21.4 MN). Figure 15.16 gives the detailing of the shear wall longitudinal and boundary element confining reinforcement.

SELECTED REFERENCES

15.1. ACI Committee 318, *Building Code Requirements for Reinforced Concrete*, American Concrete Institute, Detroit, 1995, 350 p.

15.2. International Conference of Building Officials, *Uniform Building Code (UBC)*, Vol. 2, ICBO, Whittier, Calif., 1994, 1335 p.

15.3. Norris, H. C., Hansen, R. J., Holley, M. J., Biggs, J. M., Namyet, S., and Minami, K., *Structural Design for Dynamic Loads*, McGraw-Hill, New York, 1959, 453 p.

15.4. Wakabayashi, M., *Design of Earthquake-resistant Buildings*, McGraw-Hill, New York, 1986, 308 p.

15.5. Engelkirk, R. E., and Hart, G. C., *Earthquake Design of Concrete Masonry Buildings*, Prentice Hall, Englewood Cliffs, N.J., Vol. 1, 144 p., Vol. 2, 268 p., 1982.

15.6. Schneider, R. R., and Dickey, W. L., *Reinforced Masonry Design*, Prentice Hall, Englewood Cliffs, N.J., 1994, 729 p.

15.7. Clough R. W., "Dynamic Effects of Earthquakes," *Proc. ASCE*, New York, Vol. 86, ST4, April 1960, pp. 49–65.

15.8. Ghosh, S. K., "Special Provisions for Seismic Design," PCA Publication, *Notes on ACI318-89 Code*, Chapter 31, Portland Cement Association, Skokie, Ill., 1990, pp. 31-1 to 31-81.

15.9. Derecho, A. T., Fintel, M., and Ghosh, S. K., "Earthquake-resistant Structures," Ch. 12 in *Handbook of Concrete Engineering*, 2nd ed., Van Nostrand Reinhold, New York, 1985, pp. 411–513.

15.10. Borg, S. E., *Earthquake Engineering–Damage Assessment and Structural Design*, Wiley, New York, 1983, 110 p.

PROBLEMS FOR SOLUTION

15.4. A 3×18 panel, ductile, moment-resistant building has a ground story 15 ft high (4.6 m) and 10 upper stories of equal height of 11-6" (3.5 m). Calculate the base shear V and the overturning moment at each story level in terms of the weight W_s of each floor. Given:

$$Z = 1.0, \quad I = 1.2, \quad S = 1.3$$
$$T_n = 0.12 \text{ sec.}, \quad R_w = 12$$
$$W_s \text{ per floor} = 2,400,000 \text{ lb (9560 kN)}$$

15.2. If the building in Problem 15.1 is constructed with components having the dimensions and data listed below, design the confining transverse reinforcement for the spandral exterior beams and columns (three faces confined) joint for the bottom floor. Given:

floors have slabs of thickness $h_f = 7$ in. (178 mm)

all beams: 18 in. $\times$ 24 in. (457 mm $\times$ 610 mm)

exterior columns: 20 in. $\times$ 20 in. (508 mm $\times$ 508 mm)

clear beam spans in both longitudinal and transverse

directions $= 20'$-$0''$ (6.1 m)

shear wall base length $\ell_w = 25$ ft (7.6 m)

shear wall height $h_w = 130$ ft (39.6 m)

$f'_c = 5000$ psi, normal weight (34.5 MPa)

$f_y = f_{yh} = 60,000$ psi (414 MPa)

Sketch the joint reinforcement.

15.3. Design the confining reinforcement for the joint at an interior column of the bottom floor in Problem 15.2 and sketch the joint reinforcement.

15.4. Design the shear wall in Problem 15.2 and the boundary elements for the shear wall, assuming that the magnitude of loads, forces, and moments are 110% of the values used in Ex. 15.3.

COMPUTER PROGRAMS IN BASIC

A

The programs presented in this appendix are furnished as a guide to the development of a user's own programs for the design of reinforced concrete members. While every effort has been made to utilize the existing state of the art and to assure accuracy of the analytical solution and design techniques, neither the author nor the publisher make any warranty, either expressed or implied, regarding the use of these programs for other than informational purposes. The user of the program is responsible for the final evaluation as to the validity, accuracy, and applicability of any results obtained. The programs listed can be purchased on $5\frac{1}{4}$-in. or $3\frac{1}{2}$-in. diskette from NC SOFTWARE, Box 161, East Brunswick, New Jersey 08816.

Photo 97 The Corinthian, East Side Airlines Terminal, New York: reinforced concrete 570-ft-high building built with 8000-psi-strength concrete in columns and 6000-psi concrete in the floors. (Courtesy of New York Concrete Construction Institute and the Concrete Industry Board, Inc., New York.)

A.1 COMPUTER PROGRAM EGNAWY1 FOR RECTANGULAR BEAMS IN FLEXURE, SHEAR, AND TORSION

A computer program in BASIC for IBM PC AT, or PS/2 and compatibles, EGNAWY1, is presented for the analysis (design) of beams. It is divided into three major segments or operations. The *first* covers flexure for singly reinforced and doubly reinforced beams and can be applied to T beams. The *second* deals with diagonal tension and the *third* with combined diagonal tension and torsion for normal beams, including the design of web steel spacing. In effect, the program can be considered as three separate programs in BASIC language for proportioning rectangular beams, but combined for efficiency in programming the input and output steps.

In order to use the program, one has to key in (enter) the program steps into the computer. Once these steps are entered, they can be executed by entering the command RUN.

The program will respond with "ENTER YOUR SELECTION". If the input is operation 1, the program will go to the flexural analysis part. An input 2 takes the program to the diagonal tension operation, and an input 3 takes it to the combined diagonal tension and torsion operation.

Once the program selects the appropriate analysis section, the necessary input variables are requested.

The following symbols for the variables are used in the program.

$$FC' = \text{compressive strength of concrete, psi}$$
$$FY = \text{yield strength of steel, psi}$$
$$B \text{ or } b = \text{width of the beam, in.}$$
$$D \text{ or } d = \text{depth of the beam, in.}$$
$$AS \text{ or } As = \text{area of tension steel, in.}^2$$
$$AS' = \text{area of compression steel, in.}^2$$
$$MU \text{ or } Mu = \text{design moment, in.-lb}$$
$$A = \text{depth of the rectangular stress block, in.}$$
$$Vu = \text{design shear force, lb}$$
$$At, Av = \text{area of two legs of the stirrup, in.}^2$$
$$x1, y1 = \text{shorter and longer dimensions of the stirrup, in.}$$
$$Tu = \text{torsional moment, in.-lb}$$

In flexural analysis, program operation 1 calculates and prints the design moment M_u if the given reinforcement satisfies the maximum and minimum permissible reinforcement ratios. Otherwise, the program prints the appropriate error message: "Section is overreinforced" or "Area of steel is less than minimum."

For diagonal tension in beams, program operation 2 calculates and prints the spacing of stirrups. If necessary, it will also print one of the two following messages:

```
list
1 REM -- EGNAWY1  "REINFORCED CONCRETE BEAMS IN FLEXURE, SHEAR AND TORSION"
2 REM --"***********************************************************************"
3 REM -- COPYRIGHT 1988 BY DR. EDWARD G. NAWY
4 REM -- ALL RIGHTS RESERVED
6 CLS: LOCATE 5,33: PRINT "PROGRAM EGNAWY1"
7 LOCATE 7,16: PRINT "REINFORCED CONCRETE BEAMS IN FLEXURE, SHEAR AND TORSION"
8 LOCATE 11,33: PRINT "COPYRIGHT 1988"
9 LOCATE 13,32: PRINT "DR. EDWARD G. NAWY"
10 LOCATE 15,32: PRINT "ALL RIGHTS RESERVED"
14 J=1: FOR Q=1 TO 1000: J=J+1: NEXT Q
20 SCREEN 1
30 PAINT (15,15), 2
35 SCREEN 2
40 PRINT
41 PRINT "***********************************************************************"
42 PRINT "*                                                                     *"
43 PRINT "*     REINFORCED CONCRETE BEAMS IN FLEXURE,SHEAR AND TORSION          *"
44 PRINT "*                                                                     *"
45 PRINT "***********************************************************************"
47 PRINT : PRINT
48 INPUT "PRESS (RETURN) KEY TO CONTINUE"; PPPPPPP
49 SCREEN 1
50 PAINT (15,15), 2
51 PRINT: PRINT : PRINT
60 SCREEN 2
70 PRINT T2$: PRINT "1- RECTANGULAR BEAMS IN FLEXURE" : PRINT "2- BEAMS UNDER DI
AGONAL TENSION"
75 PRINT T2$: PRINT "1-RECTANGULAR BEAMS IN FLEXURE": PRINT "2-BEAMS UNDER DIAGO
NAL TENSION"
80 PRINT " 3- BEAMS UNDER SHEAR AND TORSION"
85 LPRINT "3-BEAMS UNDER SHEAR AND TORSION"
150 INPUT "ENTER YOUR SELECTION >";AN$
152 AN = VAL (AN$)
155 IF AN < 1 OR AN > 3 THEN GOTO 150
200 IF AN = 1 THEN GOSUB 1000
210 IF AN = 2 THEN GOTO 2000
220 IF AN = 3 THEN GOTO 3000
1000 REM TRIAL PROG. **BEAM**
1005 REM  -- ANALYSIS OF DOUBLY REINFORCED CONCRETE BEAM  --
1205 LPRINT T2$: LPRINT : LPRINT "**INPUT DATA:**" : LPRINT "---------------"
1206 PRINT T2$: PRINT : PRINT "**INPUT DATA:**": PRINT "---------------"
1210 INPUT "Fc'=";F: INPUT "Fy=";S
1211 PRINT
1212 INPUT "b=";B: INPUT "d=";D: INPUT "d'=";D1
1214 INPUT "As=";A1 :INPUT "As'=";A2
1220 PRINT : PRINT "** VERIFICATION OF INPUT DATA **": PRINT
1221 LPRINT: LPRINT "** VERIFICATION OF INPUT DATA **": LPRINT
1225 LPRINT "Fc'=";F : LPRINT "Fy=";S
1226 PRINT "Fc'=";F : PRINT "Fy=";S
1228 LPRINT "b=";B : LPRINT "d=";D : LPRINT "d'=";D1
1229 PRINT "b=";B: PRINT "d=";D : PRINT "d'=";D1
1230 LPRINT : LPRINT "As=";A1 : LPRINT "As'=";A2
1231 PRINT : PRINT "As=";A1: PRINT "As'=";A2
1232 LPRINT
1233 PRINT
1234 INPUT "ARE THE DATA CORRECT ? Y/N ";Z$
1235 IF Z$ = "N" THEN GOTO 1210
1238 LPRINT T2$ : LPRINT "PLEASE WAIT FOR CALCULATIONS" : LPRINT
1239 PRINT T2$: PRINT "PLEASE WAIT FOR CALCULATIONS" : PRINT
1240 IF F < 4000! THEN 1270
1250 IF F > 8000! THEN 1280
1260 BETA=.85 - (.05 * (F - 4000!) / 1000) : GOTO 1290
1270 BETA = .85: GOTO 1290
1280 BETA = .65: GOTO 1290
1290 REM CONTINUE
1300 PM = 200 / S
1310 R1 = A1 / (B*D)
1325 F1 = S
1330 A = ((A1*S)-(A2*F1))/(.85*F*B)
```

Figure A.1 Excerpts of EGNAWY1 listing for analysis of rectangular beams in flexure, shear, and torsion.

```
1332 E1 = .003 * (A - (BETA * D1))/A
1334 IF E1 > = S/29000000# THEN F1 = S: GOTO 1350
1336 F1 = 29000000# *E1
1340 IF F1 < 0 THEN F1 =0!: GOTO 1350
1344 A= (A1*S-A2*F1)/(.85*F*B)
1350 E2 = .003 * ((BETA *D) - A) / A
1355 M1$ =""
1365 PB = .75 * ((.85 * F * BETA / S) * (.003 * 29000000# / ((.003 * 29000000#)
+ S))) + (A2 * F1 / (S * B * D))
1366 IF E2 < = (S/29000000#) THEN M1$="SECTION IS OVER REINFORCED": GOTO 1450
1370 R1=A1/(B*D):PM=200/S
1375 IF R1 > = PB THEN M1$ = "SECTION IS OVER REINFORCED": GOTO 1450
1380 IF R1 <= PM THEN M1$ = "AREA OF STEEL IS LESS THAN MINIMUM": GOTO 1450
1395 LPRINT T2$: LPRINT : LPRINT "**FINAL RESULTS**" :LPRINT
1396 PRINT T2$: PRINT : PRINT "**FINAL RESULTS**": PRINT
1430 M=.9*(.85*F*A*B*(D-A/2)+A2*F1*(D-D1))
1440 LPRINT "Mu (IN-LB)=      ";M
1441  PRINT "Mu (IN-LB)=      ";M
1450 PRINT "BETA1=          ";BETA
1451 LPRINT "BETA1=          ";BETA
1452  PRINT "b=              ";B
1453 LPRINT "b=              ";B
1454  PRINT "d=              ";D
1455 LPRINT "d=              ";D
1456  PRINT "As=             ";A1
1457 LPRINT "As=             ";A1
1458  PRINT "As'=            ";A2
1459 LPRINT "As'=            ";A2
1460 LPRINT "a=              ";A
1461  PRINT "a=              ";A
1470 LPRINT ".75*ROW.BAL=    ";PB
1471  PRINT ".75*ROW.BAL=    ";PB
1472 LPRINT "ROW=            ";R1
1473  PRINT "ROW=            ";R1
1474 LPRINT "ROW MINIMUM=    ";PM
1475  PRINT "ROW MINIMUM=    ";PM
1476 LPRINT
1477 PRINT
1480 PRINT "** MESSAGES **"
1481 LPRINT "** MESSAGES **"
1490 PRINT M1$: PRINT
1491 LPRINT M1$: LPRINT
1500 INPUT "ANOTHER SECTION ? Y/N";AN$
1510 IF AN$ = "Y" THEN GOTO 1000
1515 IF AN$ = "N" THEN GOTO 40000
2000 REM DIAGONAL TENSION
2100 LPRINT T1$: LPRINT T2$: LPRINT "***DIAGONAL TENSION***"
2110 PRINT T1$: PRINT T2$: PRINT "***DIAGONAL TENSION***"
2300 LPRINT : LPRINT
2305 CLEAR
2310 INPUT "Fc'        =";FC
2315 INPUT "Fy         =";FY
2317 INPUT "b          =";B
2319 INPUT "d          =";D
2322 INPUT "Vu         =";VU
2325 INPUT "Mu         =";M
2328 INPUT "As         =";A
2350 INPUT "At         =";AO
4400 INPUT "HIT (RETURN) TO CONTINUE"; X$  : GOTO 48
4500 REM DIAGONAL TENSION AND TORSION MODULE #3
4510 S = AV / A7
4520 VS = 4 * BW * D * (SQR(FC))
4530 SZ = D / 2
4540 IF (VN - VC) > VS THEN SZ = D / 4
4550 IF SZ > 24 THEN SZ = 24
4560 IF S > SZ THEN S = SZ
4570 LPRINT "FINAL SPACING =";S
4571 PRINT "FINAL SPACING =";S
4580 LPRINT
4581 PRINT
4600 INPUT "HIT (RETURN) TO CONTINUE"; X$ : GOTO 48
40000 END
0
```

Figure A.1   *(continued)*

"The cross-section is too small, change the dimensions" or "Shear reinforcement is not required."

For beams subjected to combined diagonal tension and torsion, program operation 3 calculates and prints the spacing of stirrups and the area of the longitudinal reinforcement. It will also print the following messages in the appropriate cases: "Enlarge section due to torsional shear" or "No torsion bars required."

When the beam section is not rectangular, such as a T-beam or an L-beam, with the neutral axis falling outside the flange, flexural analysis operation 1 of program EGNAWY 1 can be used for the analysis (design). Minor modifications have to be made in the input values of b, A_s', and d' such that b = width b_w of the web, $A_s' = A_{sf}$ and $d' = h_f/2$. A check has to be made that $\rho \leq 0.75\rho_b$ of the flanged section. If the neutral axis falls within the flange, the beam can be treated in the flexural analysis as a singly reinforced rectangular section using input values for b = width of the compression flange and $A_s' = 0$.

Excerpts from this program are given in Fig. A.1. The full program, together with the other seven programs subsequently discussed, can be procured from the author as indicated in the introductory statement.

Photo 98 St. Joseph's Hospital, Tacoma, Washington, Tower Details. Photo by Richards Studios. (Courtesy Bertrand Goldberg Associates, Inc., Architects and Engineers, Chicago, Illinois.)

Example A.1.: Analysis of a Doubly Reinforced Beam in Flexure

Calculate the design moment $M_u = \phi M_n$ of the doubly reinforced section using the following beam properties. Given:

$$f'_c = 5000 \text{ psi, normal-weight concrete}$$
$$f_y = 60,000 \text{ psi}$$
$$b = 14 \text{ in.}$$
$$d = 18.5 \text{ in.}$$
$$d' = 2.5 \text{ in.}$$
$$A_s = 5.08 \text{ in.}^2$$
$$A_s' = 1.20 \text{ in.}^2$$

Solution

```
**INPUT DATA:**
-----------------
Fc'= 5000
          Fy= 60000

B  = 14
          D = 21
                    D' = 2.5

AS = 5.08
          AS'= 1.2

** VERIFICATION OF INPUT DATA **

FC'= 5000
          FY= 60000

B  = 14
          D = 21
                    D' = 2.5

AS = 5.08
          AS'= 1.2

ARE THE DATA CORRECT  ? Y/N Y
*************************************
```

```
                    PLEASE`WAIT`FOR`CALCULATIONS
          ****************************************

          ** FINAL RESULTS **

          MU (LB-IN)=     5158856.01
          BETA1=          .8
          A=              4.26497444
          .75*ROW.BAL=    .0280461463
          ROW=            .0172789116
          ROW MINIMUM=    3.33333333E-03
```

Example A.2: Design of Web Stirrups

A rectangular beam has an effective span of 25 ft and carries a design live load of 8000 lb per linear foot and no external dead load except its self-weight. Design the necessary shear reinforcement.

$$f_c' = 4000 \text{ psi, normal-weight concrete}$$
$$f_y = 60,000 \text{ psi}$$
$$b_w = 14 \text{ in.}$$
$$d = 28 \text{ in.}$$
$$h = 30 \text{ in.}$$
$$\sum x^2 y = (14^2)(30) = 5880 \text{ in.}^3$$

Longitudinal tension steel is six No. 9 bars.   No axial force acts on the beam.

Solution   The input variables are $f_c' = 4000$ psi, $f_y = 60,000$ psi, $b = 14$ in., $d = 28$ in., $V_u = 144,499$ lb and 111,331 lb, $M_u = 0$, $A_s = 6$ in.^{2}, and $A_v = 0.4$ in.^{2}.

```
***** SHEAR AND DIAGONAL TENSION *****


Fc'     = 4000
FY      = 60000
b       = 14
d       = 28
Vu      = 144499
Mu      = 0
As      = 6
At      = 0.4
ARE DATA CORRECT Y/N              ? Y
SPACING OF STIRRUP= 5.58073207
```

```
*******************************************
***** SHEAR AND DIAGONAL TENSION *****


Fc'     = 4000
FY      = 60000
b       = 14
d       = 28
Vu      = 111331
Mu      = 0
As      = 6
At      = .4
ARE DATA CORRECT Y/N              ? Y
SPACING OF STIRRUP= 8.25622473
```

A.2 COMPUTER PROGRAMS EGNAWY2, EGNAWY3, AND EGNAWY4 FOR COMPRESSION MEMBERS

Three separate computer programs in BASIC for IBM or PS/2 and compatibles have been written with subroutines to enable analyzing and designing columns and constructing load-moment strength interaction diagrams.

The *first* program, EGNAWY2, is for the analysis and design of rectangular columns, the *second*, EGNAWY3, is for the analysis and design of circular columns, and the *third*, EGNAWY4, is for the design of biaxially loaded rectangular columns.

Uniaxial loading.   In order to use the programs, we must key in (enter) all the program steps into the computer.   Once the program steps are entered, they can be executed by entering the command RUN.

The program executes two types of operations: type 1 is for analyzing a particular column sections by carrying out the analysis for a given neutral-axis depth c.   This value of c is assumed by the user at the outset for the first input cycle, to be adjusted subsequently.   Type 2 is for evaluating coordinate values for constructing load–moment strength interaction diagrams.

Operation input 1: Analysis of column sections.   The program asks for the necessary input variables.   The following symbols for the variables are used in the program.

FC' = compressive strength of concrete, psi

FY = yield strength of steel, psi

B   = width of the column, in.

H   = total thickness of the column or diameter of the column

β_1 = adjustment factor β_1 for f_c'

C   = depth of neutral axis, in.

PU   = specified factored axial load

N   = number of rows of bars in a rectangular column section or number of bars in a circular column section

DM   = distance of the tension bar closest to the tension face measured from the extreme compression fiber

D   = depth of a specific layer of bars from the compression fibers

A   = area of the bars located at depth D

When analyzing a section for the entered neutral-axis depth c, the program calculates and prints the values of c_b, P_{ub}, M_{ub}, and e_b, the depth of neutral axis, design axial load, design moment, and eccentricity, respectively, for the balanced condition and the maximum permissible axial load P_{u0}.

For the first cycle, the computer calculates the strength reduction factor ϕ value corresponding to the factored axial load input. If the design load strength P_u differs from the applied factored load P_u, the computer goes through additional computational cycles using the *design* P_u value in the strength reduction factor ϕ equation. Convergence results in applying the correct ϕ value for the computation of P_u and moment strength M_u of the particular section that is being analyzed. This is necessary since ϕ increases in the tension-type failure from $\phi = 0.70$ for tied columns and $\phi = 0.75$ for spirally reinforced columns up to a maximum $\phi = 0.90$ for the pure bending state.

The output printout lists the actual permissible design load P_u, permissible design moment M_u, and the corresponding eccentricity e. Based on the output values of these three last parameters and the c values used in the input transfer, the user adjusts the geometrical properties of the first section. The adjustment involves reducing or increasing the dimensions and/or the reinforcement area. This step can also involve using another trial c value in order to arrive at the proper column section where the output eccentricity is *almost equal* to the eccentricity of the actual factored load P_u applied to the column.

Operation input 2: Coordinate values for load–moment strength interaction plots. In order to execute the program, a choice of the dimensions B and H of a hypothetical cross section for input is to be based on a predetermined γ value for the chart's set of diagrams and the appropriate β_1. Hypothetical reasonable values for a reinforcement area A_s, a design load P_u, and a neutral axis depth c are used in the input in order to maintain the flow of calculations for the $(\phi P_n/A_g)/(\phi M_n/A_g h)$ coordinates.

The program calculates and prints the coordinate values of $\phi P_n/A_g$ (psi) and $\phi M_n/A_g h$ (psi) for a range of gross reinforcement ratios from $\rho_g = 1\%$ to $\rho_g = 8\%$ for the particular input value of β_1 corresponding to the concrete compressive strength f'_c used.   It gives the maximum permissible axial load value $\phi P_{no(max)}/A_g$ and indicates the coordinates of the point for $f_s = 0$, that is when the stress in the extreme tension bar changes from compression to tension and $f_s = f_y$, the balanced condition.

It also prints the following input data values preceding the tabulation of the coordinate values:

$$f'_c = \text{compressive strength of concrete, psi}$$

$$f_y = \text{reinforcement yield strength, psi}$$

$$\rho_g = \text{gross reinforcement ratio } A_{st}/A_g$$

$$\gamma = \text{ratio } \frac{\text{center-line distance between extreme bars}}{\text{total thickness } h \text{ of the rectangular column}}$$

or

$$\gamma = \text{ratio } \frac{\text{center-line distance between extreme bars}}{\text{external diameter } h \text{ of the circular column}}$$

The program checks whether the minimum and maximum limits of the total permissible reinforcement ratio ρ_g are followed and then prints the appropriate error messages: "Reinforcement ratio is less than 1%" or "Reinforcement ratio is greater than 8%."

Biaxial loading.   The third computer program, EGNAWY4, presents steps for the analysis of rectangular columns subjected to biaxial bending.   The applicable equations are given in Section 9.16.2.   A contour β-factor chart for moment ratios in columns subjected to biaxial bending is given in Fig. 9.36.

1. *Operation type 1.*   The input consists of the required axial load P_u/ϕ designated as P_n (lb) and the required uniaxial moments M_{ux}/ϕ and M_{uy}/ϕ (in.-lb) designated as M_{nx} and M_{ny}, respectively, in the program, a trial shorter total dimension b (in.) and longer total dimension h (in.) of the column section, and a trial contour β-factor for moment ratios.   The computer evaluates and prints out the *required controlling* equivalent uniaxial moment strength M_{ox} and M_{oy} and the moment ratios M_{nx}/M_{ox} and M_{ny}/M_{oy}. M_{ox} controls in designing the column in operation type 2 when $M_{ny}/M_{nx} < b/h$, and M_{oy} controls when $M_{ny}/M_{oy} > b/h$, as discussed in Sec. 9.16.2 and shown in the flow chart.

2. *Operation type 2.*   The input consists of the concrete compressive strength f'_c (psi), the trial column cross-sectional dimensions b (in.) and h (in.), an

assumed neutral-axis depth c value, the required axial load P_n, the number of rows of bars, and their area (in.^{2}) and distance (in.) from the compression face of the column section. The computer evaluates and prints out the *actual* axial load strength P_n, the actual nominal moment strength M_n (in.-lb), and the corresponding eccentricity e (in.) of the assumed section. The output nominal moment strength M_n represents M_{oxn} or M_{oyn} based on the input cross-sectional column dimensions, that is, whether the larger dimension is entered as perpendicular to the bending axis of the equivalent uniaxially loaded column (H in the program) or the shorter dimension is entered as such. Computation of both M_{oxn} and M_{oyn} is needed for the solution. The resulting moment ratios M_{nx}/M_{oxn} and M_{ny}/M_{oyn} are entered into the β-factor chart of Fig. 9.36 in order to verify if the initially assumed β value for the trial section is close to the β value obtained on the basis of the computer calculated M_{oxn} and M_{oyn}.

If not, another trial and adjustment cycle is executed since a new eccentricity results from each new β-factor value used, as outlined in the step-by-step operational procedure in Section 9.16.3. The process is repeated until the β-factor value converges for the final nominal equivalent moment strengths M_{oxn} and M_{oyn} about the x and y axes, respectively. These moment values have to be at least equal to the *required* controlling equivalent moment strengths M_{ox} or M_{oy} about the respective axes. The programs on a diskette can be procured from the author as indicated in the introductory statement.

Example A.3: Analysis of a Rectangular Tied Column

Select the appropriate section for a rectangular tied column to carry a service axial dead load $P_{wd} = 120,000$ lb and a service axial live load $P_{wl} = 360,000$ lb acting at an eccentricity (a) $e = 19.5$ in. and (b) $e = 4.0$ in. Given:

$$f'_c = 4000 \text{ psi}$$
$$f_y = 60,000 \text{ psi}$$
$$d' = 3.0 \text{ in.}$$

Solution (a) *Eccentricity $e = 19.5$ in.:*

factored axial load $P_u = 1.4 \times 120,000 + 1.7 \times 360,000 = 780,000$ lb

Assume a section 27 in. × 32 in. ($d = 29$). Assume that $\rho_g \approx 0.015$.

$$A_s = A'_s = 0.0075 \times 27 \times 29$$
$$= 5.87 \text{ in.}^2$$

Try six No. 9 bars on each face $= 6.00$ in.^{2}. Assume that neutral-axis depth $c = 13$ in. and input all the data using the Problem Type 1 Program.

The input and output for this cycle are as follows:

```
]PR#1
]RUN
*******ENGINEERING PROGRAM*******
**INPUT DATA:**
---------------
TYPE OF PROBLEM=1
FC'=4000
FY=60000
B=27
H=32
B1=0.85
C=13
PU=780000
NO. OF ROWS N=2
INPUT DEPTH AND AREA OF ROW NO.1
D=3
A=6
INPUT DEPTH AND AREA OF ROW NO.2
D=29
A=6
**VERIFICATION OF INPUT DATA**
FC'=4000
FY=60000
B=27
H=32
```

```
B1=.85
C=13
PU=780000
N=2
D AND A OF ROW NO. 1   ARE   3   AND   6
D AND A OF ROW NO. 2   ARE   29 AND   6
ARE THE DATA CORRECT? Y/NY


              PLEASE`WAIT`FOR`CALCULATION

         **FINAL RESULTS**
         CB (IN)      =17.1632653
         PUB (LB)     =923194.714
         MUB (IN-LB)  =14527651.4
         EB (IN)      =15.7362809
         PUO (LB)     =2025408
         PI           =.7
         PU (LB)      =695793
         MU (IN-LB)   =13804374.7
         E (IN)       =19.8397723
```

Output $e = 19.84$ in. is approximately equal to the given eccentricity of 19.5 in., while output $P_u = 695,793$ lb is less than 780,000 lb. Hence a new trial c value for the same section or same reinforcement area would not suffice.

Try in the next cycle a reinforcement area $A_s = A'_s = 7.62$ (six No. 10 bars) and a trial value of $c = 14.9$ in.

The input and output values are as follows:

```
*******ENGINEERING PROGRAM*******
**INPUT DATA:**
---------------
TYPE OF PROBLEM=1
FC'=4000
FY=60000
B=27
H=32
B1=.85
C=14.9
PU=780000
NO. OF ROWS N=2
INPUT DEPTH AND AREA OF ROW NO.1
D=3
A=7.62
INPUT DEPTH AND AREA OF ROW NO.2
D=29
A=7.62
**VERIFICATION OF INPUT DATA**
FC'=4000
FY=60000
B=27
H=32
```

```
B1=.85
C=14.9
PU=780000
N=2
D AND A OF ROW NO. 1   ARE   3   AND   7.62
D AND A OF ROW NO. 2   ARE   29 AND   7.62
ARE THE DATA CORRECT? Y/NY
*******ENGINEERING PROGRAM*******


              PLEASE`WAIT`FOR`CALCULATION

         **FINAL RESULTS**
         CB (IN)      =17.1632653
         PUB (LB)     =919339.114
         MUB (IN-LB)  =16246568.6
         EB (IN)      =17.6720084
         PUO (LB)     =2128103.04
         PI           =.7
         PU (LB)      =795717.3
         MU (IN-LB)   =15953200.1
         E (IN)       =20.048829
```

The computer output gives a design $P_u = 795,717$ lb $\approx$ factored $P_u = 780,000$ lb and output eccentricity $e = 20.05$ in., which is close to the actual eccentricity of 19.5 in. Adopt the design.

(b) *Eccentricity $e = 4.0$ in.:* Assume a section 18 in. $\times$ 24 in. ($d = 21$ in.). Try $A_s = A'_s = 5.0$ in.^{2}. Assume that $c = 18$ in. in the first cycle.

Call the Problem Type 1 Program and input all the data. The computer input and output data are as follows:

```
*******ENGINEERING PROGRAM*******
**INPUT DATA:**
---------------
TYPE OF PROBLEM=1
FC'=4000
FY=60000
B=18
H=24
B1=.85
C=18
PU=780000
NO. OF ROWS N=2
INPUT DEPTH AND AREA OF ROW NO.1
D=3
A=5
INPUT DEPTH AND AREA OF ROW NO.2
D=21
A=5
**VERIFICATION OF INPUT DATA**
FC'=4000
FY=60000
B=18
H=24
B1= 85
```

```
C=18
PU=780000
N=2
D AND A OF ROW NO. 1   ARE   3   AND   5
D AND A OF ROW NO. 2   ARE   21 AND   5
ARE THE DATA CORRECT? Y/NY
*******ENGINEERING PROGRAM*******


              PLEASE`WAIT`FOR`CALCULATION

**FINAL RESULTS**
CB (IN)      =12.4285714
PUB (LB)     =440674
MUB (IN-LB)  =6713227.48
EB (IN)      =15.2339995
PUO (LB)     =1139488
PI           =.7
PU (LB)      =802802
MU (IN-LB)   =5090866.2
E (IN)       =6.3413721
```

The output gives $e = 6.34$ in. and $P_u = 802,802$ lb. Since both e and P_u are larger than the actual values, decrease the size of the section.

Assume a section size 18 in. $\times$ 20 in. and a value of $c = 16.5$ in for the new input. The computer output gives design $P_u = 789,703$ lb $\sim$ factored $P_u = 780,000$ lb and output eccentricity $e = 4.11$ in., which is close to the actual eccentricity. Adopt the design.

```
*******ENGINEERING PROGRAM*******
**INPUT DATA:**
---------------
TYPE OF PROBLEM=1
FC'=4000
FY=60000
B=18
H=20
B1=.85
C=16.5
PU=780000
NO. OF ROWS N=2
INPUT DEPTH AND AREA OF ROW NO.1
D=3
A=5
INPUT DEPTH AND AREA OF ROW NO.2
D=17
A=5
**VERIFICATION OF INPUT DATA**
FC'=4000
FY=60000
B=18
```

```
H=20
B1=.85
C=16.5
PU=780000
N=2
D AND A OF ROW NO. 1   ARE   3   AND   5
D AND A OF ROW NO. 2   ARE   17 AND   5
ARE THE DATA CORRECT? Y/NY
*******ENGINEERING PROGRAM*******


              PLEASE`WAIT`FOR`CALCULATION

**FINAL RESULTS**
CB (IN)      =10.0612245
PUB (LB)     =354469.429
MUB (IN-LB)  =4953791.13
EB (IN)      =13.9752281
PUO (LB)     =1002400
PI           =.7
PU (LB)      =789703.728
MU (IN-LB)   =3246273.52
E (IN)       =4.11074864
```

Example A.4: Analysis of a Spirally Reinforced Circular Column

Solve Ex. A.4 using a circular spirally reinforced column with a load eccentricity $e = 19.5$ in.

Solution   The factored axial load $P_u = 780,000$ lb.   Assume a 30.0-in.-diameter circular section reinforced with 12 No. 10 bars, giving a total reinforcement area of 15.24 in.^{2} and a trial neutral-axis depth $c = 18$ in.

The computer input and output for this cycle are as follows:

```
*******ENGINEERING PROGRAM*******
**INPUT DATA:**
--------------
INPUT TYPE OF COLUMN SPIRAL OR TIED
?SPIRAL
TYPE OF PROBLEM=1
FC=4000
FY=60000
H=30
DS=24
B1=.85
AS=15.24
NO. OF BARS N=12
C=18
PU=780000
DMAX=27
**VERIFICATION OF INPUT DATA**

THE COLUMN IS                  SPIRAL
FC'=4000
FY=60000
H=30
DS=24
B1=.85
AS=15.24
NO. OF BARS N=12
C=18
PU=780000
DMAX=27
ARE THE DATA CORRECT ? Y/NY
*******ENGINEERING PROGRAM*******



        PLEASE WAIT FOR CALCULATIONS

   **FINAL RESULTS**
   CB (IN)    =15.9795918
   PUB (LB)   =817463.413
   MUB (IN-LB)=9836549.82
   EB (IN)    =12.0330154
   PUO (LB)   =
   2082629.44
   PI         =.75
   PU (LB)    =1046339
   MU(IN-LB)  =9330154.81
   E (IN)     =8.91695218
```

The output gives a design axial load $P_u = 1{,}046{,}399$ lb, which is larger than the factored $P_u = 780{,}000$ lb.   But the output eccentricity of 8.91 in. is too low compared to given eccentricity of 19.5 in. Hence revise the section.

In the last trial cycle, a circular column section having a diameter of 36.0 in. and reinforced with 14 No. 10 bars is used as adequate.   Following are the computer input and output details.

```
*******ENGINEERING PROGRAM*******       C=16
**INPUT DATA:**                         PU=780000
--------------                          DMAX=33
INPUT TYPE OF COLUMN SPIRAL OR TIED     ARE THE DATA CORRECT ? Y/NY
?SPIRAL                                  *******ENGINEERING PROGRAM*******
TYPE OF PROBLEM=1
FC=4000
FY=60000
H=36
DS=30                                              PLEASE·WAIT·FOR·CALCULATIONS
B1=.85
AS=17.78
NO. OF BARS N=14                        **FINAL RESULTS**
C=16                                    CB (IN)    =19.5306123
PU=780000                               PUB (LB)   =1211384.98
DMAX=33                                 MUB (IN-LB)=15913504.6
**VERIFICATION OF INPUT DATA**          EB (IN)    =13.1366204
                                        PUO (LB)   =
THE COLUMN IS              SPIRAL       2848681.14
FC'=4000                                PI         =.75
FY=60000                                PU (LB)    =795686.296
H=36                                     MU(IN-LB)  =15636843.1
DS=30                                    E (IN)     =19.6520201
B1=.85
AS=17.78
NO. OF BARS N=14
```

Example A.5: Load–Moment Design Strength Interaction Plots for Rectangular Columns

Construct a set of load–moment strength interaction diagrams for rectangular columns for the two cases (a) $\gamma = 0.75$ and (b) $\gamma = 0.90$ from the sets of printout coordinate values $\phi P_n/A_g$ and $\phi M_n/A_g h$.   Give also an output set of

$$\frac{\phi P_n/A_g}{\phi M_n/A_g h}$$

values for $\rho_g = 0.04$ and $\gamma = 0.90$.   Given:

$$f_c' = 4000 \text{ psi}$$

$$f_y = 60{,}000 \text{ psi}$$

Solution   (a) *Case $\gamma = 0.75$:* The input variables are type of problem $= 2$, $f_c' = 4000$ psi, $f_y = 60{,}000$ psi, $b = 12$ in., $h = 20$ in., $\beta_1 = 0.85$, $c = 10$ in., $P_u = 200{,}000$ lb, number of rows $= 2$, $d_1 = 2.5$ in., $A_1 = 3.0$ in.^{2}, $d_2 = 17.5$ in., and $A_2 = 3.0$ in.^{2}.   The output is given in five plots in Fig. A.2 for values of $\rho_g = 0.01$, 0.02, 0.04, 0.06, and 0.08.

Coordinates of the first point on each plot define the maximum allowable axial load strength ϕP_n; that is, $\phi P_{n(\text{max})}$ or $P_{u(\text{max})} = 0.80\phi[0.85f_c'(A_g - A_{st}) + f_y A_{st}$ and

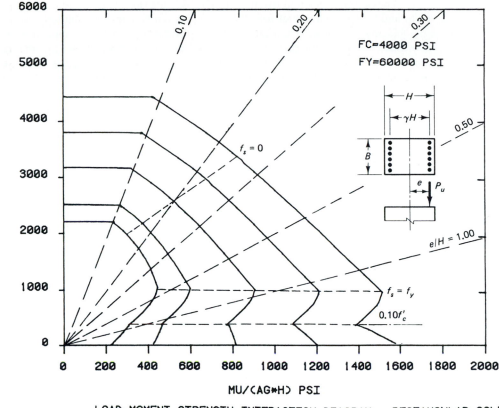

LOAD–MOMENT STRENGTH INTERACTION DIAGRAM – RECTANGULAR COLUMNS

GAMMA=0.75

Figure A.2 Load–moment strength interaction diagram: rectangular columns. $\gamma = 0.75$.

$\phi M_n = 0$.   The output plots were obtained by Tektronix Easygraphing with a peripheral interactive digital plotter.

The second point on the interaction in Fig. A.2 gives the load–moment strength
coordinate values for reinforcement tensile stress $f_s = 0$ in the bar closest to the
extreme tension fibers, that is, at the load level where the stress in the tension bar
changes from compression to tension.   The horizontal plateau of the diagrams at
$\phi P_{n(max)}$ is manually drawn to intersect with the load–moment strength interaction
curve extended upward beyond the point at $f_s = 0$.

The coordinate values in the plots at $f_s = f_y$ represent the balanced condition,
that is, the $\phi P_{nb} - \phi M_{nb}$ strength levels.   The line $e/H = 0.10$ denotes the minimum
eccentricity ratio allowed in rectangular columns in previous codes.   The eccentricity
ratio lines are hand-drawn to aid in speedy use of the charts as a design tool.

(b) *Case $\gamma = 0.90$:* The input variables are type of problem = 2, $f'_c = 4000$
psi, $f_y = 60,000$ psi, $b = 36$ in., $h = 50$ in., $\beta_1 = 0.85$, $c = 15$ in., $P_u = 600,000$ lb,
number of rows = 2, $d_1 = 2.5$ in., $A_1 = 18.0$ in.^{2}, $d_2 = 47.5$ in., and $A_2 = 18.0$ in.^{2}.

The output is given in five plots in Fig. A.3 for values of ρ_g = 0.01, 0.02, 0.04, 0.06, and 0.08.   Coordinates for other ρ_g values can be executed similarly.

Figure A.4 gives a typical computer printout of the load–moment strength coordinate values for the case of ρ_g = 0.04 and γ = 0.90.   Note the starting maximum permissible axial load value $P_{n(\text{max})}/A_g$ at $M_u/A_g h$ = 0, which is the first point on the horizontal plateau of the diagram in Fig. A.2 for the gross reinforcement percentage ρ_g = 0.04.

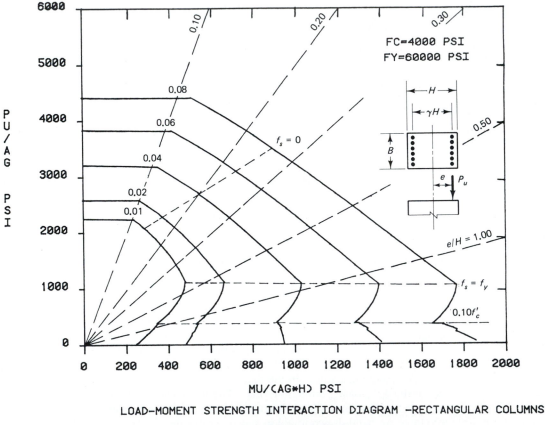

Figure A.3   Load–moment strength diagram: rectangular columns. γ = 0.90.

```
**********************************************
FC'      =4000
FY        =60000
RG       =.04
GAMMA      =.9


********COORDINATES OF INTERACTION DIAGRAM FOR RG=.04********
PU/AG(PSI)                    MU/(AG*H)(PSI)
3171.84                       0.0000
3427.88372                    241.612325

3382.3166                     262.117529

3267.71266                    316.034791

3090.41922                    395.51339

2906.57793                    471.748139

2714.96707                    545.288816

2666.65                       562.978062FS=0.0

2613.84261                    577.319265

2510.75764                    612.737784

2404.71488                    647.939712

2295.39362                    683.069376

2182.425                      718.292749

2065.38273                    753.801675

1943.77135                    789.819101

1817.01169                    826.605661

1684.42235                    864.467981

1545.19615                    903.769293

1398.36968                    944.943135

1242.78356                    988.51122

1089.82143                    1031.49922FS=FY

1085.28                       1031.39456

1024.59                       1029.1642

963.9                         1025.38625

903.21                        1020.0607

842.52                        1013.18756

781.83                        1004.76682
```

Figure A.4 Set of interaction diagram coordinates printout values for $\rho_g = 0.04$ and $\gamma = 0.90$.

721.14	994.798489
660.45	983.282562
599.76	970.219039
539.07	955.607922
478.38	939.44921
417.69	921.742902
357	902.489
357.539325	926.745678
346.983915	922.893678
339.024978	926.071187
330.326453	927.503425
321.674798	929.318559
312.887529	931.014896
304.002967	932.706453
292.205621	933.105185
270.392523	930.524481
248.943617	933.453956
225.71748	935.365361
200.85596	937.472208
174.071271	939.5692
145.089125	941.688108
113.568259	943.823469
79.0936488	945.972854
41.153979	948.130438
-.887417711	950.286111

Figure A.4   (*cont.*)

Example A.6: Load–Moment Design Strength Interaction Plots for Circular Columns

Construct a set of load–moment strength interaction diagrams for spiral circular columns for the two cases (a) $\gamma = 0.75$ and (b) $\gamma = 0.90$ from the sets of printout coordinate values $\phi P_n/A_g$ and $\phi M_n/A_g h$.   Given:

$$f_c' = 4000 \text{ psi}$$
$$f_y = 60,000 \text{ psi}$$

Solution   (a) *Case* $\gamma = 0.75$: The input variables are type of column = spiral, type of problem = 2, $f'_c = 4000$ psi, $f_y = 60,000$ psi, $h = 20$ in., $d_s = 15$ in., $\beta_1 = 0.85$, $A_s = 4.74$ in.², number of bars = 12, $c = 7.0$ in., $P_u = 120,000$ lb, and $d_{max} = 17.5$ in.

The output is given in five plots in Fig. A.5 for values of $\rho_g = 0.01, 0.02, 0.04, 0.06$, and 0.08 in a similar manner to that described in Ex. A.5 for rectangular columns. The value $f_s = 0$ in these interaction plots denotes the point at which the stress in the tension steel closest to the extreme concrete tension fibers changes from compression to tension. The value $f_s = f_y$ represents the balanced condition. The line $e/H = 0.05$ represents the minimum eccentricity ratio allowed in circular columns in previous codes. The set of eccentricity ratio lines is hand-drawn to facilitate speedy use of the charts as a design tool.

(b) *Case* $\gamma = 0.90$: The input variables are type of column = spiral, type of problem = 2, $f'_c = 4000$ psi, $f_y = 60,000$ psi, $h = 50$ in., $d_s = 45$ in., $\beta_1 = 0.85$, $A_s = 40.0$ in.², number of bars = 12, $c = 20$ in., $P_u = 800,000$ lb, and $d_{max} = 47.5$ in.

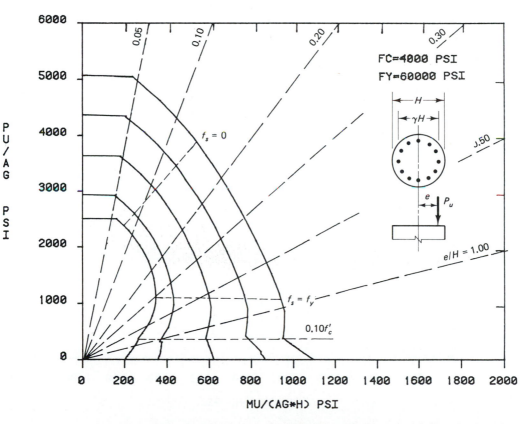

Figure A.5   Load–moment strength interaction diagram: circular columns. $\gamma = 0.75$.

The output is given in five plots in Fig. A.6 for values for $\rho_g = 0.01, 0.02, 0.04, 0.06,$ and 0.08. These plots were constructed in the same manner as in part (a) of this example using a peripheral interactive digital plotter. The horizontal plateau of this diagram at $\phi P_{n(max)} = P_{u(max)}$ has been manually drawn to intersect with the interaction curve extension upward beyond the point at $f_s = 0$.

Example A.7: Analysis of a Biaxially Loaded Column

A corner column is subjected to a factored compression axial load $P_u = 210,000$ lb, a factored bending moment $M_{ux} = 1,680,000$ in.-lb about the x axis, and a factored bending moment $M_{uy} = 980,000$ in.-lb about the y axis. Design a rectangular tied column section to resist this factored load and factored biaxial moments, given $f'_c = 4000$ psi and $f_y = 60,000$ psi.

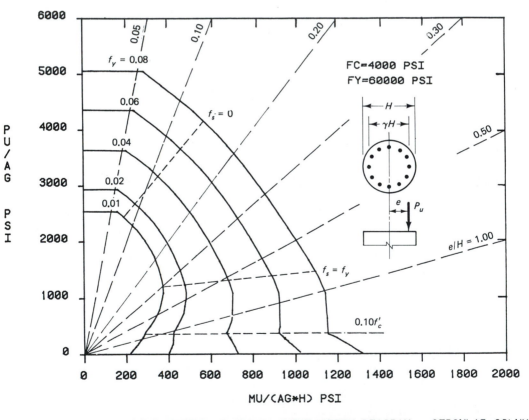

LOAD-MOMENT STRENGTH INTERACTION DIAGRAM - CIRCULAR COLUMNS

GAMMA=0.90

Figure A.6 Load–moment strength interaction diagram: circular columns. $\gamma = 0.90$.

Solution: Assume a 12 in. $\times$ 20 in. cross section with three No. 8 bars on each of the four faces of the column.   Assume that $\phi = 0.7$ as sufficiently reasonable for the biaxial loading.

$$P_n = \frac{P_n}{\phi} = \frac{210,000}{0.7} = 300,000 \text{ lb}$$

$$M_{nx} = \frac{M_{ux}}{\phi} = \frac{1,680,000}{0.7} = 2,400,000 \text{ in.-lb}$$

$$M_{ny} = \frac{M_{uy}}{\phi} = \frac{980,000}{0.7} = 1,400,000 \text{ in.-lb}$$

Operation type 1: Assume moment ratios β factor $= 0.64$.   The computer input data are $P_n = 300,000$ lb, $M_{nx} = 2,400,000$ in.-lb, $M_{ny} = 1,400,000$ in.-lb, $b = 12$ in., $h = 20$ in., and $\beta = 0.64$.

```
*******ENGINEERING PROGRAM*******
**INPUT DATA:**
-------------
TYPE OF PROBLEM=1
PN=300000
MNX=2400000
MNY=1400000
B=12
H=20
BE=.64
**VERIFICATION OF INPUT DATA**

PN=300000
MNX=2400000
MNY=1400000
B=12
H=20
BE=.64
ARE THE DATA CORRECT? Y/NY
*******ENGINEERING PROGRAM*******



          PLEASE`WAIT`FOR`CALCULATION
MNX/MOX =.646464647
MNY/MOY =.633484163
MOX=3712500
```

Therefore, controlling *required* moment strength in this example is $M_{ox} = 3,712,500$ in.-lb.

Operation type 2

(a) *Nominal moment strength* M_{oxn}: The computer input data are $f'_c = 4000$ psi, $f_y = 60,000$ psi, $b = 12$ in., $h = 20$ in., $\beta_1 = 0.85$, trial axis depth $c = 9.67$ in., $P_n = 300,000$ lb, $d_1 = 2.5$ in., $A_1 = 2.37$ in.^{2}, $d_2 = 17.5$ in., and $A_2 = 2.37$ in.^{2}.

The iterated final value of $c = 9.67$ in. for the case where the larger column cross-sectional dimension assumed perpendicular to the axis of bending was arrived at after three trials.   This c value gave the actual controlling nominal moment strength M_{oxn} almost equal in value to the required controlling moment strength M_{ox} and a correct iterated contour β-factor value.

```
******ENGINEERING PROGRAM*******          B=12
**INPUT DATA:**                           H=20
-------------                             B1=.85
TYPE OF PROBLEM=2                          C=9.67
FC'=4000                                   PU=300000
FY=60000                                   N=2
B=12                                       D AND A OF ROW NO. 1  ARE  2.5AND  2.37
H=20                                       D AND A OF ROW NO. 2  ARE  17.5AND 2.37
B1=.85                                     ARE THE DATA CORRECT? Y/NY
C=9.67                                     ******ENGINEERING PROGRAM*******
PN=300000
NO.OF ROWS N=2
INPUT DEPTH AND AREA OF ROW NO.1
D=2.5
A=2.37
INPUT DEPTH AND AREA OF ROW NO.2              PLEASE·WAIT·FOR·CALCULATION
D=17.5
A=2.37                                     **FINAL RESULTS**
**VERIFICATION OF INPUT DATA**             PN(LB)        =327297.6
FC'=4000                                   MN(IN-LB)     =4047893.32
FY=60000                                   E(IN)         =12.367623
```

Therefore, actual nominal M_{oxn} = 4,047,893 in.-lb > required controlling M_{ox} = 3,712,508 in.-lb; O.K.

    (a) *Nominal moment strength M_{oyn}:* The computer input data are f'_c = 4000 psi, f_y = 60,000 psi, b = 20 in., h = 12 in., trial axis depth c = 5.40 in., β_1 = 0.85, d_1 = 2.5 in., A_1 = 2.37 in.², d_2 = 9.5 in., and A_2 = 2.37 in.².

    The iterated final value of c = 5.40 for the case where the shorter column dimension was assumed perpendicular to the axis of bending was arrived at after three trials.

```
*******ENGINEERING PROGRAM*******         B=20
**INPUT DATA:**                           H=12
-------------                             B1=.85
TYPE OF PROBLEM=2                          C=5.4
FC'=4000                                   PU=300000
FY=60000                                   N=2
B=20                                       D AND A OF ROW NO. 1  ARE  2.5AND  2.37
H=12                                       D AND A OF ROW NO. 2  ARE  9.5AND  2.37
B1=.85                                     ARE THE DATA CORRECT? Y/NY
C=5.4                                      *******ENGINEERING PROGRAM*******
PN=300000
NO.OF ROWS N=2
INPUT DEPTH AND AREA OF ROW NO.1
D=2.5
A=2.37                                        PLEASE·WAIT·FOR·CALCULATION
INPUT DEPTH AND AREA OF ROW NO.2
D=9.5
A=2.37                                     **FINAL RESULTS**
**VERIFICATION OF INPUT DATA**             PN(LB)        =272593.667
FC'=4000                                   MN(IN-LB)     =2013462.43
FY=60000                                   E(IN)         =7.38631406
```

Therefore, actual moment M_{oyn} = 2,013,462 in.-lb.

$$\frac{M_{nx}}{M_{oxn}} = \frac{2,400,000}{4,047,893} = 0.593$$

$$\frac{M_{ny}}{M_{oyn}} = \frac{1,400,000}{2,013,462} = 0.695$$

The chart in Fig. 9.35 gives a contour β-factor value ≅ 0.64. Hence, adopt the design.

A.3 COMPUTER PROGRAMS EGNAWY5 AND EGNAWY6 FOR FLEXURAL ANALYSIS AND DESIGN OF FLANGED SECTIONS

Two separate computer programs in BASIC for IBM PC AT, PS/2, and compatibles are given for the flexural analysis and design of T beams and L beams. EGNAWY5 analyzes the input data comprising of f'_c, f_y, b, b_w, h_f, d, and A_s by first determining whether the neutral axis falls within or outside the flange. Then it proceeds to solve for the nominal moment strength.

EGNAWY6 (Fig. A.7) is a design program with input data M_u and trial input d. The program also computes β_1, $0.75\rho_b$, a, c, $0.75\bar{\rho}_b$, ρ_{min}, and ρ and then checks to see if $\rho_{min} < \rho < 0.75\rho_b$ if $c > h_f$ (i.e., a flanged section). If any of these parameters are not satisfied, an error message is displayed. The program solves for increments of effective depth d of 0.5 in. until it iterates to solve and output a depth d necessary to give the required nominal moment strength.

```
1 REM -- EGNAWY6 "DESIGN OF T AND L BEAMS IN FLEXURE"
2 REM -- ******************************************
3 REM -- COPYRIGHT BY DR. EDWARD G. NAWY
4 REM -- ALL RIGHTS RESERVED
5 CLS: LOCATE 6,33: PRINT "PROGRAM EGNAWY6"
7 LOCATE 8,24: PRINT "DESIGN OF T AND L BEAMS IN FLEXURE"
8 LOCATE 11,33: PRINT "(C) COPYRIGHT 1988"
9 LOCATE 13,32: PRINT "DR. EDWARD G. NAWY"
10 LOCATE 15,32: PRINT "ALL RIGHTS RESERVED"
11 J=0: FOR Q=1 TO 1000: J=J+1: NEXT Q
12 SCREEN 1
13 PAINT (15,15), 2
15 SCREEN 2
22 PRINT "**************************************************"
23 PRINT "*                                                *"
24 PRINT "*         DESIGN OF T AND L BEAMS IN FLEXURE      *"
25 PRINT "*                                                *"
26 PRINT "**************************************************"
27 REM
28 PRINT "**************    INPUT  DATA    ****************"
29 INPUT "MUI= ";MI
30 INPUT "Fc'= ";FC  :  INPUT "Fy= ";FY  :  INPUT "b= ";B
31 INPUT "bw= ";BW  :  INPUT "hf= ";HF  :  INPUT "d= ";D
32 INPUT "As = ";AS
35 PRINT "******    VERIFICATION OF INPUT DATA    *******"
38 PRINT "**************************************************"
40 REM
42 PRINT "MUI= ";MI
45 PRINT "Fc'= ";FC  :  PRINT "Fy= ";FY  :  PRINT "b= ";B
46 PRINT "bw= ";BW  :  PRINT "hf= ";HF  :  PRINT "d= ";D
47 PRINT "As = ";AS
50 INPUT "IS INPUT DATA CORRECT (Y/N)";  Z$
51 IF Z$ = "N" THEN GOTO 30
200 IF FC < 4000 THEN GOTO 800
300 IF FC > 8000 THEN GOTO 600
400 BE=.85 - .05*(FC-4000)/1000
500 GOTO 810
600 BE=.65
700 GOTO 810
800 BE=.85
810 FLAG1=0
820 FLAG2=0
```

Figure A.7 Program EGNAWY6 listing for the design of flanged beams in flexure.

```
830 INCR=1
840 INCRM=.5
900 RB=.85*FC*BE*87000!/(FY*(87000!+FY))
1000 ASF = .85 * FC * HF * (B - BW) / FY
1100 RF = ASF / (BW * D)
1200 NRB=.75*BW*(RB+RF)/B
1300 A=AS*FY/(.85*FC*B)
1400 C=A/BE
1410 R=AS/(B*D)
1500 IF C > HF AND A > HF THEN GOTO 2300
1700 IF R > .75*RB THEN GOTO 3400
1800 RW=AS/(BW*D)
1900 RMIN = 200 /FY
2000 IF RW < RMIN THEN GOTO 3391
2100 MU=.9*(AS*FY*(D-A/2))
2110 GOTO 2720
2300 A = (AS - ASF) * FY / (.85 * FC *BW)
2320 C=A/BE
2500 IF R > NRB THEN GOTO 3400
2600 RMIN = 200 / FY
2700 IF R < RMIN THEN GOTO 3391
2710 MU = .9 * (FY * (AS - ASF) * (D - A / 2) + ASF * FY * (D - HF / 2 ))
2715 GOTO 2720
2720 IF ABS (MU-MI) < = .01*MI THEN GOTO 2801
2721 IF FLAG1 = 1 AND FLAG2 = 1 AND ABS(D1 - D) < = INCRM THEN GOTO 3350
2730 IF MU > MI THEN GOTO 3330
2740 GOTO 3300
2801 PRINT : PRINT "**FINAL RESULT**": PRINT
2802 LPRINT : LPRINT "**FINAL RESULT**" : LPRINT
2806 PRINT "bw(IN)     =";BW
2807 LPRINT "bw        =";BW
2808 PRINT "d(IN)      =";D
2809 LPRINT "d(IN)      =";D
2810 PRINT "c(IN)      =";C
2820 LPRINT "c(IN)      =";C
2830 PRINT "a(IN)      =";A
2840 LPRINT "a(IN)      =";A
2850 PRINT "BE         =";BE
2855 LPRINT "BE         =";BE
2900 PRINT "0.75RB     =";NRB
2910 LPRINT "0.75RB     =";NRB
2920 PRINT "R          =";R
2930 LPRINT "R          =";R
2940 PRINT "RW         =";RW
2950 LPRINT "RW         =";RW
2960 PRINT "RMIN       =";RMIN
2970 LPRINT "RMIN       =";RMIN
3100 PRINT "MU(IN-LB)  =";MU
3110 LPRINT "MU(IN-LB)  =";MU
3200 GOTO 3500
3300 D1=D
3301 D=D+.25 * INCR
3302 FLAG1=1
3305 PRINT "d=";D
3306 PRINT "INCR=";INCR
3310 GOTO 1100
3330 D1=D
3331 D=D - .25 * INCR
3332 FLAG2=1
3335 PRINT "d=";D
3336 PRINT "INCR=";INCR
3340 GOTO 1100
3350 WW=WW/2
3360 W=W/2
3370 FLAG1=0
```

Figure A.7   *(continued)*

```
3380 FLAG2=0
3390 GOTO 2730
3391 PRINT "INCREASE REINFORCEMENT "
3392 LPRINT "INCREASE REINFORCEMENT "
3393 INPUT "As=";AS
3395 LPRINT "As=";AS
3398 GOTO 22
3400 PRINT "DECREASE REINFORCEMENT"
3405 LPRINT "DECREASE REINFORCEMENT"
3410 INPUT "As=";AS
3415 PRINT "As=";AS
3420 LPRINT "As=";AS
3430 GOTO 22
3500 END
```

Figure A.7   (*continued*)

Example A.8: Analysis of a T Beam for Moment Capacity

Calculate the design moment capacity of the precast T beam simply supported over a span of 30 ft and with a distance between webs of 10 ft.   Given:

$$f'_c = 4000 \text{ psi, normal-weight concrete}$$

$$f_y = 60,000 \text{ psi}$$

$$b = 40 \text{ in.}$$

$$b_w = 10 \text{ in.}$$

$$d = 18 \text{ in.}$$

$$h_f = 2.5 \text{ in.}$$

reinforcement area at the tension side: $A_s = 6.0$ in.^{2}

Computer Solution   A flange-width check is not necessary for a precast beam since the precast section can act independently depending on the construction system.

```
*********************************************
*                                           *
*     T AND L BEAM ANALYSIS IN FLEXURE      *
*                                           *
*********************************************
******   INPUT DATA    ********
Fc'= ? 4000
Fy= ? 60000
b= ? 40
bw=   ? 10
hf= ? 2.5
d= ? 18
As=   ? 6
*****  VERIFICATION OF DATA    *****


  **FINAL RESULTS**
```

```
*******      FINAL RESULTS    *********
c(IN)       = 3.633218
a(IN)       = 3.088236
0.75RB      = 9.772109E-03
R           = 8.333334E-03
RW          = 0
RMIN        = 3.333334E-03
MU (IN-LB) = 5399206
Ok
[
```

Example A.9: Design of an End-span L Beam

Design the end-span spandrel beam for a roof-garden floor composed of a monolithic one-way slab system on beams. Details of the floor system are given in Ex. 5.8 and Fig. 5.23.

Computer Solution   From Ex. 5.8, use a factored external moment MUI 8,499,000 in.-lb.   Given:

$$f'_c = 3000 \text{ psi}$$

$$f_y = 60,000 \text{ psi}$$

$$b = 38 \text{ in.}$$

$$b_w = 14 \text{ in.}$$

$$h_f = 4 \text{ in.}$$

$$A_s = 8.0 \text{ in.}^2$$

The input to the computer for effective depth d can be any value.   The computer program by iteration will determine the depth required to resist a design moment of 8,499,000 in.-lb.

```
*********************************************
*                                           *
*      DESIGN OF T AND L BEAMS IN FLEXURE    *
*                                           *
*********************************************
***************    INPUT   DATA   ****************
MUI= ? 8499000
Fc'= ? 3000
Fy= ? 60000
b= ? 38
bw= ? 14
hf= ? 4
d= ?
As = ? 8
******    VERIFICATION OF INPUT DATA    *******
*********************************************
```

```
**FINAL RESULT**

bw          = 14
d(IN)       = 22.5
c(IN)       = 7.750865
a(IN)       = 6.588236
BE          = .85
0.75RB      = 9.486608E-03
R           = 9.356725E-03
RW          = 0
RMIN        = 3.333334E-03
MU(IN-LB)   = 8582061
```

A.4 COMPUTER PROGRAM EGNAWY7 FOR CORBELS

A computer program in BASIC for IBM PC AT, PS/2, and compatibles for the proportioning of reinforced concrete brackets or corbels is given in Fig. A.8. The program computes the reinforcement required for a bracket or corbel with a shear span/depth ratio a/d not greater than unity, and subject to a horizontal tensile force N_{uc} not larger than V_u, the factored shear force. The program is based on the provisions of the ACI building code. Because of the small a/d ratio for corbels, there is a tendency for a pure shear failure to occur through essentially vertical planes; ACI recommends a shear-fraction approach.

By using the input values of f_c', f_y, b_w, d, h, V_u, N_{uc}, μ, a, and λ, the program checks the allowable shear capacity. If this capacity is exceeded, the program displays "Enlarge section." It then proceeds to calculate A_s, the top horizontal reinforcement, and A_h, the horizontal reinforcement in the side faces of the bracket or corbel. The program ensures that N_{uc}, the horizontal tensile force, is greater than $0.2V_u$. If the input value is less than $0.2V_u$, the program uses $0.2V_u$.

```
LOAD"egnawy7
0
list
1 REM -- EGNAWY7 "CONCRETE CORBEL DESIGN"
2 REM -- ********************************
3 REM -- COPYRIGHT BY DR. EDWARD G. NAWY
4 REM -- ALL RIGHTS RESERVED
5 CLS: LOCATE 6,33: PRINT "PROGRAM EGNAWY7"
6 LOCATE 8,31: PRINT "CONCRETE CORBEL DESIGN"
7 LOCATE 11,33: PRINT "(C) COPYRIGHT 1988"
8 LOCATE 13,32: PRINT "DR. EDWARD G. NAWY"
9 LOCATE 15,32: PRINT "ALL RIGHTS RESERVED"
10 J=0: FOR Q=1 TO 1000: J=J+1: NEXT Q
20 SCREEN 1
30 PAINT (15,15) ,2
35 SCREEN 2
40 PRINT
41 PRINT "*************************************************************"
42 PRINT "*     PROGRAM TO SELECT SIZE AND REINFORCEMENT FOR      *"
44 PRINT "*           CONCRETE BRACKETS AND CORBELS               *"
45 PRINT "*                                                       *"
46 PRINT "*************************************************************"
```

Figure A.8 Program EGNAWY7 for proportioning brackets and corbels.

```
47 PRINT :PRINT :PRINT
48 INPUT "PRESS (RETURN) KEY TO CONTINUE ";PPPPPP
49 SCREEN 1
50 PAINT (15,15),2
60 SCREEN 2
65 PRINT TAB(5); "***   TYPE OF CONCRETE   ***"
70 PRINT TAB(5); "*********************" : PRINT : PRINT
80 PRINT TAB(2); "1- Lambda = 1.0, Normalweight  Concrete " : PRINT
85 PRINT TAB(2); "2- Lambda = .85, Sand-Lightweight Concrete" : PRINT
90 PRINT TAB(2); "3- Lambda = .75, All-Lightweight  Concrete" : PRINT
95 INPUT "WHAT IS YOUR CHOICE ? "; WW : PRINT
96 INPUT "WW =1, .85, .75 "; WW :PRINT
100 PRINT : PRINT"INPUT DATA" :PRINT "---------------------" : PRINT
110 ON WW GOTO 120, 142, 150
120 PRINT TAB(5); "****   FRICTION COEFFICIENT  u   ****" : PRINT
122 PRINT TAB(5); "******************************************" : PRINT : PRINT
125 PRINT TAB(2); "1- u = 1.4*Lambda, Concrete Placed Monolithically" :PRINT
127 PRINT TAB(2); "2- u = 1.0*Lambda, Concrete Against Roughened Conc":PRINT
129 PRINT TAB(2); "3- u = 0.6*Lambda, Concrete Against Unroughened Conc":PRINT
132 PRINT TAB(2); "4- u = 0.7*Lambda, Concrete Anchored to Steel Studs ":PRINT
135 INPUT "WHAT IS YOUR CHOICE ? ";ZZ : PRINT
136 INPUT "ZZ =1.4, 1, .6, .7 ";  ZZ : PRINT
137 IF WW =   1  GOTO 140, 150, 500
138 IF WW = .85 GOTO 140, 150, 512
139 IF WW = .75 GOTO 140, 150, 512
140 ON ZZ GOTO 142, 150
142 INPUT "WW ="; WW : PRINT
150 U = WW * ZZ
160 PRINT : PRINT "INPUT DATA" : PRINT "----------------------------- ":PRINT
300 REM "LINES 310 - 380 INPUT MATERIAL AND SECTION PROPERTIES" : PRINT
310 INPUT "CONCRETE COMPRESSIVE STRENGTH f'c (PSI) ?"; FC: PRINT
320 INPUT "REINFORCEMENT YIELD  STRENGTH fy  (PSI) ?"; FY: PRINT
330 INPUT "VERTICAL LOAD REACTION  Vu (LB)       ?"; VU: PRINT
340 INPUT "HORIZONTAL FRICTION FORCE Nuc(LB)      ?"; NUC: PRINT
350 INPUT "DO YOU WANT TO CORRECT DATA (Y/N)"; AS : IF AS="Y" THEN  310
360 GOTO 370, 470
370 INPUT "TOTAL DEPTH h (INCH), EFFECTIVE DEPTH  d (INCH)"; H, D :PRINT
380 INPUT "MOMENT ARM  a (INCH), CORBEL WIDTH  b (INCH) "; A, B :PRINT
390 INPUT "DO YOU WANT TO CORRECT DATA ? (Y/N)"; AS : IF AS="Y" THEN 370
400 GOTO 470, 480, 500, 520, 526, 540
470 IF (A/D)>1 THEN 480 ELSE 490
480 PRINT "(a/d)>1 NOT A CORBEL."
485 GOTO 720
490 IF (A/D)<1 THEN VN=VU/.85
500 IF VN<=.2*FC*B*D AND VN<=800*B*D    THEN  520 ELSE 510
510 PRINT "Vn>=.2*Fc*b*d  AND  Vn>=800*b*d  ,   ENLARGE SECTION"
512 IF VN<=(.2-.07*A/D)*FC*B*D AND VN<=(800-280*A/D)*B*D THEN 520 ELSE 515
515 PRINT "VN>=(.2-.07*a/d)*FC*B*D AND VN>=(800-280*a/d)*B*D, ENLARGE SECTION"
517 GOTO 720
520 AVF = VN/(FY*U)
530 IF   NUC<.2*VU THEN NUC=.2*VU
540 AF=(VU*A+(NUC*(H-D)))/(.9*FY*.85*D)
550 AN=NUC/(.85*FY)
560 AS1=(2/3)*AVF+AN
570 AS2=AF+AN
580 IF AS1<AS2 THEN AS=AS2 :GOTO 590
590 AH=(1/2)*AF
600 IF AS1>AS2 THEN  AS=AS1 :GOTO 610
610 AH=(1/3)*AVF
620 IF AS<.04*(FC/FY)*B*D  THEN  AS=.04*(FC/FY)*B*D
630 IF AH>.5*(AS-AN) THEN  AH=.5*(AS-AN)
640 PRINT
650 PRINT
660 PRINT "AS =" , AS
670 PRINT "AH =" , AH
710 PRINT
720 INPUT "DO YOU WANT TO ENTER NEW DATA (Y/N)" ; XXS: IF XXS = "Y" GOTO 80
750 END
O
```

Example A.10: Design of a Bracket or Corbel

Design a corbel to support a factored vertical load $V_u = 90,000$ lb acting at a distance of $a = 5$ in. from the face of the column. It has a width $b = 10$ in., a total thickness $h = 18$ in., and an effective depth $d = 14$ in. Given:

$$f'_c = 5000 \text{ psi, normal-weight concrete}$$

$$f_y = 60,000 \text{ psi}$$

Assume the corbel to be either cast after thé supporting column was constructed or both simultaneously cast. Neglect the weight of the corbel.

Computer Solution

```
***************************************************
*    PROGRAM TO SELECT SIZE AND REINFORCEMENT FOR   *
*  ,         CONCRETE BRACKETS AND CORBELS           *
*                                                    *
***************************************************



PRESS (RETURN) KEY TO CONTINUE ? [

INPUT DATA
-------------------------------------

CONCRETE COMPRESSIVE STRENGTH f'c (PSI) ?? 5000

REINFORCEMENT YIELD  STRENGTH fy  (PSI) ?? 60000

VERTICAL LOAD REACTION   Vu (LB)          ?? 90000

HORIZONTAL FRICTION FORCE   Nuc(LB)       ?? 18000

DO YOU WANT TO CORRECT DATA (Y/N)? N
TOTAL DEPTH h (INCH), EFFECTIVE DEPTH   d (INCH)? 18, 14

MOMENT ARM  a (INCH), CORBEL WIDTH   b (INCH) ? 5, 10

DO YOU WANT TO CORRECT DATA ? (Y/N)? N


AS =          1.529412
AH =          .5882353

DO YOU WANT TO ENTER NEW DATA (Y/N)? N
```

A.5 COMPUTER PROGRAM EGNAWY8 FOR DEEP BEAMS

A computer program in BASIC for IBM PC AT, PS/2, and compatibles is given for the proportioning of deep beams. It computes the necessary reinforcement and its spacing for deep beams ($a/d < 2.5$ for concentrated loading and $l_n/d < 5.0$ for uniform loading) either simply supported or continuous and either uniformly loaded or subjected to concentrated loading. The program designs both flexural and shear reinforcement and uses the same criteria as those discussed in Chapter 6. Because plane sections before bending do not necessarily remain plane after bending in deep beams, the resulting strain distribution is no longer considered as linear, and shear deformations that are neglected in normal beams become significant compared to pure flexure.

While the ACI code does specify a rather simple approach to the design of shear reinforcement in deep beams, it does not specify a design procedure for flexural analysis but does require a rigorous nonlinear analysis. As in Chapter 6 for flexure, this program is based on the simplified provisions recommended by the Euro-International Concrete Committee (CEB).

By using the input values of f_c', f_y, l_n, l, b, d, h, V_u, and M_u at $0.5a \le d$ for concentrated loading or $0.15l_n < d$ for uniform loading, $M_{u(\max)}$, and twice the area of one vertical and one horizontal shear reinforcing bar, the program computes the total area, A_s, of the flexural reinforcement required (both negative and positive for continuous beams); the vertical distance over which it should be distributed, Y_{H1} and Y_{H2} for negative moment regions and Y_{H3} for positive moment regions; the allowable concrete shear capacity, V_c, in psi; and the spacing of both the vertical and horizontal shear reinforcement.

For continuous beams, use the program with the maximum negative moment for the section at the support and then run the program with maximum positive moment for the section at or near the center of the span. The positive flexural reinforcement placed in the bottom of the beam (within Y_{H3}) should be extended through the support region.

A summary of the equations applied to proportioning the beam based on Section 6.9 are as follows:

simply supported: $jd = 0.2(l + 2h)$ if $l/h \ge 1.0$

$jd = 0.6l$ if $l/h < 1.0$

continuous: $jd = 0.2(l + 1.5h)$ if $l/h \ge 1.0$

$jd = 0.5l$ if $l/h < 1.0$

$$A_s = \frac{M_{u(\max)}}{\phi f_y jd} \ge \frac{200bd}{f_y} \qquad \phi = 0.9$$

simply supported: A_s should be placed within:

$$Y_{H3} = (0.25h - 0.05l) \le 0.2h$$

continuous: $A_{s1} = 0.5(l/h - 1)A_s$ if $l < h$ $A_{s1} = \dfrac{200bd}{f_y} = \dfrac{3\sqrt{f_c'}}{f_y}$

$A_{s2} = A_s - A_{s1}$ if $l < h$ $A_{s2} = A_s$

$Y_{H1} = 0.2h$ $Y_{H2} = 0.6h$

If $l_n/d < 2.0$,

$$V_u \le \phi(8\sqrt{f_c'}\, bd)$$

If $l_n/d \ge 2.0$,

$$V_u \le \phi\left[\frac{2}{3}\left(10 + \frac{l_n}{d}\right)\sqrt{f_c'}\, bd\right]$$

$$V_c = \left(3.5 - 2.5\frac{M_u}{V_u d}\right)\left(1.9\sqrt{f_c'} + 2500\rho_w \frac{V_u d}{M_u}\right)bd \le 6\sqrt{f_c'}\, bd$$

$$1.0 \le 3.5 - 2.5\frac{M_u}{V_u d} \le 2.5$$

$$V_s = \frac{V_u}{\phi} - V_c \qquad V_s = \left(\frac{A_v}{s_v}\frac{1 + l_n/d}{12} + \frac{A_{vh}}{s_h}\frac{11 - l_n/d}{12}\right)f_y d$$

The program solves the equation above assuming that $s_v = s_h$. If $s_v \ne s_h$, the program does not recompute s_v and s_h. The result of having $s_v = s_h$ will be slightly conservative.

Example A.11: Simply Supported Uniformly Loaded Deep Beam

Design the flexural and shear reinforcement for a deep beam given the following details:

$$f_c' = 5000 \text{ psi}$$
$$f_y = 60{,}000 \text{ psi}$$
$$l_n = 130 \text{ in.}$$
$$l = 155 \text{ in.}$$
$$b = 16 \text{ in.}$$
$$h = 110 \text{ in.}$$
$$V_u = 700{,}000 \text{ lb at } 0.15l_n$$
$$M = 11{,}000{,}000 \text{ in.-lb at } 0.15l_n$$
$$M_{u(max)} = 20{,}000{,}000 \text{ in.-lb}$$

Computer Solution   Assume that $d = 0.9h$.   Try No. 3 bars for shear reinforcement. $A_v = 2 \times 0.11 = 0.22$ in.^{2}.

```
*******************************************
*                                         *
*   REINFORCED CONCRETE DEEP BEAMS        *
*                                         *
*******************************************
PRESS (RETURN) KEY TO CONTINUE ? [

CONCRETE COMPRESSIVE STRENGTH  f'c (psi)     ?? 5000

STEEL YIELD STRENGTH   fy (psi)              ?? 60000

BEAM CLEAR SPAN  Ln (in.)                     ?? 130

CENTER TO CENTER SPAN DIMENSION L (in.)      ?? 155

BEAM WIDTH   b (in.)                          ?? 16

SHEAR SPAN FOR CONCENTRATED LOAD a (in.)     ??

EFFECTIVE DEPTH d, TOTAL DEPTH h (in.)      ?? 99,110

VERTICAL SHEAR FORCE Vu (lb)               ?? 700000

FACTORED MOMENT AT CRITICAL SECTION Mu(in.-lb)  ?? 11000000

MAXIMUM FACTORED MOMENT VALUE Mum (in.-lb)    ?? 20000000

SHEAR STEEL BARS AREA Av, Avh (sq. in)         ?? 0.22,.34[
DO YOU WANT TO CORRECT DATA (Y/N) ?? N


S=1 FOR SIMPLE SUPPORT, S=2 FOR CONTINUOUS, INPUT S =    ? 1

*****************     FINAL  RESULTS   *******************

As     = 5.28

YH3    = 20.025

Vc     = 672034.3

Sv     = 9.166667

Sh     = 8.5


Ok
```

TABLES AND NOMOGRAMS

B

B.1   Selected Conversion Factors to SI Units
B.2a   Geometrical Properties of Reinforcing Bars
B.2b   ASTM Standard Metric Reinforcing Bars
B.3   Cross-sectional Area of Bars for Various Bar Combinations
B.4   Area of Bars in a 1-Foot-Wide Slab Strip
B.5   Gross Moment of Inertia of T Sections
B.6 to B.11   Design Moment Strength for Slab Sections 10 In. and 12 In. Wide
B.12   Design Shear Strength of U Stirrups; f_y = 400,000 psi
B.13   Design Shear Strength of U stirrups; f_y = 60,000 psi
B.14 to B.21   Rectangular Columns Load–Moment Strength Interaction Diagrams
B.22   Moment Distribution Factors for Slabs without Drop Panels
B.23   Moment Distribution Factors for Slabs with Drop Panels; h_1 = 1.25h
B.24   Moment Distribution Factors for Slabs with Drop Panels; h_1 = 1.5h
B.25   Values of Torsion Constant C
B.26   Stiffness and Carry-over Factors for Columns

Photo 99   City Spire, one of the tallest concrete buildings in New York City. (Courtesy The Concrete Industry Board, Inc., New York.)

To convert from	to	multiply by
	Length	
inch	millimeter (mm)	25.4E*
foot	meter (m)	0.3048E
yard	meter (m)	0.9144E
mile (statute)	kilometer (km)	1.609
	Area	
square inch	square centimeter (cm^2)	6.452
square foot	square meter (m^2)	0.09290
square yard	square meter (m^2)	0.8361
	Volume (Capacity)	
ounce	cubic centimeter (cm^3)	29.57
gallon	cubic meter (m^3)	0.003785
cubic inch	cubic centimeter (cm^3)	16.4
cubic foot	cubic meter (m^3)	0.02832
cubic yard	cubic meter (m^3)	0.765
	Force	
kilogram-force	newton (N)	9.807
kip-force	kilonewton (kN)	4.448
pound-force	newton (N)	4.448
	Pressure or Stress (Force per Area)	
kilogram-force/square meter	pascal (Pa)	9.807
kip-force/square inch (ksi)	megapascal (MPa)	6.895
newton/square meter (N/m^2)	pascal (Pa)	1.000E
pound-force/square foot	pascal (Pa)	47.88
pound-force/square inch (psi)	pascal (Pa)	6895
	Bending Moment or Torque	
inch-pound-force	newton-meter (Nm)	0.1130
foot-pound-force	newton-meter (Nm)	1.356
meter-kilogram-force	newton-meter (Nm)	9.807
	Mass	
ounce-mass (avoirdupois)	gram (g)	28.35
pound-mass (avoirdupois)	kilogram (kg)	0.4536
ton (metric)	megagram (Mg)	1.000E
ton (short 2000 lbm)	megagram (Mg)	0.9072
	Mass per Volume	
pound-mass/cubic foot	kilogram/cubic meter (kg/m^3)	16.02
pound-mass/cubic yard	kilogram/cubic meter (kg/m^3)	0.5933
pound-mass/gallon	kilogram/cubic meter (kg/m^3)	119.8
	Temperature	
deg Fahrenheit (F)	deg Celsius (C)	$t_C = (t_F - 32)/1.8$
deg Celsius (C)	deg Fehrenheit (F)	$t_F = 1.8t_C + 32$

*E = English Unit

Figure B.1 Selected conversion factors to SI units.

Bar size designa-tion	Nominal cross section area, sq. in.	Weight, lb per ft	Nominal diameter, in.
#3	0.11	0.376	0.375
#4	0.20	0.668	0.500
#5	0.31	1.043	0.625
#6	0.44	1.502	0.750
#7	0.60	2.044	0.875
#8	0.79	2.670	1.000
#9	1.00	3.400	1.128
#10	1.27	4.303	1.270
#11	1.56	5.313	1.410
#14	2.25	7.650	1.693
#18	4.00	13.600	2.257

Figure B.2a Geometrical properties of reinforcing bars.

Bar Size Designation	Nominal Dimensions		
	Mass (Kg/m)	Diameter (mm)	Area (mm^2)
10 M	0.785	11.3	100
15 M	1.570	16.0	200
20 M	2.355	19.5	300
25 M	3.925	25.2	500
30 M	5.495	29.9	700
35 M	7.850	35.7	1000
45 M	11.775	43.7	1500
55 M	19.625	56.4	2500

Note: ASTM A615M Grade 300 is limited to size Nos. 10 M through 20 M; otherwise grades 400 or 500 MPa for all the sizes. Check availability with local suppliers for Nos. 45 M and 55 M.

Figure B.2b   ASTM standard netruc reinforcing bars

Areas, A_s (or A'_s) in sq. in.
Columns headed **0 5** contain data for bars of one size in groups of one to ten.
Columns headed **1 2 3 4 5** contain data for bars of two sizes with from one to five of each size.

#4 / #4 (one size) and **#3** (second size)

n	0	5	#3: 1	2	3	4	5
1	0.20	1.20	0.31	0.42	0.53	0.64	0.75
2	0.40	1.40	0.51	0.62	0.73	0.84	0.95
3	0.60	1.60	0.71	0.82	0.93	1.04	1.15
4	0.80	1.80	0.91	1.02	1.13	1.24	1.35
5	1.00	2.00	1.11	1.22	1.33	1.44	1.55

#5 / #5 (one size), **#4**, and **#3**

n	0	5	#4: 1	2	3	4	5	#3: 1	2	3	4	5
1	0.31	1.86	0.51	0.71	0.91	1.11	1.31	0.42	0.53	0.64	0.75	0.86
2	0.62	2.17	0.82	1.02	1.22	1.42	1.62	0.73	0.84	0.95	1.06	1.17
3	0.93	2.48	1.13	1.33	1.53	1.73	1.93	1.04	1.15	1.26	1.37	1.48
4	1.24	2.79	1.44	1.64	1.84	2.04	2.24	1.35	1.46	1.57	1.68	1.79
5	1.55	3.10	1.75	1.95	2.15	2.35	2.55	1.66	1.77	1.88	1.99	2.10

#6 / #6 (one size), **#5**, **#4**, and **#3**

n	0	5	#5: 1	2	3	4	5	#4: 1	2	3	4	5	#3: 1	2	3	4	5
1	0.44	2.64	0.75	1.06	1.37	1.68	1.99	0.64	0.84	1.04	1.24	1.44	0.55	0.66	0.77	0.88	0.99
2	0.88	3.08	1.19	1.50	1.81	2.12	2.43	1.08	1.28	1.48	1.68	1.88	0.99	1.10	1.21	1.32	1.43
3	1.32	3.52	1.63	1.94	2.25	2.56	2.87	1.52	1.72	1.92	2.12	2.32	1.43	1.54	1.65	1.76	1.87
4	1.76	3.96	2.07	2.38	2.69	3.00	3.31	1.96	2.16	2.36	2.56	2.76	1.87	1.98	2.09	2.20	2.31
5	2.20	4.40	2.51	2.82	3.13	3.44	3.75	2.40	2.60	2.80	3.00	3.20	2.31	2.42	2.53	2.64	2.75

#7 / #7 (one size), **#6**, **#5**, and **#4**

n	0	5	#6: 1	2	3	4	5	#5: 1	2	3	4	5	#4: 1	2	3	4	5
1	0.60	3.60	1.04	1.48	1.92	2.36	2.80	0.91	1.22	1.53	1.84	2.15	0.80	1.00	1.20	1.40	1.60
2	1.20	4.20	1.64	2.08	2.52	2.96	3.40	1.51	1.82	2.13	2.44	2.75	1.40	1.60	1.80	2.00	2.20
3	1.80	4.80	2.24	2.68	3.12	3.56	4.00	2.11	2.42	2.73	3.04	3.35	2.00	2.20	2.40	2.60	2.80
4	2.40	5.40	2.84	3.28	3.72	4.16	4.60	2.71	3.02	3.33	3.64	3.95	2.60	2.80	3.00	3.20	3.40
5	3.00	6.00	3.44	3.88	4.32	4.76	5.20	3.31	3.62	3.93	4.24	4.55	3.20	3.40	3.60	3.80	4.00

#8 / #8 (one size), **#7**, **#6**, and **#5**

n	0	5	#7: 1	2	3	4	5	#6: 1	2	3	4	5	#5: 1	2	3	4	5
1	0.79	4.74	1.39	1.99	2.59	3.19	3.79	1.23	1.67	2.11	2.55	2.99	1.10	1.41	1.72	2.03	2.34
2	1.58	5.53	2.18	2.78	3.38	3.98	4.58	2.02	2.46	2.90	3.34	3.78	1.89	2.20	2.51	2.82	3.13
3	2.37	6.32	2.97	3.57	4.17	4.77	5.37	2.81	3.25	3.69	4.13	4.57	2.68	2.99	3.30	3.61	3.92
4	3.16	7.11	3.76	4.36	4.96	5.56	6.16	3.60	4.04	4.48	4.92	5.36	3.47	3.78	4.09	4.40	4.71
5	3.95	7.90	4.55	5.15	5.75	6.35	6.95	4.39	4.83	5.27	5.71	6.15	4.26	4.57	4.88	5.19	5.50

#9 / #9 (one size), **#8**, **#7**, and **#6**

n	0	5	#8: 1	2	3	4	5	#7: 1	2	3	4	5	#6: 1	2	3	4	5
1	1.00	6.00	1.79	2.58	3.37	4.16	4.95	1.60	2.20	2.80	3.40	4.00	1.44	1.88	2.32	2.76	3.20
2	2.00	7.00	2.79	3.58	4.37	5.16	5.95	2.60	3.20	3.80	4.40	5.00	2.44	2.88	3.32	3.76	4.20
3	3.00	8.00	3.79	4.58	5.37	6.16	6.95	3.60	4.20	4.80	5.40	6.00	3.44	3.88	4.32	4.76	5.20
4	4.00	9.00	4.79	5.58	6.37	7.16	7.95	4.60	5.20	5.80	6.40	7.00	4.44	4.88	5.32	5.76	6.20
5	5.00	10.00	5.79	6.58	7.37	8.16	8.95	5.60	6.20	6.80	7.40	8.00	5.44	5.88	6.32	6.76	7.20

#10 / #10 (one size), **#9**, **#8**, and **#7**

n	0	5	#9: 1	2	3	4	5	#8: 1	2	3	4	5	#7: 1	2	3	4	5
1	1.27	7.62	2.27	3.27	4.27	5.27	6.27	2.06	2.85	3.64	4.43	5.22	1.87	2.47	3.07	3.67	4.27
2	2.54	8.89	3.54	4.54	5.54	6.54	7.54	3.33	4.12	4.91	5.70	6.49	3.14	3.74	4.34	4.94	5.54
3	3.81	10.16	4.81	5.81	6.81	7.81	8.81	4.60	5.39	6.18	6.97	7.76	4.41	5.01	5.61	6.21	6.81
4	5.08	11.43	6.08	7.08	8.08	9.08	10.08	5.87	6.66	7.45	8.24	9.03	5.68	6.28	6.88	7.48	8.08
5	6.35	12.70	7.35	8.35	9.35	10.35	11.35	7.14	7.93	8.72	9.51	10.30	6.95	7.55	8.15	8.75	9.35

#11 / #11 (one size), **#10**, **#9**, and **#8**

n	0	5	#10: 1	2	3	4	5	#9: 1	2	3	4	5	#8: 1	2	3	4	5
1	1.56	9.36	2.83	4.10	5.37	6.64	7.91	2.56	3.56	4.56	5.56	6.56	2.35	3.14	3.93	4.72	5.51
2	3.12	10.92	4.39	5.66	6.93	8.20	9.47	4.12	5.12	6.12	7.12	8.12	3.91	4.70	5.49	6.28	7.07
3	4.68	12.48	5.95	7.22	8.49	9.76	11.03	5.68	6.68	7.68	8.68	9.68	5.47	6.26	7.05	7.84	8.63
4	6.24	14.04	7.51	8.78	10.05	11.32	12.59	7.24	8.24	9.24	10.24	11.24	7.03	7.82	8.61	9.40	10.19
5	7.80	15.60	9.07	10.34	11.61	12.88	14.15	8.80	9.80	10.80	11.80	12.80	8.59	9.38	10.17	10.96	11.75

#14 / #14 (one size), **#11**, **#10**, and **#9**

n	0	5	#11: 1	2	3	4	5	#10: 1	2	3	4	5	#9: 1	2	3	4	5
1	2.25	13.50	3.81	5.37	6.93	8.49	10.05	3.52	4.79	6.06	7.33	8.60	3.25	4.25	5.25	6.25	7.25
2	4.50	15.75	6.06	7.62	9.18	10.74	12.30	5.77	7.04	8.31	9.58	10.85	5.50	6.50	7.50	8.50	9.50
3	6.75	18.00	8.31	9.87	11.43	12.99	14.55	8.02	9.29	10.56	11.83	13.10	7.75	8.75	9.75	10.75	11.75
4	9.00	20.25	10.56	12.12	13.68	15.24	16.80	10.27	11.54	12.81	14.08	15.35	10.00	11.00	12.00	13.00	14.00
5	11.25	22.50	12.81	14.37	15.93	17.49	19.05	12.52	13.79	15.06	16.33	17.60	12.25	13.25	14.25	15.25	16.25

#18 / #18 (one size), **#14**, **#11**, and **#10**

n	0	5	#14: 1	2	3	4	5	#11: 1	2	3	4	5	#10: 1	2	3	4	5
1	4.00	24.00	6.25	8.50	10.75	13.00	15.25	5.56	7.12	8.68	10.24	11.80	5.27	6.54	7.81	9.08	10.35
2	8.00	28.00	10.25	12.50	14.75	17.00	19.25	9.56	11.12	12.68	14.24	15.80	9.27	10.54	11.81	13.08	14.35
3	12.00	32.00	14.25	16.50	18.75	21.00	23.25	13.56	15.12	16.68	18.24	19.80	13.27	14.54	15.81	17.08	18.35
4	16.00	36.00	18.25	20.50	22.75	25.00	27.25	17.56	19.12	20.68	22.24	23.80	17.27	18.54	19.81	21.08	22.35
5	20.00	40.00	22.25	24.50	26.75	29.00	31.25	21.56	23.12	24.68	26.24	27.80	21.27	22.54	23.81	25.08	26.35

Figure B.3 Cross-sectional area of bars for various bar combinations.

Spacing, in.	Cross section area of bar, A_s (or A_s'), in.^{2}												Spacing, in.
	Bar size												
	#3	#4	#5	#6	#7	#8	#9	#10	#11	#14	#18		
4	0.33	0.60	0.93	1.32	1.80	2.37	3.00	3.81	4.68			4
4½	0.29	0.53	0.83	1.17	1.60	2.11	2.67	3.39	4.16	6.00		4½
5	0.26	0.48	0.74	1.06	1.44	1.90	2.40	3.05	3.74	5.40	9.60	5
5½	0.24	0.44	0.68	0.96	1.31	1.72	2.18	2.77	3.40	4.91	8.73	5½
6	0.22	0.40	0.62	0.88	1.20	1.58	2.00	2.54	3.12	4.50	8.00	6
6½	0.20	0.37	0.57	0.81	1.11	1.46	1.85	2.34	2.88	4.15	7.38	6½
7	0.19	0.34	0.53	0.75	1.03	1.35	1.71	2.18	2.67	3.86	6.86	7
7½	0.18	0.32	0.50	0.70	0.96	1.26	1.60	2.03	2.50	3.60	6.40	7½
8	0.17	0.30	0.47	0.66	0.90	1.19	1.50	1.91	2.34	3.38	6.00	8
8½	0.16	0.28	0.44	0.62	0.85	1.12	1.41	1.79	2.20	3.18	5.65	8½
9	0.15	0.27	0.41	0.59	0.80	1.05	1.33	1.69	2.08	3.00	5.33	9
9½	0.14	0.25	0.39	0.56	0.76	1.00	1.26	1.60	1.97	2.84	5.05	9½
10	0.13	0.24	0.37	0.53	0.72	0.95	1.20	1.52	1.87	2.70	4.80	10
10½	0.13	0.23	0.35	0.50	0.69	0.90	1.14	1.45	1.78	2.57	4.57	10½
11	0.12	0.22	0.34	0.48	0.65	0.86	1.09	1.39	1.70	2.45	4.36	11
11½	0.11	0.21	0.32	0.46	0.63	0.82	1.04	1.33	1.63	2.35	4.17	11½
12	0.11	0.20	0.31	0.44	0.60	0.79	1.00	1.27	1.56	2.25	4.00	12
13	0.10	0.18	0.29	0.41	0.55	0.73	0.92	1.17	1.44	2.08	3.69	13
14	0.09	0.17	0.27	0.38	0.51	0.68	0.86	1.09	1.34	1.93	3.43	14
15	0.09	0.16	0.25	0.35	0.48	0.63	0.80	1.02	1.25	1.80	3.20	15
16	0.08	0.15	0.23	0.33	0.45	0.59	0.75	0.95	1.17	1.69	3.00	16
17	0.08	0.14	0.22	0.31	0.42	0.56	0.71	0.90	1.10	1.59	2.82	17
18	0.07	0.13	0.21	0.29	0.40	0.53	0.67	0.85	1.04	1.50	2.67	18

Figure B.4   Area of bars in a 1-foot-wide slab strip.

$$I_g = K_{i4}\left(\frac{1}{12}\,b_w h^3\right)$$

$$K_{i4} = 1 + (\alpha_b - 1)\beta_h^3 + \frac{3(1 - \beta_h)^2(\beta_h)(\alpha_b - 1)}{1 + \beta_h(\alpha_b - 1)}$$

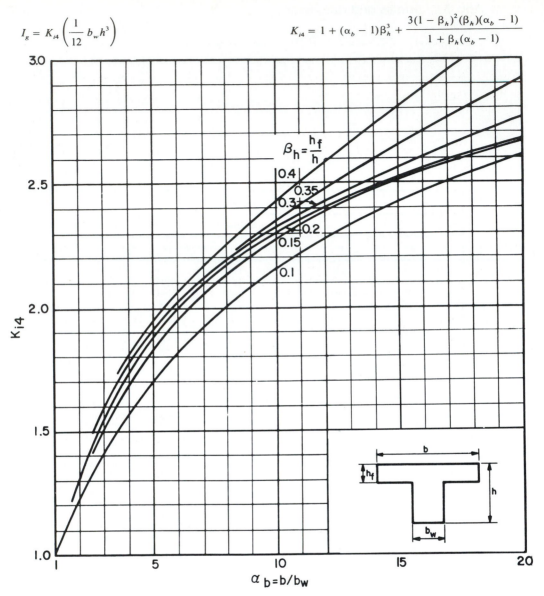

Example: For the T-beam shown, find the moment of inertia I_g:

$$\alpha_b = b/b_w = 143/15 = 9.53$$

$$\beta_h = h_f/h = 8/36 = 0.22$$

Interpolating between the curves for $\beta_h = 0.2$ and 0.3, read $K_{i4} = 2.28$.

$$I_g = K_{i4}\,\frac{b_w h^3}{12} = 2.28\,\frac{15(36)^3}{12} = 133{,}000 \text{ in.}^4$$

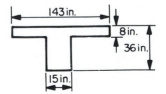

Figure B.5 Gross moment of inertia of T sections.

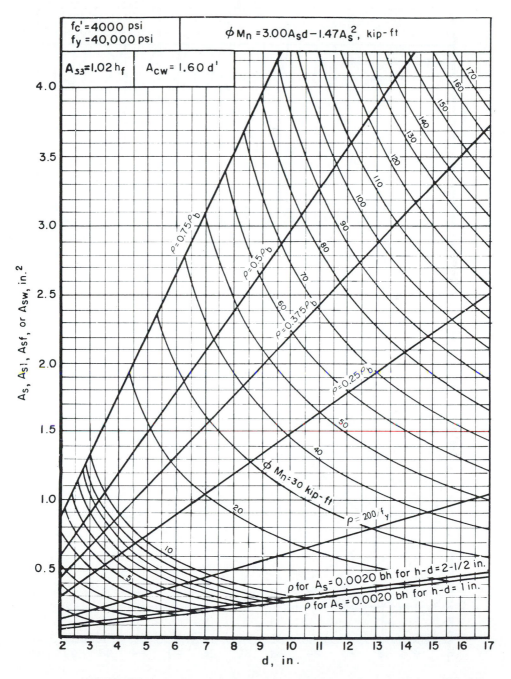

Figure B.6 Design moment strength ϕM_n for slab sections 12 in. wide, $f'_c = 4000$ psi, $f_y = 40,000$ psi (ACI-SP17). (From E. G. Nawy, *Simplified Reinforced Concrete*, Prentice Hall, Englewood Cliffs, N.J., 1986.)

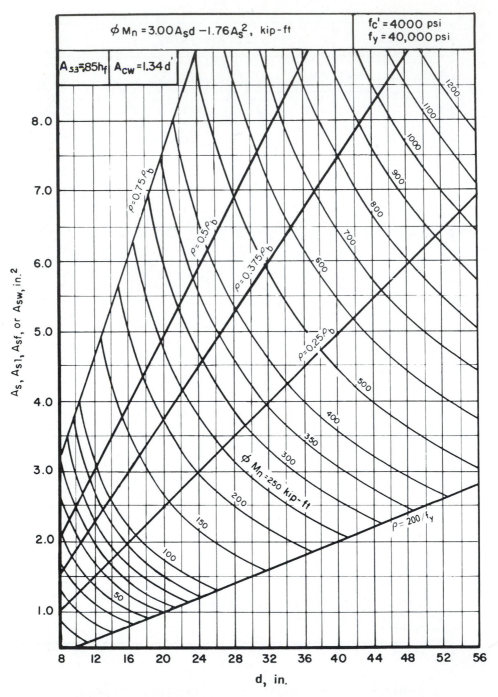

Figure B.7 Design moment strength ϕM_n for beam sections 10 in. wide, $f'_c = 4000$ psi, $f_y = 40,000$ psi (ACI-SP17). (From E. G. Nawy, *Simplified Reinforced Concrete*, Prentice Hall, Englewood Cliffs, N.J., 1986.)

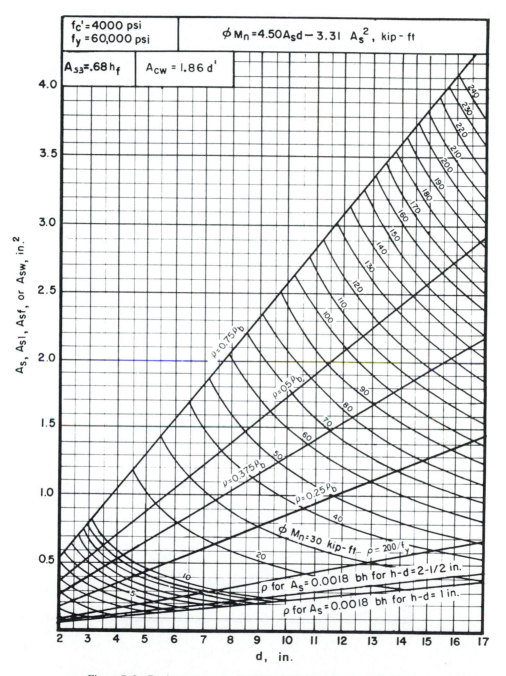

Figure B.8 Design moment strength ϕM_n for slab sections 12 in. wide, $f'_c = 4000$ psi, $f_y = 60,000$ psi (ACI-SP17). (From E. G. Nawy, *Simplified Reinforced Concrete*, Prentice Hall, Englewood Cliffs, N.J., 1986.)

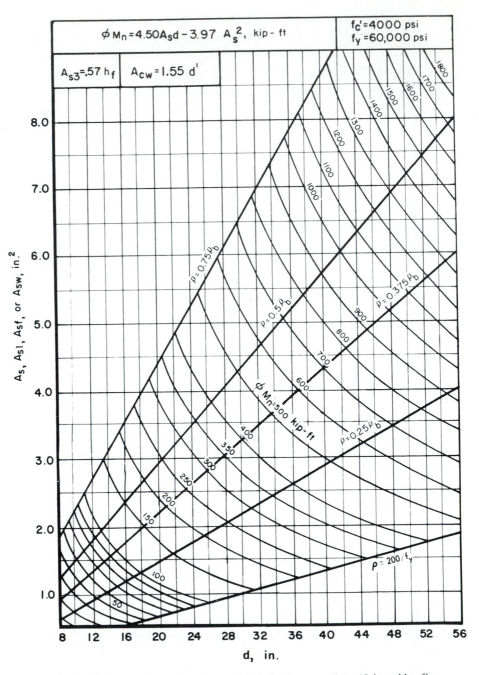

Figure B.9 Design moment strength ϕM_n for beam sections 10 in. wide, $f_c' = 4000$ psi, $f_y = 60,000$ psi (ACI-SP17). (From E. G. Nawy, *Simplified Reinforced Concrete*, Prentice Hall, Englewood Cliffs, N.J., 1986.)

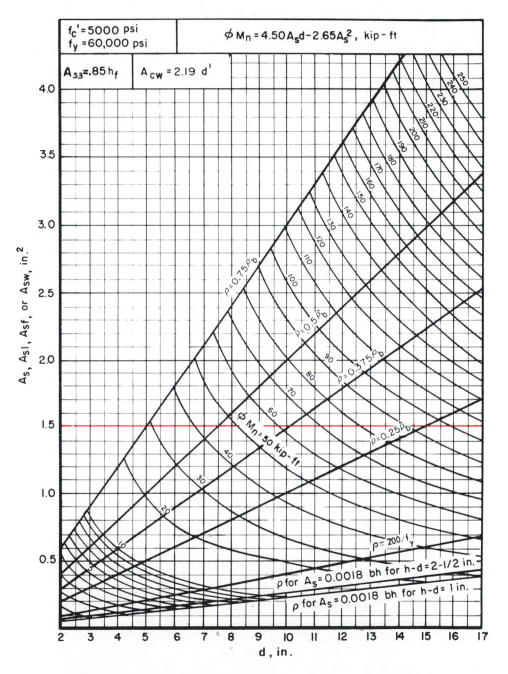

Figure B.10 Design moment strength ϕM_n for slab sections 12 in. wide, $f'_c =$ 5000 psi, $f_y = 60,000$ psi (ACI-SP17). (From E. G. Nawy, *Simplified Reinforced Concrete*, Prentice Hall, Englewood Cliffs, N.J., 1986.)

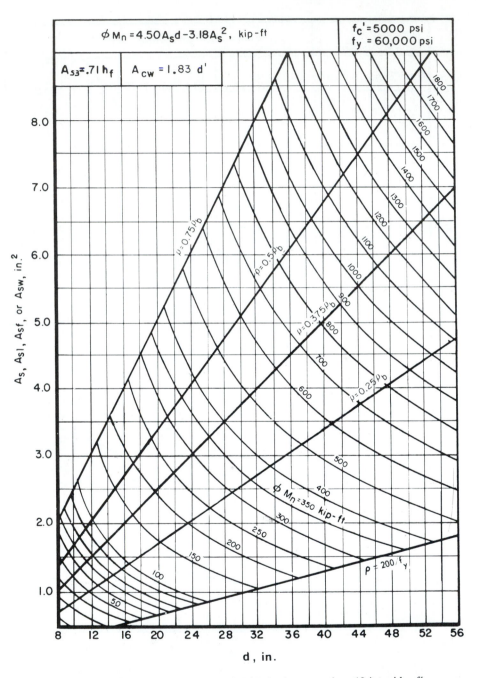

Figure B.11 Design moment strength ϕM_n for beam sections 10 in. wide, $f'_c =$ 5000 psi, $f_y =$ 60,000 psi (ACI-SP17). (From E. G. Nawy, *Simplified Reinforced Concrete*, Prentice Hall, Englewood Cliffs, N.J., 1986.)

$$\phi V_s = V_u - \phi V_c = \phi A_v f_y \frac{d}{s}$$

$$\text{Maximum } b_w = \frac{A_v f_y}{50s}$$

Stirrup size	s, in. / d, in.	2	3	4	5	6	7	8	9	10	11	12	14	16	18	20
											ϕV_s, kips					
	8	30	20	15												
	10	37	25	19	15											
	12	45	30	22	18	15										
	14	52	35	26	21	17	15									
	16	60	40	30	24	20	17	15								
	18	67	45	34	27	22	19	17	15			Maximum spacing = d/2				
	20	75	50	37	30	25	21	19	17	15						
	22	82	55	41	33	27	24	21	18	16	15					
#3 stirrups	24	90	60	45	36	30	26	22	20	18	16	15				
	26	97	65	49	39	32	28	24	22	19	18	16				
	28	105	70	52	42	35	30	26	23	21	19	17	15			
	30	112	75	56	45	37	32	28	25	22	20	19	16			
	32	120	80	60	48	40	34	30	27	24	22	20	17	15		
	34	127	85	64	51	42	36	32	28	25	23	21	18	16		
	36	135	90	67	54	45	38	34	30	27	24	22	19	17	15	
	38	142	95	71	57	47	41	36	32	28	26	24	20	18	16	
	40	150	100	75	60	50	43	37	33	30	27	25	21	19	17	15
	Max b_w, in.	88	59	44	35	29	25	22	19	17	16	15	13	11	10	9
	8	54	36	27												
	10	68	45	34	27											
	12	82	54	41	33	27										
	14	95	63	48	38	32	27									
	16	109	73	54	44	36	31	27								
	18	122	82	61	49	41	35	31	27			Maximum spacing = d/2				
	20	136	91	68	54	45	39	34	30	27						
	22	150	100	75	60	50	43	37	33	30	27					
#4 stirrups*	24	163	109	82	65	54	47	41	36	33	30	27				
	26	177	118	88	71	59	51	44	39	35	32	29				
	28	190	127	95	76	63	54	48	42	38	35	32	27			
	30	204	136	102	82	68	58	51	45	41	37	34	29			
	32	218	145	109	87	73	62	54	48	44	40	36	31	27		
	34	231	154	116	92	77	66	58	51	46	42	38	33	29		
	36	245	163	122	98	82	70	61	54	49	45	41	35	31	27	
	38	258	172	129	103	86	74	65	57	52	47	43	37	32	29	
	40	272	181	136	109	91	78	68	60	54	49	45	39	34	30	27
	Max b_w, in.	160	107	80	64	53	46	40	36	32	29	27	23	20	18	16

Figure B.12   Design shear strength ϕV_s of U stirrups; $f_y = 40,000$ psi (ACI-SP17).   (From E. G. Nawy, *Simplified Reinforced Concrete*, Prentice Hall, Englewood Cliffs, N.J., 1986.)

$$\phi V_s = V_u - \phi V_c = \phi A_v f_y \frac{d}{s}$$

$$\text{Maximum } b_w = \frac{A_v f_y}{50s}$$

ϕV_s, kips

Stirrup size	s, in. / d, in.	2	3	4	5	6	7	8	9	10	11	12	14	16	18	20
	8	45	30	22												
	10	56	37	28	22											
	12	67	45	34	27	22										
	14	79	52	39	31	26	22									
	16	90	60	45	36	30	26	22								
	18	101	67	50	40	34	29	25	22		Maximum spacing = $d/2$					
	20	112	75	56	45	37	32	28	25	22						
	22	123	82	62	49	41	35	31	27	25	22					
#3 stirrups	24	135	90	67	54	45	38	34	30	27	24	22				
	26	146	97	73	58	49	42	36	32	29	27	24				
	28	157	105	79	63	52	45	39	35	31	29	26	22			
	30	168	112	84	67	56	48	42	37	34	31	28	24			
	32	180	120	90	72	60	51	45	40	36	33	30	26	22		
	34	191	127	95	76	64	54	48	42	38	35	32	27	24		
	36	202	135	101	81	67	58	50	45	40	37	34	29	25	22	
	38	213	142	107	85	71	61	53	47	43	39	36	30	27	24	
	40	224	150	112	90	75	64	56	50	45	41	37	32	28	25	22
Max b_w, in.		32	88	66	52	44	37	33	29	26	24	22	18	16	15	13
	8	82	54	41												
	10	102	68	51	41											
	12	122	82	61	49	41										
	14	143	95	71	57	48	41									
	16	163	109	82	65	54	47	41								
	18	184	122	92	73	61	52	46	41		Maximum spacing = $d/2$					
	20	204	136	102	82	68	58	51	45	41						
	22	224	150	112	90	75	64	56	50	45	41					
#4 stirrups*	24	245	163	122	98	82	70	61	54	49	45	41				
	26	265	177	133	106	88	76	66	59	53	48	44				
	28	286	190	143	114	95	82	71	63	57	52	48	41			
	30	306	204	153	122	102	87	77	68	61	56	51	44			
	32	326	207	163	131	109	93	82	73	65	59	54	47	41		
	34	347	231	173	139	116	99	87	77	69	63	58	50	43		
	36	367	245	184	147	122	105	92	82	73	67	61	52	46	41	
	38	388	258	194	155	129	111	97	86	78	70	65	55	48	43	
	40	408	272	204	163	136	117	102	91	82	74	68	58	51	45	41
Max b_w, in.		240	160	120	96	80	68	60	53	48	44	40	34	30	27	24

Figure B.13 Design shear strength ϕV_s of U stirrups; $f_y = 60,000$ psi (ACI-SP17). (From E. G. Nawy, *Simplified Reinforced Concrete*, Prentice Hall, Englewood Cliffs, N.J., 1986.)

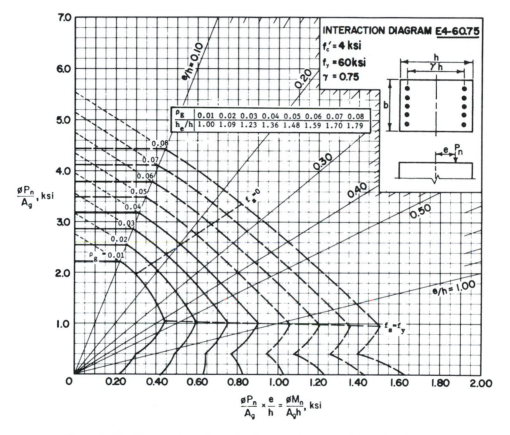

Figure B.14 Rectangular columns load–moment strength interaction diagrams; f'_c = 4000 psi, f_y = 60,000 psi, γ = 0.75 (ACI-SP17). (From E. G. Nawy, *Simplified Reinforced Concrete*, Prentice Hall, Englewood Cliffs, N.J., 1986.)

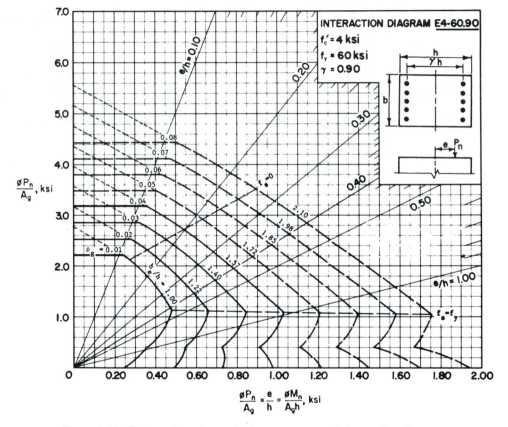

Figure B.15 Rectangular columns load–moment strength interaction diagrams, $f'_c = 4000$ psi, $f_y = 60,000$ psi, $\gamma = 0.90$ (ACI-SP17). (From E. G. Nawy, *Simplified Reinforced Concrete*, Prentice Hall, Englewood Cliffs, N.J., 1986.)

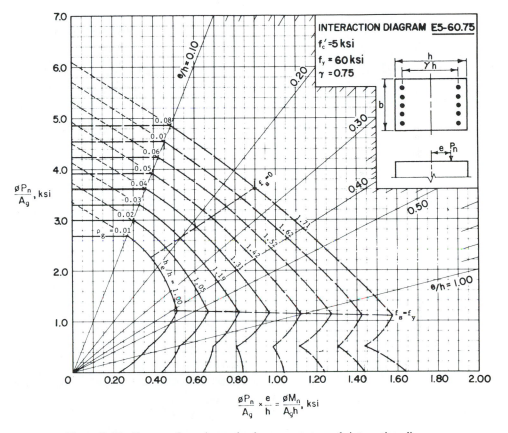

Figure B.16 Rectangular columns load–moment strength interaction diagrams, $f'_c = 5000$ psi, $f_y = 60,000$ psi, $\gamma = 0.75$ (ACI-SP17). (From E. G. Nawy, *Simplified Reinforced Concrete*, Prentice Hall, Englewood Cliffs, N.J., 1986.)

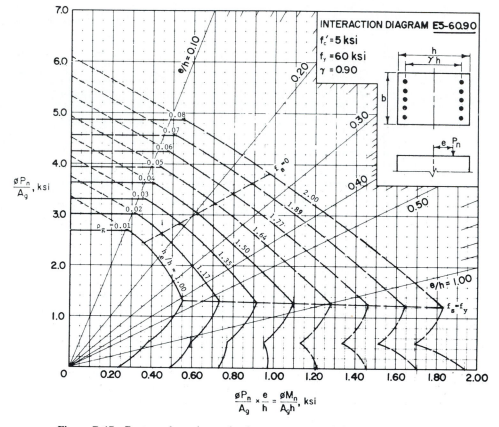

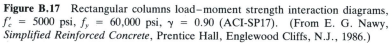

Figure B.17 Rectangular columns load–moment strength interaction diagrams,
f'_c = 5000 psi, f_y = 60,000 psi, γ = 0.90 (ACI-SP17). (From E. G. Nawy,
Simplified Reinforced Concrete, Prentice Hall, Englewood Cliffs, N.J., 1986.)

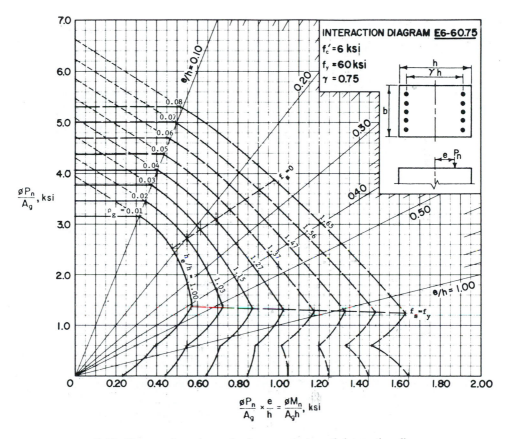

Figure B.18 Rectangular columns load–moment strength interaction diagrams, $f'_c = 6000$ psi, $f_y = 60,000$ psi, $\gamma = 0.75$ (ACI-SP17). (From E. G. Nawy, *Simplified Reinforced Concrete*, Prentice Hall, Englewood Cliffs, N.J., 1986.)

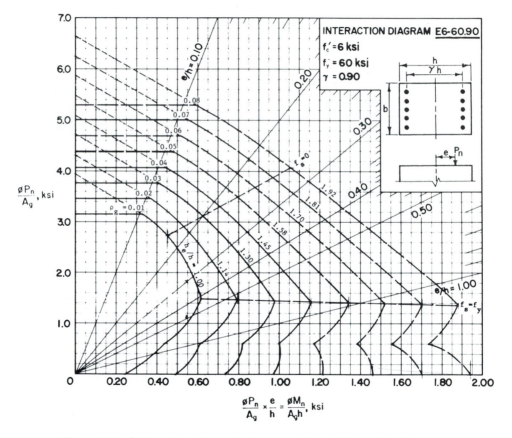

Figure B.19 Rectangular columns load–moment strength interaction diagrams, $f'_c = 6000$ psi, $f_y = 60{,}000$ psi, $\gamma = 0.90$ (ACI-SP17). (From E. G. Nawy, *Simplified Reinforced Concrete*, Prentice Hall, Englewood Cliffs, N.J., 1986.)

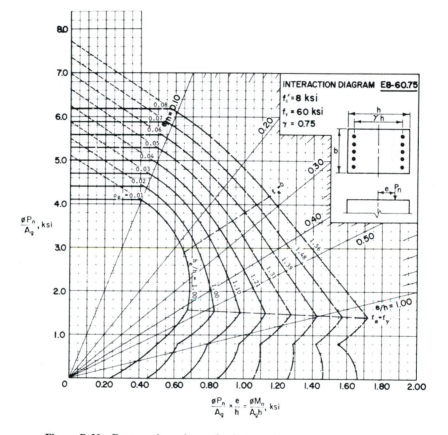

Figure B.20 Rectangular columns load–moment strength interaction diagrams, f'_c = 8000 psi, f_y = 60,000 psi, γ = 0.75 (ACI-SP17). (From E. G. Nawy, *Simplified Reinforced Concrete*, Prentice Hall, Englewood Cliffs, N.J., 1986.)

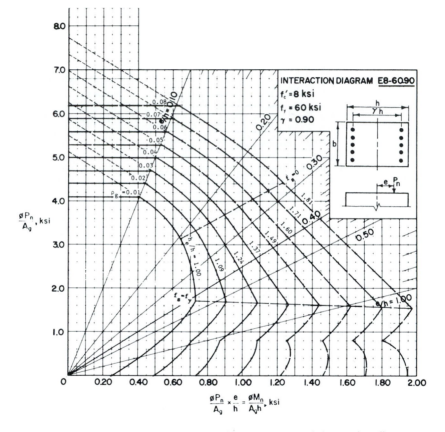

Figure B.21 Rectangular columns load–moment strength interaction diagrams, $f_c' = 8000$ psi, $f_y = 60,000$ psi, $\gamma = 0.90$ (ACI-SP17). (From E. G. Nawy, *Simplified Reinforced Concrete*, Prentice Hall, Englewood Cliffs, N.J., 1986.)

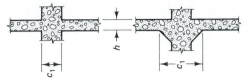

| FEM (uniform load w) = $Mw\ell_2\ell_1^2$ | | K (stiffness) = $kE\ell_2h^3/12\ell_1$ | | | | |

Carryover factor = COF

c_1/ℓ_1 \ c_2/ℓ_2		0.00	0.05	0.10	0.15	0.20	0.25	0.30
0.00	M	0.083	0.083	0.083	0.083	0.083	0.083	0.083
	k	4.000	4.000	4.000	4.000	4.000	4.000	4.000
	COF	0.500	0.500	0.500	0.500	0.500	0.500	0.500
0.05	M	0.083	0.084	0.084	0.084	0.085	0.085	0.085
	k	4.000	4.047	4.093	4.138	4.181	4.222	4.261
	COF	0.500	0.503	0.507	0.510	0.513	0.516	0.518
0.10	M	0.083	0.084	0.085	0.085	0.086	0.087	0.087
	k	4.000	4.091	4.182	4.272	4.362	4.449	4.535
	COF	0.500	0.506	0.513	0.519	0.524	0.530	0.535
0.15	M	0.083	0.084	0.085	0.086	0.087	0.088	0.089
	k	4.000	4.132	4.267	4.403	4.541	4.680	4.818
	COF	0.500	0.509	0.517	0.526	0.534	0.543	0.550
0.20	M	0.083	0.085	0.086	0.087	0.088	0.089	0.090
	k	4.000	4.170	4.346	4.529	4.717	4.910	5.108
	COF	0.500	0.511	0.522	0.532	0.543	0.554	0.564
0.25	M	0.083	0.085	0.086	0.087	0.089	0.090	0.091
	k	4.000	4.204	4.420	4.648	4.887	5.138	5.401
	COF	0.500	0.512	0.525	0.538	0.550	0.563	0.576
0.30	M	0.083	0.085	0.086	0.088	0.089	0.091	0.092
	k	4.000	4.235	4.488	4.760	5.050	5.361	5.692
	COF	0.500	0.514	0.527	0.542	0.566	0.571	0.585
$x = (1 - c_2/\ell_2^3)$		1.000	0.856	0.729	0.613	0.512	0.421	0.343

c_1 and c_2 are the widths of the column measured parallel to ℓ_1 and ℓ_2.

Figure B.22 Moment distribution factors for slabs without drop panels. (From S. H. Simmonds and M. Janko, "Design Factors for Equivalent Frame Method," *Journal of the American Concrete Institute*, Vol. 68, No. 11, November 1971. Figures B.23–B.26 are from this same source.)

		c_2/ℓ_2						
c_1/ℓ_1		0.00	0.05	0.10	0.15	0.20	0.25	0.30
0.00	M	0.088	0.088	0.088	0.088	0.088	0.088	0.088
	k	4.795	4.795	4.795	4.795	4.795	4.795	4.795
	COF	0.542	0.542	0.542	0.542	0.542	0.542	0.542
0.05	M	0.088	0.088	0.089	0.089	0.089	0.089	0.090
	k	4.795	4.846	4.896	4.944	4.990	5.035	5.077
	COF	0.542	0.545	0.548	0.551	0.553	0.556	0.558
0.10	M	0.088	0.088	0.089	0.090	0.090	0.091	0.091
	k	4.795	4.894	4.992	5.039	5.184	5.278	5.368
	COF	0.542	0.548	0.553	0.559	0.564	0.569	0.573
0.15	M	0.088	0.089	0.090	0.090	0.091	0.092	0.092
	k	4.795	4.938	5.082	5.228	5.374	5.520	5.665
	COF	0.542	0.550	0.558	0.565	0.573	0.580	0.587
0.20	M	0.088	0.089	0.090	0.091	0.092	0.093	0.094
	k	4.795	4.978	5.167	5.361	5.558	5.760	5.962
	COF	0.542	0.552	0.562	0.571	0.581	0.590	0.590
0.25	M	0.088	0.089	0.090	0.091	0.092	0.094	0.095
	k	4.795	5.015	5.245	5.485	5.735	5.994	6.261
	COF	0.542	0.553	0.565	0.576	0.587	0.598	0.600
0.30	M	0.088	0.089	0.090	0.092	0.093	0.094	0.095
	k	4.795	5.048	5.317	5.601	5.902	6.219	6.550
	COF	0.542	0.554	0.567	0.580	0.593	0.605	0.618

FEM (uniform load w) $= Mw\ell_2\ell_1^2$ K (stiffness) $= kE\ell_2h^3/12\ell_1$
Carryover factor $= COF$

h, Slab thickness; h_1, total thickness in drop panel.

Figure B.23 Moment distribution factors for slabs with drop panels; $h_1 = 1.25h$.

FEM (uniform load w) $= M w \ell_2 \ell_1^2$ K (stiffness) $= k E \ell_2 h^3 / 12 \ell_1$
Carryover factor $=$ COF

c_1/ℓ_1	c_2/ℓ_2	0.00	0.05	0.10	0.15	0.20	0.25	0.30
0.00	M	0.093	0.093	0.093	0.093	0.093	0.093	0.093
	k	5.837	5.837	5.837	5.837	5.837	5.837	5.837
	COF	0.589	0.589	0.589	0.589	0.589	0.589	0.589
0.05	M	0.093	0.093	0.093	0.093	0.094	0.094	0.094
	k	5.837	5.890	5.942	5.993	6.041	6.087	6.131
	COF	0.589	0.591	0.594	0.596	0.598	0.600	0.602
0.10	M	0.093	0.093	0.094	0.094	0.094	0.095	0.095
	k	5.837	5.940	6.042	6.142	6.240	6.335	6.427
	COF	0.589	0.593	0.598	0.602	0.607	0.611	0.615
0.15	M	0.093	0.093	0.094	0.095	0.095	0.096	0.096
	k	5.837	5.986	6.135	6.284	6.432	6.579	6.723
	COF	0.589	0.595	0.602	0.608	0.614	0.620	0.626
0.20	M	0.093	0.093	0.094	0.095	0.096	0.096	0.097
	k	5.837	6.027	6.221	6.418	6.616	6.816	7.015
	COF	0.589	0.597	0.605	0.613	0.621	0.628	0.635
0.25	M	0.093	0.094	0.094	0.095	0.096	0.097	0.098
	k	5.837	6.065	6.300	6.543	6.790	7.043	7.298
	COF	0.589	0.598	0.608	0.617	0.626	0.635	0.644
0.30	M	0.093	0.094	0.095	0.096	0.097	0.098	0.090
	k	5.837	6.099	6.372	6.657	6.953	7.258	7.571
	COF	0.589	0.599	0.610	0.620	0.631	0.641	0.651

h, Slab thickness; h_1, total thickness in drop panel.

Figure B.24 Moment distribution factors for slabs with drop panels; $h_1 = 1.5h$.

y \ x	4	5	6	7	8	9	10	12	14	16
12	202	369	592	868	1,188	1,538	1,900	2,557	—	—
14	245	452	736	1,096	1,529	2,024	2,566	3,709	4,738	—
16	388	534	880	1,325	1,871	2,510	3,233	4,861	6,567	8,083
18	330	619	1,024	1,554	2,212	2,996	3,900	6,013	8,397	10,813
20	373	702	1,167	1,782	2,553	3,482	4,567	7,165	10,226	13,544
22	416	785	1,312	2,011	2,895	3,968	5,233	8,317	12,055	16,275
24	548	869	1,456	2,240	3,236	4,454	5,900	9,469	13,885	19,005
27	522	994	1,672	2,583	3,748	5,183	6,900	11,197	16,628	23,101
30	586	1,119	1,888	2,926	4,260	5,912	7,900	12,925	19,373	27,197
33	650	1,243	2,104	3,269	4,772	6,641	8,900	14,653	22,117	31,293
36	714	1,369	2,320	3,612	5,284	7,370	9,900	16,381	24,860	35,389
42	842	1,619	2,752	4,298	6,308	8,828	11,900	19,837	30,349	43,581
48	970	1,869	3,184	4,984	7,332	10,286	13,900	23,293	35,836	51,773
54	1,098	2,119	3,616	5,670	8,356	11,744	15,900	26,749	41,325	59,965
60	1,226	2,369	4,048	6,356	9,380	13,202	17,900	30,205	46,813	68,157

$$C = (1 - 0.63x/y)\frac{x^3 y}{3}$$

x is smaller dimension of rectangular cross section.

Figure B.25   Values of torsion constant C.

$$K_c = k\frac{E\ell_c}{\ell_c}$$

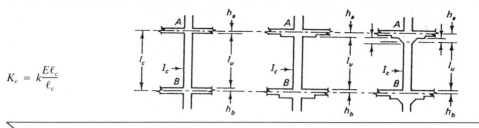

h_d/h_b	ℓ_c/ℓ_u	1.05	1.10	1.15	1.20	1.25	1.30	1.35	1.40	1.45
0.00	k_{AB}	4.20	4.40	4.60	4.80	5.00	5.20	5.40	5.60	5.80
	C_{AB}	0.57	0.65	0.73	0.80	0.87	0.95	1.03	1.10	1.17
0.2	k_{AB}	4.31	4.62	4.95	5.30	5.65	6.02	6.40	6.79	7.20
	C_{AB}	0.56	0.62	0.68	0.74	0.80	0.85	0.91	0.96	1.01
0.4	k_{AB}	4.38	4.79	5.22	5.67	6.15	6.65	7.18	7.74	8.32
	C_{AB}	0.55	0.60	0.65	0.70	0.74	0.79	0.83	0.87	0.91
0.6	k_{AB}	4.44	4.91	5.42	5.96	6.54	7.15	7.81	8.50	9.23
	C_{AB}	0.55	0.59	0.63	0.67	0.70	0.74	0.77	0.80	0.83
0.8	k_{AB}	4.49	5.01	5.58	6.19	6.85	7.56	8.31	9.12	9.98
	C_{AB}	0.54	0.58	0.61	0.64	0.67	0.70	0.72	0.75	0.77
1.0	k_{AB}	4.52	5.09	5.71	6.38	7.11	7.89	8.73	9.63	10.60
	C_{AB}	0.54	0.57	0.60	0.62	0.65	0.67	0.69	0.71	0.73
1.2	k_{AB}	4.55	5.16	5.82	6.54	7.32	8.17	9.08	10.07	11.12
	C_{AB}	0.53	0.56	0.59	0.61	0.63	0.65	0.66	0.68	0.69
1.4	k_{AB}	4.58	5.21	5.91	6.68	7.51	8.41	9.38	10.43	11.57
	C_{AB}	0.53	0.55	0.58	0.60	0.61	0.63	0.64	0.65	0.66
1.6	k_{AB}	4.60	5.26	5.99	6.79	7.66	8.61	9.64	10.75	11.95
	C_{AB}	0.53	0.55	0.57	0.59	0.60	0.61	0.62	0.63	0.64
1.8	k_{AB}	4.62	5.30	6.06	6.89	7.80	8.79	9.87	11.03	12.29
	C_{AB}	0.52	0.55	0.56	0.58	0.59	0.60	0.61	0.61	0.62
2.0	k_{AB}	4.63	5.34	6.12	6.98	7.92	8.94	10.06	11.27	12.59
	C_{AB}	0.52	0.54	0.56	0.57	0.58	0.59	0.59	0.60	0.60
2.2	k_{AB}	4.65	5.37	6.17	7.05	8.02	9.08	10.24	11.49	12.85
	C_{AB}	0.52	0.54	0.55	0.56	0.57	0.58	0.58	0.59	0.59
2.4	k_{AB}	4.66	5.40	6.22	7.12	8.11	9.20	10.39	11.68	13.08
	C_{AB}	0.52	0.53	0.55	0.56	0.56	0.57	0.57	0.58	0.58
2.6	k_{AB}	4.67	5.42	6.26	7.18	8.20	9.31	10.53	11.86	13.29
	C_{AB}	0.52	0.53	0.54	0.55	0.56	0.56	0.56	0.57	0.57
2.8	k_{AB}	4.68	5.44	6.29	7.23	8.27	9.41	10.66	12.01	13.48
	C_{AB}	0.52	0.53	0.54	0.55	0.55	0.55	0.56	0.56	0.56
3.0	k_{AB}	4.69	5.46	6.33	7.28	8.34	9.50	10.77	12.15	13.65
	C_{AB}	0.52	0.53	0.54	0.54	0.55	0.55	0.55	0.55	0.55
3.5	k_{AB}	4.71	5.50	6.40	7.39	8.48	9.69	11.01	12.46	14.02
	C_{AB}	0.51	0.52	0.53	0.53	0.54	0.54	0.54	0.53	0.53
4.0	k_{AB}	4.72	5.54	6.45	7.47	8.60	9.84	11.21	12.70	14.32
	C_{AB}	0.51	0.52	0.52	0.53	0.53	0.52	0.52	0.52	0.52
4.5	k_{AB}	4.73	5.56	6.50	7.54	8.69	9.97	11.37	12.89	14.57
	C_{AB}	0.51	0.52	0.52	0.52	0.52	0.52	0.51	0.51	0.51
5.0	k_{AB}	4.75	5.59	6.54	7.60	8.78	10.07	11.50	13.07	14.77
	C_{AB}	0.51	0.51	0.52	0.52	0.51	0.51	0.51	0.50	0.49
6.0	k_{AB}	4.76	5.63	6.60	7.69	8.90	10.24	11.72	13.33	15.10
	C_{AB}	0.51	0.51	0.51	0.51	0.50	0.50	0.49	0.49	0.48
7.0	k_{AB}	4.78	5.66	6.65	7.76	9.00	10.37	11.88	13.54	15.34
	C_{AB}	0.51	0.51	0.51	0.50	0.50	0.49	0.48	0.48	0.47
8.0	k_{AB}	4.78	5.68	6.69	7.82	9.07	10.47	12.01	13.70	15.54
	C_{AB}	0.51	0.51	0.50	0.50	0.49	0.49	0.48	0.47	0.46

Figure B.26   Stiffness and carry-over factors for columns.

INDEX

V_j = factored shear force at section due to externally applied loads occurring simultaneously with M_{max}.

V_n = nominal shear strength.

w_u = factored load per unit length of beam or per unit area of slab.

x = shorter overall dimension of rectangular part of cross section.

x_1 = shorter center-to-center dimension of closed rectangular stirrup.

y = longer overall dimension of rectangular part of cross section.

y_t = distance from centroidal axis of gross section, neglecting reinforcement, to extreme fiber in tension.

y_1 = longer center-to-center dimension of closed rectangular stirrup.

α = total angular change of prestressing tendon profile in radians from tendon jacking end to any point x.

α = ratio of flexural stiffness of beam section to flexural stiffness of a width of slab bounded laterally by centerlines of adjacent panels (if any) on each side of the beam.

$$= \frac{E_{cb}I_b}{E_{ca}I_s}$$

α_m = average value of α for all beams on edges of a panel.

β_a = ratio of dead load per unit area to live load per unit area (in each case without load factors).

β_d = ratio of maximum factored dead load moment to maximum factored total load moment, always positive.

β = a ratio of clear spans in long to short direction of two-way slabs.

γ_f = fraction of unbalanced moment transferred by flexure at slab-column connections.

γ_p = factor for type of prestressing tendon.
= 0.55 for f_{py}/f_{pu} not less than 0.80
= 0.40 for f_{py}/f_{pu} not less than 0.85
= 0.28 for f_{py}/f_{pu} not less than 0.90

γ_v = fraction of unbalanced moment transferred by eccentricity of shear at slab-column connections.
= $1 - \gamma_f$

δ_{ns} = moment magnification factor for frames braced against sidesway, to reflect effects of member curvature between ends of compression member.

δ_s = moment magnification factor for frames not braced against sidesway, to reflect lateral drift resulting from lateral and gravity loads.

μ = curvature friction coefficient.

ξ = time-dependent facor for sustained load.
(xi)

ρ = ratio of nonprestressed tension reinforcement.
(rho)
= A_s/bd

ρ' = ratio of nonprestressed compression reinforcement.
= A_s'/bd

ρ_b = reinforcement ratio producing balanced strain conditions.

ρ_p = ratio of prestressed reinforcement.
= A_{ps}/bd_p

ρ_v = A_{sv}/A_{cv}; where A_{av} is the projection on A_{cv} of area of distributed shear reinforcement crossing the plane of A_{cv}.

θ = angle of compression diagonals in truss analogy for torsion.

ϕ = strength reduction factor.

ω = pf_y/f_c'.

ω' = $p'f_y/f_c'$.

ω_p = $\rho_p f_{pe}/f_c'$.

$\omega_{pw}, \omega_w, \omega'_w$
= reinforcement indices for flanged sections computed as for ω, ω_p, and ω' except that b shall be the web width, and reinforcement area shall be that required to develop compressive strength of web only.